The Special Theory of Relativity

Helmut Günther · Volker Müller

The Special Theory of Relativity

Einstein's World in New Axiomatics

Springer

Helmut Günther
Berlin, Germany

Volker Müller
Leibniz-Institut fuer Astrophysik
University Potsdam
Potsdam, Germany

Revised English edition of: H. Günther "Die Spezielle Relativitätstheorie - Einsteins Welt in einer neuen Axiomatik" Springer Fachmedien Wiesbaden 2013.
Extension to General Relativity Theory and translation: Volker Müller.

95 illustrations including 13 portraits, 54 completely solved problems.

Picture references:

Albert Einstein, p. 24, Einstein tower, p. 355, Leonhard Euler, p. 88, Carl Friedrich Gauss, p. 518, Werner Karl Heisenberg, p. 269, David Hilbert, p. 266, Hendrik Antoon Lorentz, p. 47, James Clerk Maxwell, p. 206, Albert Abraham Michelson, p. 45, Hermann Minkowski, p. 227, Isaac Newton, p. 86, Emmy Noether, p. 56, Georg Friedrich Bernhard Riemann, p. 339, Erwin Schrödinger, p. 276 by Christina Günther (Berlin) mixed technique, 2004–2012.

Atomic clock, p. 3, with permission of the Physikalisch-technische Bundesanstalt Braunschweig, 2014.

Michelson interferometer, p. 49, Angela Müller, 2015.

Einstein cross, CASTLES-survey, p. 382, with permission of the Space Telescope Science Institute, 2014.

Cosmic Microwave Background fluctuations, Planck-Satellite, p. 385, with permission of the European Space Agency, 2014.

Other illustrations by the authors.

Additional material to this book can be downloaded from ftp.springernature.com.

ISBN 978-981-13-7782-2 ISBN 978-981-13-7783-9 (eBook)
https://doi.org/10.1007/978-981-13-7783-9

This Springer imprint is published by the registered company Springer Nature Singapore Pte Ltd.
The registered company address is: 152 Beach Road, #21-01/04 Gateway East, Singapore 189721, Singapore

Preface

The Special Theory of Relativity (SRT) is frequently wrongly classed as scientific terrain, in which very few scientists really feel competent. However, the consequences of SRT are widespread, and the concepts and derivations are significant for many physical disciplines today. It has to be underlined that the SRT was developed as a result of a flash of genius, thereby solving a vast range of problems simultaneously. However, we do not have to take the most difficult route to get an intimate knowledge and understanding.

Can one really comprehend the Special Theory of Relativity (SRT) the same way we understand the mechanics of automobiles and bicycles?

Teaching Special Relativity today generally follows EINSTEIN's original method. The unprepared reader is firstly confronted with the amasing claim of the universal constancy of the speed of light, i.e. EINSTEIN's famous principle of relativity from the year 1905. The reader is left all on his own to brood over the no less amasing consequences of the behaviour of moving measuring rods and clocks.

However, already in 1910, W. v. IGNATOWSKI demonstrated that the universal constancy of the speed of light has not to be claimed in order to derive the LORENTZ transformation. In a similar way, P. FRANK AND H. ROTHE demanded in 1912 the equivalence of all inertial frames instead of universal constancy of the light velocity.

In contrast to all other approaches to derive LORENTZ transformation, we do not postulate a principle of relativity at all. In our exposition, the principle of relativity including the universal constancy of the speed of light will be a result at the end.

This book aims to make possible a real understanding of Special Relativity without including the entire instruments of theoretical physics. Based on the aforementioned papers, we intend to work our way up gradually to the momentous EINSTEIN principle. Our method will be less abstract than traditional approaches, yet without being less exact. We aim to develop an approach, whereby the Special Theory of Relativity will certainly be no more difficult to understand than any other fundamental physical issue. Why does my bike fall over when it is stationary and not when I am riding it? What keeps a Boeing, weighing so many tones, in the air?

We intend to impart to readers, who do not intend to become experts in theoretical physics, independent competence in Special Relativity. Before addressing the issue of the relativity of physical laws, we start with an accurate description of motion in space and time, Chap. 1. The pioneering work of this approach goes back to LANGE 1885. Further contributions were reviewed by DISALLE in 1990. In the first place, we have to discuss in this connection the definition of simultaneity with particular care, Chap. 1, Sect. 4.

Then we will be confronted with two issues, which cannot be ignored. Namely, when I move a rod, does it have to have the same length as when it is stationary? When I move a clock, does it has the same oscillating period as when it is stationary? Is there a logical answer or do we have to let answer nature—via experiments? To avoid complex premises at this stage, we will first describe motions, leaving the answer to these issues open, Chap. 1, Sect. 5.

EINSTEIN's famous principle of relativity is discussed in Chap. 2, Sect. 1. Nevertheless, we base our approach neither on this fundamental law of nature nor on a principle of relativity at all. Once simultaneity is introduced in one special inertial system, we formulate the so-called elementary principle of relativity—which is nothing but the so-called reciprocity principle—for the definition of simultaneity in all other inertial systems. The meaning of simultaneity in curved space-time is discussed in the overview of General Relativity, Chap. 13.

With the knowledge of the principle of relativity, everything begins to fall into place by itself. To some extend as preliminary practice for our main problem, we start with the elementary structure of classical space-time, with which we are familiar, based on the GALILEAN transformation, Chap. 3. Subsequently, the reader will virtually be in a position to calculate without any trouble the famous LORENTZ transformation, the key to relativistic space-time.

In Chap. 2, Sect. 3, a certain principle of relativity is discussed which is somewhere between EINSTEIN's and the elementary principle of relativity. As it was already discussed in 1910 by IGNATOWSKI and more detailed in 1912 by FRANK and ROTHE, our new approach does not need a discussion of electrodynamics.

We report on the experimental situation in Special Relativity and present key modern test experiments in Chap. 4, Sect. 6. This chapter is intended as a first suggestion for the experimentally interested reader.

The relativistic corrections of the classical mechanics are performed with the aid of the known TOLMAN thought experiment, thereby enabling us to understand EINSTEIN's famous energy mass equivalence, Chap. 6, Sect. 3. We then present the main relativistic phenomena and paradoxes, which are easy to understand at a basic level, Chap. 8, Sects. 1–9.

Additionally to the traditional solution of measuring rod paradox, Chap. 8, Sect. 4, and twin paradox, Chap. 8, Sect. 8, we present a surprising explanation: both paradoxes disappear on its own, if we make use, in an appropriate manner, of the freedom in the definition of simultaneity, Chap. 8, Sect. 9. In this case, the transformation formula using absolute simultaneity applies, which at first was written down by THIRRING in 1992.

In Chap. 12, Sect. 1, we discuss a lattice model of relativistic space-time. In this context, we present a physical curiosity, namely a mini version of Special Relativity as presented by nature with real existing crystalline solids. Philosophical questions concerning Special Relativity are also included.

In Chap. 13, we provide a short, but concise introduction to the General Theory of Relativity. The Chap. 13, Sects. 3–5 contain some interesting modern applications of the theory on astronomical observations, as to black holes, gravitational waves and cosmology.

Finally, we have collected a set of exercises, including hints on their solutions, Appendix A. Mathematical tools are at the readers' disposal in Appendix B. The numerous illustrations serve to clarify the text and the solutions to the exercises.

The book benefits strongly from lectures and exercises, delivered by the authors on Special and General Relativity Theory over years at Humboldt University of Berlin, University Potsdam and University of Applied Sciences Bielefeld. We thank warmly for critical discussions and remarks.

The references are intended to encourage particularly interested readers in a more in-depth study of the subject. In particular, we mention the textbooks of Arzeliès (1966), where the author gives detailed bibliographical information up to 1966, French (1968), Liebscher (2005), Rindler (1977), Misner, Thorne, Wheeler (1973).

SI units are applied throughout.

We express our gratitude to Drs. Christoph Baumann and Loyola d'Silva for cooperation to create out of the German version an extended Englisch textbook.

Berlin, Germany — Helmut Günther
Potsdam, Germany — Volker Müller
January 2019

Contents

Chapter 1
Space, Time and Motion

The measuring process in mechanics and the description of the motion of bodies in space and time cannot be separated. The careful formulation of motion is of fundamental importance. The roots of Special Relativity Theory are already present in the problem of the measuring process.

1 Measuring Rods and Clocks

An event is described by the location in space, where it took place, and by the time, when it happened. We need measuring rods and clocks in order to describe distances and time intervals. And we must guarantee that we have sufficiently many identically built normal measuring rods L_N for the length measurement, and normal clocks with a period T_N for time measurement.

For precision measurements, it is reasonable to refer to comparison measures for lengths and times provided by nature. One avails oneself of the spectra of electromagnetic radiation of characteristic wavelengths and frequencies from the atoms or molecules that are supposed to be unchanged by space and time.[1]

The metre L_N is defined as the 1 650 763.73 multiple of the wavelength of a certain orange spectral line of the Krypton isotope $^{86}\mathrm{Kr}$.

The time interval T_N of one second is the duration of 9 192 631 770 oscillations of a certain spectral line of the Cesium isotope $^{133}\mathrm{Cs}$.

The quantitative description of the measurement of each physical quantity consists always in two information, the measuring unit, which provides a comparison

[1]The metre was originally understood as the 40-millionth part of the circumference of the Earth. A definition of the metre with the help of the speed of light is avoided in the beginning for avoiding confusion with the logic of our exposition.

H. Günther and V. Müller, *The Special Theory of Relativity*,
https://doi.org/10.1007/978-981-13-7783-9_1

quantity, and the measured value, which indicates, how often one must take the comparison quantity in order to compose the quantity to be measured. Today always the SI-system of units is used.

1 m and 1 s are the units for length and time in the SI-system of units.

The distance of 100 m = 100 L_N means, one puts 100 times 1 650 763.73 wavelengths from the above-mentioned spectral line of the Krypton isotope. One writes $l = 100$ m and likewise for the numerical measure $l = 100$.

A time interval lasts 2.5 s = 2.5 T_N, if it coincides with the duration of 2.5 times 9 192 631 770 oscillations from the above-mentioned spectral line of the Cesium isotope (Fig. 1). One writes t =2.5 s and likewise for the numerical measure $t = 2.5$.

The metre and the second are thereby no abstract concepts, but physical characteristics of atoms and molecules. Thus, we find we can consider the instruments of our measurements themselves as objects of measurements. In particular, it will be of fundamental importance to determine the behaviour of resting and moving rods and/or clocks.

2 Inertial Systems

Motion of a body is always motion with respect to another body. The order of events in space and time requires the distinction of a system of tightly connected bodies, a reference system, in which we can measure the occurrence of an arbitrary event. In principle, we can take any reference system. But only in a certain class of reference systems, the description of the motion will follow simple rules that allow a deeper understanding of physical laws. Historical examples represent the Copernican revolution, the transition of the geocentric picture of PTOLEMÄUS to the heliocentric word model of COPERNICUS. We can define

Those reference systems, where bodies remain at rest or in uniform motion so long as no physical forces are acting on them are called after GALILEI *inertial systems*.

Do they really exist?[2]

In a sufficient approximation for many purposes, a laboratory on Earth surface, or simply a lecture hall, represents such an inertial system. For more precise measurements, we will notice the influence of the rotation of the Earth, at first during its daily circulation around its north–south axis, second its yearly motion around the Sun, and

[2]Soon after establishing his Special Relativity Theory, EINSTEIN (1955) has shown, that the universal mass attraction, the gravitation requires a substantial extension of the theoretical framework of space and time. The understanding of Special Relativity, a logically closed theory, will become more difficult by the inclusion of gravitational effects. Therefore, in this book, all gravitational effects are excluded besides in the last Chap. 13, where the fundamentals of this extension called General Relativity Theory are introduced.

Fig. 1 The atomic clock CSF1 of the Physikalisch-Technische Bundesanstalt Braunschweig. Built from 1996 to 1999, it represents like its successor CSF2 from 2008 a high-quality standard of time measurements by counting the oscillations of Cs-atoms (the name means Caesium Fountain) and providing the standard second as given in our definition. The accuracy is about $\delta t/t \approx 10^{-15}$, i.e. the clock may fail by one second in 30 million years. Since the time measurement represents a very exact metrological standard, we get the value for the definition of the metre of the last page via the speed of light in vacuum as given by the value of Eq. (4). By the way, the accuracy of the atomic clock is comparable to the time standard as provided by the millisecond pulsar, c.p. the discussion of neutron stars in Chap. 13, Sect. 3.3 and the attempts to use them for measuring a gravitational wave background in Chap. 13, Sect. 4.

sometimes in astronomical observations, the rotation of the solar system around the centre of the Milky Way. For a sufficient approximation, we imagine to construct a reference system with origin in the centre of the solar system and pointing to the

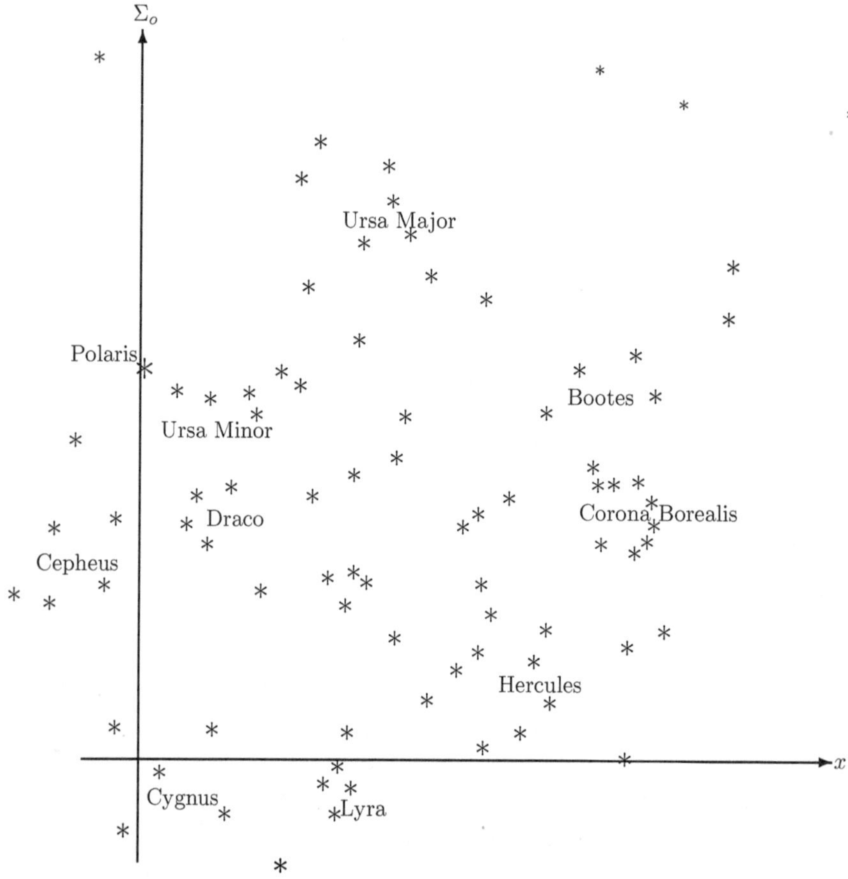

Fig. 2 Since thousands of years, the sky positions of fixed stars of our nearby Milky Way remain almost unchanged. They define a references system that in the following is designated as Σ_o.

system of fixed stars, called Σ_o, s. Fig. 2.[3] We do not intend to further proceed with the precise definition of an inertial system Σ_o, cp. Fig. 3.

> The reference system of fixed connected bodies, that are resting with respect to the system of fixed stars, represents an inertial system Σ_o.

[3]Well known is FOUCAULT's pendulum, a long, well-suspended pendulum on the roof of high room of building, that shows a continuous change of its oscillation plane without any force impact. The room does not represent an inertial system, the Earth is in rotation with respect to the inertial system Σ_o. Considered from this system Σ_o, the oscillation plane does not move at all. The property of our rotating Earth, not to represent an exact inertial system, will become especially significant in precision experiments to the relativistic time dilatation, s. Problem 4.

If a force-free body, observed from system Σ_o, keeps in straight uniform motion, it will stay at rest with respect to a system Σ', that is attached to this body. This connection is sufficient for passing over from one inertial system Σ_o to all others:

> The inertial systems are realised by the union of all with respect to Σ_o uniformly moving reference systems.

The standard form of introducing inertial systems has a long historical evolution. Its starting point was the distinguished paper of LANGE (1885). The interested reader is further recommended to follow the review by DISALLE (1990).

3 Coordinates and Velocities

3.1 One *Inertial System*

We use in the following an arbitrary, but then fixed inertial system as defined in the last Section, Σ_o. An event will be described by four numbers, three for space and one for time.

3.1.1 Spatial Coordinates

In space, we choose an arbitrary zero point, the origin of the coordinates, $O_3(0, 0, 0)$, and three orthogonal spatial directions as x-, y- and z-axes of a Cartesian coordinate system. By laying our length normals L_N one after the other, we assign to each point $P(\mathbf{x})$, three Cartesian coordinates $(\mathbf{x}) = (x, y, z)$. For example, we come to the point $P(2, -3, 5)$ starting from the origin in 2 steps with the measure L_N in x-direction, then 3 in negative y-direction and 5 in z-direction. In this way, we can practice geometry in this inertial system, we can measure distances and angles and compare them in all the space.[4] For one and the same point P, one can also define any other three numbers as coordinates. The only constraint is the unique attribution for the doubtless identification of points by its coordinates. We want to specialise on the following definition of the spatial coordinates of a point:

> The Cartesian coordinates $(\mathbf{x})$ of a point $P(\mathbf{x})$ in Σ_o are defined as the measuring numbers of the distance from the coordinate origin $O_3(0, 0, 0)$.

[4]Thereby, we suppose that the sum of inner angles in a triangle measures 180° or π in arc scale. For the three-space holds the EUCLIDean geometry. This basic assumption will only be modified in EINSTEIN's General Relativity Theory.

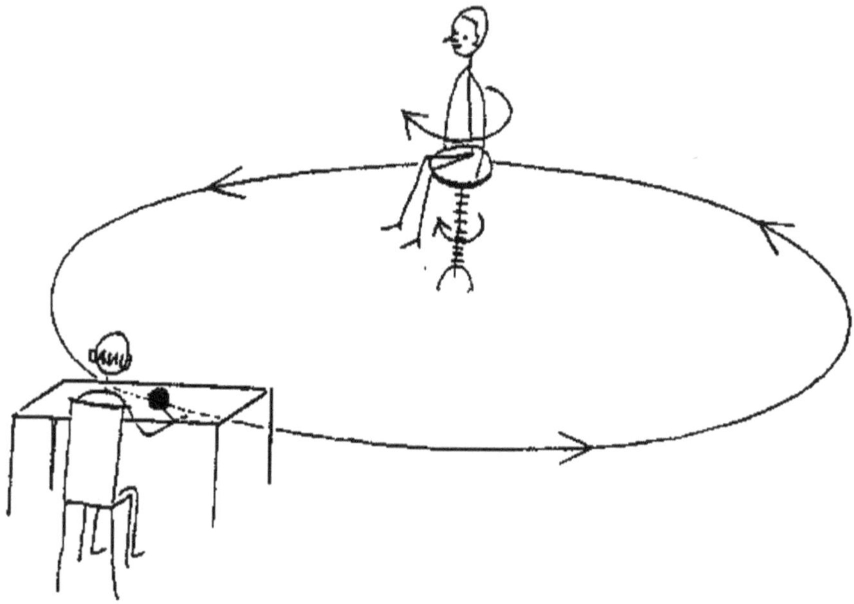

Fig. 3 An observer on a desk and an observer on a rotating chair. The last one states that the mass resting on the desk performs a circular motion around himself. Motion is always motion with respect to something, a reference system, where the observer is sitting and measuring space and time. A mass m is moving with respect to the reference system of the observer. The observer is assumed to rest in his system by definition. Taken we have two observers with their respective reference systems, that are moving against each other. They will describe the motion of mass m quite differently. I am sitting on my desk with the resting mass m on it. A colleague is circulating on his chair around his own axis. He is seeing the mass and the desk with me on my chair moving on a circular orbit around himself.

Have both reference systems equal rights, me at my desk and he on his rotation chair? — Now, he will become amiss after a while, me not. If I would be very sensitive, I should become a little bit amiss because of the rotation of the Earth around its axis, and if still more sensitive, because of the rotation of the Earth around the Sun, and then, even by the accelerated motion of the solar system. Special reference systems, where one is absolutely not becoming amiss, are the inertial systems.

These inertial systems are the reference systems, with respect to them EINSTEIN has formulated his Special Relativity Theory. Accelerated systems will become the starting point for his theory of gravitation, his General Relativity Theory, that is sketched in Chap. 13.

3.1.2 The Problem of Time Measurements

We must be quite careful with the time ordering of events. We distribute normal clocks sufficiently dense, so that everywhere clocks are available. However, we have to *synchronise* the clocks at different spatial points, i.e. we have to make them show the 'same time'.

Now we arrive at a problem: Should we first set up the clocks at the origin, and then distribute them over the space, or should we first distribute them over space before setting them into run. How do we know, when we should set in run the clock,

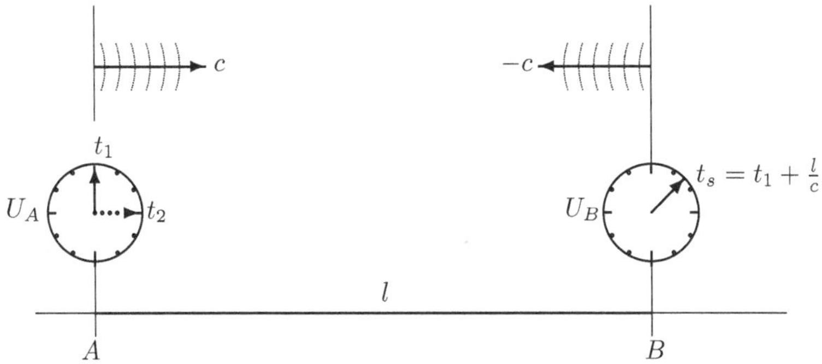

Fig. 4 If we suppose for the inertial system Σ_o homogeneity and isotropy, then we can measure the speed of light with a single clock U_A, and this then allows the synchronisation of both clockes U_A and U_B in the system Σ_o.

e.g. at $P(2, -3, 5)$, that it is synchronous with a clock at the origin $O_3(0, 0, 0)$? On the other hand, we may ask with EINSTEIN (1923), how we can be sure '... that the state of motion of a clock had no influence on its rate ...", so that putting in run the clocks before its distribution in space will later have no value anymore.

Already in 1898 POINCARÉ (1898, 1913), presents the following remarkable analysis: 'It is difficult to separate the qualitative problems of simultaneity from the quantitative problem of the measurement of time; either one uses a chronometer, or one takes into account a transmission velocity such as the one of light, since one cannot measure such a velocity without *measuring* a time ... we have no direct intuition about the equality of two time intervals'. POINCARÉ concludes: 'The simultaneity of two events or the order of their succession, as well as the equality of two time intervals, must be defined in such a way that the statements of the natural laws be as simple as possible. In other words, all rules and definitions are but the result of an unconscious opportunism'.

For synchronising two clocks U_A and U_B at the end points A and B of a line of length l we need a velocity. The pointer of clock U_A will be fixed at t_1, when a light signal, a photon, with velocity c is passing. When the signal reaches the clock U_B at end point B of the line, the pointer of this clock will be fixed at $t_s = t_1 + l/c$, it now runs synchronously with the clock U_A, Fig. 4.

Where do we know the speed of light from, the velocity of the photons? We need to measure the time $t_s - t_1$, taken by the light to travel the length l, and we form the velocity $c = l/(t_s - t_1)$. Here we need the difference of pointer positions $t_s - t_1$ at both end points of the line, Fig. 4. For this procedure, we must first synchronise both clocks, what had to be done with light. We are in a circular reasoning, just restating POINCARÉ[5]:

[5]For a deeper discussion of the time definition we refer to BARBOUR (1999, 2001).

The knowledge of a velocity allows a definition of simultaneity.
The definition of simultaneity allows the measurement of velocities. (1)

How to proceed? For the system Σ_o we postulate a basic experience, the *homogeneity* and *isotropy* of space and time:

It is possible to synchronise the clocks in Σ_o in such a way that at each point and in each direction the same physical properties are measured. For light, we will then determine at each place and in each direction the same velocity. (2)

If the light signal, emitted at the beginning A of our length l at time t_1 gets to the end point B, an observer at this point will send back a light signal, that will arrive at the first point A at time t_2. Because of the supposed isotropy of space, we are sure that the speed of light c in both directions has the same value, and we find for this velocity c,

$$c = \frac{2l}{t_2 - t_1} \ . \tag{3}$$

The time values t_1 and t_2 are measured with one and the same clock U_A at the starting point A of the length.[6] For the numerical value of the speed of light c, we get

$c = 299\,792\,458\,\mathrm{ms}^{-1}$. Speed of light in vacuum (4)

This photon velocity c is independent of the state of motion of the emitting source
This property is also explicitely formulated in EINSTEIN's axiomatics, s. Eq. (38), and it is physically justified, since light waves are solutions of the homogeneous, i.e. source free MAXWELL equations in vacuum, s. Chap. 9, Sect. 3.1.5. (5)

Using in this way the determined velocity c of photons in vacuum, all clocks in space can now be synchronised. Thereby, we consider only the inertial system Σ_o. The clock U_B is synchronous with the clock U_A, when its pointer at the arrival of the light signal shows the time t_s,

$t_s = t_1 + \frac{l}{c}$. Rule for synchronisation of clocks in system Σ_o (6)

If all clocks are synchronised, we denote the time t as fourth coordinate of an event $E(\mathbf{x}, t)$ at place $P(\mathbf{x})$:

---

[6] If one replaces the photons by bodies K, that are given identical velocities v by some precision machine, then from practical points of view we will never get the same precision as with light. Furthermore, there we will not have the independence of the velocity according to statement (5).

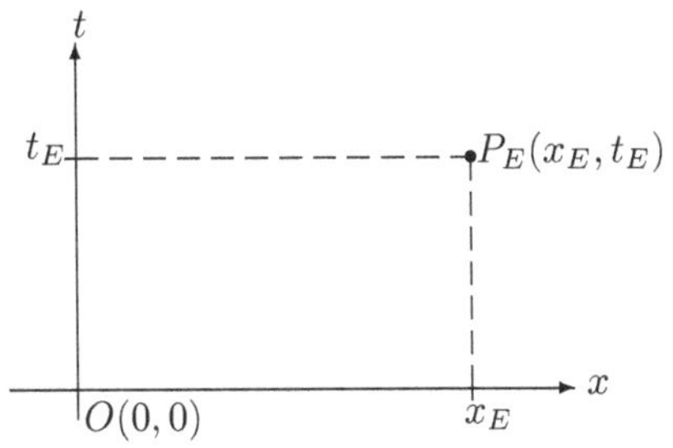

Fig. 5 Each event E is in the x-t-plane a point P_E with coordinates x_E and t_E.

The time coordinate t of an event $E(\mathbf{x}, t)$ is determined in system Σ_o as pointer position of the clock at place $P(\mathbf{x})$.

Each *event* $E(\mathbf{x}, t)$ is characterised by four numbers, three for the place and one for the time. All events get both a spatial and a chronological order. For presenting many problems, it is sufficient and practical to suppress two spatial dimensions, and to present the inertial system in a two-dimensional space-time diagram. The event $E(x_E, t_E)$ is then a point P_E in the x-t-plane, Fig. 5.

3.1.3 The Relative Velocity

We are now in a position to measure and to compare velocities of arbitrary objects in the inertial system Σ_o. The position of an object K on the x-axis should be described by $x_1 = x_1(t)$. This object can be an arbitrary body, e.g. a steel ball, or the front of a light wave. It is characteristic, that both objects transport energy, i.e. they can transfer signals.

Alternatively, we can imagine that we take a very long ruler, that is inclined by a small angle α against the x-axis. As 'object', we take the crossing point of the ruler with the x-axis. When we shift the ruler in the direction of the negative y-axis with a velocity g, the position of the crossing point with the x-axis has a velocity $v = g/\tan\alpha$ in x-direction. Here no energy is transferred, and so also no signal can be transmitted, Fig. 6.

Independent from its physical nature, we determine the velocity v of an 'object' following a path $x_1 = x_1(t)$ according to equation

$$v = \text{passed way/time difference, i.e. } v = \Delta x_1/\Delta t \text{ or as differential } v = dx_1/dt.$$

Considering the crossing point of the ruler with the x-axis, we can get *arbitrary* large velocities by decreasing the inclination angle α. Essentially these large velocities have no physical relevance. With respect to the steel ball or the front of the light wave, the velocity v is a velocity of energy tranfer. A possible limit for its amount represents a physical statement, and this has to be investigated, s. Problem 7.

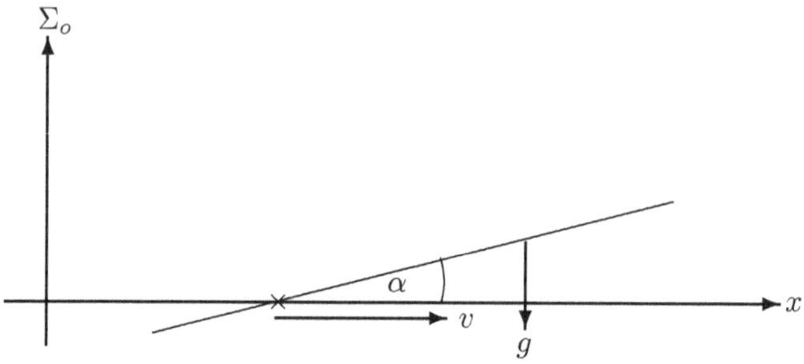

Fig. 6 Demonstrating the existence of arbitrarily large velocities (explanations in the text).

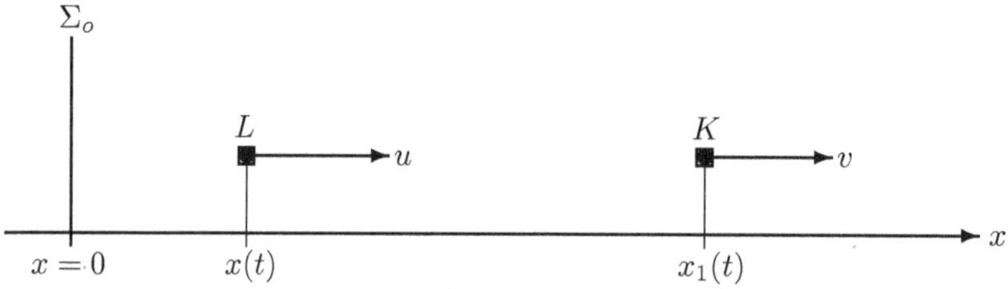

Fig. 7 Relative velocity. Let the objects L and K in inertial system Σ_o have velocities $u = dx/dt$ and $v = dx_1/dt$, respectively. Then the object L approaches in Σ_o to object K with relative velocity $w = u - v$. Is e.g. $v = 0.8\,c$ and $u = 0.9\,c$, then the object L approaches with relative velocity $w = 0.1\,c$ to the object K.

Now we add a second object L, that is moving as $x = x(t)$ on the x-axis with velocity $u = dx(t)/dt$. We may ask for the *relative velocity* w, that we define as[7]

$$w := \frac{d}{dt}\left[x(t) - x_1(t)\right] , \tag{7}$$

that describes the approach or recession rate of object L to the object K on the x-axis, Fig. 7,

$w = u - v \quad \longleftrightarrow \quad u = w + v\ .$	Relative velocity w of two objects in *one* inertial system	(8)

This relative velocity w is according to the definition (7) nothing else than the *definition of the temporal change of a coordinate difference.*

Two steel balls approaching against each other with velocities $v = 0.9\,c$ and $u = -0.9\,c$ have a relative velocity $w = -0.9\,c - 0.9\,c = -1.8\,c$ (negative since they are approaching). Two approaching light waves in a system Σ_o lead to a relative velocity of $w = -c - c = -2\,c$, that means an amount of twice the speed of light.

[7]The meaning of the symbol ':=' is 'equal by definition'.

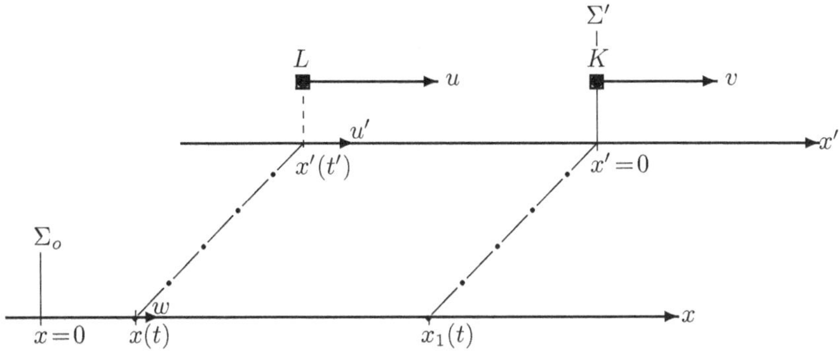

Fig. 8 Addition theorem of velocities. The body K and an "object" L might have velocities $v = dx_1/dt$ and $u = dx/dt$, respectively, in the inertial system Σ_o. Then the body L approaches to the body K in Σ_o with relative velocity $w = u - v$. This velocity w is in general different from the velocity u', that an observer sitting on the body K measures for the approaching body L. The dash-dotted lines connect points representing the same event.

This represents not a strange peculiarity. The relative velocity w is *not* the speed of an energy (or signal) transfer.

We must strictly separate the relative velocity w from the velocity u' of object L, that an observer measures who is sitting on a body K, i.e. resting relative to K, where the body K may have a velocity v relative to the system Σ_o. The connection between the velocities u' and u is subject of the addition theorems of velocities, that will be discussed in Fig. 8. This is always a statement including two inertial systems, therefore the laws of coordinate transformations are required. This is the subject of the next Section.

3.2 Two *Inertial Systems*

Now we add a second inertial system Σ' that has a constant velocity $\mathbf{v}$ with respect to Σ_o.

In Σ', we also take advantage of the there resting normal lengths and normal clocks as in Σ_o. There exist the same atoms and molecules as measuring devices. For defining the spatial coordinates $(\mathbf{x}')$ of points $P(\mathbf{x}')$ we proceed as in Σ_o, and we fix the zero point $O'_3(0, 0, 0)$:

> The Cartesian coordinates $(\mathbf{x}')$ of a point $P(\mathbf{x}')$ in Σ' are determined as measured values of its distance from the coordinate origin $O'_3(0, 0, 0)$.

Furthermore, we assume that in Σ' sufficient many resting clocks are tightly distributed. Subject to a still missing synchronisation, we will define as time coordinates t' the measured value of these clocks:

The time coordinate t' of an event $E(\mathbf{x}', t')$ is determined in Σ' as measured value of the time measurement at the place $P(\mathbf{x}')$.

3.2.1 Coordinate Transformations

Each event E can be described both in Σ_o as well as in Σ',

$$E(\mathbf{x}, t) = E(\mathbf{x}', t') \ . \tag{9}$$

The relation between primed and unprimed coordinates is called *coordinate transformation.* We take into account the relative velocity $\mathbf{v}$ of Σ' with respect to Σ_o as parameter, and we also write down the inverse of the relation,

$$\left.\begin{array}{lcl} x' = f_1(\mathbf{x}, t, \mathbf{v}) \,, & & x = \varphi_1(\mathbf{x}', t', \mathbf{v}) \,, \\ y' = f_2(\mathbf{x}, t, \mathbf{v}) \,, & \longleftrightarrow & y = \varphi_2(\mathbf{x}', t', \mathbf{v}) \,, \\ z' = f_3(\mathbf{x}, t, \mathbf{v}) \,, & & z = \varphi_3(\mathbf{x}', t', \mathbf{v}) \,, \\ t' = f_4(\mathbf{x}, t, \mathbf{v}) \,, & & t = \varphi_4(\mathbf{x}', t', \mathbf{v}) \ . \end{array}\right\} \quad \text{General coordinate transformation} \tag{10}$$

As *initial condition* we take an event O, so that the coordinate origin O_3' at time $t' = 0$ coincides with O_3 at time $t = 0$,

$$\text{Event } O: \quad \left.\begin{array}{lcl} x' = 0 \,, & & x = 0 \,, \\ y' = 0 \,, & \longleftrightarrow & y = 0 \,, \\ z' = 0 \,, & & z = 0 \,, \\ t' = 0 \,, & & t = 0 \ . \end{array}\right\} \quad \text{Initial condition} \tag{11}$$

3.2.2 Addition Theorem of Velocities

The general form of the coordinate transformation (10) has an immediate consequence, an *addition theorem of velocities*, that will be discussed at a few further occasions.

For a body K, that moves in Σ_o according to $x_1 = x_1(t)$ in x-direction, we measure there the velocity $v = dx_1(t)/dt$. This body may realise an inertial system Σ'. In Σ_o, we measure for a second "object" L the path $x = x(t)$ in x-direction with velocity $u = dx(t)/dt$. For the same object L, one finds a path $x' = x'(t')$ in the system Σ' and therefore a velocity $u' = dx'(t')/dt'$, cp. Fig. 8,

$$\begin{array}{lll} & \text{Body } K & \text{Body } L \\ \Sigma_o: & v = \dfrac{dx_1}{dt}\,, & u = \dfrac{dx}{dt}\,, \\ \Sigma': & v' = \dfrac{dx_1'}{dt'} := 0\,, & u' = \dfrac{dx'}{dt'}\,. \end{array}$$

The connection between u' and u is now represented by transformation formulas (10). We restrict ourselves on a motion in the x-direction and find using the chain rule of differentiation

$$\begin{aligned} u' = \frac{dx'}{dt'} = \frac{dx'}{dt}\left(\frac{dt'}{dt}\right)^{-1} &= \frac{d}{dt}\, f_1\big(x(t), t, \mathbf{v}\big)\left(\frac{d}{dt}\, f_4\big(x(t), t, \mathbf{v}\big)\right)^{-1}\,, \\ &= \left(\frac{\partial f_1}{\partial x}\frac{dx}{dt} + \frac{\partial f_1}{\partial t}\right)\left(\frac{\partial f_4}{\partial x}\frac{dx}{dt} + \frac{\partial f_4}{\partial t}\right)^{-1}\,, \end{aligned}$$

and with $u = dx/dt$,

$$u' = \left(\frac{\partial f_1}{\partial x}u + \frac{\partial f_1}{\partial t}\right)\left(\frac{\partial f_4}{\partial x}u + \frac{\partial f_4}{\partial t}\right)^{-1}. \qquad \text{Addition theorem of velocities for transformation (10)} \qquad (12)$$

This Eq. (12) represents the general form of an addition theorem of velocities, which is required by the general transformation (10). Only with EINSTEIN's (1905a, b, s. 1955) Special Relativity Theory the notion of an addition theorem of velocities was introduced in physics. This theorem will play a significant role in the following.

4 Special Coordinate Transformations

In the following, the connections of two inertial systems are formulated in such a way that we can cover both the classical as well as the relativistic notion of space-time.

We specialise on Cartesian coordinates oriented with parallel axes, and we consider only motions of the system Σ' with constant velocity v in one axis direction of Σ_o, here taken as x-direction.

We call them *special coordinate transformations*.[8]

The unprimed coordinates $(\mathbf{x}, t)$ are used for the system Σ_o. Inertial systems with velocities v and u in x-direction of Σ_o are called Σ' with primed coordinates $(\mathbf{x}', t')$ and Σ'' with $(\mathbf{x}'', t'')$, resp.

In the next Section, we derive precisely which conclusions we get from these coordinate transformations on the measured lengths and oscillation periods of moving rulers and clocks. The following first simple experience will be now presented, that was originally formulated by H. A. LORENTZ:

[8]The notion 'special' is used in this sense for 'special LORENTZ transformations'. In the destination 'Special Relativity Theory' it has another meaning. There we distinguish it from the 'General Relativity Theory', the theory of space, time and gravitation, s Chap. 13.

For a bar that is moving *transverse* to its linear extension, we measure the same length in the moving as in the rest system. Then it follows: (13)

$$y' = y \ , \quad z' = z \ . \tag{14}$$

Indeed, a bar should be resting on the y'-axis of Σ' with end coordinates 0 and y'. These coordinates are determined as measured values of multiples of our length standard. Therefore, in system Σ' a length y' is measured. With Eqs. (11) and (14), the bar in Σ_o has the same end coordinates 0 and $y = y'$ in fact independently on the time, at which these coordinates are measured. Therefore, the bar will have the same length $y = y'$. There are no observations known that put into question these observations.

Between the primed and unprimed coordinates of an event E, it holds

$$E(x, t) = E(x', t') \ . \tag{15}$$

We have to specify the connections

$$\left.\begin{array}{lcl} x' = f_1(x, t, v) \ , & \longleftrightarrow & x = \varphi_1(x', t', v) \ , \\ t' = f_4(x, t, v) \ , & & t = \varphi_4(x', t', v) \ . \end{array}\right\} \tag{16}$$

A point P_o' resting in Σ' should have the unchanged coordinate x_o'. At time t_o in Σ_o we observe, that it should be found at position x_o, i.e.

$$x_o' = f_1(x_o, t_o, v) \ .$$

We take into account, that the relative velocity v is constant. In Σ_o the point P_o' will be found at time $t_o + t$ at position $x_o + v\,t$, while its coordinate x_o' in Σ' remains unchanged,

$$x_o' = f_1(x_o + v\,t, t_o + t, v) = f_1(x_o, t_o, v) \ . \tag{17}$$

For an arbitrary x_o and t_o, we solve this equation by the ansatz

$$x' = f_1(x, t, v) = f(x - v\,t, v) \ . \qquad \text{for } v = \text{const.} \tag{18}$$

4.1 *Definition of Simultaneity*

Now we note that the function $f_4(x, t, v)$ of transformation (16) includes a rule, when the clocks of Σ' are initialised. Starting from a synchronisation of clocks in system Σ_o a clock resting in Σ', that is found in system Σ_o at time $t = 0$ at position x, has according to Eq. (16) a time coordinate $t' = f_4(x, 0, v)$. This relation is denoted as *general synchronisation function* Ω,

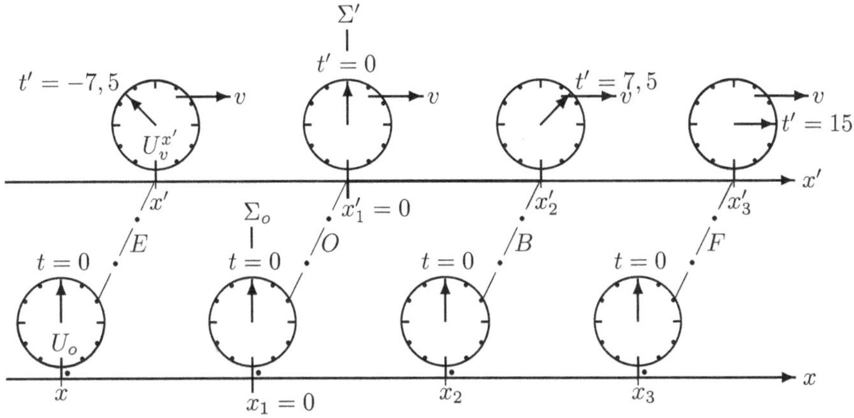

Fig. 9 The definition of simultaneity in Σ' by the fixation of an arbitrary, linear synchronisation function $\tau(x, v) = f_4(x, 0, v) = \theta(v)\,x$. The dash-dotted lines connect again points in the figure, that represent identical events, here the events E, O, B, F.

$$\Omega(x, v) := f_4(x, 0, v) \ . \qquad \text{General synchronisation function} \tag{19}$$

We have chosen the coordinates as measured values of space and time measurements. The coordinate transformations (16) contain two very different statements: at first a *definition*, namely, the just introduced, in principle freely eligible synchronisation function (19) for clocks in Σ', and otherwise a *physical statement* about the behaviour of moving and resting rulers and clocks, that we shall discuss in Sect. 5. There the following should be noted: As long as we use local measurements, so that only the time at a special place is of interest, all measurements are independent on the definition of simultaneity. We will discuss this extensively in Chap. 4, Sect. 7.

For keeping the coordinate transformations simple, we restrict ourselves on synchronisation functions, that are linear in the x-coordinate.

Among all possible clock adjustments $\Omega(x, v)$ in systems Σ' we consider in the following only in x linear functions. Therefore, we write instead of $\Omega(x, v)$ the function $\tau(x, v)$ with a *synchronisation parameter* $\theta(v)$, s. Fig. 9. Similar presentations for the comparison of clocks in different inertial systems can be found, e.g. at Arzeliès (1966),

$$f_4(x, 0, v) = \tau(x, v) = \theta(v)x \ . \qquad \text{Linear synchronisation function} \tag{20}$$

4.2 The Linear Transformation Formulas

If we admit the possibility that the length of a bar and the period of a clock depend on its velocities, then we will assume that these properties satisfy in the Sect. 3, rule (2), formulated general experience of the *homogeneity* and *isotropy* of space and time, again only assumed in the system Σ_o:

> In system Σ_o holds: The quotient of a length of a bar in motion and at rest and of a clock period in motion and at rest does not depend on the coordinates x and on time t, and not on the bar length and the period itself, but only on its relative velocity. (21)

When also the *definition* of simultaneity with the linear synchronisation function (20) is supposed to be subject of the homogeneity requirement, then it follows from this statement (21), that the transformations (16) should be linear in the coordinates (x, t), s. Problem 2.

With (14) and (18), we then find

$$\left.\begin{array}{ll} x' = k\,(x - v\,t)\,, & x = \dfrac{q}{\Delta}\,x' + \dfrac{v\,k}{\Delta}\,t'\,, \\ & \longleftrightarrow \\ t' = \theta\,x + q\,t\,, & t = -\dfrac{\theta}{\Delta}\,x' + \dfrac{k}{\Delta}\,t'\,, \\ \multicolumn{2}{c}{y' = y\;,\quad z' = z\,,} \\ \multicolumn{2}{l}{\Delta = k\,(v\,\theta + q)\;,} \\ \multicolumn{2}{l}{k = k(v)\;,\quad q = q(v)\;,\quad \theta = \theta(v)\;.} \end{array}\right\} \quad \text{Special coordinate transformations} \qquad (22)$$

The Eqs. (22) still contain 3 parameters, $k(v)$, $q(v)$ and $\theta(v)$ that include all the secrets of classical and relativistic physics.[9]

We yet remark: The necessity to justify the linearity of the transformation formulae were already indicated by Frank and Rothe (1912). There it was formulated purely geometrical by postulating that parallel lines should transform into parallel lines.

4.3 The Addition Theorem of Velocities

Using the linear transformations (22), i.e. $f_1 = k(x - v\,t)$ and $f_4 = \theta\,x + q\,t$, we get with the general form (12) of the addition theorem of velocities the equation

---

[9]As complement to Eq. (18), we also add
$f_1(x_o + v\,t, t_o + t, v) = f(x_o + v\,t - v(t_o + t), v) = f(x_o + v\,t - v\,t_o - v\,t), v) = f(x_o - vt_o, v) = f(x_o, t_o, v)$.

$$u' = k\,\frac{u - v}{\theta\,u + q} \quad \longleftrightarrow \quad u = \frac{q\,u' + v\,k}{-\theta\,u' + k}\,. \qquad \text{Addition theorem of velocities for transformation (22)} \tag{23}$$

A special case is instructive: The body L is assumed to rest in Σ_o, i.e. its velocity is $u = u_o = 0$. The observer resting in system Σ' measures for L a velocity u'_o that is the velocity of system Σ_o,

$$u'_o = \frac{-k\,v}{q}\,. \qquad \text{In } \Sigma' \text{ measured velocity of system } \Sigma_o \tag{24}$$

The observer resting in system Σ_o measures a velocity v for the system Σ', but the observer resting in system Σ' measures a velocity $u'_o = \dfrac{-k\,v}{q}$ of Σ_o.

We shall see how this apparent asymmetry between systems Σ_o and Σ' will be regulated by the *definition* of simultaneity in system Σ'.

5 Moving Rulers and Clocks

The results of the considerations in this Section play a key role for the further exposition of the theory.

Here we transform the instruments of our measurements, the rulers and clocks, to objects of measurements. We compare the length l_v of a moving bar with the length l_o of the same bar at rest. We furthermore compare the period T_v of a moving clock with the period T_o of the same clock at rest. We shall explore how this relation is enclosed in the properties of the coordinate transformations (22).

For our calculations, we take into account the initial conditions in Eq. (11), i.e., for $(x = 0,\ t = 0)$ it holds also $(x' = 0,\ t' = 0)$.

5.1 Moving and Stationary Rulers

For measuring the length of a bar in a reference system, we need the coordinates of its end points. But there is an essential difference whether the bar is at rest with respect to this system or whether it is in motion with respect to the measuring observer. In the first case it does not play any role, at which time we determine the coordinates of its end points, they remain constant.

We take as *rest length* l_o of the bar the length measured by an observer relative to which the bar is at rest. It represents the difference of the coordinates of its end points. We establish that we can use in all inertial systems the same length normals, their rest length is independent from the reference system used:

The rest length l_o of a bar is an invariable quantity characterising the material. (25)

Especially in the relativistic space-time, we define

The rest length l_o of a bar is denoted as eigen length. (26)

A different consideration is necessary for determining the length of a moving bar. Its end points on the x-axis, x_1 and x_2, are changing with time, $x_1 = x_1(t)$ and $x_2 = x_2(t)$. It makes no sense to measure the left end point at a time t_1, and the right one some times later, at a time t_2, and then to define the difference $x_2(t_2) - x_1(t_1)$ with $t_2 \neq t_1$ as its length. We shall follow the generally accepted *definition*:

The length l_v of a bar moving with velocity v is defined as difference $x_2(t) - x_1(t) = l_v$ of the coordinates of its end points at one and the same time t. (27)

This definition implies a *definition of simultaneity*!

The 'moving ruler' l_v is in no way a fixed concept.

In system Σ_o, we have based the simultaneity on the hypothesis of isotropy. With this statement also the moving length in Σ_o is well defined, s. Problem 1. For all other systems Σ', we have connected the simultaneity with the linear synchronisation function $\tau(x, v) = \theta(v)\, x$ to the synchronisation of clocks in Σ_o.

A length l_v measured in the moving system Σ' will in general depend on the choice of this function $\theta(v)$. With the statement that a moving bar is contracted with respect to his eigen length, we should be very careful.

A bar might be resting with his left end point x_1' at the origin of the x'-axis of system Σ', so that his eigen length l_o can be measured as coordinate of his right end point, $x_1' = 0$, $x_2' = l_o$. In system Σ_o, we observe the end points in uniform motion. We write $x_1 = x_1(t)$ and $x_2 = x_2(t)$.

We intend to determine the length $l_v = x_2(t) - x_1(t)$ of the bar that is moving in system Σ_o with velocity v. To this end, we need the position of the end points $x_1 = x_1(t)$ and $x_2 = x_2(t)$ at one and the same time t in Σ_o, e.g. at time $t = 0$, s. Fig. 10.

The measured values for the left end point ($x_1' = 0$, $t' = 0$) in Σ' and ($x_1(0) = 0, t = 0$) in Σ_o are just given by the initial conditions (11).

The right end point has by definition at the time t' in Σ' the coordinate $x_2' = l_o$. We measure for the right end point in Σ_o at time $t = 0$ the coordinate $x_2(0)$, then it holds because of the transformation $x' = k(x - v\,t)$ in (22) the equation $l_o = k\, x_2(0)$.

For the positions of the bar at the same time Σ_o, we get, therefore,

$$\Sigma_o:\ t = 0\,,\quad \begin{array}{l} x_1(0) = 0\,, \\ x_2(0) = \dfrac{l_o}{k}\,, \end{array} \longrightarrow\ l_v = x_2(0) - x_1(0) = \frac{1}{k}\, l_o\,, \quad \Big\}\ \text{Length } l_v \text{ of a moving bar in } \Sigma_o \tag{28}$$

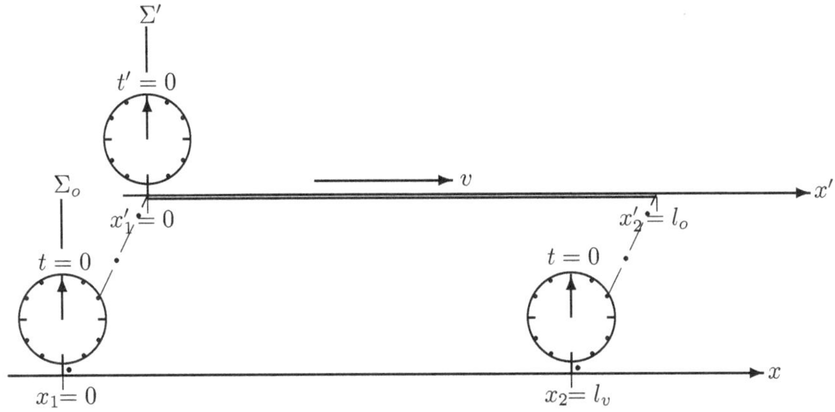

Fig. 10 The measurement of the length l_v of a moving bar (explanation in the text). The dash-dotted lines connect again points in the figure that represent the same event.

$$\Sigma_o : \quad \frac{\text{length of a moving bar } \Sigma_o}{\text{eigen length of the bar}} = \frac{l_v}{l_o} = \frac{1}{k} . \tag{29}$$

Equation (29) derived in Σ_o represents a physical interpretation of the parameter k in the coordinate transformation (22). The shape of the function $k = k(v)$ can be found according to Eq. (29) by precision measurements in system Σ_o.

The inverse question is also interesting, when the bar with eigen length l_o, resting on the x-axis of the system Σ_o will be now observed from the system Σ'. Since the following considerations are not influenced by this question, we shall consider this case in Chap. 2, Sect. 3.

5.2 *Moving and Stationary Clocks*

As clock we understand a system in periodic motion. The pointer of the time measurement t counts oscillations.

To avoid misunderstanding, we first state the following:

The clocks resting in a certain inertial system are all gauged, i.e. they show the same time t. The number of oscillations connected with the advance by Δt depends on the construction of the clock. We remind on the discussion in Sect. 1: The time interval of $\Delta t = 1$ s requires 9 192 631 770 oscillations of a certain spectral line of the Caesium isotope ^{133}Cs. The spring-driven pocket watch will need maybe 2 or else, 20 oscillations for $\Delta t = 1$ s.

We understand under the *eigen period* T_o of a clock that oscillation period that is measured with a normal clock resting relative to this clock. We assume, that in an inertial system we have a sufficient number of these normal clocks. The eigen period is independent from the reference system, where it is measured:

The eigen period T_o of a clock represents an invariant quantity. (30)

Especially, in the relativistic space-time it is defined:

The time of a clock in its own rest system is denoted as *eigen time.* (31)

In the following, we must admit that the oscillation period of a clock may change if it is moving.

We ask, whether a moving clock U_v shows a different time as the stationary clock.[10]

We want to measure the period T of a clock U_v moving with respect to a reference system Σ_o, i.e. we want to determine its time t with our normal clocks, s. Fig. 11.

For determining the time of a clock U_v in motion in an inertial system Σ_o, we need two in Σ_o stationary normal clocks, where the clock U_v is passing by, s. Fig. 11.

When the period of the clock U_v changes, then there changes the displayed or the 'running' time. When the period becomes longer, then the pointer runs more slowly. When we compare the measured time of a moving clock with the same clock at rest, we measure times t' and t, and on the other hand the periods of these clocks are T_v and T_o, then it holds:

The quotient of two time intervals $\Delta t'$ and Δt is inversely proportional to the corresponding quotient of its periods T_v and T_o,

$$\frac{\Delta t'}{\Delta t} = \frac{T_o}{T_v} . \tag{32}$$

The measured time is inversely proportional to the periods. There it is assumed that the layout of the clock remains unchanged.

A clock U_v is assumed to be stationary at the origin of system Σ', i.e. it shows at the position $x' = 0$ the time t'. We observe this clock from system Σ_o.

The initial conditions (11) tell us that for $(x = 0,\ t = 0)$ it holds also $(x' = 0,\ t' = 0)$, i.e. the clock U_v resting at the origin of Σ' shows the same pointer position as the clock U_o^0 resting at the origin of Σ_o, when it just passes by s. Fig. 11,

$$\textbf{First time measurement} \quad E_o : \left. \begin{array}{l} \Sigma_o : x = 0\,,\ t = 0\,, \\ \Sigma' : x' = 0\,,\ t' = 0\,. \end{array} \right\} \tag{33}$$

The clock U_v is located after the time t at the position $x = v\,t$ in Σ_o. We compare the pointer position t' of U_v with the clock resting at $x = v\,t$ in Σ_o. According to Eq. (22), it holds $t' = \theta\,x + q\,t$, and with $x = v\,t$,

[10] Indeed, there exists an additional effect: The period of a clock depends on the strength of the gravitational field at its place. However, here we neglect all gravitational effects.

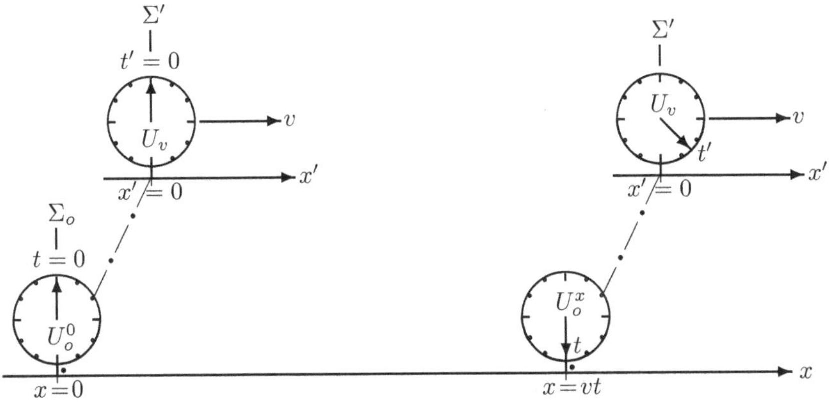

Fig. 11 The pointer position t' of a clock U_v resting in Σ' is compared with the pointer position t of a clock U_v resting in Σ_o and just passing by. The left part of the figure represents the initial conditions (11). Dash-dotted lines again connect points in the illustration that represent the same event.

Second time measurement $$E : \left.\begin{array}{l} \Sigma_o : x = v\,t\,,\ t\,, \\ \Sigma' : x' = 0\,,\ \ t' = (v\,\theta + q)\,t\,. \end{array}\right\} \tag{34}$$

It follows

$$\Sigma_o : \frac{\text{difference of pointer positions of } one \text{ moving clock in } \Sigma_o}{\text{difference of pointer positions of } two \text{ in } \Sigma_o \text{ resting clocks}} = \frac{t'}{t} = v\,\theta + q\,. \tag{35}$$

The period T is inverse to the pointer position t, s. Eq. (32). In terms of the proper period T_o of a clock and the period T_v of the same moving clock we can write instead of Eq. (35) also

$$\Sigma_o : \frac{\text{period of a moving clock in } \Sigma_o}{\text{proper period}} = \frac{T_v}{T_o} = \frac{1}{v\,\theta + q}\,. \tag{36}$$

According to Eqs. (35) and (36), respectively, the parameter combination $v\,\theta + q$ has a physical meaning and can be measured in precision experiments in Σ_o.

Once again we remark that up to this point, we do not require an equal status of both inertial systems *from the outset*, but an imaginable special role of system Σ_o is a priori admissible.

In Problem 2, we conclude from the postulate of homogeneity and isotropy in the inertial system Σ_o, s. Sect. 4.2, the linearity of the coordinate transformations[11], i.e. transformations of type (22). Here we conclude from this linearity of transformations

[11]The standard proof for the linearity of coordinate transformations is shown in Chap. 9, Sect. 1, cp. e.g. also RINDLER (1977), FOCK (1964), WEYL (1952). It follows from EINSTEIN's relativity principle, s. Chap. 2, Sect. 1. With the postulate of a universal constancy of the speed of light EINSTEIN introduced also a definition of simultaneity, and indeed with a linear synchronisation

(22) for the quotient of the moving to the eigen length and for the quotient of moving to proper periods expressions of the form (29) and (36), that only depend on the velocity of this motion:

> The coordinate transformations are exactly then linear, when for a linear synchronisation (20) the length ratios of moving and resting rulers and the periods of clocks fulfil the postulate of homogeneity and isotropy (21). (37)

Until now we have checked which statements can be made for the parameter $k(v)$, $q(v)$ and $\theta(v)$ describing the transformations (22) to moving lengths and clocks.

How we get these parameters?

Can we assume a priori a specific behaviour of moving lengths and clocks? Do we have a fundamental principle that allows a definitive answer. When such a principle leads to consequences, that contradict other evidence, what can be our basis?

The last authority in physics is the experiment. As long as we stay on the level of classical physics with its relative low precision measurements, we will not find contradictions with simple assumptions, Sects. 1 and 2.

There is a remarkable thought experiment, the so-called light clock. A change of lengths and periods of moving systems is there naturally suggested. This will be discussed Chap. 4, Sect. 2, when we interpret the law of time dilatation, EINSTEIN's experimentum crucis of the Special Relativity Theory (SRT). In Chap. 4, Sect. 1, we shall discuss the key experiment of the SRT, the MICHELSON experiment. In Chap. 4, Sect. 6, we will further review more recent precision experiments. These observations will confront us directly with the peculiarities of the space-time that is called relativistic, and it will require more reflections. The interested reader can furthermore consult Chap. 12, Sect. 1 for a conceivable advanced justification, cp. also GÜNTHER (2000).

EINSTEIN (1905a, b) has justified the determination of parameters $k(v)$, $q(v)$ and $\theta(v)$ by a unique principle, his relativity principle, s. Chap. 2, Sect. 1. This has caused a complete reformulation of the classical physics. We will try to follow a less abstract way to come to the same goal, s. Chap. 2, Sect. 2.

---

recipe with the prescription of Eq. (20) for the function θ in Eq. (103) as will be shown in Chap. 4, Sect. 4.

Chapter 2
The Relativity Principle

1 EINSTEIN's Relativity Principle

The principle of the indistinguishability of inertial systems, or the equivalence of all of them, stands at the beginning of physics when it became an exact science.

The deep crisis of physics at the beginning of the twentieth century that arose from contradictions between classical mechanics and at this time new electrodynamics could be solved by EINSTEIN (1905a, b) with a new formulation of the whole physics that is based on a single postulate, his *relativity principle*[1]:

> "1. The laws by which the states of physical systems undergo change are not affected, whether these changes of state be referred to the one or the other of two coordinate systems in translatory motion.
> 2. Any ray of light moves in the 'stationary' system of coordinates with a determined velocity V whether the ray be emitted by a stationary or a moving body. Hence (38)
> $$\text{velocity} = \frac{\text{light path}}{\text{time interval}}\,,$$
> where 'time interval' is to be taken in the sense of the definition in Chap. 1, Sect. 1."

EINSTEIN's (Fig. 1) 'paragraph one', Sect. 1, the 'definition of simultaneity' contains the regulation for synchonisation of the clocks with the help of the speed of light, as we described this in Chap. 1, Sect. 3, Eq. (6), for the system Σ_o, see also Fig. 4.

The distinctive feature of EINSTEIN's procedure consists of the use of one and the same light signal for synchronising all clocks in all inertial systems. In Chap. 1, Sect. 3, we have applied EINSTEIN's synchronisation algorithm only for one specific system Σ_o that is distinguished in this way.

[1] EINSTEIN uses the term 'coordinate system' instead of the term preferred in later years, that we also use, 'reference system' or 'frame'.

H. Günther and V. Müller, *The Special Theory of Relativity*,
https://doi.org/10.1007/978-981-13-7783-9_2

Fig. 1 Albert Einstein, * Ulm 14.3.1879, † Princeton 18.4.1955.

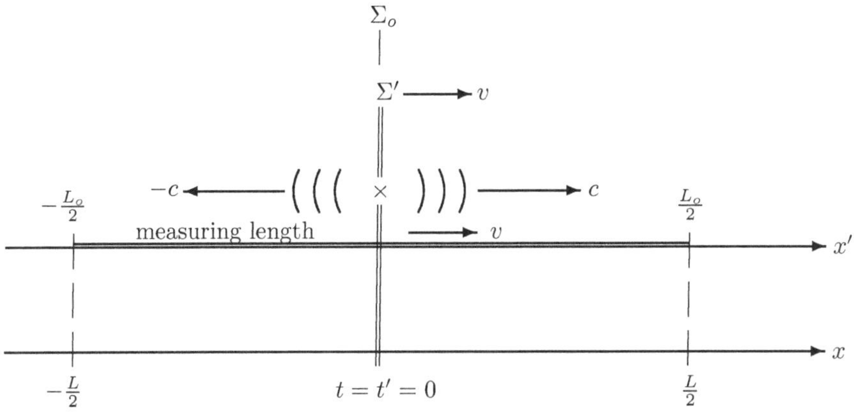

Fig. 2 The relativity of EINSTEIN's simultaneity is a consequence of the synchronisation of clocks in all inertial systems by means of the principle of the universal constancy and therefore also of the isotropy of the speed of light. The same signal can synchronise in principle the clocks in all inertial systems.

By our experience, the strictly deductive path to derive the Special Relativity Theory according to EINSTEIN, the conventional way to present SRT, leads for the unprepared reader quite often to irritations in applying the theory. The reason is obviously that the *definition* of simultaneity by means of the invariant speed of light in all inertial systems is immediately built into the axiomatic exposition of the theory. The relativity of simultaneity is then fixed from the beginning *per definitionem*—a brilliant anticipation of the LORENTZ group.

In distinction to EINSTEIN's (1923, 1955) own presentation, the relativity of simultaneity is time and again *erroneously* taken as a physical law, although this relativity *is fixed* in EINSTEIN's axiomatics obviously *before* all considerations by means of a *definition* of simultaneity, cp. however Chap. 4, Sects. 3 and 4 as well as Chap. 8, Sect. 9, s. also REICHENBACH (2003), THIRRING (1992), GÜNTHER (2000, 2007, 2010). In Fig. 2 we have shown an example for this.

A measuring rod might be resting in the system Σ' that has a velocity v in x-direction as seen from Σ_o. In the picture, we sketch the emission of a light signal, the event $(t = t' = 0, x = x' = 0)$.

According to EINSTEIN one postulates such a synchronisation of clocks in all inertial systems Σ', so that the measured speed of light has one and the same value. In other words, if the clocks are synchronised in all inertial systems with the speed of light, then we measure in all inertial systems a speed of light with the same numerical value of $c = 299\,792\,458\,\mathrm{ms}^{-1}$, independent from the inertial system in which the light signal is emitted.

This must appear quite curious to the beginner. If we replace in Fig. 2 the atomic light by balls expelled with a mechanical precision catapult, then we can also synchronise the clocks with these balls, cp. footnote on Chap. 1, Sect. 3.1.2. The velocity that we measure for these balls depends not only on the catapults, but also whether the catapult is stationary in this frame or not. Therefore, in order to synchronise the clocks we must use identical catapults that are resting in the corresponding inertial systems. The only exception is light. There in principle only one light signal in all inertial systems is sufficient since, according to EINSTEIN's principle, its velocity does not depend on the state of motion of the emitting source.

Still we should remark: The propagation speed of sound waves is independent on the state of motion of the source, if the observer is at rest with respect to the propagation medium of the sound. The motion of this medium can now be observed by frequency measurements, s. Problem 19. Because of the time dilatation to be discussed extensively in Chap. 4, Sect. 2, this is not possible for light. This is the difference. The fictive notion of an aether[2] as medium for light propagation had to be rejected. The only remaining of aether is nothing else than the physical vacuum.

Considered in Σ' the light gets to both end points of the measuring rod per definition at the same time—the Σ'-clocks are just set up in this way. Observed in system Σ_o, the light will approach to the right end point with velocity $c - v$ because of the motion of the measuring track, and to the left one with $c + v$, since both velocities are measured in the same system Σ_o, and therefore their velocities must be added according to Eq. (8). When the length of the moving measuring rod in Σ_o might be denoted by L, so the light signal arrives at $t_1 = L/[2(c + v)]$ to the left end and at $t_2 = L/[2(c - v)]$ to the right end.

Therefore in Σ_o holds

$$t_2 - t_1 = \frac{L\,v}{c^2 - v^2} \ . \tag{39}$$

While the light signals get at the same time to the end points of the measuring track as observed in Σ', there the track is at rest, and this equality is a consequence of the isotropy of the light propagation, the observer in Σ_o finds that the light arrives at first at the left end point at time t_1, and later at the right end point at time t_2, retarded according to Eq. (39), i.e. at $t_2 = t_1 + L\,v/(c^2 - v^2)$. In both the events, arrival of the signal at the end points of the measuring rod is simultaneous in Σ', but not in Σ_o.

We summarise the results of EINSTEIN's principle:

1. It implies a *selection principle*: Among all possible definitions of simultaneity it fixes *one* synchronisation rule.
2. By means of this definition the *law* of light propagation is formulated, and it determines the mathematical structure of space-time.
3. The mathematical form of physical laws are subject to this space-time structure.

The prospective theoretical physicist has to study carefully this reformulation of the fundamental equations of physics on the basis of EINSTEIN's postulate. The three-dimensional space and the time are connected due to the famous work of

---

[2]The former spelling used by FITZGERALD, LORENTZ and EINSTEIN was 'ether'.

MINKOWSKI, s. LORENTZ (1952), into a four-dimensional space-time continuum. This leads to a mathematically elegant and highly effective formulation of the relativity principle that will be presented in Chap. 9, Sect. 1.6. The new mathematical form of mechanics and electrodynamics is discussed in Chap. 9, Sects. 2 and 3. Without giving any evaluation, we only mention among the many excellent presentations of the SRT the textbooks of JOOS (1987), TAYLOR and WHEELER (1966), NOLTING (2016).

In the following chapters, we will follow a less abstract way for solving the relativity problems. Therefore, we will discuss at first only the characterisation of the space-time and in the following the equations of physics. Here we still study only the mechanics and exclude preliminarily electrodynamics. In Chap. 9, Sect. 3 we will see that the inclusion of electrodynamics will not require additional considerations to the relativity problem, since already the classical electrodynamics fulfils the relativity principle, as EINSTEIN (1905a, b) has shown in his exceptional paper.

2 Elementary Relativity

From EINSTEIN's relativity principle now we shall select the part that only regulates the definition of simultaneity in inertial systems.

To this aim, we select at first one special homogeneous and isotropic inertial system Σ_o, where we know the speed of light and where we can synchronise all clocks, Chap. 1, Sect. 3. As we have seen, the structure of space-time is fixed by three parameters, $k(v)$, $q(v)$ and $\theta(v)$, with quite different physical meaning to be underlined:

$$\Sigma_o: \frac{l_v}{l_o} = \frac{1}{k(v)} . \tag{40}$$

The quotient of a moving length and the length at rest of a bar in Σ_o defines the parameter $k(v)$.

$$\Sigma_o: \frac{T_v}{T_o} = \frac{1}{v\,\theta(v) + q(v)} . \tag{41}$$

The quotient of the period of a moving clock and its proper period in Σ_o defines the combination $v\,\theta(v) + q(v)$ of the parameters $\theta(v)$ and $q(v)$.

With the parameter $\theta(v)$ we *define* the synchronisation of clocks in systems Σ'. (42)

In principle, we can freely choose—and this represents the logical simplicity of our approach—which convention we are taking for the synchronisation of clocks in systems Σ'. Besides from special cases, that we will especially remark, we follow an excellent advice of POINCARÉ (1898) to take this freedom of choice to fix the parameter $\theta(v)$ in such a way that the transformation laws become especially simple and symmetric. To this aim, we formulate a simple symmetry principle that alone fixes the procedure for synchronisation of clocks in inertial systems Σ', s. GÜNTHER (2000), it is called:

The elementary relativity principle[3]

When an observer in the initially distinguished reference system Σ_o for the inertial system Σ' has measured the velocity v, then he should in Σ' set into operation stationary normal clocks, so that an observer stationary in Σ' should measure a velocity $-v$ for the reference system Σ_o. (43)

If the clocks in system Σ_o are started up, then the elementary relativity principle is only a *selection principle* for the synchronisation of clocks in systems Σ'.

The observer resting in Σ_o measures for the system Σ' a velocity v. In Eq. (24), we have calculated the velocity u'_o that an observer resting in Σ' measures for the system Σ_o, namely, $u'_o = -k\,v/q$.

The elementary relativity principle requires simply $u'_o = -v$, i.e. with Eq. (24)

$$q = k\ . \qquad \text{Elementary relativity principle} \qquad (44)$$

In this way, we have devised the method by which we can establish the classical structure of space-time in Chap. 3, Sects. 1–2. In the following Chap. 4, Sects. 1–8, we will consider the relativistic case. This will be done in three steps:

We assume that because of sufficient accurate measurements in system Σ_o the quotient of lengths in motion and at rest l_v/l_o, and the quotient of periods T_v/T_o is known for our space-time. This means:

1. The parameter $k(v) = \dfrac{l_o}{l_v}$ is determined from precision measurements in the reference system Σ_o.

2. The combination of parameters $v\,\theta(v) + q(v) = \dfrac{T_o}{T_v}$ is determined by precission measurements in the reference system Σ_o. (45)

In the coordinate transformations (22), there remains one free parameter θ, for which we get from Eq. (41) $v\,\theta = T_o/T_v - q$. The parameter θ defining the simultaneity in systems Σ', is then fixed by the elementary relativity principle, i.e. with $q = k$ according to Eq. (44), it follows with $k = l_o/l_v$ from (40) the relation

[3]In literature, it is conventionally called 'reciprocity principle'. Since we will formulate in Sect. 3, a reciprocity theorem, that has nothing to do with this reciprocity principle, we use here the notion 'elementary relativity'. Also IGNATOWSKI (1910) used this term when he writes: 'So it must apparently hold: $q' = -q, \ldots$'. Furthermore, it has to be respected that the reciprocity principle is derived as consequence from the relativity principle, cp. BERZI, GORINI (1969), while our elementary relativity principle is independently only a synchronisation recipe.

$$3.\ \theta(v) = \frac{T_o/T_v - l_o/l_v}{v}\,. \qquad \text{Synchronisation, Elementary relativity principle} \tag{46}$$

The synchronisation of clocks following from the elementary relativity principle according to Eq. (46) is called *conventional simultaneity*. (47)
Each synchronisation different from this form is called non-conventional.

From Eq. (46), it follows that the definition of the conventional simultaneity in the systems Σ' for ensuring the elementary relativity depends on the behaviour of moving measuring rods and clocks.

With precision measurements for the behaviour of moving rods and clocks in a single inertial system Σ_o, the complete space-time structure is fixed by the convention of the elementary relativity. We get in this manner all knowledge of the parameters $k(v)$, $q(v)$ and $\theta(v)$.

We conclude

With exception of a selection principle for fixing the synchronisation of all clocks we do not establisch any a priori postulate concerning the properties of our space-time. We only respect the results of experiments in an initially distinguished inertial system Σ_o. The thought experiment with a light clock in Chap. 4, Sect. 2 will provide hints for the expected experimental results.

In Chap. 3, Sect. 1 and Chap. 4, Sect. 3, we will formulate postulates concerning the behaviour of measuring rods and clocks on the basis of experimental results.

Will we come in this way also to the relativity of reference systems, the expected indistinguishability of all inertial systems? Furthermore, can we follow our method really without coming into contradictions? Let us assume that the clocks in systems Σ' and Σ'' are synchronised with respect to Σ_o according to the elementary relativity. Do systems Σ' and Σ'' also fulfil among themselves the elementary relativity?

We shall see: If we introduce for l_v/l_o and T_v/T_o the functions from Eqs. (45) and (46) that are experimentally determined in the classical or relativistic space-time, Chap. 3, Sect. 1, respectively, Chap. 4, Sect. 3, then we can proof with the so determined coordinate transformations (22) the relativity of all inertial systems directly by simple algebra. This holds true both for the long-known classical case according to GALILEI as well as for the here especially interesting relativistic case according to EINSTEIN.

In a more general view, we understand these answers by means of a stricter relativity principal that postulates in addition to the elementary relativity some simple properties of space-time, that are, however, weaker than EINSTEIN's relativity principle. We shall discuss this in the following Sect. 3.

Here, we have to remark the following: The question of the independence of the relativity principle from the universal constancy of the speed of light was already posed by IGNATOWSKI (1910). Our definition of simultaneity, introduced as elementary relativity (43), is there treated as an obvious fact, s. there the Eq. (5), $q' = -q$.

A rigorous axiomatics of IGNATOWSKI's ansatz was presented by FRANK and ROTHE (1912). In their postulate II, they demand that parallel straight lines again transform into straight lines with the consequence of the linearity of the transformation rules. In Chap. 1, Sect. 4.2, we required the linearity from the homogeneity and isotropy of space-time in the linear synchronisation prescription, where the true proof is contained in Problem 2.

The meaning of simultaneity on large scales is discussed in the overview of General Relativity, Chap. 13.

For a deeply interested reader, we want to advertise the book *Histoire du principe de relativité* by TONNELAT (1971), a truly encyclopedic presentation, that covers the subject in a historical and cultural overview of thousands of years, from the antiquity to the present time. To the French understanding reader, the book is a true pleasure and unputdownable.

3 A Metric Relativity Principle

In this Section, we intend to discuss a relativity principle that equals in its logical structure to EINSTEIN's principle. It connects the claimed physical postulate inseparable with the definition of a certain simultaneity. We have not based our presentation on this principle. But it gives us interesting insights into the space-time structure. We call it

The metric relativity principle:

> It is possible to synchronise the clocks in all inertial systems in such a way, that we find in all inertial systems the same formulae, when we compare moving and stationary measuring rods and clocks. (48)

While EINSTEIN's relativity principle postulates the equivalence of inertial systems for all physical laws, s. Sect. 1, the statement (48) requires only the equivalence for lengths and time measurements, a *metrical equivalence* of inertial systems. It is much weaker than EINSTEIN's principle, i.e. it is much easier to grasp:

$$\left.\begin{array}{l}\text{By the quotients}\\ \dfrac{\text{length of a moving bar}}{\text{proper length}}=\dfrac{l_v}{l_o},\\ \dfrac{\text{period of a moving clock}}{\text{eigen period}}=\dfrac{T_v}{T_o}\\ \text{no inertial system will be distinguished.}\end{array}\right\}\quad \begin{array}{l}\text{Metric}\\ \text{relativity principle}\end{array} \tag{49}$$

For discussing this relativity principle, we want to resume the discussion of Chap. 1, Sect. 5 and consider the following situation: A rod with the eigenlength l_o is taken to rest on the x-axis of the system Σ_o with end point coordinates $x_1 = 0$ and $x_2 = l_o$.

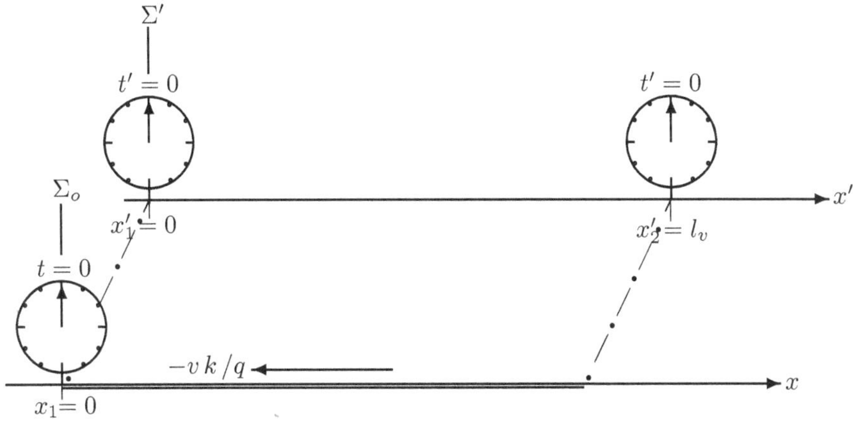

Fig. 3 Measuring the length l_v of a moving rod. It rests in system Σ_o. For a given relative velocity v between Σ_o and Σ', its length will be measured in Σ'. For the in Σ' determined velocity of Σ_o it holds Eq. (24). The dash-dotted lines connect again points in the picture that represent one and the same event.

We will determine the length l_v of this rod moving in Σ' with velocity v. To this end, we must determine the position of its end points at one and the same time in Σ', so, e.g. for $t' = 0$, Fig. 3.

We insert into the equation $x = (x'\,q + t'\,v\,k)/\Delta$ from Eq. (22), $t' = 0$ and $x = x_1 = 0$ and get for the left end point $x'_1 = 0$. With $t' = 0$ and $x = x_2 = l_o$ it follows for the right end point $l_o = x'_2\,q/\Delta$. Therefore, we have for the simultaneous end point positions of the rod in Σ'

$$\Sigma' : \; t' = 0 \,, \quad \begin{array}{l} x'_1 = 0 \,, \\ x'_2 = \dfrac{l_o\,\Delta}{q} \,, \end{array} \quad \longrightarrow \quad l_v = x'_2 - x'_1 = \frac{\Delta}{q}\,l_o \,. \quad \begin{array}{l}\text{Length } l_v \text{ of a with } \Sigma' \\ \text{moving rod}\end{array} \tag{50}$$

It follows, when we take into account $\Delta = k(v\,\theta + q)$,

$$\Sigma' : \quad \frac{\text{length of the in } \Sigma' \text{ moving rod}}{\text{rest length of the rod}} = \frac{l_v}{l_o} = \frac{k(v\,\theta + q)}{q} \,. \tag{51}$$

Equation (51) is different from Eq. (29). The *description* of our space-time may in general become asymmetric because of the principle arbitrary synchronous function.

In addition we consider a clock U^* that rests in the coordinate origin of Σ_o, and it shows there at the position $x = 0$ the time t.

We observe this clock in the system Σ'. For $x = 0,\ t = 0$ it holds because of the initial condition (11), also $x' = 0,\ t' = 0$. That is, the resting clock U^* at origin of system Σ_o shows the same pointer position as the clock resting at the origin of Σ' when it just passes by Fig. 4.

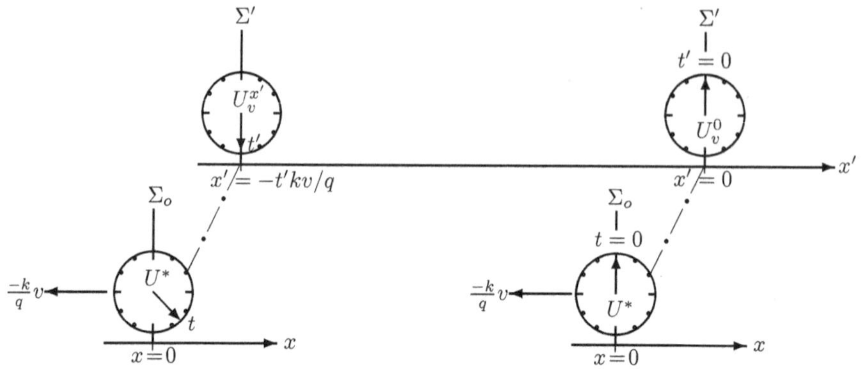

Fig. 4 For a given relative velocity between systems Σ_o and Σ', the pointer positions t of Σ_o resting clock U^* are compared with the time values t' of the in Σ' resting clock, when it just passes by. It may be remarked that for the in Σ' measured velocity of Σ_o at first it holds the general Eq. (24). The dash-dotted lines again connect points in the figure representing the same event.

$$\textbf{First time determination} \quad E_o : \left.\begin{array}{l} \Sigma' : x' = 0\,,\ t' = 0\,, \\ \Sigma_o : x = 0\,,\ \ t = 0\,. \end{array}\right\} \tag{52}$$

The in Σ' resting observer measures the velocity $u'_o = -k\,v/q$ for the in Σ_o stationary clock U^* according to Eq. (24). Then the clock U^* in Σ' sits after the time t' at the position $x' = u'_o\,t' = -t'\,k\,v/q$. Now we compare the pointer position t of U^* with the clock resting at $x' = -t'\,k\,v/q$ in Σ'. With $\Delta = (v\,\theta + q)\,k$ we find

$$t = -\frac{\theta}{\Delta}\,x' + \frac{k}{\Delta}\,t' = \frac{\theta}{\Delta}\,\frac{k\,v}{q}\,t' + \frac{k}{\Delta}\,t' = \frac{v\,\theta + q}{\Delta}\,\frac{k}{q}\,t' = \frac{1}{q}\,t'\,,$$

so

$$\textbf{Second time measurement} \quad E : \left.\begin{array}{ll} \Sigma' : x' = -\dfrac{k\,v}{q}\,t'\,, & t'\,, \\ \Sigma_o : x = 0\,, & t = \dfrac{1}{q}\,t'\,. \end{array}\right\} \tag{53}$$

$$\Sigma' : \frac{\text{difference of pointer positions of } \textit{one} \text{ in } \Sigma' \text{ moving clock}}{\text{difference of pointer positions of } \textit{two} \text{ in } \Sigma' \text{ resting clocks}} = \frac{t}{t'} = \frac{1}{q}\,. \tag{54}$$

Equation (53) is different from Eq. (35). The *description* of space-time can in general become asymmetrical also because of the principle of freely selectable synchronsation function.

In terms of the oscillation periods T_o and T_v, the relation of the, with respect to Σ', resting and moving clocks can be written instead of Eq. (53) as

$$\Sigma' : \quad \frac{\text{period of a in } \Sigma' \text{ moving clock}}{\text{eigen period}} = \frac{T_v}{T_o} = q \; . \qquad (55)$$

From Eqs. (29) and (36) as well as (51) and (55), we read that the determinant Δ of the coordinate transformation (22) is determined by lengths and time measurements:

$$\Sigma_o : \quad \frac{T_v}{T_o}\frac{l_v}{l_o} = \frac{1}{\Delta} = \frac{1}{k(v\,\theta + q)} \,, \qquad (56)$$

$$\Sigma' : \quad \frac{T_v}{T_o}\frac{l_v}{l_o} = \Delta = k(v\,\theta + q) \, . \qquad (57)$$

Now we come back to our metric relativity principle (49). It requires that Eqs. (29) and (51) as well as Eqs. (36) and (55) coincide. Therefore, with $k(v\,\theta + q) = \Delta$ it follows:

$$\frac{1}{k} = \frac{\Delta}{q} \quad \text{and} \quad \frac{k}{\Delta} = q \; . \qquad \text{Metric relativity principle} \qquad (58)$$

From Eq. (58) we get

$$\Delta = \frac{q}{k} = \frac{k}{q} \quad \longrightarrow \quad q^2 = k^2 \; . \qquad (59)$$

We restrict our consideration on the case $\Delta > 0$,[4] and we find from Eq. (59)

$$\left.\begin{array}{l} q = k \\ \text{and} \\ \Delta = 1 \; . \end{array}\right\} \qquad \text{Metric relativity principle} \qquad (60)$$

The first equation (60) reproduces the elementary relativity principle, Sect. 2. The second equation (60) provides a remarkable *reciprocity theorem* by taking into account Eqs. (56) and (57):

$$\Delta = 1 \quad \longrightarrow \quad \frac{T_v}{T_o} = \frac{l_o}{l_v} \; . \qquad \text{Reciprocity} \qquad (61)$$

From the metric equivalence of all inertial systems it follows, that the period changes of clocks is inverse to the lengths changes of measuring rulers.

The experimental results of Eqs. (65) and (66) in Chap. 3, Sect. 1 and Eqs. (99) and (100) in Chap. 4, Sect. 3 for the classical and relativistic space-time, resp., just fulfil this reciprocity (61).

---

[4]That means that the orientation of axes is conserved, i.e. for $v \longrightarrow 0$ the space and time axes of inertial systems should have the same direction.

From Eq. (61) it follows: With the lengths ratio of moving and stationary measuring rods we also measure the periods ratio of moving and stationary clocks, and inversely.

According to Eq. (60), it follows from $\Delta = k\,(v\,\theta + q) = 1$ with $k = q$ directly

$$\theta(v) = \frac{1-k^2}{v\,k}\,. \qquad \text{Synchronisation for the metric relativity principle} \qquad (62)$$

Therefore, when we postulate the metric relativity then the simultaneity in systems Σ' is already defined by the parameter k, e.g. by the ratio of the rest length to the length of a bar in motion in Σ_o.

The transformation (22) now reads[5]

$$\left.\begin{array}{ll} x' = k\,(x - v\,t)\,, & x = k(x' + v\,t')\,, \\ t' = \dfrac{1-k^2}{v\,k}\,x + k\,t\,, \quad \longleftrightarrow & t = -\dfrac{1-k^2}{v\,k}\,x' + k\,t'\,, \\ k = k(v)\,. & \end{array}\right\} \quad \text{Coordinate transformation for metric relativity principle} \qquad (63)$$

With Eqs. (23), (60) and (62) there follows the addition theorem of velocities

$$u' = \frac{u - v}{1 + \dfrac{1-k^2}{v\,k^2}\,u}\,. \qquad \text{Addition theorem of velocities for metric relativity principle} \qquad (64)$$

Using Eqs. (64) and (62), we conclude from a simple calculation:

For the metric relativity follows exactly GALILEI's addition theorem (70), when the absolute simultaneity according to Eq. (3) is fulfilled, and EINSTEIN's addition theorem (106) follows exactly, when LORENTZ's simultaneity as given in Eq. (103) is fulfilled.

We summarise

The metric relativity principle leaves among the whole space-time problem just one single parameter undetermined, namely $k(v) = l_o/l_v$, the quotient of the resting and moving lengths of a rod.
The parameter $k = k(v)$ already determines the coordinate transformations, and also the definition of simultaneity and the addition theorem of velocities.
The classical and relativistic space-time differ only by this parameter $k = k(v)$.

[5]From an arithmetic point of view, the entirety of transformations (63) represents a mathematical group, both for classical space-time with $k = 1$, and for the relativistic case with $k = 1/\sqrt{1 - v^2/c^2}$, s. Chap. 3, Sect. 2 and Chap. 4, Sect. 4. In this way, the equivalence of inertial systems is secured.

Since the precision measurements of length contraction (or rather their consequences) and of time dilatation have got a tremendous improvement, it has been possible for us to take them as the starting point of our axiomatics of SRT, without postulation a relativity principle. We shall discuss this in the following extensively. We shall return back to R. Feynman's thought experiment with the so-called light clock in Chap. 4, Sect. 2.1. (Sands, Feynman, Leighton 1970, cp. also Lewis, Tolman 1909).

Chapter 3
Elementary Structure of Classical Space-Time

Now, we want to follow the simplest thinkable way for exploring the space-time structure. We start with an empirical determination of the ratios of moving and resting measuring rods an clocks in an at first fixed reference system Σ_o, and for the definition of simultaneity in all other systems Σ', we require the elementary relativity.

1 The Physical Postulates of Classical Space-Time

Since in the beginning, the classical mechanics was based tacitly on the assumption of an invariability of lengths and periods, when measuring devices and clocks are in motion. Within the bounds of measuring accuracy, this was also constantly verified. We will formulate this hypothesis as postulates of classical space-time. In our procedure, we have to suppose this property only for the at first distinguished system Σ_o:

The physical postulates of classical space-time:

$\Sigma_o: \quad \dfrac{l_v}{l_o} = \dfrac{1}{k} = 1 \; ,$	Moving and resting rulers have in Σ_o the same lengths.	(65)
$\Sigma_o: \quad \dfrac{T_v}{T_o} = \dfrac{1}{v\,\theta(v) + q(v)} = 1 \; .$	Moving and resting clocks have in Σ_o the same periods.	(66)

Our experience tells us that for length and time measurements the *reciprocity* $T_v/T_o = l_o/l_v$ holds true in a trivial sense. In Sect. 3, Chap. 2, this relation was derived quite general from a relativity postulate, it stands in his range of coverage between EINSTEIN's postulate and the elementary relativity, s. Eq. (61).

H. Günther and V. Müller, *The Special Theory of Relativity*,
https://doi.org/10.1007/978-981-13-7783-9_3

In Sect. 1, Chap. 4, we will see how we will come in conflict with an experiment of higher precision using this classical hypothesis (65), cp. also Problem 3.

We remark again that we postulate Eqs. (65) and (66) only for the as isotropical declared system Σ_o. The behaviour of these quotients of moving and resting lengths and periods in other inertial systems Σ' is then a consequence of the there *defined* synchronisation of clocks.

2 Elementary Relativity—The GALILEI Transformation

We now assume the validity of the elementary relativity principle according to Eqs. (43) and (44), Sect. 2, Chap. 2. In systems Σ', the clocks should, therefore, be set up according to this principle. Our physical postulates (65) and (66) for the distinguished reference system Σ_o now require, according to Eq. (46) for a conventional synchronisation, a so-called absolute synchronisation parameter θ_a as

$$\theta_a = \frac{T_o/T_v - l_o/l_v}{v} = \frac{1-1}{v} = 0\,. \qquad \text{Absolute synchronisation parameter} \qquad (67)$$

This synchronisation is illustrated in Fig. 1. We summarise

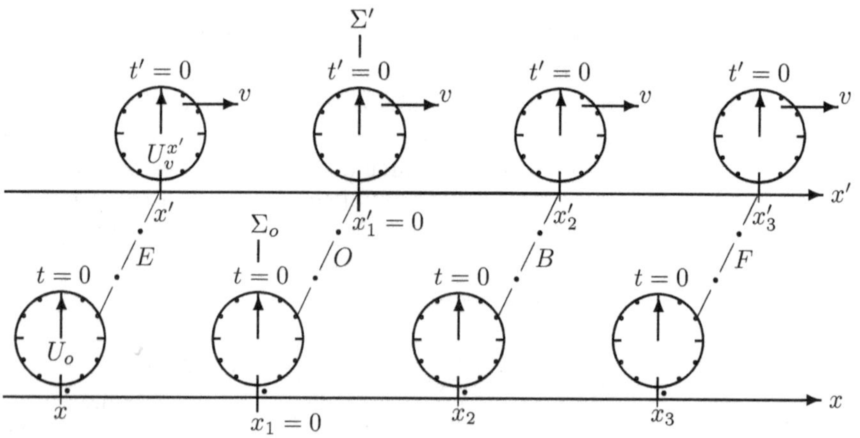

Fig. 1 Realisation of elementary relativity principle in classical space-time by a synchronisation of systems Σ' with parameter $\theta_a(v) = 0$, that introduces an absolute simultaneity. At time $t = 0$ in Σ_o, all Σ'-clocks are started with $t' = 0$. The dash-dotted lines connect points representing one and the same event, here the events E, O, B and F.

$$\left.\begin{array}{lll}\text{The physical postulates} & k = 1\,, & \\ \text{of classical space-time} & q = 1\,, & \\ \quad + & & \\ \text{Elementary relativity principle} & \theta_a(v) = 0\,. & \text{Absolute synchronisation parameter}\end{array}\right. \tag{68}$$

With Eq. (68) we get for the coordinate transformation (22) the famous GALILEI *transformation* of classical space-time,

$$\left.\begin{array}{ll} x' = x - v\,t\,, & x = x' + v\,t'\,, \\ t' = t\,, \quad \longleftrightarrow & t = t'\,. \end{array}\right\} \quad \text{GALILEI transformation} \tag{69}$$

For $\theta = 0$ and $k = q = 1$, we get from the addition theorem (23)

$$u' = u - v \quad \longleftrightarrow \quad u = u' + v\,. \qquad \text{GALILEI's addition theorem of velocities} \tag{70}$$

Therefore, the velocity u' fulfills the same equation as the relative velocity w in Sect. 3, Chap. 1, Eq. (8), but we should keep in mind the conceptual difference of these two quantities.

Equation (70) is also easily derived from the GALILEI transformation (69): We assume a bit more general, the body L is in motion in Σ_o as

$$\mathbf{x}(t) = \big(x(t),\; y(t),\; z(t)\big)$$

with a velocity

$$\mathbf{u} = \left(\frac{dx}{dt},\; \frac{dy}{dt},\; \frac{dz}{dt}\right) = (u_x,\; u_y,\; u_z)\,.$$

For its motion in Σ'

$$\mathbf{x}'(t') = \big(x'(t'),\; y'(t'),\; z'(t')\big)$$

with the velocity

$$\mathbf{u}' = \left(\frac{dx'}{dt'},\; \frac{dy'}{dt'},\; \frac{dz'}{dt'}\right) = (u'_x,\; u'_y,\; u'_z)$$

it holds because of $t' = t, x' = x - vt$ and $y' = y,\; z' = z$ according to Eqs. (69) and (22),

$$\frac{d}{dt'} = \frac{d}{dt}\,, \quad \text{so that}$$

$$\frac{dx'}{dt'} = \frac{dx'}{dt} = \frac{dx}{dt} - v\,, \qquad \frac{dy'}{dt'} = \frac{dy}{dt}\,, \quad \frac{dz'}{dt'} = \frac{dz}{dt}$$

and therefore again the theorem (70) with the conservation of the other two velocity components,

$$u'_x = u_x - v\,, \quad u'_y = u_y\,, \quad u'_z = u_z\,. \tag{71}$$

In distinction to the formally identical relation (8), the theorem (70) or (71) now has far reaching physical consequences. We consider an example:

A spaceship, that represents a system Σ', might have a velocity $v = 200\,000\,\text{km s}^{-1}$ relative to the Earth. From the spaceship there might be started a second spaceship with the same velocity, but now in the system Σ', and therefore with $u'_1 = 200\,000\,\text{km s}^{-1}$. According to Eq. (70), this second spaceship will have a velocity measured from Earth of $u_1 = u'_1 + v = 400\,000\,\text{km s}^{-1}$. This process can be continued from the second space ship can be started a third one, and then a fourth one, etc. The velocities measured from each will be $u_2 = u'_2 + u'_1 = 600\,000\,\text{km s}^{-1}$, $u_3 = 800\,000\,\text{km s}^{-1}$, etc. The addition theorem (70) allows arbitrary large velocities.

Signals with arbitrary high velocities can then be transmitted by means of spaceships. In agreement with this possibility, Newton has based his theory of gravitation on the assumption of an instantaneous interaction. This is also called an action on the distance theory. The gravitational forces of a mass are assumed to propagate with infinite velocity. Each change of the position of the mass will then be instantaneous, without any retardation, be felt in the whole world. An important consequence of the theorem (70) will be met in Problem 14. We shall show, that in Newton's mechanics, the inertial mass of a body should be a fixed constant, independent on the inertial system, i. e. the mass of a body cannot depend on its velocity.

When the velocities v and u of $\Sigma'(x', t')$ and $\Sigma''(x'', t'')$ are in x-direction of $\Sigma_o(x, t)$, then we find from the corresponding Galilei transformation

$$\left.\begin{array}{ll} x'' = x - u\,t\,, & x' = x - v\,t\,, \\ t'' = t\,, & t' = t\,. \end{array}\right\} \longrightarrow \begin{array}{l} x'' = x' - u'\,t'\,, \\ t'' = t'\,, \end{array} \quad u' = u - v\,. \tag{72}$$

That is, $\Sigma''(x'', t'')$ and $\Sigma'(x', t')$ also are connected by a Galilei transformation, where $u' = u - v$ is now the velocity measured of Σ'' in system Σ'.

According to the Galilei transformation (69), two events $E_1(x_1, t)$ and $E_2(x_2, t)$ in the classical space-time, that are simultaneous in a system $\Sigma_o(x, t)$, are also simultaneous in each other system $\Sigma'(x', t')$. Furthermore, according to Eq. (69), it holds not only $\Delta t = 0$ exactly then, when $\Delta t' = 0$, but also the times itself coincide.

This is NEWTON's famous absolute time:

$$t' = t\,. \qquad \text{NEWTON's absolute time} \qquad (73)$$

Therefore, when we require for the synchronisation of clocks in the systems Σ' the principle of elementary relativity, we will find that two events are simultaneous either in all inertial systems or in no one.

The time perception of our daily experience is based on this construction, cp. on this point also the discussion on I. KANT (1998) in Chap. 10, Sect. 7. With Eqs. (46), (47) and (67) we summarise

With the GALILEI transformation the *conventional simultaneity of classical space-time* is realised by the absolute synchronisation parameter θ_a. (74)

Obviously, from the GALILEI transformation there follows again the invariability of moving rulers and clocks:

An in Σ' a resting bar has there a length l_o, that is given by the difference of its end point coordinates,

$$l_o = x_2' - x_1'\,.$$

The bar with velocity v in Σ_o has there the length l_v, given also by the coordinate difference of its end points at the same time t, i. e. with Eq. (69),

$$l_v = x_2(t) - x_1(t) = x_2'(t') - v\,t' - x_1'(t') + v\,t' = x_2' - x_1' = l_o\,.$$

It is a trivial consequence that the moving clock shows the same time $t' = t$ as the stationary clock. Equation (72) demonstrates that all inertial systems are connected by the same form of coordinate transformations.

In mathematical terms, the GALILEI transformations represent a group. In this way, we have found the mathematical simplest form for the equivalence of all inertial systems. It follows that the invariability of moving rulers and clocks, Eqs. (65) and (66) are also measured in all inertial systems in the same way. We remark especially that this is not obvious, and it is connected with the conventional definition of simultaneity.

Indeed, one can get to a description of classical space-time with a definition of a non-conventional simultaneity and therefore with a deviation of the GALILEI transformation, as we shall discuss in Sect. 7, Chap. 4 with the v/c linearised LORENTZ transformation, cp. also Problem 12.

We summarise once again

GALILEI's relativity principle, the physical equality of all inertial systems, can be in the mathematical simplest way be realised with the GALILEI transformation in the conventional simultaneity of classical space-time.

For a historical assessment of the relativity principles of classical physics, we refer to the comprehensive presentation of M.- A. TONNELAT (1971).

Chapter 4
Elementary Structure of Relativistic Space-Time

When we now discuss the two historical key experiments for Special Relativity, the MICHELSON experiment and the observation of the red H_α-line in fast channel rays, we want to start from an empirical determination of the ratio of moving and stationary rulers and clocks, at first in the distinguished system Σ_o. We remark again explicitly that EINSTEIN's universal constancy of the speed of light is not postulated in our procedure, but follows from the fully formulated theory, when we require for the definition of simultaneity in all other systems Σ' only the elementary relativity, cp. Sect. 4.

Modern precision experiments to the relativity theory are reviewed in Sect. 6.

1 The Moving Rod is Shortened—The MICHELSON Experiment

Here, we describe the schematic experimental set-up of the MICHELSON–MORLEY experiment, shown in Figs. 1 and 3. A light source L emits a wave segment with a stable phase relation (a coherent wave segment) that hits a semi-transparent plate P inclined by a half right angle. There it splits off in two coherent wave segments that are propagating along the two arms l_1 and l_2 of the so-called MICHELSON (Fig. 2) interferometer. At their ends, both waves are reflected by mirrors S_1 and S_2, returning back and interfere with each other at B forming an interference picture.

According to the statement (2), the speed of light c has in the distinguished reference system Σ_o in each direction one and the same value, Eq. (4). The reference system Σ' might have the velocity v with respect to Σ_o.

At first we consider the case, where the MICHELSON interferometer rests in system Σ_o at its starting position, Fig. 1a.

Because of the isotropy of light propagation in Σ_o, the time span t_1^o of light is the same for the forward and backward ways along l_1, hence $t_1^o = 2l_1/c$, and likewise the time span along l_2 is $t_2^o = 2l_2/c$. The interference picture after unification of both

H. Günther and V. Müller, *The Special Theory of Relativity*,
https://doi.org/10.1007/978-981-13-7783-9_4

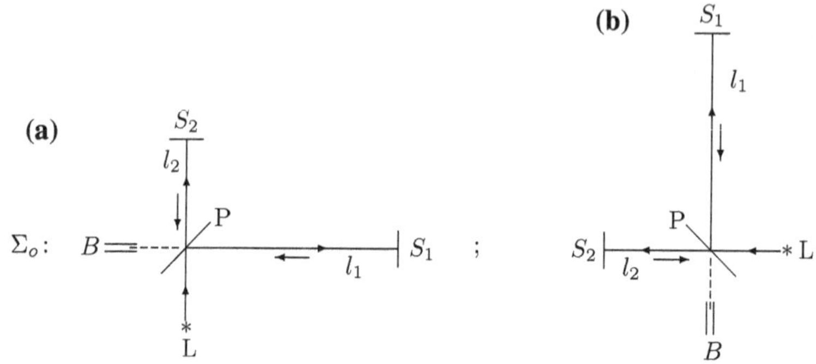

Fig. 1 Schema of the MICHELSON interferometer, **a** at first resting in system Σ_o, and **b** rotated by an angle of $\pi/2$.

wave segments at P is determined by the difference in the run times Δt^o,

$$\Delta t^o = t_2^o - t_1^o = \frac{2l_2}{c} - \frac{2l_1}{c}. \tag{75}$$

The idea of the experiment, going back to J. C. MAXWELL, consists of rotating the interferometer by an angle of $\pi/2$, Fig. 1b.[1] Because of the isotropy of the light propagation in Σ_o, the run time difference $\Delta t^o_{\frac{\pi}{2}} = t^o_{2,\frac{\pi}{2}} - t^o_{1,\frac{\pi}{2}}$ along both arms l_2 and l_1 of the interferometer has clearly not changed by this rotation,

$$\Delta t^o_{\frac{\pi}{2}} = t^o_{2,\frac{\pi}{2}} - t^o_{1,\frac{\pi}{2}} = \frac{2l_2}{c} - \frac{2l_1}{c} = \Delta t^o \,. \tag{76}$$

We denote with $\delta := \Delta t_{\frac{\pi}{2}} - \Delta t$ the difference of the run times before and after the rotation. This quantity δ is a measure for the change of the interference picture because of the rotation.

If the interferometer is at rest in the distinguished reference system Σ_o, it holds

$$\delta^o = \Delta t^o_{\frac{\pi}{2}} - \Delta t^o = 0 \,. \qquad \text{Stationary interferometer} \tag{77}$$

From Eq. (77), we read that the interference picture does not change during the rotation, when the interferometer is at rest in Σ_o.

Now the interferometer might be at rest in the reference system Σ' that has the velocity v in the distinguished system Σ_o. We now derive how an observer resting in the system Σ_o describes the experiment, Fig. 3.

[1] MAXWELL had all of his life the opinion, that there exists a mechanical medium, a so-called aether, cp. Chap. 2, Sect. 1 and Chap. 8, Sect. 3, that can be used for the description of light propagation in a similar way as the propagation of sound waves through the air. A. EINSTEIN believed probably until 1901 in such an aether.

Fig. 2 ALBERT ABRAHAM MICHELSON, *Strelno (near Hohensalza) 19.12.1852, † Pasadena 9.5.1931.

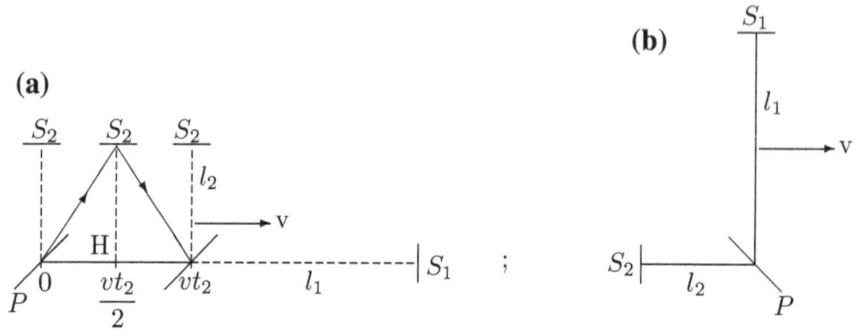

Fig. 3 The interferometer has a velocity v with respect to the distinguished reference system Σ_o. **a** Initial position. **b** The interferometer moving with the velocity v is rotated by an angle $\pi/2$.

According to our Eq. (8) in Chap. 1, Sect. 3, the wavefront approaches to the mirror S_1 on the way from O to S_1 with the relative velocity $c - v$, on the way back it approaches to the starting position O with a relative velocity $c + v$.

Therefore, the observer in Σ_o measures a total run time t_1 along l_1

$$t_1 = \frac{l_1}{c-v} + \frac{l_1}{c+v} = \frac{l_1}{c}\left(\frac{1}{1-v/c} + \frac{1}{1+v/c}\right),$$

$$t_1 = \frac{2l_1}{c}\frac{1}{1-v^2/c^2}. \tag{78}$$

The run time along l_2 has the same value $t_2/2$ for the forward and backward way, s. Fig. 3a. The wavefront has the velocity c, the interferometer the velocity v. From the triangle $O S_2 H$ then it follows:

$$\left(\frac{c\,t_2}{2}\right)^2 = \left(\frac{v\,t_2}{2}\right)^2 + l_2^2 \longrightarrow \frac{t_2^2}{4}(c^2 - v^2) = l_2^2 \longrightarrow t_2^2 = \frac{4\,l_2^2}{c^2 - v^2} = \frac{4\,l_2^2}{c^2}\frac{1}{1-v^2/c^2},$$

hence

$$t_2 = \frac{2\,l_2}{c}\frac{1}{\sqrt{1-v^2/c^2}}. \tag{79}$$

For the difference Δt of run times t_1 and t_2, we get then

$$\Delta t = t_2 - t_1 = \frac{2\,l_2}{c}\frac{1}{\sqrt{1-v^2/c^2}} - \frac{2l_1}{c}\frac{1}{1-v^2/c^2}. \tag{80}$$

Now, we rotate the interferometer again by the angle $\pi/2$, Fig. 3b. The run time to the mirrors S_1 and S_2 are denoted by $t_{1,\frac{\pi}{2}}$ and $t_{2,\frac{\pi}{2}}$. Now we need to replace in Eq. (78) only l_1 by l_2 and in Eq. (79) l_2 by l_1 to get $t_{2,\frac{\pi}{2}}$ and $t_{1,\frac{\pi}{2}}$. Then it holds

$$\Delta t_{\frac{\pi}{2}} = t_{2,\frac{\pi}{2}} - t_{1,\frac{\pi}{2}} = \frac{2\,l_2}{c}\frac{1}{1-v^2/c^2} - \frac{2l_1}{c}\frac{1}{\sqrt{1-v^2/c^2}}. \tag{81}$$

Fig. 4 HENDRIK ANTOON LORENTZ, *Arnheim 18.7.1853, † Haarlem 4.2.1928.

The difference $\delta = \Delta t_{\frac{\pi}{2}} - \Delta t$ is a measure for a possible change of the interference picture because of the rotation. With Eqs. (80) and (81), we get for it

$$\begin{aligned}\Delta t_{\frac{\pi}{2}} - \Delta t &= \frac{2l_2}{c}\frac{1}{1-v^2/c^2} - \frac{2l_1}{c}\frac{1}{\sqrt{1-v^2/c^2}} - \left(\frac{2l_2}{c}\frac{1}{\sqrt{1-v^2/c^2}} - \frac{2l_1}{c}\frac{1}{1-v^2/c^2}\right)\\ &= \frac{2l_2}{c}\left(\frac{1}{1-v^2/c^2} - \frac{1}{\sqrt{1-v^2/c^2}}\right) + \frac{2l_1}{c}\left(\frac{1}{1-v^2/c^2} - \frac{1}{\sqrt{1-v^2/c^2}}\right),\end{aligned}$$

that means

$$\delta = \Delta t_{\frac{\pi}{2}} - \Delta t = \left(\frac{2\,l_1}{c} + \frac{2\,l_2}{c}\right)\left(\frac{1}{1-v^2/c^2} - \frac{1}{\sqrt{1-v^2/c^2}}\right). \quad \text{Moving interferometer} \tag{82}$$

If a number $a \ll 1$, i.e. much smaller than 1, then there follow (as can be illustrated by inserting a small fixed value, e.g. $a = 10^{-4}$) approximation formulae

$$\left.\begin{aligned}\frac{1}{1-a} &\approx 1+a\,, & \frac{1}{1+a} &\approx 1-a\,,\\ \sqrt{1-a} &\approx 1-\frac{1}{2}a\,, & \frac{1}{\sqrt{1-a}} &\approx 1+\frac{1}{2}a\end{aligned}\right\}\quad \text{for} \quad a \ll 1\,. \tag{83}$$

These relations follow from the TAYLOR expansion that are often used in this book. It becomes more and more accurately, as a gets smaller. For $a = v^2/c^2$, we obtain from Eqs. (83) and (82)

$$\delta \approx \frac{2}{c}\,(l_1 + l_2)\left(1 + \frac{v^2}{c^2} - 1 - \frac{1}{2}\frac{v^2}{c^2}\right) = \frac{2}{c}\,(l_1 + l_2)\,\frac{1}{2}\frac{v^2}{c^2}$$

and therefore

$$\delta = \Delta t_{\frac{\pi}{2}} - \Delta t \approx \frac{l_1 + l_2}{c}\frac{v^2}{c^2}\,. \quad \text{Moving interferometer} \tag{84}$$

The quantity δ determines the shift of the interference fringes because of the rotation of the interferometer by $\pi/2$.

For analysing the experiment, we assume that the interferometer is disposed in a laboratorium fixed on Earth. The Earth represents our reference system Σ'. The Earth velocity around the Sun amounts about $v = 30\,000$ m/s. This is the velocity of Σ' with respect to a distinguished reference system Σ_o, the rest system of the Sun approximately realising an inertial system.

In the historical experiment of A. A. MICHELSON in the basement of the Astrophysikalisches Observatorium Potsdam in 1881, cp. Bleyer et al. (1979), s. Fig. 5, the total length of the light path $l_1 + l_2$ amounted about $l_1 + l_2 = 30$ m, even if the set-up has an extension of only 1.20 m. When we consider the interference of yellow line of sodium (Na) with $\lambda = 6 \cdot 10^{-7}$ m, we get with $c = 3 \cdot 10^8$ m/s a period of

Fig. 5 MICHELSON experiment in the basement of the Astrophysikalisches Observatorium Potsdam. The set-up was remodeled after original drafting in 1979. At present, the model belongs to the Michelson-house of the Potsdam Institute for Climate Impact Research in the Science Park ALBERT EINSTEIN on the Telegrafenberg, Potsdam. The spark lamp is seen on the left, the half-permeable mirrors in the centre of the picture, the two arms leading to the background and to the left and the observing optics in the foreground left. The experimental set-up lies below the eastern dome of the observatory for protection against vibrations of the ground. Previous attempts by MICHELSON to perform the experiment in Berlin have been unsuccessful due to the vibrations by traffic. The accuracy of the experiment war about 10 km/s, to be compared with the aether effect of 30 km/s by the motion of the Earth around the Sun.

$$\tau = \frac{\lambda}{c} = \frac{6 \cdot 10^{-7}}{3 \cdot 10^8}\,\mathrm{s} = 2 \cdot 10^{-15}\,\mathrm{s}\ .$$

On the other hand, we get from Eq. (84) the following value for the light time difference δ of interfering wave sequences due to the rotation of the interferometer

$$\delta \approx \frac{30}{3 \cdot 10^8}\left(\frac{3 \cdot 10^4}{3 \cdot 10^8}\right)^2 \mathrm{s} = 10^{-15}\,\mathrm{s}\ ,$$

and therefore

$$\delta = \Delta t_{\frac{\pi}{2}} - \Delta t \approx \frac{1}{2}\tau\ . \qquad (85)$$

Moving interferometer

In result of the rotation, the quantity δ should change by the travel time $\tau/2$, i.e. a half wave length! Therefore, the interference picture should change during the rotation by one half of a strip, the dark stripes had to become bright, and the bright ones in reverse dark.

However, not the slightest change in the interference picture was observed, neither in 1881 in Potsdam neither in one of the many following larger experiments by MICHELSON and MORLEY, and also not in the modern follow set-ups, s. Sect. 6.

During the calculation of the run times we tacitly assumed that the in Σ_o measured lengths l_1 and l_2 of the interferometer legs are independent on the velocity of Σ_o. This is the error.

Already FitzGerald (1889) presented an explanation of the outcome of the MICHELSON–MORLEY experiments with the hypothesis, 'that the length of material bodies changes, according as they are moving through the ether or across it, by an amount depending on the square of the ratio of their velocities to that of light'. Lorentz (1892), s. Fig. 4, had independently established such an hypothesis. LORENTZ's efforts were to find an expression for the required contraction. Lorentz (1904) wrote: 'We are therefore led to suppose that the influence of a translation on the dimensions (of the separate electrons and of a ponderable body as a whole) is confined to those that have the direction of the motion, these becoming k times smaller than they are in the state of rest'. Here $k = \sqrt{1 - v^2/c^2}$ and 'ponderable' is an older expression for heavy.

For the interested reader, we refer to the reprints of the most important papers leading to the emergence of the relativity theory in the book 'The Principle of Relativity' by Lorentz (1958, 1952).

The length change for explanation of the MICHELSON experiments is now denoted as FITZGERALD–LORENTZ contraction or short LORENTZ contraction, cp. also Fig. 10:

If in a distinguished system Σ_o for a resting rod the length l_o is observed, then for the same rod, that is in motion relative to Σ_o with a velocity v, in Σ_o we will measure the length l_v :

$$\Sigma_o: \quad l_v = l_o \sqrt{1 - \frac{v^2}{c^2}} \,. \qquad \text{LORENTZ contraction} \qquad (86)$$

Indeed, when we replace in Eq. (80) the length l_1 of the interferometer leg in the direction of motion by the moving length $l_1 \sqrt{1 - v^2/c^2}$ and after rotation by $\pi/2$ in Eq. (81) now in the direction of motion lying leg l_2 by the moving length $l_2 \sqrt{1 - v^2/c^2}$, then it follows $\Delta t = \Delta t_{\frac{\pi}{2}}$ and therefore according to Eq. (85) $\delta = 0$, i.e. the same value as in Eq. (77). We keep up our hypothesis (13) in agreement with LORENTZ that orthogonal to the direction of motion, a rod does not suffer any length change.

The difference δ of the run times remains unchanged during the rotation, when the LORENTZ contraction (86) of moving lengths is taken into account. Then we will not observe a change in the interference picture, cp. also Problem 3. Nevertheless, the presented scenario for the experimental justification of the LORENTZ contraction remains a thought experiment (likewise the light clock discussed in Sect. 2.1), since we do not have the situation of distinguished reference systems Σ_o, where we measure the rest length, and a relative to this system moving one Σ', where the interferometer is at rest. For our exposition of the SRT, this is indeed irrelevant, since we will take the LORENTZ contraction in Σ_o as an axiomatic basis for the theory, Sect. 3. Furthermore, it should be denoted that we made only a statement on the speed of light in system Σ_o.

2 The Moving Clock Goes Behind—EINSTEIN's Experimentum Crucis of the Special Relativity Theory

The caption means an experiment that compares the period T_v of a moving clock with the eigenperiod T_o of identical resting clocks.

In the historical experiment, the oscillating system of a moving clock was a hydrogen atom that produces in its rest system a red spectral line H_α of the Paschen series with an eigenperiod $T_o = 2.1876 \cdot 10^{-15}$ s. When the H-atoms are observed in channel rays of high velocity v, the frequency is, according to EINSTEIN's Special Relativity Theory, the period to $T_v = T_o \,/\sqrt{1 - v^2/c^2}$. The corresponding frequency change suffers a relativistic correction to the classical theory of the DOPPLER-effect, s. Chap. 8, Sect. 5. These precision experiments were successfully performed first in the years 1938/39, and a redshift of the spectral lines was confirmed. EINSTEIN has denoted this effect as the 'experimentum crucis' of the SRT, cp. the textbook of Sommerfeld (1952) §27 D, p. 227, i.e. as the deciding experiment for his postulate of an universal constancy of the speed of light, that leads to this effect. Further precision experiments to the time dilatation will be presented in Sect. 6.

Different from EINSTEIN's theoretical foundation of his Special Relativity Theory, we do not assume an a priori guess for the behaviour of moving clocks.[2] Here we take advantage of the so-called light clock, extensively described by FEYNMAN's thought experiment (Sands et al. (1970), cp. also Lewis and Tolman (1909).

2.1 *The Light Clock*

Between two mirrors S_1 and S_2, that are arranged with a fixed separation l_o, there should run a light signal forth and back. The number of incoming periodic light signals arriving at mirror S_1 is counted and indicated as positions of the clock pointer. This arrangement is called a light clock. We consider the case, that it rests in system Σ_o s. Fig. 6. The time between two signals arriving at S_1 represents the period T_o, i.e. it follows with the speed of light c in Σ_o

[2]In Chap. 2, Sect. 3, we get the time dilatation (97) as consequence from the LORENTZ contraction (86), when we postulated a somewhat stronger relativity principle as elementary principle, and when we require more than only the definition of simultaneity.

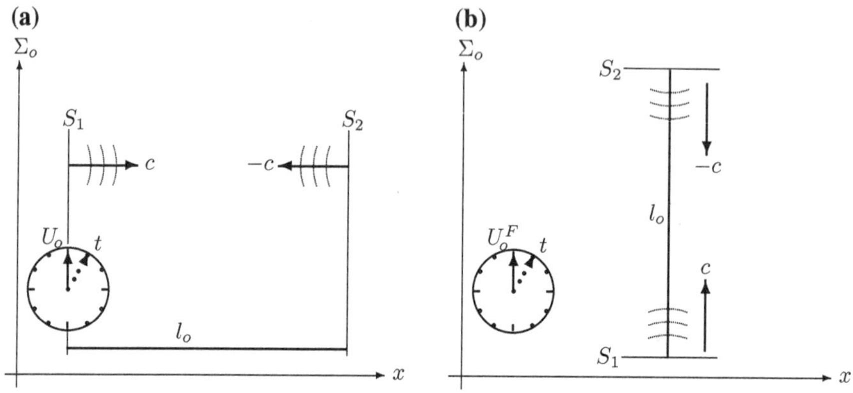

Fig. 6 **a** The light clock U_o with a separation l_o between two mirrors S_1 and S_2 resting in system Σ_o. The light propagates along the x-axis. The period amounts $T_o = 2l_o/c$. **b** In FEYNMAN's light clock U_o^F with a light propagation along the y-axis we will measure the same period $T_o^F = 2l_o/c$ because of the isotropy of the light propagation in Σ_o, if the clock also rests in the system Σ_o.

$$T_o = \frac{2l_o}{c} \,. \qquad \text{Period of a in } \Sigma_o \text{ resting light clock} \tag{87}$$

In the left part (a) of Fig. 6, the light signal runs in x-direction. In part (b), we consider additionally the originally by FEYNMAN discussed light clock, where the light signal propagate in y direction forth and back, but this plays only a role when both clocks are moving in x-direction.

Now we consider the cases that one after another both light clocks are moving both in x-direction.

At first we consider the clock U_o in Fig. 6a in motion in x-direction. This means it rests in a system Σ', that has a velocity v in x-direction with respect to Σ_o, s. Fig. 7. We evaluate the period T_v of the clock now moving with respect to Σ_o.

To this aim, we take into account the possibility, that the observer in Σ_o would measure a separation between the mirrors l_v for the case of motion, that can be different from the separation l_o of an in Σ_o resting system. But we will not take into account the results of the LORENTZ contraction from Sect. 1.

According to the addition of velocities in one reference system (8), the light will travel the forward way of the distance l_v from S_1 to S_2 with the relative velocity $c - v$, and back from S_1 to S_2 with a velocity $c + v$. Then it follows the total time needed for both ways, the period T_v of the moving clock,

$$\begin{aligned} T_v &= \frac{l_v}{c - v} + \frac{l_v}{c + v} \,, \\ &= l_v \frac{c + v + c - v}{(c - v)(c + v)} \,, \\ &= 2\,l_v \frac{c}{c^2 - v^2} \,, \\ &= \frac{2\,l_v}{c} \frac{c^2}{c^2 - v^2} \,, \end{aligned}$$

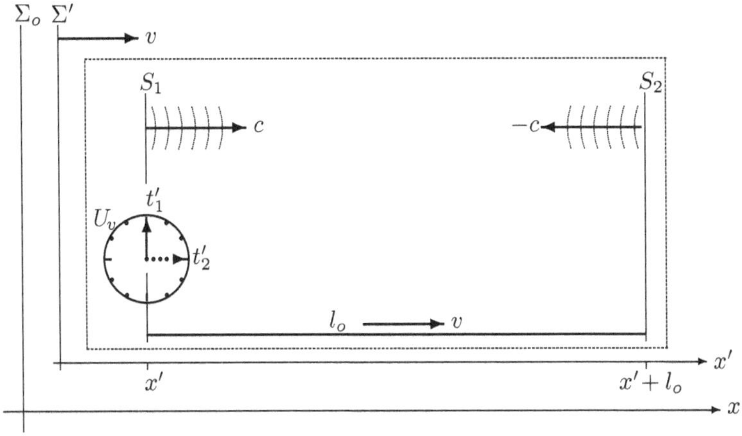

Fig. 7 The moving light clock U_v. The enframed part, the length l_o with the mirrors S_1 and S_2, resting in the system Σ', is in motion with a velocity v in x-direction in the system Σ_o. The clock hand t' counts the between S_1 and S_2 forth and back travelling light signals. In the text we have provisionally denoted the clock hands by t_v. As time coordinate in Σ' we write again $t' = t_v$.

therefore

$$T_v = \frac{2l_v}{c} \frac{1}{1 - v^2/c^2} \, . \qquad \text{Period of the in } \Sigma_o \text{ moving light clock } U_v \qquad (88)$$

The connection between the period T_o of an in Σ_o resting light clock and its period T_v in a system moving with velocity v with respect to Σ_o now follows from a consideration of FEYNMAN's light clock, where the light signal propagates in the y-direction forth and back. For evaluating the period T_v^F of FEYNMAN's light clock U_v^F, we consider Fig. 8 with a system Σ' connected with the clock.

From the orthogonal triangle OS_2H, we find

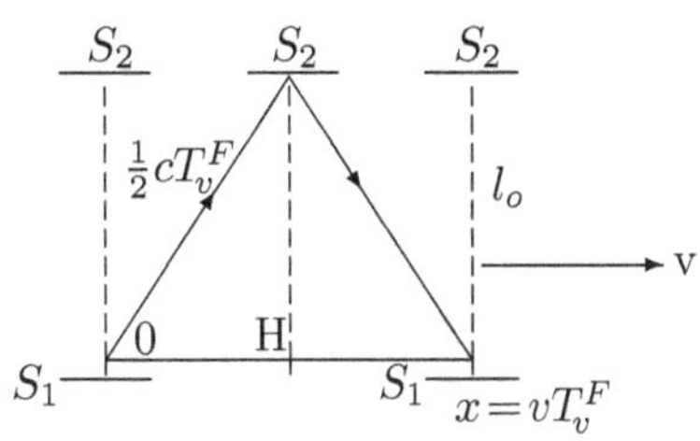

Fig. 8 Schematic presentation for the evaluation of the period of FEYNMAN's moving light clock U_v^F.

$$
\begin{aligned}
\left(\frac{c\,T_v^F}{2}\right)^2 &= \left(\frac{v\,T_v^F}{2}\right)^2 + l_o^2\,,\\
\frac{\left(T_v^F\right)^2}{4}\,(c^2 - v^2) &= l_o^2\,,\\
\left(T_v^F\right)^2 &= \frac{4\,l_o^2}{c^2 - v^2} = \frac{4\,l_o^2}{c^2}\,\frac{1}{1 - v^2/c^2}\,,
\end{aligned}
$$

and therefore

$$T_v^F = \frac{2\,l_o}{c}\,\frac{1}{\sqrt{1 - v^2/c^2}}\,. \tag{89}$$

When we take into account Eq. (87), we note the astonishing fact, that the period T_v^F of moving FEYNMAN's light clock depends on its velocity v and is bigger than the rest period T_o,

$$T_v^F = \frac{T_o}{\sqrt{1 - v^2/c^2}}\,. \tag{90}$$

Is this behaviour universal for all clocks? With our physical understanding it would be unbelievable, when we would have different laws for different clocks. We will call this effect time dilatation. We make the assumption that the clocks U_o and U_v behave identical to FEYNMAN's clocks. For the period T_v, there must follow $T_v = T_v^F$, and consequently with Eqs. (88) and (89)

$$\frac{2l_v}{c}\,\frac{1}{1 - v^2/c^2} = \frac{2l_o}{c}\,\frac{1}{\sqrt{1 - v^2/c^2}}\,.$$

There it follows a remarkable second consequence,[3]

$$l_v = l_o\,\sqrt{1 - v^2/c^2}\,. \tag{91}$$

The distance between both mirrors that are, for e.g. fixed on an iron rod has in case of motion with velocity v in the reference system Σ_o a smaller length l_v compared with the length l_o in a system at rest. Also here we must raise the question whether this property (91), the so-called length contraction, holds true for all lengths.

In Sect. 1, we have already discussed the length contraction (91) with the historical key experiment to the SRT, the MICHELSON experiment. We shall further explore the time dilatation (90) at the end of this section.

---

[3]One could come to the idea that this experiment could also be done with sound waves. As long as such a 'sound clock' is resting in system Σ_o as in Fig. 6, there is nothing wrong with this attempt. However, the sound clock might be running in a medium like air or water. For a moving sound clock, the transmitting medium must be moving with the clock. Different from the case with light in Fig. 7, we do not get the in Σ_o -observed velocity of sound waves without an additional hypothesis for the addition of the velocity v of the medium and the velocity of sound waves with respect to this medium.

Now we will show that both the time dilatation (90) and also the length contraction (91) follow alone from the behaviour of light clocks with parallel propagation in the direction of motion, when we use arguments of a fundamental symmetry in nature.

Starting points are the formulas (87) and (88) for the periods of resting and moving light clocks with light propagation parallel to the direction of motion.

For getting a connection between the period T_o of Σ_o resting light clock U_o and its period T_v for a motion with velocity v with respect to Σ_o, we have to make an assumption whether and how the length l_v of in Σ_o, moving bar is distinct from the length l_o of the same bar resting in Σ_o.

1. assumption: The length l_v between the mirrors moving in Σ_o is not different from the length l_o at rest. We insert $l_v = l_o$ in (88) and respect (87), then it follows:

$$\text{1. assumption} \quad \Sigma_o: \quad l_v = l_o \quad \longrightarrow \quad T_v = T_o \frac{1}{1 - v^2/c^2} \,. \tag{92}$$

2. assumption: We observe, that the separation l_v between both mirrors of the in Σ_o with velocity v moving light clock is shorter than its separation l_o at rest, i.e. it should hold $l_v = l_o\,(1 - v^2/c^2)$. We insert this assumption into (88) and use (87), then it follows:

$$\text{2. assumption} \quad \Sigma_o: \quad l_v = l_o\,(1 - v^2/c^2) \quad \longrightarrow \quad T_v = T_o \,. \tag{93}$$

In the first case, the length of the bar remains unchanged and the period alters. In the second case it is inverse, the length of the bar changes, and the period remains constant. Both cases are logically allowed, and likewise all laws in between. Since it is a pure thought experiment, we cannot determine what is the truth. But it is very important to start from the most realistic assumption since only then we have a chance to get an experimental verification.

At this stage, a general experience with our natural laws can help. In physics, symmetries are playing a fundamental role. After preparation and still incomplete insights by A. EINSTEIN and D. HILBERT, a deciding break through was delivered by E. NOETHER (cp. Fig. 9) in 1918. The corresponding fundamental law is called to her honor NOETHER's theorem. It tells us that symmetries in physical equations lead to conservation laws. The most important example is the conservation law of energy and momentum as a consequence of the translation invariance of the space-time, cp. Pais (1982) and Günther (2012).

We remark, that a remarkable asymmetry in the *mathematical description* of the induction law of electrodynamics could not be accepted by A. EINSTEIN, and gave him the deciding stimulus for the formulation of his Special Relativity Theory, s. Chap. 9, Sect. 3 and the there in the beginning cited statement.

It does not fit in our understanding of symmetrical connections in nature that the periods does change during a motion, but the length of a bar not, or inverse the length changes but the period not.

It is our desire to uncover symmetries in nature. This drives us to the assumption, that both the length of a rod and also the period of a clock do change, when they are

Fig. 9 AMALIE EMMY NOETHER, * Erlangen 23.3.1882, † Bryn Mawr (Pennsylvania) 14.4.1935.

in motion. We will show: The connections between T_v and T_o on one hand, as well as l_v and l_o on the other hand, take a symmetrical shape, if we assume

$$l_v = l_o \sqrt{1 - v^2/c^2}\,, \qquad \text{Length changes of an in } \Sigma_o \text{ moving rod} \tag{94}$$

since then we get from Eq. (88)

$$T_v = \frac{2l_o}{c} \frac{\sqrt{1 - v^2/c^2}}{1 - v^2/c^2}\,,$$

i.e. with Eq. (87)

$$T_v = T_o \frac{1}{\sqrt{1 - v^2/c^2}}\,. \qquad \text{Period of an in } \Sigma_o \text{ moving light clock} \tag{95}$$

If we construct a clock with the period T, than the time shown by the pointer of the clock t changes inverse to the change of T, s. Eq. (32). We call the pointer position of a moving light clock t_v, then it follows from Eq. (95) for the connection between t_v and t_o of the moving and resting light clock

$$t_v = t_o \sqrt{1 - v^2/c^2}\,. \qquad \text{Pointer position of an in } \Sigma_o \text{ moving light clock} \tag{96}$$

The formulas (94) and (96) show the intended symmetry, and we have guessed both formulas for the time dilatation (90) as well as the length contraction (91).

Now we will not attempt to construct light clocks for control. But we have fixed expectations on the results, and we could already confirm the length contraction with the MICHELSON experiment.

2.2 *The General Law of Time Dilatation*

Now we formulate the generalisation of our results with the light clocks:

The period change (95) must hold for each periodic system, and today, it could be directly verified with excellent precision by Cäsium atomic clocks, cp. Problem 4.

For the pointer positions of the moving light clock with the period T_v we use now instead of t_v the denotation $t' = t_v$, as was already done in Fig. 7, since (x', t') are space and time coordinates in Σ'.

We explore the behaviour of moving and resting clocks once again in Fig. 10: The clock resting in Σ_o at position x is denoted by U_o^x. A normal clock, say U_v, is set into motion with velocity v, so that the positions of U_v are given by $x = vt$. The pointer positions on this clock are denoted as t'. At time $t = 0$ in Σ_o, the pointer of U_v might be put to $t' = 0$. The clock U_v is then situated just at $x = 0$, it shows the

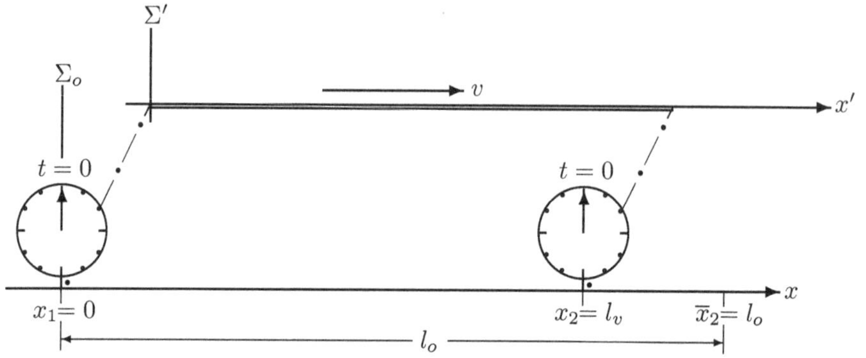

Fig. 10 Schematic illustration of the LORENTZ contraction observed in the distinguished system Σ_o. For the in the moving system Σ' resting rod, there should hold at time $t = 0$ in Σ_o the end point coordinates $x_1 = 0$ and $x_2 = l_v$. When the same rod rests in Σ_o, we measure for the coordinates of its end points $x_1 = 0$ and $\overline{x}_2 = l_o$. Dash-dotted lines connect points representing the same event.

same pointer position as the clock U_o^0 resting at the origin O of U_v. When the clock U_v arrived at the in Σ_o resting clock U_o^x at position $x = v\,t$ and showing the pointer position t, the pointer of U_v shows the time t'.

Both pointer positions are different.

The pointer position t' of a moving clock stays behind the pointer positions t of the resting clocks. Here we formulate this effect in the same way as at the LORENTZ contraction (86) of moving lengths, only for the at first distinguished reference system Σ_o. Each moving clock in Σ_o stays behind,

$$\Sigma_o:\quad t' = t\sqrt{1 - \frac{v^2}{c^2}}\,. \qquad \text{Time dilatation} \qquad (97)$$

The pointer of the clock counts oscillations. The period T_v of a with respect to Σ_o moving clock will be stretched, i.e. it becomes larger than the period T_o of the in Σ_o resting clocks. Therefore, we call this effect *time dilatation* (*from late latin dilatatio = extension*).

When in a distinguished system Σ_o for a there resting clock the eigenperiod T_o is measured, then the same clock, when it moves relativ to Σ_o with a velocity v, will show in Σ_o a streched periode T_v:

$$\Sigma_o:\quad T_v = \frac{T_o}{\sqrt{1 - v^2/c^2}}\,. \qquad \text{Time dilatation} \qquad (98)$$

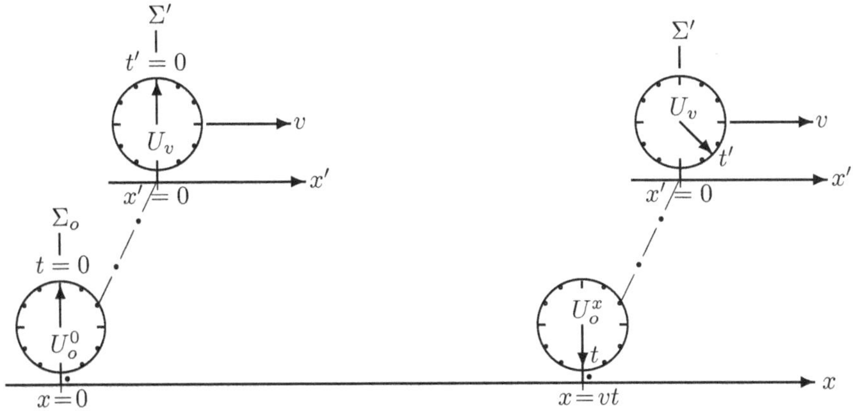

Fig. 11 Schematic illustration of EINSTEIN's time dilatation observed in the distinguished system Σ_o. The pointer position t' of the moving clock U_v stays behind the pointer positions t of the in Σ_o resting clocks, where the clock U_v passes by. Dash-dotted lines connect again points in the figure representing the same event.

3 The Physical Postulates of Relativistic Space-Time

With the two following figures, we sketch once again the facts concerning the LORENTZ contraction and time dilatation.[4] We always suppose the initial conditions (11), i.e. for $(x = 0,\ t = 0)$ it holds also $(x' = 0,\ t' = 0)$ (Fig. 11).

The high level of measuring accuracy leads to the consequence that we must abandon the hypothesis of classical space-time of an invariability of moving lengths and periods. Instead of Eqs. (65) and (66) for the classical space-time, we come to Eqs. (86) and (98) in a distinguished reference system Σ_o because of experimental results, 'the physical content, free of convention', according to Einstein (1923), the relativistic space-time.[5]

The physical postulates of relativistic space-time:

$$\Sigma_o: \quad \frac{l_v}{l_o} = \frac{1}{k} = \sqrt{1 - \frac{v^2}{c^2}}\,, \qquad \text{In } \Sigma_o \text{ the moving rod is contracted} \tag{99}$$

[4]For the geometrically interested reader we refer to the presentation of SRT with all its phenomenena and paradoxa in space-time diagrams by LIEBSCHER (2005).

[5]For those readers, that still are in doubt whether we must admit the logical *possibility* of length changes of moving rulers, and period changes of moving clocks, the nature provides an extra phenomenon, a miniature version of the Special Relativity Theory in solid states. We will discuss these issues in Chap. 12, Sects. 1 and 2, cp. also GÜNTHER (1994, 2000, 2006).

$$\Sigma_o: \quad \frac{T_v}{T_o} = \frac{1}{v\,\theta(v) + q(v)} = \frac{1}{\sqrt{1 - v^2/c^2}}\,. \tag{100}$$

When we express the periods in Eq. (100) by the pointer positions of clocks and write t_v for t' and t_o for t, then we get

$$\Sigma_o: \quad \frac{t_v}{t_o} = \sqrt{1 - \frac{v^2}{c^2}}\,. \qquad \text{In } \Sigma_o \text{ the moving clock stays behind} \tag{101}$$

The direct experience shows here again for length and time measurements a *reciprocity* $T_v/T_o = l_o/l_v$. The so-called metrical relativity principle that stands in its range of applicability between EINSTEIN's postulate and the elementary relativity principle, shows the theoretical origin of this relation, Chap. 2, Sect. 3.

As in the case of classical space-time it holds for the relativistic space-time:

We postulate Eqs. (99) and (100) indeed only for the as isotropical declared system Σ_o. In other inertial systems Σ', the quotient of moving and resting lengths and periods is then a consequence of the there *defined* synchronisation of clocks.

For later applications, we introduce the following abbreviations:

$$\left.\begin{aligned} &\gamma := \frac{1}{\sqrt{1 - v^2/c^2}}\,, \quad \beta := \frac{v}{c}\,, \quad \text{i.e.} \quad \gamma = \frac{1}{\sqrt{1-\beta^2}} \\ &\gamma \text{ is called LORENTZ factor. For differentiation we also use} \\ &\gamma_u := \frac{1}{\sqrt{1 - u^2/c^2}}\,, \quad \gamma_v := \frac{1}{\sqrt{1 - v^2/c^2}}\,, \quad \gamma_1 := \frac{1}{\sqrt{1 - v_1^2/c^2}}\,. \end{aligned}\right\} \tag{102}$$

4 Elementary Relativity—The LORENTZ Transformation

As required in Chap. 3, Sect. 2, we now introduce the synchronisation of clocks in systems Σ', again using the principle of elementary relativity (43) with $q = k$ according to Eq. (44). From the physical postulates (99) and (100) for the distinguished system Σ_o there follows according to Eq. (46)

$$\theta = \frac{T_o/T_v - l_o/l_v}{v} = \frac{\sqrt{1 - v^2/c^2} - 1/\sqrt{1 - v^2/c^2}}{v} = \frac{1 - v^2/c^2 - 1}{v\,\sqrt{1 - v^2/c^2}}\,.$$

In distinction to the classical space-time with an absolute simultaneity as consequence of the elementary relativity principle, we now require a conventional synchronisation in agreement with the same principle for the relativistic space-time, leading to LORENTZ's synchronisation parameter θ_L and then to EINSTEIN's definition of simultaneity,

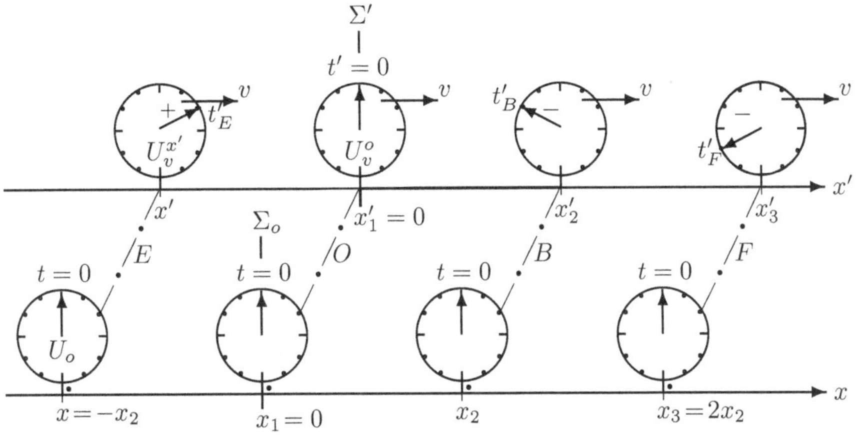

Fig. 12 The realisation of the elementary relativity principle in the relativistic space-time by means of LORENTZ's synchronisation function $\tau_L(x, v) = -v\,x/(c^2\gamma)$. At time $t = 0$ in Σ_o, the pointer positions of Σ'-clocks are fixed as $t' = -v\,x/(c^2\gamma)$. In the figure we have chosen $v = 0.8\,c$, and $\gamma = 1/\sqrt{1 - v^2/c^2} = 1.67$, and we have gauged the clocks in such a way, that the time $\Delta t_o := 2\,x_2/c$ corresponds to a pointer position of 'one quarter', i.e. $2\,x_2/c = 15$ for 60 minutes on the face plate. With it there follow the sketched pointer positions $t'_E = t'(x, 0) = -x\,v\gamma/c^2 = x_2 \cdot 0.8\,c \cdot 1.67/c^2 = 7.5 \cdot 0.8 \cdot 1.67 = 10$, $t'_B = t'(x_2, 0) = -10$, $t'_F = t'(x_3, 0) = -20$. The dash-dotted lines connect again points representing the same events, namely E, O, B, F.

$$\theta = \theta_L = \frac{-v/c^2}{\sqrt{1 - v^2/c^2}} \,. \qquad \text{LORENTZ's synchronisation parameter} \qquad (103)$$

This synchronisation is illustrated in Fig. 12. We summarise

The physical postulates of relativistic space-time	$k = \dfrac{1}{\sqrt{1 - v^2/c^2}}\,,\quad q = \dfrac{1}{\sqrt{1 - v^2/c^2}}\,,$	(104)
+		
elementary relativity principle	$\theta_L(v) = \dfrac{-v/c^2}{\sqrt{1 - v^2/c^2}}\,.$ LORENTZ's synchronisation parameter	

With Eq. (104), we get for the special coordinate transformation (22) the famous *Special* LORENTZ *transformation*

$$\left.\begin{aligned} x' &= \frac{x - v\,t}{\sqrt{1 - v^2/c^2}}\,, & \longleftrightarrow \quad x &= \frac{x' + v\,t'}{\sqrt{1 - v^2/c^2}}\,, \\ t' &= \frac{t - xv/c^2}{\sqrt{1 - v^2/c^2}}\,, & t &= \frac{t' + x'v/c^2}{\sqrt{1 - v^2/c^2}}\,. \end{aligned}\right\} \quad \text{Special LORENTZ transformation} \qquad (105)$$

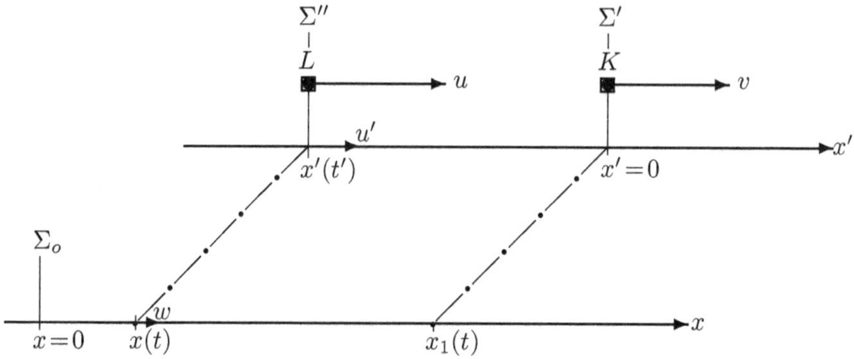

Fig. 13 EINSTEIN's addition theorem of velocities. The body K, being in motion with velocity v with respect to Σ_o, represents the reference system Σ'. The on K sitting observer looks at a body or any object L at positions $x' = x'(t')$, which moves with respect to it with velocity $u' = dx'/dt'$. The object L has the velocity $u = dx/dt$ in the reference system Σ_o, while the body K (the reference system Σ') has the velocity $v = dx_1/dt$ in system Σ_o. The observer in Σ_o measures, that L approaches to the body K with a relative velocity $w = u - v$. This velocity w is now different from the velocity u', with which the object L approaches to the body K according to the conclusion of the observer resting in Σ'. We chose as example again $v = 0.8c$. Furthermore, the velocity of object L in Σ_o might be $u = 0.9c$, so that L, observed from Σ_o, has again a relative velocity $w = u - v = 0.1c$ approaching to the body K. In the contrary, for the velocity u' one gets from the addition theorem (106) $u' = (u - v)\big/\big(1 - (uv/c^2)\big) = (0.9c - 0.8c)\big/\big(1 - (0.9c \cdot 0.8c/c^2)\big) = 0.36\,c$. Therefore, the point $x'(t')$ approaches on the x'-axis with the velocity $u' = 0.36\,c$ to the point $x' = 0$, and the point $x(t)$ approaches on the x-axis with the velocity $w = 0.1c$ to the point $x_1(t)$. The dash-dotted lines connect again points representing the same event.

For $\theta = \theta_L = -v\big/(c^2\,\gamma)$ and $k = q = 1/\gamma$, we get from the addition theorem (23) EINSTEIN's famous *addition theorem of velocities*, s. also Fig. 13,

$$u' = \frac{u - v}{1 - u\,v/c^2} \longleftrightarrow u = \frac{u' + v}{1 + u'\,v/c^2}\,. \qquad \text{EINSTEIN's addition theorem of velocities} \qquad (106)$$

The velocity u' of an object L determined in a system Σ' now differs from the in Eq. (8) used relative velocity w, which only denotes the temporal change of Σ_o measured coordinate differences of bodies L and K. The system Σ' is realised by the rest system of the body K, Fig. 13.

The case of an arbitrarily oriented velocity $\mathbf{u}$, we shall consider in Sect. 5. If measured in Σ_o, the front of a light wave has a velocity $u = c$. The velocity of the inertial systems Σ' might be v. The in Σ' measured velocity of the front is denoted by $u' = c'$. The three velocities c, c' and v are connected with EINSTEIN's addition theorem of velocities (106),

$$c' = \frac{c - v}{1 - c\,v/c^2} = \frac{c - v}{1 - v/c} = \frac{c(1 - v/c)}{1 - v/c}$$

and therefore

$$c' = c \, . \tag{107}$$

The front of a light wave has in each inertial system one and the same value, $c = 299\,792\,458 \text{ m s}^{-1}$. (108)

This statement is EINSTEIN's famous principle of the universal constancy of the speed of light. Hence EINSTEIN's relativity postulate is reproduced, s. also Problems 5 and 7.

The enormous relevance of this principle let to many new precision experiments for its verification, s. Sect. 6.

Now we came to our goal.

The system Σ_o cannot be distinguished from any other inertial system.
The LORENTZ transformation (105) holds true between every two inertial systems.

Therefore, any two inertial systems Σ' and Σ'' must be connected by a LORENTZ transformation. With velocities v and u of systems $\Sigma'(x', t')$ and $\Sigma''(x'', t'')$ in x-direction of system $\Sigma_o(x, t)$, we find from the LORENTZ transformations between Σ_o and Σ' as well as between Σ_o and Σ'' that indeed Σ' and Σ'' are connected with a LORENTZ transformation, when we apply EINSTEIN's addition theorem of velocities:

$$\left.\begin{array}{lll} x'' = \gamma_u\,(x - u\,t)\,, \quad x' = \gamma_v\,(x - v\,t)\,, & \longrightarrow & x'' = \gamma_{u'}\,(x' - u'\,t')\,, \\ t'' = \gamma_u\,(t - x\,u/c^2)\,, t' = \gamma_v\,(t - x\,u/c^2)\,, & & t'' = \gamma_{u'}/,\,(t' - x'\,u'/c^2), \\ & & u' = \dfrac{u - v}{1 - u\,v/c^2}\,, \end{array}\right\} \tag{109}$$

and in the transformation between Σ' and Σ'' appears u', also the 'right' parameter, namely, the in Σ' measured velocity $u' = (u - v)/(1 - u\,v/c^2)$ of Σ''. We will proof Eq. (109) explicitly in Problem 6.

For two events $E_1(x_1, t_1)$ and $E_2(x_2, t_1)$, that are simultaneously in one system $\Sigma_o(x, t)$ and there occur at different positions ($x_2 \neq x_1$), one reads off immediately from the LORENTZ transformation (105), that they occur in each other system $\Sigma'(x', t')$, in motion with respect to Σ_o, not any more at the same time,

$$t_1' = \gamma\,(t_1 - \frac{v\,x_1}{c^2}) \neq \gamma\,(t_1 - \frac{v\,x_2}{c^2}) = t_2' \;\text{ for }\; x_1 \neq x_2 \; .$$

The elementary relativity as applied to the relativistic space-time leads to EINSTEIN's famous *relativity of simultaneity*:

$$\Sigma_o : t_1 = t_2 \;\text{ and }\; x_1 \neq x_2 \quad \longrightarrow \quad \Sigma' : t_1' \neq t_2' \, . \tag{110}$$

Relativity of simultaneity

The connection between the temporal order of two events in different inertial systems and the causality are discussed in Problem 7.

We conclude

> The LORENTZ transformation with LORENTZ's synchronisation parameter θ_L realises the *conventional simultaneity of relativistic space-time*. (111)

All inertial systems are connected by the same form of coordinate transformation (109). Mathematically the coordinate transformations represent a group. The mathematical properties of LORENTZ transformations will be discussed in Chap. 9, Sect. 1 and Chap. 10, Sect. 2. Now it is sufficient to state that we found the mathematically simplest form with Eq. (105) that shows the equivalence of all inertial systems. In particular, we read off directly from Eqs. (105) or (109) that the inverse of a LORENTZ transformation is once again a LORENTZ transformation, in particular with velocity $-v$, as required by the elementary relativity principle.

For every inertial system Σ, there are now defined the same parameters k, q and θ.

In consequence, EINSTEIN's addition theorem of velocities (106) holds true between any two inertial systems.

Likewise the contraction of moving rods and the time dilatation of moving clocks is measured in every system Σ, since Eqs. (29) and (35) now are true in every inertial system:

$$l_v = l_o\sqrt{1-\frac{v^2}{c^2}}\,. \qquad \text{In arbitrary system } \Sigma \text{ the bar in motion is contracted} \qquad (112)$$

$$t_v = t_o\sqrt{1-\frac{v^2}{c^2}}\,. \qquad \text{In every system } \Sigma \text{ the moving clock is retarded} \qquad (113)$$

This is by no means obvious. If we abandon the elementary relativity and synchronise the clocks in Σ' in a different way from Eq. (103), an *non-conventional simultaneity* is introduced, e.g. by $\theta = 0$, a moving rod and a period of a moving clock *by definition* behave with other formulas, i.e. with other dependencies from the velocity in Σ' as given by Eqs. (112) and (113). In Sect. 7 we will discuss a problem, whose solution is best found with an non-conventional simultaneity, cp. also Problem 11.

In EINSTEIN's axiomatics, the procedure is quite different. According to EINSTEIN's relativity principle, Chap. 2, Sect. 1, the *conventional definition* of simultaneity is fixed from the beginning by means of the constant speed of light in all inertial systems as basis of the axiomatic exposition of the theory. The relativity of simultaneity follows then *by definition* for the relativistic space-time. A deviating definition of synchronisation is not possible anymore. This represents a conceptual difficulty that is often underestimated. It may lead to long discussions in resolving the relativistic paradoxa.

Our axiomatics fixes the synchronisation only at the end, so that we can in principle freely chose how to put the clocks into motion in systems Σ'. The definition of an non-conventional simultaneity leads, however, to an *asymmetric description* of space-time, from which the truly physical equivalence of inertial systems is much harder to follow. In Chap. 8, Sect. 9 we discuss two examples of non-conventional definitions of simultaneity of relativistic space-time, cp. also THIRRING (1992), Vol.1, Ch. 6.4, and GÜNTHER (1996, 2000).

In THIRRING's (1992) book, the mathematically interested reader can find a detailed analysis of motions in the classical and relativistic space-time including a discussion of the definition of simultaneity. Later these questions are again treated by GIULINI (2006) in the Lecture Notes in Physics entitled 'Algebraic and Geometric Structures in Special Relativity' with an detailed treatment of the LIE algebra of the GALILEI and LORENTZ groups, not covered in this book.

Both presentations are far beyond the here used mathematical level. Our textbook is supposed to be suitable in its basic parts for undergraduate students, where Special Relativity Theory is normally on the teaching schedule.

5 EINSTEIN's Addition Theorem for Arbitrary Oriented Velocities

EINSTEIN's addition theorem of velocities is necessary for the explanation of various relativistic effects, s. Chap. 8. Therefore, we shall consider the general case of an object that moves in Σ_o with an arbitrarily oriented velocity $\mathbf{u} = (u_x,\, u_y,\, u_z)$,

$$\Sigma_o : \mathbf{u} = (u_x,\, u_y,\, u_z) = \left(\frac{dx}{dt},\, \frac{dy}{dt},\, \frac{dz}{dt}\right). \tag{114}$$

We derive the velocity $\mathbf{u}' = (u'_{x'},\, u'_{y'},\, u'_{z'})$ of this object in the system Σ',

$$\Sigma' : \mathbf{u}' = (u'_{x'},\, u'_{y'},\, u'_{z'}) = \left(\frac{dx'}{dt'},\, \frac{dy'}{dt'}\, .\frac{dz'}{dt'}\right). \tag{115}$$

At start we assume again that Σ' has a velocity $\mathbf{v} = (v_1, 0, 0)$ with respect to Σ_o, and we use the LORENTZ factor $\gamma_1 = 1/\sqrt{1 - v_1^2/c^2}$. We insert the motion

$$x = x(t),\ y = y(t),\ z = z(t) \text{ in } \Sigma_o \text{ or } x' = x'(t'),\ y' = y'(t'),\ z' = z'(t') \text{ in } \Sigma'$$

in the LORENTZ transformation (105),

$$\left.\begin{array}{lcl} x' = \gamma_1\,(x - v_1\,t)\,, & & x = \gamma_1\,(x' + v_1\,t')\,, \\ y' = y, & \longleftrightarrow & y = y'\,, \\ z' = z\,, & & z = z'\,, \\ t' = \gamma_1\,(t - y\,v_1/c^2)\,, & & t = \gamma_1\,(t' + x'\,v_2/c^2)\,, \end{array}\right\} \begin{array}{l}\text{Motion in } x \text{ -direction} \\ \text{special} \\ \text{LORENTZ transformation}\end{array} \tag{116}$$

and we find

$$u'_{x'} = \frac{dx'}{dt'} = \frac{dx'}{dt}\left[\frac{dt'}{dt}\right]^{-1} = \gamma_1\left(u_x - v_1\right)\left[\gamma_1\left(1 - \frac{u_x v_1}{c^2}\right)\right]^{-1},$$
$$u'_{y'} = \frac{dy'}{dt'} = \frac{dy}{dt}\left[\frac{dt'}{dt}\right]^{-1} = u_y\left[\gamma_1\left(1 - \frac{u_x v_1}{c^2}\right)\right]^{-1},$$
$$u'_{z'} = \frac{dz'}{dt'} = \frac{dz}{dt}\left[\frac{dt'}{dt}\right]^{-1} = u_z\left[\gamma_1\left(1 - \frac{u_x v_1}{c^2}\right)\right]^{-1}.$$

Consequentially we get the general form of EINSTEIN's addition theorem

$$\left.\begin{aligned} u'_{x'} &= \frac{u_x - v_1}{1 - u_x v_1/c^2}, & & u_x = \frac{u'_{x'} + v_1}{1 + u'_{x'} v_1/c^2}, \\ u'_{y'} &= \frac{u_y/\gamma_1}{1 - u_x v_1/c^2}, & \longleftrightarrow\quad & u_y = \frac{u'_{y'}/\gamma_1}{1 + u'_{x'} v_1/c^2}, \\ u'_{z'} &= \frac{u_z/\gamma_1}{1 - u_x v_1/c^2}, & & u_z = \frac{u'_{z'}/\gamma_1}{1 + u'_{x'} v_1/c^2}. \end{aligned}\right\} \begin{array}{l}\text{Addition theorem for}\\ \text{velocity } (v_1, 0, 0)\\ \text{of } \Sigma' \text{ with respect to } \Sigma_o,\\ \gamma_1 = 1/\sqrt{1-\beta_1^2},\ \beta_1 = v_1/c\end{array} \quad (117)$$

Now we still consider the case that the system Σ' is in motion along the y-axis of Σ_o, i.e. $\mathbf{v} = (0, v_2, 0)$. With $\gamma_2 = 1/\sqrt{1 - v_2^2/c^2}$ the special LORENTZ transformation reads

$$\left.\begin{aligned} x' &= x, & & x = x', \\ y' &= \gamma_2\,(y - v_2\,t), & \longleftrightarrow\quad & y = \gamma_2\,(y' + v_2\,t'), \\ z' &= z, & & z = z', \\ t' &= \gamma_2\,(t - y\,v_2/c^2), & & t = \gamma_2\,(t' + y'\,v_2/c^2). \end{aligned}\right\} \begin{array}{l}\text{Motion in } y\text{-direction}\\ \text{special}\\ \text{LORENTZ transformation}\end{array} \quad (118)$$

The randomly oriented motion of an object is now seen again from Σ_o and Σ' as given in (114) and (115), resp. Now it follows

$$u'_{x'} = \frac{dx'}{dt'} = \frac{dx'}{dt}\left[\frac{dt'}{dt}\right]^{-1} = u_x\left[\gamma_2\left(1 - \frac{u_y v_2}{c^2}\right)\right]^{-1},$$
$$u'_{y'} = \frac{dy'}{dt'} = \frac{dy}{dt}\left[\frac{dt'}{dt}\right]^{-1} = \gamma_2\left(u_y - v_2\right)\left[\gamma_2\left(1 - \frac{u_y v_2}{c^2}\right)\right]^{-1},$$
$$u'_{z'} = \frac{dz'}{dt'} = \frac{dz}{dt}\left[\frac{dt'}{dt}\right]^{-1} = u_z\left[\gamma_2\left(1 - \frac{u_y v_2}{c^2}\right)\right]^{-1},$$

and then the following addition theorem,

$$\left.\begin{aligned} u'_{x'} &= \frac{u_x/\gamma_2}{1-u_y\, v_2/c^2}\,, & & & u_x &= \frac{u'_{x'}/\gamma_2}{1+u'_{y'}\, v_2/c^2}\,, \\ u'_{y'} &= \frac{u_y - v_2}{1-u_y\, v_2/c^2}\,, & &\longleftrightarrow & u_y &= \frac{u'_{y'}+v_2}{1+u'_{y'}\, v_2/c^2}\,, \\ u'_{z'} &= \frac{u_z/\gamma_2}{1-u_y\, v_2/c^2}\,, & & & u_z &= \frac{u'_{z'}/\gamma_2}{1+u'_{y'}\, v_2/c^2}\,. \end{aligned}\right\} \begin{array}{l} \text{Addition theorem for a} \\ \text{velocity } (0, v_2, 0) \\ \text{of } \Sigma' \text{ with respect to } \Sigma_o, \\ \gamma_2 = 1/\sqrt{1-\beta_2^2}\,,\ \beta_2 = \frac{v_2}{c} \end{array} \qquad (119)$$

6 Test Experiments of Special Relativity Theory

A physical theory can never be verified, but only falsified. We can never prove that a physical theory is true. We can only show, where its range of applicability ends, where it becomes wrong.

In its axiomatic structure, the Special Relativity Theory is equally consistent as the basic laws of geometry. This was explicitly shown in LIEBSCHER (1999, 2005). However it is important to check, whether the axiomatic assumptions of the Special Relativity Theory and all its consequences are in agreement with our experiences. To what degree of precision coincide the experimental results with our theoretical predictions? Here the following has to be observed:

The predictions of the Special Relativity Theory about space and time can only be sustained so long the influence of gravitating masses can be neglected. In 1915 A. EINSTEIN (s. 1923) has shown the inclusion of gravitation leads to a new superior theory, his General Relativity Theory, see Chap. 13, Sect. 1. All special relativistic effects, time dilatation, length contraction, constancy of the speed of light, addition of velocities and others, will be modified, since all masses, appearing in the equations of motion as inertial masses, they are also heavy masses and act also as sources of gravitational fields, and therefore they modify special relativistic effects. Conceptual and also in a practical sense, this is unproblematic because we already have this better theory that overcomes these restrictions, cp. Chap. 13. The importance of heavy masses for time dilatation will be discussed in Problem 4.

The actual test experiments of Special Relativity Theory ask for experimental consequences of the theory under the condition of a negligible gravity, or for a corresponding one if the gravitational influences are theoretically estimated and then eliminated. The resulting range of experiments is huge and widespread, since the entire field of theoretical physics with the single exception of gravitation is based on the Special Relativity Theory. Therefore, we have to prove not only the statements on the propagation of light and the behavior of moving measuring rulers and clocks, but also so-called secondary relativistic effects as, e.g. the pair production and the vacuum polarisation as predicted by the relativistic quantum theory, s. Gabrielse et al. (1995) . We intend to describe a few important experiments and refer to special literature for a more complete survey, e.g. Haughan and Will (1987). Also the detailed description of the experimental set-up and the discussion of its results exceed the range of our presentation.

Two elementary effects, 'which hold, *mutatis mutandis* for every system of reference, form the physical content, free of convention, of the LORENTZ transformation', EINSTEIN (1923, p. 35), the LORENTZ contraction and the time dilatation. In Sect. 4 we have shown: The Special Relativity Theory is just correct so long, or experimentally spoken, so good confirmed, as we can verify both these effects in a single reference system. The experimental precision, with which we can measure both effects determines the accuracy of the proof of the non-existence of an 'aether drift', and in this way, the universal constancy of the speed of light. It is clear that every consequence of the Relativity Theory has to be checked once again at each time.

The traditional MICHELSON–MORLEY experiment was schematically described in Sect. 1. It has been repeated from 1881 until 1930 at different places of the Earth (Potsdam, Cleveland, Mt. Wilson, Heidelberg, Pasadena, Mt. Rigi, Jena) with increasing level of experimental refinements.

To the putative proof of an aether drift in 1921 by Miller (1933) at Mt. Wilson Observatory, EINSTEIN's comment became known, 'Subtle is the Lord, but malicious He is not'. Later he added the wonderful remark: 'Nature hides its secret because of its essential loftiness, but not by means of ruse'. Shankland et al. (1955) reanalysed these experimental data and come to the expected null result, that is explained by the FITZGERALD- LORENTZ contraction hypothesis, Sect. 1. We mention the further development of these experiments by Kennedy and Thorndike (1932), as well as by Brillet and Hall (1979) and Hils and Hall (1990) in using modern Laser technology.

As we will see in Chap. 8, Sect. 4, the transversal DOPPLER effect, s. Eq. (243), $\overline{\nu} = \nu\sqrt{1 - v^2/c^2}$, is a direct expression of the time dilatation. From present point of view, it is a bit curious that the first attempts of Ives and Stilwell (1938) were performed with the aim to verify the non-existence of the transversal DOPPLER effect, i.e. the non-existence of the relativistic redshift of spectral lines, however not with the wanted result. In contrast the successful experiments of Otting (1939) have been oriented from the beginning in the confirmation of this effect.

In checking the time dilatation with the transversal DOPPLER effect, the measuring accuracy is bounded by the precision of frequency measurements. Here γ quanta emitted from excited atomic nuclei play an important role. However the energy $E_\gamma = h\nu$ of the emitted quanta cannot be simply equated to the excitation energy E_o of the atomic nucleus. At first it must be remarked that we never can use a strictly monochromatic, infinitely long wave, but always a finite wave train, that represents a finite bandwidth as seen by a FOURIER transformation. This is the *natural line width* with a so-called half width $\Delta\nu$: There the intensity dropped off from its maximum I_o at the frequency ν_o to the half maximum $I_o/2$ at the frequency $\nu_o \pm \Delta\nu$. In quantum mechanics the natural line width is explained by the finite lifetime of the emitting or absorbing quantum states. The natural line width is the basic frequency uncertainty of emitted quanta, that cannot be diminished. The relative line width $\Delta\nu/\nu_o$ can however be, e.g.[6]

$$E^o_{Fe} = 14.4 \text{ keV}$$

for

$$^{57}Fe - \text{nuclei}$$

and then

$$\Delta\nu/\nu_{Fe} = 3 \cdot 10^{-13}\,.$$

The energy E_γ of an emitted (or absorbed) quant is subject to further influences. The emission (or absorption) of a γ quant can be considered as an elementary process subject to the conservation laws of energy and momentum. The nucleus suffers a change of velocity during the emission of the γ quants, that follows from a classical, i.e. non-relativistic form of the conservation law. The nucleus might have a momentum $\mathbf{p}_K = m\,\mathbf{v}_K$ different from zero, i.e. a kinetic energy $E_1 = p_K^2/(2m)$. The direction of the emitted γ quant of momentum $\mathbf{p} = \hbar\,\mathbf{k}$ is in general not in the direction of $\mathbf{p}_K$, so that the momentum conservation requires a vector addition.

The energy of the emitted quant is denoted as $E_\gamma = h\,\nu = h\,c/\lambda = \hbar\,c\,2\pi/\lambda = \hbar\,k\,c = p\,c$ with $p = |\mathbf{p}|, k = |\mathbf{k}|$. The nucleus gets a *repulsion momentum* $\mathbf{p}_r$ during the emission. The conservation laws of energy and momentum in this elementary process then require

$$\left.\begin{aligned} \mathbf{p}_K &= \hbar\mathbf{k} + (\mathbf{p}_K + \mathbf{p}_r)\,, \\ E_o + \frac{\mathbf{p}_K^2}{2m} &= E_\gamma + \frac{(\mathbf{p}_K + \mathbf{p}_r)^2}{2m}\,. \end{aligned}\right\} \tag{120}$$

From the first equation, we get the repulsion momentum $\mathbf{p}_r = -\hbar\,\mathbf{k}$, and we find an energy shift $\Delta E = E_o - E_\gamma$ between the excitation energy of the nucleus E_o and the energie E_γ of the emittierted γ quant of

$$\Delta E = E_o - E_\gamma = \frac{\hbar^2\,k^2}{2\,m} - \frac{\hbar\,\mathbf{k}\cdot\mathbf{p}_K}{m} = \frac{E_\gamma^2}{2\,m\,c^2} - \frac{p_K\,\hbar k\,\cos\Theta}{m} = \frac{E_\gamma^2}{2\,m\,c^2} - E_\gamma\,\frac{p_K\,\cos\Theta}{m\,c}\,,$$

where Θ is the angle between vectors $\mathbf{p}_K$ and $\mathbf{k}$, therefore

$$\Delta E = \frac{E_\gamma^2}{2\,m\,c^2} - \frac{E_\gamma\,v_K\,\cos\Theta}{c}\,. \tag{121}$$

The energy shift ΔE is equal to the difference of the *repulsion energy* $E_r = E_\gamma^2/(2mc^2)$ taken up by the nucleus, and the contribution from a DOPPLER effect $p_r v_K \cos\Theta = -E_\gamma v_K\,\cos\Theta/c$. For the assumed freely moving nuclei with thermally distributed velocities, the energy of the emitted γ quanta becomes in the vast majority of cases so far shifted away, that it cannot be absorbed by other nuclei

---

[6] For the unit: If an elementary charge runs through a potential difference of one Volt, it takes up an energy of $1\,\text{eV} = 1.602 \cdot e^{-19}\,J$.

because of the small natural line width. Therefore in the overwhelming number of emission processes, no *resonance absorption* is possible, but it may happen in rare cases.

Other circumstances are met in a crystal lattice that is only movable as a whole. The velocity v_K is now the velocity of the crystal. Compared to the atomic mass, here we have a very large mass of the crystal, this cannot take up any repulsion energy E_r. It is $m \longrightarrow \infty$ therefore $E_r = E_\gamma^2/(2mc^2) \longrightarrow 0$. The energy E_r remains at the emitted γ quant if it is not absorbed from the phonons of the lattice vibrations that are multiples of a ground level $\hbar\omega$. This is only possible if $E_r > \hbar\omega$.[7] Therefore, one looks for such crystal lattices where $E_r < \hbar\omega$. Then the repulsion energy E_r cannot be submitted by the γ quanta. This happens already for the above-mentioned ^{57}Fe at room temperatures. The γ quanta of the $E_o = E^o_{Fe} = 14.4$ keV line are repulsion-free emitted in over 90% of cases and then again absorbed. For other elements, this can be achieved by cooling. The repulsion free resonance absorption is called MÖSSBAUER effect, found in 1957 by R. L. MÖSSBAUER. With the freely choosable velocity v_K of the crystal one can shift the absorption line arbitrarily as seen from the second term in Eq. (121), i.e. one allows or prevents the resonance absorption. This characterises the metrological significance of the MÖSSBAUER effect, enabling measurements with an up to this time unknown precision.

The isotop ^{57}Co has the property to decay in an excited state of the ^{57}Fe nucleus, that then emits the above mentioned 14.4 keV γ quanta. The ^{57}Co atoms are fixed in a crystal grid, also called a matrix. The emitted γ quanta should now be absorbed by ^{57}Fe nuclei in the ground level. They are bound in another matrix, a different crystal grid. The binding in a grid means a slight shift of the energy level of the 14.4 keV line. This difference in the energy levels of source and absorber by the imbedding in different crystal grids is extremely small. The measurement became only possible after the discovery of the MÖSSBAUER effect with the help of the classical longitudinal DOPPLER effect.

The frequencies ν_{Co} and ν_{Fe} have a frequency difference $\delta\nu$ in the rest state of source and observer, that prohibits an immediate resonance because of the small line width,

$$\delta\nu := \nu_{Fe} - \nu_{Co} \,. \tag{122}$$

The absorber are set in motion away from the source with a velocity $v \ll c$, i.e. away from the matrix with the ^{57}Co-atoms. According to the classical longitudinal DOPPLER effect (233), where source and absorber are approaching, we only have to replace v by $-v$, and we find that the absorber can be received at the frequency ${\nu_{Fe}}'$

$${\nu_{Fe}}' = \nu_{Fe}(1 - v/c) \,. \tag{123}$$

We write ${\nu_{Fe}}' = \nu_{Fe} - \delta\nu' = \nu_{Fe}(1 - v/c)$, so that

[7]If the grid takes up the repulsion energy E_r, the γ quant looses this amount of energy and can no longer be absorbed from the grid as it would require an energy E_r larger than E_o.

$$\delta\nu'/\nu_{Fe} = v/c\,. \tag{124}$$

If the velocity v is so long varied until $\delta\nu' = \delta\nu$, the emitted γ quanta are absorbed from the ^{57}Co matrix, i. e. source and absorber come into resonance by the classical DOPPLER shift. The required velocities v lie in the range of some millimetres per second. The measurement of the frequency shift of the DOPPLER effect, Eq. (124), represents a direct test of the time dilatation. This was realised in the rotator experiments of CHAMPENEY, ISAAK, KHAN (1963, 1965).

In the set-up of Champeney et al. (1965) the resonance was found at $v = 1.88 \cdot 10^{-4}\,\mathrm{m\,s^{-1}}$, less than two-tenth of a millimetre per second, so that we get an extremely small relative frequency shift (124) of

$$\delta\nu'/\nu_{Fe} = \delta\nu/\nu_{Fe} = v/c \approx 6 \cdot 10^{-13}\,. \tag{125}$$

As illustrated in Fig. 14, the source is placed on the centre of a rotator, so that its velocity is practically zero. The observer is put at a radius $R = 4\,\mathrm{cm}$, and has a pure transversal velocity $v_{Fe} = R\,\Omega$ with respect to the arriving γ quanta, Ω denotes the angular frequency; the term with velocity v_K in Eq. (121) is zero because of the angle $\Theta = \pi/2$.

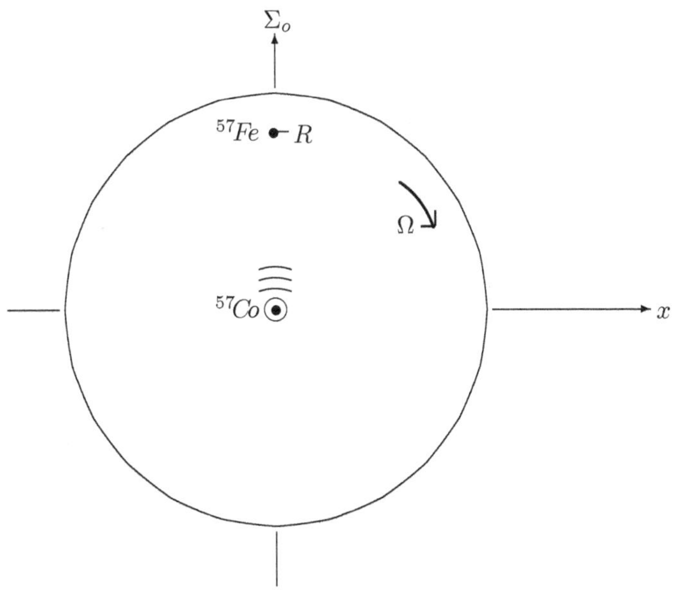

Fig. 14 Schema of the experiment of Champeney et al. (1965) for measuring the time dilatation by means of a high-frequency rotator. The source of the γ quanta sits in the centre and the absorber near to the edge. Because of the transversal DOPPLER effect, the rotation velocity Ω leads to a reduction of the absorbed frequency.

For getting again resonance absorption as for the longitudinal measurement, the authors determine a velocity of the rotor of 1313 revolutions per second, so that $\Omega = 2\pi \cdot 1313 \approx 8250\,\mathrm{s}^{-1}$, Fig. 14.

Due to the transversal DOPPLER effect (243), we can write for the frequency $\nu_{FeT}{}'$ of the absorber atoms, moving transversal to the incoming waves on the edge of the rotors[8]

$$\nu_{FeT}{}' = \nu_{Fe}\sqrt{1 - \frac{v_{Fe}^2}{c^2}} = \nu_{Fe}\sqrt{1 - \frac{R^2\,\Omega^2}{c^2}} \approx \nu_{Fe}\Big(1 - \frac{1}{2}\,\frac{R^2\Omega^2}{c^2}\Big)\,. \tag{126}$$

With $\nu_{FeT}{}' = \nu_{Fe} - \delta\nu_T$ it follows (for $R = 4 \cdot 10^{-2}$ m)

$$\frac{\delta\nu_T}{\nu_{Fe}} = \frac{1}{2}\,\frac{8250^2 \cdot 4^2 \cdot 10^{-4}}{3^2 \cdot 10^{16}} \approx 6 \cdot 10^{-13}\,, \tag{127}$$

and therefore with Eq. (125) $\delta\nu_T = \delta\nu' = \delta\nu$. On this way, the formula (243) for the transversal DOPPLER effect is confirmed, and therefore also the time dilatation. The IVES–STILWELL experiments confirmed Eq. (243) with 1% accuracy. By means of MÖSSBAUER experiment, the accuracy became about 0.001%, and later even about 0.00001%, cp. Grieser et al. (1994).

A further test of the time dilatation is presented by the observation of unstable elementary particles at high energies, as is discussed in Problem 16, cp. the experiments of Durbin et al. (1952) with π^+ mesons and of Burrowes et al. (1959) with K mesons. Here, the rest life-time in the laboratory T_o is compared with the life-time T of mesons moving with high velocities, i.e. the equation $T = T_o\sqrt{1 - v^2/c^2}$. The mesons decay according to the equation $N = N_o \exp[-t/T]$. When they are traversing a length L with a velocity v, i.e. $N = N_o \exp[-L/(vT)]$, one counts the particle numbers at the start and at the end of the distance and gets in this way the life time T, that is compared with the known rest lifetime T_o.

A very high accuracy in measuring time dilatation can also be achieved with the help of accelerator storage ring experiments. There charged elementary particles circulate in vacuum circular torus rings let by strong magnetic coils, which comprise this torus.

FARLEY et al. (1966, 1968) observed fast μ^- mesons that run and decay with a life-time T in the storage ring. This is again compared to the known lifetime T_o of μ^- mesons at rest.

GRIESER et al. (1994; 1996) put Li^+ ions in a storage ring on a very high velocity. By means of Laser-spectroscopy a precise measurement of excitation emission lines are compared with the corresponding rest wave length. Like in the historical IVES-STILWELL-OTTING experiments here the time dilatation is tested with the transversale DOPPLER effect $\overline{\nu} = \nu\sqrt{1 - v^2/c^2}$. The old precision of about 1% accuracy of

[8] The condition (234) for application of the pure transversal DOPPLER effect is here fulfilled. In our case we use in Eq. (234) $\Omega R/c \ll R\,\nu/c$, hence $\Omega \ll \nu$. Ω has an order of magnitude of about 1000 Hz, while 14.4 keV quanta correspond to a frequency of $3.5 \cdot 10^{18}$ Hz. This follows from $h\,\nu = 14.4$ keV with $h = 4.14 \cdot 10^{-15}$ eV s.

the confirmation of the time dilatation is then reduced to a remarkable maximum deviation of 0.00008 %.

A focus of experimental research has been since long time the exclusion of the existence of an *aether*, — or even the verification of its existence. The latter would distinguish a reference system, the rest system of the aether, i.e. it would immediately exclude any relativity principle. The fundamental of SRT is the universal constancy of the speed of light and thus the nonexistence of an aether. For the experimental test of this statement, one asks two different questions:

1. If protons with an energy of $19.2 \cdot 10^9$ eV are shot on a beryllium target, there are produced π^o-mesons that propagate with the extremely high velocity of $v = 0.99975\, c$ with respect to the laboratory system and then decay in two γ-quanta, $\pi^o \longrightarrow \gamma_{\rightarrow} + \gamma_{\leftarrow}$. In the rest system of the pion, these photons run in opposite direction with the speed of light c. The velocity c' of these photons is now measured in the laboratory system, i.e. the addition theorem of velocities is tested. If there would be deviations from EINSTEIN's result of Eqs. (106) and (107), $c' = c$, one would get an indication of an distinguished reference system, the rest system of a supposed aether. Conceivable deviations from the equation $c' = c$ are below 0.013 %, cp. Alvänger et al. (1964).

2. An other test is a possible *dispersion*, a dependency of the speed of light from its frequency. If one supposes that the rest mass of photons is exactly zero, then the reasons for such a dispersion could be a discrete structure of our vacuum, cp. Chap. 12, Sect. 1. Using astronomical observations on pulsars, no dispersion was found up to frequencies of $2.5 \cdot 10^{20}$ Hz, corresponding to MeV- γ-quanta. Therefore, the relative deviation of the speed of light is $\Delta c/c < 10^{-14}$, cp. Rawls (1972). In terrestrial measurements up to an energy of γ-Quanten of 7 GeV a precision of $\Delta c/c < 10^{-5}$ was observed, cp. Brown et al. (1973).

We conclude:

With an up to now possible precision the vacuum speed of light c is a universal constant in each inertial system and for each frequency.

7 Linear Approximation of the Special Relativity Theory

The inertial system Σ' might be in motion with velocity v with respect to Σ_o. We want to describe the same phenomenon from both systems Σ_o and Σ'. As linear approximation of the SRT, we understand the case that the velocity v is small compared to the speed of light c,

$$\frac{v}{c} \ll 1 \longrightarrow \frac{v^2}{c^2} \approx 0 \,. \qquad \text{Linear approximation of Special Relativity Theory} \qquad (128)$$

Equation (128) means that in derivations we keep linear terms in v/c and neglect higher powers of this ratio. Or one supposes that our measuring accuracy is not sufficient to detect higher orders in v/c.[9]

We consider the following TAYLOR expansions, where the points mean higher order terms in v/c,

$$\left.\begin{aligned} \sqrt{1-\frac{v^2}{c^2}} &= 1-\frac{1}{2}\frac{v^2}{c^2}+\ldots, \\ \frac{1}{\sqrt{1-v^2/c^2}} &= 1+\frac{1}{2}\frac{v^2}{c^2}+\ldots, \\ \frac{1}{1-v/c} &= 1+\frac{v}{c}+\ldots. \end{aligned}\right\} \tag{129}$$

From a comparison of the physical postulates (65) and (66) of the classical space-time with the corresponding relativistic formulas (99) and (100) it follows immediately that the relativistic space-time in the linear approximation goes over to the classical space-time:

> The in v/c linear approximation of the relativistic space-time is physically equivalent with the classical space-time.
> All in v/c linear effects can be basically understood in the framework of the classical space-time. (130)

In addition to the classical, in v/c linear effects, the Special Relativity Theory provides nonlinear corrections, that we will investigate in Chap. 8, Sects. 5 and 6, e.g. the DOPPLER effect and the aberration. These are genuine relativistic effects, beginning in the order v^2/c^2, and absent at the classical level. There one must either measure very accurately, or the velocity v must become high, one tells this relativistic speeds. Examples are the THOMAS precession discussed in Chap. 8, Sect. 3, and the transversal DOPPLER effect in Chap. 8, Sect. 5.

The situation looks apparently different, when we apply the linearisation in v/c to the coordinate transformation. From Eqs. (128) and (129), it follows the linear approximation of the LORENTZ transformation (105)

$$\begin{aligned} x' &= x - v\,t, \\ t' &= t - \frac{v}{c}\frac{x}{c}, \end{aligned} \quad\longleftrightarrow\quad \left.\begin{aligned} x &= x' + v\,t', \\ t &= t' + \frac{v}{c}\frac{x'}{c}. \end{aligned}\right\} \quad \text{Linear approximation of the LORENTZ transformation} \tag{131}$$

The transformation (131) now looks different from the GALILEI transformation (69). This difference remains puzzling, when the conventional character of simultaneity is forgotten.

[9] It should be observed that the linearity of space-time in the coordinates x and t, agreed upon in Chap. 1, Sect. 4, has nothing to do with the linearisation in v/c here discussed.

The classical and the relativistic space-time are characterised by the results of measurements, described by Eqs. (65) and (66) in the classical case, and by Eqs. (99) and (100) in the relativistic case.

If one abandons the important symmetric mathematical structure of coordinate transformations, suggested by the principle of elementary relativity, then we can choose an arbitrary synchronisation parameter for fixing the clocks in systems Σ', e.g. θ_L for the classical space-time and θ_a for the relativistic case.

We remind the reader

> The LORENTZ synchronisation parameter $\theta_L = -\gamma\, v/c^2$ leads to the conventional simultaneity in the relativistic space-time and an non-conventional simultaneity for the classical space-time.
> Likewise the absolute synchronisation parameter $\theta_a = 0$ leads to the conventional simulaneity in the classical space-time and an non-conventional simulataneity for the relativistic space-time. (132)

With the choice $\theta_L = -\gamma\, v/c^2$ for the classical space-time it follows from Eq. (66) that $q = 1 - v\,\theta_L = 1 + \gamma\, v^2/c^2$. Together with $k = 1$ according to Eq. (65), it follows from the coordinate transformation (22) instead of the GALILEI transformation (69) at first

$$x' = x - v\,t\,, \quad t' = -\gamma\frac{v}{c^2}\,x + \left(1 + \gamma\frac{v^2}{c^2}\right)\,t\,. \qquad \text{Classical space-time with an non-conventional simultaneity} \qquad (133)$$

The second formula in (133) contains terms of order v^2/c^2 not measurable in classical precision. If we consequently disregard in Eq. (133) the nonlinear terms in v/c that means we set the LORENTZ factor $\gamma = 1$, then we get instead of Eq. (133) for the classical space-time the transformation formula

$$x' = x - v\,t\;, \quad t' = t - \frac{v}{c}\frac{x}{c}\,. \qquad \text{Classical space-time with an non-conventional simultaneity} \qquad (134)$$

A comparison of Eqs. (134) and (131) now shows:
In v/c linearised LORENTZ *transformation leads to a description of classical space-time with an non-conventional simultaneity, by using the synchronisation parameter $\theta_L = -\gamma\, v/c^2$ instead of $\theta_a = 0$ and a following linearisation in v/c.*

The linearised LORENTZ transformation and the GALILEI transformation are different only in the definition of simultaneity.

There are applications of this procedure, s. LIEBSCHER (1998), Günther (2001). Especially for experiments with light it can be mathematically advantageous first to evaluate relativistic expressions, and then to derive effects of classical space-time by a following linearisation in v/c, cp. Chap. 8, Sects. 1 and 6. Exactly as in the relativistic space-time, there follows the relativity of simultaneity (110) from the LORENTZ transformation (105), now we get also for the classical space-time the relativity of simultaneity from the in v/c linearised LORENTZ transformation (134) or (131). The reason is simply that for the classical space-time the clocks are not synchronised as in Chap. 3, Fig. 1, but according to the agreement in Chap. 4, Fig. 12 and the following linear approximation in v/c. In Problem 12, we shall proof this explicitly.

Here we remark that all experiments in physics so far are carried out with a conventional regulation of clocks: the experiments of classical physics with the absolute simultaneity and precision experiments that take us to the relativistic case with a synchronisation that realises by definition a constant speed of light. An experimental verification of formulas that are based on an non-conventional definition of simultaneity, requires a careful examination of the corresponding setting of the clocks.

But all physically measurable effects are affected by a change in the definition of simultaneity only if we need to use two clocks at different places to determine some experimental quantities, so the synchronisation of these two clocks are entering directly into the measurement, as we have shown in Chap. 1, Sect. 3 in measurements of velocities, s. statement (1).

We will show this again with the speed of light.

Let be $c' = c$ the speeds of light measured in systems Σ' and Σ_o in the relativistic space-time with conventional definition of simultaneity according to EINSTEIN. The system Σ' might be in motion with velocity v in x-direction of Σ_o.

Now we consider the classical space-time with the conventional definition of simultaneity, and the speed of light in Σ_o might be c. With the addition theorem (70) of the GALILEI transformation (69) we get then for the classical value c'_{cl} of the speed of light in Σ'

$$c'_{cl} = c - v = c\left(1 - \frac{v}{c}\right). \tag{135}$$

Therefore the classical value for c'_{cl} differs clearly by an effect of first order in v/c from the relativistic value c, while we expect the difference between classical and relativistic measurement to be of second order in v/c!

The velocities measured in system Σ' can only be compared, when the same synchronisation of clocks is taken, s. Chap. 1, Sect. 3, statement (1).

A light signal, emitted from $x'_1 = 0$ at time $t'_1 = 0$, will arrive to the Σ'-clock at position x'_2 for time t'_2. Then there follows a speed of light c' in Σ'

$$c' = \frac{x'_2 - x'_1}{t'_2 - t'_1} = \frac{x'_2}{t'_2},$$

naturally it depends on the synchronisation, whether the Σ'-clock x'_2 was set into operation according to Chap. 3, Fig. 1 or to Chap. 4, Fig. 12.

In order to compare the classical value of the speed of light with the relativistic one, we must take also for the classical space-time the same synchronisation for the Σ'-clocks as for the relativistic one. If we consider the conventional simultaneity for the relativistic space-time, hence taking the synchronisation parameter $\theta_L = -v/(c^2\gamma)$, then we have to take the same parameter θ_L also for the classical calculation.

As we have explicated above, we have for $\theta = \theta_L = -\gamma\, v/c^2$ in the classical space-time $q = 1 + \gamma\, v^2/c^2$ and $k = 1$.

From the addition theorem (23) we get with $u = c$ in Σ_o the classically measured speed of light $u' = \widetilde{c}'_{cl}$ in Σ'. Here we use a tilde for quantities resulting from an non-conventional synchronisation of clocks,

$$\widetilde{c}'_{cl} = \frac{c - v}{1 - \gamma\, v/c + \gamma\, v^2/c^2} \approx c\big(1 - \frac{v}{c}\big)\big(1 + \gamma\, \frac{v}{c} - \frac{\gamma^2}{2}\frac{v^2}{c^2} - \gamma\, \frac{v^2}{c^2}\big)\,,$$

so that

$$\widetilde{c}'_{cl} \approx c\,(1 - \frac{v}{c})\,(1 + \frac{v}{c})\,,$$

$$\widetilde{c}'_{cl} \approx c\,(1 - \frac{v^2}{c^2})\,,$$

$$\widetilde{c}'_{cl} \approx c\,\Big(1 + O\big(\frac{v^2}{c^2}\big)\Big)\,. \tag{136}$$

This equation means that we get the classical value of the speed of light $\widetilde{c}'_{cl}$ in Σ' also for an assumed non-conventional synchronisation, i.e. simply c, since the in v/c nonlinear terms are classically not measurable.

This classical value for the speed of light $\widetilde{c}'_{cl}$ in Σ' only differs from the relativistic value c by nonlinear terms in v/c in agreement with Eq. (130). This is expressed by the notation $O\big(v^2/c^2\big)$. In Problem 11, we discuss the same problem using the absolute simultaneity, i.e. the conventional one for the classical space-time, and the non-conventional one for the relativistic case.

Still we remark:

Clearly the GALILEI transformation can also be considered as limiting case of the LORENTZ transformation, namely in considering the limit $c \longrightarrow \infty$.

8 Overview Over the Axiomatic Structure of Special Relativity Theory

As preparation for the later elaboration of the Special Relativity Theory, Lange (1885) presented an investigation of the concept of an inertial system. This was further extended by DiSalle (1990).

It remains indisputable that the EINSTEIN-MINKOWSKI method for the derivation of the SRT is indispensable for the theoretical physicist. The EINSTEIN postulate from 1905 of the universal constancy of the speed of light, irrespective of the state of motion of the emitting source and irrespective of the observers inertial system, we have cited in statement (38). We will continue to elaborate in Chap. 9, Sect. 1 how the LORENTZ transformation can be derived from Eq. (318). Part 1 of EINSTEIN's postulate also contains the general condition to be applied to all physical equations that no inertial system is distinguished from all others. The MINKOWSKI method provides the required mathematical apparatus for finding the basic physical laws. This is the traditional method for representing SRT. Besides the books from Einstein already cited, further textbooks are listed in the bibliography.

Einstein's fundamental paper from 1905 was based on a discussion of electrodynamics and mechanics with the aim to put it into a uniform theoretical concept. Soon later there appeared a number of analyses and axiomatic foundations of the SRT which, among other objectives, aimed to avoid its justification by electrodynamics.

Already Ignatowski (1910) has shown that the principle of the universal constancy of the speed of light must not be assumed from the beginning, but can be replaced by a number of 'relativity properties'.

This idea was further explored by Frank and Rothe (1912). They replaced Einstein's postulate by three axioms for the transition from one inertial system to an other one that can be summarised as follows: 1. Parallel straight lines go again into straight lines. 2. Two uniform and parallel velocities in one system go over to the same velocities in an other system. 3. The inverse transformation follows, when one replaces the velocity v by $-v$.

On an advanced mathematical level, Giulini (2006) discusses the algebraic and geometric structure of the SRT as basis for the derivation of the Lorentz transformation.

In distinction to all previous presentations of the SRT, we have here presented a method that will be summarized on the following page, with which we could base the Special Relativity Theory, without postulating any relativity principle. The Lorentz transformation follows exclusively from the two key experiments, the length contraction of moving rods and the time dilatation of moving clocks. This must be observed *only in one inertial system* Σ_o, when we realise the *definition* of simultaneity in all other inertial systems by means of the elementary relativity, i.e. with the reciprocity principle on the other hand. As consequence of the Lorentz-transformation this principle is discussed by Berzi and Gorini (1969).

A special reference for the foundation of the SRT should be given to the thought experiment with the light clock of R. Feynman that we discussed in Sect. 2. There we use only the second part of Einstein's postulates (38), the requirement that the speed of light in one *single* inertial system Σ_o does not depend on the state of motion of the emitting source. This was formulated in statement (5). The postulates of the symmetry of the basic laws and the homogeneity and isotropy of our space-time then require already the Lorentz transformation—without any further experiment.

Chapter 5
The Complete Theory on Two Pages

We start with a *temporarily distinguished* **initial system** $\Sigma_o(\mathbf{x}, t)$.
The system $\Sigma'(\mathbf{x}', t')$ might have a velocity $(v, 0, 0)$ with respect to Σ_o.

The coordinates should be the values of spatial and time measurements.

l_v and l_o are the lengths of the same rod moving and resting, respectively, in Σ_o.

T_v and T_o are the periods of the same clock moving and resting, respectively, in Σ_o.

The homogeneity and isotropy of space-time is postulated in Σ_o.
For a linear synchronisation in Σ' it follows: The coordinate transformations are linear,

$$\left.\begin{aligned} x' &= k(x - v\,t)\,, \quad y' = y\,, \quad z' = z\,, \\ t' &= \theta\,x + q\,t\,. \end{aligned}\right\} \tag{I}$$

For a velocity u in Σ_o, there follows a value u' in Σ' according to the addition theorem

$$u' = \frac{k(u - v)}{\theta\,u + q}$$

$$\text{with the special case} : u = 0 \quad \longrightarrow \quad u' = -\frac{k}{q}\,v\,. \tag{II}$$

From (I) we get the result

$$\frac{l_o}{l_v} = k(v)\,, \quad \frac{T_o}{T_v} = v\,\theta(v) + q(v)\,. \tag{III}$$

H. Günther and V. Müller, *The Special Theory of Relativity*,
https://doi.org/10.1007/978-981-13-7783-9_5

Elementary relativity principle: $u' = -v$ for $u = 0$. From (II) follows then

$$k = q\,,$$

and with (III) we get the parameter θ for the synchronisation of clocks in Σ',

$$\theta = \frac{T_o/T_v - l_o/l_v}{v}\,.$$

Experimental determination of l_v/l_o and T_v/T_o in Σ_o:

classical space-time

$$l_v = l_o\,, \quad T_v = T_o\,.$$

$$\Downarrow$$

$$k = q = 1\,, \quad \theta = 0\,.$$

relativistic space-time

$$l_v = l_o\sqrt{1 - v^2/c^2}\,, \quad T_o = T_v\sqrt{1 - v^2/c^2}\,.$$

$$\Downarrow$$

$$k = q = \frac{1}{\sqrt{1 - v^2/c^2}}\,, \quad \theta = \frac{-v/c^2}{\sqrt{1 - v^2/c^2}}\,.$$

Substituting this in (I) provides:

GALILEI *transformation*

$$x' = x - v\,t\,, \quad y' = y\,, \quad z' = z\,,$$
$$t' = t\,.$$

LORENTZ *transformation*

$$x' = \frac{x - v\,t}{\sqrt{1 - v^2/c^2}}\,, \quad y' = y\,, \quad z' = z\,,$$
$$t' = \frac{t - v\,x/c^2}{\sqrt{1 - v^2/c^2}}\,.$$

From the mathematical structure of these transformations, there follows the equalitz of all inertial systems. It holds GALILEI's or EINSTEIN's relativity principle.

Chapter 6
NEWTONian Mechanics

Until now, we have seen that the structure of our space-time provides the same picture in each inertial system. The principle of relativity now requires more. Also, the physical laws should stay the same in each inertial system. This physical experience was made at first in mechanics. This is now our subject. Electrodynamics will be studied in Chap. 9, Sect. 3.

The formulation of the NEWTONian mechanics has *not* the classical space-time as preposition. The NEWTONian laws assume, however, the validity of a relativity principle in mechanics:

It is not posible to perform an experiment in mechanics, by which one inertial system is distinguished among all others.	(137)

Now, we are concerned with the extension of the relativity postulate to the physical laws, here to the laws of mechanics.

Starting from the elementary relativity in the description of space-time, we have to formulate an empirical insight, that the laws of motion in mechanics have to conform with the principle (137). This statement reads as follows[1]:

If in physics a force is measured in two inertial systems, then the results of the two measurements coincide.	(138)

If in a system Σ_o there might, for example, act a force $F = 1\,\text{N}$, on a body K, we consider only one dimension, so one would measure also a force of $F' = 1\,\text{N}$ in Σ'. (Here, N is an abbreviation for 'Newton', the unit of force introduced below in the SI-system.)

[1]For the LORENTZ force in the relativistic mechanics of charged particles, this statement can be explicitly verified with the help of electrodynamics, s. Chap. 9, Sects. 2 and 3

H. Günther and V. Müller, *The Special Theory of Relativity*,
https://doi.org/10.1007/978-981-13-7783-9_6

This statement is not trivial. With its help, the NEWTONian axioms of mechanics can be adapted either on the classical or on the relativistic space-time, so the equivalence of all inertial system is assured.

1 NEWTONian Axioms

The first NEWTON*ian axiom* asserts, that there exists a reference system, where bodies that are not suspended to physical forces, remain in a state of rest or of uniform motion. This axiom is also called GALILEI's law of inertia. The corresponding reference systems are called *inertial systems*, introduced in Chap. 1, Sect. 2.

The second NEWTON*ian axiom* describes the action of a force **F** on a mass m in motion $\mathbf{x} = \mathbf{x}(t)$ with velocity $\mathbf{u} = \mathbf{dx/dt}$. In an inertial system, the temporal change (the time derivative) of the momentum $\mathbf{p} = m\mathbf{u} = md\mathbf{x}/dt$ is proportional to the force,

$$\frac{d}{dt}\,\mathbf{p} \equiv \frac{d}{dt}\,(m\mathbf{u}) \sim \mathbf{F} \quad \longrightarrow \quad \frac{d}{dt}\,(m\mathbf{u}) = k\,\mathbf{F}\,. \qquad \text{Second NEWTONian axiom} \qquad (139)$$

The constant of proportionality k is determined by the unit of force.

At first, we remind on the definition of the unit of mass. As already defined about 200 years ago, it holds: The kilogram prototype is reposited in Paris and copied to different national metrological centres. It defines the mass unit in the SI-convention[2]:

1 kg is the unit for the mass in the SI-system.

Then we define: One *Newton* is the force, that a resting mass of one kilogram gives an acceleration of one metre per second squared. Since we do not know whether the mass is changing with its velocity, we make this definition unique by specifying a resting mass:

A Newton, $1\,\mathrm{N} = 1\,\mathrm{kg} \cdot 1\,\mathrm{m} \cdot 1\,\mathrm{s}^{-2}$, is the unit of force in the SI-convention.

The Newton is therefore a secondary unit for the force, derived from the basic units metre, kilogram and second of the SI-system. Then the constant in Eq. (139) is $k = 1$. The unit dyn $\equiv 10^{-5}\,\mathrm{N}$ is not in use anymore.

Also in the so-called absolute unit system, that reduces all physical units on the basic quantities length, mass and time, remains the Newton the force unit, if instead of the old cgs-units centimetre and gram the metre and kilogram are used. So in

[2]With this definition, the unit for the quantity of a material, a *mole* is $L = 6.0221367 \cdot 10^{23}$ atoms. For a carbon isotope ^{12}C, one mole has a mass of 12 g. Because of the uncertainty in the determination of the LOSCHMIDT number L (also called AVOGADRO number) this statement is still not more accurate than the definition by the Paris mass prototype.

mechanics, there is no difference between the SI-units and the modern absolute unit system.[3]

The second NEWTONian axiom (139) can now be written as

$$\frac{d}{dt}\mathbf{p} = \frac{d}{dt}(m\mathbf{u}) = \frac{d}{dt}(m\frac{d}{dt}\mathbf{x}) = \mathbf{F}\ . \qquad \text{Second NEWTONian axiom} \qquad (140)$$

If we now set $\mathbf{F} = 0$, so follows the first axiom

$$\mathbf{p} = m\mathbf{u} = \mathbf{const} \quad \text{for} \quad \mathbf{F} = 0\ . \qquad \text{First NEWTONian axiom} \qquad (141)$$

The third NEWTON*ian axiom*, the so-called reaction axiom *actio* = *reactio*, states a general property of all interaction forces. In particular, it tells us that for the force of a mass m_b acting on a mass m_a, $\mathbf{F}_{ba}$ fulfils the equation

$$\mathbf{F}_{ba} = -\mathbf{F}_{ab}\ . \qquad \text{Third NEWTONian axiom} \qquad (142)$$

According to this Eq. (142), the force a mass m_b exerts on a mass m_a is of equal amount, but of opposite directions as the force, the mass m_b exerts on the mass m_a. With Eq. (142), it is also expressed that a mass cannot exert a force (142) on itself, it holds $\mathbf{F}_{aa} = 0$.

To be more general than this formulation (142), we derive a property of the inner forces of a system of particles, that we also denote as the third axiom of mechanics.

For n particles with masses m_a at positions $\mathbf{x}_a$ there hold Eq. (140) at first for each single particle,

$$\frac{d}{dt}\mathbf{p}_a = \frac{d}{dt}(m_a\frac{d}{dt}\mathbf{x}_a) = \sum_{b=1}^{n}\mathbf{F}_{ba} + \mathbf{F}_a\ . \qquad (143)$$

Here, $\mathbf{F}_a$ are the external forces acting on particles with masses m_a.

From Eq. (142) it follows for the interaction forces $\mathbf{F}_{ba}$ the general property

$$\sum_{a=1}^{n}\sum_{b=1}^{n}\mathbf{F}_{ba} = 0\ . \qquad (144)$$

For a system of n particles that are only influenced by inner interaction forces, it follows by summation from Eqs. (143) and (144) the conservation of the total momentum $\mathbf{P} := \sum_{a=1}^{n}\mathbf{p}_a$. This form of the axiom goes back to C. HUYGENS,

[3]Not in use is the unit kilopond, 1 kp = 9.80665 N, as unit of the weight of the *kilogram of the archive* in Paris, equal to the weight of one litre water at 4 °C.

$$\mathbf{F}_a = 0: \quad \frac{d}{dt}\sum_{a=1}^{n} \mathbf{p}_a = \frac{d}{dt}\sum_{a=1}^{n} m_a \mathbf{u}_a = \frac{d}{dt}\mathbf{P} = 0. \qquad \text{Third NEWTONian axiom conservation of total momentum} \quad (145)$$

This conservation law for a closed system, with only inner forces acting on the particles, is a fundamental property. In the relativistic formulation of mechanics, we will replace the NEWTONian formulation (142) by the law (145). With Eqs. (140), (141) and (145) instead of Eq. (142), the formulation of the NEWTONian mechanics is so general that we must not distinguish between classical and relativistic mechanics.

NEWTON (Fig. 1) has allowed in his law (140), that the masses m_a can change during the motion. As simple case, one can assume that the mass m of a body depends on the amount u of its velocity $\mathbf{u}$. This, in general, allowed dependence of a mass from its velocity will be written with a curly bracket,

$$m = m\{u\}. \qquad \text{The inertial mass } m \text{ of a body may be in general a function of its velocity } u. \quad (146)$$

As long as we do not know this function $m\{u\}$, in principle we cannot operate with the NEWTONian equations.

With a thought experiment going back to R. C. TOLMAN, one can generally derive the function $m = m\{u\}$, if one starts from the third NEWTONian axiom (145) and the GALILEI transformation (69), for the classical, and the LORENTZ transformation (105), for the relativistic space-time on the momenta $m\,\mathbf{u}$ of colliding particles in an inertial system Σ_o. We will see, that it is, in particular, the addition theorem of velocities that leads to different properties of masses in the classical and relativistic space-time.

2 Classical Mechanics

In the next chapter, we will derive the function $m = m\{u\}$, the dependence of the inertial mass on its velocity for the relativistic space-time within the framework of the LORENTZ transformation by means of TOLMAN's thought experiment. The linear approximation in v/c provides the known result, the independence of the mass on the velocity within the validity of the GALILEI transformation,[4] s. also the direct proof of the following equation in Problem 14:

$$\frac{dm\{u\}}{du} = 0 \longrightarrow \frac{dm}{dt} = \frac{dm\{u\}}{du}\frac{du}{dt} = 0. \qquad \text{Constancy of the mass Classical space-time} \quad (147)$$

The mass m of a body is independent from its velocity u within the range of validity of the GALILEI transformation (105).

[4]Here, we consider masses of bodies with unchanging constitution, i.e. with unchanging numbers of atoms or elementary particles. The mass of a spaceship stays constant only as long as the mass of the expelled power fuel is included.

The motion of a body might be described from systems Σ' and Σ_o by the functions $x' = x'(t')$ and $x = x(t)$, resp. We insert these motions in the GALILEI transformation (69) and find by using two time derivatives because of $t = t'$,

$$\frac{dx'}{dt'} = \frac{dx'}{dt}\frac{dt}{dt'} = \frac{dx'}{dt} = \frac{d}{dt}(x - vt) = \frac{dx}{dt} - v \ ,$$
$$\frac{d^2x'}{dt'^2} = \frac{d}{dt}\Big(\frac{dx}{dt} - v\Big)\frac{dt}{dt'} = \frac{d^2x}{dt^2} \ .$$

Now, we also use Eq. (147) and get the characteristic acceleration terms of NEWTONian mechanics in all inertial systems Σ of classical space-time

$$\Sigma : \frac{d}{dt}\mathbf{p} = \frac{d}{dt}(m\,\mathbf{u}) = m\frac{d^2}{dt^2}\mathbf{x} = m\frac{d^2}{dt'^2}\mathbf{x}' \ . \quad \text{Momentum change, Classical space-time} \tag{148}$$

According to the statement (138), we now assume that in all inertial systems, the identical forces $\mathbf{F}$, $\mathbf{F}_a$, and $\mathbf{F}_{ba}$ are measured. Then in the classical space-time, the NEWTONian fundamental laws of mechanics get the following uniform shape for all inertial systems Σ:

GALILEI transformation: In each inertial system Σ holds

$$\frac{d}{dt}\mathbf{p} = m\frac{d^2}{dt^2}\mathbf{x} = \mathbf{F} \ . \quad \text{Second NEWTONian axiom}$$

$$\mathbf{p} = m\mathbf{u} = \mathbf{const} \quad \text{for} \quad \mathbf{F} = 0 \ . \quad \text{First NEWTONian axiom} \tag{149}$$

$$\mathbf{F}_{ba} = -\mathbf{F}_{ab} \ . \quad \text{Third NEWTONian axiom}$$

For n particles with positions $\mathbf{x}_a$ and constant masses m_a it holds

$$m_a\frac{d^2}{dt^2}\mathbf{x}_a = \sum_{b=1}^{n}\mathbf{F}_{ba} + \mathbf{F}_a \ . \tag{150}$$

For vanishing outer forces $\mathbf{F}_a$, we can express the third axiom also by means of momentum conservation law (145):

$$\frac{d}{dt}\sum_{a=1}^{n}\mathbf{p}_a = \sum_{a=1}^{n}m_a\frac{d\,\mathbf{u}_a}{dt} = \frac{d}{dt}\mathbf{P} = 0 \ . \quad \text{Third NEWTONian axiom, GALILEI transformation} \tag{151}$$

If the interaction of particles occurs only on a very small time interval δt, and the particles are in free motion before and after the interaction, we speak of an *impact* or a *collision*. The explicit evaluations of the impact process of two particles on the basis of the GALILEI transformation are discussed in Problem 13.

Fig. 1 SIR ISAAC NEWTON, * Woolsthorpe (near Grantham) 4.1.1643, † Kensington (today London) 31.3.1727.

As long as we consider bodies with velocities u, very small compared to the speed of light c ($u \ll c$), so long the NEWTONian equations (149) hold true with unchanging masses in all inertial systems, that are again realised by such bodies with small velocities. Each motion that is possible in one inertial system, is also possible in each other inertial system.

The relativity principle of mechanics (137) is realised with Eqs. (149) of the classical space-time.

As long as we cannot detect effects of the order of magnitude of v^2/c^2, so long we can be sure, that this relativity principle is experimentally confirmed in classical mechanics. Physically, we can therefore also characterise inertial system as those reference systems, where in the limit of small velocities, the NEWTONian equations in the form (149) are true.

When we have solved the Eqs. (149), we are able to predict the position of a certain body of mass m_a at a given time. Such a statement is a result of the so-called LAGRANGE *equations of motion*. However, all field theories are of another type, where one speaks of EULER*'s* (Fig. 2) *equations*. There one is interested in the temporal change of physical fields at a fixed place in space. The forces resulting from these fields represent force densities acting on a fixed point. For the connection of mechanics with field theory as, e.g. electrodynamics, one needs a formulation of the basic equations for a continuous mass density ϱ and for a force density $\mathbf{f} = d\mathbf{F}/dV$, e.g. the LORENTZ force density. This question is discussed in Problem 38, and furthermore we recommend, e.g. the book of A. Papapetrou (1974).

3 TOLMAN's Thought Experiment—Relativistic Mechanics

The application of NEWTONian axioms (145) in a single inertial system, say in Σ_o, on an ideal collision between two masses leads to the determination of the function $m = m\{\mathbf{u}\}$, the dependence of the mass m of a body on its velocity $\mathbf{u}$. This function will now be derived for the LORENTZ transformation (105) characterising the relativistic space-time.

3.1 Relativistic Mass Formula

R. C. TOLMAN presented the following thought experiment. We consider an ideal elastic collision of two ideal smooth balls A and B as shown in Fig. 3. Both balls are physically identical, with the same masses m. The ball A should have in the reference system Σ_o only a velocity component in y-direction,

$$\Sigma_o : \quad \mathbf{u}_A = (dx/dt,\ dy/dt) = (u_{Ax},\ u_{Ay}) = (0,\ w)$$

Fig. 2 LEONHARD EULER, * Basel 15.04.1707, † Petersburg 18.09.1783.

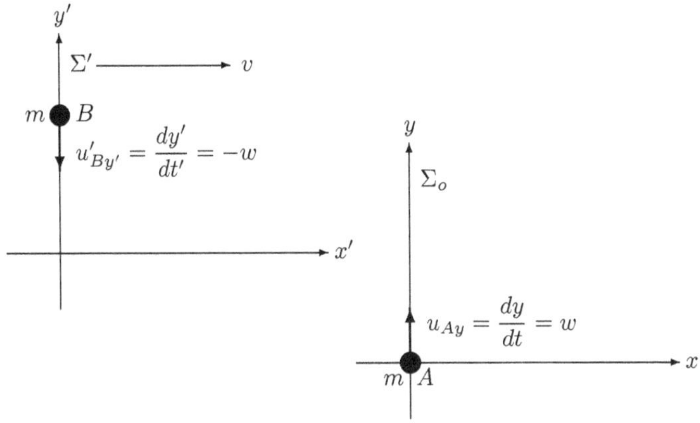

Fig. 3 Schematic illustration of TOLMAN's thought experiment.

The reference system Σ' has a velocity v in x-direction as measured from Σ_o. The ball B has only a velocity component in direction of the negative y'-axis as measured from Σ',

$$\Sigma' : \quad \mathbf{u}'_B = (dx'/dt',\ dy'/dt') = (u'_{Bx'},\ u'_{By'}) = (0,\ -w).$$

As observed from Σ_o, we get for $\mathbf{u}_B$ with (105) and the chain rule of differentiation

$$\begin{aligned} u_{Bx} &= dx/dt = dx/dt' \cdot (dt/dt')^{-1} = d(\gamma\,(x' + vt'))\Big/dt' \cdot (dt/dt')^{-1} \quad \text{and with } u'_{Bx'} = 0 \\ &= (\gamma\,(u'_{Bx'} + v)) \cdot (\gamma\,(1 + u'_{Bx'}\,v/c^2))^{-1} = (\gamma\,v)/\gamma = v \qquad \text{and} \\ u_{By} &= dy/dt = dy'/dt = dy'/dt' \cdot (dt/dt')^{-1}) = u'_{By'}\,/\,\gamma = -w\,/\,\gamma\,, \end{aligned}$$

(s. also Eq. (117)). Overall, it holds then from the viewpoint of Σ_o,

$$\Sigma_o : \left.\begin{aligned} \mathbf{u}_A &= (u_{Ax},\ u_{Ay}) = (0,\ w)\ , \\ \mathbf{u}_B &= (u_{Bx},\ u_{By}) = (v,\ -w\,/\,\gamma)\ . \end{aligned}\right\} \quad \begin{array}{c}\text{Velocity components} \\ \text{before the collision}\end{array} \tag{152}$$

The velocities v and w are chosen in such a way, and the balls are so positioned, that they will collide at the moment, when the y'-axis just matches with the y-axis, so that the balls are sitting vertically one over the other.

The assumption of ideal slippery balls means that at the collision no tangential forces, i.e. forces in x-direction are transferred. In y-direction, only forces occur satisfying the counteraction axiom, so that we can apply the conservation law of momenta (145) for the balls. With an overbar for the momenta and velocities after the impact, the conservation law for the total momentum during the impact reads

$$\mathbf{p}_A + \mathbf{p}_B = \overline{\mathbf{p}}_A + \overline{\mathbf{p}}_B \qquad \text{Conservation of momenta in}\,\Sigma_o \tag{153}$$

or in components

$$\left.\begin{aligned} m\{u_A\}\, u_{Ax} + m\{u_B\}\, u_{Bx} &= m\{\overline{u}_A\}\, \overline{u}_{Ax} + m\{\overline{u}_B\}\, \overline{u}_{Bx}\,, \\ m\{u_A\}\, u_{Ay} + m\{u_B\}\, u_{By} &= m\{\overline{u}_A\}\, \overline{u}_{Ay} + m\{\overline{u}_B\}\, \overline{u}_{By}\,. \end{aligned}\right\} \quad \text{Conservation of momenta in } \Sigma_o \qquad (154)$$

Here, we took into consideration that the masses m can depend on their velocities, and such a dependence is always denoted by curly brackets. The masses m therefore cannot be extracted from the equations.

Since no tangential forces occur, the velocities in x- and x'-direction remain unchanged after the impact: $\overline{u}_{Ax} = u_{Ax} = 0, \overline{u}_{Bx} = u_{Bx} = v$ and $\overline{u}'_{Bx'} = u'_{Bx'} = 0$. The velocity components of balls A and B after the impact can be written as

$$\begin{aligned} \Sigma_o: &\quad \overline{\mathbf{u}}_A = (0,\ \overline{w}_A)\,, \\ \Sigma': &\quad \overline{\mathbf{u}}'_B = (\overline{u}'_{Bx'},\ \overline{u}'_{By'}) = (0,\ \overline{w}_B). \end{aligned}$$

The components $\overline{u}_{Ay} = \overline{w}_A$ and $\overline{u}'_{By'} = \overline{w}_B$ are still unknown.

Observed in Σ_o it follows for $\overline{u}_{By}$ as above,

$$\overline{u}_{By} = dy/dt = dy'/dt = dy'/dt' \cdot \left(dt/dt'\right)^{-1} = \overline{u}'_{By'} \,/\, \gamma = \overline{w}_B \,/\, \gamma\,.$$

Altogether we can write for the velocity components in Σ_o after the impact

$$\Sigma_o: \left.\begin{aligned} \overline{\mathbf{u}}_A &= (\overline{u}_{Ax},\ \overline{u}_{Ay}) = (0,\ \overline{w}_A)\ , \\ \overline{\mathbf{u}}_B &= (\overline{u}_{Bx},\ \overline{u}_{By}) = (v,\ \overline{w}_B\,/\gamma)\ . \end{aligned}\right\} \quad \text{Velocity components after the impact} \qquad (155)$$

In the relativistic space-time it is $|v| < c < \infty$ and therefore $\gamma \neq 1$ for $v \neq 0$.

At first, we show with an indirect argumentation, that in this case, the mass m of a body *must* depend on its velocity.

Suppose the bodies collided ideally elastic, so that both bodies change their velocities in the result of the impact under conservation of their translationary energy. Now suppose that the mass m is a velocity independent constant, then we can reduce the fraction of m in Eq. (154), and we receive with Eqs. (152) and (155) from the second equation (154) with arbitrary v

$$w - w\,\gamma = \overline{w}_A + \overline{w}_B \,/\, \gamma\,. \qquad (156)$$

Then it follows for $v \longrightarrow 0$, i.e. $\gamma \longrightarrow 1$, that $\overline{w}_A = -\overline{w}_B$.

The same holds true also for arbitrary v, since for vanishing tangential forces $\overline{w}_A$ and $\overline{w}_B$ cannot depend on v. Equation (156) then reads

$$w(1 - 1\,/\,\gamma) = -(1 - 1\,/\,\gamma)\,\overline{w}_B\ . \qquad (157)$$

since $\gamma \neq 1$ for $v \neq 0$ we can divide by the factor $(1 - 1\,/\,\gamma)$, so that

$$\overline{w}_B = -w \quad \text{and} \quad \overline{w}_A = w\,. \qquad (158)$$

This means both balls continue running with unchanged velocities, identical as without collision, this contradicts our assumption of a collision.

The velocity independence of masses is incompatible with the LORENTZ-transformation.

In case of the GALILEI transformation, the relativistic factor γ is equal to 1 for arbitrary v, and from Eq. (157), we cannot derive the relation (158).

As simplest case, we now assume that the mass m in Σ_o is strictly monotonic, and therefore invertible unambiguous dependent on the amount of its velocity $|\mathbf{u}|$, or what is equivalent, strictly monotonic dependent on the square of the velocity,

$$\Sigma_o : \; m = m\{|\mathbf{u}|^2\} = m\{u_x^2 + u_y^2\} \; . \tag{159}$$

With (152) and (155), we get the x-component of the momentum relation (154)

$$m\{\mathbf{u}_B^2\}\, u_{Bx} = m\{\overline{\mathbf{u}}_B^2\}\, \overline{u}_{Bx} ,$$

and therefore

$$\Sigma_o : m\left\{v^2 + w^2\left(1 - \frac{v^2}{c^2}\right)\right\} v = m\left\{v^2 + \overline{w}_B^2\left(1 - \frac{v^2}{c^2}\right)\right\} v \, . \quad \begin{matrix}x\text{-component}\\ \text{of the momentum balance}\end{matrix} \tag{160}$$

For arbitrary v, this equation can only be satisfied for $\overline{w}_B^2 = w^2$. When an impact took place, what we will now assume, then the B-ball must move back in positive y-direction. The solution $\overline{w}_B = -w$ is excluded,[5]

$$\Sigma_o : \; \overline{w}_B = +w \; . \tag{161}$$

The y-component of the momentum relation (154) now reads with Eqs. (152), (155) and (161)

$$m\{\mathbf{u}_A^2\}\, u_{Ay} + m\{\mathbf{u}_B^2\}\, u_{By} = m\{\overline{\mathbf{u}}_A^2\}\, \overline{u}_{Ay} + m\{\overline{\mathbf{u}}_B^2\}\, \overline{u}_{By} ,$$

and therefore

$$\Sigma_o : \left.\begin{aligned} & m\{w^2\}\, w - m\left\{v^2 + w^2\left(1 - \frac{v^2}{c^2}\right)\right\} w / \gamma \\ &= m\{\overline{w}_A^2\}\, \overline{w}_A + m\left\{v^2 + w^2\left(1 - \frac{v^2}{c^2}\right)\right\} w / \gamma \, . \end{aligned}\right\} \quad \begin{matrix}y\text{-component}\\ \text{of momentum balance}\end{matrix} \tag{162}$$

[5] In literature, there is found sometimes the remark on classical mechanics for justification, namely the limiting case of small velocities, i.e. $\gamma \approx 1$. But with this approximation, the selection of the solution has no connection. Both in the relativistic and also the classical mechanics, there exist both solutions. Only one needs to replace the mass m of one of the two balls by the centre of mass of a system of two distant bodies, and this system will in most cases simply move apart from the other ball.

Equation (162) must hold true for arbitrary velocities v and w. At first, we consider again the limiting case $v \longrightarrow 0$, hence $\gamma \longrightarrow 1$,

$$0 = m\{\overline{w}_A^2\}\ \overline{w}_A + m\{w^2\}\ w\ , \quad \text{when} \quad v = 0\ . \tag{163}$$

For $w \longrightarrow 0$, it follows from Eq. (163) $m\{\overline{w}_A^2\}\ \overline{w}_A \longrightarrow 0$, and then, since the mass is not zero, also $\overline{w}_A \longrightarrow 0$. In the limit

$$\lim_{w\to 0} \frac{\overline{w}_A}{w} = -\lim_{w\to 0} \frac{m\{w^2\}}{m\{\overline{w}_A^2\}} = -\frac{m\{0\}}{m\{0\}} = -1$$

there follows from Eq. (163)

$$\Sigma_o : \overline{w}_A = -w + O(w^2)\ . \tag{164}$$

Here, we denote with $O(w^2)$ non linear terms in w, that we do not need for our further considerations.

We are interested in the limit $w \longrightarrow 0$ in Eq. (162). To this aim we assume at first $w \neq 0$ and we insert Eq. (164) into (162), where all non linear terms $O(w^2)$ are omitted, and we find

$$2\,m\left\{v^2 + w^2(1 - \frac{v^2}{c^2})\right\} = 2\,\gamma\, m\{w^2\}\ . \tag{165}$$

We take the limit $w \longrightarrow 0$ and denote

$$m\{0\} := m_o\ , \quad m\{v^2\} := m\ , \tag{166}$$

then it follows a dependency of the mass m on its velocity, when we denote the particle velocity again with u, while we denote, in general, the velocity of the reference system with v,

$$m = \frac{m_o}{\sqrt{1 - u^2/c^2}}\ . \qquad \text{Relativistic mass formula} \tag{167}$$

We see:

The inertial mass of a body depends according to Eq. (167) on the velocity in the same way as the period of a moving clock, Eq. (100).

We summarise:

In relativistic space-time with the LORENTZ-transformation (105), the momenta must be defined with the velocity dependent masses from Eq. (167) according to

$\mathbf{p}_a = m_a\,\mathbf{u}_a$ in order to guarantee the conservation of the total momentum $\mathbf{P} = \sum_a^n \mathbf{p}_a$ in Eq. (153). Here $m\{0\} := m_o$ defines the *rest mass* of a body.[6]

3.2 Relativistic Basic Equations of Mechanics

With Eq. (167), we have found the amendment that we need for the first only in the inertial system Σ_o formulated NEWTONian equations, when we want to fulfil both the principle of relativity (137) and also the physical postulate (99).

However with the third NEWTONian axiom, we must be careful. For two particles, sitting at time t in system Σ_o at positions $P_1(x_1, y_1, z_1)$ and $P_2(x_2, y_2, z_2)$, the third axiom (142) requires for the force $\mathbf{F}_{12}$ of particle 2 on particle 1 at position P_1 and for $\mathbf{F}_{21}$ of particle 1 on particle 2 at position P_2, that $\mathbf{F}_{12}(x_1, y_1, z_1, t) = -\mathbf{F}_{21}(x_2, y_2, z_2, t)$. Such a statement implies the simultaneity of these forces at different positions, and it is therefore not negotiable without special considerations on arbitrary inertial systems.

We will solve this problem by generalising not the original third NEWTONian axiom (142), but its consequence (145), the conservation of the total momentum when no other forces are present, now taken as part of the basic laws of mechanics.

Instead of classical Eqs. (149)–(151), there hold the following equations of motion of relativistic mechanics, and indeed in the same form in every inertial system Σ, what will be verified in the following,

LORENTZ transformation: In each inertial system Σ, it holds

$$\frac{d}{dt}\,\mathbf{p} = \frac{d}{dt}\Big(\frac{m_o}{\sqrt{1-u^2/c^2}}\,\mathbf{u}\Big) = \mathbf{F}\,.$$

Second axiom of relativistic mechanics

$$\mathbf{p} = \frac{m_o}{\sqrt{1-u^2/c^2}}\,\mathbf{u} = \mathbf{const} \quad \text{for} \quad \mathbf{F} = 0\,. \tag{168}$$

First axiom of relativistic mechanics

If only inner forces act, then it holds

$$\frac{d}{dt}\sum_{a=1}^{n} \mathbf{p}_a = \sum_{a=1}^{n} \frac{d(m_a\,\mathbf{u}_a)}{dt} = \frac{d}{dt}\,\mathbf{P} = 0\,.$$

Third axiom of relativistic mechanics

[6]The rest mass m_o is a physical parameter of a body or of a particle. The masses m in Eq. (167) are correspondingly denoted as momentum masses. In relativistic mechanics, it will be the proportionality factor between force and acceleration, defined by NEWTON's law, the inertia of a mass is in general different from the rest mass, as can be seen from the second axiom (168) taking into account Eq. (170), namely when the particle has a finite velocity. We will explicitly discuss this on Chap. 9, Sect. 2.2.

For n particles with velocities $\mathbf{u}_a = (d/dt)\mathbf{x}_a$ and rest masses m_{oa}, it holds

$$\frac{d}{dt}\mathbf{p}_a = \frac{d}{dt}(\frac{m_{oa}}{\sqrt{1-u_a^2/c^2}}\mathbf{u}_a) = \sum_{b=1}^{n}\mathbf{F}_{ba} + \mathbf{F}_a \ . \tag{169}$$

Equation (167) replaces the independence of mass from velocity following from the GALILEI transformation. This is the only change in the classical NEWTONian equations of motion. It is required that it holds in this form in all inertial systems in the relativistic space-time. Now we shall verify it. To this aim, we substitute the LORENTZ transformation (105) in Eq. (168). For simplicity, we consider only motion along the x-axis.

With $\mathbf{u} = (u, 0, 0)$, $\mathbf{p} = m\,\mathbf{u} = (p, 0, 0)$ and $\mathbf{a} = (a, 0, 0)$ for the acceleration, it follows:

$$\begin{aligned}\frac{dp}{dt} &= \frac{d}{dt}(\frac{m_o\,u}{\sqrt{1-u^2/c^2}}) = m_o\,\frac{a\sqrt{1-u^2/c^2} + au^2/(c^2\sqrt{1-u^2/c^2})}{1-u^2/c^2}\\ &= m_o\,\frac{a - au^2/c^2 + au^2/c^2}{\sqrt{1-u^2/c^2}^{\,3}} = m_o\,\frac{a}{\sqrt{1-u^2/c^2}^{\,3}} \ .\end{aligned}$$

This expression will be formulated in two systems Σ_o and Σ',

$$\left.\begin{aligned} &1.\ \Sigma_o: \quad \frac{dp}{dt} = m_o\,\frac{a}{\sqrt{1-u^2/c^2}^{\,3}}, \quad u := \frac{dx}{dt}\ , \ a := \frac{du}{dt}\ ,\\ &2.\ \Sigma': \quad \frac{dp'}{dt'} = m_o\,\frac{a'}{\sqrt{1-u'^2/c^2}^{\,3}}, \quad u' := \frac{dx'}{dt'},\ a' := \frac{du'}{dt'}\ . \end{aligned}\right\} \tag{170}$$

Now we shall show that the first and second expression in Eq. (170) are identical, when we substitute the LORENTZ transformation (105). There the undashed quantities concern the reference system Σ_o, and the dashed ones the system Σ', that is in motion in x-direction with velocity v with respect to Σ_o. The velocity v of the reference system is a constant, while the independent particle velocity u will, in general, depend on time.

We use the addition theorem (106) and take the notations γ_u, γ_v and $\gamma_{u'}$ of Eq. (102). By simply taking the square we verify the formulas

$$\left.\begin{aligned} \gamma_{u'} &= \left(1 - \frac{u\,v}{c^2}\right)\gamma_u\,\gamma_v \ , \quad u' = \frac{u - v}{1 - u\,v/c^2}\ ,\\ \gamma_u &= \left(1 + \frac{u'\,v}{c^2}\right)\gamma_{u'}\gamma_v \ , \quad u = \frac{u' + v}{1 + u'\,v/c^2}\ . \end{aligned}\right\} \tag{171}$$

Then we get

$$\frac{dp}{dt} = \frac{dp}{dt'}\left(\frac{dt}{dt'}\right)^{-1} = \frac{dp}{dt'}\left(\frac{d}{dt'}\gamma_v\,(t' + vx'/c^2)\right)^{-1} = m_o\,\frac{d}{dt'}\,(\gamma_u\,u)\,\left(\gamma_v\,(1 + vu'/c^2)\right)^{-1}$$
$$= m_o\,\left(\gamma_v\,(1 + vu'/c^2)\right)^{-1}\,\frac{d}{dt'}\,(\gamma_u\,u)$$
$$= m_o\,\left(\gamma_v\,(1 + vu'/c^2)\right)^{-1}\,\frac{d}{dt'}\left(\frac{u' + v}{(1 + u'v/c^2)}\,\gamma_{u'}\,\gamma_v\,(1 + u'v/c^2)\right).$$

Here, we resolve both parentheses in the fraction, and we take out the time-independent factor γ_v, getting

$$\frac{dp}{dt} = m_o\,\frac{1}{1 + vu'/c^2}\,\frac{d}{dt'}\,\frac{u' + v}{\sqrt{1 - u'^2/c^2}}$$
$$= m_o\,\frac{1}{1 + vu'/c^2}\,\frac{a'\sqrt{1 - u'^2/c^2} + (u' + v)\,u'a'/\left(c^2\sqrt{1 - u'^2/c^2}\,\right)}{1 - u'^2/c^2}$$
$$= m_o\frac{1}{1 + vu'/c^2}\,\frac{a'(1 - u'^2/c^2) + (u' + v)u'a'/c^2}{\sqrt{1 - u'^2/c^2}^{\,3}}$$
$$= m_o\frac{1}{1 + vu'/c^2}\,\frac{a' + vu'a'/c^2}{\sqrt{1 - u'^2/c^2}^{\,3}} = m_o\frac{1}{1 + vu'/c^2}\,\frac{a'(1 + vu'/c^2)}{\sqrt{1 - u'^2/c^2}^{\,3}}\,.$$

With Eq. (170) it holds as claimed

$$\frac{dp}{dt} = m_o\,\frac{a}{\sqrt{1 - u^2/c^2}^{\,3}} = m_o\,\frac{a'}{\sqrt{1 - u'^2/c^2}^{\,3}} = \frac{dp'}{dt'}\,. \tag{172}$$

From the validity of Eq. (168) in reference system Σ_o, it follows that the basic equation of mechanics holds true in each other inertial system Σ', if we assume in agreement with Eq. (138), that in each reference system the same forces $\mathbf{F}' = \mathbf{F}$ occur in the equations of motion.

The relativity principle of mechanics (137) is realised with Eqs. (168) for the relativistic space-time. (173)

Chapter 7
EINSTEIN's Energy–Mass Equivalence

1 The Inertia of Energy

The dependency of mass m of a particle on its velocity u from Eq. (167) now leads to very far-reaching conclusion, lying on the basis of Eq. (168) holding in each inertial system.

In classical mechanics, it is well known that one gets the conservation law of energy by scalar multiplication of the second NEWTONian axiom with velocity.[1] Therefore, we multiply here the first Eq. (168) scalar with the velocity $\mathbf{u}$ of the particle,

$$\mathbf{u}\cdot\frac{d}{dt}\Big(\frac{m_o}{\sqrt{1-u^2/c^2}}\,\mathbf{u}\Big)=\mathbf{F}\cdot\mathbf{u}\ . \tag{174}$$

With $\mathbf{a}:=\dfrac{d\mathbf{u}}{dt}$, $\dfrac{d(u^2)}{dt}=\dfrac{d}{dt}(\mathbf{u}\cdot\mathbf{u})=2\,\mathbf{u}\cdot\dfrac{d\mathbf{u}}{dt}$ holds

$$\frac{d}{dt}\frac{m_o}{\sqrt{1-u^2/c^2}}=\frac{\mathbf{a}\cdot\mathbf{u}/c^2}{\sqrt{1-u^2/c^2}^{\,3}}\,m_o$$

and therefore

$$\begin{aligned}\mathbf{u}\cdot\frac{d}{dt}(m\,\mathbf{u})&=\frac{dm}{dt}\,\mathbf{u}\cdot\mathbf{u}+m\,\mathbf{u}\cdot\frac{d\mathbf{u}}{dt}=\mathbf{u}\cdot\mathbf{u}\,\frac{d}{dt}\frac{m_o}{\sqrt{1-u^2/c^2}}+\frac{1}{\sqrt{1-u^2/c^2}}\,m_o\,\mathbf{u}\cdot\mathbf{a}\\&=\mathbf{u}\cdot\mathbf{u}\,\frac{\mathbf{u}\cdot\mathbf{a}/c^2}{\sqrt{1-u^2/c^2}^{\,3}}\,m_o+\frac{1}{\sqrt{1-u^2/c^2}}\,m_o\,\mathbf{u}\cdot\mathbf{a}=\frac{u^2/c^2+1-u^2/c^2}{\sqrt{1-u^2/c^2}^{\,3}}\,m_o\,\mathbf{u}\cdot\mathbf{a}\\&=\mathbf{u}\cdot\mathbf{a}\,\frac{m_o}{\sqrt{1-u^2/c^2}^{\,3}}=\frac{d}{dt}\,\frac{m_o\,c^2}{\sqrt{1-u^2/c^2}}=\frac{d}{dt}\,(mc^2)\,,\end{aligned}$$

---

[1] For simplicity, we formulate this in one dimension, assume a conservative force $F=-\,dV/dx$ with a potential V, and therefore $m\,\ddot{x}=-\,dV/dx$. Multiplication with $\dot{x}$ leads to $m\,\ddot{x}\,\dot{x}=-\,dV/dx\,\dot{x}$ and $d(\dot{x}^2/2)/dt=-dV/dt$, providing the conservation of energy $d(m\,\dot{x}^2/2+V)/dt=0$ where V is the potential energy.

H. Günther and V. Müller, *The Special Theory of Relativity*,
https://doi.org/10.1007/978-981-13-7783-9_7

and for Eq. (174) we can write

$$\mathbf{u} \cdot \frac{d}{dt}(m\,\mathbf{u}) = \frac{d}{dt}\frac{m_o\,c^2}{\sqrt{1-u^2/c^2}} = \frac{d}{dt}\,(mc^2) = \mathbf{F}\cdot\mathbf{u}\ . \tag{175}$$

On the right-hand side of Eq. (175) appears the power exerted by the force **F**, i.e. the work done per unit time on the with velocity **u** moving particle.

For understanding this equation, we use the TAYLOR expansion $1/\sqrt{1-x^2} = 1 + \frac{1}{2}x^2 + \frac{3}{8}x^4 + \frac{5}{16}x^6 + \cdots$ for γ

$$m\,c^2 = \frac{m_o\,c^2}{\sqrt{1-u^2/c^2}} = m_o\,c^2\left[1 + \frac{1}{2}\frac{u^2}{c^2} + \frac{3}{8}\left(\frac{u^2}{c^2}\right)^2 + \frac{5}{16}\left(\frac{u^2}{c^2}\right)^3 + \cdots\right]. \tag{176}$$

We consider a free particle, e.g. an electron in an electric field, that is resting at time t_o, so $\mathbf{u}(t_o) = 0$, and under the action of a force **F** it gets at time t a velocity u. Then integration of Eq. (175) provides in taking care of Eq. (176)

$$\int_{t_o}^{t} \frac{d}{d\tilde{t}}\,(mc^2)\,d\tilde{t} = m\,c^2\big|_{t_o}^{t} = m\,c^2 - m_o\,c^2$$

$$= m_o\,c^2\left[1 + \frac{1}{2}\frac{u^2}{c^2} + \frac{3}{8}\left(\frac{u^2}{c^2}\right)^2 + \frac{5}{16}\left(\frac{u^2}{c^2}\right)^3 + \cdots\right] - m_o\,c^2 = \int_{t_o}^{t} \mathbf{F}\cdot\mathbf{u}\,d\tilde{t}\,,$$

hence

$$\int_{t_o}^{t} \mathbf{F}\cdot\mathbf{u}\,d\tilde{t} = \int_{x_o}^{x} \mathbf{F}\cdot\,d\tilde{\mathbf{x}} = \frac{1}{2}\,m_o\,u^2 + \frac{3}{8}\,m_o\,c^2\left(\frac{u^2}{c^2}\right)^2 + \frac{5}{16}\,m_o\,c^2\left(\frac{u^2}{c^2}\right)^3 + \cdots\ . \tag{177}$$

The work on the free particle $\int_{x_o}^{x} \mathbf{F}\cdot\,d\tilde{\mathbf{x}}$ equals the increase of its kinetic energy.

In nonrelativistic mechanics with its velocity independent mass, only the term $\frac{1}{2}m_o u^2$ appears. The power $\mathbf{F}\cdot\mathbf{u}$ appears in the temporal change of the classical kinetic energy $E_{kin}^{cl} = \frac{1}{2}m_o u^2$ of the particle.

The higher powers of u^2/c^2 in Eq. (177) can therefore be considered as relativistic correction to the kinetic energy of a particle,

$$E_{kin}^{rel} = mc^2 - m_o c^2\ . \tag{178}$$

Relativistic kinetic energy

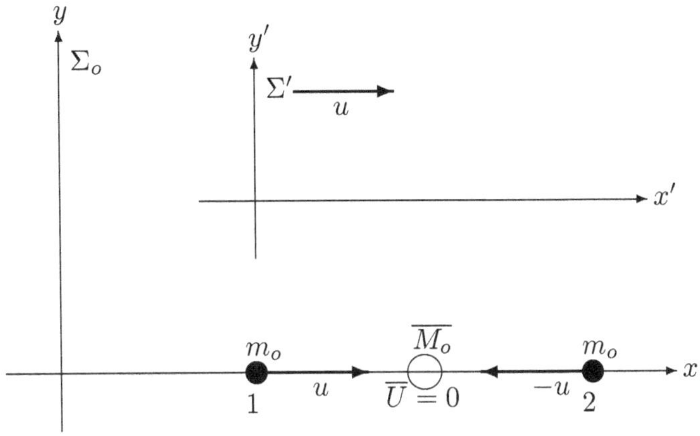

Fig. 1 Both particles 1 and 2 are colliding inelastically. After the impact, the quantities are denoted with an overbar.

What is the meaning of the term $m_o c^2$?

For answering this question, we consider a completely inelastic collision of two particles without the influence of outer forces, $\mathbf{F}_a = 0$, Fig. 1. Again it holds the conservation law of momenta, the third Eq. (168), indeed in each inertial system as proved in Chap. 6, Sect. 3.2 with the law (173).

The quantities after the impact are again denoted with an overbar.

Both particles should have identical rest masses $m_{o1} = m_{o2} = m_o$, and might have, observed in the reference system Σ_o, opposite-oriented velocities of equal amounts u along the x-axis, i.e. $\mathbf{u}_1 = (u, 0, 0)$, $\mathbf{u}_2 = (-u, 0, 0)$, and therefore $\mathbf{p}_1 = (mu, 0, 0)$, $\mathbf{p}_2 = (-mu, 0, 0)$. They collide inelastically, i.e. so that after the collision they move together as one single particle with the rest mass $\overline{M}_o$, the velocity $\overline{\mathbf{U}} = (\overline{U}, 0, 0)$ and the momentum $\overline{\mathbf{P}} = (\overline{M}\,\overline{U}, 0, 0)$.

By the third axiom in (168), the total momentum cannot change in the collision. In reference system Σ_o, this reads as $\overline{P} = P$ with $\overline{P} = \overline{M}\,\overline{U}$ and $P = p_1 + p_2$, i.e.

$$\Sigma_o\colon\ \overline{M}\,\overline{U} = \frac{m_o\,u}{\sqrt{1-u^2/c^2}} + \frac{m_o\,(-u)}{\sqrt{1-u^2/c^2}} = 0\ . \qquad \text{Conservation of momentum in } \Sigma_o \tag{179}$$

Therefore, it follows as in the classical mechanics,

$$\overline{U} = 0 \quad \longrightarrow \quad \overline{M}\{0\} = \overline{M}_o\ . \tag{180}$$

Because of the validity of the mechanical laws (168) in each inertial system, the conservation of the total momentum holds true also for the inertial system Σ', being in motion with respect to Σ_o with velocity $v = u$. The first particle is then at rest in Σ', so $u_1' = 0$. The combined particle after the collision is at rest in Σ_o, therefore, it holds $\overline{U}' = -u$ in Σ' according to the principle of elementary relativity. The velocity u_2' of the second particle before the collision is derived from the addition theorem (106),

$$u' = \frac{u - v}{1 - u\,v/c^2} \;,$$

where we take u for the velocity v of Σ', and we use instead of u the velocity $-u$ of the second particle in Σ_o. In Σ', we observe then the velocities

$$\Sigma' : \quad \left. \begin{aligned} & u_1' = 0 \;, \qquad u_2' = \frac{-2u}{1 + u^2/c^2} \;, \\ & \overline{U}\,' = -u \;. \end{aligned} \right\} \tag{181}$$

With the mass formula (167), the third axiom in (168), the conservation law of momenta, is written as $\overline{P}' = P'$ in Σ', with $\overline{P}' = \overline{M}\,\overline{U}'$ and $P' = m_1 u_1' + m_2 u_2'$, so with Eq. (181) it follows

$$\Sigma' : \; \overline{P}' = \gamma_{\overline{U}'}\,\overline{M}_o\,\overline{U}\,' = P' = \gamma_{u_2'}\,m_o\,u_2' \;. \qquad \text{Conservation of momentum in } \Sigma' \tag{182}$$

Now we still need the γ-factors. Obviously it holds $\gamma_{\overline{U}'} = \gamma_u$, and for $\gamma_{u_2'}$ we find from Eq. (181)

$$\gamma_{u_2'} = \frac{1}{\sqrt{1 - u_2'^2/c^2}} = \frac{1}{\sqrt{1 - \frac{1}{c^2}\frac{4u^2}{(1+u^2/c^2)^2}}} = \frac{c^2 + u^2}{\sqrt{(c^2 + u^2)^2 - 4c^2u^2}}$$
$$= \frac{c^2 + u^2}{\sqrt{(c^2 - u^2)^2}} \;,$$

so all together

$$\gamma_{\overline{U}'} = \frac{1}{\sqrt{1 - u^2/c^2}} \;, \quad \gamma_{u_2'} = \frac{c^2 + u^2}{c^2 - u^2} \;. \tag{183}$$

With Eqs. (183) and (182), we find in applying the third axiom in Eq. (168) for Σ'

$$\frac{-\overline{M}_o u}{\sqrt{1 - u^2/c^2}} = \frac{c^2 + u^2}{c^2 - u^2}\,\frac{-2m_o u}{1 + u^2/c^2} \;, \quad \text{hence} \quad \frac{\overline{M}_o}{\sqrt{1 - u^2/c^2}} = \frac{2m_o}{1 - u^2/c^2} \;,$$

so that

$$\overline{M}_o = \frac{2m_o}{\sqrt{1 - u^2/c^2}} = 2m \;. \tag{184}$$

With (178) we can then formulate

$$\left.\begin{aligned} 2\,(m_o c^2 + E_{kin}^{rel}) &= \overline{M}_o c^2 \\ &\text{or} \\ 2\left(m_o + \frac{E_{kin}^{rel}}{c^2}\right) &= \overline{M}_o\,. \end{aligned}\right\} \quad \text{Energy conservation in } \Sigma_o \qquad (185)$$

This is indeed the relativistic form of the energy conservation law.

Before the collision, both particles together have the relativistic kinetic energy $E_{kin}^{rel} = 2(mc^2 - m_o c^2)$. In addition, there is still for each particle the term $m_o c^2$.

The combined particle after the collision is at rest in system Σ_o and has therefore no kinetic energy. But its rest mass has increased with respect to the sum of rest masses $2m_o c^2$ of the particles before the collision by the amount E_{kin}^{rel}/c^2.

The conservation law holds true for the sum:

relativistic kinetic energy + rest mass · square of the speed of light.

After the collision, the total energy of the incoming particles is preserved in the term $\overline{M}_o\, c^2$, the relativistic energy of a particle at rest with rest mass $\overline{M}_o$. In this way, we got an interpretation of the term $m_o\, c^2$ in Eq. (178):

Each mass at rest m_o contains an energy, the rest energy $E_o = m_o\, c^2$.

The quantity mc^2 is the total energy of a particle, it contains the kinetic energy E_{kin}^{rel} and the rest energy $m_o c^2$.

It holds EINSTEIN*'s energy–mass equivalence*:

Each mass m is equivalent to an energy E.
Each energy has an inertial mass.

The conversion factor is the square of the speed of light,

$$E = m\,c^2\,. \qquad \text{Energy-mass equivalence} \qquad (186)$$

For a body in motion with velocity u, one has to use a mass $m = m\{u\}$ according to Eq. (167),

$$E = \frac{m_o}{\sqrt{1 - u^2/c^2}}\, c^2\,. \qquad (187)$$

It is important to observe that, in general, also the rest energy $m_o c^2$ is included in energy conversions, and that it cannot be omitted as an insignificant energy constant.

According to Eq. (185), energy is not removed from the system of the two colliding particles. The inner forces can only cause a transformation of one form, here of the kinetic energy of the incoming particles, into an another form of energy, the rest energy of the combined particle after the collision, cp. Problem 18.

If in a system act only inner forces, than the energy of the system remains conserved.

In classical space-time, there results also for the inelastic collision instead of Eq. (185) the conservation law of rest masses, known from classical physics, cp. Problem 14. On the contrary, a conservation of rest masses does not exist in Special Relativity Theory. Instead, the conservation of masses is equivalent with the conservation of energy.

If the inelastic collision, described by Eq. (185) occurs with macroscopic particles, the sum of the kinetic energies of these bodies E^{cl}_{kin} can be detected after the collision as heat energy Q. For satisfying the general principle of energy conservation, one formulates an conservation law for the sum of mechanical and heat energy for the energy balance of inelastic collisions. In relativistic physics, the conservation of energy is already a consequence of the equations of motion also for the inelastic collisions.

We formulate a first approximation of Eq. (185) that corrects the classical conservation of rest masses,

$$\overline{M}_o c^2 \approx 2\,m_o\,c^2 + 2\,\frac{m_o}{2}u^2 = 2\,m_o\,c^2 + E^{cl}_{kin} = 2\,m_o\,c^2 + \Delta M_o c^2 \ . \qquad (188)$$

Because of $Q = E^{cl}_{kin}$ the heat energy Q increases the inertial mass by the amount

$$\Delta M_o = \frac{Q}{c^2} = \frac{m_o u^2}{c^2} \ . \qquad (189)$$

By the warming, we increase the rest mass of a body. Here, we have shown the conversion of kinetic energy into rest energy using the example of an inelastic collision.

The most famous law of Special Relativity Theory, the energy–mass equivalence, we have derived by using the equivalence of inertial systems in the mechanical laws, and we got it only discussing mechanics. Electrodynamic was not necessary, just as in the derivation of the LORENTZ transformation.

Historically, the procedure went different. EINSTEIN (1905a, b) first discovered the equivalence of energy and mass for the energy of the electromagnetic field. EINSTEIN's very instructive thought chain are presented in the next Section, mathematically complemented by Problem 35.

In relativistic mechanics, creation and annihilation processes of elementary particles can be treated as collision processes. If the rest masses of all particles remain unchanged, one speaks of an *elastic collision*. If the rest masses of particles change, or if particles are destroyed or new particles are created, we call it *inelastic collisions*.[2]

---

[2]In Chap. 9, Sect. 2, Eq. (449), we show that energy and momentum of a particle in Special Relativity Theory are connected. In different inertial systems, both quantities decompose in different constituents. Independent physical quantities in classical physics as energy and momentum are tightly connected by the LORENTZ transformation. This connection plays a fundamental role in the mathematical formulation of relativistic theories.

The most famous examples for such processes represent the *nuclear fission* and the *nuclear fusion*, cp. Problem 17. In both cases, a part of the rest energy of the initial masses is released as kinetic energy of the reaction products, or as energy of electromagnetic radiation.

In principle, the rest energy is available for various energy transformations, if the physical conditions of the reactions are fulfilled, e.g. if the conservation laws of the various charges permit the reactions.

According to the energy–mass equivalence, mass and also rest mass cannot be transformed into energy. This is not possible simply from dimensional reasons. The sum of masses stays constant in the same way as the sum of energies. However, rest mass can be destroyed, e.g. in favour of the mass of the electromagnetic radiation or the kinetic energy. When a form of energy is transformed into an another one, then this means at the same time, a transformation of the corresponding masses. Each mass can be *converted* in an equivalent energy, with the transformation factor c^2 (in the same way as we can conclude from the number of cows on the number of hooves, without transforming a cow into a hoof). Each energy has a corresponding inertia, resulting from the equivalent mass. From EINSTEIN's formulation in the year 1905, 'Does the inertia of a body depend upon its energy content', results the shortened parlance of an *inertia of energy*.

2 EINSTEIN's Idea of the Energy–Mass Equivalence

Only a few months after his famous paper (Einstein 1905a) on the Special Relativity Theory, Einstein (1905b) presented a surprisingly simple chain of thoughts leading to the equivalence of mass and energy. In distinction to all other statements of the Special Relativity Theory, for this assertion there was no precursor. We intend to present here EINSTEIN's idea, and we will shift the complex mathematical details requiring some electrodynamics to Problem 35.

A body B might be at rest in the system Σ_o, and it has an energy U_o. In a certain finite time interval, the body might emit an amount of light of energy $\Delta E/2$ in a direction $\mathbf{k}$, and at the same time the same amount of light also of the same energy in opposite direction. We denote the quantities after the emission process with an overbar. The body B remains at rest after the light emission. Its energy is denoted as $\overline{U}_o$. The energy conservation requires, that the energy U_o of the body before the emission coincides with the sum of the energies after the emission, i.e. with both radiation energies and the energy $\overline{U}_o$ of the remaining body, so

$$U_o = \overline{U}_o + \left[\frac{\Delta E}{2} + \frac{\Delta E}{2}\right] . \qquad \text{Energy conservation in } \Sigma_o \qquad (190)$$

The system Σ' might be moving in x-direction of Σ_o with velocity v. As observed from Σ', the body has the energies U_v' and $\overline{U}_v'$ before and after the emission. Einstein (1905a) had been proven that the equations of electrodynamics remain unchanged in each inertial system. The theory behind this statement will be discussed in Chap. 9,

Sect. 3. In Problem 35, it will than be shown how the energy of the emitted light as seen from Σ' can be calculated on this basis. For this we get the amount $\gamma\,(\Delta E/2 + \Delta E/2)$, and then with Eq. (1471) the energy conservation law in Σ' reads

$$U_v' = \overline{U}_v' + \gamma\,(\Delta E/2 + \Delta E/2) \ . \qquad \text{Energy conservation law in } \Sigma' \tag{191}$$

From Eqs. (190) and (191), we get

$$\left(U_v' - U_o\right) - \left(\overline{U}_v' - \overline{U}_o\right) = (\gamma - 1)\,\Delta E \ . \tag{192}$$

The energy U_o of the in Σ_o resting body can differ from the energy U_o', when it is at rest in Σ', only by an arbitrary constant, that may be agreed upon only different for different inertial systems. So it holds before and after the collision

$$U_o = U_o' + C \ , \quad \overline{U}_o = \overline{U}_o' + C \ . \tag{193}$$

Inserting Eq. (193) into Eq. (192) one finds

$$\left(U_v' - U_o' - C\right) - \left(\overline{U}_v' - \overline{U}_o' - C\right) = (\gamma - 1)\ \Delta E \ . \tag{194}$$

The difference $\left(U_v' - U_o'\right)$ is equal to the kinetic energy U_{kin}' of the body in Σ' before the emission, and likewise the difference $\left(\overline{U}_v' - \overline{U}_o'\right)$ is equal to its kinetic energy $\overline{U}_{kin}'$ after the emission. For Eq. (194), we can then write

$$U_{kin}' - \overline{U}_{kin}' = \Delta U_{kin}' = (\gamma - 1)\ \Delta E \ . \tag{195}$$

In system Σ', it is observed that the kinetic energy of the body decreases due to the emission of electromagnetic waves, even if its velocity does not change. It must be the mass of the body that has changed due to the emission. The emitted energy of the waves is equivalent to an emitted mass. The transformation factor can be read off from the first post-classical approximation.

For the classical approximation of the kinetic energy, we write with $\Delta v = 0$

$$\Delta U_{kin}' \approx \Delta\left(\frac{m}{2}\,v^2\right) = \frac{1}{2}\,v^2 \Delta m \ . \tag{196}$$

With the TAYLOR expansion in v/c

$$\gamma = \left(1 - v^2/c^2\right)^{-1/2} \approx 1 + (1/2)v^2/c^2$$

we get the classical approximation of the right-hand side of Eq. (195)

$$(\gamma - 1)\,\Delta E \approx \left(1 + \frac{1}{2}\frac{v^2}{c^2} - 1\right)\Delta E = \frac{1}{2}\frac{v^2}{c^2}\,\Delta E \ . \tag{197}$$

The right sides of Eqs. (196) and (197) coincide, when

$$\Delta m = \Delta E/c^2 \ . \tag{198}$$

If one accepts the validity of the equivalence for all energy transformations, so that without exception each mass element Δm is equivalent to an energy ΔE, then the summation of all term (198) provides

$$E = m\,c^2 \; . \qquad \text{EINSTEIN's energy-masse equivalence} \qquad (199)$$

Chapter 8
Relativistic Phenomena and Paradoxa

1 FRESNEL's Dragging Coefficient

We consider a transparent medium (e.g. air or water) with a refraction index n that is at first at rest in an inertial system Σ_o. Observed in Σ_o the front of a light wave propagates with the velocity $u = c/n$ s. Chap. 9, Sect. 3.1.

The medium might now be at rest in the system Σ' that has a velocity v in x-direction with respect to Σ_o. In Σ_o we then observe v as constant streaming velocity of the medium.

We ask for the velocity u of propagation of light wave fronts in the streaming medium measured in Σ_o. There we consider for the streaming the classical case of slow velocities with respect to the speed of light.

$$v \ll c \,. \qquad \text{Slowly moving matter} \qquad (200)$$

The simplest solution of the problem is found with a relativistic derivation and the following linearisation in v/c. According to EINSTEIN's relativity principle, we start from the equivalence of all inertial systems also with respect to electrodynamics. This will be presented extensively in Chap. 9, Sect. 3. The light propagation in the medium, measured from system Σ' is then $u' = c/n$. For our considerations, we can abandon from physical particularities of the light propagation in a medium. We only need the composition of velocities, i.e. EINSTEIN's addition theorem (106). There we take the velocity v of the streaming medium, that defined the inertial system Σ' and the in Σ' determined light velocity $u' = c/n$. We linearise in v/c and get

$$u = \frac{u' + v}{1 + \frac{v}{c}\frac{u'}{c}} \approx (u' + v)\left(1 - \frac{v}{c}\frac{u'}{c}\right) . \qquad (201)$$

H. Günther and V. Müller, *The Special Theory of Relativity*,
https://doi.org/10.1007/978-981-13-7783-9_8

With $u' = c/n$ and under repeated negligence of the in v/c quadratic terms we get immediately FRESNEL's solution of the problem:

$$u = \frac{c}{n} + v\left(1 - \frac{1}{n^2}\right) . \qquad \text{The factor } \left(1 - 1/n^2\right) \text{ is called FRESNEL's dragging coefficient.} \tag{202}$$

The light wave front will be *carried* with the streaming medium with a reduced velocity $v\left(1 - 1/n^2\right)$.

This formula is in agreement with the classical experiment of FIZEAU measuring the light propagation in a moving medium.

The formula (202) can also be understood fully in the framework of classical physics. To this aim, one needs MAXWELL's equations for slowly moving media in order to derive the velocity u of light waves propagating in them. However, a considerable effort is needed, cf. e.g. the textbook of Becker (1964), Chap. DIV, Sect. 72, pp. 296, 304, 308.

2 A Paradox of the Dragging Coefficient

The following argumentation in the derivation of Eq. (202) for FRESNEL's dragging coefficient leads to a paradox.

With the restriction on in v/c linear terms, we remain according to Chap. 4, Sect. 7 in the range of classical space-time. We now start from the GALILEI transformation (69) with the corresponding addition theorem of velocities (70), $u = u' + v$ and there we insert $u' = c/n$. Then it follows a velocity u expected for the propagation velocity of the light wave front in Σ_o

$$u = u' + v = \frac{c}{n} + v \ , \tag{203}$$

that contradicts Eq. (202), but is also physically wrong by a simple consideration:

In the limit of an extreme dilution of the mediums, we come to the refraction index $n \approx 1$. Then we are practically in a vacuum, so that we have to take $u \approx c$ while according to Eq. (203) it follows $u = c + v$ in this limit, an obvious contradiction.

With the refraction index n as material constant and the speed of light c in Σ_o, the light propagation within an in Σ_o resting medium proceeds with velocity c/n. On the basis of the LORENTZ-transformation and the included definition of simultaneity in all inertial systems, the speed of light is also c in Σ'. Then the material constant n delivers the velocity $u' = c/n$ for the light propagation within an in Σ' resting medium. When we observe this light propagation in system Σ_o, we have to compose the velocities c/n and v according to EINSTEIN's addition theorem (106) following from the LORENTZ-transformation. Then we get for $v \ll c$ the result (202).

For the classical space-time, we cannot argue in this way. From the speed of light c in Σ_o, we find with the help of the GALILEI transformation for the propagation of light in Σ' a velocity $c'_{cl} = c - v$ so in no way again the same speed of light as in Σ_o s. Chap. 4, Sect. 7. Concerning the velocity u' of light in a medium resting in Σ' we have at first no answer at all in the framework of classical physics. Since in classical physics, we cannot claim that MAXWELL's equations are valid in equal form in all inertial systems as basis for the light propagation. For the derivation of Eq. (202), we assumed a velocity of light $u' = c/n$ in Σ' but this cannot be assumed within the range of the validity of the GALILEI transformation. Simply in the framework of classical space-time we do not know the propagation velocity u' of electromagnetic waves in the system Σ' for applying GALILEI's addition theorem (70). In so far we have shown, that the conclusion above, a velocity $u = u' + v = c/n + v$ with paradox consequences is not acceptable.

As already remarked at the end of the previous section, we can also explain the formula (202) completely within classical physics.

Basically, it remains to be added that the laws of light propagation in different inertial systems can always be understood only approximately within classical space-time, s. Chap. 4, Sect. 7.

3 THOMAS Precession

We consider three reference systems $\Sigma_o(x_i)$, $\Sigma'(x_i')$ and $\Sigma''(x_i'')$. The velocities of its coordinate origins are denoted by the three characters u g and w. A body L might have in Σ_o the velocity $\mathbf{v} = (v_1,\ v_2,\ 0)$, s. Fig. 1,

$$\mathbf{v} = (v_1,\ v_2,\ 0)\ . \qquad \text{In } \Sigma_o \text{ measured velocity of body } L \tag{204}$$

We use the abbreviation

$$\gamma_1 = \frac{1}{\sqrt{1-\frac{v_1^2}{c^2}}}\ ,\quad \gamma = \frac{1}{\sqrt{1-\frac{v_1^2}{c^2}-\frac{v_2^2}{c^2}}}\ ,\quad \beta_1 = \frac{v_1}{c}\ ,\quad \beta_2 = \frac{v_2}{c}\ . \tag{205}$$

The connecting line from the coordinate origin of Σ_o to the body L forms an angle φ with the x-axis of Σ_o,

$$\tan\varphi = \frac{v_2}{v_1}\ . \tag{206}$$

In Σ_o, we assume that the system Σ' moves *with parallel axes* against Σ_o with a velocity $\mathbf{g} = (v_1,\ 0,\ 0)$. For deriving the velocity $\mathbf{v}'$ of L in Σ' we use EINSTEIN's addition theorem of velocities (117), where we replace the velocity $\mathbf{u}$ by the velocity

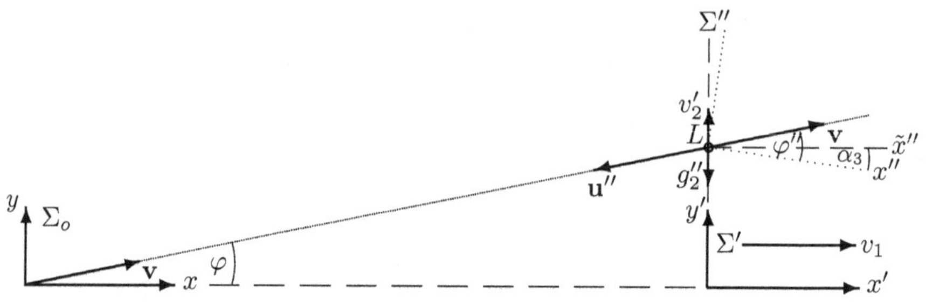

Fig. 1 Scheme of the THOMAS precession. The velocity vectors of system Σ_o, Σ' and Σ'' are denoted with **u**, **g** and **w**. For the body L, one measures in Σ_o the velocity **v**. The body L is at rest in Σ''. Hence, it holds $\mathbf{w} = \mathbf{v}$. The system Σ' is in axis parallel motion with respect to Σ_o with velocity v_1 in x-direction, hence $\mathbf{g} = (v_1, 0, 0)$ and Σ'' is in axis parallel motion with respect to Σ' with velocity v_2' in y'-direction, hence according to Eq. (207) $\mathbf{w}' = \mathbf{v}' = (0, \gamma_1 v_2, 0)$. For Σ', there is measured a velocity $\mathbf{g}'' = (0, -\gamma_1 v_2, 0)$ of Σ'' according to Eq. (209). Using the elementary relativity, we have $|\mathbf{u}''| = |\mathbf{v}|$ Eq. (211), and $|\mathbf{g}''| = |\mathbf{w}'|$ hence $|g_2''| = |v_2'|$ Eqs. (209) and (207), it follows: An observer in Σ'' states, that the connecting line of the body L tightly fixed with Σ'' and the system Σ_o, i.e. the vector $-\mathbf{u}''$ forms an angle φ'' with the x''-axis. The axes of Σ'' are represented by dotted lines. The orientation of the with small dashes drawn Σ''-axes are observed in Σ'. For the x''-axis we have indicated this with a tilde. It should be observed: The angle φ'' is measured in Σ''. For example, the orientation of the x''-axis does not coincide with the direction of this axis as measured in system Σ_o cp. Chap. 4, Sect. 5. We have drawn the relations for velocities $v_1 = 0.8\,c$, $v_2 = 0.16\,c$ and therefore $v_2' = \gamma_1 v_2 = 0.27c$. Then it follows with Eqs. (206) and (213) that the angle $\varphi \approx 11.31°$ and $\varphi'' \approx 19.08°$ and therefore $\alpha_3 := \varphi - \varphi'' = -7.77°$. Since $v_1 = 0.8\,c$, $v_2 = 0.16\,c$ the approximation (215) is not applicable anymore.

$\mathbf{v} = (v_1, v_2, 0)$ of L in Σ_o. The velocity v_1 keeps up its meaning as velocity of Σ' in Eq. (117). Therefore, we obtain for the velocity $\mathbf{u}'$ in Eq. (117) the components $\mathbf{v}' = (v_1', v_2', v_3')$ von L in Σ'[1] (cp. also the direct evaluation of $\mathbf{v}'$ in Chap. 4, Sect. 1),

$$\mathbf{v}' = (v_1', v_2', v_3') = (0, \gamma_1 v_2, 0) \ . \qquad \text{In } \Sigma' \text{ measured velocity of body } L \qquad (207)$$

The system $\Sigma''(x_i'')$ might be defined as follows: As measured from $\Sigma'(x_i')$, the system Σ'' is moving *axis parallel* along the y' axis with velocity $\mathbf{w}' = (0, v_2', 0) = \mathbf{v}'$. Hence, the body L is at rest in Σ'', Fig. 1. (The explicit transformation formulae between Σ_o and Σ'' are derived in the following Section). In system Σ'', there is measured a velocity $\mathbf{g}''$ of system Σ'. We use the elementary relativity for systems Σ' and Σ'', s. (43),

$$|\mathbf{g}''| = |\mathbf{w}'| \ , \qquad \text{Elementary relativity} \qquad (208)$$

---

[1]The following relation (207) can lead to wrong conclusions: Since γ_1 can become arbitrary large, one could conclude that then also $v_2' = v_2\,\gamma_1$ could become arbitrary large—however each velocity of a material body is limited by the speed of light c. The contradiction is resolved, when one takes into account that $v_1^2 + v_2^2 < c^2$. With increasing γ_1, i.e. with approaching of v_1 to c, the velocity v_2 must become accordingly smaller.

so with $\mathbf{w}' = \mathbf{v}'$ according to Eq. (207) taking into account the velocity directions,

$$\mathbf{g}'' = (g_1'',\, g_2'',\, g_3'') = (0,\, -\gamma_1\, v_2,\, 0)\ . \qquad \text{In } \Sigma'' \text{ measured velocity of system } \Sigma' \qquad (209)$$

Now we look for the in Σ'' measured velocity $\mathbf{u}'' = (u_1'',\, u_2'',\, u_3'')$ of system Σ_o. Since Σ_o has the same positions with respect to the z''- and y''-coordinates as Σ', it holds with (209)

$$u_2'' = g_2'' = -\gamma_1\, v_2\ , \quad u_3'' = g_3'' = 0\ . \qquad (210)$$

For evaluating the missing component u_1'', we use the elementary relativity principle in systems Σ_o and Σ'', hence $|\mathbf{u}''| = |\mathbf{w}|$. The body L is at rest in Σ'', so that $\mathbf{w} = \mathbf{v}$ and therefore

$$|\mathbf{u}''| = |\mathbf{v}| \qquad \text{Elementary relativity} \qquad (211)$$

i.e. $u''^2 = v^2$ and with Eqs. (204) and (210)

$$u_1''^2 + u_2''^2 + u_3''^2 = v_1^2 + v_2^2 + v_3^2\ ,$$
$$u_1''^2 + \gamma_1^2\, v_2^2 = v_1^2 + v_2^2\ ,$$
$$u_1''^2 = v_1^2 + v_2^2 - \gamma_1^2\, v_2^2 = \gamma_1^2\left(v_1^2 + v_2^2 - v_1^2(v_1^2 + v_2^2)/c^2 - v_2^2\right) = \gamma_1^2\left(v_1^2\left(1 - (v_1^2 + v_2^2)/c^2\right)\right)\ .$$

When we take into account the negative sign of the velocity of Σ_o we find with Eqs. (205) and (210)

$$-\mathbf{u}'' = (-u_1'',\, -u_2'',\, -u_3'') = \left(\frac{v_1\gamma_1}{\gamma},\, \gamma_1\, v_2,\, 0\right)\ . \qquad \text{In } \Sigma'' \text{ measured velocity of systems } \Sigma_o \qquad (212)$$

Hence, the angle φ'' between the vector $-\mathbf{u}''$ and the x''-axis of Σ'' differs from the angle φ between the vector $\mathbf{v}$ and the x-axis of Σ_o. From Eq. (212), it follows for the angle φ''

$$\tan\varphi'' = \frac{u_2''}{u_1''} = \frac{\gamma\, v_2}{v_1}\ . \qquad (213)$$

The system Σ' is moving with parallel axes with respect to Σ_o, and Σ'' is moving with parallel axis with respect to Σ'. However, the vector of the relative velocity between the systems Σ_o and Σ'' of unchanging amount according to Eq. (211), decomposes according to Eqs. (204) and (212) with respect to the coordinate axes of Σ_o and Σ'' in different components. The x''-axis of Σ'' is rotated against the x-axis of Σ_o:

> The velocity vector between the coordinate origins of Σ_o and Σ'' has different direction with respect to the coordinate axes of these systems.
> We describe this difference by a rotation angle α_3 given by $\alpha_3 := \varphi - \varphi''$.

Since $\tan(\varphi - \varphi'') = (\tan\varphi - \tan\varphi'')/(1 + \tan\varphi\tan\varphi'')$ we get the angle α_3 from Eqs. (206) and (213),

$$\tan\alpha_3 = \tan(\varphi - \varphi'') = \frac{(v_2/v_1) - (\gamma\, v_2/v_1)}{1 + \gamma\, v_2^2/(v_1^2)} = \frac{(1-\gamma)\, v_2/v_1}{1 + \gamma\, v_2^2/v_1^2} = \frac{(1-\gamma)\,\beta_2/\beta_1}{1 + \gamma\,\beta_2^2/\beta_1^2}\,,$$

$$\tan\alpha_3 = \frac{(1-\gamma)\,\beta_1\,\beta_2}{\beta_1^2 + \gamma\,\beta_2^2}\,. \tag{214}$$

We derive the first non-vanishing approximation of Eq. (214) for the case $v_1 \ll c\, v_2 \ll c$, and $v_2 \ll v_1$:
With $\tan x \approx x$ and $1 - \gamma \approx -(1/2)(\beta_1^2 + \beta_2^2)$ hence
$(1-\gamma)\,\beta_1^2 \approx -(1/2)(\beta_1^2\beta_1^2 + \beta_2^2\beta_1^2) \approx 0$ hence $\gamma\beta_1^2 \approx \beta_1^2$,
it follows

$$\alpha_3 \approx \tan\alpha_3 = \frac{(1-\gamma)\beta_1\beta_2}{\beta_1^2 + \gamma\beta_2^2} \approx -\frac{1}{2}\frac{(\beta_1^2 + \beta_2^2)\beta_1\beta_2}{\gamma\beta_1^2 + \gamma\beta_2^2} \approx -\frac{1}{2\gamma}\,\beta_1\,\beta_2 \approx -\frac{1}{2}\,\frac{v_1 v_2}{c^2}\,,$$

$$\alpha_3 \approx -\frac{v_1\, v_2}{2\, c^2}\,. \tag{215}$$

In the laboratory system Σ_o, we observe the motion of a body on a circular path, e.g. the classical orbit of an electron in an atom. To each circular motion of a body there corresponds an angular momentum vector **L** standing orthogonal on the plane of the orbit with a direction, so that the vector forms a right-handed system with the orbit. The length L of the vector is equal to the product of the separation of the body to the orbital centre and the orthogonal component of its momentum. If the body is in rotation about itself, one calls this eigen angular momentum. However, in case of elementary particles, this quantity cannot be reduced on the rotation around an axis, but must be understood as quantum phenomenon. The eigen angular momentum is called *spin* of a particle. More details on these connections are discussed in Chap. 10, Sects. 1–7.

We assume the system $\Sigma''(x_i'')$ might be a tightly with the electron connected axis system. The central forces from the atomic nucleus causes no change of the eigen angular momentum, the spin **S** of the electron, that keeps a fixed direction with respect to the axis (x_i'') of Σ''. Might $\mathbf{v} = (v_1,\, 0,\, 0)$ be the temporal orbital velocity of the electron, that changes because of the central acceleration $\mathbf{a} = (0,\, a,\, 0)$ during the time Δt by the amount $\Delta\mathbf{v} = (0,\, \Delta v = v_2,\, 0) = \mathbf{a}\Delta t$. According to Eq. (215) it holds

$$\Delta\alpha_3 = -v_1\Delta v/(2c^2),$$

that is a rotation of the with the electron connected coordinate system Σ'' with respect to Σ_o during the time Δt. In Σ_o we observe therefore the THOMAS *precession*.[2]: Since the eigen angular momentum vector **S** of the electron in Σ'' remains fixed, it is in rotation as seen from Σ_o with the angular velocity $\omega_T = \Delta\alpha_3/\Delta t$ around the z-axis, or in vector notation[3] given by the known formula, cp. e.g. Arzeliès (1960), p. 179,

$$\omega_T = -\frac{\mathbf{v} \times \mathbf{a}}{2\,c^2}\,. \qquad \text{Angular velocity of the } Thomas \text{ precession} \qquad (216)$$

4 The Measuring Rod Paradox

Both the THOMAS precession as well as the measuring rod paradox are connected with the peculiarities occurring by the superposition of two special LORENTZ transformations with two orthogonal velocities. Now, we consider the following experimental situation.

Parallel to the x-axis there might be arranged a series of point-like barriers with equal separations l_o in the inertial system Σ_o. In system Σ' in axis parallel motion relative to Σ_o with velocity v_1 in x-direction, there might rest a rod on the x'-axis of the same length l_o as measured in Σ'. Observed in Σ_o the rod is parallel oriented to the x-axis and has therefore a length of a moving object $l_v = l_o\,/\,\gamma_1 = l_o\sqrt{1 - v_1^2/c^2}$, Fig. 2.

The observer in Σ_o now notices that the rod with fixed orientation gets an additional velocity v_2 in y-direction. It moves against the chain of barriers. The observer in Σ_o concludes: The rod has a smaller length l_v because of the LORENTZ contraction. The barriers have the rest separation $l_o > l_v$. Therefore the rod can pass the barriers without touching them, if it fits a gap.

For the observer in Σ', the barriers are moving with a velocity $-v$. Therefore the barriers have a LORENTZ-contracted separation $l_v = l_o/\gamma$ while the rod at rest has a larger length $l_o > l_v$. Considered from Σ' the rod should collide in any case with the barriers. Both statements together form the measuring rod paradox:

The rod has passed without resistance, and the rod was stopped—this is paradox!

Where is the error?

In our derivation, we have not been carefully enough with the orientation of the rod. As in Sect. 3 we define a system Σ'' that moves, as measured in $\Sigma'(x_i')$, *axis parallel* to Σ' with the velocity $\mathbf{v}' = (0,\, v_2',\, 0)$.

---

[2] As a purely kinematic effect, the THOMAS precession is also denoted simply as THOMAS effect.

[3] The special LORENTZ transformations (105), (109), are restricted on the direction of motion and form a group $\mathcal{L}$. Because of the rotation (214), the reference systems Σ'' and Σ_o are now not connected by a special LORENTZ transformation. The group property of the special LORENTZ transformations is lost, when we add different directions of motion. This is the general mathematical origin of the THOMAS precession. In Chap. 9, Sect. 1.4 we will discuss this extensively.

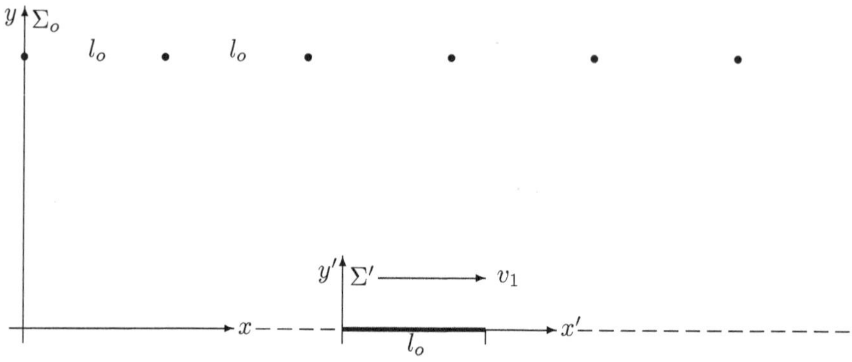

Fig. 2 The initial situation to the measuring rod paradox. The in Σ_o resting and parallel to the x-axis lined up barriers have separations l_o. The system Σ' is moving axis parallel to Σ_o with a velocity v_1 in x-direction. A rod resting on the x'-axis of Σ' has there also a rest length l_o. For the observer in Σ_o the rod is LORENTZ-contracted, for an observer in Σ' this concerns the separations of the barriers. Can a rod pass through the barriers when he gets an additional velocity component in y-direction?.

As seen from Σ_o the coordinate origin of Σ'' has a velocity $\mathbf{v}$,

$$\mathbf{v} = (v_1,\ v_2,\ 0)\ . \qquad \text{In } \Sigma_o \text{ measured velocity of } \Sigma'' \tag{217}$$

The amount of v_2' is determined by a LORENTZ transformation (105) using the chain rule of differentiation, when we take into account, that the motion of Σ'' is per definition $dx'/dt' = 0$ hence

$$\frac{dy'}{dt'} = \frac{dy}{dt}\frac{dt}{dt'} = v_2 \cdot \gamma_1$$

and therefore as in Sect. 3

$$\mathbf{v}' = (v_1',\ v_2',\ v_3') = (0,\ v_2\,\gamma_1,\ 0)\ . \qquad \text{In } \Sigma' \text{ measured velocity of } \Sigma'' \tag{218}$$

Here, we use the abbreviations

$$\gamma := \frac{1}{\sqrt{1 - v_1^2/c^2 - v_2^2/c^2}}\ ,\quad \gamma_1 := \frac{1}{\sqrt{1 - v_1^2/c^2}} \quad \longrightarrow \quad \gamma_2' := \frac{1}{\sqrt{1 - v_2'^2/c^2}} = \frac{\gamma_1}{\gamma}\ . \tag{219}$$

We use the LORENTZ transformations (105) and (118). For Σ' and Σ'' we find with the used γ-factors

$$x'' = x'\ ,\quad y'' = (y' - v_2'\,t')\,\gamma_2'\ ,\quad t'' = (t' - y'\,v_2'/c^2)\,\gamma_2'$$

and using Eqs. (218) and (219)

$$\left.\begin{array}{ll} x'' = x' \,, & x' = x'' \,, \\ y'' = \gamma\,(y'/\gamma_1 - v_2\,t') \,, & y' = \gamma\,(y''/\gamma_1 + v_2\,t'') \,, \\ t'' = \gamma\,(t'/\gamma_1 - y'\,v_2/c^2) \,, & t' = \gamma\,(t''/\gamma_1 + y''\,v_2/c^2) \,. \end{array}\right\} \tag{220}$$

For Σ' and Σ_o it holds

$$\left.\begin{array}{ll} x' = \gamma_1\,(x - v_1\,t) \,, & x = \gamma_1\,(x' + v_1\,t') \,, \\ y' = y \,, & y = y' \,, \\ t' = \gamma_1\,(t - x\,v_1/c^2) \,, & t = \gamma_1\,(t' + x'\,v_1/c^2) \,. \end{array}\right\} \tag{221}$$

Taken together Eqs. (220) and (221) one gets

$$\left.\begin{array}{ll} x'' = \gamma_1\,(x - v_1\,t) \,, & x = \gamma_1 x'' + \gamma\gamma_1 \dfrac{v_1\,v_2}{c^2} y'' + \gamma v_1\,t'' \,, \\ y'' = \gamma\,\gamma_1\,(\dfrac{v_1 v_2}{c^2}\,x + (1 - \dfrac{v_1^2}{c^2})\,y - v_2\,t) \,, & y = \gamma\,(y''/\gamma_1 + v_2\,t'') \,, \\ t'' = \gamma\,(t - \dfrac{v_1}{c^2}\,x - \dfrac{v_2}{c^2}\,y) \,, & t = \gamma t'' + \gamma_1 \dfrac{v_1}{c^2}\,x'' + \gamma\gamma_1 \dfrac{v_2}{c^2} y'' \,. \end{array}\right\} \tag{222}$$

The rod should be at rest in Σ''. Now we have to distinguish two cases, Fig. 3.

1. case: The rod is at rest on the x''-axis of system Σ'' and moves axis parallel to system Σ'.

The observer in Σ' measures the separations between the barriers LORENTZ contracted, hence it is shorter than the parallel to the x'-axis oriented rod, that consequently must collide with the barriers. What is the statement of an observer in Σ_o? We shall now show, that seen from Σ_o the rod will collide, since it is observed to be inclined against the x-axis.

In Σ'' we observe two events, the event $O(0, 0, 0, 0)$ connected with the left endpoint of the in Σ'' resting rod at time $t'' = 0$, and an event $E(x''{=}l_o, 0, 0, t_o'')$ representing the right endpoint at another time $t'' = t_o''$. The origin of coordinates should have both systems together. For the event E we find in Σ_o with help of the right-hand side of Eq. (222)

$$E\Big(x_E = \gamma_1 l_o + \gamma_1\,v_1\,t_o'', y_E = \gamma\,v_2\,t_o'', 0, t_E = \gamma\,t_o'' + \gamma_1\,(v_1/c^2)\,l_o\Big) \,. \tag{223}$$

Since the origin of Σ_o lies at $O(0, 0, 0, 0)$ from the sight of Σ_o the right endpoint of the rod will be at $t_E = 0$ simultaneously with $O(0, 0, 0, 0)$ hence $t_E = 0$ in Eq. (223) for

$$t_o'' = -\frac{v_1\,\gamma_1\,l_o}{c^2\,\gamma} \,. \tag{224}$$

Inserting in Eq. (223) we find the right endpoint of the rod in Σ_o with some evaluations

$$x_E = l_o/\gamma_1 \ , \quad y_E = -\gamma_1 \frac{v_1 v_2}{c^2} l_o \ . \tag{225}$$

Then in Σ_o there is observed that the rod has an inclination angle κ against the x-axis,

$$\tan\kappa = \frac{y_E}{x_E} = -\gamma_1^2 \frac{v_1 v_2}{c^2} \ . \tag{226}$$

Equation (226) shows that the rod is so inclined that it hits the barriers also for the observer in Σ_o s. Fig. 3. The paradox is solved.[4]

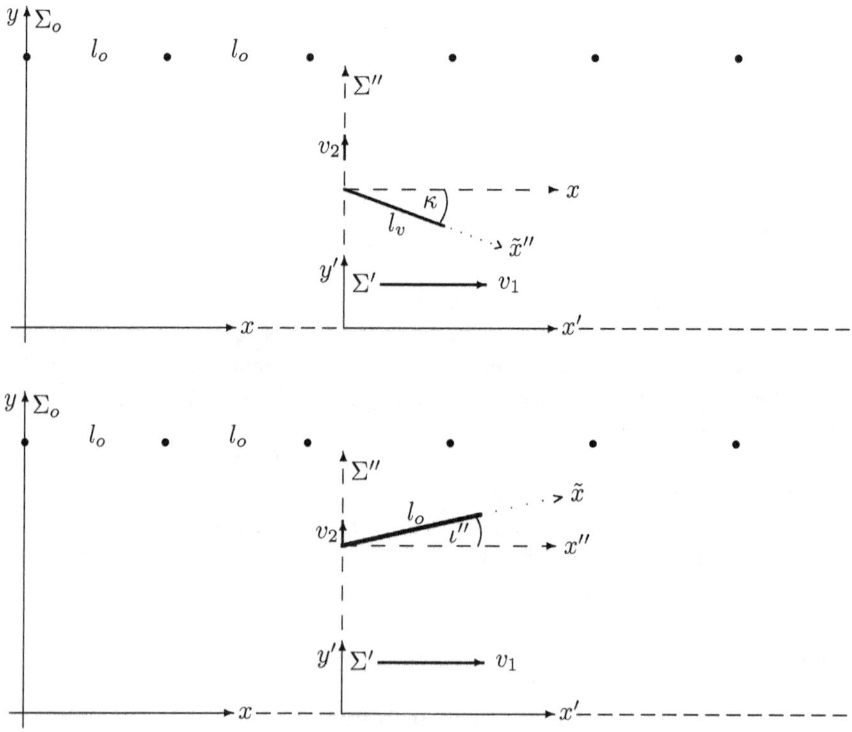

Fig. 3 Resolution of the measuring rod paradox. The system Σ' is moving axis parallel against Σ_o with velocity v_1 in x-direction, and Σ'' is in axis parallel motion to Σ' with velocity component v_2' in y'-direction. In both pictures, it is seen that the orientation of moving rods is determined by the conventional definition of simultaneity in the relativistic space-time.

[4]For simplicity we have assumed that the rod lies with the left end at the origin. It is easy to verify that the time $t^* = (l_o - x_E)/v_1$ that the right end of the rod takes to get to $x = l_o$ is still smaller

Upper picture (**1. case**): As seen from Σ' and Σ'' the separations of the barriers are LORENTZ contracted, so that the rod cannot pass through the barriers. The rod on the x''-axis of Σ'' is inclined by the negative angle κ against the x-axis as seen from Σ_o. Because of this inclination, the rod will collide with the barriers also from the sight of Σ_o during its motion. The orientation of the dotted Σ''-axis is determined in system Σ_o. It is indicated by the tilde. If we choose $v_1 = 0.8\,c$ and $v_2 = 0.16\,c$ as in Fig. 1, then $\kappa = -19.57°$.

Lower picture (**2. case**): Now we suppose that the in Σ'' resting rod is seen from Σ_o as axis parallel to the x-axis. As seen from Σ_o the rod will therefore be LORENTZ contracted and can pass, when it fits a gap. The observer in Σ'' now states that the rod is inclined by an angle ι'' against the x''-axis of system Σ'' according to Eq. (228). Because of this inclination, the rod can also pass through the barriers as seen from Σ''. The orientation of the dotted Σ''-axis is determined in system Σ''. It is denoted by a tilde. If we choose $v_1 = 0.8\,c$ and $v_2 = 0.16\,c$ as shown in Fig. 1, then we find $\iota'' = +12.48°$.

2. case: Again the rod might be at rest in system Σ'' but now so oriented that it is axis parallel to the x-axis of Σ_o as seen from Σ_o.

The observer in Σ_o concludes: The rod is LORENTZ contracted with respect to the separation of the barriers and can pass them when it fits a gap. The rod is in motion with a velocity v_1 axis parallel in x-direction. As seen from Σ_o, the rod has the LORENTZ contracted length l_o/γ_1. In Σ_o we observe two events, $O(0, 0, 0, 0)$ and $E(l_o/\gamma_1, 0, 0, 0)$ hence the simultaneous positions of the end points of the rod at time $t = 0$ in Σ_o. The origin is common to both systems. For the event E we find in Σ'' according to the left-hand side of Eq. (222)

$$E\Big(x''_E = l_o,\ y''_E = \gamma\,(v_1\,v_2/c^2)\,l_o,\ 0,\ t''_E = -\gamma\,(v_1/c^2)\,(l_o\,/\gamma_1)\Big)\,. \tag{227}$$

The rod is at rest in Σ'', therefore, the t''_E-coordinate of the right endpoint of the rod does not matter. Therefore in Σ'', it is observed that the rod has an inclination angle ι'' against the x''-axis with

$$\tan\iota'' = \frac{y''_E}{x''_E} = \gamma\,\frac{v_1\,v_2}{c^2}\;. \tag{228}$$

Consequently, the rod is inclined so that it will also pass the barriers as seen from Σ'' if it finds a gap, s. Fig. 3. The paradox is solved.

We have seen that always the incorrect treatment of the relativity of simultaneity leads us to wrong conclusions. In Chap. 8, Sect. 9.1, we will show how the measuring rod paradox is solved quasi by itself, when we introduce an absolute simultaneity in the relativistic space-time, especially for the discussion of this experiment.

---

than the time $t^\dagger = y_E/v_2$ that the right end of the rod needs to get to $y = 0$. The rod hits therefore in any case the barrier.

5 DOPPLER Effect

The DOPPLER effect concerns a change in the measured frequency of waves due to the relative motion between sender and receptor. The derived frequency shift depends on the assumption whether the classical space-time with an invariability of periods is used as in Eq. (66), or the relativistic postulate of time dilatation (100) is supposed. In the latter case it is to be respected, in which ratio the signal velocity of the considered waves, denoted by C stands to the in Eq. (100) occurring speed of light c. The formulas of the optical DOPPLER effect with $C = c$ will be different from the acoustic frequency shift, where $C = c_a$ is the sound velocity. We will see that the DOPPLER effect is suited to decide whether we can ascribe a state of motion to the medium of the wave propagation or not, Problem 19.

In an inertial system Σ_o, we consider a standardised sender S emitting by construction a well defined harmonic wave with period T_S hence with frequency ν_S like a tuning fork in music,

$$\nu_S = \frac{1}{T_S} \ . \tag{229}$$

In the surrounding space (or for an in Σ_o resting medium) a sender might produce a monochromatic wave of frequency ν_S that propagates through space (or the medium) with a velocity C, s. Fig. 4.

A receptor E might measure the frequency ν_E of this wave. When the receptor E is also at rest in the system Σ_o then it holds

$$\nu_E = \nu_S \ . \qquad \text{Sender and receptor are at rest in } \Sigma_o \tag{230}$$

As an example, we can imagine as monochromatic light wave of the yellow sodium line with $\nu_{Na} = 5.0847416 \cdot 10^{14}$ Hz, hence $\lambda_{Na} = 589.5923 \cdot 10^{-9}$ m since here we have the propagation with $C = c = 299\,792\,458\,\text{ms}^{-1}$, the speed of light. The period of the sender is then $T_{Na} = 1/\nu_{Na} = 1.9666683 \cdot 10^{-15}$ s.

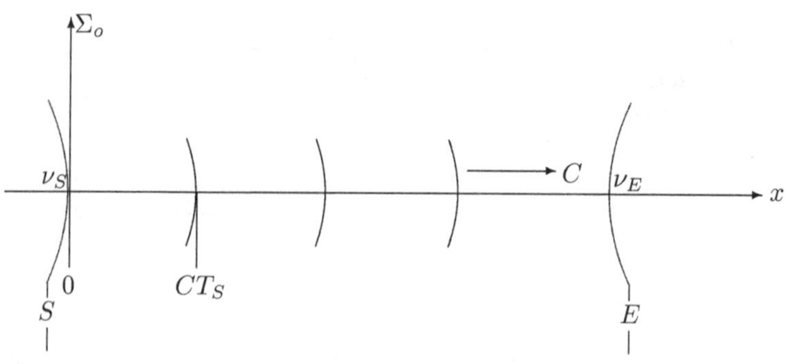

Fig. 4 Receptor E and sender S are both at rest in the reference system Σ_o.

Now we will discuss the DOPPLER effect at first under the assumption of the classical space-time (66), and then we will derive the exact, relativistic formulae taking into account the time dilatation (100).

5.1 *Classical Theory of the* DOPPLER *Effect*

Here, we start from the classical space-time with the GALILEI transformation (69), and we distinguish two cases:

5.1.1 Longitudinal Motion

Here we assume that sender and receptor are in motion on its connecting line. This is called *longitudinal* DOPPLER effect.

(a) The receptor E is at rest in the reference system $\Sigma_o(x, t)$ at position $x > 0$ while the sender S is at rest in a system $\Sigma'(x', t')$ at $x' = 0$ that is in motion with velocity v in the direction of the receptor.

The initial conditions of systems Σ_o and Σ' are chosen as in Eq. (11). At time $t = t' = 0$ there is emitted the first wave maximum, when the sender is at $x = x' = 0$ left from the receptor. After time $t = t' = T_S$, the first wave crest is at $x_2 = CT_S$, and the sender reaches $x_1 = v\,T_S$ where it emits a second wave crest in the same direction. Both crests are separated by a wavelength λ as observed from Σ_o, cp. Fig. 5,

$$\lambda = x_2 - x_1 = (C - v)\,T_S \ . \tag{231}$$

The wave crests at the separation λ run with velocity C onto the receptor. The second wave hits the receptor later by a time $T_E = \lambda/C$ than the first wave crest. The receptor now measures a frequency ν_E with

$$\nu_E = \frac{1}{T_E} = \frac{C}{\lambda} = \frac{C}{(C - v)\,T_S} \ ,$$

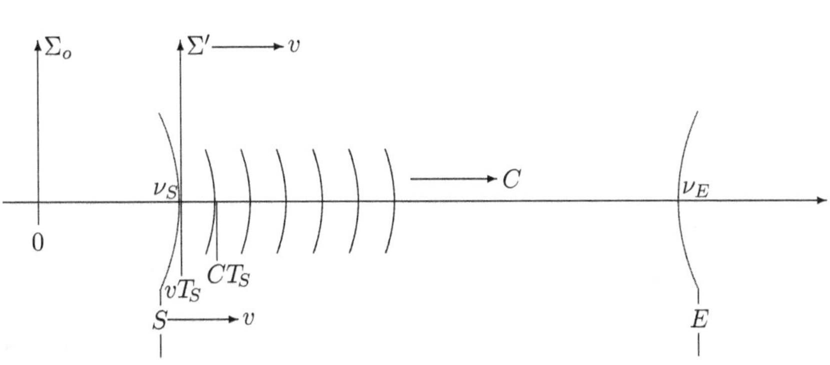

Fig. 5 The receptor E is at rest in Σ_o and the sender S at rest in the system Σ'. In the figure we have $v = 0.8\,C$.

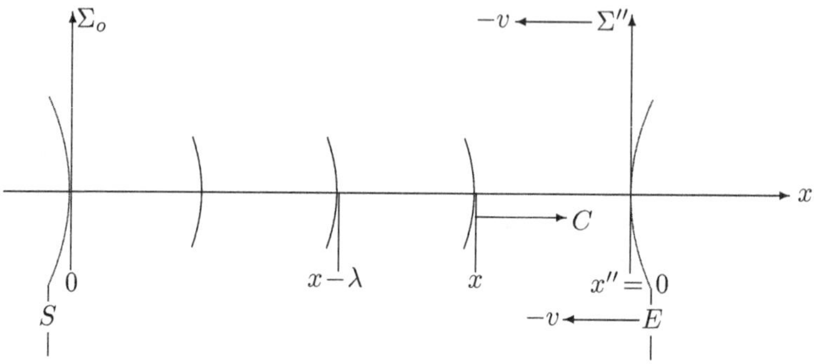

Fig. 6 The sender S is at rest in Σ_o and the receptor E rests in the system Σ''. The figure is again for the case $|v| = 0.8\,C$.

hence

$$\nu_E = \nu_S \frac{1}{(1 - v/C)} \;. \qquad \textit{Doppler} \text{ effect for a moving sender, Classical space-time} \qquad (232)$$

(b) Now the sender S is at rest in the reference system $\Sigma_o(x, t)$ while the receptor E is at rest in a system $\Sigma''(x'', t'')$ at $x'' = 0$ that has a velocity $-v$ with respect to Σ_o. The receptor E is therefore in motion with a velocity amount v, i.e. it is coming from right onto the sender, s. Fig. 6.

As seen in Σ_o the sender produces a wave in space with frequency ν_S hence with wavelength $\lambda = C/\nu_S$ propagating in positive x-direction. The receptor approaches to this wave with velocity v according to Eq. (8) it covers therefore a wavelength with the added velocity $C + v$. The corresponding time T_E for this covering is then

$$T_E = \frac{\lambda}{C + v} = \frac{C}{\nu_S} \frac{1}{C + v} = \frac{C}{\nu_S} \frac{1}{C} \frac{1}{1 + v/C} \;,$$

and the frequency $\nu_E = 1/T_E$ measured by the receptor is

$$\nu_E = \nu_S \left(1 + \frac{v}{C}\right) . \qquad \text{DOPPLER effect for a moving receptor, Classical space-time} \qquad (233)$$

Only for $v \ll C$ the frequency shifts (232) and (233) coincide because of the TAYLOR expansion (83), $1/(1 - x) \approx 1 + x$ for $x \ll 1$. Beyond this approximation, the frequency shift (232) differs for the case of a moving sender from that for a moving receptor, Eq. (233).

With this difference it would be possible to measure the state of motion of the carrier medium of the waves with the frequency shift. In this way, there is determined the distinguished inertial system Σ_o where the carrier medium is at rest. The DOPPLER effect concerns all waves, both electromagnetic and sound waves. For the sound velocity $C = c_a$ in Eqs. (232) and (233), cp. Problem 19, we determine by means of the acoustic DOPPLER shift the reference system, in which the carrier of these waves is at rest, so, for example, with respect to the air.

If the GALILEI transformation would be valid without restrictions, i.e. not only approximately for small velocities v so that we could apply Eqs. (232) and (233) without the restriction $v \ll c$ on light wave propagating with the speed of light c for C then we could determine the state of motion of the carrier of electromagnetic waves relative to the sender or receptor by optical measurements. The corresponding inertial system Σ_o with the resting carrier would then be *physically* distinguished. Since the classical mechanics holds true in each inertial system, s. Chap. 6, Sect. 2, then only in this inertial system Σ_o it would hold true both MAXWELL's equations of electrodynamics as well as the equations of classical mechanics. Then we would have found an absolute space that was historically introduced in expectation of such measurements as *aether*, but it was vainly searched for.

5.1.2 Transversal Observation

The receptor might be at rest in Σ_o and the sender at rest in a system Σ' that is in motion along the positive y-axis as measured from Σ_o. The sender might pass the receptor at a large separation R. The receptor measures the waves that are emitted at the time of smallest separation of receptor and sender. This situation describes the so-called *transversal* DOPPLER effect, there are observed waves that are emitted orthogonal to the direction of motion of the sender, Fig. 7.

A large separation means that the rate of change of this separation at the time of smallest separation R_o s. Fig. 8, for the duration of an eigen period of the sender can be neglected. The sender with velocity v transverses during the period $T_S = 1/\nu_S$ a length $v\,T_S = v/\nu_S$. According to the notion introduced in Fig. 8, the condition

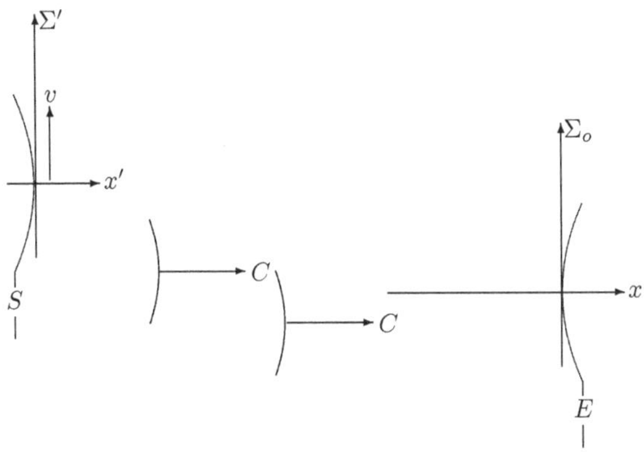

Fig. 7 Experiment for the purely transversal DOPPLER effect.

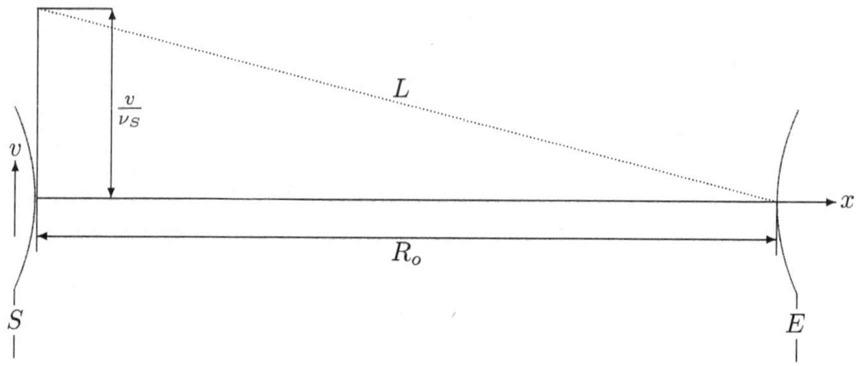

Fig. 8 The condition for the observation of the purely transversal DOPPLER effect. The condition $L \approx R_o$ is realised for $v/\nu_S \ll R_o$.

$L \approx R_o$ is realised if $v/\nu_S \ll R_o$ since according to Eq. (83) we have $\sqrt{1+a} \approx 1 + a/2$ for $a \ll 1$, hence $a = v^2/\nu_S^2$

$$L = \sqrt{R_o^2 + \frac{v^2}{\nu_S^2}} = R_o\sqrt{1 + \frac{v^2}{R_o^2\,\nu_S^2}} \approx R_o\bigl(1 + \frac{v^2}{2R_o^2\,\nu_S^2}\bigr) = R_o + \frac{v^2}{2R_o\,\nu_S^2} \approx R_o$$

for

$$\frac{v^2}{2R_o\,\nu_S^2} \ll R_o \quad , \text{ hence} \quad \frac{v}{\nu_S} \ll \sqrt{2}R_o$$

and then also

$$L \approx R_o \quad \text{for} \quad \frac{v}{\nu_S} \ll R_o \quad \text{or for} \quad \frac{v}{C} \ll \frac{R_o}{\lambda_S}. \qquad \text{Large separation between sender and receptor} \qquad (234)$$

After a time T_S, a second wave crest is emitted from the sender. Both move with the same velocity C, and they must travel the same distance R_o as given by Eq. (234). As we still assume the validity of the GALILEI transformation, both sender and receptor measure for all events the same time. Therefore the receptor also measures the same time T_S when the second wave crest will arrive after the first one, and it finds then also the same frequency for the arriving waves as emitted by the sender. Therefore the GALILEI transformation does not lead to a transversal DOPPLER effect, s. also Problem 20,

$$\nu_E = \nu_S \ . \qquad \text{Transversal observation Classical space-time} \qquad (235)$$

5.2 *Exact Theory of the* DOPPLER *Effect*

Until now we have disregared the in Chap. 4, Sect. 3 formulated relativistic properties of our measuring devices. The approximations applied in classical space-time are now corrected. To this aim, we have to take into account only the postulate (100) of time dilatation, $T_v = T_o/\sqrt{1 - v^2/c^2}$. This equation is valid for arbitrary oscillatory systems, both for the balance spring of a mechanical clock as well as for the oscillations of a tuning fork or the atomic oscillations leading to the emission of electromagnetic waves. The corresponding frequency then fulfils the relation

$$\nu_v = \nu_o \sqrt{1 - \frac{v^2}{c^2}} \,. \tag{236}$$

In Eq. (236) *the symbol c means in each case the speed of light.*

Now we consider the two in Sect. 5.1 considered situations.

5.2.1 Longitudinal Observation

(a) The receptor E is at rest in Σ_o. The received frequency, measured, for example, by resonance absorption, is denoted as ν_E. The sender S is at rest in Σ' which is in motion with velocity v against Σ_o s. Fig. 5. In the formula (232), we have to replace the frequency of the sender ν_S by the frequency ν_v according to Eq. (236), i.e. replacing ν_S for ν_o or $\nu_S \rightarrow \nu_S \sqrt{1 - v^2/c^2}$. This provides the exact relativistic formula for the frequency shift, when the sender, in motion with velocity v emits a wave with the signal velocity C which remains for the moment unspecified.

$$\nu_E = \nu_S \frac{\sqrt{1 - v^2/c^2}}{1 - v/C} \,. \quad \text{Exact theory of the } \textit{Doppler} \text{ effect} \quad \text{Moving sender, arbitrary waves} \tag{237}$$

Considering sound waves with $C = c_a$ and supposing subsonic velocities of the sender $v < c_a$. Under conditions of a laboratory on Earth we have then because of $c_a \ll c$ also $v \ll c$. Then we get from Eq. (237) the relation (232) for $C = c_a$ as non-relativistic approximation since we have $\sqrt{1 - v^2/c^2} \approx 1$

$$\nu_E = \nu_S \frac{\sqrt{1 - v^2/c^2}}{1 - v/c_a} \approx \nu_S \frac{1}{1 - v/c_a} \,, \quad \text{for } v < c_a \ll c \,. \quad \text{Sound waves} \quad \text{Moving sender} \tag{238}$$

However in case of highly compressed stellar matter, there the sound velocity c_a can come in the order of magnitude of the speed of light c, $c_a \approx c$ however with $c_a < c$. In this case we have to keep the square root in Eq. (238) without approximation.

In Eq. (237), we now consider the case $C = c$, a propagation with the speed of light. Then we find

$$\nu_E = \nu_S \frac{\sqrt{1 - v^2/c^2}}{1 - v/c} = \nu_S \frac{\sqrt{(1 - v/c)(1 + v/c)}}{1 - v/c} = \nu_S \sqrt{\frac{1 + v/c}{1 - v/c}} \; ,$$

$$\nu_E = \nu_S \sqrt{\frac{c + v}{c - v}} \; . \qquad \text{Electromagnetic waves, Moving sender} \qquad (239)$$

(b) Now the sender S might be at rest in Σ_o and we can keep the frequency of the sender ν_S in Eq. (233). However, the receptor now has a velocity of magnitude v with respect to Σ_o and it determines the frequency ν_E in its reference system Σ'', s. Fig. 6. As seen from Σ_o there follows the frequency ν_v from Eq. (236), where we replace ν_o by ν_E. In Eq. (233) we have therefore to replace the reception frequency ν_E by ν_v, i.e. $\nu_E \rightarrow \nu_E \sqrt{1 - v^2/c^2}$ and then

$$\nu_E \sqrt{1 - \frac{v^2}{c^2}} = \nu_S \, (1 + \frac{v}{C}) \; ,$$

hence

$$\nu_E = \nu_S \frac{1 + v/C}{\sqrt{1 - v^2/c^2}} \; . \qquad \text{Exact theory of the DOPPLER effect, Moving receptor, arbitrary waves} \qquad (240)$$

For sound velocity $C = c_a$ we again restrict ourself to velocities $v < c_a$. For conditions on the Earth, there is then always $c_a \ll c$ and therefore also $v \ll c$. We get from Eq. (240) the relation above (233) for $C = c_a$ as non-relativistic approximation,

$$\nu_E = \nu_S \frac{1 + v/c_a}{\sqrt{1 - v^2/c^2}} \approx \nu_S (1 + \frac{v}{c_a}) \;\; \text{for } v < c_a \ll c \, . \qquad \text{Sound waves, Moving receptor} \qquad (241)$$

For extremely dense stellar matter with $c_a \approx c$ and $c_a < c$ we have to keep the square root in Eq. (241).

In each case for a sufficient precise measurement, the difference in the formulae for the acoustic DOPPLER effect allows us to determine the state of motion of the carrier of the sound waves, s. Problem 19.

In Eq. (240) we now assume $C = c$, the speed of light. Then we get

$$\nu_E = \nu_S \frac{1 + v/c}{\sqrt{1 - v^2/c^2}} = \nu_S \frac{1 + v/c}{\sqrt{(1 - v/c)(1 + v/c)}} = \nu_S \sqrt{\frac{1 + v/c}{1 - v/c}} \; ,$$

$$\nu_E = \nu_S \sqrt{\frac{c+v}{c-v}} \,. \qquad \text{Electromagnetic waves, Moving receptor} \qquad (242)$$

Equations (239) and (242) are now identically. For electromagnetic waves, there is no difference between a moving sender and a moving receptor. Since in this situation, sender and receptor are in motion in the direction of observations, Eq. (239) describes the *longitudinal* DOPPLER effect.

Simply from the consideration of the postulate of time dilatation (100), we are led to the conclusion, that it is *impossible* to determine the state of motion of the carrier of electromagnetic waves with the DOPPLER shifts, since now only the relative velocity v between sender and receptor enters into the description of the effect.

5.2.2 Transversal Observation

As in case (**a**) the receptor E might be at rest in Σ_o. The receptor frequency is ν_E. The sender S is at rest in Σ' which moves with the velocity v against Σ_o. In the formula (235) we must replace the frequency ν_S by ν_v according to Eq. (236) with $\nu_o = \nu_S$. This leads to the *transversal* DOPPLER-effect, s. also Problem 20,

$$\nu_E = \nu_S \sqrt{1 - \frac{v^2}{c^2}} \,. \qquad \text{Transversal DOPPLER effect, Exact theory, arbitrary waves} \qquad (243)$$

Due to the time dilatation (100), we find in distinction to the classical approximation (235) a frequency shift for transversal observations. It is significant that Eq. (243) is independent of the physical nature of the considered waves and its signal velocity C. Eq. (243) is nothing else than the time dilatation of a moving clock expressed in terms of the frequencies. The physical nature of an oscillatory system, the nature of the observed waves has no influence. Under laboratory conditions on Earth with $v < c_a \ll c$ for acoustic waves, there is no chance to experimentally measure this effect. For highly compressed stellar matter with $c_a \approx c$, $c_a < c$, one could detect the transversal DOPPLER effect both for sound and for electromagnetic waves.

For transversal motions, the factor $\sqrt{1 - v^2/c^2}$ determines the total effect. For longitudinal observations, the same factor determines the difference between the classical approximations (232) or (233) to the exact Eq. (239). Since $\sqrt{1 - v^2/c^2} \approx 1 - v^2/(2c^2)$ each measurement of the DOPPLER-shift with a precision of quadratic terms in v/c provides a test of Eq. (100) for the time dilatation. The first experiment from 1938/39 of the time dilatation used the oscillating system of a hydrogen atom. The red spectral line H_α has in its rest system a period $T_o = 2.1876 \cdot 10^{-15}$ s.

When the H-atoms are channel rays in motion with high velocity v with respect to the receptor, one observes the period $T_v = T_o/\sqrt{1 - v^2/c^2}$. For transversal observation, orthogonal to the direction of motion of the canal rays, the DOPPLER-shift from Eq. (243) is a direct test of the time dilatation. Due to experimental circumstances, the measurements were done with an inclined canal ray, and the experimental precision was sufficient to measure the in v/c quadratic terms. In this way, the formula (100) for the time dilatation was verified. Newer precision experiments to prove the time dilatation by means of the transversal DOPPLER effect have been discussed in Chap. 4, Sect. 6.

6 Aberration

In the neighbourhood of the coordinate origin of system Σ_o, light is observed arriving from a distant source, say from a star S. We consider here the case of light coming from the zenith, we observe the star in Σ_o in y-direction orthogonal to the horizon, as seen in Figs. 9 and 10.

Now we consider the case that our telescope is in motion in x-direction with a velocity v or $-v$ with respect to Σ_o, i.e. our telescope is at rest in a reference

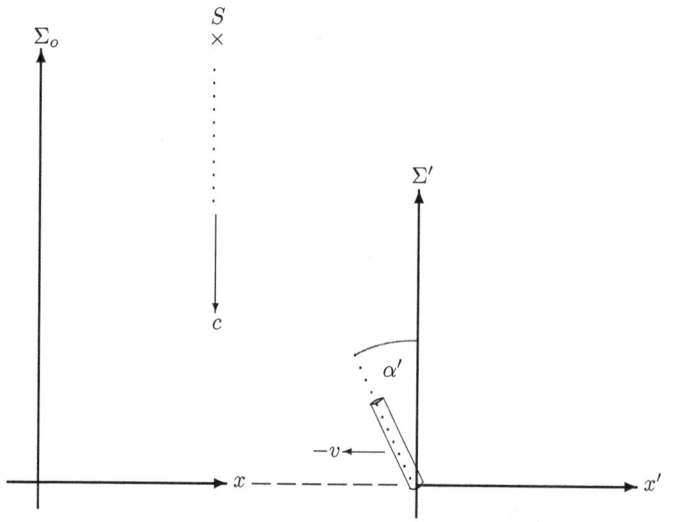

Fig. 9 Aberration in the particle picture. As seen from Σ_o photons arrive from the vertical y-direction with the speed of light c. In system Σ_o there is measured, for example, a velocity $v = 0.448c$ in negative x-direction of system Σ'. In order that the photons pass through the tube, not hitting the sides, the instrument must be tilted by an angle α with $\tan\alpha = v/c = \Delta x/\Delta y$ as seen from Σ_o. Here Δx and Δy denote the path length of the tube, or of the photons during time Δt. For $v \ll c$ this is equal to the angle α' that the observer in Σ' has to tilt the telescope. In Σ' photons travel a path $\Delta x'$ in x'-direction during the time $\Delta t'$ and in negative y'-direction the path $\Delta y'$ so that $\tan\alpha' = \Delta y'/\Delta x'$. For relativistic velocities, we have to respect the LORENTZ contraction, hence $\Delta x' = \Delta x/\gamma$. Since $\Delta y' = \Delta y$ we get then $\tan\alpha' = \gamma\tan\alpha$ that can be written as $\sin\alpha' = v/c$. As seen from Σ' we have to tilt the instrument in our example then by an angle α' against the y'-axis. $\alpha' = \arcsin 0.448 \approx 26.6°$ is different from the angle expected by the classical approximation (251) $\alpha = \arctan 0.448 \approx 24.1°$ by about 2.5°.

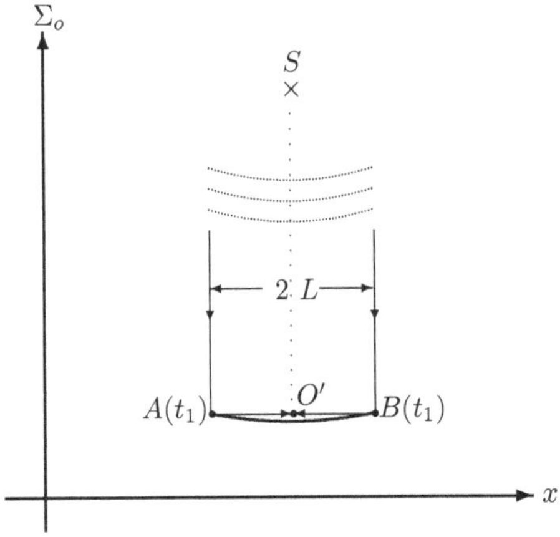

Fig. 10 The observation of a star with a telescope in the waves picture.

system Σ'. Then we will see the star not vertically above us, but in a slightly inclined direction. In astronomy this effect is called *aberration*. We notice the important point: The state of motion of the sender, so, for example, of the star that emits light waves or a particle flux, does not play any role. It is only important, that the waves or the particle flux are observed from two different inertial systems.

The arguments in the explanation of the observed phenomenon of aberration depends on the supposed assumption on the physical nature of light.

6.1 Aberration in Particle Picture

We first assume that the light consists of photons, i.e. of particles with rest mass zero, that are propagating with the speed of light c and have a momentum $\mathbf{p}$. In Σ_o we observe these particles coming vertically from above along the y-direction, so that $\mathbf{p} = (0, p, 0)$, Fig. 9.

Σ' might be the reference system, that is in motion in negative x-direction of Σ_o with a velocity amount v. When we orient a telescope, resting in Σ', parallel to the y'-direction, then a photon entering the telescope cannot reach the ocular because of the motion in the negative x-direction. The photon hits the tube of the telescope and is there absorbed. The picture remains dark.

As seen from Σ_o the photons can pass through the tube, when the telescope is tilted by an angle α so that $\tan\alpha$ equals the ratio of the horizontal velocity v of the telescope in system Σ_o and the vertical velocity c of photons in Σ_o.

As seen from Σ' we have to tilt the telescope by an angle α' against the y'-axis, so that $\tan\alpha'$ equals the ratio of the horizontal velocity v of the telescope in system Σ_o and the vertical velocity $u'_{y'}$ of photons in Σ', s. Fig. 9,

$$\tan\alpha = \frac{v}{c}\,, \quad \tan\alpha' = \frac{v}{|u'_{y'}|}\,. \tag{244}$$

For evaluation of α' we describe the motion of photons in system Σ_o with the path $\big(x = x(t),\ y = y(t),\ z = z(t)\big)$ and therefore a velocity

$$(u_x,\ u_y,\ u_z) = (dx/dt,\ dy/dtdz/dt) = (0,\ -c,\ 0)\,. \tag{245}$$

In system Σ' we observe a motion $\big(x' = x'(t'),\ y' = y'(t'),\ z' = z'(t')\big)$ with the velocity

$$(u'_{x'},\ u'_{y'},\ u'_{z'}) = (dx'/dt',\ dy'/dt'dz'/dt')\,.$$

Here $u'_{x'} = v$ is simply the in Σ' observed velocity of system Σ_o. For evaluation of the component in y'-direction, we replace in the LORENTZ transformation (105) the parameter v for the velocity of Σ' by $-v$ and find by application of the chain rule of derivatives

$$u'_{y'} = \frac{dy'}{dt'} = \frac{dy}{dt}\cdot\left(\frac{dt'}{dt}\right)^{-1} = -c\sqrt{1-\frac{v^2}{c^2}}\,.$$

The velocity component in z'-direction remains zero. Hence it holds

$$(u'_{x'},\ u'_{y'},\ u'_{z'}) = (dx'/dt',\ dy'/dt'dz'/dt') = (v,\ -\sqrt{c^2-v^2}\,0)\,. \tag{246}$$

This expression also follows simply from the application of EINSTEIN's addition theorem of velocities (117).

The amount of this velocity is again c. Also in Σ' the photons have the speed of light. From Eqs. (244) and (246) we get therefore an aberration angle α' with $\tan\alpha' = v\gamma/c$. A simpler expression follows using the simple calculation:

$$\sin\alpha' = \frac{\tan\alpha'}{\sqrt{1+\tan^2\alpha'}} = \frac{v/\sqrt{c^2-v^2}}{\sqrt{1+v^2/(c^2-v^2)}} = \frac{v}{\sqrt{c^2-v^2+v^2}} = \frac{v}{c}\,,$$

so that we can summarise

$$\begin{aligned} \Sigma_o:&\quad \tan\alpha = \frac{v}{c}\,,\\ \Sigma':&\quad \sin\alpha' = \frac{v}{c}\,. \end{aligned} \qquad \text{Exact theory of aberration} \tag{247}$$

In classical space-time, the same phenomenon is written as follows: With the velocity vector (245) of photons in Σ_o its motion is described by

$$\big(x(t),\ y(t),\ z(t)\big) = (0,\ -c\,t,\ 0)\,. \tag{248}$$

With Eq. (248) and the Galilei transformation

$$x' = x + v\,t\,,\quad y' = y\,,\quad z' = z\,,\quad t' = t$$

it follows a motion of photons in Σ'

$$\big(x'(t'),\, y'(t'),\, z'(t')\big) = (v\,t',\, -c\,t',\, 0) \tag{249}$$

and therefore a velocity vector

$$(u'_{x'},\, u'_{y'},\, u'_{z'}) = (v,\, -c,\, 0)\ . \tag{250}$$

Considered from Σ' we find that the photon velocity now seems to be not equal to c but $\sqrt{c^2 + v^2} \approx c\,(1 + \frac{1}{2}v^2/c^2)$. If quadratic terms in v/c are measurable, the classical description of light propagation becomes inapplicable, s. Chap. 4, Sect. 7.

With Eqs. (244) and (250) the tilt angles α' and α in systems Σ' and Σ_o, respectively, would coincide,

$$\tan\alpha = \tan\alpha' = \frac{v}{c}\ . \qquad \text{Classical approximation for aberration} \tag{251}$$

6.2 *Aberration in Wave Picture*

Now we start with the picture, that a plane parallel electromagnetic wave, emitted from the star, produces a picture of the star in the telescope. The telescope has a diameter or aperture of $2L$, so that a wave train of this extension (across to the propagation direction) can be observed.[5] In a wave train, a fixed relative phase relation is needed for creation of an interference picture. Two wave crests entering the objective of the telescope might be directed to the middle of the detector for superposition. The symmetry axis of the telescope gives the direction, in which we observe the star. We will assume that the star S observed in system Σ_o stands in the zenith in the direction of the y-axis of our coordinates. Rays propagating parallel to the y-axis would hit the telescope and create a picture, when the plane of the telescope is orthogonal oriented to the y-direction, and the telescope is at rest in Σ_o, s. Fig. 10.

Now we go over again to a reference system Σ' moving in negative x-direction of Σ_o with a velocity amount v, and we consider the cutout of rays of the waves, Fig. 11.

We ask: By which angle α' we have to tilt the instrument, so that both wave crests are meeting at the middle point of the detector?

---

[5] With respect to the modern radio observations, the so-called very long baseline interferometry (VLBI) we take $2L$ as the maximum separation of the observation stations.

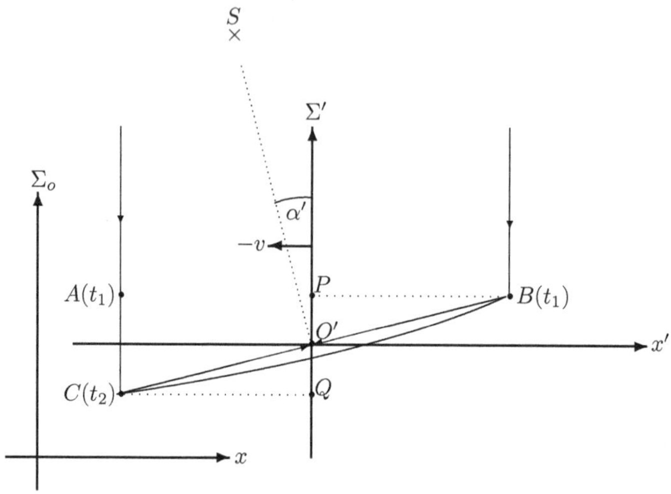

Fig. 11 Aberration in the wave picture. The boundary rays might arrive at the telescope at points A and B at the same time t_1 in Σ_o and in the same phase relation, e.g. both at wave crests. In Σ' the telescope must be inclined by an aberration angle α' in order that both wave crests arrive at the same time at O'. The aperture of the telescope is $2L = CB$ or $L = CO' = O'B$.

The coordinate origin O' of Σ' is put into the symmetry point of the telescope, where the waves should positively interfere. Both bounding rays at A and B might have the same phase relation at time t_1 of system Σ_o. When the in Σ_o evaluated run times of the wave crests for distances ACO' and BO' coincide, then they interfere positively at O' and we will observe the star in Σ' under the angle α'.

The telescope is at rest in Σ'. The aperture is there $2L = BC$. Then it holds

$$\Sigma' : \; AP = BP = CQ = L\cos\alpha' \;, \quad \frac{1}{2}AC = QO' = O'P = L\,\sin\alpha' \;. \tag{252}$$

The lengths AP, BP and CQ are in motion with velocity amount v along the directions of its extension as observed from system Σ_o and they are seen LORENTZ-contracted according to Eq. (99). However, the lengths QO' and $O'P$ are transverse to the direction of motion and remain unchanged. In Σ_o it holds

$$\Sigma_o : \; AP = BP = CQ = \frac{1}{\gamma}\,L\,\cos\alpha' \,, \quad QO' = O'P = L\,\sin\alpha' \,. \tag{253}$$

We simplify the computation without changing the interference property, when we replace the path of the left boundary ray from C to O' by the path through point Q and the path of the right boundary ray from B to O' by the path over the point P. The evaluation for the direct path of the boundary rays are discussed in Problem 46. The rays interfere positively at O', when the run times coincide,

$$\Delta t_{AC} + \Delta t_{CQ} + \Delta t_{QO'} = \Delta t_{BP} + \Delta t_{PO'} \,.$$

Always in Σ_o observed, the wave crest transverses the distance from A to C with the speed of light c. Since the telescope approaches to the wave crest at C with the velocity v then in system Σ_o the crest has a velocity $c + v$ according to Eq. (8) for the distance from C to Q and correspondingly a velocity $c - v$ for the path from B to P since the telescope runs away from the wave crest with velocity v. The run time is the distance divided by the velocity. For the equality of both run times, we can therefore write, using $\Delta t_{QO'} = \Delta t_{PO'}$ and $AC = 2QO'$

$$\frac{2L\sin\alpha'}{c} + \frac{L\cos\alpha'}{\gamma(c+v)} = \frac{L\cos\alpha'}{\gamma(c-v)} ,$$

hence

$$\frac{2}{c}\tan\alpha' = \frac{1}{\gamma\,(c-v)} - \frac{1}{\gamma\,(c+v)} = \frac{2\,v}{\gamma\,(c-v)(c+v)} = 2\,\frac{v}{c}\,\frac{1}{\gamma\,(1-v^2/c^2)} = 2\,\gamma\,\frac{v}{c^2} .$$

For the angle α' by which the observer must tilt the telescope in Σ' we get in Σ_o the relation

$$\tan\alpha' = \gamma\,\frac{v}{c} .$$

In Σ_o the in the direction of motion lying length is smaller by the factor $1/\gamma$ than the rest length in Σ' so that $\tan\alpha = \tan\alpha'/\gamma$, and we get again the formula (247) for the Σ' and Σ_o measured aberration angles α' and α, respectively,

$$\left.\begin{aligned} \tan\alpha &= \frac{v}{c} , \\ \sin\alpha' &= \frac{v}{c} . \end{aligned}\right\} \qquad \text{Exact theory} \qquad (254)$$

For the classical approximation we can repeat all arguments from above, and we have only to set always $\gamma = 1$. So we get again[6]

$$\tan\alpha' = \tan\alpha = \frac{v}{c} . \qquad \text{Classical approximation} \qquad (255)$$

7 A Paradox for the Aberration of Waves

Again and again the following consideration of aberration of waves in the classical approximation leads to irritations.

---

[6]Here, we have evaluated the aberration according to the arguments of FRESNEL concerning energy streamings. Within the framework of VLBI one can measure the aberration of wave fronts directly. The angle change of the wave fronts determines the time differences of signals. For a separation of the telescopes of 3000 km, one gets an aberration of 300 m corresponding to a time difference of one microsecond. Here we follow the arguments of Liebscher and Brosche (1998).

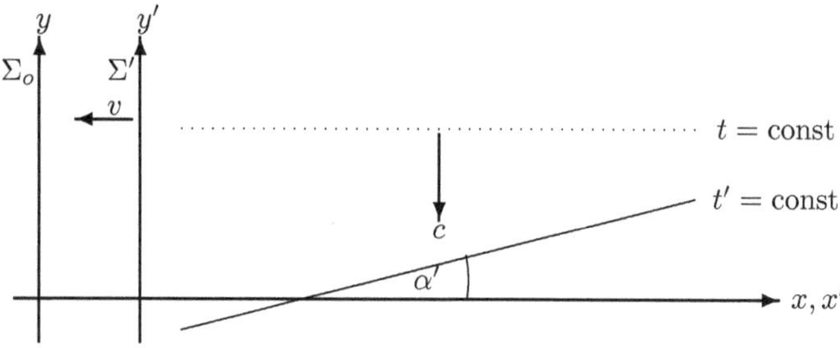

Fig. 12 Aberration and wave front.

The equation of the considered plane wave from the star light S reads in Σ_o, s. the dotted line in Fig. 12,

$$\Sigma_o : \; y = -c\,t \; . \qquad \text{Plane of constant phase in } \Sigma_o \tag{256}$$

The system Σ' might be in motion with velocity amount v in negative x-direction of Σ_o. We insert into Eq. (256) the LORENTZ transformation (105) with $-v$ instead of v hence $t = \gamma\,(t' - vx'/c^2)$ and $y = y'$ so that $y' = -c\,\gamma\,(t' - vx'/c^2)$, and we get in Σ' the relation

$$\Sigma' : \; y' = \frac{\gamma\,v}{c}\,(x' - \frac{c^2}{v}\,t') \, . \qquad \text{Plane of constant phases in } \Sigma' \tag{257}$$

From Eq. (257) we immediately read off the tilt of the phase plane in agreement with our result (247), $\tan\alpha' = \gamma\,v/c$.

However if we insert into Eq. (256) the GALILEI transformation (69), then a paradox follows,

$$\Sigma' : \; y' = -c\,t' \; . \tag{258}$$

This plane is not tilted—in contradiction to the in Eq. (254) given tilt angle $\tan\alpha = v/c$ of the in Σ' observed wave normal in the framework of classical space-time. Before we clarify this contradiction, we scrutinise our argumentation.

Independent from its physical nature, one can represent waves as superposition of plane waves

$$A = A_o \cos(\phi) = A_o \cos(\omega\,t - \mathbf{k}\cdot\mathbf{x}) \qquad \text{Plane wave} \tag{259}$$

with an amplitude A_o and a phase ϕ

$$\phi = \omega\,t - \mathbf{k}\cdot\mathbf{x} \, . \qquad \text{Phase of a plane wave} \tag{260}$$

We set $k := |\mathbf{k}|$ and $\mathbf{k_o} = (\mathbf{k}/k)$ so that $|\mathbf{k_o}| = 1$ and we write $\mathbf{x} = (x, y, z)$, $\mathbf{k} = (k_1, k_2, k_3)$. Points of constant phase define planes in space, i.e. planes of the same oscillation state. For a space-independent vector $\mathbf{k}$ these planes $\phi = \omega t - \mathbf{k} \cdot \mathbf{x} = \text{const}$ run through space in direction $\mathbf{k}$ with velocity $u = \omega/k$. For a fixed time t there result phase planes $\mathbf{k} \cdot \mathbf{x} = \mathrm{C}$ with a normal vector $\mathbf{k}$.

When we proceed in the direction of the unit vector $\mathbf{k_o}$ by $\Delta\mathbf{x} = (2\pi/k)\mathbf{k_o}$, we have $\mathbf{k} \cdot \Delta\mathbf{x} = 2\pi$. Because of the periodicity of the cos-function, we have the same oscillation state, i.e. we proceed by one waves length λ. Hence it holds $k = 2\pi/\lambda$. Likewise, the oscillation state must repeat at a fixed point after time $\Delta t = T$, i.e. when we wait for one period T. From $\omega\,\Delta t = 2\pi$ we get $\omega = 2\pi/T$.

If we stay at a fixed point $\mathbf{x}$ and wait for a time $\Delta t = n\,T$, then n wave crests past, and the phase changes by $\Delta\phi = n\,2\,\pi$. That means: The by $2\,\pi$ divided phase ϕ equals the *number* n of passing wave crests. Such a natural number cannot change, when one counts in another inertial system

The phase ϕ is invariant at a change of the reference system.
This invariance is independent of the coordinate transformation, by which we describe the change of the reference system, this holds true both for a LORENTZ- as well as for a GALILEI transformation.

The significance of the phase invariance for the DOPPLER effect and the aberration are examined in Problem 20.

For the exploration of the paradox of aberration of waves, we must distinguish two directions:

1. The normal vector $\mathbf{k}_o$ is defined as the direction orthogonal to the planes of constant phases.

2. We define a vector $\mathbf{n}_o$ as the direction, in which energy is transported with the waves through space.

A plane of constant phase is fixed by all the points in space, that at a fixed time t have the same phase. This plane and consequently its normal vector $\mathbf{k}_o$ is fixed by the *definition* of simultaneity.

The direction vector $\mathbf{n}_o$ of the energy flux is fixed in the experiment through the direction of the telescope tube, with which we observe the star. We have to orient the telescope in such a direction, that the energy, transported from the waves, reaches the detector at the end of the tubes. For a wrong orientation, the energy will be absorbed from the boundaries of the tube.

The directions $\mathbf{k}_o$ and $\mathbf{n}_o$ can in general be different. Now we will discuss this statement more in detail.

In our at first selected system Σ_o, we start with the assumption, that the normal vectors $\mathbf{k}_o$ of phases determine the direction vector $\mathbf{n}_o$ of the energy flux. The definition of simultaneity in Σ_o is simply done in such a way. The paradox occurs in the transition to a moving reference system Σ'.

Again the planes of constant phase and therefore the direction of the corresponding normal vectors $\mathbf{k}'_o$ are fixed by the *definition* of simultaneity, but now in system Σ'.

The phase planes in Σ' resulting from the substitution of coordinates in Eq. (256), are then determined by the in the transformation contained definition of simultaneity. Then we substitute the general linear coordinate transformation (26) into Eq. (256),

$$\Sigma' : \; y' = \frac{\theta c}{\Delta}\,(x' - \frac{k}{\theta}\,t') \,. \tag{261}$$

From Eq. (261) we read off the tilt of the phase planes

$$\tan\alpha' = \frac{\theta c}{\Delta} = \theta c \quad \text{for} \quad \Delta = 1 \,, \tag{262}$$

when we restrict ourself for simplicity on transformations with $\Delta = 1$, cp. Chap. 2, Sect. 3.

The parameter θ regulating the synchronisation of clocks in systems Σ' is in principle arbitrary. When we chose θ arbitrarily, one can produce an arbitrary tilt of the phase planes in Σ' with the substitution method using Eq. (256). In general, this has however nothing to do with the observed aberration.

We get a paradox, when we confuse the fixed normal vector of wave planes by the definition of simultaneity, that we get by the substitution of coordinate transformations in Eq. (256) for the dashed coordinates, with the observation direction of waves in system Σ'.[7] The direction $\mathbf{n}'_o$ of the in Σ' observed energy flux is independent from the definition of simultaneity in Σ' but fixed by experiment. The tube of our telescope is tilted until the incoming waves reach the detector at the end of the tube, cp. Fig. 9. If the energy arrives vertically in system Σ_o as assumed, then the tilt of the tube defines the aberration angle.

The missing tilt of the phase planes in Σ' according to Eq. (258) are therefore only produced by the definition of an absolute simultaneity in Σ' in using the Galilei transformation. Then the vectors $\mathbf{n}'_o$ and $\mathbf{k}'_o$ will not coincide anymore. With our paradox, we are captured in the long-known trouble of absolute simultaneity.

We convince ourself, that we get the correct aberration angle when using the direction of the energy flux of the waves. As discussed above, in aberration we observe the propagation direction of energy contained for example in a finite wave packet. Only energy can excite a detector or our retina. In system Σ_o, these wave packets might move with a velocity amount c in direction of the negative y-axis. Seen from system Σ', we measure for Σ_o a negative velocity of amount v the wave packet has then in the Galilei space-time of Eqs. (70) or (71) a velocity component $u'_x = v$ in x'-direction, while $u'_y = u_y = -c$ remains unchanged, so that the wave packet, as seen from Σ' again has an aberration angle α with $\tan\alpha = v/c$. Hence,

---

[7] In Sect. 9, we discuss an application of the transformation (298) already given by W. Thirring for the relativistic space-time with absolute simultaneity. Inserting this transformation (298) in Eq. (256) leads to the same confusion in the question of the aberration as the substitution of the Galilei transformation. On the other hand, one gets the correct equation (255) for the aberration in the classical space-time, when one inserts in Eq. (256) not the Galilei transformation, but the linear approximation (131) of the Lorentz transformation, hence the classical space-time with non-conventional simultaneity, cf. Thirring (1992), Vol. 1, Sect. 6.4.

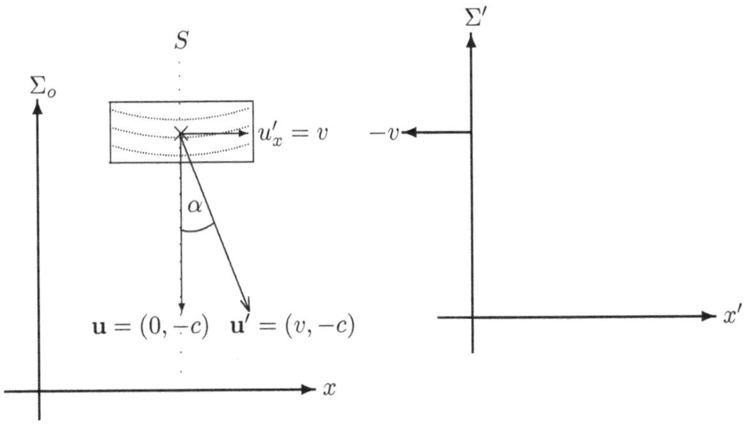

Fig. 13 The wave packet limited by a frame moves in system Σ_o with the vector components of the velocity $(u_x, u_y) = (0, -c)$. System Σ' has a negative velocity of amount v as measured from Σ_o. After a GALILEI transformation (69) the wave packet is observed in Σ' according to Eq. (70) or (71) with a velocity $(u'_x, u'_y) = (v, -c)$. Again the tilt angle α is given by $\tan\alpha = v/c$.

we find the correct observed tilt of the wave front in Σ', s. Fig. 13. We remark that the physical nature of the waves has no influence on this argumentation.[8] When we substitute the LORENTZ transformation in Eq. (256) for the phase plane in Σ_o, we get in the dashed coordinates in system Σ' another plane with another normal vector as for the substitution of the GALILEI transformation in Eq. (256), since in Σ', we get by definition with the LORENTZ transformation other spatial points as simultaneously as with the GALILEI transformation.

Still it remains the question, what is the reason that the substitution of the LORENTZ transformation in Eq. (256) delivers the tilt angle of the normal of the phase planes in Σ' in agreement with the direction of the energy flux of electromagnetic waves. In Chap. 9, Sect. 3 we will discuss that the equations of the electromagnetic field, MAXWELL's equations, are invariant against LORENTZ transformation. This is the reason that the vector of the energy flux in the transition to system Σ' will transform exactly as the normal vector of wave planes when substituting the LORENTZ transformation in Eq. (256). The invariance of MAXWELL's equations under LORENTZ transformations is also the reason that both the DOPPLER effect as well as the aberration of electromagnetic waves can be derived from the invariance of the phases ϕ with respect to LORENTZ transformations, cp. Problem 20 and Problem 33.

[8]In principle also sound waves suffer from aberration. The physical properties of sound waves must be especially respected. When we assume that we use the sound speed in Eq. (256), then the substitution of the LORENTZ transformation as well as the substitution of the GALILEI transformation in Eq. (256) lead to a wrong results for the in Σ' expected tilt of the wave normal. For the correct tilt angle we have to substitute in Fig. 13 the quantity c by the sound speed c_a and for the in Σ' measurable tilt of the front of sound waves there follows the correct result $\tan\alpha = v/c_a$.

8 The Twin Paradox

We base our discussion on the relativistic space-time with EINSTEIN's synchronisation of clocks and therefore the LORENTZ transformation for the coordinates of events in different inertial systems.

The twin paradox concerns the following dispute:

A pair of twins, say for distinction twin A and B undertake a travel. Precisely, twin B stays in a reference system Σ_o and twin A stays in system Σ'. But the system of A moves with constant velocity v as seen from B and inversely, B travels with constant velocity $-v$ as seen from A. At start at the common coordinate origin, the twins put their respective clocks, say clock U^A of twin A and clock U^B of twin B at position 0.

Now if twin B compares the clock U^A of twin A with clocks U_o^x at positions x resting in his reference systems Σ_o so he states because of the time dilatation (113), that the pointer of clock U^A remains behind the pointer positions of clocks U_o^x that U^A just passes, Fig. 14.

Twin A uses the same arguments. He also states, that according to Eq. (113) the pointer of clock U^B of twin B remains behind the pointer positions of those clocks $U_v^{x'}$ that rest at positions x' in the reference systems Σ' where clock U^B passes by, and just by the same factor, since the time dilatation depends only on the square of the velocity, Fig. 15.

This sounds really strange. However, a truly paradox situation does not appear since different clocks are compared with another. A logical contradiction cannot be constructed.

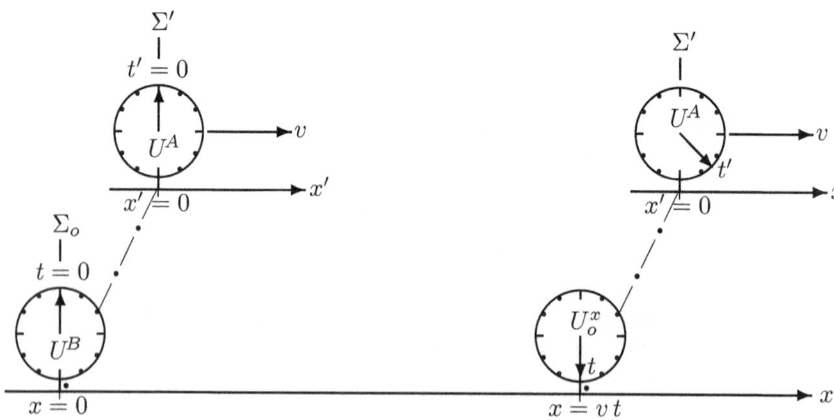

Fig. 14 Twin A is with his clock U^A at time t in Σ_o at position $x = vt$. By Eq. (113) on clock U^A a pointer position $t' = t\sqrt{1 - v^2/c^2}$ is found because of motion of U^A relative to Σ_o. The dash-dotted lines connect in the following always points in the figures belonging to the same event.

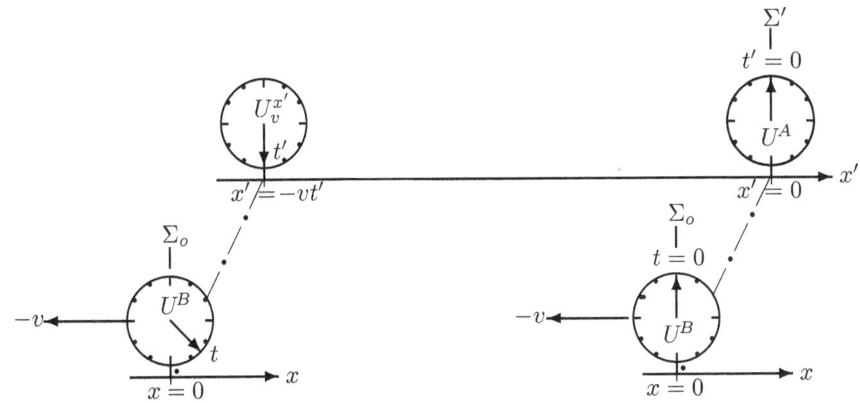

Fig. 15 Twin B is with his clock U^B at time t' in Σ' at position $x' = -v\,t'$. Now the clock U^B is moving relative to Σ'. According to Eq. (113), in this case holds $t = t'\sqrt{1 - v^2/c^2}$. Hence take, for example, $v = -\frac{2}{3}\,c$ on the moving clock U^B, twin B will be measure a time $t \approx 22.4$ while the clock at rest shows $t' = 30$. The dashed-dotted lines connect again the same space-time points.

Now twin A might return his velocity, or twin B might follow twin A with higher velocity, so that the twins approach to one another with opposite velocities but equal amount. Again the two will argue that the pointer of each other's clock stays behind because of the time dilation, and the effect shown in Figs. 14 and 15 should occur twice.

At their encounter twin B tells that twin A's pointer of clock U^A remained behind his own clock U^B while twin A claims, that his pointer should be in progress against that of clock U^B just by the same reason.

One pointer at two different positions—this would be paradox!

To solve this paradox we consider three inertial systems, $\Sigma_o(x, t)$, $\Sigma'(x', t')$ and $\Sigma''(x'', t'')$ with a common coordinate origin $O(0, 0)$.

At the event O the twins separate from another. Later twin B decides to follow twin A in an inertial system Σ'' so that both meet again in an event $Z(x_Z, t_Z) = Z(0, t'_Z) = Z(x''_Z, t''_Z)$, s. Figs. 16 and 17.

Twin A sits all time at point $x'_Z = 0$ in his inertial system Σ' that is in motion with velocity v with respect to Σ_o. His travel time t_A from the separation until the meeting point, from event O until event Z, can be read off from his clock U^A,

$$t_A = t'_Z \;. \qquad \text{Travel time of twin } A \qquad (263)$$

For specifying a definite situation, we assume that the inertial system Σ'' in which twin B follows twin A, might have the velocity $u' = v$ as seen from Σ', hence Σ' would move with velocity $-v$ with respect to Σ''. Twin A notes that twin B approaches with velocity v, while B approaches to A with $-v$.

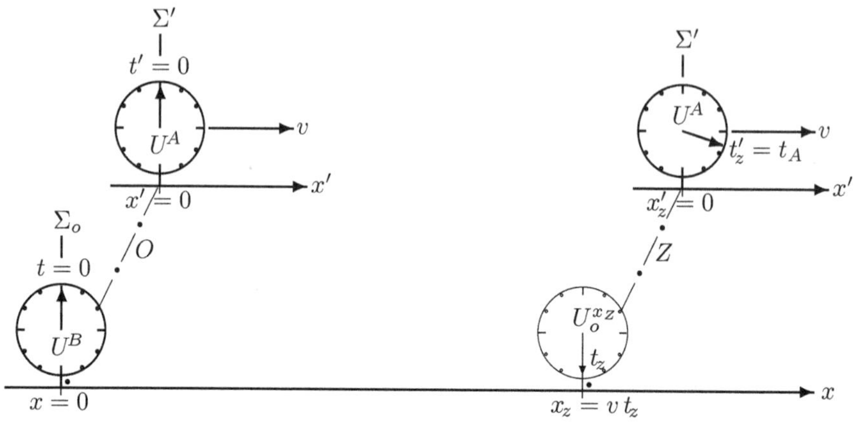

Fig. 16 At event O (left side) the twins separate. At the event Z of meeting again (right side) there only the clock U^A of twin A in system Σ' is shown, and for comparison a clock of system Σ_o. For $v = 0.8\,c$ and therefore $\gamma_v = 1.67$, it follows a travel time of twin A of $t_A = t'_Z = t_Z/\gamma_v = 18$ when $t_Z = 30$. The dash-dotted lines again connect the same space-time points.

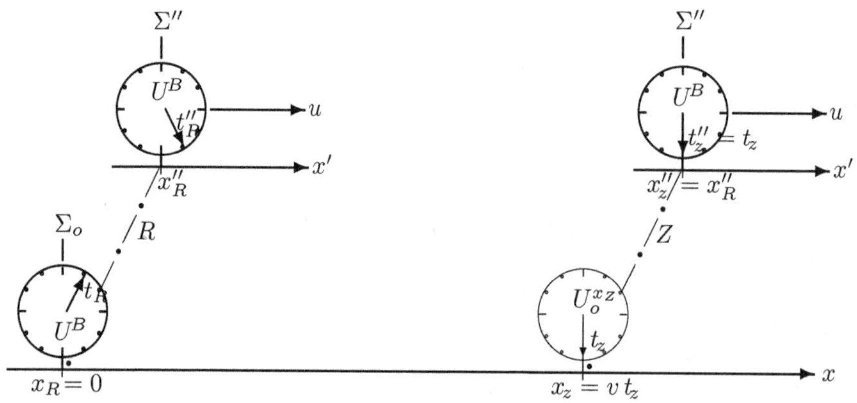

Fig. 17 Twin B returns at time t_R of system Σ_o, i.e. it changes to the system Σ'' that has the velocity u with respect to Σ_o. This is event R (left part of the figure). As an example, we assume that this velocity u is chosen so, that twin A in his system Σ' measures a velocity $u' = v$ with which twin B and his system Σ'' approaches to him. Then we have formulas (264)–(273). For the event Z of the meeting (right panel) we only draw clock U^B of twin B and a comparison clock of Σ_o. For $v = 0.8\,c$ hence $\gamma_v = 1.67$ it follow for $t_Z = 30$ from Eqs. (268) and (273) the times $t_R = 0.36 \cdot 15 = 5.4$ $t''_R = \gamma_v'^2\, t_R\,(1 + v^2/c^2) = 2.8 \cdot 5.4 \cdot 1.64 = 24.6$ and $t''_Z = t_Z = 30$. The dash-dotted lines connect the same space-time points.

With the initial condition of a common coordinate origin $O(0, 0)$ the LORENTZ transformation between Σ' and Σ'' reads

$$x'' = \gamma_v\,(x' - v\,t')\ , \qquad t'' = \gamma_v\,(t' - x'\,v/c^2)\ . \tag{264}$$

With the velocities v of Σ' with respect to Σ_o, and $u' = v$ of Σ'' with respect to Σ' we get from the addition theorem (106) a velocity u of Σ'' as measured in Σ_o,

$$u = \frac{u' + v}{1 + u'\,v/c^2} = \frac{2vc^2}{c^2 + v^2} \quad \longrightarrow \quad \gamma_u = \frac{1}{\sqrt{1 - \frac{u^2}{c^2}}} = \frac{c^2 + v^2}{c^2 - v^2} \; . \tag{265}$$

The reference systems Σ' and Σ'' are in motion with velocities v and u, with respect to Σ_o, if always a common coordinate origin is assumed. Then the corresponding LORENTZ transformations read

$$\begin{aligned} x' &= \gamma_v\,(x - v\,t)\;, & &\longleftrightarrow & x &= \gamma_v\,(x' + v\,t')\;, \\ t' &= \gamma_v\,(t - x\,v/c^2)\;, & & & t &= \gamma_v\,(t' + x'\,v/c^2)\;, \end{aligned} \tag{266}$$

$$\begin{aligned} x'' &= \gamma_u\,(x - u\,t)\;, & &\longleftrightarrow & x &= \gamma_u(x'' + u\,t'')\;, \\ t'' &= \gamma_u\,(t - x\,u/c^2)\;, & & & t &= \gamma_u\,(t'' + x''\,u/c^2)\;. \end{aligned} \tag{267}$$

Twin B is initially at $x = 0$ in Σ_o.

At the event $R(0, t_R) = R(x'_R, t'_R) = R(x''_R, t''_R)$ he might return by entering the moving system Σ''.

Until that event on his clock U^B it elapsed the time t_R.

From formulas (266) and (267), we find with relation (265) from the coordinates of event R in Σ_o the coordinates in Σ' and Σ'',

$$\left.\begin{aligned} \Sigma_o &: x_R = 0\,, & & t_R\,, \\ \Sigma' &: x'_R = -\gamma_v\,v\,t_R\,, & t'_R &= \gamma_v\,t_R\;, \\ \Sigma'' &: x''_R = -\frac{2vc^2}{c^2 - v^2}\,t_R\,, & t''_R &= \frac{c^2 + v^2}{c^2 - v^2}\,t_R\,. \end{aligned}\right\} \quad \text{Return event } R \tag{268}$$

The time t_R is a parameter in the sequence of events that can be freely chosen.

For the event of final meeting Z, cp. Fig. 16, it follows at first by the time dilatation or from $t_Z = \gamma_v\,(t'_Z + x'_Z v/c^2)$ using the right-hand side of Eq. (266) because of $x'_Z = 0$ and with Eq. (263),

$$t_A \equiv t'_Z = t_Z/\gamma_v \;. \tag{269}$$

Twin A observes that twin B initially removes with velocity v, and it returns after time t'_R with the same velocity. For his travel time $t_A = t'_Z$ it holds

$$t_A = t'_Z = 2t'_R \,. \tag{270}$$

With Eq. (268) it follows

$$t_A = t'_Z = 2\,\gamma_v\,t_R \,. \qquad \text{Travel time of twin } A \tag{271}$$

From Eq. (264) follows for $t''_Z = \gamma_v (t'_Z - x'_Z v/c^2)$ with $x'_Z = 0$ and using Eqs. (269) and (271)

$$t''_Z = \gamma_v t'_Z = \gamma_v t_A , \qquad (272)$$

hence[9]

$$t''_Z = t_Z = 2\gamma_v^2 t_R . \qquad (273)$$

The complete twin history is then characterised by three events:

$$\left.\begin{array}{llll} O(0,0) = & O(0,0) = & O(0,0) , & \text{Start time of twins} \\ R(0,t_R) = & R(x'_R, t'_R) = & R(x''_R, t''_R) , & \text{Return of twin } B \\ Z(x_Z, t_Z) = & Z(0, t'_Z) = & Z(x''_Z, t''_Z) . & \text{Meeting of the twins} \end{array}\right\} \qquad (274)$$

The travel time of twin B consists of the time t_R in the system Σ_o and the time, that he takes after the transfer into the system Σ''. After the return, twin B measures the time $t''_Z - t''_R$ on his clock U^B, Fig. 17.

The travel time t_B of twin B amounts then in total

$$t_B = t_R + t''_Z - t''_R . \qquad \text{Travel time of twin } B \qquad (275)$$

With Eqs. (273) and (268) it follows

$$t_B = t_R + 2\gamma_v^2 t_R - \frac{c^2 + v^2}{c^2 - v^2} t_R = \frac{c^2 - v^2 + 2c^2 - c^2 - v^2}{c^2 - v^2} t_R ,$$

hence

$$t_B = 2t_R . \qquad \text{Travel time of twin } B \qquad (276)$$

With Eq. (271) we find

$$t_B = t_A/\gamma_v , \quad \gamma_v > 1 \quad \longrightarrow \quad t_B < t_A . \qquad (277)$$

From the sight of A, the returning twin B is at the meeting younger than the twin A. In other words: Younger stays the twin that has changed his velocity.

This is now remarkable. The actual paradox results from the following argumentation of twin B:

"On the first part of the travel the pointer of my clock U^B proceeded by t_R. Twin A removes himself from me with velocity v so that the pointer of his clock then stands at $t'_1 = t_R/\gamma_v$ because of the time dilatation.

[9]The equality $t''_Z = t_Z$ is accidental in our example with $u' = v$.

When I transfer into system Σ'', he approaches me with velocity v. I stay there during the time $t''_Z - t''_R$. By this amount the pointer of my clock moves forward. The pointer of his clock U^A stays behind again due to the time dilatation $1/\gamma_v$, i.e. it proceeds only by $t'_2 = (t''_Z - t''_R)/\gamma_v$.

Then at the end the pointer of my clock U^B shows the time

$$t_B = t_R + t''_Z - t''_R\,, \qquad \text{Correct evaluation of the pointer position of clock } U^B \text{ by twin } B \tag{278}$$

whereas the pointer of his clock U^A at the meeting in total shows

$$t'_1 + t'_2 = (t_R + t''_Z - t''_R)/\gamma_v = t_B/\gamma_v \qquad \text{Incorrect evaluation of the pointer position of clock } U^A \text{ by twin } B \tag{279}$$

and not inverse as in Eq. (277)."

Where is the error?

Again we suffer from the pitfall of the relativity of simultaneity.

We remember at first the common initial conditions for all three systems Σ_o Σ', and Σ'', cp. Eq. (11). In Σ_o the clock U^A moves as $x = v\,t$, and for Σ'' we have chosen the velocity in such a way that the in Σ' resting clock U^A moves in Σ'' as $x'' = -v\,t''$

$$\left.\begin{array}{ll} x = v\,t\,, & \text{Clock } U^A \text{ in } \Sigma_o \\ x'' = -v\,t''\,. & \text{Clock } U^A \text{ in } \Sigma'' \end{array}\right\} \tag{280}$$

The simultaneous position of clock U^A at the transfer event $R(0, t_R)$ in Σ_o is then $x''_q = -v\,t''_R$, hence

$$x_p = v\,t_R\,, \qquad \text{Simultaneous position of clock } U^A \text{ at transfer-event } R \text{ in } \Sigma_o \tag{281}$$

$$x''_q = -v\,t''_R\,. \qquad \text{Simultaneous position of clock } U^A \text{ at transfer-event } R \text{ in } \Sigma'' \tag{282}$$

From the right-hand side of Eq. (267), $x = \gamma_u\,(x'' + u\,t'')$ we find with Eqs. (268) and (265) for the coordinate x_q in Σ_o at time t''_R the position

$$x_q = \gamma_u\,(x''_q + u t''_R) = \gamma_u\,(u - v)\,t''_R = \gamma_u \left(\frac{2vc^2}{c^2 + v^2} - v\right) t''_R$$
$$= \gamma_u\,\frac{2vc^2 - v^3 - vc^2}{c^2 + v^2}\,t''_R = \frac{v(c^2 - v^2)}{c^2 + v^2}\,\frac{c^2 + v^2}{c^2 - v^2}\,t''_R\,,$$

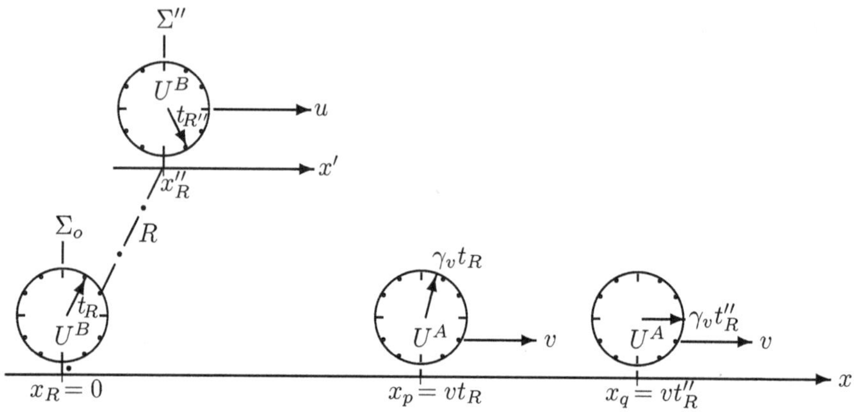

Fig. 18 Positions x_p and x_q of clock U^A according to Eqs. (281) and (283) in Σ_o at return R. The dash-dotted lines again connect the same space-time-points.

hence with Eq. (268)

$$x_q = v\, t''_R = \frac{c^2 + v^2}{c^2 - v^2}\, v\, t_R \,. \quad \begin{array}{l}\Sigma_o\text{-coordinate of simultaneous position of clock } U^A \\ \text{at return } R \text{ in } \Sigma''\end{array} \tag{283}$$

From Eqs. (281) and (283) we read off[10]

$$x_p < x_q \,. \tag{284}$$

When twin B leaves the system Σ_o, then he took for the last pointer position of clock U^A of twin A his position x_p in Σ_o. As soon as twin B sits in Σ'', he takes as first pointer position of clock U^A its position x_q in Σ_o. Hence he arrives at the summation $t_1 + t_2 = (t_R + t''_Z - t''_R)/\gamma_v = t_B/\gamma_v$ for the travel time of twin A. He has overlooked the continuing run of the pointer of clock U^A during its motion from x_p to x_q, since he disregarded the relativity of simultaneity, Fig. 18.

During the motion of clock U^A from x_p to x_q in Σ_o, a time T_{pq} is running with

$$T_{pq} = \frac{1}{v}\,(x_q - x_p) = \frac{1}{v}(v\, t''_R - v\, t_R) = t''_R - t_R \,.$$

On clock U^A the pointer then proceeds because of the time dilatation by an amount $\Delta t'_A = T_{pq}/\gamma_v$,

$$\Delta t'_A = (t''_R - t_R)/\gamma_v \,. \tag{285}$$

[10]We remark that the simplicity of Eq. (283) is again a consequence of our example with $u' = v$.

Twin B has forgotten this time in his evaluation (279) of the pointer position of clock U^A. When we add the time $(t''_R - t_R)/\gamma_v$ to the right-hand side of Eq. (279), then we get indeed the correct pointer position t_A of twin A at their encounter in agreement with our evaluations above, Eqs. (272) and (277), since

$$t_A = t'_1 + t'_2 + \Delta t'_A = (t_R + t''_Z - t''_R)/\gamma_v + (t''_R - t_R)/\gamma_v = t''_Z/\gamma_v = 2\,\gamma_v\,t_r = \gamma_v\,t_B\ . \tag{286}$$

The paradox is solved.

Finally we intend to show that the full twin story has a simple algebraic explanation. To this aim, we follow the passing times t_B and t_A between start and encounter of clocks U^B and U^A in the reference system Σ_o. Here we consider the general case:

In Σ_o, there is measured a time t_Z for the travelling of the twins.

Twin A is the complete time in motion with system Σ', at velocity v with respect to Σ_o. Because of the time dilatation it follows Eq. (269),

$$t_A = t_Z/\gamma_v\ . \tag{287}$$

Twin B might be at rest in Σ_o until time t_u so that the pointer of his clock proceeds by time t_u. Now we observe in Σ_o, that twin B transfers at time t_u into a system Σ'' which has a velocity u with respect to Σ_o. This velocity u is chosen so that he just catches twin A at time t_Z. Because of the time dilatation, the pointer of his clock proceeds by $(t_Z - t_u)/\gamma_u$ so finally its position is

$$t_B = t_u + (t_Z - t_u)/\gamma_u\ . \tag{288}$$

Now we show

$$t_B < t_A \tag{289}$$

for an arbitrary transfer time t_u which has only to be chosen, so that twin B can catch up twin A with velocity $u < c$ at time t_Z.

Hence it is only required that

$$0 < v < u < c\ . \tag{290}$$

The same way $v\,t_Z$ of twin A in Σ_o must be transversed by twin B during the time $t_Z - t_u$, $u\,(t_Z - t_u) = v\,t_Z$ hence

$$t_Z = \frac{u}{u - v}\,t_u\ . \tag{291}$$

The claimed inequality (289) now reads with Eqs. (287), (288) and (291)

$$t_B = \left[1 + \frac{v}{\gamma\,(u - v)}\right] t_u < \left[\frac{u}{\gamma_v\,(u - v)}\right] t_u = t_A\ . \tag{292}$$

Taking advantage from Eq. (290) one proofs that this inequality is connected with the inequality between the geometrical and arithmetical mean for arbitrary positive velocities u and v s. GÜNTHER (2000), or somewhat simplified using a proposal of FREUND (2009):

$$0 < (u-v)^2 c^2 \ , \tag{293}$$

$$c^4 - u^2c^2 - v^2c^2 + u^2v^2 < c^4 - 2c^2uv + u^2v^2 \ ,$$

$$(c^2 - v^2)(c^2 - u^2) < (c^2 - uv)^2 \ ,$$

$$c^2\sqrt{1-\frac{v^2}{c^2}}\sqrt{1-\frac{u^2}{c^2}} < c^2 - uv \ ,$$

$$\frac{c^2}{\gamma_u\,\gamma_v} < c^2 - uv \ ,$$

$$\frac{2uv}{\gamma_u\,\gamma_v} < 2uv - \frac{2u^2v^2}{c^2} \ ,$$

$$-2uv < -\frac{2u^2v^2}{c^2} - \frac{2\,u\,v}{\gamma_u\,\gamma_v} \ ,$$

$$u^2 - 2uv + v^2 < u^2 - \frac{u^2\,v^2}{c^2} + v^2 - \frac{u^2v^2}{c^2} - \frac{2\,u\,v}{\gamma_u\,\gamma_v} \ ,$$

$$(u-v)^2 < \left(\frac{u}{\gamma_v} - \frac{v}{\gamma_u}\right)^2 \ .$$

It is $0 < u - v$ and since $\gamma_v < \gamma_u$ also $0 < u\gamma_u - v\gamma_v$ and therefore

$$u - v < \frac{u}{\gamma_v} - \frac{v}{\gamma_u} \ ,$$

$$\frac{u - v + v/\gamma_u}{u - v} < \frac{u}{(u-v)\,\gamma_v} \ , \quad \text{hence}$$

$$1 + \frac{v}{(u-v)\,\gamma_u} < \frac{u}{(u-v)\,\gamma_v} \ . \qquad \text{Twin inequality} \tag{294}$$

This is indeed the claimed twin inequality (292) or (289). Obviously also in the limit of late return with $u \longrightarrow c$ and $\gamma_u \longrightarrow \infty$ the inequality can be fulfilled because of $1 < \sqrt{(c+v)/(c-v)} = c/[\gamma_v(c-v)]$. It holds always

$$t_B < t_A \ . \qquad \text{The following up twin } B \text{ stays younger.} \tag{295}$$

We see:

> The twin will stay younger that changed the reference system for returning or catching up the other twin.

With other words, when there exists *one* inertial system, where one twin is at rest during the whole story, than this is finally the older twin. Who rests, rusts.

It follows: We did not come into trouble with the twin story, when we consider it in one single inertial system. However when we describe the story from the sight of a reversing twin, or after the first part of the travel by the catching up of twin B we get easily in the pitfall of the problem of simultaneity, since with the change of the inertial systems, there also changes the simultaneity of events.

In the following section, we will show that the twin paradox dissolves by itself when we introduce as an exception an absolute simultaneity in the relativistic space-time for this application.

9 Measuring Rod and Twin Paradox for Non-conventional Simultaneity

The troubles, we encounter especially in the measuring rod and the twin paradox, are a consequence of the definition of conventional simultaneity in relativistic space-time. But only this EINSTEIN simultaneity allows to formulate mathematically the equivalence of all inertial systems in such a way, that each inertial system is connected to any other one with the same transformation, a LORENTZ transformation. However we again and again meet the pitfalls of the resulting relativity of simultaneity. The further exposition of relativistic physics without this formulation is however practically not feasible. For the explanation of relativistic paradoxa, i.e. for mistakes in our thinking with the relativity of simultaneity, it can be indeed sometimes helpful to use another definition of simultaneity. This will be shown in the following:

The relativistic space-time is defined by the physical postulates (99) and (100). If we now introduce instead of the by Eq. (46) defined conventional simultaneity in relativistic space-time an absolute simultaneity,

$$\theta = \theta_a = 0, \qquad \text{Absolute simultaneity} \qquad (296)$$

then it follows from Eq. (100) instead of Eq. (104)

$$k = \frac{1}{\sqrt{1 - v^2/c^2}}, \quad q = \sqrt{1 - v^2/c^2}, \quad \theta = 0. \qquad \text{Relativistic space-time with absolute simultaneity} \qquad (297)$$

From the general transformation formulas (26), we get instead of the LORENTZ transformation (105) formulas given by (THIRRING 1992, s. Chap. 2, Sect. 1) and independently introduced by GÜNTHER (2000) as REICHENBACH transformation,

$$\left.\begin{aligned} x' &= \frac{x - v\,t}{\sqrt{1-v^2/c^2}}\,, & \longleftrightarrow \quad x &= \frac{(1-v^2/c^2)\,x' + v\,t'}{\sqrt{1-v^2/c^2}}\,, \\ t' &= t\sqrt{1-v^2/c^2}\,, & t &= \frac{t'}{\sqrt{1-v^2/c^2}}\,. \end{aligned}\right\} \begin{array}{l}\text{REICHENBACH–}\\ \text{transformation}\\ \Sigma_o(x,t)\text{ distinguished}\end{array} \tag{298}$$

Already from the form of the inverse transformation, one sees the asymmetry in the description of inertial systems. The system $\Sigma_o(x, t)$ is here distinguished by the *mathematical description*:

The validity of the relativity principle, i.e. the physical equivalence of all inertial systems, means in this formulation, that we can select each arbitrary inertial system for this pure mathematical exceptional position.

Deciding for the simplicity in the discussion of the paradoxa on the basis of this transformation is the circumstance, that in the second line of Eq. (298), in the time transformation, the coordinates x and x' do not appear. We shall only need these equations.

An advantage can be immediately read off from the REICHENBACH transformation (298):

When we consider the in v/c linear approximation of the relativistic space-time, so it follows with $\gamma \longrightarrow 1$ directly the GALILEI transformation (69).

9.1 Measuring Rod Paradox

Again we consider the in Sect. 4, Fig. 2, described situation.

The system Σ_o with the parallel to the x-axis placed barriers with separation l_o is now in the sense of the transformation formulae (298) distinguished. In system Σ' moving axis parallel to Σ_o with velocity v_1 in x-direction, the bar is at rest on the x'-axis, and the length in Σ' has also the value l_o. Using Eq. (297), it follows from Eq. (29), that from sight of Σ_o the bar has a LORENTZ-contracted moving length l_v^S

$$l_v^S = \frac{l_o}{k} = l_o\sqrt{1 - v_1^2/c^2} < l_o\,. \tag{299}$$

The observer in Σ_o now notices that the bar with the constant orientation has an additional velocity v_2 in y-direction, so that the bar moves towards the queue of barriers.

During the discussion in Sect. 4 we came in a discrepancy with this statement. The reason was that the statement of a 'constant orientation' depends in general on the used inertial system, in which the orientation is measured. For setting the bar in motion without change of the orientation, the velocity v_2 in y-direction must be awarded at both end points *simultaneously*. For conventional or EINSTEIN simultaneity in the framework of relativistic space-time, the simultaneous start of the end points in Σ_o looks differently from sight of system Σ'. At first the right endpoint starts, and later the left one. The bar gets a tilt as seen from Σ' shown in the lower drawing of Fig. 3. If conversely the start of the motion of the bar takes place simultaneously in the sight of frame Σ' then it starts, as observed in Σ_o at first the left endpoint and later the right one, as seen in the upper drawing of Fig. 3.

Now this is else, when we—exceptional—use the absolute simultaneity and hence the transformation formulas (298). Then the two events are just simultaneous in Σ' when they are simultaneous in system Σ_o. If the end points of the bar start simultaneously in sight of Σ_o then also from sight of Σ'. The orientation of our bar remains independent from the reference system parallel to the x-axis. The description of the experiment becomes quite simple:

The observer in Σ_o states: The bar has due to Eq. (298) a moving length l_v as given in Eq. (299). The barriers have larger separations l_o with $l_v^S < l_o$. In consequence the bar can pass without resistance the series of barriers, if it fits a gap.

The observer in Σ' states: The bar has length l_o. He observes the barriers in a moving separation l_v^H. This follows from the right-hand side of the REICHENBACH transformation (298) by means of the relation between x and x'. The separation of the first barrier from the zero point, $x = l_o$ in Σ_o he finds a separation $x' = l_v^H$ from the common origin, hence at $t' = 0$ when we resolve for $x' = x/\sqrt{1 - v_1^2/c^2}$ hence

$$l_v^H = l_o k = \frac{l_o}{\sqrt{1 - v_1^2/c^2}} \quad \text{and therefore} \quad l_o < l_v^H \,. \tag{300}$$

From his sight, the separation of the barriers is larger than the rest length l_o of the bar, $l_o < l_v^H$ and consequently it can pass the barriers for a velocity component in y-direction, when it fits a gap.

Both observers come to the same conclusion. There is no paradox.

9.2 Twin Paradox

We consider the in Sect. 8 discussed twin story, where one twin A remains all time in one inertial system Σ' at rest, while twin B changes during the travel the inertial systems. Now when we use again the definition of an absolute simultaneity, then only system Σ' can take over the role as distinguished system.

Twin B is at first at rest in the inertial system Σ_o. Twin A measures in his system Σ' for twin B, as long as he sits in system Σ_o, a velocity $-v$. For the REICHENBACH transformation we chose Σ' as distinguished system. In Eq. (298), we have to change the dashed and undashed variables. We need only the formulas for the in $\Sigma'(x', t')$ and $\Sigma_o(x, t)$ measured times t' and t. With $\gamma_v = 1/\sqrt{1 - v^2/c^2}$ it holds

$$t = \frac{t'}{\gamma_v}\,, \longleftrightarrow t' = \gamma_v\, t\,. \tag{301}$$

In the inertial system $\Sigma''(x'', t'')$ twin B follows twin A with such a velocity u so that he can catch up the first twin, i.e. with $0 < v < u < c$. In the distinguished system Σ' we get that twin B travels for times t' and t'' in $\Sigma'(x', t')$ and $\Sigma''(x'', t'')$, with $\gamma_u = 1/\sqrt{1 - u^2/c^2}$

$$t'' = \frac{t'}{\gamma_u} \quad \longleftrightarrow \quad t' = \gamma_u\, t''\,. \tag{302}$$

In distinction to the situation on the basis of EINSTEIN's simultaneity in the relativistic space-time, s. the analysis of Sect. 8, the correct description of the time span under consideration of the transfer of twin B from Σ_o to Σ'' is without any difficulty, since the spatial coordinates x and x' do not appear in the conversion of time lapses from one reference system to the other.

We write for the elapsed time on the clock U^A before the return $\Delta t_1'$ and for the time after the return $\Delta t_2'$ so that clock U^A shows a time Δt_A

$$t_A = \Delta t_1' + \Delta t_2'\,. \tag{303}$$

With Eqs. (301) and (302) both twins consistently state, that the clock U^B of twin B shows the time t_B

$$t_B = \Delta t + \Delta t'' = \Delta t_1'/\gamma_v + \Delta t_1'/\gamma_u\,. \tag{304}$$

The following statement

$$t_B < t_A \tag{305}$$

holds true for any u and v with $0 < v < u < c$ and this without any dispute.

There remains no paradox.

Chapter 9
Mathematical Formalism of Special Relativity

We shall present the mathematically more sophisticated content of this chapter as simple as possible. We refer to GIULINI (2006) for an exposure 'on an advanced level …(of) the algebraic and geometric structures that underly the theory of Special Relativity'.

Mathematical tools are presented in the Appendix.

1 The LORENTZ Group

1.1 *Special* LORENTZ *Transformation*

The significance of EINSTEIN's relativity principle, s. citation in Chap. 2, Sect. 1, consists of the unification of facts because of its physical nature that are however unexpected and quite different from for our classical perception.

In the first part of his relativity principle, EINSTEIN postulates the equivalence of inertial systems both for electrodynamics and for mechanics. In the second part, he added the postulate of the universal constancy of the speed of light, and with it he distinguished the definition of simultaneity that we introduced with LORENTZ's synchronisation parameter (103). The coordinate transformations between inertial systems form a group $\mathcal{A}$, the general LORENTZ group, as we shall see.

In the inertial system Σ_o with space-time coordinates (x, y, z, t), we consider D'ALEMBERT's wave equation for a scalar field $f = f(x, y, z, t)$,

$$\frac{1}{c^2}\frac{\partial^2 f}{\partial t^2} - \frac{\partial^2 f}{\partial x^2} - \frac{\partial^2 f}{\partial y^2} - \frac{\partial^2 f}{\partial z^2} = 0 \ , \qquad (306)$$

D'ALEMBERT's wave equation

H. Günther and V. Müller, *The Special Theory of Relativity*,
https://doi.org/10.1007/978-981-13-7783-9_9

or for ease of expressions, with two space dimension suppressed,

$$\frac{1}{c^2}\frac{\partial^2 f(x,t)}{\partial t^2} - \frac{\partial^2 f(x,t)}{\partial x^2} = 0 \ . \tag{307}$$

The general solutions $f(x,t) = f(x-ct)$ and $f(x,t) = f(x+ct)$ describe the propagation of plane waves in positive or negative x-direction with the speed of light c.

The system Σ' with coordinates (x', y', z', t') might have a velocity v in x-direction with respect to Σ_o. If the dashed coordinates would be connected with the undashed ones by a Galilei transformation (69), so it would follow

$$f = f(x-ct) = f(x' + vt' - ct') = f[x' - (c-v)t'] \ ,$$

that the light wave observed in Σ' fulfilled the equation

$$\frac{1}{(c-v)^2}\frac{\partial^2 f}{\partial t'^2} - \frac{\partial^2 f}{\partial x'^2} = 0 \, , \tag{308}$$

and therefore it would propagate with a velocity $(c-v)$, in contradiction to the principle of a universal constancy of the speed of light. The Galilei transformation (69) is consequently insufficient for Einstein's relativity principle, as we already know.

Consequently, we look for such transformations that lead to the same wave equation (306) or (307) in all inertial systems, i.e. that leave the wave equation invariant.

Minkowski recognised that the postulate of the invariance of the speed of light can be formulated independently from electrodynamics as a pure space-time problem.[1] Starting from an inertial reference system, there the three-dimensional space is connected with the time to a four-dimensional space. A point P in this space has then four coordinates, i.e. an event E takes place at place (x, y, z) in time t. In order to have also the additional fourth coordinate with the dimension of a length, one takes not t but ct.

In this four-dimensional space for an event $E(ct, x, y, z,)$, there is defined a quantity s^2, the separation from the coordinate origin $O(0, 0, 0, 0)$ by

$$s^2 := c^2t^2 - x^2 - y^2 - z^2 \ , \tag{309}$$

or for two arbitrary events $E(ct, x, y, z,)$ and $F(c(t+\Delta t), x+\Delta x, y+\Delta y, z+\Delta z)$

$$\Delta s^2 := c^2\Delta t^2 - \Delta x^2 - \Delta y^2 - \Delta z^2 \ . \tag{310}$$

For $\Delta x = \Delta y = \Delta z = 0$, Δs^2 is equal to the square of the light travel time Δt, and for $\Delta t = 0$, Δs^2 is equal to the (-1) multiplied square of the three-dimensional Euclidean separation.

[1] We have presented this procedure, based on the results of precision experiments, with an elementary procedure in Chap. 4, Sects. 1–4.

For mathematical convenience, it is reasonable to enumerate the coordinates,

$$ct,\ x,\ y,\ z \equiv x^0,\ x^1,\ x^2,\ x^3\ . \tag{311}$$

With the in Eq. (354) defined metric tensor η with elements η_{ik}, cp. also Eq. (1623),

$$\eta_{ik} = \eta^{ik} \quad \text{with} \quad \eta_{00} = 1,\ \eta_{11} = \eta_{22} = \eta_{33} = -1 \ \text{and elsewhere zero}\ , \tag{312}$$

we introduce for Δs^2 (or for ds^2 meaning infinitesimally neighbouring points) a covariant notation and call this quantity *line element*,

$$\left.\begin{array}{l} \Delta s^2 = \eta_{ik}\Delta x^i \Delta x^k\ , \\ ds^2 = \eta_{ik} dx^i dx^k\ . \end{array}\right\} \qquad \begin{array}{l}\text{Line element}\\ \text{Covariant notation}\end{array} \tag{313}$$

Here we use EINSTEIN's summation convention, according to which we sum over equal lower and upper indices. The indices run over the values 0, 1, 2, 3, s. also Appendix, Sect. B.1.

From the definition (310), it follows:

The line element can have both negative as well as positive values, or it can be zero, even if the two events do not coincide.

Physically, one calls two events as related by a time-like, space-like or a light-like separation:

$$\begin{array}{lll} \text{Two events} & \textit{spacelike} & \qquad\qquad\qquad\quad < 0 \\ E(x^i) \text{ and } F(x^i + \Delta x^i) & \textit{timelike}\ , & \text{if} \quad \Delta s^2 = \eta_{ik}\Delta x^i \Delta x^k > 0\ . \\ \text{are related} & \textit{lightlike} & \qquad\qquad\qquad\quad = 0 \end{array} \tag{314}$$

With this definition, we introduce the MINKOWSKI space that has an indefinite metric according to the relations (314). One denotes this as signature $(+, -, -, -)$ of the MINKOWSKI space.[2]

[2]The MINKOWSKI space is a physically very important abstract mathematical construction, but it is beyond our practical perception. This concerns both its four-dimensionality and its indefinite metric. This should be kept in mind in all illustrations of examples in this space. For this reason, there also exist different but completely equivalent definitions for the MINKOWSKI space, with different pros and cons, but they can easily be transformed from one to another. For the signature $(+, -, -, -)$ used here, the invariant line element has a direct meaning: $ds = c\,d\tau$, where $d\tau$ denotes the eigentime of a clock moving from E to F, cp. also Sect. 2.1.1. An other signature used in literature is $(-, +, +, +)$ and furthermore, if the time is used as fourth coordinate and written at the last place $(+, +, +, -)$ and $(-, -, -, +)$. The last signature is different from our convention only in the kind of numbering. In the covariant representation of electrodynamics, it is important to take care of the counting convention. We shall see further below that also the use of an imaginary fourth coordinate as in (331) can be helpful. However, for the transition to the General Relativity Theory, this construction is unsuitable and will not be used in this book.

In the following, we use the so-called LAPLACE operator Δ, cp. also Eq. (1665),

$$\Delta := \frac{\partial^2}{\partial x^2} + \frac{\partial^2}{\partial y^2} + \frac{\partial^2}{\partial z^2} \tag{315}$$

as well as D'ALEMBERT's wave operator $\Box$, cp. also Eq. (1666),

$$\Box := \frac{1}{c^2}\frac{\partial^2}{\partial t^2} - \frac{\partial^2}{\partial x^2} - \frac{\partial^2}{\partial y^2} - \frac{\partial^2}{\partial z^2} . \tag{316}$$

The wave equation (306) then reads in covariant notation

$$\eta^{ik}\frac{\partial}{\partial x^i}\frac{\partial}{\partial x^k} f(x^r) = \Box f = 0 . \qquad \text{D'ALEMBERT's wave equation, Covariant notation} \tag{317}$$

The transition from one inertial system Σ_o with coordinates (x^0, x^1, x^2, x^3) to an other one Σ' with coordinates $(x^{0'}, x^{1'}, x^{2'}, x^{3'})$ is a coordinate transformation $x^{i'} = x^{i'}(x^i)$ in MINKOWSKI space.

MINKOWSKI has recognised that the transformations (10), leaving invariant the wave equation, are identical with the transformations that leave the mathematical form of the line element invariant.[3]

The axiomatic principle for the foundation of the Special Relativity Theory can be formulated after MINKOWSKI:

One is looking for transformations of reference systems $x^{i'} = x^{i'}(x^i)$, that leave the line element form-invariant:

$$\eta_{ik}dx^i dx^k = \eta_{i'k'}dx^{i'}dx^{k'} . \qquad \text{MINKOWSKI's postulate} \tag{318}$$

The tensor η_{ik} must be numerically invariant for these transformations, i.e. we are looking for such transformations (10), for which with Eq. (312) also holds

$$\eta_{i'k'} = \eta^{i'k'} \quad \text{with} \quad \eta_{0'0'} = 1, \;\; \eta_{1'1'} = \eta_{2'2'} = \eta_{3'3'} = -1 \text{ and else zero} . \tag{319}$$

When we have determined these transformations $x^{i'} = x^{i'}(x^i)$ fulfilling Eq. (318), one can easily prove that with these transformations, from the validity of the wave equation (317) in Σ_o, there follows also its validity in Σ',

$$\eta^{i'k'}\frac{\partial}{\partial x^{i'}}\frac{\partial}{\partial x^{k'}} f(x^r(x^{r'})) = 0 . \tag{320}$$

We restrict ourself on transformations fulfilling the initial conditions (11). All inertial systems have therefore the same coordinate origin $O(0, 0, 0, 0)$.

At first, we show that the transformations fulfilling Eq. (318) are linear.

[3] EINSTEIN (1905a) has already mentioned this statement in a footnote of his most famous paper.

We insert on the right-hand side of (318) the looking for transformation

$$dx^{i'} = \frac{\partial x^{i'}}{\partial x^i} dx^i , \quad dx^{k'} = \frac{\partial x^{k'}}{\partial x^k} dx^k ,$$

and find

$$\eta_{ik} dx^i dx^k = \eta_{i'k'} \frac{\partial x^{i'}}{\partial x^i} \frac{\partial x^{k'}}{\partial x^k} dx^i \, dx^k .$$

Since dx^i and dx^k are arbitrary, it follows

$$\eta_{ik} = \eta_{i'k'} \frac{\partial x^{i'}}{\partial x^i} \frac{\partial x^{k'}}{\partial x^k} . \tag{321}$$

We differentiate Eq. (321) for x^j,

$$0 = \eta_{i'k'} \left(\frac{\partial x^{i'}}{\partial x^i} \frac{\partial^2 x^{k'}}{\partial x^j \partial x^k} + \frac{\partial^2 x^{i'}}{\partial x^j \partial x^i} \frac{\partial x^{k'}}{\partial x^k} \right) . \tag{322}$$

We repeat this equation still two times with cyclic interchange of indices i, j, k,

$$0 = \eta_{i'k'} \left(\frac{\partial x^{i'}}{\partial x^j} \frac{\partial^2 x^{k'}}{\partial x^k \partial x^i} + \frac{\partial^2 x^{i'}}{\partial x^k \partial x^j} \frac{\partial x^{k'}}{\partial x^i} \right) , \tag{323}$$

$$0 = \eta_{i'k'} \left(\frac{\partial x^{i'}}{\partial x^k} \frac{\partial^2 x^{k'}}{\partial x^i \partial x^j} + \frac{\partial^2 x^{i'}}{\partial x^i \partial x^k} \frac{\partial x^{k'}}{\partial x^j} \right) . \tag{324}$$

We add these last two equations and subtract (322) finding

$$0 = \eta_{i'k'} \left(\frac{\partial x^{i'}}{\partial x^j} \frac{\partial^2 x^{k'}}{\partial x^k \partial x^i} + \frac{\partial^2 x^{i'}}{\partial x^i \partial x^k} \frac{\partial x^{k'}}{\partial x^j} \right) . \tag{325}$$

Since $\eta_{i'k'}$ vanishes for different indices, the two terms in the parenthesis of (322) coincide, so that

$$\eta_{i'k'} \frac{\partial x^{i'}}{\partial x^j} \frac{\partial^2 x^{k'}}{\partial x^i \partial x^k} = 0 . \tag{326}$$

We multiply Eq. (326) with $\partial x^j / \partial x^{l'}$ and sum over j. So we get

$$\frac{\partial x^{i'}}{\partial x^j} \frac{\partial x^j}{\partial x^{l'}} = \frac{\partial x^{i'}}{\partial x^{l'}} .$$

Since the coordinates of our space are independent from each another, the partial derivative $\partial x^{i'}/\partial x^{l'}$ vanishes for $i' \neq l'$, and it is equal to 1 for $i' = l'$. With the KRONECKER symbol δ^i_k, that has the value 1 for equal indices, and else is zero, s. Appendix B.1, we find from Eq. (326),

$$\eta_{i'k'} \frac{\partial x^{i'}}{\partial x^j} \frac{\partial x^j}{\partial x^{l'}} \frac{\partial^2 x^{k'}}{\partial x^i \partial x^k} = \eta_{i'k'} \, \delta^{i'}_{l'} \frac{\partial^2 x^{k'}}{\partial x^i \partial x^k} = \eta_{l'k'} \frac{\partial^2 x^{k'}}{\partial x^i \partial x^k} = 0 \longrightarrow \frac{\partial^2 x^{l'}}{\partial x^i \partial x^k} = 0 \, , \tag{327}$$

since also $\eta_{i'k'}$ is different from zero only for equal indices.

The second derivatives are zero. The looking for transformations are linear.

Now we derive at first special transformations, which leave the y- and z-coordinates invariant, hence $y' = y$, $z' = z$. Then MINKOWSKI's postulate (318) is written as $c^2t^2 - x^2 = c^2t'^2 - x'^2$, where we have again replaced x^0 and x^1 by ct and x. This equation is still multiplied with -1,

$$x^2 - c^2t^2 = x'^2 - c^2t'^2 \; . \tag{328}$$

For determining the linear transformations $x' = a \cdot x + b \cdot t$, $t' = c \cdot x + d \cdot t$ fulfilling Eq. (328), we remember on the normal geometry on a plane.

In Cartesian coordinates, the square of the separation r of a point $P(x, y)$ from the coordinate origin vom $O(0, 0)$ is determined by PYTHAGORAS rule, $r^2 = x^2 + y^2$. For a rotation of the coordinate axes in the x-y-plane by an angle α this formula for the separation of the point $P(x', y') = P(x, y)$ from the origin is preserved,

$$x^2 + y^2 = x'^2 + y'^2 \, , \tag{329}$$

and the new Cartesian coordinates (x', y') of the point P follow from the old Cartesian coordinates (x, y) by

$$\left.\begin{aligned} x' &= x \cos\alpha + y \sin\alpha \, , \\ y' &= -x \sin\alpha + y \cos\alpha \, . \end{aligned}\right\} \tag{330}$$

Equation (329) is different from MINKOWSKI's postulate (328) only by the sign of the second summand. We compensate this difference formally by introducing an imaginary coordinate τ,

$$\tau := ict \; , \quad i^2 = -1 \; . \tag{331}$$

Then Eq. (328) has the same mathematical form as Eq. (329),

$$x^2 + \tau^2 = x'^2 + \tau'^2 \; . \tag{332}$$

Therefore, also the transformation formulas fulfilling Eq. (332) must have the same mathematical form as Eq. (330),

$$\left.\begin{aligned} x' &= x \cos\alpha + \tau \sin\alpha \, , \\ \tau' &= -x \sin\alpha + \tau \cos\alpha \, . \end{aligned}\right\} \tag{333}$$

Equation (333) follows immediately the looking for solution of Eq. (332). Since x and x' are real, but τ and τ' are imaginary, there must be $\cos\alpha$ real and $\sin\alpha$ imaginary, simply fulfilled by an imaginary angle $\alpha = i\varphi$. Using the relations

$$\cos\alpha = \cos(i\varphi) = \cosh\varphi \quad \text{and} \quad \sin\alpha = \sin(i\varphi) = i\,\sinh\varphi$$

between trigonometric and hyperbolic functions, there follows from Eq. (333) by replacing τ again with ict and cancel the imaginary unit i,

$$\left.\begin{aligned} x' &= x\cosh\varphi - ct\sinh\varphi\;, \\ ct' &= -x\sinh\varphi + ct\cosh\varphi\;. \end{aligned}\right\} \tag{334}$$

Using the relation $\cosh^2\varphi - \sinh^2\varphi = 1$ one easily verify that Eq. (334) is indeed the mathematical solution of the transformation problem (328). With some thoughts this solution could be simply guessed.

For understanding the physical content of Eq. (334), we replace the still unspecified parameter φ by

$$\tanh\varphi := \frac{v}{c}\;, \tag{335}$$

where here the parameter v is unspecified. Since

$$\cosh\varphi = \frac{1}{\sqrt{1-\tanh^2\varphi}} \quad \text{and} \quad \sinh\varphi = \frac{\tanh\varphi}{\sqrt{1-\tanh^2\varphi}}, \tag{336}$$

we find

$$\cosh\varphi = \frac{1}{\sqrt{1-v^2/c^2}} \quad \text{and} \quad \sinh\varphi = \frac{v/c}{\sqrt{1-v^2/c^2}}\;, \tag{337}$$

and from Eq. (334) we get

$$x' = \frac{x - vt}{\sqrt{1-v^2/c^2}} \quad \text{and} \quad t' = \frac{t - xv/c^2}{\sqrt{1-v^2/c^2}}\;. \tag{338}$$

The interpretation of the parameter v now follows from the requirement that these transformation formulas should turn into the GALILEI transformation (69) in the limit $c \longrightarrow \infty$. Then we see immediately that v is the velocity of the inertial system Σ' with respect to Σ_o.

Our problem is solved. Equations (338) are the special LORENTZ transformations (105) that let invariant the line element (328) and the wave equation (307). The parameter v is the velocity of system Σ' in x-direction of Σ_o, and the x'-, y'-, z'-axes of Σ' remain parallel to the x-, y-, z-axes of Σ_o.

From the geometry of the plane or from the three space we know, that exactly the rotations of Cartesian coordinates, i.e. the orthogonal transformations fulfil Eq. (329) or $x^2 + y^2 + z^2 = x'^2 + y'^2 + z'^2$. Therefore, one denotes the transformations leaving the line element (309) invariant, i.e. fulfilling Eq. (318) as *pseudo-orthogonal transformations* or *pseudo-rotations*. The special LORENTZ transformation (338) is then a special pseudo-rotation, namely, in the x^0-x^1-plane of MINKOWSKI space. We write for Eq. (338) the tensor equation of transformations in MINKOWSKI space, also adding the easily verified inverse transformation,

$$\left.\begin{aligned} \begin{pmatrix} x^{0'} \\ x^{1'} \\ x^{2'} \\ x^{3'} \end{pmatrix} &= \begin{pmatrix} \gamma & -\beta\gamma & 0 & 0 \\ -\beta\gamma & \gamma & 0 & 0 \\ 0 & 0 & 1 & 0 \\ 0 & 0 & 0 & 1 \end{pmatrix} \begin{pmatrix} x^0 \\ x^1 \\ x^2 \\ x^3 \end{pmatrix}, \qquad \gamma = \frac{1}{\sqrt{1-\beta^2}}, \\ \begin{pmatrix} x^0 \\ x^1 \\ x^2 \\ x^3 \end{pmatrix} &= \begin{pmatrix} \gamma & \beta\gamma & 0 & 0 \\ \beta\gamma & \gamma & 0 & 0 \\ 0 & 0 & 1 & 0 \\ 0 & 0 & 0 & 1 \end{pmatrix} \begin{pmatrix} x^{0'} \\ x^{1'} \\ x^{2'} \\ x^{3'} \end{pmatrix}, \qquad \beta := \frac{v}{c}. \end{aligned}\right\} \text{Special LORENTZ-transformation} \tag{339}$$

With EINSTEIN's summation convention this reads

$$x^{i'} = L^{i'}_{i}\, x^i \quad \longleftrightarrow \quad x^i = L^{i}_{i'}\, x^{i'} \tag{340}$$

with the matrices $L^{i'}_{i}$ and $L^{i}_{i'}$ in Eq. (339).

1.2 *General* LORENTZ *Transformation*

When we stay in an inertial system and only the orthogonal x-, y-, z-axes rotate, then it is $dx^{0'} = dx^0$, and the condition (318) of pseudo-orthogonality in the four-dimensional MINKOWSKI space becomes identically with the usual orthogonality in our three-dimensional space. With a general rotation in the three-dimensional Euclidean space, we get a further special case **D** of pseudo-rotations of the MINKOWSKI space (in the following, the indices $a, b, \ldots$ run over 1, 2, 3),

$$\left.\begin{aligned} &\begin{pmatrix} x^{0'} \\ x^{1'} \\ x^{2'} \\ x^{3'} \end{pmatrix} = \begin{pmatrix} 1 & 0 & 0 & 0 \\ 0 & D^{1'}_1 & D^{1'}_2 & D^{1'}_3 \\ 0 & D^{2'}_1 & D^{2'}_2 & D^{2'}_3 \\ 0 & D^{3'}_1 & D^{3'}_2 & D^{3'}_3 \end{pmatrix} \begin{pmatrix} x^0 \\ x^1 \\ x^2 \\ x^3 \end{pmatrix} \\ &\text{with} \sum_{a'=1}^{3} D^{a'}_{a} D^{a'}_{b} = \delta_{ab}\,. \end{aligned}\right\} \text{General rotation} \tag{341}$$

For the matrix multiplication, we also write shorter

$$x^{a'} = D^{a'}_{a}\, x^a\,. \tag{342}$$

The general LORENTZ transformation with fixed coordinate origin $O(0,0,0,0)$ is written as matrix $\mathbf{A}$,

$$\begin{pmatrix} x^{0'} \\ x^{1'} \\ x^{2'} \\ x^{3'} \end{pmatrix} = \begin{pmatrix} A_0^{0'} & A_1^{0'} & A_2^{0'} & A_3^{0'} \\ A_0^{1'} & A_1^{1'} & A_2^{1'} & A_3^{1'} \\ A_0^{1'} & A_1^{2'} & A_2^{2'} & A_3^{2'} \\ A_0^{1'} & A_1^{3'} & A_2^{3'} & A_3^{3'} \end{pmatrix} \begin{pmatrix} x^0 \\ x^1 \\ x^2 \\ x^3 \end{pmatrix} \tag{343}$$

or shorter

$$x^{i'} = A_i^{i'} \, x^i \; . \tag{344}$$

When we insert the matrix $\mathbf{A}$ in the condition of the invariance of the line element $\eta_{ik} x^i x^k = \eta_{ik} x^{i'} x^{k'}$, hence $\eta_{ik} x^i x^k = \eta_{ik} A_i^{i'} \, x^i A_i^{k'} \, x^k$, we get a condition for $\mathbf{A}$, the condition of pseudo-orthogonality,

$$A_i^{i'} A_k^{j'} \eta_{i'j'} = \eta_{ik} \; . \qquad \text{Pseudo-orthogonality} \tag{345}$$

According to Eq. (345), one gets the inverse matrix of $\mathbf{A}$, when one exchanges rows and columns, and in addition if one multiplies all matrix elements with exactly one index 0 by (-1). The 16 components of a pseudo-orthogonal matrix $\mathbf{A}$ must then fulfil 10 symmetry conditions Eq. (345). Consequently, one has for the general LORENTZ transformation $\mathbf{A}$ with fixed coordinate origin six independent parameters. Thereby, three parameters characterise the general rotation matrix of three-dimensional space, e.g. by EULER's angles, and three additional independent components of velocity $\mathbf{v} = (v_1, v_2, v_3)$, for the relative motion of an inertial system with respect to another one.

The spatial rotation (330) in the x-y-plane, i.e. a rotation around the z-axis, is denoted by $\mathbf{D}_3$, the other rotations around the x- and y-axes by $\mathbf{D}_1$ and $\mathbf{D}_2$, resp. The special LORENTZ transformations, where an inertial system moves with velocities v_1, v_2 or v_3 in x-, y- or z-direction are denoted by $\mathbf{L}_1$, $\mathbf{L}_2$ and $\mathbf{L}_3$, respectively. The corresponding matrices are

$$\left.\begin{aligned}
\mathbf{D}_1 &= \begin{pmatrix} 1 & 0 & 0 & 0 \\ 0 & 1 & 0 & 0 \\ 0 & 0 & \cos\alpha_1 & \sin\alpha_1 \\ 0 & 0 & -\sin\alpha_1 & \cos\alpha_1 \end{pmatrix}, \quad
\mathbf{L}_1 = \begin{pmatrix} \gamma_1 & -\beta_1\gamma_1 & 0 & 0 \\ -\beta_1\gamma_1 & \gamma_1 & 0 & 0 \\ 0 & 0 & 1 & 0 \\ 0 & 0 & 0 & 1 \end{pmatrix}, \\
\mathbf{D}_2 &= \begin{pmatrix} 1 & 0 & 0 & 0 \\ 0 & \cos\alpha_2 & 0 & -\sin\alpha_2 \\ 0 & 0 & 1 & 0 \\ 0 & \sin\alpha_2 & 0 & \cos\alpha_2 \end{pmatrix}, \quad
\mathbf{L}_2 = \begin{pmatrix} \gamma_2 & 0 & -\beta_2\gamma_2 & 0 \\ 0 & 1 & 0 & 0 \\ -\beta_2\gamma_2 & 0 & \gamma_2 & 0 \\ 0 & 0 & 0 & 1 \end{pmatrix}, \\
\mathbf{D}_3 &= \begin{pmatrix} 1 & 0 & 0 & 0 \\ 0 & \cos\alpha_3 & \sin\alpha_3 & 0 \\ 0 & -\sin\alpha_3 & \cos\alpha_3 & 0 \\ 0 & 0 & 0 & 1 \end{pmatrix}, \quad
\mathbf{L}_3 = \begin{pmatrix} \gamma_3 & 0 & 0 & -\beta_3\gamma_3 \\ 0 & 1 & 0 & 0 \\ 0 & 0 & 1 & 0 \\ -\beta_3\gamma_3 & 0 & 0 & \gamma_3 \end{pmatrix}.
\end{aligned}\right\} \tag{346}$$

We have introduced the following abbreviations, when an arbitrary-oriented velocity $\mathbf{v} = (v_1, v_2, v_3)$ occurs:

$$\begin{aligned} \gamma_1 &= \frac{1}{\sqrt{1-\beta_1^2}} , \quad \beta_1 = \frac{v_1}{c} , \\ \gamma_2 &= \frac{1}{\sqrt{1-\beta_2^2}} , \quad \beta_2 = \frac{v_2}{c} , \\ \gamma_3 &= \frac{1}{\sqrt{1-\beta_3^2}} , \quad \beta_3 = \frac{v_3}{c} , \end{aligned} \qquad \begin{aligned} \gamma &= \frac{1}{\sqrt{1-\beta^2}} , \\ \beta &= \sqrt{\beta_1^2 + \beta_2^2 + \beta_3^2} . \end{aligned} \tag{347}$$

The quantities $(l_1, l_2, l_3, l_4, l_5, l_6) := (\alpha_1, \alpha_2, \alpha_3, \beta_1, \beta_2, \beta_3)$ are the six independent parameters of the general LORENTZ transformation $\mathbf{A}$. Each matrix $\mathbf{A}$ can be decomposed in a product of these matrices. There it must be observed that the multiplication by these matrices can in general not be permuted. It makes a difference whether one first performs, e.g. a rotation with the matrix $\mathbf{D}_3$ and then a special LORENTZ transformation $\mathbf{L}_1$ or inversely. We shall come back to these properties.

Only for infinitesimal transformations, when all nonlinear terms in the parameters of the LORENTZ transformations can be neglected, the sequence of the transformations does not play any role.

By means of the unity matrix $\mathbf{1}$ and the six basic matrices $\mathbf{a}_1, \mathbf{a}_2, \mathbf{a}_3, \mathbf{b}_1, \mathbf{b}_2, \mathbf{b}_3$,

$$\left. \begin{aligned} \mathbf{a}_1 &= \begin{pmatrix} 0&0&0&0\\ 0&0&0&0\\ 0&0&0&-1\\ 0&0&1&0 \end{pmatrix} , \quad \mathbf{a}_2 = \begin{pmatrix} 0&0&0&0\\ 0&0&0&1\\ 0&0&0&0\\ 0&-1&0&0 \end{pmatrix} , \quad \mathbf{a}_3 = \begin{pmatrix} 0&0&0&0\\ 0&0&-1&0\\ 0&1&0&0\\ 0&0&0&0 \end{pmatrix} , \\ \mathbf{b}_1 &= \begin{pmatrix} 0&1&0&0\\ 1&0&0&0\\ 0&0&0&0\\ 0&0&0&0 \end{pmatrix} , \quad \mathbf{b}_2 = \begin{pmatrix} 0&0&1&0\\ 0&0&0&0\\ 1&0&0&0\\ 0&0&0&0 \end{pmatrix} , \quad \mathbf{b}_3 = \begin{pmatrix} 0&0&0&1\\ 0&0&0&0\\ 0&0&0&0\\ 1&0&0&0 \end{pmatrix} , \end{aligned} \right\} \tag{348}$$

one defines with the six parameters $(\alpha_1, \alpha_2, \alpha_3, \beta_1, \beta_2, \beta_3)$ the six matrices $\mathbf{d}_1, \mathbf{d}_2, \mathbf{d}_3, \mathbf{l}_1, \mathbf{l}_2, \mathbf{l}_3$,

$$\mathbf{d}_a = \mathbf{1} - \alpha_{(a)}\, \mathbf{a}_{(a)} , \quad \mathbf{l}_a = \mathbf{1} - \beta_{(a)} \mathbf{b}_{(a)} , \quad a = 1, 2, 3 , \tag{349}$$

(here the parentheses indicate that we do not sum over the repeated indices). By summation over $a = 1, 2$ and 3 we form a matrix $\mathbf{\Lambda}$,

$$\mathbf{\Lambda} = \mathbf{1} - \alpha_a\, \mathbf{a}_a - \beta_a\, \mathbf{b}_a . \tag{350}$$

one shows that the general, six-parametric finite LORENTZ transformation $\mathbf{A}$ is represented by means of the matrix $\mathbf{\Lambda}$ as

$$\mathbf{A} = \exp[\, \Lambda - \mathbf{1}] . \qquad \text{General LORENTZ transformation} \tag{351}$$

The entirety of pseudo-orthogonal transformations (351) form the group $\mathcal{A}$ of general LORENTZ transformations.

Here the exponential function of a matrix $\mathbf{B}$ is understood by means of a TAYLOR series,

$$\exp[\mathbf{B}] \equiv e^{\mathbf{B}} := \mathbf{1} + \mathbf{B} + \frac{1}{2}\mathbf{B}^2 + \frac{1}{3!}\mathbf{B}^3 + \frac{1}{4!}\mathbf{B}^4 + \cdots + \frac{1}{n!}\mathbf{B}^n + \cdots \ , \tag{352}$$

and because of $e^{\mathbf{B}}\, e^{-\mathbf{B}} = e^{\mathbf{B}-\mathbf{B}} = e^0 = \mathbf{1}$ then it holds

$$\left(e^{\mathbf{B}}\right)^{-1} = e^{-\mathbf{B}} \ . \tag{353}$$

For a compact notation, we also use

$$\mathbf{x} = \begin{pmatrix} x^0 \\ x^1 \\ x^2 \\ x^3 \end{pmatrix}, \quad \mathbf{x}^T = (x^0,\, x^1,\, x^2,\, x^3)\,, \quad \eta = \eta^T = \eta^{-1} = \begin{pmatrix} 1 & 0 & 0 & 0 \\ 0 & -1 & 0 & 0 \\ 0 & 0 & -1 & 0 \\ 0 & 0 & 0 & -1 \end{pmatrix}. \tag{354}$$

$\mathbf{B}^T$ is the to $\mathbf{B}$ transposed matrix. Then the line element (313) simply reads

$$ds^2 = d\mathbf{x}^T\, \eta\, d\mathbf{x} \ , \tag{355}$$

and the defining relation for the pseudo-orthogonality (345) of the general LORENTZ transformations $\mathbf{A}$ is written as

$$\mathbf{A}\, \eta\, \mathbf{A}^T = \eta \ . \qquad \text{Pseudo-orthogonality} \tag{356}$$

Then there follows an equivalent property to Eq. (356) by multiplication with η from the right side, and then with $\mathbf{A}^{-1}$ from the left side

$$\eta\, \mathbf{A}^T\, \eta = \mathbf{A}^{-1} \ . \qquad \text{Pseudo-orthogonality} \tag{357}$$

We write for abbreviation

$$\mathbf{C} := \mathbf{\Lambda} - \mathbf{1} = -\alpha_a\, \mathbf{a}_a - \beta_a\, \mathbf{b}_a \ . \tag{358}$$

By tedious matrix multiplication, we find

$$\eta\, \mathbf{C}^T\, \eta := -\mathbf{C} \ . \tag{359}$$

Because of $\eta^2 = \mathbf{1}$ it follows $\eta\, \mathbf{B}^2\, \eta = \eta\, \mathbf{B}\, \mathbf{B}\, \eta = \mathbf{B}\, \eta^2\, \mathbf{B} = \eta\, \mathbf{B}\, \eta\, \eta\, \mathbf{B}\eta$. Similar transcriptions hold true for the higher powers.

Furthermore, we have $\left(\mathbf{C}^2\right)^T = (\mathbf{C}\,\mathbf{C})^T = \mathbf{C}^T\,\mathbf{C}^T = \left(\mathbf{C}^T\right)^2$ with the corresponding transcriptions for the higher powers. Therefore, we have always

$$\left.\begin{array}{l}\eta\, e^{\mathbf{B}}\, \eta = e^{\eta\, \mathbf{B}\, \eta}\ , \\ \left(e^{\mathbf{B}}\right)^T = e^{(\mathbf{B}^T)}\ . \end{array}\right\} \tag{360}$$

For the matrix $e^{\mathbf{C}}$, we can now easily prove condition (357) of pseudo-orthogonality, and therefore also formula (351) for $\mathbf{A}$. From the equations above, we infer directly

$$\eta\, \left(e^{\mathbf{C}}\right)^T \eta = \eta\, e^{(\mathbf{C}^T)}\, \eta = e^{\eta\, \mathbf{C}^T \eta} = e^{-\mathbf{C}} = \left(e^{\mathbf{C}}\right)^{-1}\ , \tag{361}$$

what had to be demonstrated.

When we now form from the six parameters α_a and β_a six parameters a_a and b_a with the natural numbers n

$$a_a := \frac{\alpha_a}{n}\ , \quad b_a := \frac{\beta_a}{n}\ , \quad 1 \ll n\ , \tag{362}$$

so the a_a and b_a will become arbitrary small for sufficient large n,

$$a_a \ll 1\ , \quad b_a \ll 1\ , \tag{363}$$

and from the matrix $\mathbf{\Lambda}$ we get the general, i.e. six-parametric infinitesimal LORENTZ transformation $\mathbf{A}_{inf}$,

$$\mathbf{A}_{inf} = \mathbf{1} - a_a\, \mathbf{a}_a - b_a\, \mathbf{b}_a \tag{364}$$

General infinitesimal LORENTZ transformation

with the general infinitesimal rotation matrix $\mathbf{d} = \mathbf{1} - a_a\, \mathbf{a}_a$ and the general infinitesimal proper LORENTZ transformation $\mathbf{l} = \mathbf{1} - b_a\, \mathbf{b}_a$

$$\mathbf{d} = \begin{pmatrix} 1 & 0 & 0 & 0 \\ 0 & 1 & a_3 & -a_2 \\ 0 & -a_3 & 1 & a_1 \\ 0 & a_2 & -a_1 & 1 \end{pmatrix}\ , \quad \mathbf{l} = \begin{pmatrix} 1 & -b_1 & -b_2 & -b_3 \\ -b_1 & 1 & 0 & 0 \\ -b_2 & 0 & 1 & 0 \\ -b_3 & 0 & 0 & 1 \end{pmatrix}\ . \tag{365}$$

The application of these infinitesimal matrices are now commutable.

Still we consider the limit of an indefinite product of infinitesimal LORENTZ transformations $\mathbf{A}_{inf}$ and write again $a_a = \alpha_a/n\ ,\quad b_a = \beta_a/n$,

$$\lim_{n\to\infty} \left(\mathbf{A}_{inf}\right)^n = \lim_{n\to\infty} \left(\mathbf{1} - \frac{1}{n}\left(\alpha_a\, \mathbf{a}_a - \beta_a\, \mathbf{b}_a\right)\right)^n\ . \tag{366}$$

With the identity $\lim_{n\to\infty}\left(1 - \frac{x}{n}\right)^n = e^{-x}$ then it follows from Eq. (366)

$$\lim_{n\to\infty} \left(\mathbf{A}_{inf}\right)^n = e^{-\alpha_a\, \mathbf{a}_a - \beta_a\, \mathbf{b}_a} = e^{\mathbf{\Lambda}-\mathbf{1}} = \mathbf{A}\ . \tag{367}$$

The non-commutativity of two (finite) LORENTZ transformations follows from the commutation relations of the basic matrices $\mathbf{a}_a$ and $\mathbf{b}_a$.
With the abbreviation for a commutator, $[\mathbf{A}, \mathbf{B}] := \mathbf{AB} - \mathbf{BA}$, one finds for the commutators of the in Eq. (348) defined matrices $\mathbf{a}_a$ and $\mathbf{b}_a$ by simple evaluations

$$[\mathbf{a}_a, \mathbf{a}_b] = \epsilon_{abc}\,\mathbf{a}_c\ , \quad [\mathbf{a}_a, \mathbf{b}_b] = \epsilon_{abc}\,\mathbf{b}_c\ , \quad [\mathbf{b}_a, \mathbf{b}_b] = -\epsilon_{abc}\,\mathbf{a}_c\ . \tag{368}$$

Here we used the antisymmetric tensor with components $\epsilon_{123} = \epsilon_{231} = \epsilon_{312} = 1$, $\epsilon_{132} = \epsilon_{321} = \epsilon_{213} = -1$ and else zero, cp. Appendix B.1.
The squares of these matrices are

$$\left.\begin{aligned}
\mathbf{a}_1^2 &= \begin{pmatrix} 0&0&0&0\\ 0&0&0&0\\ 0&0&-1&0\\ 0&0&0&-1 \end{pmatrix}, \quad \mathbf{a}_2^2 = \begin{pmatrix} 0&0&0&0\\ 0&-1&0&0\\ 0&0&0&0\\ 0&0&0&-1 \end{pmatrix}, \quad \mathbf{a}_3^2 = \begin{pmatrix} 0&0&0&0\\ 0&-1&0&0\\ 0&0&-1&0\\ 0&0&0&0 \end{pmatrix},\\
\mathbf{b}_1^2 &= \begin{pmatrix} 1&0&0&0\\ 0&1&0&0\\ 0&0&0&0\\ 0&0&0&0 \end{pmatrix}, \quad \mathbf{b}_2^2 = \begin{pmatrix} 1&0&0&0\\ 0&0&0&0\\ 0&0&1&0\\ 0&0&0&0 \end{pmatrix}, \quad \mathbf{b}_3^2 = \begin{pmatrix} 1&0&0&0\\ 0&0&0&0\\ 0&0&1&0\\ 0&0&0&0 \end{pmatrix}.
\end{aligned}\right\} \tag{369}$$

For $\beta_a = 0$, we get the general rotation matrices $\mathbf{D}$ of the three-dimensional space,

$$\mathbf{D} = \exp[-\alpha_a\,\mathbf{a}_a]\ . \qquad \text{General rotation} \tag{370}$$

The rotation matrices (370) of the three-dimensional position space form a non-commutative subgroup $\mathcal{D}$ of the general LORENTZ-group $\mathcal{A}$.

For $\alpha_a = 0$, we get from $\mathbf{A}$ the general proper, i.e. rotation-free LORENTZ transformation $\mathbf{L}$,

$$\mathbf{L} = \exp[-\beta_a\,\mathbf{b}_a]\ . \qquad \text{General proper LORENTZ transformation} \tag{371}$$

Restricting on special LORENTZ transformations along only one coordinate axis, one finds that they form a simple commutative group $\mathcal{L}_a$, s. Chap. 4, Sect. 4. In Problem 6, we show: If v and u are the velocities of $\Sigma'(x', t')$ and $\Sigma''(x'', t'')$ in x-direction of $\Sigma_o(x, t)$, we perform the special LORENTZ transformations

$$\mathbf{x}'' = \mathbf{L}_1(u)\,\mathbf{x}\ , \quad \mathbf{x}' = \mathbf{L}_1(v)\,\mathbf{x} \quad \longrightarrow \quad \mathbf{x} = \mathbf{L}_1(-v)\,\mathbf{x}'\ , \tag{372}$$

then the systems $\Sigma''(x'', t'')$ and $\Sigma'(x', t')$ are connected by a special LORENTZ transformation $\mathbf{L}_1(u')$,

$$\mathbf{x}'' = \mathbf{L}_1(u')\,\mathbf{x}' = \mathbf{L}_1(u)\,\mathbf{L}_1(-v)\,\mathbf{x}\ , \tag{373}$$

where the group parameter u' is the velocity of system Σ' as measured from Σ'', hence just the velocity calculated from EINSTEIN's addition theorem from v and u,

$$\left.\begin{aligned} &\mathbf{L}_1(u') = \mathbf{L}_1(u)\,\mathbf{L}_1(-v) = \mathbf{L}_1(-v)\,\mathbf{L}_1(u) \quad \text{with} \quad u' = \frac{u-v}{1-u\,v/c^2} \\ &\text{and} \\ &\mathbf{L}_1(v)\,\mathbf{L}_1(-v) = \mathbf{1}\,, \quad \text{hence} \quad \mathbf{L}_1(-v) = \Big(\mathbf{L}_1(v)\Big)^{-1}. \end{aligned}\right\} \tag{374}$$

This proves the claimed group property. The group $\mathbf{L}_1(v)$ is called simple, since it has only the unity element as trivial subgroup. We call commutative groups as ABEL groups, hence we can formulate for $a = 1, 2, 3$:

The special LORENTZ transformations $\mathbf{L}_a(v)$ form simple ABEL groups $\mathcal{L}_a$.

The follow-up of a sequence of special LORENTZ transformations $\mathbf{L}_1$ was the reason of the twin paradox, s. Chap. 8, Sect. 8.

The group properties of the proper LORENTZ transformations is lost, when we allow arbitrary relative velocities.

The set L of proper LORENTZ transformations (371) for arbitrary directed relative velocities does not form a group.

We consider, e.g. a transformation $\mathbf{L}_1$ from the inertial system Σ_o to the inertial system Σ' moving along the x-axis of Σ_o, and in the following a transformation $\mathbf{L}_2$ to the inertial system Σ'' moving along the y'-axis of Σ', so the resulting transformation $\mathbf{L}_2 \cdot \mathbf{L}_1$ will be described again by a unique pseudo-orthogonal transformation $\mathbf{A}$ according to Eqs. (351) or (343) with the conditions (345). However, this transformation will not be anymore from type $\mathbf{L}$ according to Eq. (371),

$$\mathbf{L}_2\,\mathbf{L}_1 = \mathbf{A} \neq \mathbf{L}\,. \tag{375}$$

By elementary considerations, we have discussed this situation already in Chap. 8, Sects. 3 and 4. There we have found that the axes of systems Σ'' after the transformations $\mathbf{L}_1$ and $\mathbf{L}_2$ are rotated with respect to the axes of Σ_o, Chap. 8 Figs. 1 and 3. We will return to this point in Sect. 1.4.

Now we shall determine the general proper LORENTZ transformation $\mathbf{L}$, connecting two inertial systems Σ_o and Σ''.

1.3 *General Proper* LORENTZ *Transformation*

In Problem 22, we show that the coordinate axes of a system Σ'', emerging from a system Σ_o by a transformation of type $\mathbf{L}$, will, in general, be seen from Σ_o not as orthogonal. The property of orthogonality of the axes as seen from an other reference system is preserved only in the case of an axis-parallel velocity, hence, e.g. for $\mathbf{L}_1$. There exists however always a common rotation matrix $\mathbf{D}$, so that after its application both on Σ_o and on Σ'' the axes of the, respectively, other reference systems are seen as orthogonal. One of the axes of Σ_o and Σ'' is then oriented in the direction of the relative motion.[4]

Inertial systems, connected by proper LORENTZ transformations $\mathbf{L}$, are often denoted as *axis parallel*, and this parallelity can always be generated. The operator $\mathbf{L}$ is denoted as *boost*.

The general proper LORENTZ transformations $\mathbf{L}$ of Eq. (371) and also the general rotations $\mathbf{D}$ of Eq. (370) are defined by the TAYLOR series (352) of the exponential function. The matrices $\mathbf{D}_1 \ldots \mathbf{L}_3$ defined in Eq. (346) and also the general matrices $\mathbf{L} = \exp[-\beta_a\, \mathbf{b}_a]$ and $\mathbf{D} = \exp[-\alpha_a\, \mathbf{a}_a]$ can be directly derived from Eq. (352). The relations $\mathbf{D}_1 = \exp[-\alpha_1\, \mathbf{a}_1]$ and $\mathbf{L}_1 = \exp[-\beta_1\, \mathbf{b}_1]$ are derived as examples in Problem 21.

Now the matrix for a general boost $\mathbf{L} = \exp[-\beta_a\, \mathbf{b}_a]$ will be evaluated not from the TAYLOR series (352), but by elementary considerations.

We consider a body K moving with constant velocity $\mathbf{v} = (v_1, v_2, 0)$ in the inertial system Σ_o. We call the inertial system, in which the body is at rest Σ''. We want to determine the proper, hence, rotation-free LORENTZ transformation $\mathbf{L}_{1,2}(\beta_1, \beta_2) = \exp[-\beta_1\, \mathbf{b}_1 - \beta_2\, \mathbf{b}_2]$ that converts the coordinates x^i in Σ_o of an arbitrary event E into coordinates $\tilde{x}^{i''}$ of system Σ''.

In both inertial systems Σ_o and Σ'', we consider at first those coordinate systems $\bar{x}^i$ and $\bar{x}^{i''}$, for which Σ'' is moving along the common $\bar{x}$-$\bar{x}''$-axis with velocity $v = \sqrt{v_1^2 + v_2^2}$ as observed from Σ_o. Then the coordinates $\bar{x}^{i''}$ and $\bar{x}^i$ are connected by a special LORENTZ transformation (339), here called $\mathbf{L}_o$. When we rotate in the following the coordinate system, then the amount of the velocity v remains constant. The index '1' for β and γ is now omitted for avoiding later confusions. We write

$$\bar{\mathbf{x}}'' = \mathbf{L}_o\, \bar{\mathbf{x}} \; . \tag{376}$$

Now we rotate both coordinate systems around the z- and z''-axis, respectively, by an angle φ as follows:

$$\left.\begin{aligned} &\tan\varphi = \frac{\beta_2}{\beta_1}\,, \quad \text{hence} \;\; \beta_1 = \beta\cos\varphi\,, \;\; \beta_2 = \beta\sin\varphi\,, \\ &\beta = \sqrt{\beta_1^2 + \beta_2^2} \quad \text{and} \;\; \gamma := \frac{1}{\sqrt{1 - \beta_1^2 - \beta_2^2}} \end{aligned}\right\} \tag{377}$$

[4]The shear of the axes of Σ'' as seen from Σ_o after the application of the proper LORENTZ transformation $\mathbf{L}$ on Σ_o is connected with the formulas for the rotation (399) evaluated elementary in Chap. 8, Sects. 3 and 4, s. Problem 22.

and we write

$$\mathbf{D}_3 = \mathbf{D}_3(\varphi)\ . \tag{378}$$

Then we get the coordinate systems $x^{i''}$ of Σ'' and x^i of Σ_o,

$$\bar{\mathbf{x}}'' = \mathbf{D}_3(\varphi)\,\tilde{\mathbf{x}}''\,,\quad \bar{\mathbf{x}} = \mathbf{D}_3(\varphi)\,\mathbf{x}\ . \tag{379}$$

Equations (376) and (379) then lead to

$$\tilde{\mathbf{x}}'' = \mathbf{D}_3^{-1}\,\mathbf{L}_o\,\mathbf{D}_3\,\mathbf{x} := \mathbf{L}_{1,2}\,\mathbf{x}\ . \tag{380}$$

Now we use the rotation matrix $\mathbf{D}_3$ with the angle φ and the boost $\mathbf{L}_1$ according to Eq. (346) with matrix elements without the index 1, and we find

$$\left.\begin{array}{l}\mathbf{D}_3^{-1}\,\mathbf{L}_o\,\mathbf{D}_3 = \\ \begin{pmatrix}1 & 0 & 0 & 0\\ 0 & \cos\varphi & -\sin\varphi & 0\\ 0 & \sin\varphi & \cos\varphi & 0\\ 0 & 0 & 0 & 1\end{pmatrix}\begin{pmatrix}\gamma & -\beta\gamma & 0 & 0\\ -\beta\gamma & \gamma & 0 & 0\\ 0 & 0 & 1 & 0\\ 0 & 0 & 0 & 1\end{pmatrix}\begin{pmatrix}1 & 0 & 0 & 0\\ 0 & \cos\varphi & \sin\varphi & 0\\ 0 & -\sin\varphi & \cos\varphi & 0\\ 0 & 0 & 0 & 1\end{pmatrix},\end{array}\right\} \tag{381}$$

hence after multiplication $\mathbf{L}_{1,2} = \exp[-\beta_1\,\mathbf{b}_1 - \beta_2\,\mathbf{b}_2]$,

$$\mathbf{L}_{1,2}(\beta_1,\beta_2) = \begin{pmatrix}\gamma & -\beta\gamma\cos\varphi & -\beta\gamma\sin\varphi & 0\\ -\beta\gamma\cos\varphi & \gamma\cos^2\varphi + \sin^2\varphi & \gamma\sin\varphi\cos\varphi - \sin\varphi\cos\varphi & 0\\ -\gamma\beta\sin\varphi & \gamma\sin\varphi\cos\varphi - \sin\varphi\cos\varphi & \gamma\sin^2\varphi + \cos^2\varphi & 0\\ 0 & 0 & 0 & 1\end{pmatrix},$$

and we get after some simplifications

$$\mathbf{L}_{1,2}(\beta_1,\beta_2) = \begin{pmatrix}\gamma & -\beta_1\gamma & -\beta_2\gamma & 0\\ -\beta_1\gamma & 1+(\gamma-1)\beta_1^2/\beta^2 & (\gamma-1)\beta_1\beta_2/\beta^2 & 0\\ -\beta_2\gamma & (\gamma-1)\beta_2\beta_1/\beta^2 & 1+(\gamma-1)\beta_2^2/\beta^2 & 0\\ 0 & 0 & 0 & 1\end{pmatrix} \tag{382}$$

with

$$\mathbf{L}_{1,2}(\beta_1,\beta_2) = \mathbf{D}_3^{-1}(\varphi)\,\mathbf{L}_o(\beta)\,\mathbf{D}_3(\varphi)\ . \tag{383}$$

We see that we need for the decomposition of the boost-transformation $\mathbf{L}_{1,2}(\beta_1,\beta_2)$ both a special Lorentz transformation $\mathbf{L}_o$ and a rotation $\mathbf{D}_3$. The rotation matrix $\mathbf{D}_3$ arises since in the Taylor series (352) for $\mathbf{L}_{1,2}(\beta_1,\beta_2) = \exp[-\beta_1\,\mathbf{b}_1 - \beta_2\,\mathbf{b}_2]$ there appear products of matrices $\mathbf{b}_1$ and $\mathbf{b}_2$, where the order of factors is fixed. The commutator of $\mathbf{b}_1$ and $\mathbf{b}_2$, namely, $[\mathbf{b}_1,\mathbf{b}_2] = -\epsilon_{123}\,\mathbf{a}_3 = -\mathbf{a}_3$ leads according to (368) to the needed rotation $\mathbf{D}_3$.

The generalisation on an arbitrary-oriented velocity $\mathbf{v} = (v_1, v_2, v_3)$ can be easily given from symmetry considerations.

The general, hence, three-parametric proper LORENTZ transformation $\mathbf{L}$ reads

$$\mathbf{L} = \begin{pmatrix} \gamma_t & -\beta_1\gamma_t & -\beta_2\gamma_t & -\beta_3\gamma_t \\ -\beta_1\gamma_t & 1+\frac{(\gamma_t-1)\beta_1^2}{\beta_t^2} & \frac{(\gamma_t-1)\beta_1\beta_2}{\beta_t^2} & \frac{(\gamma_t-1)\beta_1\beta_3}{\beta_t^2} \\ -\beta_2\gamma_t & \frac{(\gamma_t-1)\beta_2\beta_1}{\beta_t^2} & 1+\frac{(\gamma_t-1)\beta_2^2}{\beta_t^2} & \frac{(\gamma_t-1)\beta_2\beta_3}{\beta_t^2} \\ -\beta_3\gamma_t & \frac{(\gamma_t-1)\beta_3\beta_1}{\beta_t^2} & \frac{(\gamma_t-1)\beta_3\beta_2}{\beta_t^2} & 1+\frac{(\gamma_t-1)\beta_3^2}{\beta_t^2} \end{pmatrix} . \tag{384}$$

General proper LORENTZ-transformation *Boost*

Here we denote

$$\beta_t := \sqrt{\beta_1^2 + \beta_2^2 + \beta_3^2} \ , \quad \gamma_t := \frac{1}{\sqrt{1 - \beta_1^2 - \beta_2^2 - \beta_3^2}} \ . \tag{385}$$

Now we consider a system Σ' moving with velocity $\mathbf{v}_1 = (v_1, 0, 0)$ with respect to Σ_o,

$$\begin{pmatrix} x^{0'} \\ x^{1'} \\ x^{2'} \\ x^{3'} \end{pmatrix} = \begin{pmatrix} \gamma_1 & -\beta_1\gamma_1 & 0 & 0 \\ -\beta_1\gamma_1 & \gamma_1 & 0 & 0 \\ 0 & 0 & 1 & 0 \\ 0 & 0 & 0 & 1 \end{pmatrix} \begin{pmatrix} x^0 \\ x^1 \\ x^2 \\ x^3 \end{pmatrix} , \tag{386}$$

or in a more compact form

$$\mathbf{x}' = \mathbf{L}_1(\beta_1)\,\mathbf{x} \ . \tag{387}$$

In Σ_o, the velocity $\mathbf{v} = (v_1, v_2, 0)$ of our body K is measured from Σ' as $\mathbf{v}' = (0, v_2', 0) = (0, \gamma_1 v_2, 0)$, as seen from Eq. (117) with components $(u_x, u_y, u_z) = (v_1, v_2, 0)$, s. Chap. 8, Sect. 3, Eq. (207),

$$\Sigma_o : \mathbf{v} = (v_1, v_2, 0) \quad \longrightarrow \quad \Sigma' : \mathbf{v}' = (0, v_2', 0) = (0, \gamma_1 v_2, 0) \ . \tag{388}$$

In the rest system Σ'' of body K, we consider now the coordinate system $x^{i''}$, obtained from the coordinates $x^{i'}$ of Σ' by the special LORENTZ transformation $\mathbf{L}_2$ in y'-direction of Σ' with the velocity $v_2' = \gamma_1 v_2$. With

$$\sqrt{1 - \frac{v_2'^2}{c^2}} = \sqrt{\frac{c^2 - (v_1^2 + v_2^2)}{c^2 - v_1^2}} = \frac{\gamma_1}{\gamma} , \tag{389}$$

it follows

$$\begin{pmatrix} x^{0''} \\ x^{1''} \\ x^{2''} \\ x^{3''} \end{pmatrix} = \begin{pmatrix} \frac{\gamma}{\gamma_1} & 0 & -\beta_2\gamma & 0 \\ 0 & 1 & 0 & 0 \\ -\beta_2\gamma & 0 & \frac{\gamma}{\gamma_1} & 0 \\ 0 & 0 & 0 & 1 \end{pmatrix} \begin{pmatrix} x^{0'} \\ x^{1'} \\ x^{2'} \\ x^{3'} \end{pmatrix} \tag{390}$$

or in compact form

$$\mathbf{x}'' = \mathbf{L}_2(\gamma_1\,\beta_2)\,\mathbf{x}' \ . \tag{391}$$

Now the connection between $x^{i''}$ and x^i is given by the product of both special LORENTZ transformations as

$$\mathbf{x}'' = \mathbf{L}_2\,\mathbf{L}_1\,\mathbf{x} \ , \tag{392}$$

hence

$$\begin{pmatrix} x^{0''} \\ x^{1''} \\ x^{2''} \\ x^{3''} \end{pmatrix} = \begin{pmatrix} \frac{\gamma}{\gamma_1} & 0 & -\beta_2\,\gamma & 0 \\ 0 & 1 & 0 & 0 \\ -\beta_2\,\gamma & 0 & \frac{\gamma}{\gamma_1} & 0 \\ 0 & 0 & 0 & 1 \end{pmatrix} \begin{pmatrix} \gamma_1 & -\beta_1\,\gamma_1 & 0 & 0 \\ -\beta_1\,\gamma_1 & \gamma_1 & 0 & 0 \\ 0 & 0 & 1 & 0 \\ 0 & 0 & 0 & 1 \end{pmatrix} \begin{pmatrix} x^0 \\ x^1 \\ x^2 \\ x^3 \end{pmatrix} \ . \tag{393}$$

Multiplication of the matrices delivers

$$\mathbf{L}_2\,\mathbf{L}_1 = \begin{pmatrix} \gamma & -\beta_1\,\gamma & -\beta_2\,\gamma & 0 \\ -\beta_1\,\gamma_1 & \gamma_1 & 0 & 0 \\ \beta_2\,\gamma\,\gamma_1 & \beta_1\,\beta_2\,\gamma\,\gamma_1 & \gamma/\gamma_1 & 0 \\ 0 & 0 & 0 & 1 \end{pmatrix} \ . \tag{394}$$

The matrix (394) of the pseudo-orthogonal transformation $\mathbf{L}_2\,\mathbf{L}_1$ is not identical to the matrix (382) of the proper LORENTZ transformation $\mathbf{L}_{1,2}$, a boost. Both matrices differ by a rotation matrix of type $\mathbf{D}_3$, what contains the key for understanding the THOMAS precession as we shall see next. For avoiding confusion with the rotation matrix $\mathbf{D}_3(\varphi)$ of Eq. (377), we decorate this matrix with a star, *. We define the matrix $\mathbf{D}^*$ by

$$\mathbf{L}_2\,\mathbf{L}_1 = \mathbf{D}^*\,\mathbf{L}_{1,2} \quad \text{mit} \quad \mathbf{D}^* \neq \mathbf{1} \ . \tag{395}$$

1.4 General Theory of the THOMAS *Precession*

The matrix $\mathbf{D}^*$ of Eq. (395) is then

$$\mathbf{D}^* = \mathbf{L}_2\,\mathbf{L}_1\,\mathbf{L}_{1,2}^{-1} \tag{396}$$

with

$$\mathbf{x}'' = \mathbf{D}^*\,\tilde{\mathbf{x}}'' \ . \tag{397}$$

In Eq. (396), we use $\mathbf{L}_2\,\mathbf{L}_1$ from Eq. (394). For $\mathbf{L}_{1,2}^{-1}$, we take into account the pseudo-orthogonality of LORENTZ transformations. From Eqs. (345) and (357), we get the inverse matrix $\mathbf{L}_{1,2}^{-1}$ to $\mathbf{L}_{1,2}$, when we exchange rows and columns, and in addition multiply the matrix elements containing exactly one index "0" with (-1). The matrix $\mathbf{D}^*$ is then

$$\mathbf{D}^* = \begin{pmatrix} \gamma & -\beta_1\gamma & -\beta_2\gamma & 0 \\ -\beta_1\gamma_1 & \gamma_1 & 0 & 0 \\ -\beta_2\gamma\gamma_1 & \beta_1\beta_2\gamma\gamma_1 & \gamma/\gamma_1 & 0 \\ 0 & 0 & 0 & 1 \end{pmatrix} \begin{pmatrix} \gamma & \beta_1\gamma & \beta_2\gamma & 0 \\ \beta_1\gamma & 1+\frac{(\gamma-1)\beta_1^2}{\beta^2} & \frac{(\gamma-1)\beta_1\beta_2}{\beta^2} & 0 \\ \beta_2\gamma & \frac{(\gamma-1)\beta_2\beta_1}{\beta^2} & 1+\frac{(\gamma-1)\beta_2^2}{\beta^2} & 0 \\ 0 & 0 & 0 & 1 \end{pmatrix} . \tag{398}$$

Matrix multiplication delivers

$$\mathbf{D}^* = \begin{pmatrix} 1 & 0 & 0 & 0 \\ 0 & \frac{\gamma_1}{\gamma}\frac{\beta_1^2+\gamma\beta_2^2}{\beta^2} & (1-\gamma)\beta_1\beta_2\frac{\gamma_1}{\gamma} & 0 \\ 0 & (\gamma-1)\beta_1\beta_2\frac{\gamma_1}{\gamma} & \frac{\gamma_1}{\gamma}\frac{\beta_1^2+\gamma\beta_2^2}{\beta^2} & 0 \\ 0 & 0 & 0 & 1 \end{pmatrix} . \tag{399}$$

We check that this is indeed a rotation matrix of type (346). We use again

$$\gamma_1 = \frac{1}{\sqrt{1-\beta_1^2}} , \quad \gamma = \frac{1}{\sqrt{1-\beta_1^2-\beta_2^2}} , \quad \beta^2 = \beta_1^2 + \beta_2^2 .$$

Because of the following inequalities:

$$\begin{aligned} 0 &< (1-1/\gamma)^2 , \\ 2/\gamma &< 1+1/\gamma^2 , \\ \beta_1^2 + 2/\gamma &< 1+\beta_1^2+1/\gamma^2 , \\ \beta_1^2\,\beta_2^2(\beta_1^2+2/\gamma) &< \beta_1^2\,\beta_2^2\,(1+\beta_1^2+1/\gamma^2) , \\ \beta_1^4\,\beta_2^2 + 2\,\beta_1^2\,\beta_2^2/\gamma + \beta_2^4 &< \beta_1^2\,\beta_2^2\,(2-\beta_2^2) + \beta_2^4 , \\ -\beta_1^4\,\beta_2^2 + 2\,\beta_1^2\,\beta_2^2/\gamma + \beta_2^4 &< -2\beta_1^4\,\beta_2^2 + \beta_1^2\,\beta_2^2\,(2-\beta_2^2) + \beta_2^4 , \\ (1-\beta_1^2)\,\beta_1^4 - \beta_1^4\,\beta_2^2 + 2\,\beta_1^2\,\beta_2^2/\gamma + \beta_2^4 &< (1-\beta_1^2)\,\beta_1^4 + (1-\beta_1^2)\,\beta_2^4 + (1-\beta_1^2)\,2\,\beta_1^2\,\beta_2^2 , \\ (1-\beta_1^2+\beta_2^2)\,\beta_1^4 + 2\sqrt{1-\beta_1^2+\beta_2^2} + \beta_2^4 &< (1-\beta_1^2)(\beta_1^4+2\,\beta_1^2\,\beta_2^2+\beta_2^4) , \\ (\beta_1^2/\gamma+\beta_2^2)^2 &< (\beta_1^2+\beta_2^2)^2/\gamma_1^2 \end{aligned}$$

it holds

$$\beta_1^2/\gamma + \beta_2^2 < (\beta_1^2+\beta_2^2)/\gamma_1 ,$$

and therefore

$$\frac{\beta_1^2/\gamma+\beta_2^2}{\beta^2/\gamma_1} < 1 . \tag{400}$$

Furthermore, it is

$$\begin{aligned}(\beta_1^2/\gamma+\beta_2^2)^2+(1-1/\gamma)^2\beta_1^2\beta_2^2 &= 2\beta_1^2\beta_2^2/\gamma+\beta_1^4/\gamma^2+\beta_2^4+\beta_1^2\beta_2^2-2\beta_1^2\beta_2^2/\gamma+\beta_1^2\beta_2^2/\gamma^2\\ &= \beta_1^2\,(\beta_1^2+\beta_2^2)/\gamma^2+\beta_2^2\,(\beta_1^2+\beta_2^2)\\ &= (\beta_1^2+\beta_2^2)\,(\beta_1^2-\beta_1^4-\beta_1^2\,\beta_2^2+\beta_2^2)\\ &= (1-\beta_1^2)\,(\beta_1^2+\beta_2^2)\,(\beta_1^2+\beta_2^2),\end{aligned}$$

and therefore

$$\left(\beta_1^2/\gamma+\beta_2^2\right)^2+\left((1/\gamma-1)\,\beta_1\,\beta_2\right)^2=(\beta^2/\gamma_1)^2\,. \tag{401}$$

With Eqs. (400) and (401), we can write the matrix $\mathbf{D}^*$ defined in Eq. (399) as rotation matrix $\mathbf{D}_3^*(\alpha_3)$

$$\mathbf{D}^*=\mathbf{D}_3^*(\alpha_3)=\begin{pmatrix}1&0&0&0\\0&\cos\alpha_3&\sin\alpha_3&0\\0&-\sin\alpha_3&\cos\alpha_3&0\\0&0&0&1\end{pmatrix} \tag{402}$$

with the rotation angle α_3

$$\sin\alpha_3=\frac{\gamma_1(1-\gamma)\beta_1\beta_2}{\gamma\beta^2}\;,\;\cos\alpha_3=\frac{\gamma_1(\beta_1^2+\beta_2^2)}{\gamma\beta^2}\quad\longrightarrow\;\tan\alpha_3=\frac{(1-\gamma)\beta_1\beta_2}{\beta_1^2+\gamma\beta_2^2}\;. \tag{403}$$

Therefore, we have to apply a rotation by an angle α_3 around the z''-axis in the inertial system Σ'' for bringing both coordinate systems atop of each other Fig. 1,

$$\mathbf{x}''=\mathbf{D}_3^*(\alpha_3)\,\tilde{\mathbf{x}}''\;. \tag{404}$$

We determine the first non-vanishing terms of the formula for

$$v_1\ll c\;,\;v_2\ll c\;,\;v_2\ll v_1\;.$$

With $\tan x\approx x\;,\;\sqrt{1-x}\approx 1-x/2\quad\longrightarrow\;1/\gamma\approx 1-(1/2)\left(\beta_1^2+\beta_2^2\right)\;,$

and especially $\beta_1^2/\gamma+\beta_2^2\approx\left(1-(1/2)\left(\beta_1^2+\beta_2^2\right)\right)\beta_1^2+\beta_2^2\approx\beta_1^2+\beta_2^2\;,$

we find

$$\alpha_3\approx\tan\alpha_3=\frac{(1/\gamma-1)\beta_1\beta_2}{\beta_1^2/\gamma+\beta_2^2}\approx\frac{(-1/2)\left(\beta_1^2+\beta_2^2\right)\beta_1\beta_2}{\beta_1^2+\beta_2^2}\;,$$

hence

$$\alpha_3=-\frac{v_1v_2}{2c^2}\;. \tag{405}$$

In this way, we have reproduced Eqs. (214) and (215) from Chap. 8, Sect. 3. Equations (383) and (399) lead together to the matrix relation

$$\mathbf{D}_3^*(\alpha_3) = \mathbf{L}_2(\tfrac{\beta_2}{\gamma_1})\,\mathbf{L}_1(\beta_1)\,(\mathbf{D}_3)^{-1}(\varphi)\,\mathbf{L}_o(\beta)\,\mathbf{D}_3(\varphi) \tag{406}$$

with $\tan\varphi = \beta_1/\beta_2$ and $\tan\alpha_3$ as in Eq. (403), and the matrix $\mathbf{L}_o(\beta)$ from Eq. (381). Equations (399) and (406) contain the general theory of the THOMAS precession.

There appear two rotation matrices of type $\mathbf{D}_3$ that should not be mixed. The matrix $\mathbf{D}_3(\varphi)$ is needed for decomposing the general boost $\mathbf{L}_{1,2}$ into a product of special LORENTZ transformations and a rotation $\mathbf{D}_3(\varphi)$, Eq. (383). The other rotation matrix $\mathbf{D}_3^*(\alpha_3)$ describes the measurable effect of the THOMAS precession. For composing boosts, we refer to GIULINI (2006).

Now we return to our arguments from Chap. 8, Sect. 3. In the laboratory system Σ_o, we observe the motion of a body on a circular orbit, e.g. the classical path of an electron in an atom. $\Sigma''(x^{i''})$ might be the axis system connected with the electron. The atomic nucleus is the origin of electric central forces not influencing the proper angular momentum vector $\mathbf{S}$ of the electron, and this vector keeps constant its direction with respect to the axes $(x^{i''})$ of Σ''.

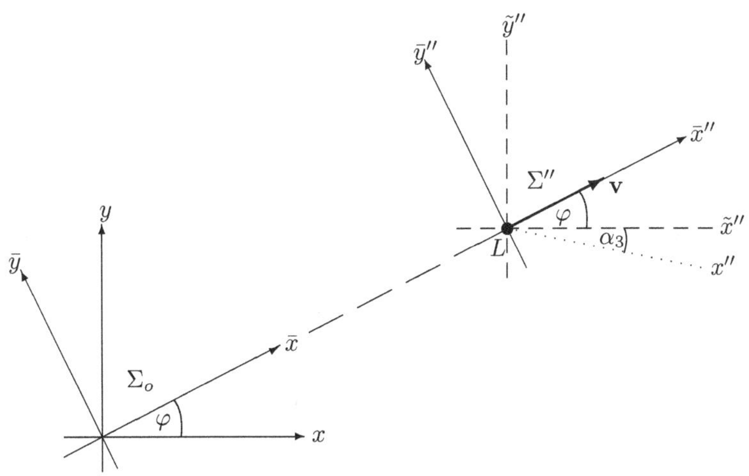

Fig. 1 Proper LORENTZ transformation (boost) and THOMAS precession. Starting point is the special LORENTZ transformation $\mathbf{L}_o$ between axis-parallel coordinates with overbar, Eq. (376), of Σ_o and Σ''. Here, $\mathbf{v}$ is an arbitrary velocity within the x-y-plane of Σ_o. Then there follows the proper LORENTZ transformation (380) with the matrix (382) that the coordinate system (x^i) of Σ_o transforms into $(\tilde{x}^{i''})$ of Σ''.

We perform the special LORENTZ transformation $\mathbf{L}_1$ with Eq. (386) and then the special LORENTZ transformation $\mathbf{L}_2$ with Eq. (390), and we come to the coordinate system $(x^{i''})$ of Σ''. The angle α_3 between the axes $\tilde{x}''$ and x'' in Σ'' defines the THOMAS precession, described by the rotation matrix $\mathbf{D}_3^*$ with Eqs. (399) and (404). It has be noted that the directions of axes $\tilde{x}''$ and x'', and therefore also the angle α_3, refer to system Σ''.

Let $\mathbf{v} = (v_1, 0, 0)$ be the current orbital velocity of the electron that changes due to the centripetal acceleration of the electron $\mathbf{a} = (0, a, 0)$ during the time interval Δt by the velocity $\Delta_v = \Delta v_2 = a\,\Delta t$. Then the angle $\Delta\alpha_3 = -v_1 \Delta v/(2c^2)$ describes the rotation of the coordinate axes of Σ'' against the axes $(\tilde{x}^{i''})$—determined in the labour system Σ_o by the proper LORENTZ transformation $\mathbf{L}_{1,2}$—during the time Δt. Since the eigen angular momentum $\mathbf{S}$ of the electron in Σ'' is constant, we observe in Σ_o the THOMAS *precession*:

The vector $\mathbf{S}$ rotates with the angular velocity $\omega_T = \Delta\alpha_3/\Delta t$ around the z-axis, or written as vector, cp. also Chap. 8, Sect. 3, Eq. (216),

$$\omega_T = -\frac{\mathbf{v}\times\mathbf{a}}{2\,c^2}\ . \qquad \text{Angular velocity of THOMAS precession} \qquad (407)$$

1.5 *Geometry in* MINKOWSKI *Space*

Tensor calculus is the analytic method of geometry. Therefore, we shall deal in the following with vectors and tensors. A short summary of tensor calculus is presented in Appendix B.1.

The requirement of the form invariance of the line element (313) distinguishes those coordinate systems $(x^i), i = 0, 1, 2, 3,$ of the four-dimensional MINKOWSKI space, that can physically assigned to inertial systems, e.g. to $\Sigma_o(x^i)$. The coordinates (x^i) of a point in MINKOWSKI space are the coordinates of an event E. The three Cartesian space coordinates (x^1, x^2, x^3) determine, where the event took place, and the time coordinate $x^0 = ct$ gives the moment when it happens. The coordinate differentials dx^i are assigned to two neighbouring events (in spatial and temporal meaning) of the MINKOWSKI space. As seen from any other inertial system $\Sigma'(x^{i'})$, or in a special case also from the same inertial system, but with rotated coordinate axes, the same events define distinct differentials $dx^{i'}$, for which it holds

$$dx^{i'} = A^{i'}_i\, dx^i \quad \longleftrightarrow \quad dx^i = A^i_{i'}\, dx^{i'}\ , \quad A^{i'}_i A^{j'}_k\, \eta_{i'j'} = \eta_{ik}\ . \qquad (408)$$

Here $A^{i'}_i$ is the matrix of the general LORENTZ transformation; therefore, it fulfils the condition of the pseudo-orthogonality Eqs. (345) or (356). If we restrict us on such coordinate systems, that have a common origin $O(0, 0, 0, 0)$, then Eq. (408) holds true also for the coordinates themselves, s. Eq. (344).

The mathematical properties of physical quantities, having in general more components, are classified after the behaviour of these components at coordinate changes as known from three-dimensional vector calculus.

The prototype of a *contravariant* vector is given by the coordinate differentials dx^i according to Eq. (408). When a coordinate change leads to transformations of the quantity V^i as

$$V^{i'} = A^{i'}_i\, V^i\ , \qquad \text{Contravariant vector} \qquad (409)$$

it represents the contravariant components of a vector in MINKOWSKI space. If it represents a vector field $V^i = V^i(x^i)$, then in the new inertial system $\Sigma'(x^{i'})$ one is concerned with changed functional dependence caused by the coordinate transformation, and hence

$$V^{i'} = V^{i'}(x^{k'}) = A_i^{i'}\, V^i\left(A_k^{k'} x^k\right) ,$$

without writing this down so explicitly in each case.

A scalar function $\phi = \phi(x^k)$ is defined by the simple recipe that for a coordinate change one has only to substitute the coordinates,

$$\phi = \phi(x^k) = \phi(A_{k'}^k x^{k'}) . \qquad \text{Scalar field} \qquad (410)$$

The transformation rule of a *covariant* vector field V_i is given by the properties of the gradients of a scalar field,

$$\frac{\partial\phi(x^{k'})}{\partial x^{i'}} = A_{i'}^i \frac{\partial\phi(x^k)}{\partial x^i} \quad \longleftrightarrow \quad \frac{\partial\phi(x^k)}{\partial x^i} = A_i^{i'} \frac{\partial\phi(x^{k'})}{\partial x^{i'}} . \qquad (411)$$

When a quantity V_i transforms for a coordinate change as in Eq. (411),

$$V_{i'} = A_{i'}^i\, V_i , \qquad \text{Covariant vector} \qquad (412)$$

it represents the covariant components of vectors in MINKOWSKI space. For a vector field, one has to substitute again also the new coordinates. Scalars and vectors are also called tensors of zeroth and first rank, respectively. Tensors of second and higher ranks transform as the respective products of vectors, Appendix B.1, and hence, e.g.

$$T_{i'}{}^{k'} = A_{i'}^i\, A_k^{k'}\, T_i{}^k \quad \longleftrightarrow \quad T_i{}^k = A_i^{i'}\, A_{k'}^k\, T_{i'}{}^{k'} . \qquad (413)$$

When we form the partial derivatives of a tensor field $T_i{}^k$, so we get a new tensor field $T_{li}{}^k := \partial_l\, T_i{}^k$ with a by one higher covariance level.[5]

$$T_{l'i'}{}^{k'} = A_{l'}^l\, A_{i'}^i\, A_k^{k'}\, T_{li}{}^k \quad \longleftrightarrow \quad T_{li}{}^k = A_l^{l'}\, A_i^{i'}\, A_{k'}^k\, T_{l'i'}{}^{k'} . \qquad (414)$$

The co- and contravariant components of vectors and tensors are convertible with the in Eq. (312) defined metric tensor η_{ik} of the MINKOWSKI space,

$$\left.\begin{array}{lll} V_i = V^k \eta_{ik} & \longleftrightarrow & V^i = V_k\, \eta^{ik} , \\ T_{ik} = T^{rs}\, \eta_{ir}\eta_{ks} & \longleftrightarrow & T^{ik} = T_{rs}\, \eta^{ir}\eta^{ks} . \end{array}\right\} \qquad (415)$$

Because of the diagonal form (354) of the MINKOWSKI metric η_{ik}, the co- and contravariant components of a vector differ only by the sign. In particular, because

[5] When we also admit curvilinear coordinates, e.g. spherical polar coordinates in space, or when we even transform to accelerated reference systems, then for keeping the tensor properties, we have to replace the partial by the covariant derivative, cp. Chap. 13 and Appendix B.1.

of the chosen signature (312) the time-like component remains unchanged, and the three spatial components change its sign. For a tensor of arbitrary level, only the components with an uneven number of spatial indices containing “1”, “2” or “3” have to be multiplied with -1.

The scalar product

$$U^i\, V_i = U^i\, V^k \eta_{ik} = U_i\, V_k \eta^{ik} = U^{i'}\, V_{i'} \tag{416}$$

of two vectors U^i and V^k is an invariant quantity with respect to general LORENTZ transformations. This means physically, the quantity $U^i\, V_i$ has in each inertial system the same numerical value. Likewise, the respective tensor operations are independent on the coordinate system, where they are evaluated, e.g.

$$T_i{}^i = T_{i'}{}^{i'}\,, \quad T_{ik} T^{ik} = T_{i'k'} T^{i'k'}\,, \quad \ldots\,. \tag{417}$$

Now there holds true the following definitions:

$$\text{A vector } \; V^k \; \text{ is called} \quad \begin{matrix} \textit{space like} \\ \textit{time like} \\ \textit{null vector} \end{matrix} \;, \quad \text{when} \quad V^k\, V_k \begin{matrix} < 0 \\ > 0 \\ = 0 \end{matrix}\,. \tag{418}$$

The property of vectors to be space-like, time-like or a null vector is therefore independent on the inertial system.

In distinction to the Euclidean space, the invariant $V^k\, V_k$ can be zero in MINKOWSKI space when the vector V^k does not vanish, since the metric η_{ik} of the MINKOWSKI space is indefinite.

In particular, the line element (318) is invariant with respect to LORENTZ transformations. One says, two points have a separation in MINKOWSKI space being:

$$\begin{matrix} \text{Two points in MINKOWSKI} \\ \text{space } P(x^i) \text{ and } Q(x^i + \Delta x^i) \\ \text{lying respectively} \end{matrix} \quad \begin{matrix} \textit{space like} \\ \textit{time like} \\ \textit{light like} \end{matrix} \;, \quad \text{when} \quad \Delta s^2 = \eta_{ik}\, \Delta x^i\, \Delta x^k \begin{matrix} < 0 \\ > 0 \\ = 0 \end{matrix}\,. \tag{419}$$

This is nothing else than the definition (314), since each point is now an event.

With the three-dimensional separation $\Delta \mathsf{x}^2 := \sum_{a=1}^{3} \Delta x^a$ it follows from $\Delta s^2 = c^2 \Delta t^2 - \Delta \mathsf{x}^2 > 0$, that $\Delta \mathsf{x}^2 / \Delta t^2 < c^2$. Then there exists a body K moving with velocity $|v| = |\Delta \mathsf{x} / \Delta t| < c$, so that it has at time t the coordinates of the event P and at time $t + \Delta t$ the coordinates of event Q. Therefore, the event P can have influence on the body K at the event Q, it can cause an effect. Then we denote the events as causally connected.

This is still possible for $\Delta s^2 = 0$. In this case, $\Delta \mathsf{x}^2 / \Delta t^2 = c^2$. Then there does not exist a body travelling between both events P and Q. But the event Q can be triggered by a light signal that had at time t the coordinates of P and at time $t + \Delta t$ the coordinates of Q. In this case, one tells that Q is sitting on the light cone starting from P. The light cone is also called zero cone.

For the following consideration, we suppress again the y- and z-coordinates. In the $x - ct$-plane, the light cone is described by the equations $x = \pm ct$. All events Q lying within or on the light cone originating from P are causally connected. Therefore, we call the light cone also causal cone.

For $\Delta s^2 = c^2 \Delta t^2 - \Delta \mathrm{x}^2 < 0$ we have $\Delta \mathrm{x}^2 / \Delta t^2 > c^2$. Now there are no signals that can connect the events P and Q. The event Q is outside of the light cone originating from P. Both events are not causally connected, cp. Fig. 2. We summarise:

All events from the inner region or the boundary of the light cone originating from event P, and only these events, can be causally connected with P.

The orbit of a body in MINKOWSKI space is called world line.

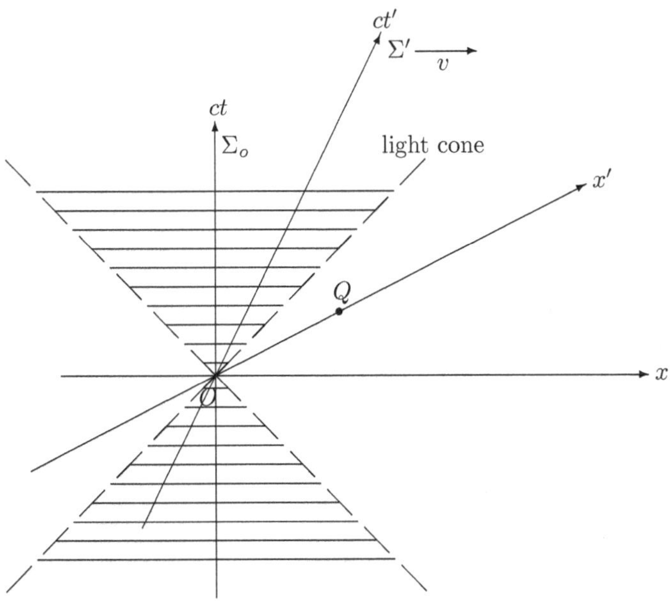

Fig. 2 The light cone from the coordinate origin O. In the figure, two space dimensions are suppressed. All events (points) lying in and on the cone around the time axis, and only these points, can be in causal connection with the coordinate origin O (dashed region). In particular, the event O can influence the events in the future cone with $t \geq 0$, and the origin can be influenced by events from the past cone with $t \leq 0$. For all points Q outside of the light cone, there is an inertial system $\Sigma'(x', t')$, in which these events Q can be simultaneously with O. Henceforth, Q cannot be causally connected with O. If the x'- and the ct'-axes are drawn symmetrically to the light cone, one gets the graphic representation of a special LORENTZ transformation with the resulting properties of the length contraction (112) and the time dilatation (113), s. Fig. 3.

When the body has a constant velocity, the world line is a straight line. The light cone is formed by the world lines of photons. The world line of an in Σ_o resting body is parallel to the ct-axis. The world line of a body moving with velocity v along the x-axis is a tilted straight line in the x-ct-diagram of Fig. 2 going through the coordinate origin, the slope is given by $\tan\varphi = c\Delta t/\Delta x = c/v$. This line is the ct'-axis of a system Σ', in which this body is at rest. All world lines of bodies with non-vanishing rest mass through the coordinate origin O lie within the light cones. Lines going through the coordinate origin O, but lying outside of the light cones, e.g. the x-axis and the x'-axis, cannot represent world lines.

For an event Q outside of the light cones of P in Σ_o,[6] there exist always inertial systems Σ', in which these events P and Q are simultaneous, and there exist inertial systems Σ'', in which these events P and Q occurring in reversed temporal order as compared to the order in Σ_o, s. Problem 7.

For simplicity, we show this for the case

$$\Sigma_o: \ P = O(0,0,0,0)\,, \ Q = (x_o, 0, 0, ct_o) \text{ with } ct_o < x_o \ \longrightarrow \ \Delta s^2 = s^2 = c^2 t_o^2 - x_o^2 < 0\,.$$

Both events P and Q are situated space-like with respect to each other.

We consider a formal velocity u

$$u := x_o/t_o \ \longrightarrow \ c < u\,.$$

With it we form velocities v being always smaller than c

$$\Sigma_o: \ v \le c^2/u < c\,. \tag{420}$$

An inertial system Σ' might be in motion with a velocity v in x-direction with respect to the inertial systems Σ_o. Then we have the special LORENTZ transformation

$$x' = \gamma\,(x - v\,t)\,, \quad t' = \gamma\,(t - xv/c^2)\,.$$

The coordinate origin O retains the time coordinate $t' = 0$, and for the event Q we find with Eq. (420) the time coordinate t_o',

$$t_o' = \gamma\,(t_o - \frac{v\,x_o}{c^2}) = \gamma\,(t_o - \frac{t_o\,v\,x_o}{c^2\,t_o}) = \gamma\,(t_o - t_o\frac{v\,u}{c^2}) = \gamma\,t_o\left(1 - \frac{v\,u}{c^2}\right) \le 0\,,$$

so that in inertial system Σ' the event P occurs earlier or simultaneously to the event Q as claimed. In inertial system Σ', it is obvious:

Two spatially separated and simultaneous events P and Q cannot be causally connected, seen especially when P happens earlier than Q. This property of two events lying spatially related to each other is independent from the reference system, as it is formulated by a LORENTZ-invariant relation,

---

[6] Really the light cone does not depend on the inertial system because of the universality of the speed of light.

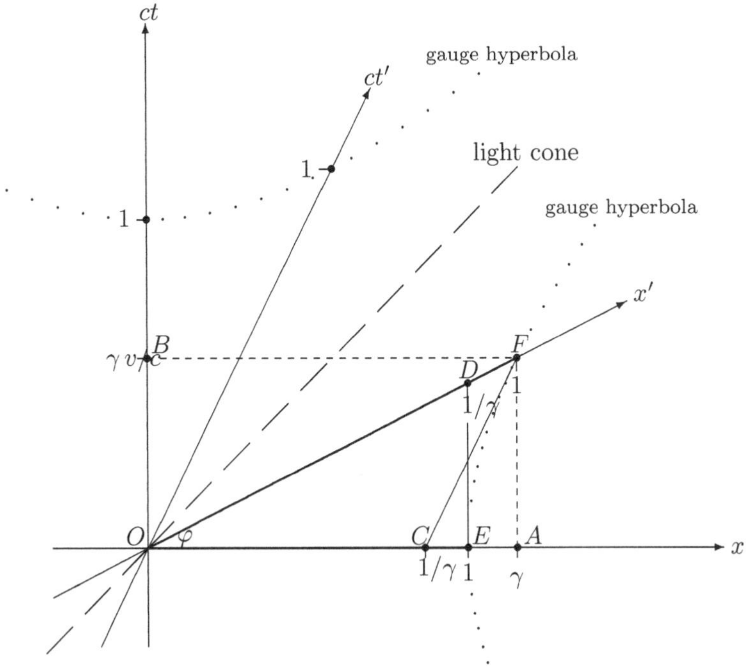

Fig. 3 Special LORENTZ transformation and LORENTZ contraction. The light cone has a tilt of 45° to the x-axis, and it is the bisection of the angle between the x'- and the ct'-axes. In the main text, we evaluate the coordinates of the five different points giving $A(x = \gamma,\ t = 0)$, $D(x' = 1/\gamma,\ t' = 0)$, $C(x = 1/\gamma,\ t = 0)$, $E(x = 1,\ t = 0)$, $F(x' = 1,\ t' = 0)$. The line $\overline{CF}$ is parallel to the ct'-axis. The points E and F lie on the unit hyperbola Eqs. (422) or (423) shown as pointed curve. For geometrical considerations, one has to respect that on the x'- and the ct'-axes one uses different length units than on the x- and the ct-axes. It is $\tan\varphi = v/c$. Here we have considered the special case $v = 0.5\,c$, and hence $\varphi \approx 26.565°$ and $\gamma = 1.155$. The Euclidean separation of point F with coordinate $x' = 1$ to the coordinate origin O is given by $\overline{OF}^2 = \overline{OA}^2 + \overline{AF}^2$, and hence $\overline{OF} = \gamma\sqrt{c^2 + v^2}/c \approx 1.29$. Since we have drawn the unit length on the x-axis as 4.5 cm, we had to draw the unit length on the x'-axis with $4.5 \cdot 1.29 = 5.8$ cm. $\overline{OC}$ are the in Σ_o simultaneously observed end points of the in Σ' resting bar $\overline{OF}$, for which we have measured a unit length 1. Similarly, $\overline{OD}$ is the in Σ' simultaneously observed position of the end points of the in Σ_o resting unit ruler $\overline{OC}$.

$$s'^2 = s^2 < 0\,.$$

The special LORENTZ transformation and the resulting relativity of LORENTZ contraction have been graphically illustrated in Fig. 3. We shall now demonstrate the origin of the relativity of length contraction. The analogous relativity of time dilatation is illustrated in Appendix A, Fig. A.10 and discussed in Problem 23.

The light cone bisects the angle between the x- and ct-axes as well as the angle between the x'- and the ct'-axes.

The x'-axis represents all points having the same time as the coordinate origin in system Σ'. It results from the LORENTZ transformation given above, when we set simply $t' = 0$ in the second equation. Likewise, the ct'-axis is described by the first equation for $x' = 0$,

$$\left.\begin{aligned} ct &= \frac{v}{c}\,x\ , \quad x'\text{-axis} \\ ct &= \frac{c}{v}\,x\ . \quad ct'\text{-axis} \end{aligned}\right\} \tag{421}$$

The x'-axis has an angle φ against the x-axis given by $\tan\varphi = v/c$ as seen directly from Fig. 3. The same angle appears between the ct-axis and the ct'-axis.

On the dashed x'- and ct'-axes now we shall define units, i.e. we define the points that have the measure 1, i.e. they have the dashed coordinate 1. This happens for all inertial systems by means of the LORENTZ-invariant relation $s^2 = -1$ and $s^2 = 1$, respectively, which define curves in the x-ct-plane, also called gauge hyperbolae in this connection,

$$s^2 = -1\ :\ c^2t^2 - x^2 = -1\ , \qquad \text{Gauge hyperbola for the } x'\text{-axes} \tag{422}$$

$$s^2 = 1\ :\ c^2t^2 - x^2 = 1\ . \qquad \text{Gauge hyperbola for the } c\,t'\text{-axes} \tag{423}$$

For measuring the same units on all axes, we use on the temporal axis the time multiplied with the speed of light, i.e. also a length. When we translate the MINKOWSKI space to the units of the SI-systems, then this requires that the number of the coordinate value of, e.g. 1 stands for a distance of 1 m from the coordinate origin, both for the x- and the ct-axis. For a measured number 1 on the ct-axis, the pointer of a clock shows than a time $(1/c)\,\text{s} \approx (1/3)\cdot 10^{-8}\,\text{s}$.

The intersection $F(x_F,\ ct_F)$ of the x'-axis (421) with the gauge hyperbola (422) is determined by

$$ct_F = \frac{v}{c}\,x_F\ , \quad \text{and} \quad x_F^2 - c^2\,t_F^2 = 1.$$

Then it follows

$$x_F = \gamma\,, \quad ct_F = \gamma\,\frac{v}{c}\ . \tag{424}$$

(Here, the second intersection for negative x' is ignored.)

We have the invariant line element. For the point F according to Eq. (424) $s^2 = c^2\,t_F^2 - x_F^2 = s^2 = c^2\,t'^2_F - x'^2_F = -1$, in Σ' the unit on the x'-axis is fixed with $t'_F = 0$ and $x'_F = 1$.

The unit length (shown with thick line) $\overline{OE}$ on the x-axis of Σ_o is assigned in Σ' the simultaneous position $\overline{OD}$, since $\overline{ED}$ is the world line of the right end point E.

The point A has the x-coordinate γ according to Eq. (424). According to the intercept theorem, it holds $\gamma/1 \equiv \overline{OA}/\overline{OE} = \overline{OF}/\overline{OD}$.

The length unit in Σ' is determined by the gauge hyperbola $\overline{OF}$. Then also its measured value, and its end coordinate x'_F in Σ' is given by 1. The intersection theorem therefore tells us that in Σ' the end coordinate is $x'_D = 1/\gamma$:

For the in system Σ_o resting unit measure $\overline{OE}$, one measures in Σ' the LORENTZ-contracted length $1/\gamma$.

On the other hand, to the (thick drawn) unit length $\overline{OF}$ on the x'-axis of Σ', there corresponds in Σ_o the simultaneous position of $\overline{OC}$, since $\overline{CF}$ is the world line of the right end point F, being in motion with velocity v of Σ'. Therefore, the coordinates (x, ct) of this world line fulfil the equation $(x - x_F)/(t - t_F) = v$. In this equation, we insert the coordinates (x_F, t_F) from Eq. (424), and we find $ct = (c/v)\,(x - \gamma + \gamma\, v^2/c^2)$, hence $ct = (c/v)\,(x - 1/\gamma)$. The intersection point C with the x-axis is given by $t = 0$, hence $x = 1/\gamma$.

Now x_E is evaluated in Σ_o as 1, so that the end coordinate of the unit ruler of Σ' in Σ_o has a measured value $x_D = 1/\gamma$:

For the unit length $\overline{OF}$ resting in system Σ' one gets in Σ_o the LORENTZ-contracted length $1/\gamma$.

1.6 EINSTEIN*'s Relativity Principle in* MINKOWSKI *Space*

If for a tensor of arbitrary rank in one single inertial system holds true $\mathbf{T} = 0$, the same equation holds true also in all other inertial systems. This can be seen immediately from the transformation law from the dashed to the undashed components, and inversely. So it follows, for example, from $T_i{}^k = 0$ according to Eq. (413) $T_{i'}{}^{k'} = 0$ and the reverse relation. Also, it follows from the equation $A_i{}^k = B_i{}^k$ in inertial system Σ_o, that then $A_{i'}{}^{k'} = B_{i'}{}^{k'}$ in each other inertial system Σ'. These simple statements have far-reaching physical consequences, and they allow a general mathematical formulation of EINSTEIN's relativity principle, already cited in Chap. 2, Sect. 1. Now we have to formulate all physical processes mathematically as equations between tensors in MINKOWSKI space. These equations then hold true in each inertial system in the same form. So we find the following general rule of EINSTEIN's relativity principle,

The laws for the behaviour of physical systems and its change are tensor equations in MINKOWSKI space. (425)

One calls this also four-dimensional *covariant* formulation of physical laws. This means that all quantities occurring in these equations behave in the same way—*covariant*—with respect to LORENTZ transformations, and hence as tensors in

MINKOWSKI space. The equations are then *form invariant* with respect to LORENTZ transformations, they preserve its mathematical form in each inertial system.

This formulation contains an immense heuristic principle for establishing physical laws. We have 'only' to invent, which mathematical formulas are possible for presenting a physical connection as tensor equation in MINKOWSKI space. One important application of this principle, going back to MINKOWSKI, will be presented in Sect. 3.2.3, Eq. (580). Laws that cannot be formulated in such a way are excluded ab initio as an exact description of physical phenomena. Now we will present tensor equations in MINKOWSKI space starting from classical physics, and we will demonstrate it for mechanics and electrodynamics in the next two sections.

From a mathematical point of view, the statement (425) must be still extended. Besides tensors, so-called integer representation of the LORENTZ group, there exists a whole class of geometrical objects, so-called spinors, half-integer representations of the LORENTZ group. The transformation law for the connection of spinor components in two different inertial systems is also given by matrices, depending on a more complex form on the LORENTZ matrix. The mathematical description of particles with spin, the internal angular momentum of electrons, protons, neutrons, π-mesons and others is provided by spinor equations, e.g. the famous DIRAC equation for particles with spin $1/2$. In full generality, the formulation of the relativity principles requires instead of 'tensor equation' the term 'tensor or spinor equations'. We shall discuss it in Chap. 10.

2 Covariant Formulation of Relativistic Mechanics

The physical content of NEWTON's second axiom (139) is the proportionality of the force $\mathbf{F}$ on a body and the temporal change of its momentum $\mathbf{p} = m\mathbf{u}$. There the equality follows from the choice of the force unit. This law even allows that the mass m, entering the definition of the momentum, depends on the velocity $\mathbf{u}$. Only the assumption of the GALILEI transformation (69) requires an invariability of the inertia $m = m_o$ as proven in Problem 14 in the interpretation of TOLMAN's experiment.

In Chap. 6, Sects. 3 and Chap. 7, Sect. 1, we have derived the relativistic corrections to Eq. (149) of classical mechanics by supposing the LORENTZ transformation, namely, the mass formula (167) and as consequence by means of the completely inelastic impact the energy–mass equivalence in Eq. (186).

The GALILEI transformation is the limiting case of the LORENTZ transformation for $c \longrightarrow \infty$. From the relativistic equations (167), it follows for $c \longrightarrow \infty$ again the constant mass m_o in the classical limit. As long as we perform experiments with bodies having velocities u much smaller than the speed of light, $|u| \ll c$, so long NEWTON's laws hold approximately true for invariable masses (149) in each inertial system.

Now we will search for such tensor equations in MINKOWSKI space by EINSTEIN's relativity principle formulated in Sect. 1.6, Eq. (425) that in the limit of small velocities go over to NEWTON's law (149) for constant masses. To this aim, we at first discuss the description of motions in MINKOWSKI space.

2.1 *Motion of a Particle in* MINKOWSKI *Space*

2.1.1 Proper Time of Particle Motion

The orbit of a body and its velocity is described by equations

$$
\begin{aligned}
x &= x(t) \longrightarrow u_x = \frac{dx(t)}{dt} , \\
y &= y(t) \longrightarrow u_y = \frac{dy(t)}{dt} , \\
z &= z(t) \longrightarrow u_z = \frac{dz(t)}{dt} .
\end{aligned}
$$

In the GALILEI-invariant NEWTONian mechanics, the time t is an invariant quantity, with the same value in each inertial system. This is different now due to the LORENTZ transformation. Instead of t we need a LORENTZ-invariant parameter for describing the motion. To this aim, we consider two neighbouring points P and Q on the orbit of a body in our three-dimensional space,

$$
\left.
\begin{aligned}
&P\Big(x(t),\ y(t),\ z(t)\Big) , \\
&Q\Big(x(t+dt), y(t+dt), z(t+dt)\Big) = Q\Big(x(t) + u_x\, dt, y(t) + u_y\, dt, z + u_z\, dt\Big) .
\end{aligned}
\right\} \tag{426}
$$

The orbit corresponds to a world line in four-dimensional MINKOWSKI space. If the body goes through the coordinate origin, and if it has constant velocity, the world lines represent straight lines within the light cone, and with the time axis ct for a body at rest, Fig. 2. In general, they are curves with slope larger than 1. If we write again

$$
x^0 = ct\ ,\ \ x^1 = x(t)\ ,\ \ x^2 = y(t)\ ,\ \ x^3 = z(t)\ ,
$$

so the space points P and Q correspond to points on the world line of the body, the events E_P and E_Q,

$$
\left.
\begin{aligned}
&E_P\Big(x^0,\ x^1,\ x^2,\ x^3\Big) , \\
&E_Q\Big(x^0 + c\, dt\ ,\ x^1 + u_x\, dt\ ,\ x^2 + u_y\, dt\ ,\ x^3 + u_z\, dt\Big) .
\end{aligned}
\right\} \tag{427}
$$

For simplicity, we assume that the world line goes through the coordinate origin. We know the invariant $ds^2 = \eta_{ik} dx^i dx^k = ds'^2 = \eta_{ik} dx^{i'} dx^{k'}$ and we form the related invariant $d\tau$,

$$
d\tau := \frac{ds}{c} = \frac{1}{c}\sqrt{c^2dt^2 - u^2dt^2} = \sqrt{1 - u^2/c^2}\, dt = \frac{dt}{\gamma_u} . \tag{428}
$$

Differential of eigentime

Here in γ_u the current velocity $\mathbf{u}$ of the body has to be inserted. The invariant τ formed by integration is called *proper time* or *eigentime* of the moving body. For the integration from $t = 0$ at event O until time t_E at an event E on the world line of the body, we get with the substitution of τ by t according to Eq. (428):

$$\tau_E = \int_0^{\tau_E} d\tau = \int_0^{t_E} \sqrt{1 - u^2(t)/c^2}\, dt \ . \qquad \text{Eigentime} \qquad (429)$$

The quantity τ is the parameter that we searched for a relativistic description of the orbit of a body replacing the classical parameter, the time t. In the non-relativistic approximation, for $u^2/c^2 \ll 1$, one can write instead of the eigentime τ again simply the time t. In the relativistic case, it is the eigentime τ that has one and the same value in each inertial system,

$$\int_0^{t_E} \sqrt{1 - u^2(t)/c^2}\, dt = \int_0^{t'_E} \sqrt{1 - u'^2(t')/c^2}\, dt' \ . \qquad \text{Independence of the eigentime from the inertial system} \qquad (430)$$

This invariance follows simply from the definition of τ from the invariant line element ds. In Problem 24, we prove this explicitly.

2.1.2 Four-Vectors of Particle Motion

The coordinate differentials dx^i corresponding to two neighbouring points E_P and E_Q form according to Eq. (408) contravariant components of a vector in Minkowski space. Since $d\tau$ is an invariant, the quantities $dx^i/d\tau$ now also form contravariant components of a four-vector, the *four-vector of velocity* u^i. It reads as follows, when we take into account Eq. (428):

$$\begin{aligned} u^i &:= \frac{dx^i}{d\tau} \\ &= \left(\frac{c}{\sqrt{1-u^2/c^2}}, \frac{u_x}{\sqrt{1-u^2/c^2}}, \frac{u_y}{\sqrt{1-u^2/c^2}}, \frac{u_z}{\sqrt{1-u^2/c^2}} \right) \end{aligned} \qquad \text{Four-vector of velocity} \qquad (431)$$

with the velocity $\mathbf{u}$ of the body with components $\mathbf{u} = \big(u_x(t)\,,\ u_y(t)\,,\ u_z(t)\big)$. We write

$$\left. \begin{aligned} &u^2 := \sum_{a=1}^{3} u^a u^a = u^a u^b\, \delta_{ab} \\ &\text{and for distinction} \\ &\mathbf{u}^2 := \sum_{i=0}^{3} u^i u_i = u^i u^k\, \eta_{ik} \ . \end{aligned} \right\} \qquad (432)$$

There the covariant components of the four-velocity are

$$u_i := \eta_{ik}u^k = \left(\frac{c}{\sqrt{1-u^2/c^2}}, \frac{-u_x}{\sqrt{1-u^2/c^2}}, \frac{-u_y}{\sqrt{1-u^2/c^2}}, \frac{-u_z}{\sqrt{1-u^2/c^2}}\right). \quad (433)$$

From Eqs. (431) and (433), it follows immediately that the four-velocity for all velocities **u** forms the invariant

$$\begin{aligned} \mathsf{u}^2 = u^i\, u^k \eta_{ik} &= \left(u^0\right)^2 - \left(u^1\right)^2 - \left(u^2\right)^2 - \left(u^3\right)^2 \\ &= \frac{c^2}{1-u^2/c^2} - \frac{(u_x)^2}{1-u^2/c^2} - \frac{\left(u_y\right)^2}{1-u^2/c^2} - \frac{(u_x)^2}{1-u^2/c^2}\,, \end{aligned}$$

hence

$$\mathsf{u}^2 = u^i\, u^k \eta_{ik} = u^i\, u_i = c^2\,. \quad (434)$$

The four-velocity is a time-like vector. The simplest way to evaluate the invariant (434) is to use the current rest frame of the bodies, where it holds $\mathbf{u} = 0$, and it follows directly Eq. (434).

From the four-velocity u^i, we can form the contravariant components of the vector of four-acceleration a^i by taking the derivative with respect to the eigentime τ. There we have to take into account that the velocity dependence appears both in numerator and denominator.

By Eq. (428), we have

$$\frac{d}{d\tau} = \frac{1}{\sqrt{1-u^2/c^2}}\frac{d}{dt}\,,$$

hence

$$\begin{aligned} a^i &= \frac{du^i}{d\tau} \\ &= \frac{1}{\sqrt{1-u^2/c^2}}\left(0, \frac{a_x}{\sqrt{1-u^2/c^2}}, \frac{a_y}{\sqrt{1-u^2/c^2}}, \frac{a_z}{\sqrt{1-u^2/c^2}}\right) \\ &\quad + \frac{1}{\sqrt{1-u^2/c^2}}\left(\frac{\mathbf{a}\cdot\mathbf{u}/c}{\sqrt{(1-u^2/c^2)^3}}, \frac{\mathbf{a}\cdot\mathbf{u}u_x/c^2}{\sqrt{(1-u^2/c^2)^3}}, \frac{\mathbf{a}\cdot\mathbf{u}u_y/c^2}{\sqrt{(1-u^2/c^2)^3}}, \frac{\mathbf{a}\cdot\mathbf{u}u_z/c^2}{\sqrt{(1-u^2/c^2)^3}},\right). \end{aligned}$$

The four-acceleration a^i therefore is composed in a complex way from the components of the three-dimensional velocity **u** and the three-dimensional acceleration **a**,

$$\left.\begin{aligned} a^i &= \frac{du^i}{d\tau} \\ &= \left(\gamma_u^4\frac{\mathbf{a}\cdot\mathbf{u}}{c}, \gamma_u^2 a_x + \gamma_u^4\frac{\mathbf{a}\cdot\mathbf{u}u_x}{c^2}, \gamma_u^2 a_y + \gamma_u^4\frac{\mathbf{a}\cdot\mathbf{u}u_y}{c^2}, \gamma_u^2 a_z + \gamma_u^4\frac{\mathbf{a}\cdot\mathbf{u}u_z}{c^2}\right) \end{aligned}\right\}\ \text{Four-vector of acceleration} \quad (435)$$

with the three-acceleration **a** of the body

$$\mathbf{a} = \big(a_x(t)\,,\, a_y(t)\,,\, a_z(t)\big) = \left(\frac{du_x}{dt}\,,\, \frac{du_y}{dt}\,,\, \frac{du_z}{dt}\right)\ ,$$

and it holds

$$\mathbf{a}\cdot\mathbf{u} = u_x\,a_x + u_y\,a_y + u_z\,a_z\,.$$

For $u \ll c$, the last three components of u^i and of a^i are identical with the components of the three-dimensional velocity **u** and the three-dimensional acceleration **a**.

2.2 *Dynamics of Particles in* MINKOWSKI *Space*

Now we look for such tensor equations in MINKOWSKI space for the motion of particles that go over to the NEWTONian equations (149) for small velocities, i.e. $v/c \longrightarrow 0$, for constant, i.e. velocity-independent masses.

The experimental parameter of a particle is its rest mass m_o, i.e. the inertial mass that we measure in the current rest system of the body. Since there the velocity $u = 0$, the inertial mass m_o is already in the framework of classical mechanics exactly defined. Using EINSTEIN's energy–mass equivalence (186), one can determine m_o also from the energy that can be potentially delivered by this mass. The quantity m_o is therefore by definition a LORENTZ-invariant quantity, a tensor of rank zero, with the same value in each inertial system:

The measureable LORENTZ-invariant parameter of a particle is its rest mass m_o, the inertial mass in the temporal rest system.

Following to classical mechanics and with the aim to form tensors in MINKOWSKI space, we define under consideration of Eq. (431) the four-vector of momentum p^i of a body,

$$p^i := m_o\,\frac{dx^i}{d\tau} = m_o\,u^i = \left(\frac{m_o\,c}{\sqrt{1-u^2/c^2}}\,,\, \frac{m_o\,u_x}{\sqrt{1-u^2/c^2}}\,,\, \frac{m_o\,u_y}{\sqrt{1-u^2/c^2}}\,,\, \frac{m_o\,u_z}{\sqrt{1-u^2/c^2}}\right). \qquad (436)$$

Four vector of momentum

For $u \ll c$, the last three components of the four-vector p^i go over into the classical momentum **p** of NEWTONian mechanics. The same holds true for the four-velocity.

The tensor equations, going over into the classical equations (149) for $u \ll c$, with Eq. (151) instead of the third law of Eq. (149), can then be written as

Covariant form of relativistic mechanics in MINKOWSKI space :

$$\frac{d}{d\tau}\,p^i = m_o\,\frac{d}{d\tau}\,u^i = m_o\,\frac{d}{d\tau}\,\frac{dx^i}{d\tau} = F^i\,. \qquad \text{Second axiom of relativistic mechanics}$$

$$p^i = \frac{m_o}{\sqrt{1-u^2/c^2}}\,u^i = \text{const} \quad \text{for} \quad F^i = 0\,. \qquad \text{First axiom of relativistic mechanics} \tag{437}$$

If there are only inner forces, it holds

$$\frac{d}{d\tau}\sum_{p=1}^{n} p_p^i = \frac{d}{d\tau}\,P^i = \frac{1}{\sqrt{1-u^2/c^2}}\,\frac{d}{dt}\,P^i = 0\,. \qquad \text{Third axiom of relativistic mechanics}$$

The physical interpretation of the 'four-vector of force' F^i will now be explained.

For n particles with rest masses m_{op} at positions x_p^i, the second law reads as follows, when they are exposed to external forces F_p^i and interaction forces F_{qp}^i,

$$\frac{d}{d\tau}\,p_p^i = m_{op}\,\frac{d}{d\tau}\,\frac{d}{d\tau}\,x_p^i = \sum_{q=1}^{n}\,F_{qp}^i + F_p^i\,. \tag{438}$$

We shall see below that Eq. (438) indeed guarantees the conservation of the total momentum $\mathbf{P}$ and of the total energy E.

The quantities F^i or F_p^i and F_{qp}^i are called MINKOWSKI force vectors. We write for the last three components of the vector F^i in (634)

$$\left(F^1\,,\;F^2\,,\;F^3\right) := \left(\frac{F_x}{\sqrt{1-u^2/c^2}}\,,\;\frac{F_y}{\sqrt{1-u^2/c^2}}\,,\;\frac{F_z}{\sqrt{1-u^2/c^2}}\right) \tag{439}$$

and we take into account Eqs. (428) and (431) for the differential of the proper time and for the four-vector of velocity. Then we rewrite the last three components in the second axiom, cancelling a factor γ_u,

$$\frac{d}{dt}\mathbf{p} = \frac{d}{dt}(m\,\mathbf{u}) = \frac{d}{dt}\,\frac{m_o\mathbf{u}}{\sqrt{1-u^2/c^2}} = \mathbf{F} \tag{440}$$

with $\mathbf{F} = \big(F_x(t)\,,\;F_y(t)\,,\;F_z(t)\big)$ and $\mathbf{u} = \big(u_x(t)\,,\;u_y(t)\,,\;u_z(t)\big)$.

As we see, with the ansatz (439) for the MINKOWSKI's force vectors, the spatial components of the covariant Eq. (636) are identical with Eq. (168) of relativistic mechanics. The covariant formulation of EINSTEIN's relativity principle replaces in one single formal step the detailed considerations in Chap. 6, Sect. 3.1.2 with TOLMAN's thought experiments for deriving relativistic mechanics. This gives insight into the theoretical significance of MINKOWSKI's formalism.

In the relativistic interaction of particles and fields, the four-dimensional, tensorial form of the equations of motions always plays a central role. MINKOWSKI's vector equations (436) always consist of four components in distinction to the three components of the three-dimensional vector equations (168). Still we have to explain what is the meaning of the fourth equation in the covariant formulation of the equations of motion of particles. To this aim, we multiply the second axiom (437) with the four-velocity u_i and contract it over i

$$u_i \frac{d}{d\tau} p^i = u_i \frac{1}{\sqrt{1-u^2/c^2}} \frac{d}{dt} dp^i = \frac{m_o}{\sqrt{1-u^2/c^2}} u_i \frac{d}{dt} u^i = u_i F^i \ . \tag{441}$$

Now it follows from Eq. (434) by differentiation

$$\frac{d}{dt} \left(u_i u^i \right) = u_i \frac{d}{dt} u^i + u^i \frac{d}{dt} u_i = 2 u_i \frac{d}{dt} u^i = 0 \ , \tag{442}$$

so that

$$u_i F^i = u_0 F^0 + u_1 F^1 + u_2 F^2 + u_3 F^3 = 0 \ . \tag{443}$$

Here we insert Eqs. (433) and (439) and get for the zero component of MINKOWSKI's force vector

$$F^0 = \frac{1}{c} \frac{\mathbf{u} \cdot \mathbf{F}}{\sqrt{1-u^2/c^2}} \ . \tag{444}$$

Equations (439) and (444) now completely determine MINKOWSKI's force vector,

$$F^i = \left(\frac{\mathbf{u} \cdot \mathbf{F}}{c\sqrt{1-u^2/c^2}} , \frac{F_x}{\sqrt{1-u^2/c^2}} , \frac{F_y}{\sqrt{1-u^2/c^2}} , \frac{F_z}{\sqrt{1-u^2/c^2}} \right) . \tag{445}$$

MINKOWSKI's force vector

The zero component of the second axiom in Eq. (437) then reads

$$\frac{d}{d\tau} p^0 = F^0 \ .$$

With Eq. (428) for $d\tau$ and Eqs. (436), (444) it follows

$$\frac{1}{\sqrt{1-u^2/c^2}} \frac{d}{dt} p^0 = \frac{1}{\sqrt{1-u^2/c^2}} \frac{d}{dt} \frac{m_o c}{\sqrt{1-u^2/c^2}} = F^0 = \frac{1}{c} \frac{\mathbf{u} \cdot \mathbf{F}}{\sqrt{1-u^2/c^2}} \ .$$

The factor γ_u can be cancelled. The zeroth component of the second axiom of relativistic mechanics in Eq. (438) then reads

$$\frac{d}{dt} \frac{m_o c^2}{\sqrt{1-u^2/c^2}} = \frac{d}{dt} E = \mathbf{u} \cdot \mathbf{F} \ . \tag{446}$$

This is nothing else than the relativistic energy balance (175) of mechanics, here contained as zeroth component in the covariant formulation of the equations of motion:

On the right-hand side of Eq. (446), there appears the work per time, i.e. the power exerted by the force **F** on the particle moving with velocity **u**. It finds its equivalent in the energy change per time, $dE/dt = d\,(mc^2)/dt$, on the left-hand side. In Chap. 7, Sect. 1, we have seen that to each energy E, there corresponds a mass $m = E/c^2$ according to Eq. (186). On the example of the total inelastic impact, we have demonstrated that the rest mass m_o of a system can change due to energy conversions. We shall evaluate this in Problem 25, for the covariant form of relativistic mechanics in MINKOWSKI space.

Now we show that according to the principle (138), in each inertial system, there acts the same three-dimensional force vector **F**, when the quantity F^i represents a four-dimensional MINKOWSKI force vector according to Eq. (445). For simplicity, we restrict the consideration on an one-dimensional motion along the x-axis. Let be $\mathbf{u} = (u, 0, 0)$ a variable velocity of a particle that is imposed to a force $\mathbf{F} = (F, 0, 0)$. The reference systems Σ' moves with constant velocity v in x-direction of Σ_o.

With $\gamma_u = 1/\sqrt{1 - u^2/c^2}$ we write for the MINKOWSKI force vector

$$\mathbf{F} \equiv \left(F^0,\, F^1,\, F^2,\, F^3\right) = \left(\frac{\gamma_u\, u\, F}{c}\,,\ \gamma_u\, F\,,\, 0\,,\, 0\right)\ . \tag{447}$$

The components $F^{i'}$ of the force in reference system Σ' are obtained from a LORENTZ transformation (339), when we replace the coordinates by the components of the four-force. We write γ_v instead of γ and employ EINSTEIN's addition theorem of velocities (106) to get for Eq. (447)

$$\begin{aligned} F^{0'} &= \gamma_v\, F^0 - \frac{\gamma_v\, F^1\, v}{c} = \frac{\gamma_u\, \gamma_v\, F\,(u - v)}{c} = \frac{\gamma_u\, \gamma_v\, u'\, F\,(1 - uv/c^2)}{c}\ , \\ F^{1'} &= \frac{-\gamma_v\, F^0\, v}{c} + \gamma_v\, F^1 = \gamma_u\, \gamma_v\, F\,(1 - uv/c^2)\ . \end{aligned}$$

From Eq. (171), it is

$$\gamma_u\, \gamma_v\left(1 - \frac{u\, v}{c^2}\right) = \gamma_{u'},\ u' = \frac{u - v}{1 - u\, v/c^2}\ ,$$

so that finally we have

$$\left(F^{0'},\, F^{1'},\, F^{2'},\, F^{3'}\right) = \left(\frac{\gamma_{u'}\, u'\, F}{c}\,,\ \gamma_{u'}\, F,\, 0\,,\, 0\right)\ . \tag{448}$$

These are the components of MINKOWSKI's force vector $F^{i'}$ in reference system Σ' for a constant force $\mathbf{F} = (F, 0, 0)$. If, e.g. in Σ_o there is measured a force $\mathbf{F} = 1\,\mathrm{N}$, then one observes also from Σ' a force $\mathbf{F}' = 1\,\mathrm{N}$.

With the energy–mass equivalence (186) and the mass formula (167) we find the four-vector p^i of momentum (436):

$$(p^i) = (m\,c\,,\,m\,u_x\,,\,m\,u_y\,,\,m\,u_z) = \left(\frac{E}{c}\,,\,p_x\,,\,p_y\,,\,p_z\right)\;. \tag{449}$$

Equation (449) represents a remarkable result. The three components of the momentum $\mathbf{p} = m\,\mathbf{u}$ form together with the energy $E = m\,c^2$ divided by c components of a vector in MINKOWSKI space. In classical mechanics, the energy of a particle or a particle system is defined up to an additive constant. The covariant formulation of mechanics however requires that the classical energy constant vanishes, otherwise momentum and energy would not form a four-vector. The four-momentum is also called energy–momentum vector.

The third axiom of mechanics in MINKOWSKI space, s. Eq. (437), therefore contains both the conservation law of energy and of momentum. At transition from one to an other inertial system the energy E/c and the momentum (p_x, p_y, p_z) of a body convert by a LORENTZ transformation as the time coordinate ct and the space coordinates (x, y, z).

For evaluating the invariant $\mathsf{p}^2 := p^i\,p_i$ of the energy–momentum vector p^i from (436), we use the independence of an invariant from the inertial system. Hence, we can evaluate p^2 simply in the rest system of the body writing Eq. (436) with $\mathbf{u} = 0$,

$$\mathbf{p}^2 := p^i\,p^k\,\eta_{ik} = p^i\,p_i = m_o^2\,c^2\;. \tag{450}$$

With (449) this means $E^2/c^2 - p^2 = m_o^2\,c^2$, and we get the relativistic relation between momentum $\mathbf{p}$ and energy E of a body in the form

$$E = \sqrt{p^2\,c^2 + m_o^2\,c^4}\;. \tag{451}$$

With Eq. $\mathbf{p} = m_o\mathbf{u}/\sqrt{1 - u^2/c^2}$ Eq. (451) contains again EINSTEIN's energy–mass equivalence,

$$E = m\,c^2\;. \qquad \text{Energy-mass equivalence} \tag{452}$$

For $v \ll c$, we get the approximation

$$\begin{aligned} E &= \sqrt{m_o^2\,c^4\left(1 + \frac{p^2}{m_o^2\,c^2}\right)} = \sqrt{m_o^2\,c^4\left(1 + \frac{m^2\,u^2}{m_o^2\,c^2}\right)} \\ &\approx \sqrt{m_o^2\,c^4\left(1 + \frac{m_o^2\,u^2}{m_o^2\,c^2}\right)} = \sqrt{m_o^2\,c^4\left(1 + \frac{u^2}{c^2}\right)} \;\approx\; m_o\,c^2\left(1 + \frac{u^2}{2\,c^2}\right)\,, \end{aligned}$$

hence an agreement with our result (188)

$$E \;\approx\; m_o\,c^2 + \frac{1}{2}\,m_o\,u^2\;. \tag{453}$$

Here we add a remark to the notion of inertial mass m of a body. This is understood as ratio $m = F/a$ of the force F to the observed acceleration a. The quantity m_o is uniquely fixed because of the supposed initial velocity $\mathbf{u} = 0$. When the ratio of force and acceleration does not change with the velocity of the body as in classical mechanics, then one can measure the inertial mass for each initial velocity in this way. If the quotient F/a changes with velocity of the body, then it will, in general, also change with the direction of the force. The inertial mass m, defined by the quotient $m = F/a$, will then depend on the direction, whether the body is accelerated parallel or vertical to its current velocity.

To illustrate this behaviour, we consider as example the case that the velocity $\mathbf{u}$ of the body has only a component in x-direction, while its acceleration $\mathbf{a}$ lies in the x-y-plane, $\mathbf{u} = (u_x = u,\ 0,\ 0)$, $\mathbf{a} = (a_x,\ a_y,\ 0)$. From the formula (435) for the four-vector a^i of acceleration, we find for the spatial components

$$\left(\gamma_u\, a_x + \gamma_u^3\, u\, a_x\, ,\ \gamma_u\, a_y\, ,\ 0\right) = \left(\gamma_u^3\, a_x\ ,\ \gamma_u\, a_y\ ,\ 0\right)\ , \tag{454}$$

and we get the relativistic equation of motion (440)

$$\left(\gamma_u^3\, m_o\, a_x\, ,\ \gamma_u\, m_o\, a_y\, ,\ 0\right) = \left(F_x\, ,\ F_y\, ,\ 0\right)\ . \tag{455}$$

This relation was earlier taken as hint to speak of an acceleration in direction of motion, here the x-direction, of a so-called longitudinal mass $m_l = \gamma_u^3\, m_0 = F_x/a_x$, and for acceleration vertical to the direction of motion, here the y-direction, of a so-called transversal mass $m_t = \gamma_u\, m_o = F_y/a_y$. However, these terms have no further meaning for the theory. Important is the mass that determines the momentum of a body, Chap. 6, Sect. 3, Eq. (167), the quantity $\gamma\, m_o$ is also denoted in this connection as momentum mass. This mass also enters into EINSTEIN's energy–mass equivalence (186).

As application of the covariant equations (438) of relativistic mechanics, we now discuss the collision of particles. An interaction of particles, occurring only during a limited and sometimes only a very small time interval δt, is called collision. Besides this interaction interval, the particles are considered as force free, while during the impact only inner forces are supposed. Without more precise knowledge on the physical nature of these interaction forces, we can apply the third axiom (437): The energy–momentum vector P^i of the total systems of particles remains constant during the time of the collision. After the collision, this vector has the same value as before. If we denote the quantities after the collision with an overline, then this property can be written as

$$P^i := \sum_{p=1}^{n} p_p^i = \sum_{p=1}^{n} \overline{p_p^i} := \overline{P^i}\,. \qquad \text{Energy–momentum conservation at collision} \tag{456}$$

Here we have considered n particles with energy–momentum vectors p_p^i, P^i for the initial state before the collision, and $\overline{p_p^i}$, $\overline{P^i}$ for the final state after the impact. Independent on the number of involved particles, Eq. (456) delivers four conditions on the collision process. A collision is called elastic, when the rest masses of the particles do not change. In other cases, we speak of an inelastic collision. A particular case concerns the creation or destruction of particles: The collision is always inelastic, when the number of particle after the impact differs from the number before. The central elastic collision of two particles with rest masses m_{o1} and m_{o2} are considered in Problem 28, two examples of an inelastic collision are discussed in Problems 25 and 26.

3 Electrodynamics—Covariant Formulation

An asymmetry in the classical formulation of electrodynamics led EINSTEIN (1905) (notice the translation in LORENTZ 1923, 1952) to a solution of the relativity problem. In his famous paper from 1905 'On the Electrodynamics of Moving Bodies', the historical origin of the revolution of all theoretical physics by the Special Relativity Theory, he directed the attention in the introduction on this fact:

"... Take, for example, the reciprocal electrodynamic action of a magnet and a conductor. The observable phenomenon here depends only on the relative motion of the conductor and the magnet, whereas the customary view draws a sharp distinction between the two cases in which either the one or the other of these bodies is in motion. For if the magnet is in motion and the conductor at rest, there arises in the neighbourhood of the magnet an electric field with a certain definite energy, producing a current at the places where parts of the conductor are situated. But if the magnet is stationary and the conductor in motion, no electric field arises in the neighbourhood of the magnet. In the conductor, however, we find an electromotive force, to which in itself there is no corresponding energy, but which gives rise—assuming equality of relative motion in the two cases discussed—the electric currents of the same path and intensity as those produced by the electric forces in the former case."

A principle conviction let EINSTEIN to the formulation of his Special Relativity Theory:

MAXWELL's equations of electrodynamics are valid without correction in each inertial system.

All further conclusions are a logical consequence. The propagation of a light wave represents an electromagnetic phenomenon. The speed of light c must therefore have in all inertial systems the same value, as EINSTEIN writes in the second part of his relativity principle, s. Chap. 2, Sect. 1. The light propagation is described by D'ALEMBERT's wave equation (306). This equation would have been correct in case of the validity of the GALILEI transformation at most in *one* distinguished reference system, cp. (308). In consequence, the GALILEI transformation cannot be true under all circumstances. One has to look for such transformations that guarantee the invariance of the wave equation (306). This exposition of the theory we have chosen in Sect. 1.

Now we have 'only' to show, how the classical MAXWELL equations can be written as tensor equations in MINKOWSKI space. With the forces on electrically charged particles, we become familiar simultaneously with an important example of the MINKOWSKI forces, cp. Eqs. (437) and (445).

At first, we shall present the traditional formulation of MAXWELL's theory with three-dimensional vectors.

3.1 MAXWELL*'s Theory*

Initially, we consider a distinguished inertial system $\Sigma_o(x, y, z, t)$, where our laboratory is at rest.

The theory of electromagnetism deals with electric charges and electric currents, and with the connected fields. The electromagnetic field is characterised by four field vectors, where each two are connected, the electric field strength **E** and the magnetic induction **B** on the one side, which are connected with the forces acting on charges in electromagnetic fields, and on the other side, the electric (also dielectric) flux density **D** and the magnetic field strength **H** providing the connection of fields with the electric charges and the electric currents.[7] Vector fields can be represented by oriented *field lines* in space or on a plane. The magnitude of the vector field is encoded in the separation of the draws lines in such a way that the density of the lines is a measure for the magnitude of the corresponding field.

For the formulation of MAXWELL's theory, it is quite important to choose the correct units. In the SI-units, we used as basis the metre (m), the second (s) and the kilogram (kg). The Newton (N) is then a secondary SI-unit with $1\ \mathrm{N} = 1\ \mathrm{m\,kg\,s^{-2}}$ introduced so that the proportionality constant in the equations of motion (139) just takes the value $k = 1$. The Newton is used also in the so-called absolute unit system, when the metre and the kilogram instead of centimetre and gram are used.[8] In this way, it is possible to reduce the number of basic units by means of physical laws.

The SI-system introduces an additional basic unit for measurements in electrodynamics, the unit Ampere for the strength of electric currents. The units of all other electromagnetic quantities are reduced on the Ampere.

Basis for definition of the unit of the electic current is AMPÈRE's law for the force per length unit $\Delta F/\Delta L$ between two straight parallel wires, which carry currents J_1 and J_2 and have a separation r that follows from OERSTEDT's law (493) and the expression of the LORENTZ force in the form of Eq. (474),

---

[7]The identic notion of electric *field strength* for **E** and magnetic *field strength* for **H** originates from a historically wrong understanding of the magnetic field quantities. Not **E** and **H**, but **E** and **B** are both physically and mathematically related. A relict from this misunderstanding is preserved in the definition of permeability. We write $\mathbf{H} = (1/\mu)\,\mathbf{B}$, but $\mathbf{D} = \varepsilon\,\mathbf{E}$, s. Eqs. (487), (491), (497) and (498).

[8]As unit for a certain amount of matter we have also introduced the mole as SI-basic unit, cp. footnote 2 of Chap. 6 and Problem 15.

$$\frac{\Delta F}{\Delta L} = \frac{\mu_o}{2\pi} \frac{J_1 J_2}{r} , \qquad \text{1. AMPÈRE's law} \qquad (457)$$

cp. also Halliday, Resnick and Walker (2008), p. 771, Eq. (29–31), and Becker (1982), p. 163, Fig. 38.

The proportionality factor μ_o in this equation is called *magnetic field constant* or also permeability of the vacuum. In the SI-system, we define

Currents of one Ampere (A) flow in two straight parallel conductors at a separation of one metre, when the force per length unit $\Delta F/\Delta L$ between them amounts $2 \cdot 10^{-7}$ Newton per metre.

Equivalent to this definition is the fixing of the numerical value and the unit of μ_o :

$$\mu_o = 4\pi \cdot 10^{-7} \frac{\text{N}}{\text{A}^2} . \qquad \text{Magnetic field constant} \qquad (458)$$

With this definition, we must accept a general asymmetry in the further exposition of electrodynamics that results from the use of SI-units. Advantageously for the SI-system is truly the introduction of different quantities for both vectors **E** and **B** on the one side, and **D** and **H** on the other one, also visible in vacuum. Since today it is in principle obligatory to use the SI-system, we follow this convention.

In books, especially on theoretical physics, and also in older presentations, the use of the so-called absolute unit system has been shown to be reasonable. There, all units, both for mechanics and for electrodynamics, are reduced to three basic units, length, mass and time. The older cgs-system, also called after GAUSS, uses the centimetre (cm), the gram (g) and the second (s), and then a derived quantity follows as unit of force, the dyn = 10^{-5} Newton. In the following, we shall refer sometimes to the modern absolute unit system with basic units metre (m), kilogram (kg) and second (s) that again leads to the force unit Newton as the SI-system. For MAXWELL's vacuum equations, and especially for its presentation in MINKOWSKI space, the symmetry of the theory is more obvious in the absolute unit system. For the theoretically interested reader, we summarise the most important equations of electrodynamics in the absolute unit system in Sect. 3.3. In particular, there in Eq. (581) we shall provide the conversion between both unit systems of electromagnetic quantities.

3.1.1 Charges and Currents—Continuity Equation

With the unit of the electric current one defines the unit of charge:

The unit of charge of one Coulomb (C) is defined by a current
of one Ampere during one second, one Coulomb := one Ampere second.
$1\,\text{C} := 1\,\text{A}\,\text{s}$.

The electric charge e of a body or particle is written as volume integral over the spatial charge density $\rho = de/dV$,

$$e = \iiint \rho\, dxdydz\,. \tag{459}$$

There the charge density ρ and with it the charge e can be positive or negative.

Charged bodies of density ρ_u in space (x, y, z) moving with velocity $\mathbf{u} = \mathbf{u}(x, y, z)$ produce consequently a vector $\mathbf{j} = (j_x, j_y, j_z)$, the density of the electric current:

The quantity $j_x\, dydzdt$ equals the amount of charges de flowing during time dt through the orthogonal to the x-axis lying area element $dydz$. Respective relations hold true for the both other components,

$$(j_x\, dt, j_y\, dt, j_z\, dt) = \frac{de}{dxdydz}(dx, dy, dz)\ . \tag{460}$$

With the velocity $\mathbf{u} = (dx/dt, dy/dt, dz/dt)$ of the charge carrier, we get for the current density

$$\mathbf{j} = \rho_u\, \mathbf{u}\ . \tag{461}$$

Such a current is also called *convection current*.

The case of an electric current, e.g. through a metallic wire, is characterised by freely moving negatively charged, so-called conduction electrons with a charge density ρ_- leading to an *conduction current* $\mathbf{j} = \rho_-\, \mathbf{u}$. The negative charge density ρ_- is everywhere compensated by a charge density ρ_+, so that the total charge density ρ is always zero, $\rho = \rho_- + \rho_+ = 0$. There is flowing a current, but the wire is uncharged. Due to the theoretical background of charged electrons, we will in the following always consider convection currents (461). However, we remark already at this stage that this assumption cannot be applied to the inner angular momentum, the so-called spin of charged particles. The magnetic field of a spinning electron can only be understood in quantum theory and not from the classical picture of a current density from the rotation of charges. Here the classical electrodynamics meets its boundaries. These connections are discussed in Chap. 10.

Through the finite cross section S there might be flowing a total current $J = de/dt$, given by the charges running the per second through the cross section S. It is given by the area integral

$$J = \frac{de}{dt} = \iint\limits_S \mathbf{j} \cdot d\mathbf{S}\ . \tag{462}$$

There it holds Eqs. (459) and (460), and the charge density ρ flows with velocity $\mathbf{u}$. An electric field $\mathbf{E}$ produces, in general, in a medium a current density $\mathbf{j}$.

If the relation between **E** and **j** is linear, one tells the relation OHM's law,

$$\mathbf{j} = \sigma\,\mathbf{E}\ . \qquad \text{OHM's law} \qquad (463)$$

The *conductivity* σ of the medium is a property of the material. The conductivity of vacuum is zero.

We consider the simplest case of a constant electric field **E** in direction of a homogeneous wire of length L with constant current j and cross section S, then integration of (463) over the length L of the wire and its cross section S provides with (462)

$$\int_0^L dl \iint_S \mathbf{j}\cdot d\mathbf{S} = \int_0^L dl \iint_S \sigma\,\mathbf{E}\cdot d\mathbf{S} \quad\longrightarrow\quad J\,L = \sigma\,S\,E\,L\ .$$

We get OHM's law in the known form with OHM's resistance $R = L/(S\,\sigma)$ and the voltage $U = EL$

$$U = R\,J\ . \qquad \text{OHM's law} \qquad (464)$$

The unit of resistance of one Ohm (Ω) is defined for a charge conductor, when a current of one Ampere through the conductor results from a voltage of one Volt (V), 1 Ohm = 1 Volt per Ampere.
1 Ω = 1 V/A.

The definition of the unit of one Volt for voltage is given later following Eq. (479).

In a medium, there exist also other causes for electric current densities **j** occurring at contact of metals (in a so-called galvanic chain), the contact of a metal with an electrolyte or else for a concentration gradient in an electrolyte. One called these causes imprinted electric forces, but we will not treat them here explicitly.

Electric currents, running through thin wires, are described by singular current densities with δ-functions, s. Appendix B.3. For example, for a vector $\mathbf{J} = (0, 0, J_o)$ of total electric current J_o, running along the z-axis, the current density is $\mathbf{j} = \big(0, 0, J_o\,\delta(x)\,\delta(y)\big)$, s. Eq. (1731).

If a single point-like particle of charge e_o moves along $\big(x_o = \xi(t), y_o = \eta(t), z_o = \zeta(t)\big)$ with velocity $\mathbf{u} = \big(d\xi/dt, d\eta/dt, d\zeta/dt\big)$, so its singular charge density is $\rho = e_o\delta\big(x - \xi(t)\big)\delta\big(y - \eta(t)\big)\delta\big(z - \zeta(t)\big)$ due to Eq. (1729), and with Eq. (461) we have a singular current density

$$\mathbf{j} = e_o\,\delta\big(x - \xi(t)\big)\,\delta\big(y - \eta(t)\big)\,\delta\big(z - \zeta(t)\big)\left(\frac{d\xi}{dt}, \frac{d\eta}{dt}, \frac{d\zeta}{dt}\right)\ . \qquad \text{Moving point charge} \qquad (465)$$

Now we consider a finite volume K that contains a certain amount of charge carriers. Due to its motion, the volume K changes its position and shape during time, hence $K = K(t)$. In K, the number of positive and the number of negatively charged particles can change, but not the sum over all charges. The total charge e in the volume K remains constant in time,

$$e = \iiint\limits_{K(t)} \rho\, dxdydz = \text{const} . \qquad \text{Charge conservation} \quad (466)$$

If we have point-like charges with charge density described by δ-functions, then integration of Eq. (466) provides

$$\sum e_p = \text{const} . \qquad \text{Charge conservation} \quad (467)$$

From Eq. (466), it follows

$$\frac{de}{dt} = \frac{d}{dt} \iiint\limits_{K} \rho\, dxdydz = 0 . \tag{468}$$

Here we only consider one kind of charge carrier with velocity $\mathbf{u}$. The surface ∂K of volume K is also in motion and changes with the variable velocity $\mathbf{u} = (dx/dt, dy/dt, dz/dt)$ of the charges. The derivative of the volume integral (468) for variable integration region K follows from Eq. (1711) in Appendix B.2,

$$\frac{de}{dt} = \iiint\limits_{K(t)} \frac{\partial}{\partial t}\rho(x, y, t)\, dxdydz + \iint\limits_{\partial K(t))} \rho(x, y, z, t)\, \mathbf{u} \cdot d\mathbf{S} = 0 .$$

We use Gauss's law (1703), for the surface integral,

$$\begin{aligned} \frac{de}{dt} &= \iiint\limits_{K(t)} \frac{\partial}{\partial t}\rho(x, y, t)\, dxdydz + \iiint\limits_{K(t)} \operatorname{div}(\rho\mathbf{u})\ dxdydz \\ &= \iiint\limits_{K(t)} \left[\frac{\partial}{\partial t}\rho + \operatorname{div}(\rho\mathbf{u})\right] dxdydz = 0 . \end{aligned}$$

This equation holds true for an arbitrary integration volume $K(t)$, and therefore, it can only be fulfilled, when the integrand vanishes. Therefore, we can write the continuity equation for ρ and $\mathbf{j} = \rho\mathbf{u}$,

$$\frac{\partial}{\partial t}\rho + \operatorname{div}\mathbf{j} = 0 . \qquad \text{Continuity equation} \quad (469)$$

If different charge carriers are included, then $\mathbf{j}$ represents the total current density. Equation (469) is the differential expression for charge conservation (466).

3.1.2 LORENTZ Force

The electric field strength **E** and the magnetic induction **B** describe the forces of electromagnetic fields. By means of these forces, one fixes the units for **E** and **B** in the SI-system[9]:

The unit of the electric field strength **E**, one Volt per metre (V/m) is defined by the force of one Newton, acting on a unit charge of one Coulomb,
one Volt per metre = one Newton per Coulomb.
$1\,\text{V/m} = 1\,\text{N/C}$.
The unit of magnetic induction **B** of one Tesla (T) is defined by the force of one Newton acting on the unit charge of one Coulomb that moves with a velocity of one metre per second,
one Tesla = one Volt second per square metre.
$1\,\text{T} = 1\,\text{V s/m}^2$.

With these definitions, we have constants of unity in the force laws that can now be formulated:

On a spatial charge density ρ and an initially independent current density **j** acts a LORENTZ force density **f**,

$$\mathbf{f} = \rho\mathbf{E} + \mathbf{j} \times \mathbf{B} \ , \qquad \text{LORENTZ force density} \qquad (470)$$

or in components, cp. Eq. (1645),

$$f_i = \rho\, E_i + \epsilon_{ikl}\, j_k\, B_l \ . \qquad \text{LORENTZ force density} \qquad (471)$$

Here we consider convection currents of charges of density ρ with the velocity field **u** and we get instead of (470)

$$\mathbf{f} = \rho\big(\mathbf{E} + \mathbf{u} \times \mathbf{B}\big) \ . \qquad \text{LORENTZ force density} \qquad (472)$$

In electrodynamics of media, Sect. 3.2.3, we also consider conduction currents.

In Eq. (472), we now take point-like charge distributions, e.g. an electron of a charge e_o at a point $P(\xi, \eta, \zeta)$ with a charge density described by $\rho = e_o\, \delta(x - \xi)\, \delta(y - \eta)\, \delta(z - \zeta)$. The force $\mathbf{F} = \iiint \mathbf{f} dxdydz$ on the charge e_o is obtained from Eq. (470) by integration over a volume containing the point P,

$$\mathbf{F} = e_o\, (\mathbf{E} + \mathbf{u} \times \mathbf{B}) \ . \qquad \text{LORENTZ force} \qquad (473)$$

Equation (473) follows from (470), if one assumes that the fields **E** and **B** do not change over the extension of the charge.

[9]In GAUSS' cgs-System the unit for the magnetic induction is one Gauß = 10^{-4} Tesla.

All fields have to be taken at the point $P(\xi, \eta, \zeta)$ of the charge e_o.

If only a line current exists, e.g. with $\mathbf{j} = \big(0, 0, J_o\,\delta(x)\,\delta(y)\big)$ and $\mathbf{f} = \mathbf{j} \times \mathbf{B}$, then it follows by integration of the force density $\mathbf{f}$ over a cylinder of length ΔL surrounding the current

$$\Delta\,\mathbf{F} = \big(\mathbf{J} \times \mathbf{B}\big)\,\Delta L\ . \qquad \text{LORENTZ force} \qquad (474)$$

If significant field gradients or continuously distributed charges are present, then we have to consider the LORENTZ force density (470) in connection with the equations of motion of a mechanical continuum. Then NEWTON's second axiom (140) or the corresponding relativistic equations (168) or (438) for the temporal change of momentum $\mathbf{p}$ of a body must be transformed into the temporal change of the momentum density $\mathbf{g} = \Delta\mathbf{p}/\Delta V$ in the continuum at a given point in space, s. Problem 38.

The LORENTZ force (473) can be used in Eq. (140) of NEWTON's point mechanics or in the first Eq. (168) of relativistic point mechanics for determining the motion of charged particles in electromagnetic fields,

$$\frac{d}{dt}\,\frac{m_o\mathbf{u}}{\sqrt{1-u^2/c^2}} = e_o\,(\mathbf{E} + \mathbf{u} \times \mathbf{B}) + \mathbf{F}_e\ . \qquad (475)$$

Here we have admitted the action of other forces $\mathbf{F}_e$ as friction or gravitational forces and others acting on a mass m_o besides the LORENTZ force.

3.1.3 Induction Flux and Induction Law

The induction flux Φ through an area S is, according to definition (1698), the flux of the vector $\mathbf{B}$ through this area,

$$\Phi := \iint_S \mathbf{B} \cdot d\mathbf{S}\ . \qquad \text{Induction flux} \qquad (476)$$

It is a basic experimental experience that there do not exist point sources for field lines of magnetic induction $\mathbf{B}$. This means: no magnetic monopoles could be detected in experiments.[10] We consider some closed area ∂K in the space that surrounds the volume K, Appendix B.2. Each field line of $\mathbf{B}$, going into this area, must also leave this area, i.e. the induction flux through a closed area must vanish,

[10]In hypothetical theories attempting to unify the electromagnetic interaction with other fundamental interactions, the possible existence of magnetic monopoles becomes again a theoretical concept. One has been looking for magnetic monopoles in cosmic rays. In accelerator experiments, the very high energies may come into the range for producing magnetic monopoles. Its inclusion into the theory of electromagnetism is quite straight forward. A hypothetical density of magnetic monopoles ϱ_m and a corresponding current density ς_m could be introduced with replacements of Eqs. (478) and (483) by $\operatorname{div}\mathbf{B} = \varrho_m$ and $\operatorname{rot}\mathbf{E} + \dfrac{\partial\mathbf{B}}{\partial t} = -\varsigma_m$. More changes are not necessary.

$$\iint\limits_{\partial K} \mathbf{B} \cdot d\mathbf{S} = 0 \ . \qquad \text{Sourceless magnetic induction} \qquad (477)$$

We apply in Eq. (477) Gauss law (1703),

$$\iint\limits_{\partial K} \mathbf{B} \cdot d\mathbf{S} = \iiint\limits_{K} \operatorname{div} \mathbf{B} \, dxdydz = 0 \ .$$

This equation holds true for an arbitrary volume K only if the integrand vanishes,

$$\operatorname{div} \mathbf{B} = 0 \ . \qquad \text{Sourceless magnetic induction} \qquad (478)$$

The next experience concerns the induction law. The curve integral of the second kind, Appendix B.2, Eqs. (1685) and (1686), performed over the electric field strength $\mathbf{E}$ is called voltage U. More precisely, let $x = x(t)$, $y = y(t)$, $z = z(t)$, $t_1 \leq t \leq t_2$, be a finite curve in space. The voltage along this curve U_{12} is given by the following integral:

Voltage U_{12} between the end points $t_1 \leq t \leq t_2$ of a curve $\mathbf{x} = \mathbf{x}(t)$:

$$U_{12} = \int\limits_{t_1}^{t_2} \left[E_x\big(x(t), y(t), z(t)\big)\frac{dx}{dt} + E_y\big(x(t), y(t), z(t)\big)\frac{dy}{dt} + E_z\big(x(t), y(t), z(t)\big)\frac{dz}{dt} \right] dt. \qquad (479)$$

The unit of voltage, one Volt (V), is given by the voltage of a constant electric field strength of one Volt per metre in direction of the curve, in a separation of one metre,
$1\,\mathrm{V} = 1\,\mathrm{m}\,1\,\mathrm{V/m} = \mathrm{N\,m/(A\,s)}$.

For a closed curve, we are concerned with a ring voltage,

$$U = \oint E_x dx + E_y dy + E_z dz = \oint \mathbf{E} \cdot d\mathbf{s} \ . \qquad (480)$$

Let ∂S represent the closed boundary curve of an area S. The induction law then reads: the temporal change of the induction flux Φ through the area S creates an electric ring voltage called induction voltage U along the closed boundary curve ∂S. Faraday found in 1831 this induction law:

$$U = -\frac{d\Phi}{dt} \ . \qquad \text{Faraday's induction law} \qquad (481)$$

Written more in detail this means

$$\oint_{\partial S} \mathbf{E} \cdot d\mathbf{s} = -\frac{d}{dt} \iint_S \mathbf{B} \cdot d\mathbf{S} \ . \qquad \text{Induction law} \qquad (482)$$

On the left-hand side, we use STOKES's integral law (1701). The time derivative can be performed under the integral sign when the area S is time independent, fixed in space, and we get

$$\iint_S \text{rot}\mathbf{E} \cdot d\mathbf{S} = -\iint_S \left(\frac{\partial}{\partial t} \mathbf{B} \right) \cdot d\mathbf{S} \ ,$$

hence

$$\iint_S \left(\text{rot}\mathbf{E} + \frac{\partial}{\partial t} \mathbf{B} \right) \cdot d\mathbf{S} = 0 \ .$$

This equation can be true for an arbitrary area S only in case that the integrand vanishes, and we get the differential form of the induction law,

$$\text{rot}\mathbf{E} + \frac{\partial \mathbf{B}}{\partial t} = 0 \ . \qquad \text{Induction law} \qquad (483)$$

We put a conductor, e.g. a wire, along the closed space curve ∂S. Then we can measure this induction voltage, or if the conductor is closed and it has OHM's electric resistance R, we can measure an induction current $J = U/R$ due to the changing induction flux Φ through the area S fixed in space. The induction law has been experimentally verified in this way. In practice, one moves a permanent magnet against the wire loop fixed in space. The induction flux Φ through the loop grows with the approaching magnet. According to Eq. (483), in space there is an excited electric field strength with a non-vanishing ring voltage, measurable as induction voltage U.

If inversely the wire loop is moving against the magnet, which is now fixed in space, then the magnetic induction $\mathbf{B}$ in space is temporally constant, and there is not created an electic field in space according to the induction law, and no induction voltage. However, the experiment shows a voltage at the ends of the wire of the same amount as in the first experiment. We get the asymmetry in the theoretical explanation of an obviously completely symmetric observation cited at the beginning of this section. We can explain the measured voltage in the second case not with the induction law but with the LORENTZ force.

If we move the wire loop in the field $\mathbf{B}$ with a velocity $\mathbf{u}$, so the free electrons with charge e_o in the wire feel a force (473) $e_o \mathbf{u} \times \mathbf{B}$. According to Eq. (475), the electrons come in motion, as if they are accelerated in the wire by an effective electric field strength $\mathbf{E}^{eff} := \mathbf{u} \times \mathbf{B}$. In the wire with OHM's resistance R, there would be an exited ring current J, without any electric field in space. The current follows from the effective ring voltage U^{eff}, and hence from the line integral over the effective field strength $\mathbf{E}^{eff}$ along the closed wire,

$$J = \frac{1}{R} U^{eff} = \frac{1}{R} \oint_{\partial S} \mathbf{E}^{eff} \cdot d\mathbf{s} \ . \tag{484}$$

During time dt the elements $d\mathbf{s}$ of the conductor loop move by $\mathbf{u}\,dt$, and fill the surface of the cylinder M with bottom and upper area S. The area element $d\mathbf{S}$ of the surface of the cylinder is $d\mathbf{S} = d\mathbf{s} \times \mathbf{u}\,dt$. Then using the identity $\mathbf{a} \cdot (\mathbf{b} \times \mathbf{c}) = \mathbf{b} \cdot (\mathbf{c} \times \mathbf{a})$,

$$dt\, U^{eff} = dt \oint_{\partial S} \mathbf{E}^{eff} \cdot d\mathbf{s} = dt \oint_{\partial S} (\mathbf{u} \times \mathbf{B}) \cdot d\mathbf{s} = \oint_{\partial S} (d\mathbf{s} \times \mathbf{u}dt) \cdot \mathbf{B} = \iint_M \mathbf{B} \cdot d\mathbf{S} = \Phi_M .$$

Now because of Eq. (477), the total induction flux Φ_Z through the cylinder must vanish, and hence $\Phi_Z = \Phi_M + \Phi_2 - \Phi_1 = 0$, where Φ_1 and Φ_2 are the fluxes going from the cylinder through the bottom and upper areas. The change $d\Phi$ of the flux (481) through the area S is then

$$d\Phi = (\Phi_2 - \Phi_1) = -\Phi_M = -U^{eff}\, dt \ ,$$

and therefore

$$U^{eff} = -\frac{d\Phi}{dt} \ . \tag{485}$$

In a temporal fixed and spatially variable field $\mathbf{B}$, such a current is excited by the LORENTZ force on the free charges that is identical to the current produced according to the induction law for a magnet in motion and a wire loop at rest. The clarification of this strange asymmetric interpretation of an experimental completely symmetric effect was the deciding stimulus for EINSTEIN in creating his Special Relativity Theory.

3.1.4 Electric Displacement and Magnetic Field

The vectors of the electric (also dielectric) displacement field $\mathbf{D}$ and the magnetic field $\mathbf{H}$ are defined by the sources of the electromagnetic field. In the SI-system, we distinguish in vacuum the vectors $\mathbf{D}$ and $\mathbf{E}$ on the one hand, and $\mathbf{H}$ and $\mathbf{B}$ on the other hand.

(a) We consider at first fields in vacuum. The area ∂K might be the boundary of the volume K containing electric charges with a charge density ρ. In the SI-units, the vector of dielectric displacement field $\mathbf{D}$ is directly defined by the charge density, so that the flux Φ_D of the vector $\mathbf{D}$ through the closed area ∂K is equal to the included charges. Therefore, $\mathbf{D}$ is also called electric flux density,

$$\Phi_D \equiv \iint\limits_{\partial K} \mathbf{D} \cdot d\mathbf{S} := \iiint\limits_{K} \rho\, dxdydz = \sum e_p \,. \qquad \text{Flux of electric displacement vector} \qquad (486)$$

This relation fixes the unit of the vector **D** as seen in this example:

On a metal ball of radius r, we insert a charge e that is distributed uniformly. Due to symmetry, we get a spherically symmetric field. We surround the ball with a concentric sphere K of radius R. We write D for the amount of the displacement field **D**, then the flux Φ_D through the sphere $4\pi R^2$ is given by

$$\Phi_D = 4\pi R^2 D = e \,.$$

The displacement D has therefore the SI-unit Coulomb per square metre, and it is 1 C/m^2 on a surface of a sphere of radius $R = 1$ m, when the included charge is $1/(4\pi)$ Coulomb.[11] A special name for this unit is not in use.

The unit for the displacement density **D** is one Coulomb per square metre (C/m^2).

Now we write still in the vacuum case

$$\mathbf{D} = \varepsilon_o \mathbf{E} \,. \qquad (487)$$

Now the units for the electric field strength **E** and the electric displacement **D** are fixed. The dielectricity constant of vacuum, also called *electric field constant* (or vacuum permittivity) ε_o, becomes in the SI-system an experimentally measurable quantity quite different from the corresponding magnetic field constant μ_o, s. Eq. (458), fixed per definition. This role of μ_o describes the already mentioned asymmetry of quantities in the SI-system that becomes especially obvious in the four-dimensional formalism. The experimental value of ε_o is given by

$$\varepsilon_o = 8.85418782 \cdot 10^{-12} \left[\frac{\mathrm{A\,s}}{\mathrm{V\,m}}\right] . \qquad \text{Electric field constant} \qquad (488)$$

We show how ε_o can be determined from a simple measurement of a capacity:

We consider a plane of a plate capacitor (without dielectric material in between) of capacity $C = Q/U$ with plates of area F and separation d. We neglect boundary effects at the capacitor and assume a constant area charge density ω. The fields have only normal components E_n and D_n assumed to be constant. Equation (479) gets simplified, reading $U = E_n\, d$. Equation (486) becomes $\iint\limits_{\partial K} \mathbf{D} \cdot d\mathbf{S} = D_n\, F = Q$, and Eq. (487) reads $D_n = \varepsilon_o\, E_n$. Then

[11] For avoiding the factor $1/(4\pi)$ in the absolute unit system one defines instead of Eq. (486) $\tilde{\Phi}_D \equiv \iint\limits_{\partial K} \tilde{\mathbf{D}} \cdot d\mathbf{S} := 4\pi \iiint\limits_{K} \tilde{\rho}\, dxdydz = 4\pi \sum \tilde{e}_i$. Here we have superscribed all quantities with a tilde. The conversion between both unit systems is given in Eqs. (581).

$$C = \frac{Q}{U} = \frac{D_n\,F}{E_n\,d} = \frac{\varepsilon_o\,E_n\,F}{E_n\,d}\left[\frac{A\,s}{V}\right]$$

leads to

$$\varepsilon_o = \frac{d}{F}\,\mathrm{C}\left[\frac{A\,s}{V\,m}\right], \tag{489}$$

the field constant ε_o can be measured with the capacitor.

We shall see below that a measurement of ε_o is equivalent to the determination of the speed of light c. The measurement of ε_o is therefore a physically remarkable method for determining the speed of light, but until now, it cannot exceed the precision of standard methods.

(b) Now we consider a medium, resting *as a whole* in our reference system Σ_o. We shall get rid of this restriction, when we treat the covariant formulation of electrodynamics in moving media, s. Sect. 3.2.3, which gives impressive conclusions.

In the medium, denoted dielectricum in this connection, the vector **D** is defined, so that Eq. (486) is preserved. The electric charges within the medium $\iiint_K \rho\, dxdydz$ are then denoted as true or free charges.

On the left-hand side of Eq. (486), we use GAUSS's law (1703). It follows

$$\iiint_K \left(\operatorname{div}\mathbf{D} - \rho\right) dxdydz = 0\ .$$

This equation, valid for an arbitrary volume K, can only be true, if the integrand vanishes, and hence the displacement vector **D** fulfils

$$\operatorname{div}\mathbf{D} = \rho\ . \tag{490}$$

Equation (490) is valid for the true charge density ρ both in vacuum and in a medium. We compare the vacuum with a dielectric neutral medium. Then in the dielectric material there will be excited a polarisation that partly compensates the genuine charges, so that the forces in the medium caused by the vector **E** become reduced with respect to the forces in vacuum. The vector **D** in the dielectric medium is then defined so that it coincides with the vector $\varepsilon_o\,\mathbf{E}^v$. One writes

$$\left.\begin{aligned} \mathbf{D} &= \varepsilon\,\mathbf{E}\ , \\ \varepsilon &:= \varepsilon_o\,\varepsilon_r\ . \end{aligned}\right\} \tag{491}$$

Here is $\varepsilon_r\,\mathbf{E} = \mathbf{E}^v$. The quantity $\varepsilon = \varepsilon_o\,\varepsilon_r$ is called dielectricity constant of the medium. The relative dielectricity constant ε_r is dimensionless due to Eqs. (487) and (491), and it is therefore independent from the used units. ε_r has the value unity in vacuum, and it is for most media not larger than 100. Water has a value 81.

Equation (491) is now supplemented by the electric polarisation **P** and the dimensionless dielectric susceptibility $\chi = (\varepsilon_r - 1)$

$$\left.\begin{array}{l}\mathbf{D} = \varepsilon_o \mathbf{E} + \mathbf{P} \,, \\ \mathbf{P} = \varepsilon_o \chi \mathbf{E} = \varepsilon_o(\varepsilon_r - 1)\, \mathbf{E}\end{array}\right\} . \tag{492}$$

In mono-crystals, the quantity ε will become a tensor. Then the connection (491) between vectors **D** and **E** will become a tensor equation $D_i = \sum \varepsilon_{ik} E_k$. In general, the directions of both vectors **D** and **E** will not coincide.

The formation of a magnetic field **H** by a stationary current distribution, described by the current density **j**, is given by OERSTEDT's law. It states that the line integral of **H** over a closed curve in vacuum ∂S is proportional to the total current J, flowing through the area S,

$$\oint_{\partial S} \mathbf{H} \cdot d\mathbf{s} \sim \iint_S \mathbf{j} \cdot d\mathbf{S} \quad \longrightarrow \quad \oint_{\partial S} \mathbf{H} \cdot d\mathbf{s} = \alpha \iint_S \mathbf{j} \cdot d\mathbf{S} = \alpha\, J \,. \qquad \text{2. AMPÈRE's law} \tag{493}$$

cp. Becker (1982), Part C, p. 172, Eq. (46.1), as well as Halliday et al. (2008), p. 772, Eq. (29.14). The closed-loop integral over the magnetic field is also called magnetomotive flux. The connection (493) is then also called *law of magnetomotive flux* or 2. AMPÈRE's law, cp. Sands, Feynman, Leighton (1970). Sometimes, this law is also called OERSTEDT law.

In SI-system, the unit of magnetic field **H** is fixed by the definition $\alpha = 1$,

$$\oint_{\partial S} \mathbf{H} \cdot d\mathbf{s} = \iint_S \mathbf{j} \cdot d\mathbf{S} = J \,. \qquad \text{2. AMPÈRE's law} \tag{494}$$

Then the unit for the magnetic field **H** is fixed,

The unit of magnetic field **H** is Ampere per metre (A/m).

There is no specific name for this unit.[12] We read off from Eq. (493), that all electric currents, each motion of electric charges is a source of magnetism.

One has to remark that a deeper understanding of magnetism, especially of ferromagnetism, requires the introduction of elementary magnetic moments that cannot be reduced on the classical motion of electric charges. The magnetic moment of the electron, representing an elementary source of magnetism, can be understood only in quantum theory. This will be discussed in Chap. 10, Sect. 5, while studying the DIRAC equation.

For illustrating the magnetic field of $H = 1$ (A/m), we consider the limiting case of an infinite long straight wire along the z-axis carrying a current J in z-direction with current density $\mathbf{j} = \big(0, 0, J\, \delta(x)\, \delta(y)\big)$. As can be easily seen, we fulfil Eq. (493) by

[12]In the cgs-system, the unit of magnetic field is Oerstedt, $1\,\text{Oe} = (1/4\pi) \cdot 10^3\,\text{A/m}$.

$$(H_x, H_y, H_z) = \frac{J}{2\pi\rho}\left(-\frac{y}{\rho}, \frac{x}{\rho}, 0\right), \quad \rho^2 = x^2 + y^2 \ . \tag{495}$$

The field surrounds the current with circular field lines, at separation ρ, it has an amount H given by,

$$H = \frac{J}{2\pi\,\rho} \ . \tag{496}$$

This means 2π Ampere in 1 m distance will create a magnetic field $H = 1$ (A/m). Now we have in vacuum for the magnetic field $\mathbf{H}$,

$$\mathbf{H} = \frac{1}{\mu_o}\mathbf{B} \ . \tag{497}$$

The magnetic field constant μ_o, the permeability of vacuum, is already fixed in Eq. (458), where we showed that the definition leads to

$$\mu_o = 4\pi \cdot 10^{-7}\left[\frac{\mathrm{V\,s}}{\mathrm{A\,m}}\right] ,$$

what determines the SI-basic unit Ampere.

With Eq. (474) the force per length unit $\Delta F/\Delta L$ between two straight currents parallel to the z-axis at separation r has the amount

$$\frac{\Delta F}{\Delta L} = J\,\mu_o\,H = \frac{\mu_o}{2\pi}\frac{J^2}{r} \ .$$

So we rediscovered 1. AMPÈRE's law (457).

Now we insert a medium into the magnetic field, where we assume that the medium is at rest *as a whole* in Σ_o. The vector $\mathbf{H}$ is then defined in such a way that Eq. (493) remains unchanged. In the medium, one expects processes of magnetisation, so that the vector $\mathbf{B}$, determining magnetic forces, will be changed with respect to the vacuum case given by B^v. The vector $\mathbf{H}$ inside the medium is defined so that it coincides with the vector $(1/\mu_o)\,\mathbf{B}^v$. Then one writes

$$\begin{aligned} \mathbf{H} &= \frac{1}{\mu}\mathbf{B} , \\ \mu &:= \mu_o\,\mu_r \ . \end{aligned} \tag{498}$$

Here it is $(1/\mu_r)\,\mathbf{B} = \mathbf{B}^v$. The quantity $\mu = \mu_o\,\mu_r$ is called permeability of the medium. The relative permeability μ_r is dimensionless and therefore independent from the unit system. For Eq. (498), one writes also

$$\left.\begin{aligned} \mathbf{B} &= \mu_o\,\mathbf{B} + \mathbf{M} , \\ \mathbf{M} &= \mu_o\,\kappa\,\mathbf{B} = \mu_o(\mu_r - 1)\,\mathbf{B} \end{aligned}\right\} \tag{499}$$

with the vector of magnetisation **M** and the dimensionless magnetic susceptibility $\kappa = (\mu_r - 1)$. The dimensionless relative permeability μ_r has the value 1 for vacuum. To the dielectric polarisation with $\varepsilon > 1$, hence $\chi > 0$, cp. Eqs. (491) and (492), there belongs the diamagnetism in the magnetic case as a general property of matter. Because of the historic definition (498) of the permeability μ, with inverted positions of the vectors **B** and **H**, we get in comparison with the electric case, Eq. (491), a negative diamagnetic susceptibility, $\kappa_D < 0$. Positive susceptibility, the so-called paramagnetismus, means $\kappa_P > 0$. The absolute value of both quantities remains, in general, very small, $|\kappa_D|, |\kappa_P| \ll 1$. Large values of $|\kappa|$ characterise the phenomenon of ferromagnetism.

In mono-crystals, the quantity μ will be a tensor, and we have to replace Eq. (498) by a tensor equation $H_i = \sum \mu_{ik} B_k$. The directions of vectors **H** and **B** will, in general, not coincide anymore.

MAXWELL detected that the temporal change of the electric displacement vector, the so-called MAXWELL *displacement current* $\partial \mathbf{D}/\partial t$, is also a source of a magnetic field **H** besides the electric current density **j**. Hence, we have to generalise OERSTEDT's law,

$$\oint_{\partial S} \mathbf{H} \cdot d\mathbf{s} = \iint_S (\mathbf{j} + \frac{\partial \mathbf{D}}{\partial t}) \cdot d\mathbf{S} \ . \tag{500}$$

We use STOKES's integral rule (1701) for the left-hand line integral and find

$$\iint_S \left(\mathrm{rot}\mathbf{H} - \frac{\partial \mathbf{D}}{\partial t} - \mathbf{j} \right) \cdot d\mathbf{S} = 0 \ .$$

This equation holds true for arbitrary areas S, which requires that the integrand itself vanishes,

$$\mathrm{rot}\mathbf{H} - \frac{\partial \mathbf{D}}{\partial t} = \mathbf{j} \ . \tag{501}$$

If we take into account that according to Eq. (1663) $\mathrm{div}\,\mathrm{rot}\,\mathbf{H} = 0$, from Eq. (501) and (490) we get immediately the continuity equation (469). This means only the introduction of MAXWELL's displacement current $\partial \mathbf{D}/\partial t$ leads to mathematically compatible Eqs. (501) and (490).

3.1.5 MAXWELL's Equations—Electromagnetic Waves

Before we present the complete MAXWELL, s. Fig. 4, equations, we write down more detailed the current densities. The total current density **J** is formed additively from the convective current density $\mathbf{j} = \rho\,\mathbf{u}$, produced by a charge density ρ streaming with a velocity field **u**, and a conductive current connected with the conductivity of the medium $\boldsymbol{\iota} = \sigma\,\mathbf{E}$, hence

$$\mathbf{J} = \rho\,\mathbf{u} + \boldsymbol{\imath}\,. \tag{502}$$

On the right-hand side of Eq. (501), there has to appear the total current density **J**, also in the LORENTZ force density (470), while we write OHM's law (463) more precisely with the conduction current $\boldsymbol{\imath}$.

Now MAXWELL's theory is complete:

$$\left.\begin{array}{l}(a)\ \ \mathrm{rot}\mathbf{E} + \dfrac{\partial \mathbf{B}}{\partial t} = 0\ ,\ \mathrm{div}\mathbf{B} = 0,\\ (b)\ \ \mathrm{rot}\mathbf{H} - \dfrac{\partial \mathbf{D}}{\partial t} = \mathbf{J},\ \mathrm{div}\mathbf{D} = \rho,\\ (c)\ \ \mathbf{f} = \rho\,\mathbf{E} + \mathbf{J}\times\mathbf{B}.\\ (d)\ \ \mathbf{D} = \varepsilon\,\mathbf{E},\quad (e)\ \ \mathbf{B} = \mu\,\mathbf{H},\quad (f)\ \ \boldsymbol{\imath} = \sigma\,\mathbf{E},\\ (g)\ \ \varepsilon = \varepsilon_o\,\varepsilon_r,\ (h)\ \ \mu = \mu_o\,\mu_r,\ (k)\ \ \mathbf{J} = \mathbf{j} + \boldsymbol{\imath}\end{array}\right\}\ \begin{array}{l}\text{Reference system } \Sigma_o\\ \text{MAXWELL equations}\\ \text{SI-units}\end{array} \tag{503}$$

These equations are formulated in a distinguished reference system $\Sigma_o(x, y, z, t)$, they contain all phenomena of electrodynamics—however with a significant restriction. We have supposed that the medium as a whole is at rest! The electrodynamics of arbitrary moving media will be discussed using the covariant formulation in Sect. 3.2. This treatment will provide a remarkable example of the application of EINSTEIN's relativity principle in MINKOWSKI space for establishing relativistic field equations, s. Sect. 1.6.

The total current can be assembled in a complex way from different parts due to the different charge carriers. The solution of MAXWELL's equations can become quite complex. One has to keep in mind, whether the current follows from a description on the basis of electron motions or phenomenologically. The distinction of total current, convection current and conductive current is also essential for the derivation of the equations of electrodynamics in moving media. The equations (580) of Sect. 3.2.3 contain the equations of this section as limiting cases.

For our treatment, the question is especially interesting, which form have the equations of electrodynamics, determined by an observer in an inertial system $\Sigma'(x', y', z', t')$ moving with velocity v relative to Σ_o. Hold true in Σ' the same MAXWELL equations (503) as in Σ_o? How are the in Σ' measured electric and magnetic fields $\mathbf{E}'$, $\mathbf{B}'$, $\mathbf{D}'$ and $\mathbf{H}'$ connected with the corresponding fields in Σ_o? Which relations between the coordinates (x', y', z', t') and (x, y, z, t) have to be supposed for preserving form invariance of equations (503) in all inertial systems? For preparing the answers to these questions, we derive at first a very important consequence from the equations (503).

To this aim, we consider the vacuum case without electric charges and currents,

$$\left.\begin{array}{l}\mu = \mu_o\ ,\ \varepsilon = \varepsilon_o\ ,\ \sigma = 0\ ,\\ \rho = 0\ ,\ \ \mathbf{j} = 0\ .\end{array}\right\}\qquad \begin{array}{l}\text{Vacuum,}\\ \text{no charges, no currents}\end{array} \tag{504}$$

The relations (1671), rot rot $\mathbf{V} = -\boldsymbol{\Delta}\,\mathbf{V} + \mathrm{grad}\,\mathrm{div}\,\mathbf{V}$, are applied on the fields **E** and **B**. We find from Eqs. (503) and (504)

$$
\begin{aligned}
&\text{rot}\,\text{rot}\mathbf{E} + \frac{\partial}{\partial t}\text{rot}\mathbf{B} = 0\ , \qquad \longrightarrow \qquad -\Delta\mathbf{E} + \text{grad}\,\text{div}\,\mathbf{E} + \frac{\partial}{\partial t}\text{rot}\mathbf{B} = 0\ ,\\
&\text{rot}\,\text{rot}\mathbf{H} - \frac{\partial}{\partial t}\text{rot}\mathbf{D} = 0\ , \qquad \longrightarrow \qquad -\Delta\mathbf{H} + \text{grad}\,\text{div}\,\mathbf{H} - \frac{\partial}{\partial t}\text{rot}\mathbf{D} = 0\ ,\\
&\mathbf{D} = \varepsilon_o\,\mathbf{E}\ , \qquad \mathbf{H} = \frac{1}{\mu_o}\mathbf{B}\ ,\\
&\text{rot}\mathbf{B} - \varepsilon_o\,\mu_o\frac{\partial\mathbf{E}}{\partial t} = 0\ , \qquad \text{div}\mathbf{E} = 0\ ,\\
&\text{rot}\mathbf{D} + \varepsilon_o\mu_o\frac{\partial\mathbf{H}}{\partial t} = 0\ , \qquad \text{div}\mathbf{B} = 0\ .
\end{aligned}
$$

So we get the wave equations for the electromagnetic fields:

$$
\left.\begin{aligned}
\frac{1}{c^2}\frac{\partial^2}{\partial t^2}\mathbf{E} - \Delta\mathbf{E} = 0\ ,\\
\frac{1}{c^2}\frac{\partial^2}{\partial t^2}\mathbf{B} - \Delta\mathbf{B} = 0\ .
\end{aligned}\right\} \quad \text{Equations for electromagnetic waves} \qquad (505)
$$

Here we have

$$
c = \frac{1}{\sqrt{\varepsilon_o\mu_o}}\ . \qquad \text{Speed of light} \qquad (506)
$$

For the quantity $1/\sqrt{\varepsilon_o\mu_o}$, we get with Eqs. (458) and (488)

$$
\frac{1}{\sqrt{\varepsilon_o\mu_o}} = \frac{1}{\sqrt{4\pi\cdot 10^{-7}8.85418782\cdot 10^{-12}}}\sqrt{\frac{\text{A m}}{\text{V s}}\frac{\text{V m}}{\text{A s}}} = 0.0299792458\cdot 10^{10}\,\frac{\text{m}}{\text{s}}.
$$

This is indeed nothing else than the *vacuum speed of light c* that now can be determined from a pure electric measurement of the permittivity ε_o s. Eq. (489). Furthermore, we notice that without MAXWELL's displacement current $\partial\mathbf{D}/\partial t$, there would not follow the wave equations for the fields. Checking the way of the derivation, we see that simply one would miss the second time derivatives.

With the LAPLACE operator (166) we can replace Eqs. (505) by[13]

$$
\left.\begin{aligned}
\Box\,\mathbf{E} = 0\ ,\\
\Box\,\mathbf{B} = 0\ .
\end{aligned}\right\} \quad \text{Electromagnetic waves} \qquad (505)
$$

The elementary solution of the homogeneous wave equation (505) for fields **E** and **B** is found in mathematically simple form in complex presentation as plane waves,

[13]Since today the signature $(+, -, -, -)$ is in use, we have also introduced the wave operator with the opposite sign as used in older literature, $(\Delta - \partial^2/\partial t^2)$. We remark that this requires at transition to the KLEIN–GORDON Gleichung with the operator $(\Box - \kappa^2)$, that κ must be imaginary for getting a real mass.

Fig. 4 JAMES CLERK MAXWELL, *Edinburgh 13.6.1831, † Cambridge 5.11.1879.

$$\mathbf{E} = \mathbf{E}_o \exp[i\phi] = \mathbf{E}_o \exp[i(\omega t - \mathbf{k}\cdot\mathbf{x})] \,, \quad \mathbf{B} = \mathbf{B}_o \exp[i\phi] = \mathbf{B}_o \exp[i(\omega t - \mathbf{k}\cdot\mathbf{x})] \,. \qquad \text{Plane waves} \qquad (507)$$

The physical real fields are given by the real part (or the imaginary part) of Eq. (507).

The quantity ϕ is already known from the aberration in wave optics, s. Eq. (260),

$$\phi(\mathbf{x}, t) = \omega t - \mathbf{k}\cdot\mathbf{x} \,. \qquad \text{Phase of a plane wave} \qquad (508)$$

It describes the current oscillation state of the plane wave and is called *phase*. From $\exp[i\phi] = \cos\phi + i\sin\phi$ in (507), we have

$$\phi(\mathbf{x}, t) = \phi\Big(\mathbf{x} + \frac{2\pi\mathbf{k}}{|\mathbf{k}|^2}\,,\, t\Big) \quad \text{and} \quad \phi(\mathbf{x}, t) = \phi\Big(\mathbf{x}\,,\, t + \frac{2\pi}{\omega}\Big) \,.$$

The phases of the waves will repeat, when we proceed for constant time t in $\mathbf{k}$-direction by one wave length, hence $\lambda = 2\pi/|\mathbf{k}|$, and also, when we wait at a fixed place $\mathbf{x}$ for one period, hence $T = 1/\nu = 2\pi/\omega$.

The quantity $\omega = 2\pi\,\nu$, i.e. 2π-times the frequency ν, is called circular frequency. Areas of constant phases, $\phi = \text{const}$,

$$\mathbf{k}\cdot\mathbf{x} = \omega t + \text{const} \,, \qquad \text{Areas of constant phase} \qquad (509)$$

are in agreement with Eq. (509) planes with normal vector $\mathbf{k}$ called *wave vector*.

In wave propagation, one speaks of *dispersion*, when the circular frequency ω depends nonlinearly on the wave vector k.

As follows immediately by insertion, the ansatz (507) delivers a solution of Eq. (505), when equation $\omega^2 = c^2\,\mathbf{k}\cdot\mathbf{k} = c^2\,k^2$ is fulfilled with a constant c independent on $\mathbf{k}$. The wave propagation is dispersion free,

$$\omega = c\,k = \text{const} \quad \text{or} \quad c = \lambda\,\nu \,. \qquad \text{Dispersion-free electromagnetic waves} \qquad (510)$$

The dispersion-free propagation of electromagnetic waves in vacuum (510) is a fundamental consequence of MAXWELL's equations:

The phase velocity of electromagnetic waves in vacuum has for all frequencies one and the same value, the speed of light $c = \omega/k$.

For a distribution of a group of waves over a range of $\mathbf{k}$-values, we introduce densities $d\mathbf{E}_o/d^3k$ and $d\mathbf{B}_o/d^3k$. The general solution of wave equation (505) is then given with help of Eq. (510) for ω by a summation (i.e. an overlay) of plane waves (507)

$$\left.\begin{aligned} \mathbf{E} &= \iiint \left(d\mathbf{E}_o/d^3k\right) \exp[i\left(\omega t - \mathbf{k}\cdot\mathbf{x}\right)]d^3k \ , \\ \mathbf{B} &= \iiint \left(d\mathbf{B}_o/d^3k\right) \exp[i\left(\omega t - \mathbf{k}\cdot\mathbf{x}\right)]d^3k \ . \end{aligned}\right\} \tag{511}$$

Areas of constant phases propagate with velocity c through space as seen from Eq. (509) by taking the derivative, when we orient $d\mathbf{x}$ in direction of the wave vector $\mathbf{k}$ and use Eq. (510),

$$\mathbf{k}\cdot d\mathbf{x} = \omega\, dt \quad \longrightarrow \quad \frac{dx}{dt} = \frac{\omega}{|\mathbf{k}|} = c \quad \text{for} \ \ d\mathbf{x} \sim \mathbf{k} \ . \tag{512}$$

Since electromagnetic waves in vacuum are dispersion free, also the velocity of each wave group of harmonic oscillations is equal to the speed of light, $c = \omega/k = d\omega/dk$.

For each plane wave, Eq. (507), it holds

$$\left.\begin{aligned} \mathrm{div}\mathbf{E} &= \frac{\partial E_x}{\partial x} + \frac{\partial E_y}{\partial y} + \frac{\partial E_z}{\partial z} = -\,i\,\left(k_x\,E_x + k_y\,E_y + k_z\,E_z\right) = -\,i\,\mathbf{k}\cdot\mathbf{E}\ , \\ \mathrm{rot}\mathbf{E} &= \left(\frac{\partial E_z}{\partial y} - \frac{\partial E_y}{\partial z},\ \frac{\partial E_x}{\partial z} - \frac{\partial E_z}{\partial x},\ \frac{\partial E_y}{\partial x} - \frac{\partial E_x}{\partial y}\right) \\ &= -\,i\,\left(\left(k_y\,E_z - k_z\,E_y,\ k_z\,E_x - k_x\,E_z,\ k_x\,E_y - k_y\,E_x\right)\right) = -\,i\,\mathbf{k}\times\mathbf{E}\ , \\ \frac{\partial\mathbf{E}}{\partial t} &= i\,\omega\,\mathbf{E}\ . \end{aligned}\right\} \tag{513}$$

The corresponding relations hold true for the vector **B**.

We insert Eq. (513) into MAXWELL's equations (503) and take into account the preposition $\rho = 0$, $\mathbf{j} = 0$, $\varepsilon = \varepsilon_o$, $\mu = \mu_o$ to get

$$\mathbf{k}\cdot\mathbf{E} = 0\ ,\quad \mathbf{k}\cdot\mathbf{B} = 0\ ,\quad \mathbf{k}\times\mathbf{E} = \omega\,\mathbf{B}\ . \tag{514}$$

Transversality of electromagnetic waves

The amplitudes of the waves in SI-units follow from Eqs. (510) and (514)

$$|\mathbf{E}_o| = c\,|\mathbf{B}_o|\ . \tag{515}$$

Equation (514) demonstrates the transversality of electromagnetic waves: The vectors **E**, **B** and **k** are all orthogonal to each other and form in this order a right-hand screw. Due to the possibility to decompose an arbitrary electromagnetic wave into plane waves (511), the transversality is a general property of all electromagnetic waves.

Electromagnetic waves propagate with the vacuum speed of light c through empty space. At frequencies between $3.8{\cdot}10^{-14}$ and $7.9{\cdot}10^{-14}$ Hz, we can see them visually. Optics becomes a branch of electrodynamics. The speed of light has then in each inertial system the unique value c, when in all inertial system the same equations of electrodynamics are valid. The universal constancy of the speed of light becomes a central perception of electrodynamics. This represents the historical starting point of Special Relativity Theory.

(a) Solution of MAXWELL's equations by electromagnetic potentials

The first group (503a) of MAXWELL's equations is always homogeneous, it can be solved by means of an ansatz of potentials. We set

$$\mathbf{B} = \text{rot}\,\mathbf{A}\ , \quad \mathbf{E} = -\text{grad}\varphi - \frac{\partial \mathbf{A}}{\partial t}\ . \qquad \text{Ansatz of potentials} \quad (516)$$

The quantities **A** and φ are called *vector potential* and *scalar potential.*

We insert Eq. (516) into the first group (503a) of MAXWELL's equations, then it holds because of div rot = 0 and rot grad = 0, s. Eqs. (1663) and (1661), for arbitrary fields **A** and φ,

$$\begin{aligned} \text{rot}\,\mathbf{E} + \frac{\partial \mathbf{B}}{\partial t} &= -\text{rot grad}\varphi - \text{rot}\,\frac{\partial \mathbf{A}}{\partial t} + \frac{\partial \text{rot}\,\mathbf{A}}{\partial t} = 0\ , \\ \text{div}\mathbf{B} &= \text{div rot}\,\mathbf{A} = 0\ . \end{aligned}$$

The first group (503a) of MAXWELL's equations is identically fulfilled with the potential ansatz (516). This is true both for vacuum and for electrodynamics in media.

Now we take into account that due to HELMHOLTZ's fundamental theorem of vector calculus, a vector field **V** is only fixed, if we know both rot **V** and div **V**. Therefore, we can impose additional conditions that are called gauges.

Here we postulate the so-called LORENZ *gauge*[14] within media

$$\varepsilon\,\mu\,\frac{\partial \varphi}{\partial t} + \text{div}\,\mathbf{A} = 0\ . \qquad \text{LORENZ gauge in a medium} \quad (517)$$

For vacuum, with $\varepsilon \longrightarrow \varepsilon_o$ and $\mu \longrightarrow \mu_o$ because of Eq. (506) this reads

$$\frac{1}{c^2}\,\frac{\partial \varphi}{\partial t} + \text{div}\,\mathbf{A} = 0\ . \qquad \text{LORENZ gauge in vacuum} \quad (518)$$

There exist also other gauges, i.e. one can impose other additional conditions on div **A**. One adapts the gauge to the problem to be solved with the potential **A** and φ. For discussion of electromagnetic waves, and with respect to the covariant formulation of electrodynamics in vacuum, the LORENZ gauge is especially useful. We mention here in addition the so-called COULOMB gauge div $\mathbf{A} = \mathbf{0}$.[15]

We consider a pure convection current $\mathbf{j} = \rho\,\mathbf{u}$. Assuming that ε and μ are constants, we insert the potential ansatz (516) into the second group (503b) of MAXWELL's equations and find at first

---

[14]The name is due to honour the danish physicist LUDVIG VALENTIN LORENZ (1829–1891) and should not be mistaken for the dutch physicist HENDRIK ANTOON LORENTZ (1853–1928). There exists even a LORENZ - LORENTZ formula, cp. SOMMERFELD (1952).

[15]The possibility to change the quantities **A** and φ by $\mathbf{A} \rightarrow \mathbf{A}' + \nabla\chi$ and $\varphi \rightarrow \varphi' = \varphi - \partial/(\partial_t)\chi$, without changing the physical fields **E** and **B**, Eq. (516), is called a gauge transformation. The deeper theory of gauge transformations goes back to WEYL, it got a huge meaning in modern physics.

$$
\begin{aligned}
\frac{1}{\mu}\,\mathrm{rot}\,\mathrm{rot}\,\mathbf{A} + \varepsilon\,\frac{\partial}{\partial t}\Big(\mathrm{grad}\varphi + \frac{\partial \mathbf{A}}{\partial t}\Big) &= \mathbf{j}\ , \\
-\varepsilon\,\mathrm{div}\,\Big(\mathrm{grad}\varphi + \frac{\partial \mathbf{A}}{\partial t}\Big) &= \rho\ .
\end{aligned}
$$

We eliminate the scalar potential in the first equation and the vector potential in the second equation by means of the LORENZ gauge (517) and get

$$
\begin{aligned}
\frac{1}{\mu}\left[\big(\mathrm{rot}\,\mathrm{rot} - \mathrm{grad}\,\mathrm{div}\big)\,\mathbf{A} + \varepsilon\,\mu\,\frac{\partial^2}{\partial t^2}\,\mathbf{A}\right] &= \mathbf{j}\ , \\
\varepsilon\left[-\mathrm{div}\,\mathrm{grad}\,\varphi + \varepsilon\,\mu\,\frac{\partial^2}{\partial t^2}\,\varphi\right] &= \rho\ .
\end{aligned}
\tag{519}
$$

With the LAPLACE operator (1665) and Eqs. (1671) and (1668), $\mathrm{div}\,\mathrm{grad}\,\varphi = \Delta\,\varphi$ and $\mathrm{rot}\,\mathrm{rot}\ \mathbf{A} = -\,\Delta\,\mathbf{A} + \mathrm{grad}\,\mathrm{div}\,\mathbf{A}$, we introduce the operator $\Box_{\varepsilon\mu}$, in these equations similar to Eq. (317).

$$
\Box_{\varepsilon\mu} := \varepsilon\,\mu\,\frac{\partial^2}{\partial t^2} - \Delta\ ,
\tag{520}
$$

describing the propagation of electromagnetic waves in media.

From Eq. (519), one finds the inhomogeneous wave equation for the components of the scalar and vector potential,

$$
\begin{aligned}
\Box_{\varepsilon\mu}\ \mathbf{A} &= \frac{1}{c_m^2}\,\frac{\partial^2}{\partial t^2}\,\mathbf{A} - \Delta\,\mathbf{A} = \mu\,\mathbf{j}\ , \\
\Box_{\varepsilon\mu}\ \varphi &= \frac{1}{c_m^2}\,\frac{\partial^2}{\partial t^2}\,\varphi - \Delta\,\varphi = \frac{\rho}{\varepsilon}\ .
\end{aligned}
\tag{521}
$$

The quantity c_m gives the propagation velocity of electromagnetic waves in media with the dielectricity constant ε and the permeability μ. It follows

$$
c_m = \frac{1}{\sqrt{\varepsilon\,\mu}} = \frac{c}{\sqrt{\varepsilon_r\,\mu_r}} = \frac{c}{n}
\tag{522}
$$

with the refraction index $n = \sqrt{\varepsilon_r\,\mu_r}$.

For vacuum with $\varepsilon = \varepsilon_o$ and $\mu = \mu_o$, we have again the vacuum speed of light c, Eq. (506). By means of Eqs. (521) one can derive that charges in accelerated motion emit electromagnetic waves, we mention especially the HERTZ dipol, s. e.g. BECKER (1982).

If there are no charges or currents, one gets the homogeneous wave equations

$$
\begin{aligned}
\frac{1}{c_m^2}\,\frac{\partial^2}{\partial t^2}\,\mathbf{A} - \Delta\,\mathbf{A} &= 0\ , \\
\frac{1}{c_m^2}\,\frac{\partial^2}{\partial t^2}\,\varphi - \Delta\,\varphi &= 0\ .
\end{aligned}
\tag{523}
$$

With **A** and φ all other components of **E** and **B** fulfil wave equations (521) and (523) due to Eq. (516).

For the proof, that Eqs. (503) are indeed already sufficient for the requirement of the universal validity in all inertial systems, we have to write down 'only' the MAXWELL equations according to EINSTEIN's relativity principle as tensor equation in MINKOWSKI space, s. Sect. 1.6. It should be remarked that the speed of light c_m in a medium is no more a LORENTZ-invariant quantity.

3.2 *Covariant Formulation of Electrodynamics*

According to the relativity principle formulated in Sect. 1.6, now we shall show that MAXWELL's equations hold true in each inertial system by writing them as tensor equations in MINKOWSKI space. We have already noted that this relativity principle is needed to generalise MAXWELL's equations (503) for application in a medium with an arbitrary flow velocity **w**.

At first, we want to show a remarkable mathematical connection between the field vectors **E** and **B**, namely, the property to form a four-dimensional tensor. Originally, this was elaborated by A. EINSTEIN (1905a, b) to resolve the unsatisfactory asymmetry of the classical explanation of the induction experiment (s. citation and the following explanations in Sect. 2.2). By means of MINKOWSKI's method it can be done quite easily.

3.2.1 Four-Dimensional Quantities of Electrodynamics

At first, we show that the quantities on the right-hand side of Eq. (503), the current density $\mathbf{j} = \rho\,\mathbf{u}$ and the charge density ρ, can be combined to a four-vector. Each electric charge can occur in nature only as a multiple of the elementary charge e_o of an electron.[16] The elementary charge e_o is therefore a natural constant, that has according to EINSTEIN's relativity principle, in all inertial systems one and the same value.

The elementary charge e_o represents an invariant in MINKOWSKI space.

We consider a certain amount of charges Δe within a volume element of the medium. The sum Δe of elementary charges, i.e. its number, is independent from the reference system. Σ_e might be the current rest system of this small quantity of charges Δe. We denote with ΔV_o the volume, measured in system Σ_e. ΔV_o is by definition independent of the reference system, and similar, the 'rest charge density' ρ_e,

[16]Elementary charges of quarks—the constituents of protons and neutron-do not exist as free particles.

$$\rho_e := \Delta e / \Delta V_o$$

is a scalar in MINKOWSKI space. Considered from Σ_o where the charges have a velocity $\mathbf{u} = (u_x, u_y, u_z)$ and occupy a volume ΔV that has a contracted value in the direction of motion $\mathbf{u}$ because of the LORENTZ contraction, it holds

$$\Delta V = \frac{\Delta V_o}{\gamma_u}, \tag{524}$$

where $u := |\mathbf{u}|$. Therefore, the charge density $\rho = \Delta e / \Delta V$, moving in Σ_o with velocity $\mathbf{u} = (u_x, u_y, u_z)$, is given by

$$\rho = \frac{\Delta e}{\Delta V_o} \frac{\Delta V_o}{\Delta V},$$

hence

$$\rho = \gamma_u \rho_e , \quad \rho_e = \frac{\Delta e}{\Delta V_o} , \quad \gamma_u = \frac{1}{\sqrt{1 - u^2/c^2}} . \tag{525}$$

Moving charge density ρ and invariant rest charge density ρ_e

We remind the definition of the four-vector u^i of velocity (431). Because of the invariance of ρ_e the quantity $\rho_e u^i = \rho_e\big(\gamma_u c, \gamma_u u_x, \gamma_u u_y, \gamma_u u_z\big)$ forms a four-vector, the four-current density j^i, the convection current of moving charges,

$$j^i = \rho_e u^i = \rho_e \left(\frac{c}{\sqrt{1-u^2/c^2}}, \frac{u_x}{\sqrt{1-u^2/c^2}}, \frac{u_y}{\sqrt{1-u^2/c^2}}, \frac{u_z}{\sqrt{1-u^2/c^2}} \right) = \rho \left(c, u_x, u_y, u_z, \right) . \tag{526}$$

Four vector of current density

With coordinates in MINKOWSKI space $x^i = (x^0, x^1, x^2, x^3) = (ct, x, y, z)$, the continuity equation (469) can be written in differential form, describing the conservation law of electric charges, in four-dimensional form, independent on the reference system,

$$\frac{\partial}{\partial x^i} j^i = 0 \quad \text{with} \quad j^i = \rho_e u^i . \tag{527}$$

Continuity equation

From both right-hand sides of the MAXWELL equations (503b), we can form with ρc and $\rho \mathbf{u}$ the four-vector j^i in MINKOWSKI space. Now we have to determine the tensorial formulation for the left-hand side of Eq. (503).

To this aim, we write down the potential ansatz (516) for $\mathbf{E}$ and $\mathbf{B}$ explicitly,

$$\left.\begin{aligned} B_x &= \frac{\partial}{\partial y} A_z - \frac{\partial}{\partial z} A_y , \quad B_y = \frac{\partial}{\partial z} A_x - \frac{\partial}{\partial x} A_z , \quad B_z = \frac{\partial}{\partial x} A_y - \frac{\partial}{\partial y} A_x , \\ E_x &= -\frac{\partial}{\partial x} \varphi - \frac{\partial}{\partial t} A_x , \quad E_y = -\frac{\partial}{\partial y} \varphi - \frac{\partial}{\partial t} A_y , \quad E_z = -\frac{\partial}{\partial z} \varphi - \frac{\partial}{\partial t} A_z . \end{aligned}\right\} \tag{528}$$

For the covariant formulation of the MAXWELL equations (503) now the following assertion is crucial:

The vector potential $\mathbf{A} = (A_x, A_y, A_z)$ forms together with the scalar potential φ a four-vector A in MINKOWSKI space

with the co- and contravariant components of the four-vector A:

$$\left.\begin{aligned} A^i &= (A^0, A^1, A^2, A^3) = \left(\frac{\varphi}{c}, A_x, A_y, A_z\right) \,, \\ A_i &= (A_0, A_1, A_2, A_3) = \eta_{ik} A^k = \left(\frac{\varphi}{c}, -A_x, -A_y, -A_z\right) \,. \end{aligned}\right\} \tag{529}$$

The potential ansatz (516) and (528) can now be written in a four-dimensional tensorial form,

$$F_{ik} = \frac{\partial}{\partial x^i} A_k - \frac{\partial}{\partial x^k} A_i \tag{530}$$

or shorter using the notation of Eq. (1650),

$$F_{ik} = \partial_i A_k - \partial_k A_i = A_{k\,,\,i} - A_{i\,,\,k} \quad , \tag{531}$$

when we define the tensor F_{ik} as follows, also providing in addition the contravariant components $F^{ik} = \eta_{ir}\,\eta_{ks}\,F_{rs}$,

$$F_{ik} = \begin{pmatrix} 0 & \frac{E_x}{c} & \frac{E_y}{c} & \frac{E_z}{c} \\ -\frac{E_x}{c} & 0 & -B_z & B_y \\ -\frac{E_y}{c} & B_z & 0 & -B_x \\ -\frac{E_z}{c} & -B_y & B_x & 0 \end{pmatrix} , \quad F^{ik} = \begin{pmatrix} 0 & -\frac{E_x}{c} & -\frac{E_y}{c} & -\frac{E_z}{c} \\ \frac{E_x}{c} & 0 & -B_z & B_y \\ \frac{E_y}{c} & B_z & 0 & -B_x \\ \frac{E_z}{c} & -B_y & B_x & 0 \end{pmatrix} . \tag{532}$$

This definition has a far-reaching conceptional consequence: The fields **E** and **B** form a unique mathematical quantity, a tensor F in MINKOWSKI space.

With MINKOWSKI's elegant mathematics we have detected the above on Sect. 3.2 announced, by EINSTEIN established mathematical connection between the vectors **E** and **B**, the central statement of the covariantly formulated electrodynamics:

The three-dimensional vectors of electric field strength **E** and the magnetic induction **B** form with Eq. (532) a two-ranked antisymmetric tensor F im MINKOWSKI space.

The coordinate transformations in MINKOWSKI space, the general LORENTZ transformations, contain both pur spatial rotations and transitions from one inertial system to an other. In Problem 29, we shall show that the three-dimensional vector properties of **E** and **B** are fulfilled also by the four-dimensional tensor F.

The proper LORENTZ transformations now deliver a completely new behaviour of the properties of fields **E** and **B**. Considering a pure electric field or a pure magnetic field in one reference system Σ_o, an observer in a reference system Σ', being in motion with respect to the first one, will measure in both cases an electric *and* a magnetic field. The four-dimensional tensor of the electromagnetic field F_{ik} decomposes in each inertial system in MINKOWSKI space in different components of magnetic induction **B** and electric field strength **E**—in the same way as a two-dimensional vector **V** for different Cartesian coordinate axes (x, y) and (x', y') in the plane decomposes in different components (V_x, V_y) and (V_x', V_y'), respectively. We shall illustrate this with the example of a special LORENTZ transformation.

We assume that Eq. (532) gives the components of the field strength tensor F_{ik} in reference system Σ_o. With $F_{i'k'}$ we denote the components in the system Σ', that moves with a velocity v along the x-axis as measured from Σ_o. The tensor F_{ik} transforms as given in Eq. (1628), hence

$$F_{i'k'} = \frac{\partial x^i}{\partial x^{i'}} \frac{\partial x^k}{\partial x^{k'}} F_{ik} \ . \tag{533}$$

With the special LORENTZ transformation (339) we can write Eq. (533) also in matrix form. We respect that in Eq. (533) the matrix of the inverse LORENTZ transformation from Eq. (339) should be used, and we get

$$F_{i'k'} = \begin{pmatrix} 0 & \frac{E_x'}{c} & \frac{E_y'}{c} & \frac{E_z'}{c} \\ -\frac{E_x'}{c} & 0 & -B_z' & B_y' \\ -\frac{E_y'}{c} & B_z' & 0 & -B_x' \\ -\frac{E_z'}{c} & -B_y' & B_x' & 0 \end{pmatrix}$$

$$= \begin{pmatrix} \gamma & \beta\gamma & 0 & 0 \\ \beta\gamma & \gamma & 0 & 0 \\ 0 & 0 & 1 & 0 \\ 0 & 0 & 0 & 1 \end{pmatrix} \begin{pmatrix} 0 & \frac{E_x}{c} & \frac{E_y}{c} & \frac{E_z}{c} \\ -\frac{E_x}{c} & 0 & -B_z & B_y \\ -\frac{E_y}{c} & B_z & 0 & -B_x \\ -\frac{E_z}{c} & -B_y & B_x & 0 \end{pmatrix} \begin{pmatrix} \gamma & \beta\gamma & 0 & 0 \\ \beta\gamma & \gamma & 0 & 0 \\ 0 & 0 & 1 & 0 \\ 0 & 0 & 0 & 1 \end{pmatrix} .$$

Evaluation of these matrices or the summation (533) provides the components $F_{i'k'}$ of the field strength tensor in Σ'

$$F_{i'k'} = \begin{pmatrix} 0 & \frac{E_x}{c} & \frac{\gamma(E_y - vB_z)}{c} & \frac{\gamma(E_z + vB_y)}{c} \\ -\frac{E_x}{c} & 0 & -\frac{\gamma(cB_z - \beta E_y)}{c} & \frac{\gamma(cB_y + \beta E_z)}{c} \\ -\frac{\gamma(E_y - vB_z)}{c} & \frac{\gamma(cB_z - \beta E_y)}{c} & 0 & -B_x \\ -\frac{\gamma(E_z + vB_y)}{c} & -\frac{\gamma(cB_y + \beta E_z)}{c} & B_x & 0 \end{pmatrix} . \tag{534}$$

Now we can easily show that the LORENTZ force density **f** in Eq. (503c) represents the last three components of the four-vector f,

$$f^i = \left(\frac{1}{c}\mathbf{f} \cdot \mathbf{u},\ f_x,\ f_y,\ f_z\right) . \tag{535}$$

The quantity f^i is a four-vector since it can be written as the tensor product of the tensor (532) and the vector (526), where we still have to respect Eq. (525),

$$f^i = F^{ik}\, j_k = \rho \begin{pmatrix} 0 & -\frac{E_x}{c} & -\frac{E_y}{c} & -\frac{E_z}{c} \\ \frac{E_x}{c} & 0 & -B_z & B_y \\ \frac{E_y}{c} & B_z & 0 & -B_x \\ \frac{E_z}{c} & -B_y & B_x & 0 \end{pmatrix} \begin{pmatrix} c \\ -u_x \\ -u_y \\ -u_z \end{pmatrix} . \quad \text{Four-vector of LORENTZ force density} \tag{536}$$

With the LORENTZ force density (472), we form $\mathbf{u} \cdot \mathbf{f} = \rho\, \mathbf{u} \cdot \big(\mathbf{E} + \mathbf{u} \times \mathbf{B}\big) = \rho\, \mathbf{u} \cdot \mathbf{E}$, where the last relation follows since $\mathbf{u} \cdot (\mathbf{u} \times \mathbf{B}) = 0$. For the zero component of $\mathbf{f}$, we can write $f^0 = (1/c)\, \mathbf{f} \cdot \mathbf{u} = (1/c)\rho\, \mathbf{E} \cdot \mathbf{u}$. This is just the density of the power divided by c done by the electromagnetic field on the charge density ρ.

For better comparing this expression with Eq. (445) of relativistic mechanics, we consider a point-like charge density $\rho = e_o\, \delta(x)\, \delta(y)\, \delta(z)$ sitting without loss of generality at the coordinate origin. The three-dimensional total force $\mathbf{F}$ on the charge e_o is obtained by integration over the LORENTZ force density $\mathbf{f}$,

$$\left.\begin{aligned} \mathbf{F} &= \iiint \mathbf{f}\, dxdydz = \iiint \rho \big(\mathbf{E} + \mathbf{u} \times \mathbf{B}\big)\, dxdydz \\ &= \iiint e_o\, \delta(x)\, \delta(y)\, \delta(z) \big(\mathbf{E} + \mathbf{u} \times \mathbf{B}\big)\, dxdydz \\ &= e_o\, (\mathbf{E} + \mathbf{u} \times \mathbf{B})\ , \end{aligned}\right\} \tag{537}$$

precisely the LORENTZ force $\mathbf{F} = (F_x,\, F_y,\, F_z)$ of Eq. (473).

In Eq. (537), we have integrated over the volume of the charge. Since the volume element $dxdydz$ is not a LORENTZ scalar, but suffers to the LORENTZ contraction with the factor $1/\gamma_u$, the three components $(F_x,\, F_y,\, F_z)$ of the LORENTZ force do not form the three last components of a four-vector. We get the MINKOWSKI force vector, the four-vector F, in agreement with Eq. (439) in Sect. 2.2, only after multiplication with the factor γ_u, and with adding the zeroth component,

$$F^i = \left(\frac{\gamma_u}{c}\mathbf{F} \cdot \mathbf{u}\, ,\ \gamma_u\, F_x\, ,\ \gamma_u\, F_y\, ,\ \gamma_u\, F_z\right) . \quad \text{Four-vector of LORENTZ force} \tag{538}$$

This is exactly the formula (445) of relativistic mechanics for the four-force. Now we use in Eq. (538) the electromagnetic field strength tensor as in Eq. (536), so we can write the MINKOWSKI force vector as

$$F^i = e_o\, F^{ik}\, u_k = e_o \begin{pmatrix} 0 & -\frac{E_x}{c} & -\frac{E_y}{c} & -\frac{E_z}{c} \\ \frac{E_x}{c} & 0 & -B_z & B_y \\ \frac{E_y}{c} & B_z & 0 & -B_x \\ \frac{E_z}{c} & -B_y & B_x & 0 \end{pmatrix} \begin{pmatrix} c \\ -u_x \\ -u_y \\ -u_z \end{pmatrix} . \tag{539}$$

In Sect. 2, cp. Eqs. (447) and (448), we have shown that F^i represents a MINKOWSKI four-vector, when the three-dimensional force **F** is a LORENTZ-invariant quantity, as postulated according to the principle (138), for the mechanical forces. In electrodynamics, we have this shown also by the four-dimensional tensor character of the electromagnetic field.

In Problem 31, we shall use the four-dimensional character of the electromagnetic field F for deriving from the static COULOMB field of a point charge at rest the field of a uniformly moving point charge by an algebraic evaluation, just a LORENTZ transformation.

Now one also notes that the tensor character of the electromagnetic field with respect to LORENTZ transformations solves according to Eqs. (532)–(534) the initially mentioned problem of the asymmetry of the explanation of induction phenomena mentioned by EINSTEIN, cp. Sect. 3:

For the current generating force on electrons, there is responsible in each case the electric field measured by the observer at rest relative to the conductor loop.

We consider the case that in system Σ_o only the components $\mathbf{B} = (B_x, B_y, B_z)$ of magnetic induction in the tensor F_{ik} are different from zero, e.g. produced by an in Σ_o resting magnet. According to the classical induction experiment, the conductor loop moves with velocity $\mathbf{u} = (-u, 0, 0)$ against the magnet. If it is left from the magnet, the conductor is at rest in a system Σ', moving in direction of the negative x-axis of Σ_o with velocity of amount u.

When we replace in Eq. (534) v by $-u$, we read off that the observer, at rest with respect to the loop conductor, measures an electric field $\mathbf{E}' = (0, \gamma_u\, u\, B_z, -\gamma_u\, u\, B_y)$ caused by the induction voltage. In Σ', it is $u'_i = (c, 0, 0, 0)$, hence due to Eq. (539)

$$F^{i'} = e_o\,(0, \gamma_u\, u\, B_z\,, -\gamma_u\, u\, B_y\,, 0)\ . \tag{540}$$

The components F^i and $F^{i'}$ of the MINKOWSKI force vectors result from the same LORENTZ transformation as the coordinates. F^2 and F^3 do not change by this transformation, and F^1 and F^4 vanish as the components $F^{1'}$ and $F^{4'}$. The force vectors remain unchanged also in Σ_o,

$$F^i = e_o(0, \gamma_u\, u\, B_z\,, -\gamma_u\, u\, B_y\,, 0)\ . \tag{541}$$

With $\mathbf{u} = (-u, 0, 0)$ and $\mathbf{B} = (B_x, B_y, B_z)$ the observer in Σ_o measures a force **F** as given in Eq. (538)

$$\mathbf{F} = e_o\, \mathbf{u} \times \mathbf{B}\ . \tag{542}$$

He will call this force LORENTZ force. As shown in Sect. 3.1.3, one can explain the induced current in the conductor loop by the LORENTZ force (542). Here we have shown in an algebraic way, by a LORENTZ transformation of the electromagnetic tensor F_{ik}, that the force on the electrons—also in the case of the motion of the conductor in system Σ_o relative to the resting magnet—is created by an electric field measured by an observer at rest relative to the conductor. In this way, we established the symmetry in the explanation of the experimentally symmetric induction effect.

From the four-dimensional vector property of the electromagnetic potentials (529), now it follows that the LORENZ gauge (518) in vacuum is a LORENTZ-invariant relation preserved in each inertial system, when it holds in one single system. This is a mathematical benefit of this gauge especially for theoretical investigations. Instead of Eq. (518), we can write

$$\frac{\partial}{\partial x^i} A^i = 0 \, . \qquad \text{LORENZ gauge in vacuum} \qquad (543)$$

The LORENZ gauge (517) in a medium is clearly not a LORENTZ-invariant relation.

3.2.2 Four-Dimensional Electrodynamics in Vacuum

We consider a vacuum with electric charges and currents, hence

$$\varepsilon = \varepsilon_o \, , \quad \mu = \mu_o \, , \quad \sigma = 0 \, , \quad \rho \, , \quad \mathbf{j} = \rho\, \mathbf{u} \, . \qquad \text{Vacuum with moving charges} \qquad (544)$$

For $\varepsilon = \varepsilon_o$ and $\mu = \mu_o$, we can use the LAPLACE operator (520)

$$\Box_{\varepsilon_o \mu_o} \equiv \Box = \frac{1}{c^2} \frac{\partial^2}{\partial t^2} - \mathbf{\Delta} = \frac{\partial}{\partial x^i} \frac{\partial}{\partial x^j} \eta^{ij} \, . \qquad (545)$$

The operator $\Box$ is also LORENTZ invariant. With the potential ansatz (530) and the LORENZ gauge (543) the MAXWELL equations reduce in vacuum to

$$\Box A^i = \mu_o \, j^i \quad \text{with} \quad \frac{\partial}{\partial x^i} A^i = 0 \, . \qquad (546)$$

The LORENTZ-invariant form of the LORENTZ force is already given in Eq. (539).

The complete system of the MAXWELL equations, written with the field strength tensor (532) formed from the fields quantities **E** and **B**, reads then

$$\left.\begin{aligned}
&(a) \quad \frac{\partial}{\partial x^i} F_{kl} + \frac{\partial}{\partial x^k} F_{li} + \frac{\partial}{\partial x^l} F_{ik} = 0 \\
&(b) \quad \frac{\partial}{\partial x^i} F^{ik} = \mu_o \, j^k \, , \\
&(c) \quad f^i = F^{ik} \, j_k = \rho_e \, F^{ik} \, u_i \, , \\
&\qquad c = \frac{1}{\sqrt{\varepsilon_o \, \mu_o}} \, .
\end{aligned}\right\} \qquad \text{Covariant form of the MAXWELL equations in vacuum} \qquad (547)$$

Here we need no further tensor for the field **D** and **H**. The MAXWELL equations in vacuum contain only *one* physical constant, the vacuum speed of light c. The SI-units may obscure a bit this issue. The constant μ_o is arbitrarily introduced, and the factor $1/c$ at **E** in the field strength tensor F shows only an apparent asymmetry of the fields **E** and **B**. Therefore, in Sect. 3.3, we show MAXWELL's theory once again in absolute units.

With a comma for the partial derivative, one writes Eq. (634a) also as

$$F_{kl},_i + F_{li},_k + F_{ik},_l = 0.$$

As seen easily, with Eq. (634a) the potential ansatz (530) or (531) is identically fulfilled, so that MAXWELL equations are reduced to Eqs. (546),

$$F_{ik},_l + F_{kl},_i + F_{li},_k = \partial_l\partial_i A_k - \partial_l\partial_k A_i + \partial_i\partial_k A_l - \partial_i\partial_l A_k + \partial_k\partial_l A_i - \partial_k\partial_i A_i A_l = 0.$$

By means of Eq. (532), we get with the notion of Eq. (1650) from Eq. (634a) for $i, k, l = 1, 2, 3$

$$F_{12},_3 + F_{23},_1 + F_{31},_2 = \partial_z B_z + \partial_x B_x + \partial_y B_y = \operatorname{div} \mathbf{B} = 0 \ .$$

The induction law follows, when in Eq. (634a) one of the indices is 0, e.g. for $i, k, l = 1, 2, 0$ with $x_0 = ct$,

$$F_{12},_0 + F_{20},_1 + F_{01},_2 = -\frac{1}{c}\,\partial_t\, B_z - \frac{1}{c}\,\partial_x\, E_y + \frac{1}{c}\,\partial_y\, E_x = -\frac{1}{c}\left[\left(\frac{\partial \mathbf{B}}{\partial t}\right)_z + (\operatorname{rot} \mathbf{E})_z\right] = 0\,.$$

Also, Eq. (634b) is identical with Eq. (503b). We use Eq. (526) and get for $k = 0$

$$\begin{aligned} F^{i0},_i &= F^{10},_1 + F^{20},_2 + F^{30},_3 \\ &= \frac{1}{c}\big(\partial_x E_x + \partial_y E_y + \partial_z E_z\big) = \mu_0\, c\, \varrho\,, \end{aligned}$$

hence

$$\frac{1}{\mu_o\, c^2}\operatorname{div} \mathbf{E} = \rho \quad \longrightarrow \quad \varepsilon_o \operatorname{div} \mathbf{E} = \rho\,.$$

and therefore

$$\operatorname{div} \mathbf{D} = \rho\,.$$

For $k = 1, 2, 3$, there follows AMPÈRE's law with inclusion of MAXWELL's displacement current, e.g. for $k = 1$,

$$\begin{aligned} F^{i1},_i &= F^{01},_0 + F^{21},_2 + F^{31},_3 \\ &= -\frac{1}{c}\,\partial_t\,\frac{E_x}{c} + \partial_y B_z - \partial_z B_y = \mu_0\, \varrho\, u_x\,, \end{aligned}$$

hence

$$\partial_y B_z - \partial_z B_y - \varepsilon_o \mu_o \partial_t E_x = \mu_0 \varrho u_x ,$$
$$\frac{1}{\mu_o}(\mathrm{rot}\,\mathbf{B})_x - \varepsilon_o \partial_t E_x = \varrho u_x \longrightarrow (\mathrm{rot}\,\mathbf{H})_x - \partial_t D_x = \varrho u_x ,$$

and therefore

$$(\mathrm{rot}\,\mathbf{H})_x - \partial_t D_x = \varrho u_x .$$

Equation (503c) for the LORENTZ force density f^i was already discussed in Eq. (536). For a charge-free vacuum, we get from Eq. (546) the homogeneous wave equation

$$\Box A^i = 0 . \tag{548}$$

With the four-vector k in MINKOWSKI space,

$$k_i = \left(\frac{\omega}{c}, -k_x, -k_y, -k_z\right) \longleftrightarrow k^i = \left(\frac{\omega}{c}, k_x, k_y, k_z\right), \quad \text{Four-vector} \tag{549}$$

the phase ϕ, s. Chap. 8, Sect. 7, and Problem 20, is LORENTZ invariant,

$$\phi = \omega t - \mathbf{k}\cdot\mathbf{x} := k_i x^i . \quad \text{Invariant phase} \tag{550}$$

Plane waves (507) are now obtained from the four-potential A^i

$$A^i = A^i_o \exp[i(\omega t - \mathbf{k}\cdot\mathbf{x})] = A^i_o \exp[i\, k_i x^i] . \quad \text{Plane waves} \tag{551}$$

For a superposition of waves over a range of **k**-values, we introduce densities dA^i_o/d^3k. With Eq. (510) one writes the general solution of Eq. (548) by a summation or superposition of plane waves

$$A^i = \iiint \left(dA^i_o/d^3k\right) \exp[i(k\,ct - \mathbf{k}\cdot\mathbf{x})]d^3k . \tag{552}$$

(a) DOPPLER effect and aberration in the covariant treatment

The relativistic theory of the DOPPLER effect and the aberration follows now in an elegant way from the simple statement above that the phase ϕ is a LORENTZ-invariant quantity, cp. also Problems 20 and 33.

From the dispersion-free nature of electromagnetic waves (510), i.e. $\omega = c\,k$, it follows immediately

$$k_i k^i = 0 , \tag{553}$$

I.e., k^i is a null vector. This vector contains the information on the propagation direction and the frequency of the wave. If k forms with the x-, y- and z-axes the angles η,

θ and ζ, then we can write for the unit vector $\mathbf{n} = (n_x, n_y, n_z) = (\cos\eta, \cos\theta, \cos\zeta)$ in propagation direction $(n_x,\ n_y,\ n_z) = k_x/k,\ k_y/k,\ k_z/k)$.

Since $k = 2\pi/\lambda = 2\pi\nu/c = \omega/c$ we get for k^i

$$\left(k^0,\ k^1,\ k^2,\ k^3\right) = \frac{2\pi}{c}\,(\nu,\ \nu\,\cos\eta,\ \nu\,\cos\theta,\ \nu\,\cos\zeta)\ . \tag{554}$$

A sender resting in reference system Σ_o might emit a plane wave with the (three-dimensional) wave vector $\mathbf{k}$ and the frequency $\nu = \nu_S$. Then the plane wave fronts propagate with the speed of light c in the direction $(n_x,\ n_y,\ n_z)$ of $\mathbf{k}$. An inertial system Σ' might be in motion along the negative x-direction of Σ_o, it has a velocity $\mathbf{v} = (-v,\ 0,\ 0)$ with respect to Σ_o. The receiver R might be at rest in Σ'. When the wave fronts propagate along the positive x-axis, the receiver will approach them with velocity v. We take an arbitrary propagation direction, s. Fig. 5. The receiver R measures a plane wave with a four-vector $k^{i'}$ and frequency $\nu' = \nu_R$ and the three-dimensional direction vector $\mathbf{n}' = (n_{x'},\ n_{y'},\ n_{z'}) = (\cos\eta',\ \cos\theta',\ \cos\zeta')$,

$$\left(k^{0'},\ k^{1'},\ k^{2'},\ k^{3'}\right) = \frac{2\pi}{c}\,(\nu',\ \nu'\cos\eta',\ \nu'\cos\theta',\ \nu'\cos\zeta')\ . \tag{555}$$

The vector components $k^{i'}$ and k^i as well as the coordinates $x^{i'}$ and x^i of the Minkowski space are connected by the special Lorentz transformation, Eqs. (105) or (339); however, we have to replace v by $-v$ and β by $-\beta$, since here the reference system Σ' moves with velocity $-v$,

$$\left.\begin{aligned} \nu'\cos\eta' &= \frac{\nu\,\cos\eta + \nu\,\beta}{\sqrt{1-\beta^2}}\ ,\\ \nu'\cos\theta' &= \nu\,\cos\theta\ ,\\ \nu'\cos\zeta' &= \nu\,\cos\zeta\ ,\\ \nu' &= \frac{\nu + \beta\,\nu\,\cos\eta}{\sqrt{1-\beta^2}}\ . \end{aligned}\right\} \tag{556}$$

These formulas contain the complete relativistic theory of the Doppler frequency shift and the aberration denoted as change of the propagation direction of the wave when it is observed from Σ'.

For the application in these formulas, one must observe that the propagation direction of the receiver R must be described by angles η', θ', ζ' in the system Σ'. Otherwise, one gets curious results, cp. Problem 39. For deriving relations of direct use for describing the relativistic Doppler effect in Eqs. (556), we express $\cos\eta$ in terms of $\cos\eta'$. To this aim, we get from the first and last Eq. (556)

$$(1 + \beta\,\cos\eta)\,\cos\eta' = \cos\eta + \beta\ . \tag{557}$$

Solving for $\cos\eta$, we get

$$\cos\eta = \frac{\beta - \cos\eta'}{\beta\,\cos\eta' - 1}\ . \tag{558}$$

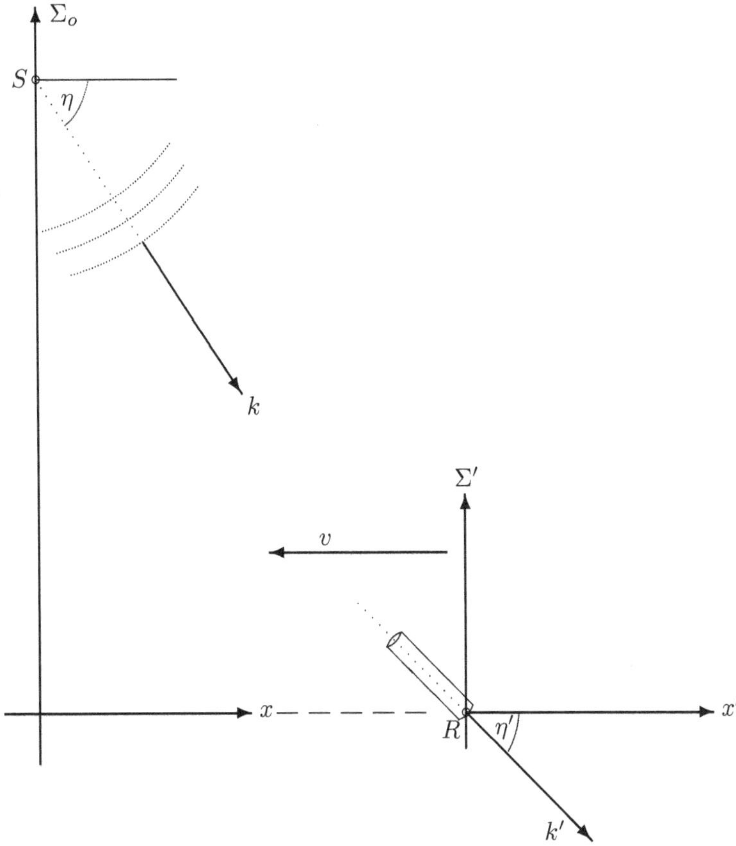

Fig. 5 Experimental scheme for the DOPPLER effect and the aberration. We use values of angle η with $\tan\eta = -1.5$ and η' with $\tan\eta' = -1$. From Eq. (562), one finds $\beta = (\cos\eta' - \cos\eta)/(1\ cos\eta'\ \cos\eta)$, hence these angles correspond to a velocity $v \approx 0.25\,c$ for system Σ' in direction of the negative x-axis of Σ_o. From Eq. (559), we get a DOPPLER shift of $\nu_R \approx 1.05\,\nu_S$.

We insert this result into the last Eq. (556) and find after short evaluation, using $\nu' = \nu_R$ and $\nu = \nu_S$,

$$\nu_R = \nu_S\,\frac{\sqrt{1-\beta^2}}{1-\beta\,\cos\eta'}\,. \qquad \text{Complete relation of the relativistic DOPPLER effect} \qquad (559)$$

Equation (559) is the complete formula for the relativistic DOPPLER effect. There the receiver moves with velocity v in negative x-direction of the rest system of the sender.

When a plane wave propagates in direction of the positive x-axis of system Σ_o, hence for $(\eta\,,\theta\,,\zeta) = (0, \pi/2, \pi/2)$, i.e. $(\cos\eta, \cos\theta, \cos\zeta) = (1, 0, 0)$, we get from Eq. (556)

$$\left.\begin{aligned} \nu'\cos\eta' &= \frac{1+\beta}{\sqrt{1-\beta^2}} \ , \\ \nu'\cos\theta' &= 0 \ , \\ \nu'\cos\zeta' &= 0 \ , \\ \nu' &= \frac{1+\beta}{\sqrt{1-\beta^2}} \ , \end{aligned}\right\}$$

where we immediately see, that in this case, the wave in Σ' has the identical direction cosinus, $(\cos\eta', \cos\theta', \cos\zeta') = (1, 0, 0)$.

We write again $\beta = v/c$, $\nu' = \nu_R$ and $\nu = \nu_S$, and we get for longitudinal observations, hence $\cos\eta' = 1$, the formula for the longitudinal DOPPLER effect, cp. Eqs. (239) and (242),

$$\nu_R = \nu_S \sqrt{\frac{c+v}{c-v}} \ . \qquad \text{Longitudinal DOPPLER effect} \qquad (560)$$

When we set $\eta' = -\pi/2$, $\cos\eta' = 0$, then from Eq. (559) it follows the formula of the transversal DOPPLER effect, cp. Eq. (243),

$$\nu_R = \nu_S \sqrt{1 - \frac{v^2}{c^2}} \ . \qquad \text{Transversal DOPPLER effect} \qquad (561)$$

When we solve Eq. (557) for $\cos\eta'$, we get the relativistic equation of aberration

$$\cos\eta' = \frac{\cos\eta + \beta}{1 + \beta\cos\eta} \ . \qquad \text{Aberration} \qquad (562)$$

With $\tan x = \sqrt{1 - \cos^2 x}/\cos x$ one finds easily the relation

$$\tan\eta' = \frac{\sin\eta}{\cos\eta + \beta}\sqrt{1-\beta^2} \ . \qquad \text{Aberration} \qquad (563)$$

For a plane wave emitted in system Σ_o in direction of the negative y-axis, we have $\eta = -\pi/2$, so $\cos\eta = 0$. According to Eq. (562), this wave will be observed from the receiver R in Σ' with an angle η' with $\cos\eta' = \beta$, cp. also Problem 39. When the position of a star is expected the direction $\tilde{\eta}$, then the telescope must not be pointed in this direction, but we have to tilt it additionally depending on our velocity by an angle in the direction η', so that $\cos\eta' = \beta$. This 'deviation' of the telescope is called aberration [*lat. aberratio = deviation*], Problem 40.

3.2.3 Four-Dimensional Electrodynamics of Moving Media

The most famous paper in the history of physics since NEWTON, EINSTEIN (1905a, cf. the English translation in LORENTZ (1952), 'On the Electrodynamics of Moving Bodies' was brought to an end by MINKOWSKI (1908). s. Fig. 6: EINSTEIN's preposition, that the body as an entity has some velocity v, will now be replaced by the assumption that the motion of the different elements of the medium is described by a velocity field $\mathbf{w}(\mathbf{x})$,

$$\mathbf{w} = (w_x,\ w_y,\ w_z)\,. \qquad \text{Velocity field of the medium} \tag{564}$$

The corresponding four-vector w is

$$w^i = \left(\frac{c}{\sqrt{1-w^2/c^2}},\frac{w_x}{\sqrt{1-w^2/c^2}},\frac{w_y}{\sqrt{1-w^2/c^2}},\frac{w_z}{\sqrt{1-w^2/c^2}}\right), \tag{565}$$

where $w^2 = w_x^2 + w_y^2 + w_z^2$.

Here we shall consider a simplified idealistic model, with three different velocities, $\mathbf{u}$, $\mathbf{v}$ and $\mathbf{w}$. The names $\mathbf{u}$ and $\mathbf{w}$ are used for the velocities of the charge carriers and the mass elements, respectively, and $\mathbf{v}$ denotes the velocity of the inertial systems. For our idealised medium, we neglect that the motion of the charge carriers is always also a motion of masses. In our model, we can consider the velocities $\mathbf{u}$ and $\mathbf{w}$ as decoupled. For example, a plasma is not covered. If the motion of the charge carriers represents purely the motion of electrons, then one can neglect the connected mass motion in good approximation. It becomes obvious that the equations of a charged and conductive medium, in general, can become very complicated.

The current density is again assumed to consist of two parts. One part represents the convective current, also possible in vacuum. In some problems, we can describe it with a predetermined current density $\mathbf{j}$ and a charge density ρ,

$$\mathbf{j} = \rho\,\mathbf{u}\ . \tag{566}$$

Due to the, in general, finite conductivity σ of the medium, we have in addition a conduction current density $\boldsymbol{\iota}$. From our experimental experience, we can determine this conduction current for a medium at rest and an electric field strength $\mathbf{E}$ by OHM's law,

$$\boldsymbol{\iota} = \sigma\,\mathbf{E}\ . \qquad \text{OHM's law} \tag{567}$$

We denote the total current density by $\mathbf{J}$, given by the sum of two parts,

$$\mathbf{J} = \mathbf{j} + \boldsymbol{\iota}\ . \tag{568}$$

In four-dimensional language

$$J^i = j^i + \iota^i \quad \text{with} \quad j^i = \rho_e\, u^i \tag{569}$$

there is still missing a four-dimensional formulation of Eq. (567) for getting a four-dimensional interpretation of the conduction current ι^i. This will be done with Eq. (577). We remark however: only in the local rest system there is no charge density connected with the conduction current, so that its fourth component vanishes. This statement is however not invariant. As seen from an arbitrary reference system, with a conductive current is also connected a charge density ϱ contributing to the total current as described in Eq. (578). Therefore, we must write the total current as

$$J^i = (\varrho c, J_x, J_y, J_z) \tag{570}$$

with a charge density ϱ, in general, different from the charge density ρ of the convective current $\mathbf{j}$ given in Eq. (566).

At first, we need a second tensor, the tensor of the electromagnetic excitation H, consisting on the source vectors **D** and **H**,

$$H_{ik} = \begin{pmatrix} 0 & c\,D_x & c\,D_y & c\,D_z \\ -c\,D_x & 0 & -H_z & H_y \\ -c\,D_y & H_z & 0 & -H_x \\ -c\,D_z & -H_y & H_x & 0 \end{pmatrix}, \quad H^{ik} = \begin{pmatrix} 0 & -c\,D_x & -c\,D_y & -c\,D_z \\ c\,D_x & 0 & -H_z & H_y \\ c\,D_y & H_z & 0 & -H_x \\ c\,D_z & -H_y & H_x & 0 \end{pmatrix}. \tag{571}$$

The source for H is the total current, and hence there follows the field equations

$$\frac{\partial}{\partial x^i} H^{ik} = J^k\,. \tag{572}$$

Now we become familiar with the powerful heuristic principle for establishing field equations that rests in the formulation of EINSTEIN's relativity principles in MINKOWSKI space. MINKOWSKI has detected in 1908 in this way the complete field equations of electrodynamics in moving media, which have been only partly known before. The simple looking principle reads:

If one knows physical equations in an inertial system, then one must only write them in covariant form, hence as tensor equation in MINKOWSKI space.

We convince us initially that we have written down with Eqs. (571) and (572) the correct equations for the tensor H. To this aim we assume that the medium is at rest is an inertial system Σ_o, so that $w^i = (c, 0, 0, 0)$. Then the contribution of the conduction current to the charge density vanishes. Equation (572) now must be identical with Eq. (503b), when we insert in Eq. (503b) for the current density $\mathbf{j}$ the total current density $\mathbf{J}$, hence $(c\,\varrho, J_x, J_y, J_z) \longrightarrow (c\,\rho, j_x, j_y, j_z)$, since we did not introduce a distinction of the current densities in Eqs. (503).

For $k = 0$, we get from Eq. (572)

$$\begin{aligned} H^{i0}{}_{,i} &= H^{10}{}_{,1} + H^{20}{}_{,2} + H^{30}{}_{,3} \\ &= c\big(\partial_x D_x + \partial_y D_y + \partial_z D_z\big) = c\,\varrho \end{aligned}$$

and then because of $\varrho \longrightarrow \rho$,

$$\operatorname{div} \mathbf{D} = \rho\,.$$

For $k = 1$ it follows

$$\begin{aligned} H^{i1}{}_{,i} &= H^{01}{}_{,0} + H^{21}{}_{,2} + H^{31}{}_{,3} \\ &= -\frac{1}{c}\,\partial_t\,(c\,D_x) + \partial_y H_z - \partial_z H_y = J^1 \end{aligned}$$

and then with $(J_x,\ J_y,\ J_z) \longrightarrow (j_x,\ j_y,\ j_z)$

$$(\mathrm{rot}\,\mathbf{H})_x - \partial_t\,D_x = j_x\ .$$

With analogous relations for $k = 2, 3$ it follows

$$\mathrm{rot}\,\mathbf{H} - \partial_t\,\mathbf{D} = \mathbf{j}\ .$$

Equations (572) are also easy to verify, they are independent from the velocity field w of the medium.

For the Lorentz force density only the total current is important. Equation (503c) now leads to the four-dimensional formulation

$$f^i = F^{ik}\,\frac{1}{c}\,J_k\ . \tag{573}$$

In vacuum, we did not need equations for the response of the medium on electromagnetic fields. Now this is different, and we must look for a tensor formulation of Eq. (503d–f) in Minkowski space.

The material parameters in Eq. (503), the quantities ε, μ and σ, are defined in the rest frame of the medium, and they are therefore per definition invariants in Minkowski space, tensors of rank zero,[17]

$$\left.\begin{array}{l} \varepsilon\ , \\ \mu\ , \\ \sigma\ . \end{array}\right\} \qquad \text{Invariants in Minkowski space} \tag{574}$$

Equations (503d–f) hold true in the local rest system of each material mass element. For transforming them into four-dimensional tensor equations, we must only note that the four-vector w of the material velocity in a local rest system reads

$$w^i = (c\,,\ 0\,,\ 0\,,\ 0)\ ,\quad w_i = (c\,,\ 0\,,\ 0\,,\ 0)\ . \qquad \text{Material velocity in the local rest system} \tag{575}$$

When we multiply in the local rest system of a mass element, the tensors F_{ik} and H_{ik} tensorial with the velocity w^k, we get with Eqs. (532) and (571)

$$F_{ik}\,w^k = (0,\ E_1,\ E_2,\ E_3)\ ,\quad H_{ik}\,w^k = (0,\ D_1,\ D_2,\ D_3)\ . \tag{576}$$

Then we can write Eqs. (503d) and (503f) in Minkowski space as

[17] Here we disregard a possible three-dimensional tensor character of these quantities in crystalline materials.

$$\left.\begin{aligned} H_{ik}\, w^k &= \varepsilon\, F_{ik}\, w^k \;, \\ \iota_i &= -\sigma\, F_{ik}\, w^k \;. \end{aligned}\right\} \tag{577}$$

The conduction current has, in general, a non-vanishing fourth component. It contributes with $\iota_0 := c\,\rho_\iota$ to the total charge density,

$$\iota_0 := c\,\rho_\iota = -\,\sigma\, F_{0\alpha}\, w^\alpha = \frac{\sigma}{c\,\gamma_w}\,\mathbf{E}\cdot\mathbf{w}\;. \tag{578}$$

The fields **B** and **H** can be represented each for itself in four dimensions under consideration of Eq. (575) for the local rest system by means of a trick. We consider the cyclic sum of expressions with indices i, j, k and form the expression $F_{ij}\,w_k + F_{jk}\,w_i + F_{ki}\,w_j$. Clearly, it represents a tensor in MINKOWSKI space. In the local rest system, *those and only those* components of this tensor are different from zero that contain exactly one index 0, and they provide just one component of the with c multiplied magnetic induction **B**, e.g. $F_{21}\,w_0 + F_{10}\,w_2 + F_{02}\,w_1 = F_{21}\,w_0 = c\,B_z$, but $F_{21}\,w_3 + F_{13}\,w_2 + F_{32}\,w_1 = 0$. The same property holds true for the tensor $H_{ij}\,w_k + H_{jk}\,w_i + H_{ki}\,w_j$. In the local rest system of the medium, we can therefore write the material equations (503e) as

$$F_{ik}\,w_l + F_{kl}\,w_i + F_{li}\,w_k = \mu\left(H_{ik}\,w_l + H_{kl}\,w_i + H_{li}\,w_k\right)\;. \tag{579}$$

Since Eq. (579) is a tensor equation in MINKOWSKI space, this equation holds true also for an arbitrary moving medium. This completes our derivation.

The full system of MAXWELL equations in moving media reads as follows:

$$\left.\begin{aligned}
&(a)\quad \frac{\partial}{\partial x^i}F_{kl} + \frac{\partial}{\partial x^k}F_{li} + \frac{\partial}{\partial x^l}F_{ik} = 0\,,\\
&(b)\quad \frac{\partial}{\partial x^i}H^{ik} = J^k\,,\\
&(c)\quad f^i = F^{ik}\,\frac{1}{c}\,J_k\,,\\
&(d)\quad H_{ik}\,w^k = \varepsilon\,F_{ik}\,w^k\,,\\
&(e)\quad F_{ik}w_l + F_{kl}w_i + F_{li}w_k = \mu\left(H_{ik}w_l + H_{kl}w_i + H_{li}w_k\right)\\
&(f)\quad \iota_i = -\sigma\,F_{ik}\,w^k\,,\\
&(g)\quad \varepsilon = \varepsilon_o\,\varepsilon_r\,,\ (h)\quad \mu = \mu_o\,\mu_r\,,\ (k)\quad J^i = j^i + \iota^i\,.
\end{aligned}\right\}\ \text{Covariant form of the MAXWELL equations of moving media} \tag{580}$$

These equations are more general than Eq. (503), since also relativistic effects in non-uniformly moving media are included.

3.3 *Electrodynamics in Absolute Units*

Since the symmetry of MAXWELL's theory becomes clearer when using the absolute units, in theoretical presentations, in general, this system with only three basic units is adopted. Instead of the old cgs-system with basic units centimetre, gram and second, we take here the modern absolute system with metre, kilogram and second. Then the

Fig. 6 HERMANN MINKOWSKI, *Aleksota (today part of Kaunas) 22.6.1864, † Göttingen 12.1.1909.

Newton remains the force unit as in the SI-system. By means of physical equations we shall now reduce all quantities on metre, kilogram and second as already done in the SI-system. For force s. Eq. (139), for the current strength, Eqs. (458), (457), for the electric field strength and the magnetic induction Eqs. (469), (471), etc.

The following table provides an easy translation between both unit systems. The quantities in absolute units are denoted by a tilde.

$$\left.\begin{array}{llll}
\tilde{\mathbf{E}} = \sqrt{4\pi\varepsilon_o}\,\mathbf{E}\ , & \tilde{\mathbf{D}} = \sqrt{\dfrac{4\pi}{\varepsilon_o}}\,\mathbf{D}\ , & \tilde{\mathbf{B}} = \sqrt{\dfrac{4\pi}{\mu_o}}\,\mathbf{B}\ , & \tilde{\mathbf{H}} = \sqrt{4\pi\mu_o}\,\mathbf{H}\ , \\
\tilde{\varphi} = \sqrt{4\pi\varepsilon_o}\,\varphi\ , & \tilde{\mathbf{A}} = \sqrt{\dfrac{4\pi}{\mu_o}}\,\mathbf{A}\ , & & \\
\tilde{\varepsilon} = \varepsilon_r = \dfrac{\varepsilon}{\varepsilon_o}\ , & \tilde{\varepsilon}_o = 1\ , & \tilde{\mu} = \mu_r = \dfrac{\mu}{\mu_o}\ , & \tilde{\mu}_o = 1\ , \\
\tilde{\sigma} = \dfrac{\sigma}{4\pi\varepsilon_o}\ , & \tilde{\rho} = \dfrac{1}{\sqrt{4\pi\varepsilon_o}}\,\rho\ , & \tilde{\mathbf{j}} = \dfrac{1}{\sqrt{4\pi\varepsilon_o}}\,\mathbf{j}\ . &
\end{array}\right\} \quad (581)$$

3.3.1 Electrodynamics in a Medium

Instead of Eq. (503), now we have

$$\left.\begin{array}{l}
(a)\ \ \mathrm{rot}\tilde{\mathbf{E}} + \dfrac{1}{c}\dfrac{\partial\tilde{\mathbf{B}}}{\partial t} = 0\ , \qquad \mathrm{div}\tilde{\mathbf{B}} = 0\ , \\
(b)\ \ \mathrm{rot}\tilde{\mathbf{H}} - \dfrac{1}{c}\dfrac{\partial\tilde{\mathbf{D}}}{\partial t} = \dfrac{4\pi}{c}\tilde{\mathbf{J}}\ , \ \mathrm{div}\tilde{\mathbf{D}} = 4\pi\tilde{\rho}\ , \\
(c)\ \ \mathbf{f} = \tilde{\rho}\tilde{\mathbf{E}} + \dfrac{1}{c}\tilde{\mathbf{J}}\times\tilde{\mathbf{B}}\ , \\
(d)\ \ \tilde{\mathbf{D}} = \tilde{\varepsilon}\tilde{\mathbf{E}}\ ,\ (e)\ \ \tilde{\mathbf{B}} = \tilde{\mu}\tilde{\mathbf{H}}\ ,\ (f)\ \ \tilde{\imath} = \tilde{\sigma}\tilde{\mathbf{E}}\ , \\
(g)\ \ \tilde{\varepsilon} = \varepsilon_r\ , \quad (h)\ \ \tilde{\mu} = \mu_r\ , \quad (k)\ \ \tilde{\mathbf{J}} = \tilde{\mathbf{j}} + \tilde{\imath}
\end{array}\right\} \ \text{MAXWELL equations, Absolute unit system}$$

(582)

3.3.2 Electrodynamics in Vacuum—Four-Dimensional Formulation

The vacuum electrodynamics, especially its four-dimensional formulation, unveils clearly the symmetry of the equations in absolute units. From the invariant charge density $\tilde{\rho}_e$ in absolute units and its velocity **u**, we form the moving charge density $\tilde{\rho} = \tilde{\rho}_e\sqrt{1 - u^2/c^2}$ and the corresponding current density $\tilde{\mathbf{j}} = \tilde{\rho}\mathbf{u}$. There follow the MAXWELL equations:

$$\begin{array}{l}
(a)\ \mathrm{rot}\tilde{\mathbf{E}} + \dfrac{1}{c}\dfrac{\partial\tilde{\mathbf{B}}}{\partial t} = 0, \qquad \mathrm{div}\,\tilde{\mathbf{B}} = 0, \\
(b)\ \mathrm{rot}\tilde{\mathbf{B}} - \dfrac{1}{c}\dfrac{\partial\tilde{\mathbf{E}}}{\partial t} = \dfrac{4\pi}{c}\tilde{\rho}\mathbf{u}\ ,\ \mathrm{div}\,\tilde{\mathbf{E}} = 4\pi\tilde{\rho}, \\
(c)\ \mathbf{f} = \tilde{\rho}\left(\tilde{\mathbf{E}} + \dfrac{\mathbf{u}}{c}\times\tilde{\mathbf{B}}\right) .
\end{array} \quad \begin{array}{l}\text{MAXWELL equations}\\ \text{Vacuum with moving charges}\\ \text{Absolute units}\end{array} \quad (583)$$

With the four-current density $\tilde{j}^{\,i}$, cp. (526),

$$\tilde{j}^i = \tilde{\rho}_e\, u^i = \tilde{\rho}\left(c,\, u_x,\, u_y,\, u_z,\right) \;, \tag{584}$$

and the field strength tensor $\tilde{\mathsf{F}}$,

$$\tilde{F}_{ik} = \begin{pmatrix} 0 & \tilde{E}_x & \tilde{E}_y & \tilde{E}_z \\ -\tilde{E}_x & 0 & -\tilde{B}_z & \tilde{B}_y \\ -\tilde{E}_y & \tilde{B}_z & 0 & -\tilde{B}_x \\ -\tilde{E}_z & -\tilde{B}_y & \tilde{B}_x & 0 \end{pmatrix} , \quad \tilde{F}^{ik} = \begin{pmatrix} 0 & -\tilde{E}_x & -\tilde{E}_y & -\tilde{E}_z \\ \tilde{E}_x & 0 & -\tilde{B}_z & \tilde{B}_y \\ \tilde{E}_y & \tilde{B}_z & 0 & -\tilde{B}_x \\ \tilde{E}_z & -\tilde{B}_y & \tilde{B}_x & 0 \end{pmatrix} , \tag{585}$$

the MAXWELL equations read

$$\left.\begin{aligned} (a)\ & \tilde{F}_{ik},_{l} + \tilde{F}_{kl},_{i} + \tilde{F}_{li},_{k} = 0 \;, \\ (b)\ & \tilde{F}^{ik},_{i} = \frac{4\pi}{c}\,\tilde{j}^{k} \;, \\ (c)\ & f^{i} = \frac{1}{c}\,\tilde{F}^{ik}\,\tilde{j}_{k} \end{aligned}\right\} \quad \begin{array}{l}\text{Covariante form of}\\ \text{MAXWELL equations}\\ \text{in vacuum}\\ \text{Absolute units}\end{array} \tag{586}$$

With the potential ansatz for solving the first group (586a) of the MAXWELL equations

$$\tilde{F}_{ik} = \tilde{A}_{k},_{i} - \tilde{A}_{i},_{k} \tag{587}$$

and the LORENZ-gauge

$$\tilde{A}^{i},_{i} = 0 \qquad \text{LORENZ-gauge} \tag{588}$$

for the four-vector $\tilde{\mathsf{A}}$,

$$\left.\begin{aligned} \tilde{A}^i &= (\tilde{A}^0,\, \tilde{A}^1,\, \tilde{A}^2,\, \tilde{A}^3) = (\tilde{\varphi},\, \tilde{A}_x,\, \tilde{A}_y,\, \tilde{A}_z) \;, \\ \tilde{A}_i &= (\tilde{A}_0,\, \tilde{A}_1,\, \tilde{A}_2,\, \tilde{A}_3) = \eta_{ik}\tilde{A}^k = (\tilde{\varphi},\, -\tilde{A}_x,\, -\tilde{A}_y,\, -\tilde{A}_z) \;, \end{aligned}\right\} \tag{589}$$

there follows from Eq. (586b) the inhomogeneous wave equations

$$\Box\,\tilde{A}^i = \tfrac{4\pi}{c}\,\tilde{j}^i \;. \tag{590}$$

3.3.3 The Energy–Momentum Tensor of the MAXWELL Field

Now we shall derive a few important consequences of MAXWELL's equations and remain to this aim in the absolute unit system for preserving the obtained symmetry of equations. Without a deeper discussion we note: the action of electromagnetic forces and energy conversations by electric charges and currents, described in MAXWELL's equations, are contained in one single tensor in MINKOWSKI space, by the *energy–momentum tensor* $\tilde{T}^{ik}$.

Resulting from MAXWELL's equations (634) there exists a connection between the tensor $\tilde{T}^{ik}$ and the four-dimensional vector of LORENTZ force density,[18]

$$\frac{\partial}{\partial x^k}\tilde{T}^{ik} = -f^i \ . \tag{591}$$

The choice of the sign is convention. In charge-free space, it holds

$$\frac{\partial}{\partial x^k}\,\tilde{T}^{ik} = 0 \quad \text{for} \quad \tilde{j}^i = \tilde{\rho}\,u^i = 0 \ . \tag{592}$$

As the continuity equation (527), this is the differential form of the conservation law of the energy and momentum of the electromagnetic field as shown next. The tensor $\tilde{T}^{ik}$ is uniquely defined only when we require in addition to Eq. (592) a related differential conservation law for the angular momentum tensor of the field,[19] defined by $\tilde{M}^{lik} := x^l\tilde{T}^{ik} - x^i\tilde{T}^{lk}$, namely,

$$\frac{\partial}{\partial x^k}\,\tilde{M}^{lik} = 0 \ . \tag{593}$$

From Eq. (593), it follows

$$\begin{aligned}\frac{\partial}{\partial x^k}\left(x^l\tilde{T}^{ik} - x^i\tilde{T}^{lk}\right) &= \delta^l_k\tilde{T}^{ik} + x^l\tilde{T}^{ik}{}_{,k} - \delta^i_k\tilde{T}^{lk} + x^i\tilde{T}^{lk}{}_{,k}\\ &= \tilde{T}^{il} - \tilde{T}^{li} + x^l\tilde{T}^{ik}{}_{,k} - x^i\tilde{T}^{lk}{}_{,k} = 0 \ ,\end{aligned}$$

and therefore with Eq. (592) the symmetry of the stress-energy tensor

$$\tilde{T}^{il} - \tilde{T}^{li} = 0 \ . \tag{594}$$

Equations (591) and (594) are now fulfilled by the following *symmetric* tensor $\tilde{T}^{ik}$ denoted as *metric energy–momentum tensor* of the electromagnetic field,[20]

---

[18]For the force density f^i, we need no tilde since the pure mechanical quantities in SI-units and the absolute MKS-units are identical.

[19]This definition generalises the mechanical angular momentum $M^{li} = \epsilon^{lik}L_k$ with $\mathbf{L} = \mathbf{x}\times\mathbf{p}$, hence $M^{li} = \epsilon^{lik}\epsilon_{krs}x^r p^s = x^l p^i - x^i p^l$, cp. Eq. (1645).

[20]The source-free MAXWELL equations are invariant under four-dimensional translations and under LORENTZ transformations in the MINKOWSKI space. The NOETHER theorem provides the existence of a so-called canonical energy–momentum tensor T^{ik}_{can} and an angular momentum tensor M^{lik} fulfilling the differential conservation laws, Eqs. (592) and (593). The canonical energy–momentum tensor T^{ik}_{can} is not symmetric. The quantity $\widehat{M}^{lik} = x^l T^{ik}_{can} - x^i T^{lk}_{can}$ describes the 'orbital angular momentum', that does not fulfil a conservation law because of the asymmetry of T^{ik}_{can}. Only the total angular momentum $M^{lik} = \widehat{M}^{lik} + S^{lik}$ fulfils the conservation law (593). The existence of the eigen angular momentum S^{lik}, the spin tensor of the electromagnetic field, is the reason for the asymmetry of the canonical tensor T^{ik}_{can}. The 'spin' is discussed intensively in Chap. 10, Sects. 3–7.

$$\tilde{T}^{ik} = \frac{1}{4\pi}\left(-\tilde{F}^{ir}\,\tilde{F}^{ks}\,\eta_{rs} + \frac{1}{4}\,\eta^{ik}\,\tilde{F}^{rs}\,\tilde{F}_{rs}\right). \quad \text{Metric energy-momentum tensor of the MAXWELL field} \tag{595}$$

By differentiation (591) of tensor (595), we find with the notation (1650), taking into account the antisymmetry $\tilde{F}_{ik} = -\tilde{F}_{ki}$ and repeated renaming of the summation indices,

$$\begin{aligned}
\tilde{T}^{ik}{}_{,k} &= \frac{1}{4\pi}\left(-\tilde{F}^{ir}{}_{,k}\,\tilde{F}^{ks}\,\eta_{rs} - \tilde{F}^{ir}[\tilde{F}^{ks}{}_{,k}]\,\eta_{rs} + \frac{1}{4}\,\eta^{ik}\,\tilde{F}^{rs}{}_{,k}\,\tilde{F}_{rs} + \frac{1}{4}\,\eta^{ik}\,\tilde{F}^{rs}\,\tilde{F}_{rs,k}\right)\\
&= \frac{1}{4\pi}\left([\tilde{F}^{sk}{}_{,k}]\tilde{F}^{ir}\,\eta_{rs} - \eta^{is}\,\tilde{F}^{kr}\,\tilde{F}_{sr,k} + \frac{1}{2}\,\eta^{ik}\,\tilde{F}^{rs}\,\tilde{F}_{rs,k}\right)\\
&= \frac{1}{4\pi}\,[\tilde{F}^{sk}{}_{,k}]\tilde{F}^{ir}\,\eta_{rs} - \frac{1}{8\pi}\,\eta^{ik}\,\tilde{F}^{rs}[2\tilde{F}_{ks,r} + \tilde{F}_{rs,k}]\\
&= \frac{1}{4\pi}\,[\tilde{F}^{sk}{}_{,k}]\tilde{F}^{ir}\,\eta_{rs} - \frac{1}{8\pi}\,\eta^{ik}\,\tilde{F}^{rs}[\tilde{F}_{ks,r} + \tilde{F}_{kr,s} + \tilde{F}_{rs,k}]\\
&= \frac{1}{4\pi}\,[\tilde{F}^{sk}{}_{,k}]\tilde{F}^{ir}\,\eta_{rs} + \frac{1}{8\pi}\,\eta^{ik}\,\tilde{F}^{rs}[\tilde{F}_{sk,r} - \tilde{F}_{kr,s} - \tilde{F}_{rs,k}]\ .
\end{aligned}$$

Then with Eq. (536) and use of MAXWELL's equations (634b) and (634a) on the square bracket, we get Eq. (591),

$$\tilde{T}^{ik}{}_{,k} = \frac{1}{4\pi}\left[-\frac{4\pi}{c}\,\tilde{j}^{s}\right]\tilde{F}^{ir}\,\eta_{rs} = -f^{i}\ .$$

By means of the matrices (585), we determine the components of tensors $\tilde{T}^{ik}$ in terms of the field strengths,

$$\tilde{T}^{ik} = \frac{1}{4\pi}\begin{pmatrix}
\frac{1}{2}(\tilde{E}^2+\tilde{B}^2) & \tilde{E}_y\tilde{B}_z-\tilde{E}_z\tilde{B}_y & \tilde{E}_z\tilde{B}_x-\tilde{E}_x\tilde{B}_z & \tilde{E}_x\tilde{B}_y-\tilde{E}_y\tilde{B}_x\\
\tilde{E}_y\tilde{B}_z-\tilde{E}_z\tilde{B}_y & \frac{1}{2}(\tilde{E}^2+\tilde{B}^2)-\tilde{E}_x^2-\tilde{B}_x^2 & -\tilde{B}_x\tilde{B}_y-\tilde{E}_x\tilde{E}_y & -\tilde{B}_x\tilde{B}_z-\tilde{E}_x\tilde{E}_z\\
\tilde{E}_z\tilde{B}_x-\tilde{E}_x\tilde{B}_z & -\tilde{B}_x\tilde{B}_y-\tilde{E}_x\tilde{E}_y & \frac{1}{2}(\tilde{E}^2+\tilde{B}^2)-\tilde{E}_y^2-\tilde{B}_y^2 & -\tilde{B}_y\tilde{B}_z-\tilde{E}_y\tilde{E}_z\\
\tilde{E}_x\tilde{B}_y-\tilde{E}_y\tilde{B}_x & -\tilde{B}_x\tilde{B}_z-\tilde{E}_x\tilde{E}_z & -\tilde{B}_y\tilde{B}_z-\tilde{E}_y\tilde{E}_z & \frac{1}{2}(\tilde{E}^2+\tilde{B}^2)-\tilde{E}_z^2-\tilde{B}_z^2
\end{pmatrix}. \tag{596}$$

For the physical interpretation of the components of the energy–momentum tensors $\tilde{T}^{ik}$, we set

$$\tilde{\mathbf{T}} = \begin{pmatrix} \tilde{\upsilon} & \frac{1}{c}\tilde{\mathbf{S}}\\ c\,\tilde{\mathbf{g}} & \tilde{\mathbf{t}} \end{pmatrix} \tag{597}$$

with

$$\left.\begin{aligned}
\tilde{\upsilon} &= \tilde{T}^{00}\ , \qquad \frac{1}{c}\tilde{\mathbf{S}} = (\,\tilde{T}^{01},\ \tilde{T}^{02},\ \tilde{T}^{03}\,)\ ,\\
c\,\tilde{\mathbf{g}} &= \begin{pmatrix}\tilde{T}^{10}\\ \tilde{T}^{20}\\ \tilde{T}^{30}\end{pmatrix},\quad \tilde{\mathbf{t}} = \begin{pmatrix}\tilde{T}^{11} & \tilde{T}^{12} & \tilde{T}^{13}\\ \tilde{T}^{21} & \tilde{T}^{22} & \tilde{T}^{23}\\ \tilde{T}^{31} & \tilde{T}^{32} & \tilde{T}^{33}\end{pmatrix}.
\end{aligned}\right\} \tag{598}$$

By use of Eqs. (535) and (597), we get from Eq. (591) for $i = 0$

$$\frac{\partial \tilde{\upsilon}}{\partial t} + \operatorname{div} \tilde{\mathbf{S}} = -\mathbf{f} \cdot \mathbf{u} \ , \tag{599}$$

and for $i = 1, 2, 3$

$$\frac{\partial \tilde{\mathbf{g}}}{\partial t} + \operatorname{Div}_2 \tilde{\mathbf{t}} = -\mathbf{f} \ , \tag{600}$$

where the operator Div_2 means action on the second index, but for the symmetric tensor $\tilde{T}_{ik}$ this does not matter.

In vacuum and without charges and currents, the free electromagnetic field fulfils the balance equations

$$\frac{\partial \tilde{\upsilon}}{\partial t} + \operatorname{div} \tilde{\mathbf{S}} = 0 \ . \qquad \text{Energy conservation of a free electromagnetic field} \tag{601}$$

and

$$\frac{\partial \tilde{\mathbf{g}}}{\partial t} + \operatorname{Div}_2 \tilde{\mathbf{t}} = 0 \ . \qquad \text{Momentum conservation of a free electromagnetic field} \tag{602}$$

By integration of Eq. (599) over a volume K, we get work $\iiint\limits_K \mathbf{f} \cdot \mathbf{u} \, dxdydz$ done by the LORENTZ-force density $\mathbf{f}$ on the charges in volume K per unit time. It is compensated by the temporal change of the field energy $\iiint\limits_K \tilde{\upsilon} \, dxdydz$ and the in- or outstreaming field energy $\iiint\limits_K \operatorname{div} \tilde{\mathbf{S}} \, dxdydz$.

We evaluate in an elementary way an expression for the energy density of the electromagnetic field $\tilde{\upsilon}$ in Problems 34 and 35.

From the integration of Eq. (600) over a volume K, we read off directly MAXWELL's detection of the near-field action of the electromagnetic field:

In a volume K, there is measured action of the LORENTZ force $\iiint\limits_K \mathbf{f} \, dxdydz$. This is equal to the temporal change of the momentum content $\iiint\limits_K \tilde{\mathbf{g}} \, dxdydz$ and the force acting in the field, $\iiint\limits_K \operatorname{Div}_2 \tilde{\mathbf{t}} \, dxdydz = \oiint\limits_{\partial K} \tilde{\mathbf{t}} \cdot d\mathbf{A}$:

Through the area element $d\mathbf{A}$ in space between charges, there are transmitted forces $\tilde{\mathbf{t}} \cdot d\mathbf{A}$ that act onto well separated charges. As we will see in Problem 38, Eq. (1498) in mechanics of continua, the surface integral of type $\oiint\limits_{\partial K} \boldsymbol{\sigma} \cdot d\mathbf{A}$ defines the stress tensor $\boldsymbol{\sigma}$. Equation (602) describes a momentum balance in the same way as Eq. (1503) in mechanics of continua. The quantity $\tilde{\mathbf{t}}$ forms the three-dimensional stress tensor of the electromagnetic field. We summarise

$$
\begin{aligned}
\tilde{v} &= \frac{1}{8\pi}\left(\tilde{\mathbf{E}}^2 + \tilde{\mathbf{B}}^2\right), && \text{Energy density of the MAXWELL field} \\
\tilde{\mathbf{S}} &= \frac{c}{4\pi}\,\tilde{\mathbf{E}} \times \tilde{\mathbf{B}}, && \text{Energy current density of the MAXWELL field} \\
\tilde{\mathbf{g}} &= \frac{1}{4\pi c}\,\tilde{\mathbf{E}} \times \tilde{\mathbf{B}}, && \text{Momentum density of the MAXWELL field} \\
\tilde{\mathbf{t}} &= \frac{1}{4\pi}\left(\frac{1}{2}(\tilde{\mathbf{E}}^2 + \tilde{\mathbf{B}}^2)\,\mathbf{1} - (\tilde{\mathbf{E}}\tilde{\mathbf{E}} + \tilde{\mathbf{B}}\tilde{\mathbf{B}})\right). && \text{MAXWELL's stress tensor}
\end{aligned}
\tag{603}
$$

The physical interpretation of the energy–momentum tensor $\tilde{T}^{ik}$ of the MAXWELL field has an important theoretical consequence. From the symmetry of the energy–momentum tensors (596), we get the general formulation of the energy–mass equivalence of the electromagnetic field,

$$
\tilde{\mathbf{S}} = \tilde{\mathbf{g}}\,c^2\,. \qquad \text{Energy-mass equivalence of the electromagnetic field} \tag{604}
$$

For a long time, the analogy between an elastic deformation field of a mechanical continuum and the electromagnetic field was the reason for the search of a mechanical continuum, the so-called aether, whose elastic deformation should describe the electromagnetic field. When one could ascribe to this aether a state of motion like to a mechanical medium, then there would exist a distinguished absolute reference system, namely, its rest system like in mechanics of continua. There field equations and balances hold true only with respect to the rest system of the mediums, e.g. of the crystal. The validity of EINSTEIN's relativity principle demonstrates that there are absolutely no remains from the physical properties of such an aether, only the properties of our physical vacuum.

For a further exposition of the electrodynamics, we refer to the monographs by HEHL and OBUKHOV (2003), cp. also HEHL, ITIN, OBUKHOV (2016).

Chapter 10
Representations of the LORENTZ Group WEYL Equation and DIRAC Equation

1 Remembering of Group Theory

We write: $\mathcal{M}$ group; $\mathcal{M}$ subgroup; M semigroup; M set; M complex;
M (transformation) matrix; $\mathbb{M}$ (MINKOWSKI) space.

The elements $a, b, c, \cdots$ of a non-empty set $\mathsf{M} = \{a, b, c, \ldots\}$ are investigated concerning the following properties:

1. It is defined a 'connection' or 'composition' $\circ$ so that two composed elements in a given order $a \in \mathsf{M}$ and $b \in \mathsf{M}$ uniquely lead to an element $c \in \mathsf{M}$,

$$a \circ b = c \quad \text{with} \quad c \in \mathsf{M} \quad \text{for all} \quad a, b \in \mathsf{M}\,. \qquad \text{Existence of a connection} \qquad (605)$$

One writes then either $a + b = c$ or $a \cdot b = c$, if this connection is read as 'plus' or 'times', or one writes simply $a\,b = c$.

2. The connection (605) is associative,

$$a \circ (b \circ c) = (a \circ b) \circ c \quad \text{for all} \quad a, b, c \in \mathsf{M}\,. \qquad \text{Associative connection} \qquad (606)$$

3. There exists a right-handed one-element e with

$$a \circ e = a \quad \text{for all} \quad a \in \mathsf{M}\,. \qquad \text{Existence of a right-handed one-element} \qquad (607)$$

If the connection is written as addition, then the one-element is called *zero element* 0.

4. There exists a right-handed inverse element a^{-1},

$$a \circ a^{-1} = e \quad \text{for all} \quad a \in \mathsf{M}\,. \qquad \text{Existence of a right-handed inverse element} \qquad (608)$$

H. Günther and V. Müller, *The Special Theory of Relativity*,
https://doi.org/10.1007/978-981-13-7783-9_10

For addition, one writes $a + (-a) = 0$.
5. The composition (605) is commutative,

$$a \circ b = b \circ a \quad \text{for all} \quad a, b \in \mathsf{M}\,. \qquad \text{Commutativity} \quad (609)$$

If the elements of a set fulfil the properties **1.** and **2.**, the set is called *semigroup* M.

If the elements of a set fulfil the properties **1.**, **2.**, **3.** and **4.**, the set is called *group* $\mathcal{M}$.

A group $\mathcal{M}$ fulfilling the additional property **5.** is called a *commutative* or ABEL *group*.

For each group $\mathcal{M}$ it holds:
There exists exactly one unity element e, and it is

$$a \circ e = e \circ a = a \quad \text{for all} \quad a \in \mathcal{M}\,. \qquad (610)$$

To each $a \in \mathcal{M}$, there exists exactly one inverse element a^{-1}, and it holds

$$a \circ a^{-1} = a^{-1} \circ a = e \quad \text{for all} \quad a \in \mathcal{M}\,. \qquad (611)$$

The equations $a \circ x = b$ and $y \circ a = b$ have each one unique solution $x = a^{-1} \circ b$ and $y = b \circ a^{-1}$. For commutative groups is $x = y$.

We speak of a *subgroup* $\mathcal{U} \subset \mathcal{M}$, when the elements of $\mathcal{U}$ consist of a non-empty subset of $\mathcal{M}$ fulfilling the properties **1.–4.**

Then it holds:
$\mathcal{U}$ is exactly then a subgroup of $\mathcal{M}$, when with $a, b \in \mathcal{U}$ also $a \circ b = c \in \mathcal{U}$ and $a^{-1} \in \mathcal{U}$ for all $a \in \mathcal{U}$.

Each group $\mathcal{M}$ has a *trivial subgroup* $\mathcal{E}$, consisting only in the unity element e.

An arbitrary subset of $\mathcal{M}$ is denoted in group theory also as *complex* $\mathtt{M}$.

Each element $a \in \mathcal{M}$ can be considered as a special complex.

The product $\mathtt{K} = \mathtt{M}\,\mathtt{N}$ of two complexes $\mathtt{M}, \mathtt{N} \subset \mathcal{M}$ consists of the set of all $c \in \mathcal{M}$ with $c = a \circ b$ for all $a \subset \mathtt{M}$ and $b \subset \mathtt{N}$.

Two complexes are called commutative, when $[\mathtt{M}, \mathtt{N}] := \mathtt{M}\,\mathtt{N} - \mathtt{N}\,\mathtt{M} = \emptyset$, where $\emptyset$ denotes the empty set it holds

The product $\mathcal{K} = \mathcal{U}\,\mathcal{V}$ of two subgroups $\mathcal{U}$, $\mathcal{V}$ is exactly then again a subgroup $\mathcal{K}$, when $[\mathcal{U}, \mathcal{V}] = \emptyset$.

For an arbitrary $a \subset \mathcal{M}$ we call the set $a\,\mathcal{U}$ of all elements $a \circ u$ for all $u \subset \mathcal{U}$ left-handed coset of $\mathcal{U}$. Correspondingly the set $\mathcal{U}\,a$ is called right-handed coset of $\mathcal{U}$.[1]

The subgroup $\mathcal{N}$ is called *normal subgroup* of group $\mathcal{M}$ or *invariant subgroup*, when a is commutable with $\mathcal{N}$,

[1] We remark: Let be $\mathcal{M} = A$ the general LORENTZ group, and $\mathcal{U} = D$ the subgroup of rotations. Then the cosets $a\,\mathcal{D}$ are transformations consisting of proper LORENTZ transformations and rotations, s. Chap. 9, Sects. 1.2 and 1.3.

$$[a, \mathcal{N}] = 0 \quad \text{for all} \quad a \in \mathcal{M}. \qquad \mathcal{N} \text{ is normal subgroup of } \mathcal{M}. \tag{612}$$

The trivial subgroup $\mathcal{E}$ is always also a trivial normal subgroup.

$\mathcal{M}$ is called *simple group*, when only the trivial normal subgroup $\mathcal{E}$ exists.

Since $a\,\mathcal{N} = \mathcal{N}\,a$ holds true for a normal subgroup $\mathcal{N}$, the product of two cosets $a\,\mathcal{N}$ and $b\,\mathcal{N}$ again are a complex, representing a coset,

$$a\,\mathcal{N}\,b\,\mathcal{N} = c\,\mathcal{N}, \text{ when } \mathcal{N} \text{ is a normal subgroup of } \mathcal{M}. \tag{613}$$

Equation (613) means: There exists an element $c \in \mathcal{M}$, so that the entirety of all compositions of elements $a \circ n$ and elements $b \circ n$ for all $n \in \mathcal{N}$ coincide with the coset $c\,\mathcal{N}$. It holds:

For a normal subgroup $\mathcal{N}$ the set of cosets $a\,\mathcal{N}$ for all $a \in \mathcal{M}$ forms a group, the *factor group* $\mathcal{M}/\mathcal{N}$ of $\mathcal{M}$.

If to each element $m \in \mathcal{M}$ of a group $\mathcal{M}$, one associates an element $n \in \mathcal{N}$ of a group $\mathcal{N}$ (mapping) with the property

$$\left.\begin{array}{l} m_1 \circ m_2 \longrightarrow n_1 \circ n_2 \;, \text{ when } m_1 \longrightarrow n_1 \text{ and } m_2 \longrightarrow n_2 \\ \text{for all } m_1, m_2 \in \mathcal{M} \text{ and } n_1, n_2 \in \mathcal{N} \;, \end{array}\right\} \quad \text{Homomorphy} \tag{614}$$

then this association is called *homomorph mapping* or a *homomorphy* of $\mathcal{M}$ *in* $\mathcal{N}$. If the mapping is surjective, i.e., to each $n \in \mathcal{N}$ there exists an image, then one calls this a homomorphy of $\mathcal{M}$ *on* $\mathcal{N}$. We write this as

$$\mathcal{M} \sim \mathcal{N}. \qquad \text{Surjective homomorphy} \tag{615}$$

Until now, we did not tell anything on the nature of the group elements. They can be numbers. For example, the natural numbers form both with the addition and the multiplication as compositions semigroups N, as can be easily verified. The integers with the addition as composition form an ABEL group $\mathcal{Z}$. Vectors and tensors of higher rank, s. Appendix B.1, also form an ABEL group with addition as composition.

The elements of a group can also be of much more general nature, e.g. transformations $\mathrm{T} = \{\alpha, \beta, \gamma, \ldots\}$. There the elements $\{\alpha, \beta, \gamma, \ldots\}$ of these groups are one-to-one maps of a set M on itself, so that to each $a \in \mathsf{M}$ it is assigned exactly one $b \in \mathsf{M}$, and each $b \in \mathsf{M}$ is exactly one times a picture. We write $b = \alpha \cdot a$. By the transformation α there must be determined an element $b \in \mathsf{M}$ for each $a \in \mathsf{M}$. Now we show that the entity of such transformations indeed forms a group.

If two transformations α and β are carried out one after another, it defines a transformation γ, hence the proposition **1.**:

$$c = \beta \cdot b \;, \quad b = \alpha \cdot a \longrightarrow c = \beta \cdot \big(\alpha \cdot a\big) = \gamma \cdot a \;, \tag{616}$$

$$\gamma = \beta \circ \alpha \,. \tag{617}$$

This composition fulfils by definition the associativity law **2.**:

$$\gamma \circ (\beta \circ \alpha) = (\gamma \circ \beta) \circ \alpha \,, \tag{618}$$

since for all $a \in M$ it holds

$$\Big(\gamma \circ (\beta \circ \alpha)\Big) \cdot a = \gamma \cdot \Big((\beta \circ \alpha) \cdot a\Big) = \gamma \cdot \Big(\beta \cdot (\alpha \cdot a)\Big)$$

and also

$$\Big((\gamma \circ \beta) \circ \alpha)\Big) \cdot a = (\gamma \circ \beta) \cdot (\alpha \cdot a) = \gamma \cdot \Big(\beta \cdot (\alpha \cdot a)\Big) .$$

Furthermore, the identic transformation ε with $a = \varepsilon \cdot a$ for all a represents the one-element, **3.**

The inverse element $(\alpha)^{-1}$ to the transformation $b = \alpha \cdot a$ is the inverse transformation $a = (\alpha)^{-1} \cdot b$, **4.**

Then the properties **1.–4.** are fulfilled, and the transformations $\{\alpha, \beta, \gamma, \ldots\}$ form a group.

Transformations in finite manifolds are called permutations $\mathcal{P} = \{\alpha, \beta, \gamma, \ldots\}$. They are well suited for demonstrating the action of a transformation group.

We consider all transformations in the set $\mathsf{M} = \{1, 2, 3\}$, hence all mappings of M on itself. This is nothing else than the set of all possible permutations of integers $1, 2, 3$, hence $\mathcal{P} = \{\varepsilon, \alpha, \beta, \gamma, \delta, \zeta\}$. Here we denote the one-element with ε. Then each element of the permutation group $\mathcal{P}$ must be assigned to an element of the set $\mathsf{M} = \{1, 2, 3\}$, all elements must appear as pictures, e.g. $\alpha \cdot 1 = 2\,,\ \alpha \cdot 2 = 3\,,\ \alpha \cdot 3 = 1\,$. When we write the pictures in a second row below the arguments (the origin) of the transformation, we can list all possible transformations of this set in the following way:

$$\left.\begin{array}{l} \varepsilon = \begin{pmatrix} 1\,2\,3 \\ 1\,2\,3 \end{pmatrix} , \quad \alpha = \begin{pmatrix} 1\,2\,3 \\ 2\,3\,1 \end{pmatrix} , \quad \beta = \begin{pmatrix} 1\,2\,3 \\ 3\,1\,2 \end{pmatrix} , \\ \gamma = \begin{pmatrix} 1\,2\,3 \\ 2\,1\,3 \end{pmatrix} , \quad \delta = \begin{pmatrix} 1\,2\,3 \\ 1\,3\,2 \end{pmatrix} , \quad \zeta = \begin{pmatrix} 1\,2\,3 \\ 3\,2\,1 \end{pmatrix} . \end{array}\right\} \tag{619}$$

With this example, one proofs easily the group laws. For example, it is

$$\beta \circ \alpha = \begin{pmatrix} 1\,2\,3 \\ 1\,2\,3 \end{pmatrix} = \varepsilon \,, \quad \zeta \circ \delta = \begin{pmatrix} 1\,2\,3 \\ 2\,3\,1 \end{pmatrix} = \alpha \,, \quad \cdots \tag{620}$$

and in the application on single elements,

$$\Big(\beta \circ (\zeta \circ \delta)\Big) \cdot 3 = (\beta \circ \alpha) \cdot 3 = \varepsilon \cdot 3 = 3 \,,$$

$$(\beta \circ \zeta) \cdot (\delta \cdot 3) = (\beta \circ \zeta) \cdot 2 = \beta \cdot (\zeta \cdot 2) = \beta \cdot 1 = 3 \ .$$

Now we consider the four-dimensional MINKOWSKI space $\mathbb{M}_4$. Each point of $\mathbb{M}_4$ is determined by four coordinates $(c\,t, x, y, z) := (x^o, x^1, x^2, x^3)$, that assign to an event in an inertial system Σ with Cartesian coordinates (x, y, z) and the with the speed of light multiplied time $c\,t$, the measuring values for space and time measurements, s. Chap. 9, Sect. 1.5. The transformations $\mathcal{T} = \{\mathbf{f}, \mathbf{g}, \ldots\}$ in MINKOWSKI space form a group, they are called coordinate transformations. We write

$$\mathbf{x} = \begin{pmatrix} x^o \\ x^1 \\ x^2 \\ x^3 \end{pmatrix}, \quad \text{hence} \quad \mathbf{x}^T = (x^o, x^1, x^2, x^3),$$

$$\mathbf{x}' = \mathbf{f}(\mathbf{x}). \qquad \text{Coordinate transformation} \qquad (621)$$

The elements $\mathbf{f}$ of group $\mathcal{T}$ are now four in general as continuous assumed functions, depending on four variables (x^o, x^1, x^2, x^3). One denotes this as a *continuous transformation group.*

The physical interesting quantity is given by the line element $ds^2 = c^2dt^2 - dx^2 - dy^2 - dz^2$, Chap. 9, Sect. 1. The physical requirement is to look for such coordinate transformations that leave the form of the line element invariant, s. Eq. (318),

$$c^2dt^2 - dx^2 - dy^2 - dz^2 = c^2dt'^2 - dx'^2 - dy'^2 - dz'^2 . \qquad (622)$$

Equation (622) defines pseudo-orthogonal transformations in MINKOWSKI space that provide a transition from one inertial system to another one, including also spatial rotations of the coordinate axes. In the following, we keep fixed the origin of coordinates. Therefore, we consider linear homogeneous transformations, that represent the general LORENTZ group $\mathcal{A}$. Furthermore, we consider transformations where the relative orientation of the axes is preserved. The elements of $\mathcal{A}$ are square four-row regular matrices that depend on six parameters:

$$\mathcal{A} = \{\mathbf{A}(\alpha_1, \alpha_2, \alpha_3, \beta_1, \beta_2, \beta_3)\} . \qquad (623)$$

The first three parameters describe spatial rotations, and the last three parameters the velocity vector for the change between inertial systems, Chap. 9, Sect. 1.2. We write

$$\mathbf{x}' = \mathbf{A}\,\mathbf{x} . \qquad \text{General homogeneous LORENTZ transformation} \qquad (624)$$

The linear coordinate transformation is a multiplication of matrices. The evaluation of the matrix $\mathbf{A}$ is described extensively in Chap. 9, Sect. 1.2.

2 Tensorial Representations of the LORENTZ Group Relativistic Mechanics and Electrodynamics

A homomorphism between a given group $\mathcal{M}$, Eq. (614), and a group $\mathcal{X}$ of regular matrices of the same rank is called a *linear representation* of the group $\mathcal{M}$. If in addition two different elements of $\mathcal{M}$ are always represented by different matrices of $\mathcal{X}$, we have the case of an *isomorphism*, also denoted as a *true linear representation* of $\mathcal{M}$.

If the elements of $\mathcal{M}$ are itself a group of matrices, one calls $\mathcal{M}$ a self-representation. The group of LORENTZ matrices $\mathbf{A}$ is therefore a self-representation of $\mathcal{A}$, that describes the linear transformations (624) in MINKOWSKI space $\mathbb{M}_4$.

One gets further representations of the LORENTZ group when the MINKOWSKI space $\mathbb{M}_4$ is described by new representation spaces.

Instead of Eq. (624), we write generally at first for contravariant vectors $\mathbf{a}$,

$$\mathbf{a}' = \mathbf{A}\,\mathbf{a}\,. \tag{625}$$

All tensorial indices run through the values $0, 1, 2, 3$, and we write with use of the sum convention, Appendix B.1,

$$a^{i'} = A^{i'}_{i}\,a^{i} \tag{626}$$

with the representation space $\mathbb{T}_4$ formed by vectors a^i.

For the metric $\mathbf{g}$, we get the to $\mathbf{a}$ belonging covariant vectors $a_i = g_{ik}a^k$, or written more compact as $(\mathbf{g}\cdot\mathbf{a})$, then we have in the representation space $\mathbb{T}_4$ the following transformations:

$$\mathbf{g}\cdot\mathbf{a}' = (\mathbf{A}^T)^{-1}\,\mathbf{g}\cdot\mathbf{a} \tag{627}$$

or in components

$$a_{i'} = A^{i}_{i'}a_i\,. \tag{628}$$

Here $A^{i}_{i'}$ is the to $A^{i'}_{i}$ inverse and transposed matrix, that exists due to the supposed regularity of the representation, i.e. the non-vanishing determinant of $A^{i'}_{i}$.

Now we consider a linear space $\mathbb{T}_{4^n}$ of 4^n dimensions as new representation space. We denote an element of this space as contravariant tensor $a^{i_1 i_2\cdots i_n} := \mathbf{a}^{(n)}$, that transforms as the product of n contravariant vectors, Eqs. (625), (626). A transformation in MINKOWSKI space, Eqs. (624) or (625), is then connected with the following transformation in the linear space $\mathbb{T}_{4^n}$:

$$\mathbf{a}^{(n)'} = (\mathbf{A})^{\mathbf{n}}\,\mathbf{a}^{(\mathbf{n})}\,, \qquad \text{A linear representation of the LORENTZ group} \tag{629}$$

or in components

$$a^{i'_1 i'_2 \cdots i'_n} = A^{i'_1}_{i_1} A^{i'_2}_{i_2} \cdots A^{i'_n}_{i_n} a^{i_1 i_2 \cdots i_n} \, . \qquad \text{A linear representation of the LORENTZ group} \tag{630}$$

The transformations (630) form a faithful linear 4^n-dimensional representation of the LORENTZ group (624). The elements $a^{i_1 i_2 \cdots i_n}$ of the representation space $\mathbb{T}_{4^n}$ are called *contravariant tensors* of rank n.

Important for understanding this representation space is the following remark:

Equations (630) describe exactly the transformation laws of products of n contravariant vectors, as seen for the case $a^{i_1 i_2 \cdots i_n} = a^{i_1} b^{i_2} c^{i_3} \cdots v^{i_n}$ directly from Eq. (626). However, the elements of the space $\mathbb{T}_{4^n}$ cannot be reduced to such products. For example, arbitrary sums of such products represent likewise elements of this space.

Likewise, one can consider a linear space $\widetilde{\mathbb{T}}_{4^n}$ of 4^n dimensions as representation space, with elements written as covariant tensors of rank n, i.e. $a^{i_1 i_2 \cdots i_n} := \mathbf{a}_{(n)}$, that transform as products of n covariant vectors according to Eqs. (627), (628). A transformation (624) or (625) in MINKOWSKI space is now connected with the following transformation in our linear space $\mathbb{T}_{4^n}$:

$$\mathbf{a}_{(n)}{}' = ((\mathbf{A}^T)^{-1})^n \, \mathbf{a}_{(n)} \, , \qquad \text{A linear representation of the LORENTZ group} \tag{631}$$

or in components

$$a_{i'_1 i'_2 \cdots i'_n} = A^{i_1}_{i'_1} A^{i_2}_{i'_2} \cdots A^{i_n}_{i'_n} a_{i_1 i_2 \cdots i_n} \, . \qquad \text{A linear representation of the LORENTZ-group} \tag{632}$$

Here the matrices $A^{i_1}_{i'_1}$ are inverse and transponed with respect to $A^{i'_1}_{i_1}$, existing due to the supposed regularity.

The transformations (632) represent also a faithful linear 4^n-dimensional representation of the LORENTZ group $\mathcal{A}$, and the remark above keeps true that the elements of the representation space $\widetilde{\mathbb{T}}_{4^n}$ cannot be reduced on the n-fold products of covariant vectors.

One can also introduce mixed-variant tensors as twofold covariant and single contravariant tensors $a^{i^3}_{i_1 i_2}$ with a 4^3-dimensional representation of the LORENTZ transformation (624)

$$a^{i'_3}_{i'_1 i'_2} = A^{i_1}_{i'_1} A^{i_2}_{i'_2} A^{i'_3}_{i_3} a^{i_3}_{i_1 i_2} \, . \qquad \text{A linear representation of the LORENTZ group} \tag{633}$$

So with the linear representations of the LORENTZ group we have introduced *tensors* in MINKOWSKI space. Due to the assumed metric, there follows clearly the representation (630) from (632) and inverse.

Using the formulation of EINSTEIN's relativity principle in MINKOWSKI space, Chap. 9, Sect. 1.6, we discussed in Chap. 9, Sects. 2 and 3 the equations of mechanics and electrodynamics as tensor equations in the MINKOWSKI space.

We summarise: The relativistic equations of mechanics, cp. Eq. (437), read as follows:

Covariant form of relativistic mechanics in MINKOWSKI *space* :

$$\frac{d}{d\tau}\,p^i = m_o\,\frac{d}{d\tau}\,u^i = m_o\,\frac{d}{d\tau}\,\frac{dx^i}{d\tau} = F^i\,,$$ Second axiom of relativistic mechanics

$$p^i = \frac{m_o}{\sqrt{1-u^2/c^2}}\,u^i = \text{const} \quad \text{for} \quad F^i = 0\,.$$ First axiom of relativistic mechanics (437)

If there exist only inner forces:

$$\frac{d}{d\tau}\sum_{p=1}^{n} p_p^i = \frac{d}{d\tau}P^i = \frac{1}{\sqrt{1-u^2/c^2}}\,\frac{d}{dt}\,P^i = 0\,.$$ Third axiom of relativistic mechanics

Eqs. (547) of electrodynamics in vacuum:

$$\left.\begin{aligned}
&(a)\ \frac{\partial}{\partial x^i}\,F_{kl} + \frac{\partial}{\partial x^k}\,F_{li} + \frac{\partial}{\partial x^l}\,F_{ik} = 0\ ,\\
&(b)\ \frac{\partial}{\partial x^i}\,F^{ik} = \mu_o\,j^k\ ,\\
&(c)\ f^i = F^{ik}\,j_k = \rho_e\,F^{ik}\,u_k\ ,\\
&\quad c = \frac{1}{\sqrt{\varepsilon_o\,\mu_o}}\ .
\end{aligned}\right\}$$ Covariant form of MAXWELL's equations in vacuum (547)

MINKOWSKI's covariant form (580) of electrodynamics of moving media,

$$\left.\begin{aligned}
&(a)\ \frac{\partial}{\partial x^i}\,F_{kl} + \frac{\partial}{\partial x^k}\,F_{li} + \frac{\partial}{\partial x^l}\,F_{ik} = 0\,,\\
&(b)\ \frac{\partial}{\partial x^i}\,H^{ik} = \mu_o\,J^k\,,\\
&(c)\ f^i = F^{ik}\,\frac{1}{c}\,J_k\ ,\\
&(d)\ H_{ik}\,w^k = \varepsilon\,F_{ik}\,w^k\,,\\
&(e)\ F_{ik}w_l + F_{kl}w_i + F_{li}w_k = \mu\left(H_{ik}w_l + H_{kl}w_i + H_{li}w_k\right),\\
&(f)\ \iota_i = -\sigma\,F_{ik}\,w^k\,,\\
&(g)\ \varepsilon = \varepsilon_o\,\varepsilon_r\ ,\ h)\ \ \mu = \mu_o\,\mu_r\ ,\ h)\ \ J^i = j^i + \iota^i\,.
\end{aligned}\right\}$$ Covariant form of MAXWELL's equations of moving media (580)

Besides the tensorial representations (630), (631), (632), ... there exist also further linear representations of the LORENTZ group $\mathcal{A}$, its representation spaces define the *spinors*, that became important for modern physics.

3 Spinorial Representations of the LORENTZ Group WEYL Equation and DIRAC Equation

3.1 The Group $\mathcal{C}_2$

The two-dimensional complex unimodular matrices **C**,

$$\mathbf{C} = \begin{pmatrix} t & q \\ r & s \end{pmatrix} \tag{634}$$

with

$$\begin{vmatrix} t & q \\ r & s \end{vmatrix} := t\,s - q\,r = 1\,, \qquad t, q, r, s \quad \text{complex numbers}\,, \tag{635}$$

form a group $\mathcal{C}_2$, that depends only on *six* real parameters because of the complex condition in Eq. (635). Also, the homogeneous LORENTZ group $\mathcal{A}$, restricted on such transformations, that keep invariant the orientation of axes in MINKOWSKI space, depends on *six* real parameters.

There exists a homomorph mapping of the LORENTZ group $\mathcal{A}$ on the complex group $\mathcal{C}_2$. We do not provide a formal proof for the existence of this homomorphism. But we shall see, cp. e.g. Eqs. (672) and (680), how we can evaluate the matrices **C**, that correspond to a certain LORENTZ matrix **A**. Therefore, it holds

$$\mathcal{A} \sim \mathcal{C}_2\,. \tag{636}$$

Each representation of group $\mathcal{C}_2$ is then also a representation of the LORENTZ group $\mathcal{A}$. Now we will investigate the representations of group $\mathcal{C}_2$.

Just as the group of LORENTZ matrices **A** is a proper representation of $\mathcal{A}$ in MINKOWSKI space $\mathbb{M}_4$, we consider at first the proper representations of group $\mathcal{C}_2$.

The matrices (634) are interpreted as transformation matrices in a two-dimensional complex space $\mathbb{U}_2$ with elements

$$\mathbf{u} = \begin{pmatrix} u_1 \\ u_2 \end{pmatrix} \quad \text{and} \quad \mathbf{u}^T = (u_1,\, u_2)\,. \tag{637}$$

Analogously to Eqs. (624), (625) in case of LORENTZ transformations, we write

$$\mathbf{u}' = \mathbf{C}\,\mathbf{u} \tag{638}$$

or in components, with a dash at the index for the transformed quantities as used for tensors,

$$\left.\begin{aligned} u_{1'} &= t\,u_1 + q\,u_2\,, \\ u_{2'} &= r\,u_1 + s\,u_2\,. \end{aligned}\right\} \tag{639}$$

The elements $(u_1, u_2) \subset \mathbb{U}_2$ are called - due to the homomorphism (615) - *elementary two-dimensional spinors* assigned to the MINKOWSKI space $\mathbb{M}_4$.

To each spinor (u_1, u_2) now we assign the complex conjugate quantity $(u_{\dot{1}}, u_{\dot{2}})$. Here we follow the convention in spinor calculus, to denote the complex conjugation for spinors, and only there, by a point over the quantity, and when a quantity has an index, by a point over the index. Hence with real $\alpha, \beta, \gamma, \delta$

$$\left.\begin{aligned} u &= \alpha + i\beta\,, \ \dot{u} = \alpha - i\beta\,, \\ u_1 &= \gamma + i\delta\,, \ u_{\dot{1}} = \gamma - i\delta\,. \end{aligned}\right\} \tag{640}$$

To the representation space $\mathbb{U}_2$ of elementary spinors, we assign the space $\mathbb{U}_{\dot{2}}$ of elementary spinors with overdot, hence $\dot{\mathbf{u}}^T = (u_{\dot{1}}, u_{\dot{2}}) \subset \mathbb{U}_{\dot{2}}$.

Each transformation (638), (639) in the space of spinors $\mathbb{U}_2$ is assigned a transformation in the space $\mathbb{U}_{\dot{2}}$ spinors with overdot,

$$\dot{\mathbf{u}}' = \bar{\mathbf{C}}\,\dot{u} \tag{641}$$

or in components, again with a dash at the indices,

$$\left.\begin{aligned} u_{\dot{1}'} &= \bar{t}\,u_{\dot{1}} + \bar{q}\,u_{\dot{2}}\,, \\ u_{\dot{2}'} &= \bar{r}\,u_{\dot{1}} + \bar{s}\,u_{\dot{2}}\,, \end{aligned}\right\} \tag{642}$$

where a complex conjugation of a matrix $\mathbf{C}$ is again indicated by a bar over the quantity, hence, e.g. $\bar{t} = \alpha - i\beta$, when $t = \alpha + i\beta$.

We remark, that Eqs. (639) and (642) describe two *different* transformations. Obviously, there result different dashed, hence differently transformed spinors for numerically identical spinors with and without overdot due to the complex elements of matrix $\mathbf{C}$.

With the transformations (639) in representation space $\mathbb{U}_2$ and Eq. (642) in representation space $\mathbb{U}_{\dot{2}}$ we have found two *different representations* of the group $\mathcal{C}_2$.

For getting at best all representations of group $\mathcal{C}_2$, we proceed similarly as for the tensorial representations of the LORENTZ group $\mathcal{A}$. We construct from spaces $\mathbb{U}_2$ and $\mathbb{U}_{\dot{2}}$ all higher dimensional representation spaces.

Analogously to tensors in Eq. (632), we form a representation space $\mathbb{P}_{\jmath\,\tilde{\jmath}}$, with spinors $\mathbf{p}_{\jmath\,\tilde{\jmath}}$ of higher degree as elements. We write

$$\mathbf{p}_{\jmath\,\tilde{\jmath}} \quad \text{or with corresponding indices} \quad p_{a_1 a_2 \ldots a_{2\jmath}\, \dot{b}_1 \dot{b}_2 \ldots \dot{b}_{2\tilde{\jmath}}}\,. \tag{643}$$

There we admit for $\jmath$ and $\tilde{\jmath}$ all half-integer values, independently one from the other, hence

$$\left.\begin{aligned} \jmath &= 0\,, \frac{1}{2}\,, 1\,, \frac{3}{2}\,, 2\,, \frac{5}{2}\,, 3\,, \ldots\,, \\ \tilde{\jmath} &= 0\,, \frac{1}{2}\,, 1\,, \frac{3}{2}\,, 2\,, \frac{5}{2}\,, 3\,, \ldots \end{aligned}\right\} \tag{644}$$

There we are concerned with spinor indices that always have values 1, 2,

$$\text{spinor indices:} \quad a_1,\, a_2,\, \ldots,\, \dot{b}_1, \dot{b}_2, \ldots = 1, 2\,. \tag{645}$$

Special spinors of type (643) are constructed from elementary spinors by means of quantities $\pi_{kk'}$, defined as

$$\pi_{kl} := u_1^{2\jmath - k} u_2^k u_{\dot 1}^{2\tilde\jmath - l} u_{\dot 2}^l \quad \text{with } k, l \text{ integer and } 0 \le k \le 2\jmath\,,\ 0 \le l \le 2\tilde\jmath. \tag{646}$$

Here the number $2\jmath$ counts how often the elementary spinors appear as factors, and $2\tilde\jmath$, how often the elementary spinors with overdot appear as factors. When the indices k, l in Eq. (646) run over the given values, so we have defined all components of $p_{a_1 a_2 \ldots a_{2\jmath}\, \dot b_1 \dot b_2 \ldots \dot b_{2\tilde\jmath}}$. We provide three examples:

$$\begin{aligned}
k = 0,\quad l = 0\ &: \pi_{11} &&= p_{11\ldots1\,\dot1\dot1\ldots\dot1} = u_1^{2\jmath}\, u_{\dot1}^{2\tilde\jmath}\,,\\
k = 2\jmath\,,\ l = 2\tilde\jmath\ &: \pi_{2\jmath\,2\tilde\jmath} &&= p_{22\ldots2\,\dot2\dot2\ldots\dot2} = u_2^{2\jmath}\, u_{\dot2}^{2\tilde\jmath}\,,\\
k = 0,\quad l = 2\tilde\jmath\ &: \pi_{12\tilde\jmath} &&= p_{11\ldots1\,\dot2\dot2\ldots\dot2} = u_1^{2\jmath}\, u_{\dot2}^{2\tilde\jmath}\,,\quad \text{etc.}
\end{aligned} \tag{647}$$

$\mathbb{P}_{\jmath\,\tilde\jmath}$ now becomes a representation space of group $\mathcal{C}_2$, when we require that the elements $p_{a_1 a_2 \ldots a_{2\jmath}\, \dot b_1 \dot b_2 \ldots \dot b_{2\tilde\jmath}} \subset \mathbb{P}_{\jmath\,\tilde\jmath}$ transform as the special elements formed from products of elementary spinors with Eq. (646). We remember that we proceeded in the same way also for the representation of tensors.

The representation spaces $\mathbb{P}_{\jmath\,\tilde\jmath}$ have dimension $(2\jmath + 1)(2\tilde\jmath + 1)$. We write down three of these representation spaces explicitly with its elements:

$$\left.\begin{aligned}
&\mathbf{1.}\ \jmath = \tfrac12\,,\ \tilde\jmath = 0\,,\ \text{hence}\quad k = 0, 1\,,\quad l = 0\,.\ \text{It follows}\\
&\mathbb{P}_{\frac12 0}\ \text{with elements}\qquad \mathbf{p}^T_{\frac12 0} = (p_{1\dot1}, p_{2\dot1}) = (u_1, u_2)\,.
\end{aligned}\right\} \tag{648}$$

Hence $\mathbb{P}_{\frac12 0} = \mathbb{U}_2$ is nothing else than the representation space of elementary spinors.

$$\left.\begin{aligned}
&\mathbf{2.}\ \jmath = 0\,,\ \tilde\jmath = \tfrac12\,,\ \text{hence}\quad k = 0\,,\quad l = 0, 1\,.\ \text{It follows}\\
&\mathbb{P}_{0\frac12}\ \text{with elements}\qquad \mathbf{p}^T_{0\frac12} = (p_{1\dot1}, p_{1\dot2}) = (u_{\dot1}, u_{\dot2})\,.
\end{aligned}\right\} \tag{649}$$

It is now also $\mathbb{P}_{0\frac12} = \mathbb{U}_{\dot2}$.

$$\left.\begin{aligned}
&\mathbf{3.}\ \jmath = \tfrac12\,,\ \tilde\jmath = \tfrac12\,,\ \text{hence}\quad k = 0, 1\,,\quad l = 0, 1\,.\ \text{It follows}\\
&\mathbb{P}_{\frac12\frac12}\ \text{with elements}\qquad \mathbf{p}^T_{\frac12\frac12} = (p_{1\dot1}, p_{1\dot2}, p_{2\dot1}, p_{2\dot2})\,.
\end{aligned}\right\} \tag{650}$$

The representation space $\mathbb{P}_{\frac12\frac12}$ now will us lead to the connection of the group $\mathcal{C}_2$ with the LORENTZ group $\mathcal{A}$.

3.2 Connection of C_2 to the LORENTZ Group $\mathcal{A}$

Since the elements $\mathbf{p}^T_{\frac{1}{2}\frac{1}{2}} = (p_{1\dot{1}}, p_{1\dot{2}}, p_{2\dot{1}}, p_{2\dot{2}})$ transform as the products of spinors with and without overdot $(u_1 u_{\dot{1}}, u_1 u_{\dot{2}}, u_2 u_{\dot{2}}, u_2 u_{\dot{2}})$, we can determine by transformations, Eqs. (639) and (642), the transformed elements $\mathbf{p}^{T\ \prime}_{\frac{1}{2}\frac{1}{2}} = (p_{1'\dot{1}'}, p_{1'\dot{2}'}, p_{2'\dot{1}'}, p_{2'\dot{2}'})$. There follows the following transformation with simple matrix multiplication:

$$\begin{pmatrix} p_{1'\dot{1}'} \\ p_{1'\dot{2}'} \\ p_{2'\dot{1}'} \\ p_{2'\dot{2}'} \end{pmatrix} = \begin{pmatrix} t\bar{t} & t\bar{q} & q\bar{t} & q\bar{q} \\ t\bar{r} & t\bar{s} & q\bar{r} & q\bar{s} \\ r\bar{t} & r\bar{q} & s\bar{t} & s\bar{q} \\ r\bar{r} & r\bar{s} & s\bar{r} & s\bar{s} \end{pmatrix} \begin{pmatrix} p_{1\dot{1}} \\ p_{1\dot{2}} \\ p_{2\dot{1}} \\ p_{2\dot{2}} \end{pmatrix} \tag{651}$$

or in matrix form, when we write in the following for simplicity instead of $\mathbf{p}_{\frac{1}{2}\frac{1}{2}}$ simply $\mathbf{p}$, and the matrix in Eq. (651), also denoted as KRONECKER product of matrices $\mathbf{C}$ and $\bar{\mathbf{C}}$, is simply written[2] as $\mathbf{M}$

$$\mathbf{p}' = \mathbf{M}\,\mathbf{p} \quad \text{where} \quad \mathbf{M} = \mathbf{C} \otimes \bar{\mathbf{C}}\,. \tag{652}$$

With Eq. (651), we derive the expression

$$\begin{aligned} p_{1'\dot{2}'}p_{2'\dot{1}'} - p_{1'\dot{1}'}p_{2'\dot{2}'} = \; & (t\bar{r}p_{1\dot{1}} + t\bar{s}p_{1\dot{2}} + q\bar{r}p_{2\dot{1}} + q\bar{s}p_{2\dot{2}})(r\bar{t}p_{1\dot{1}} + r\bar{q}p_{1\dot{2}} + s\bar{t}p_{2\dot{1}} + s\bar{q}p_{2\dot{2}}) \\ & - (t\bar{t}p_{1\dot{1}} + t\bar{q}p_{1\dot{2}} + q\bar{t}p_{2\dot{1}} + q\bar{q}p_{2\dot{2}})(r\bar{r}p_{1\dot{1}} + r\bar{s}p_{1\dot{2}} + s\bar{r}p_{2\dot{1}} + s\bar{s}p_{2\dot{2}})\,. \end{aligned}$$

The careful multiplication provides the following relation, when taking into account the unimodularity (635) of $\mathbf{C}$, $t\,s - q\,r = 1$, and the following unimodularity of $\bar{\mathbf{C}}$, $\bar{t}\,\bar{s} - \bar{q}\,\bar{r} = 1$:

$$p_{1'\dot{2}'}p_{2'\dot{1}'} - p_{1'\dot{1}'}p_{2'\dot{2}'} = p_{1\dot{1}}p_{2\dot{2}} - p_{1\dot{2}}p_{2\dot{1}}\,. \tag{653}$$

From elements $\mathbf{p}$ in representation space $\mathbb{P}_{\frac{1}{2}\frac{1}{2}}$, now we form elements $\mathbf{V}$ with a regular matrix $\mathbf{T}$. We shall see that $\mathbf{V}$ represents vectors in MINKOWSKI space. There are possible different choices of this matrix $\mathbf{T}$. The consideration is most simple for

$$\mathbf{T} = \frac{1}{\sqrt{2}} \begin{pmatrix} 1 & 0 & 0 & 1 \\ 0 & -1 & -1 & 0 \\ 0 & -i & i & 0 \\ -1 & 0 & 0 & 1 \end{pmatrix} \quad \longleftrightarrow \quad \mathbf{T}^{-1} = \frac{1}{\sqrt{2}} \begin{pmatrix} 1 & 0 & 0 & -1 \\ 0 & -1 & i & 0 \\ 0 & -1 & -i & 0 \\ 1 & 0 & 0 & 1 \end{pmatrix}, \tag{654}$$

so that

[2] Here it is possible to use the same symbol as for the tensor product, cp. Eq. (898), namely $\otimes$.

$$\mathbf{V} = \frac{1}{\sqrt{2}} \mathbf{T} \, \mathbf{p} \quad \longleftrightarrow \quad \mathbf{p} = \sqrt{2} \, \mathbf{T}^{-1} \, \mathbf{V} \,, \tag{655}$$

or in matrix language,

$$\begin{pmatrix} V^0 \\ V^1 \\ V^2 \\ V^3 \end{pmatrix} = \frac{1}{2} \begin{pmatrix} 1 & 0 & 0 & 1 \\ 0 & -1 & -1 & 0 \\ 0 & -i & i & 0 \\ -1 & 0 & 0 & 1 \end{pmatrix} \begin{pmatrix} p_{1\dot{1}} \\ p_{1\dot{2}} \\ p_{2\dot{1}} \\ p_{2\dot{2}} \end{pmatrix} \quad \longleftrightarrow \quad \begin{pmatrix} p_{1\dot{1}} \\ p_{1\dot{2}} \\ p_{2\dot{1}} \\ p_{2\dot{2}} \end{pmatrix} = \begin{pmatrix} 1 & 0 & 0 & -1 \\ 0 & -1 & i & 0 \\ 0 & -1 & -i & 0 \\ 1 & 0 & 0 & 1 \end{pmatrix} \begin{pmatrix} V^0 \\ V^1 \\ V^2 \\ V^3 \end{pmatrix} , \tag{656}$$

and there hold similar equations for elements $\mathbf{p}'$ and $\mathbf{V}'$ after a LORENTZ transformation.

Written explicitly it holds Eq. (656)

$$\left.\begin{array}{ll} V^0 = +\frac{1}{2}(p_{1\dot{1}} + p_{2\dot{2}}) \,, & p_{1\dot{1}} = +V^0 - V^3 \,, \\ V^1 = -\frac{1}{2}(p_{1\dot{2}} + p_{2\dot{1}}) \,, & p_{1\dot{2}} = -V^1 + i\,V^2 \,, \\ V^2 = -\frac{i}{2}(p_{1\dot{2}} - p_{2\dot{1}}) \,, \quad \longleftrightarrow & p_{2\dot{1}} = -V^1 - i\,V^2 \,, \\ V^3 = -\frac{1}{2}(p_{1\dot{1}} - p_{2\dot{2}}) \,, & p_{2\dot{2}} = +V^0 + V^3 \,. \end{array}\right\} \tag{657}$$

With the metric tensor in MINKOWSKI space η_{ik}, s. Eqs. (312), (313), we define the element V_i by

$$V_i := \eta_{ik} V^k \,, \quad \text{hence} \quad V_0 = V^0 \,, \;\; V_1 = -V^1 \,, \;\; V_2 = -V^2 \,, \;\; V_3 = -V^3 \,. \tag{658}$$

We insert Eq. (657) into (653) and find after simple evaluation

$$\left.\begin{array}{l} V^{i'} V^{k'} \eta_{ik} = V^i V^k \eta_{ik} \,, \quad \text{i.e.} \\ {V^{0'}}^2 - {V^{1'}}^2 - {V^{2'}}^2 - {V^{3'}}^2 = {V^0}^2 - {V^1}^2 - {V^2}^2 - {V^3}^2 \end{array}\right\} . \tag{659}$$

For a simple example, we write

$$\left.\begin{array}{llll} V^0 = dx^0 = c dt \,, & V^1 = dx^1 = dx \,, & V^2 = dx^2 = dy \,, & V^3 = dx^3 = dz \,, \\ V^{0'} = dx^{0'} = c dt' \,, & V^{1'} = dx^{1'} = dx' \,, & V^{2'} = dx^{2'} = dy' \,, & V^{3'} = dx^{3'} = dz' \,, \end{array}\right\}$$

then Eq. (659) is nothing else than the invariance of the line element, Eq. (318),

$$\left.\begin{array}{l} dx^{i'} dx^{k'} \eta_{ik} = dx^i dx^k \eta_{ik} \,, \quad \text{i.e.} \\ c dt'^2 - dx'^2 - dy'^2 - dz'^2 = c dt^2 - dx^2 - dy^2 - dz^2 \,. \end{array}\right\} \tag{660}$$

Therefore, it holds:

The elements $\mathbf{p}$ of the representation space $\mathbb{P}_{\frac{1}{2}\frac{1}{2}}$ are equivalent to vectors $\mathbf{V}$ in the MINKOWSKI space.

For the transformed elements $\mathbf{V}'$, there holds instead of Eqs. (651) or (652)

$$\mathbf{V}' = \frac{1}{\sqrt{2}}\,\mathbf{T}\,\mathbf{p}' = \frac{1}{\sqrt{2}}\,\mathbf{T}\,\mathbf{M}\,\mathbf{p} = \frac{1}{\sqrt{2}}\,\mathbf{T}\,\mathbf{M}\,\sqrt{2}\;\mathbf{T}^{-1}\,\mathbf{V}\,, \quad \text{hence}$$

$$\mathbf{V}' = \mathbf{T}\,\mathbf{M}\,\mathbf{T}^{-1}\,\mathbf{V}\,, \tag{661}$$

or written in components,

$$\begin{pmatrix} V^{0'} \\ V^{1'} \\ V^{2'} \\ V^{3'} \end{pmatrix} = \frac{1}{2}\begin{pmatrix} 1 & 0 & 0 & -1 \\ 0 & -1 & -1 & 0 \\ 0 & -i & i & 0 \\ -1 & 0 & 0 & 1 \end{pmatrix}\begin{pmatrix} t\bar{t} & t\bar{q} & q\bar{t} & q\bar{q} \\ t\bar{r} & t\bar{s} & q\bar{r} & q\bar{s} \\ r\bar{t} & r\bar{q} & s\bar{t} & s\bar{q} \\ r\bar{r} & r\bar{s} & s\bar{r} & s\bar{s} \end{pmatrix}\begin{pmatrix} 1 & 0 & 0 & -1 \\ 0 & -1 & i & 0 \\ 0 & -1 & -i & 0 \\ 1 & 0 & 0 & 1 \end{pmatrix}\begin{pmatrix} V^0 \\ V^1 \\ V^2 \\ V^3 \end{pmatrix}. \tag{662}$$

In Chap. 9, Sect. 1.2 we have derived the general LORENTZ transformation **A** from the invariance of the line element (660), s. also Eqs. (351), (371),

$$\begin{pmatrix} V^{0'} \\ V^{1'} \\ V^{2'} \\ V^{2'} \end{pmatrix} = \begin{pmatrix} A_0^{0'} & A_1^{0'} & A_2^{0'} & A_3^{0'} \\ A_0^{1'} & A_1^{1'} & A_2^{1'} & A_3^{1'} \\ A_0^{1'} & A_1^{2'} & A_2^{2'} & A_3^{2'} \\ A_0^{1'} & A_1^{3'} & A_2^{3'} & A_3^{3'} \end{pmatrix}\begin{pmatrix} V^0 \\ V^1 \\ V^2 \\ V^3 \end{pmatrix}. \tag{663}$$

From Eqs. (661) and (652), we read off

$$\mathbf{A} = \mathbf{T}\left(\mathbf{C}\otimes\bar{\mathbf{C}}\right)\mathbf{T}^{-1}\,. \tag{664}$$

Equation (664) proofs the homomorphisms between the LORENTZ group $\mathcal{A}$ and the group $\mathcal{C}_2$.

Now we shall evaluate the matrices **C** in two simple examples, on which the corresponding LORENTZ matrices **A** are depicted.

1. We consider poor spatial rotations in the x-y-plane, a rotation around the z-axis denoted with $\mathbf{D}_3$. For the matrix $\mathbf{D}_3$ we can write, s. Eq. (346),

$$\mathbf{D}_3 = \begin{pmatrix} 1 & 0 & 0 & 0 \\ 0 & \cos\alpha_3 & \sin\alpha_3 & 0 \\ 0 & -\sin\alpha_3 & \cos\alpha_3 & 0 \\ 0 & 0 & 0 & 1 \end{pmatrix}. \tag{665}$$

These transformations are

$$\left.\begin{aligned} x^{0'} &= x^0\,, \\ x^{1'} &= x^1\cos\alpha_3 + x^2\sin\alpha_3\,, \\ x^{2'} &= -x^1\sin\alpha_3 + x^2\cos\alpha_3\,, \\ x^{3'} &= x^3\,. \end{aligned}\right\} \tag{666}$$

With respect to the expressions (657), we write for Eq. (666) after a simple evaluation and using $e^{i\varphi} = \cos\varphi + i\sin\varphi$,

$$\left.\begin{aligned}(+x^{0'} - x^{3'}) &= (+x^0 - x^3)\,,\\(-x^{1'} + ix^{2'}) &= (-x^1 + ix^2)\; e^{+i\alpha_3}\,,\\(-x^{1'} - ix^{2'}) &= (-x^1 - ix^2)\; e^{-i\alpha_3}\,,\\(+x^{0'} + x^{3'}) &= (+x^0 + x^3)\,.\end{aligned}\right\} \tag{667}$$

When we now replace in Eq. (657) (V^0, V^1, V^2, V^3) by (x^0, x^1, x^2, x^3), we get for the transformation of elements $\mathbf{p}$ in the representation space $\mathbb{P}_{\frac{1}{2}\frac{1}{2}}$

$$\left.\begin{aligned}p_{1'\dot{1}'} &= p_{1\dot{1}}\,,\\p_{1'\dot{2}'} &= p_{1\dot{2}}\, e^{+i\alpha_3}\,,\\p_{2'\dot{1}'} &= p_{2\dot{1}}\, e^{-i\alpha_3}\,,\\p_{2'\dot{2}'} &= p_{2\dot{2}}\,.\end{aligned}\right\} \tag{668}$$

For getting from this equation the transformation (651), there must hold

$$\left.\begin{aligned}&t\bar{t} = 1\,,\\&t\bar{s} = e^{+i\alpha_3}\,,\\&s\bar{t} = e^{-i\alpha_3}\,,\\&s\bar{s} = 1\,,\\&\text{All other elements of matrix } \mathbf{M} \text{ have to vanish.}\end{aligned}\right\} \tag{669}$$

These equations can be fulfilled by

$$\left.\begin{aligned}t &= e^{\frac{i}{2}\alpha_3}\,,\\q &= 0\,,\\r &= 0\,,\\s &= e^{-\frac{i}{2}\alpha_3}\,.\end{aligned}\right\} \tag{670}$$

Since conditions (669) consist only of products of two elements of matrix $\mathbf{C}$, we get a second solution in inserting in each relation a minus sign:

$$\left.\begin{aligned}t &= -e^{+\frac{i}{2}\alpha_3}\,,\\q &= 0\,,\\r &= 0\,,\\s &= -e^{-\frac{i}{2}\alpha_3}\,.\end{aligned}\right\} \tag{671}$$

Therefore, the homomorphism of $\mathcal{A}$ on $\mathcal{C}_2$ does *not* represent an isomorphism. We have determined the following element of this mapping:

$$\begin{pmatrix}1 & 0 & 0 & 0\\0 & \cos\alpha_3 & \sin\alpha_3 & 0\\0 & -\sin\alpha_3 & \cos\alpha_3 & 0\\0 & 0 & 0 & 1\end{pmatrix} \longrightarrow \pm\begin{pmatrix}e^{+\frac{i}{2}\alpha_3} & 0\\0 & e^{-\frac{i}{2}\alpha_3}\end{pmatrix}. \quad \text{Mapping of } \mathbf{D}_3 \text{ onto } \mathbf{C} \tag{672}$$

2. We consider the special LORENTZ transformation $\mathbf{L}_3$, where an inertial system (dashed coordinates) move with velocity v_3 in z-direction of the initial inertial system (undashed coordinates). For the matrix $\mathbf{L}_3$, we can write, s. Eq. (346),

$$\mathbf{L}_3 = \begin{pmatrix} \gamma_3 & 0 & 0 & -\beta_3\,\gamma_3 \\ 0 & 1 & 0 & 0 \\ 0 & 0 & 1 & 0 \\ -\beta_3\,\gamma_3 & 0 & 0 & \gamma_3 \end{pmatrix} \quad \text{with} \quad \gamma_3 = \frac{1}{\sqrt{1-\beta_3^2}}\,, \quad \beta_3 = \frac{v_3}{c}\,. \tag{673}$$

It represents the transformation

$$\left.\begin{aligned} x^{0'} &= \gamma_3\,x^0 \qquad - \beta_3\,\gamma_3\,x^3\,, \\ x^{1'} &= \ x^1\,, \\ x^{2'} &= \ x^2\,, \\ x^{3'} &= -\beta_3\,\gamma_3\,x^0 + \gamma_3\,x^3\,. \end{aligned}\right\} \tag{674}$$

With respect to the expressions in Eq. (657) we write for (666) after simple evaluation

$$\left.\begin{aligned} (+x^{0'} - x^{3'}\,) &= (1+\beta_3)\,\gamma_3\,(+x^0 - x^3)\,, \\ (-x^{1'} + i x^{2'}) &= \ (-x^1 + i x^2)\,, \\ (-x^{1'} - i x^{2'}) &= \ (-x^1 - i x^2)\,, \\ (+x^{0'} + x^{3'}\,) &= (1-\beta_3)\,\gamma_3\,(+x^0 + x^3)\,. \end{aligned}\right\} \tag{675}$$

When we now replace in Eq. (657) (V^0, V^1, V^2, V^3) by (x^0, x^1, x^2, x^3), we get for the transformation of elements $\mathbf{p}$ in the representation space $\mathbb{P}_{\frac{1}{2}\frac{1}{2}}$

$$\left.\begin{aligned} p_{1'\dot{1}'} &= (1+\beta_3)\,\gamma_3\,p_{1\dot{1}}\,, \\ p_{1'\dot{2}'} &= \quad p_{1\dot{2}}\,, \\ p_{2'\dot{1}'} &= \quad p_{2\dot{1}}\,, \\ p_{2'\dot{2}'} &= (1-\beta_3)\,\gamma_3\,p_{2\dot{2}}\,. \end{aligned}\right\} \tag{676}$$

We write

$$\left.\begin{aligned} (1-\beta_3)\,\gamma_3 &= \frac{1-\beta_3}{\sqrt{1-\beta_3^2}} = \frac{\sqrt{(1-\beta_3)^2}}{\sqrt{(1-\beta_3)(1+\beta_3)}} = \sqrt{\frac{1-\beta_3}{1+\beta_3}} := \kappa\,, \\ (1+\beta_3)\,\gamma_3 &= \frac{1+\beta_3}{\sqrt{1-\beta_3^2}} = \frac{\sqrt{(1+\beta_3)^2}}{\sqrt{(1-\beta_3)(1+\beta_3)}} = \sqrt{\frac{1+\beta_3}{1-\beta_3}} = \frac{1}{\kappa}\,. \end{aligned}\right\}$$

For getting the transformation (651) from Eq. (676), there must hold

$$\left.\begin{aligned} &t\bar{t} = \frac{1}{\kappa}\,, \\ &t\bar{s} = 1\,, \\ &s\bar{t} = 1\,, \\ &s\bar{s} = \kappa\,, \\ &\text{and all other elements of the matrix } \mathbf{M} \text{ must be zero.} \end{aligned}\right\} \tag{677}$$

These equations can be fulfilled by

$$\left.\begin{aligned} t &= \frac{1}{\sqrt{\kappa}}\,, \\ q &= 0\,, \\ r &= 0\,, \\ s &= \sqrt{\kappa}\,. \end{aligned}\right\} \tag{678}$$

Since the conditions (669) contain only products of two elements of matrix **C**, we get a second solution, when we insert always a minus sign:

$$\left.\begin{aligned} t &= -\frac{1}{\sqrt{\kappa}}\,, \\ q &= \ 0\,, \\ r &= \ 0\,, \\ s &= -\sqrt{\kappa}\,. \end{aligned}\right\} \tag{679}$$

Hence, we have the following element of the homomorphisms of $\mathcal{A}$ onto $\mathcal{C}_2$ that is not an isomorphismus because of the free choice of the sign:

$$\begin{pmatrix} \gamma_3 & 0 & 0 & -\beta_3\gamma_3 \\ 0 & 1 & 0 & 0 \\ 0 & 0 & 1 & 0 \\ -\beta_3\gamma_3 & 0 & 0 & \gamma_3 \end{pmatrix} \longrightarrow \pm \begin{pmatrix} \sqrt{\gamma_3(1+\beta_3)} & 0 \\ 0 & \sqrt{\gamma_3(1-\beta_3)} \end{pmatrix} . \quad \begin{array}{l}\text{Mapping of} \\ \mathbf{L}_3 \text{ onto } \mathbf{C}\end{array} \tag{680}$$

The equations for the mappings (672) and (680) can be generalised by means of the use of the in Eq. (717) defined PAULI matrices $\boldsymbol{\sigma}_\nu\,,\ \nu = 1, 2, 3$.

At first, we consider the spatial rotations **D** that form a subgroup of the general LORENTZ transformations as known from Chap. 9, Sect. 1.2. Let be $\mathbf{n} = (n_1, n_2, n_3)$ the rotation axis and α the rotation angle. For the mapping of matrices $\mathbf{D}(\mathbf{n}, \alpha)$ from $\mathcal{A}$ on matrices $\mathbf{C}(\mathbf{n}, \alpha)$ of $\mathcal{C}_2$, we can write

$$\mathbf{D}(\mathbf{n}, \alpha) \longrightarrow \mathbf{C}(\mathbf{n}, \alpha) = \cos\frac{\alpha}{2}\ \mathbf{1} + i\ \sin\frac{\alpha}{2}\ \mathbf{n}\cdot\boldsymbol{\sigma} = \exp[\,i\ \frac{\alpha}{2}\ \mathbf{n}\cdot\boldsymbol{\sigma}]\,, \quad \begin{array}{l}\text{Mapping of} \\ \mathbf{D}(\mathbf{n}, \alpha) \text{ in } \mathbf{C}\end{array} \tag{681}$$

so that by (664)

$$\mathbf{D}(\mathbf{n}, \alpha) = \mathbf{T}\,\mathbf{C}(\mathbf{n}, \alpha) \otimes \bar{\mathbf{C}}(\mathbf{n}, \alpha)\,\mathbf{T}^{-1}\,. \tag{682}$$

Here it is $\mathbf{n}\cdot\boldsymbol{\sigma} = n_1\boldsymbol{\sigma}_1 + n_2\boldsymbol{\sigma}_2 + n_3\boldsymbol{\sigma}_3$. The exponential function with matrices as argument is defined by a TAYLOR series, cp. Eq. (352). For getting the simple expression $\left(\cos\frac{\alpha}{2}\,\mathbf{1} + i\sin\frac{\alpha}{2}\ \mathbf{n}\cdot\boldsymbol{\sigma}\right)$, one has to use the known series expansions of sin- and cos-function and the algebraic properties (726) of the PAULI matrices, that its products deliver always only the unit matrix or PAULI matrices. For the directions of coordinate axes, we verify Eq. (681) in Problem 41. In particular, our Eq. (672) is contained as a special case of Eq. (681).

Now we consider the proper LORENTZ transformations with a motion in a certain direction $\mathbf{n}$, hence the special LORENTZ transformations $\mathbf{L}(v)$, where the velocity v lies in the direction of motion. For the mapping of matrices $\mathbf{L}(v)$ of $\mathcal{A}$ onto matrices $\mathbf{C}(\mathbf{n}, \varphi)$ from $\mathcal{C}_2$, we can write

$$\mathbf{L}(v) \longrightarrow \mathbf{C}(\mathbf{n}, \alpha) = \cosh\frac{\varphi}{2}\ \mathbf{1} + \sinh\frac{\varphi}{2}\ \mathbf{n}\cdot\boldsymbol{\sigma} = \exp[\,\frac{\varphi}{2}\ \mathbf{n}\cdot\boldsymbol{\sigma}\,] \quad \begin{array}{c}\text{Mapping of}\\ \mathbf{L}(v)\text{ into }\mathbf{C}\end{array} \tag{683}$$

with $\tan\varphi = \dfrac{v}{c}$, $v =$ velocity in direction $\mathbf{n}$, and it is by Eq. (664)

$$\mathbf{L}(v), \varphi) = \mathbf{T}\,\mathbf{C}(\mathbf{n}, \varphi) \otimes \bar{\mathbf{C}}(\mathbf{n}, \alpha)\,\mathbf{T}^{-1}\,. \tag{684}$$

For the directions of coordinate axes, we verify Eq. (683) in Problem 42. In particular our Eq. (680) is a special case of (683).

3.3 Spinor Calculus

Calculations with spinors can be formalised in the same way as evaluation of tensor operations, cp. Appendix B.1. All spinor indices, for which we use normally the first small initial letters of the alphabet, should take the values 1 and 2 as assigned in Eq. (645). Also, we have to sum over equal indices.

The unimodular matrix $\mathbf{C}$ is denoted by indices as $C^a_{a'}$, and for the conjugate complex matrix $\bar{\mathbf{C}}$, we put over the indices points. The bar over the matrix C with indices is omitted. For Eqs. (639) and (642) of the transformations of an elementary spinor u_a and the spinors with point, $u_{\dot{a}}$, we get

$$u_{a'} = C^a_{a'}\, u_a\,, \tag{685}$$

$$u_{\dot{a}'} = C^{\dot{a}}_{\dot{a}'}\, u_{\dot{a}}\,. \tag{686}$$

As for tensors, we consider now objects with multiple indices, e.g. $\omega_{ab\dot{c}}$, where each single index is transformed as an elementary spinor, Eq. (639), and an index with overdot according to Eq. (642), in our example then

$$\omega_{a'b'\dot{c}'} = C^a_{a'}\, C^b_{b'}\, C^{\dot{c}}_{\dot{c}'}\, \omega_{ab\dot{c}}\,. \tag{687}$$

For two arbitrary elementary spinors u_a and v_a we form with Eq. (685) the transformed elementary spinors $u_{a'}$ and $v_{a'}$, and we find from the unimodularity (635)

$$u_{1'}v_{2'} - u_{2'}v_{1'} = u_1 v_2 - u_2 v_1\,. \tag{688}$$

Likewise, it holds

$$u_{\dot{1}'} v_{\dot{2}'} - u_{\dot{2}'} v_{\dot{1}'} = u_{\dot{1}} v_{\dot{2}} - u_{\dot{2}} v_{\dot{1}} \,. \tag{689}$$

These invariances can be reduced to a metric g_{ab} in the representation spaces $\mathbb{U}_2$ and $\mathbb{U}_{\dot{2}}$ of elementary spinors. There one uses the corresponding inverse matrix as $g^{ab} := (g_{ab})^{-1}$,

$$g_{ab} = \begin{pmatrix} 0 & -1 \\ 1 & 0 \end{pmatrix} \quad \longleftrightarrow \quad g^{ab} = \begin{pmatrix} 0 & 1 \\ -1 & 0 \end{pmatrix} \tag{690}$$

with

$$g^{bc} g_{ca} = g_{ac} g^{cb} = \delta^b_a \,, \tag{691}$$

$$\begin{pmatrix} 0 & 1 \\ -1 & 0 \end{pmatrix} \begin{pmatrix} 0 & -1 \\ 1 & 0 \end{pmatrix} = \begin{pmatrix} 0 & -1 \\ 1 & 0 \end{pmatrix} \begin{pmatrix} 0 & 1 \\ -1 & 0 \end{pmatrix} = \begin{pmatrix} 1 & 0 \\ 0 & 1 \end{pmatrix} . \tag{692}$$

The same metric applies also to the space $\mathbb{U}_{\dot{2}}$ of dotted spinors,

$$g_{\dot{a}\dot{b}} = g_{ab} \,. \tag{693}$$

This metric also fulfils the condition of numerical invariance with respect to transformations with matrices $C^a_{a'}$. This is the same property as the numerical invariance of the MINKOWSKI metric η_{ik} with respect to LORENTZ transformations. As easily verified by the unimodularity (635), it holds

$$g_{a'b'} = C^a_{a'} \, C^b_{b'} g_{ab} = \begin{pmatrix} 0 & -1 \\ 1 & 0 \end{pmatrix} = g_{ab} \,. \tag{694}$$

A significant difference to the MINKOWSKI metric η_{ik} is the property, that the spinor metric g_{ab} is antisymmetric as seen in Eq. (690). It holds

$$g_{ab} = -g_{ba} \,, \tag{695}$$

that has to be taken into account for calculations with spinors.

As in tensor calculus, we call spinors with lower indices covariant, and we define with the metric g_{ab} and g^{ab} corresponding contravariant spinors. We write at first for elementary spinors

$$u^a = g^{ab} u_b \quad \longleftrightarrow \quad u_a = g_{ab} u^b \,, \tag{696}$$

$$u^{\dot{a}} = g^{\dot{a}\dot{b}} u_{\dot{b}} \quad \longleftrightarrow \quad u_{\dot{a}} = g_{\dot{a}\dot{b}} u^{\dot{b}} \,, \tag{697}$$

where we can omit the points for the metric due to Eq. (693).

For Eqs. (688) and (689), we can now write

$$u_{a'} v^{a'} = u_a v^a \tag{698}$$

and

$$u_{\dot{a}'}\, v^{\dot{a}'} = u_{\dot{a}} v^{\dot{a}} \,. \tag{699}$$

We take into account Eqs. (694) and (690)–(692) to get

$$u_a v^a = g_{ac} u^c g^{ad} u_d = g^{ad} g_{ac} u^c u_d = -g^{da} g_{ac} u^c u_d = -\delta^d_c u^c u_d = -u^a v_a \,.$$

For arbitrary spinors u_a, v_a, $u_{\dot{a}}$, $v_{\dot{a}}$, it holds also

$$u_a v^a = -u^a v_a \,, \quad u_{\dot{a}} v^{\dot{a}} = -u^{\dot{a}} v_{\dot{a}} \,. \tag{700}$$

For two equal spinors there follows $u_a u^a = -u^a u_a = u_a u^a$ (the factors can clearly be inverted), so that the invariant of an arbitrary elementary spinor vanishes:

$$u_a u^a = 0 \,, \quad u_{\dot{a}} u^{\dot{a}} = 0 \,. \tag{701}$$

Obviously for contravariant spinors, the transformations are given by

$$u^{a'} = C^{a'}_{a}\, u^a \,, \tag{702}$$

$$u^{\dot{a}'} = \bar{C}^{\dot{a}'}_{\dot{a}}\, u^{\dot{a}} \,, \tag{703}$$

and spinors with higher rank indices transform in a similar way.

Now we write down the linear representations of the group $\mathcal{C}_2$, which are completely analogous to the equations for tensors, shown in Sect. 2. Only here there appear always additionally the indices with overdot, and all double indices are summed only over the values 1 and 2. Instead of Eqs. (630) and (632) we get

The spinorial representations of group $\mathcal{C}_2$, being due to the homomorphisms (636) also representations of the LORENTZ group $\mathcal{A}$, are

$$\left.\begin{aligned} p_{a'_1 a'_2 \ldots a'_{2J}\, \dot{b}'_1 \dot{b}'_2 \ldots \dot{b}'_{2\bar{J}}} &= C^{a_1}_{a'_1} C^{a_2}_{a'_2} \cdots C^{a_{2J}}_{a'_{2J}} C^{\dot{b}_1}_{\dot{b}'_1} C^{\dot{b}_2}_{\dot{b}'_2} \cdots C^{\dot{b}_{2\bar{J}}}_{\dot{b}'_{2\bar{J}}}\, p_{a_1 a_2 \ldots a_{2J}\, \dot{b}_1 \dot{b}_2 \ldots \dot{b}_{2\bar{J}}} \,, \\ p^{a'_1 a'_2 \ldots a'_{2J}\, \dot{b}'_1 \dot{b}'_2 \ldots \dot{b}'_{2\bar{J}}} &= C^{a'_1}_{a_1} C^{a'_2}_{a_2} \cdots C^{a'_{2J}}_{a_{2J}} C^{\dot{b}'_1}_{\dot{b}_1} C^{\dot{b}'_2}_{\dot{b}_2} \cdots C^{\dot{b}'_{2\bar{J}}}_{\dot{b}_{2\bar{J}}}\, p^{a_1 a_2 \ldots a_{2J}\, \dot{b}_1 \dot{b}_2 \ldots \dot{b}_{2\bar{J}}} \,. \end{aligned}\right\} \tag{704}$$

Since these spinors are not investigated in all its details, we have not written down the mixed-variant spinors that correspond to Eqs. (633) for tensors. The 'index operations' for spinors can be transformed completely from tensor calculus, one has only to take care of the antisymmetry of the spinor metric, Eq. (694). Considering Eq. (700), it holds, e.g.

$$\omega_{a\dot{c}} := \omega_{ab\dot{c}}{}^{b} = -\omega_{a}{}^{b}{}_{\dot{c}b} \,. \tag{705}$$

We shall not investigate whether the representations (704) are complete, and also not discuss the question, on which irreducible components they can be reduced. We note the following properties.

Already the first Eq. (704) contains *all* finite-dimensional representations of the LORENTZ group $\mathcal{A}$. If the sum $(\jmath + \tilde{\jmath})$ is an integer, we get normal tensor representations, s. e.g. Eq. (632). Explicitly we have proven this statement for an important case in Sect. 3.2. There we have shown that elements $\mathbf{p}_{\frac{1}{2}\frac{1}{2}} = (p_{1\dot{1}}, p_{1\dot{2}}, p_{2\dot{1}}, p_{2\dot{2}})$ can be reduced on vectors in MINKOWSKI-space with the conversion Eq. (657). New representations, different from tensorial representations of the LORENTZ group, we get always, when the sum $(\jmath + \tilde{\jmath})$ takes half-integer values. The elementary spinors $\mathbf{p}_{\frac{1}{2}0} = (p_{1\dot{1}}, p_{2\dot{1}}) = (u_1, u_2)$ and $\mathbf{p}_{0\frac{1}{2}} = (p_{1\dot{1}}, p_{1\dot{2}}) = (u_{\dot{1}}, u_{\dot{2}})$ are the most important examples.

From tensor calculus we know that partial derivatives $\partial/\partial x^i$, $i = 0, 1, 2, 3$, form components of a covariant vector. We write for simplicity

$$\frac{\partial}{\partial x^i} := \partial_i \,. \tag{706}$$

When we write in this case $p_{a\dot{b}} = \partial_{a\dot{b}}$, the spinorial notation for covariant vectors of partial derivatives (657) reads according to Eq. (658)

$$\begin{array}{ll} \partial_0 = \frac{1}{2}(\partial_{1\dot{1}} + \partial_{2\dot{2}})\,, & \\ \partial_1 = \frac{1}{2}(\partial_{1\dot{2}} + \partial_{2\dot{1}})\,, & \\ \partial_2 = \frac{i}{2}(\partial_{1\dot{2}} - \partial_{2\dot{1}})\,, & \\ \partial_3 = \frac{1}{2}(\partial_{1\dot{1}} - \partial_{2\dot{2}})\,, & \end{array} \longleftrightarrow \left.\begin{array}{l} \partial_{1\dot{1}} = \partial_0 + \partial_3\,, \\ \partial_{1\dot{2}} = \partial_1 - i\partial_2\,, \\ \partial_{2\dot{1}} = \partial_1 + i\partial_2\,, \\ \partial_{2\dot{2}} = \partial_0 - \partial_3\,. \end{array}\right\} \tag{707}$$

From $\partial^{a\dot{b}} = g^{ac} g^{\dot{b}\dot{d}} \partial_{c\dot{d}}$ we find

$$\partial^{1\dot{1}} = \partial_{2\dot{2}}\,, \quad \partial^{1\dot{2}} = -\partial_{2\dot{1}}\,, \quad \partial^{2\dot{1}} = -\partial_{1\dot{2}}\,, \quad \partial^{2\dot{2}} = \partial_{1\dot{1}}\,. \tag{708}$$

4 Covariant Formulation of the Principle of Relativity WEYL-Equation and DIRAC-Equation

According to EINSTEIN's relativity principle, cp. Chap. 2, Sects. 1 and Chap. 9, 1.6, we have to formulate all physical laws in all inertial systems by mathematical equations of the same form. We have fulfilled this principle by formulating the laws of mechanics and electrodynamics in the form of four-dimensional tensor equations, s. our summary in Sect. 2. When we put all terms, e.g. on the left-hand side of such an equation, and if we denote this sum by a tensor $\mathbf{T}$, that can be written by means of the metric η_{ik} by its covariant components, then we get equations of type

$$T_{i_1 i_2 \cdots i_n} = 0\,. \tag{709}$$

These tensors $T_{i_1 i_2 \cdots i_n}$ are linear representations of the LORENTZ group (632). Equation (709) holds true because of the linearity of Eq. (632) exactly when also the equations

$$T_{i'_1 i'_2 \cdots i'_n} = 0 \tag{710}$$

are fulfilled, and this is true in all inertial systems, when they are fulfilled in a single one.

Since we know with Eq. (704) all finite-dimensional linear representations of the LORENTZ group, and since we know, that they are not restricted on tensors, we must reformulate EINSTEIN's relativity principle as follows, cp. Sect. 1.6:

> The equations for the change of the states of physical systems have to be formulated as equations between spinors or tensors, that are linear representations of the LORENTZ group $\mathcal{A}$. (711)

These spinors or tensors are to be understood in the sense of Eq. (704) that may contain both spinors and tensors. Because of the linearity of Eq. (704), the spinor equations hold true according to our arguments above in all inertial systems, when they are true in a single one:

$$\omega_{a_1 a_2 \ldots a_{2j}\, \dot{b}_1 \dot{b}_2 \ldots \dot{b}_{2\bar{j}}} = 0 \quad \longleftrightarrow \quad \omega_{a'_1 a'_2 \ldots a'_{2j}\, \dot{b}'_1 \dot{b}'_2 \ldots \dot{b}'_{2\bar{j}}} = 0\ . \tag{712}$$

The formulation of physical laws as tensor- or spinor equations is also called *principle of covariance.*

4.1 WEYL *Equation*

We consider elementary spinors $\mathbf{u}^T = (u_1, u_2)$ now as functions of coordinates $x^i, i = 0, 1, 2, 3$ and we denote them as $u_1 := \varphi_1\ ,\ u_2 := \varphi_2$.
With consideration of Eq. (696), we write then

$$\varphi = \begin{pmatrix} \varphi_1 \\ \varphi_2 \end{pmatrix} = \begin{pmatrix} \varphi_1(x^0, x^1, x^2, x^3) \\ \varphi_2(x^0, x^1, x^2, x^3) \end{pmatrix}, \quad \text{and it is} \quad \left.\begin{matrix} \varphi^1 = \ \ \varphi_2\ , & \varphi^{\dot{1}} = \ \ \varphi_{\dot{2}}\ , \\ \varphi^2 = -\varphi_1\ , & \varphi^{\dot{2}} = -\varphi_{\dot{1}}\ . \end{matrix}\right\} \tag{713}$$

With the spinorial formulation $\partial_{a\dot{b}}$ for the vector of partial derivative according to Eqs. (707), (708), we find as most simple linear partial differential equation, possible with these spinors, an equation called after HERMANN WEYL,

$$\partial_{a\dot{b}}\, \varphi^a = 0\ . \qquad \text{WEYL equation} \tag{714}$$

Also, the equivalent formulation is in use,

$$\partial^{a\dot{b}}\ \varphi_a = 0\ , \qquad \text{WEYL equation} \tag{715}$$

being true in each inertial system because of its covariant form.

Under consideration of Eqs. (707) and (708) we write Eq. (715) explicitly as

$$\left.\begin{aligned} \partial^{1\dot{1}}\varphi_1 + \partial^{2\dot{1}}\varphi_2 &= -(\partial_1 - i\partial_2)\varphi_2 + (\partial_0 - \partial_3)\varphi_1 = 0\,, \\ \partial^{1\dot{2}}\varphi_1 + \partial^{2\dot{2}}\varphi_2 &= (\partial_0 + \partial_3)\varphi_2 \quad - (\partial_1 + i\partial_2)\varphi_1 = 0\,. \end{aligned}\right\} \tag{716}$$

Here we introduce the PAULI matrices $\boldsymbol{\sigma}_\alpha, \alpha = 1, 2, 3,$

$$\boldsymbol{\sigma}_1 = \begin{pmatrix} 0 & 1 \\ 1 & 0 \end{pmatrix}, \quad \boldsymbol{\sigma}_2 = \begin{pmatrix} 0 & -i \\ i & 0 \end{pmatrix}, \quad \boldsymbol{\sigma}_3 = \begin{pmatrix} 1 & 0 \\ 0 & -1 \end{pmatrix}. \qquad \text{PAULI matrices} \tag{717}$$

We denote the two-dimensional unity matrix $\mathbf{1} := \boldsymbol{\sigma}_0$, then we can summarise the PAULI matrices $\boldsymbol{\sigma}_\alpha\,,\ \alpha = 1, 2, 3$, with $\boldsymbol{\sigma}_0$ as

$$\boldsymbol{\sigma}_i \quad \text{or} \quad \boldsymbol{\sigma}^i = \eta^{ik}\,\boldsymbol{\sigma}_k\,, \quad i, k = 0, 1, 2, 3\,, \tag{718}$$

with the numerical invariant tensor $\eta^{ik} \equiv \eta_{ik}$ with $\eta_{00} = 1,\ \eta_{11} = \eta_{22} = \eta_{33} = -1$, all other components are zero, s. Appendix B.1.

In this way, one can reformulate the derivative

$$\partial^{a\dot{b}}\,\varphi_a \equiv \boldsymbol{\sigma}^i\,\partial_i\,\varphi_a\,. \tag{719}$$

Instead of Eq. (715), we can write the WEYL equation also as

$$\eta^{ik}\,\boldsymbol{\sigma}_k\,\partial_i\,\varphi_a \equiv \boldsymbol{\sigma}^i\,\partial_i\,\varphi_a = 0\,, \quad i,\ k = 0, 1, 2, 3\,, \quad a = 1, 2, \ \text{WEYL equation} \tag{720}$$

or

$$\boldsymbol{\sigma}^i\,\partial_i\,\varphi = (\mathbf{1}\partial_0 - \boldsymbol{\sigma}_1\partial_1 - \boldsymbol{\sigma}_2\partial_2 - \boldsymbol{\sigma}_3\partial_3)\begin{pmatrix}\varphi_1 \\ \varphi_2\end{pmatrix} = 0\,. \quad \text{WEYL equation} \tag{721}$$

Equation (721) follows in each inertial system with the always identical PAULI matrices, since due to its covariant formulation, Eq. (715) holds true in each inertial system.

Connecting the PAULI matrices with the unity matrix to a four-vector $\boldsymbol{\sigma}_i$ or $\boldsymbol{\sigma}^i, i = 0, 1, 2, 3$, it is still not enough to garantee, that constructions as $\eta^{ik}\,\boldsymbol{\sigma}_k\,\partial_i \equiv \boldsymbol{\sigma}^i\,\partial_i$, $i,\ k = 0, 1, 2, 3$, represent numerically invariant operations. The transformation law for the PAULI matrices that guarantees its numerical invariance is more complicated and follows from the laws of spinor calculus:

$$\left.\begin{aligned} \partial^{a\dot{b}}\,\varphi_a &= \boldsymbol{\sigma}^i\,\partial_i\,\varphi_a\,, \\ \partial^{a'\dot{b}'}\,\varphi_{a'} &= \boldsymbol{\sigma}^{i'}\,\partial_{i'}\,\varphi_{a'}\,, \\ \partial^{a'\dot{b}'}\,\varphi_a &= C^{\dot{b}'}_{\dot{b}}\,\partial^{a\dot{b}}\,\varphi_a\,, \\ \partial^{a'\dot{b}'}\,\varphi_a &= C^{\dot{b}'}_{\dot{b}}\,\boldsymbol{\sigma}^i\,\partial_i\,\varphi_a = \boldsymbol{\sigma}^{i'}\,\partial_{i'}\,\varphi_{a'}\,, \end{aligned}\right\} \tag{722}$$

hence

$$C^{b'}_{b}\,\boldsymbol{\sigma}^{i}\,\partial_i\,\varphi_a = \boldsymbol{\sigma}^{i'}\,\frac{\partial x^i}{\partial x^{i'}}\,\partial_i\,C^{a}_{a'}\,\varphi_a \tag{723}$$

and therefore as condition on the PAULI matrices

$$C^{b'}_{b}\,\boldsymbol{\sigma}^{i} = \frac{\partial x^i}{\partial x^{i'}}\,\boldsymbol{\sigma}^{i'}\,C^{a}_{a'}\,. \tag{724}$$

As we know from tensor calculus, Appendix B.1, a contravariant index is transformed by the transposed and inverse matrix, i.e. contragredient to the covariant index. In compact notation, we have to write in Eq. (722) for the matrix $C^{b'}_{b}$ a summation of the lower index then $\left(\bar{\mathbf{C}}^{T}\right)^{-1}$, when we write for $C^{a}_{a'}$ a summation over the upper index $\mathbf{C}$, hence

$$\left(\bar{\mathbf{C}}^{T}\right)^{-1}\boldsymbol{\sigma}^{i} = \frac{\partial x^i}{\partial x^{i'}}\,\boldsymbol{\sigma}^{i'}\,\mathbf{C}$$

and, when we resolve for $\boldsymbol{\sigma}^{i}$ or $\boldsymbol{\sigma}^{i'}$,

$$\left.\begin{aligned} \boldsymbol{\sigma}^{i} &= \frac{\partial x^i}{\partial x^{i'}}\,\bar{\mathbf{C}}^{T}\,\boldsymbol{\sigma}^{i'}\,\mathbf{C}\,,\\ \boldsymbol{\sigma}^{i'} &= \frac{\partial x^{i'}}{\partial x^{i}}\,\left(\bar{\mathbf{C}}^{T}\right)^{-1}\boldsymbol{\sigma}^{i}\,\mathbf{C}^{-1}\,. \end{aligned}\right\} \tag{725}$$

The transformation (725) secures the numerical invariance of PAULI matrices (717) $\boldsymbol{\sigma}^{i'} = \boldsymbol{\sigma}^{i}$. With the matrices in our examples, Eqs. (672) and (680), we can explicitly verify Eq. (725).

The covariance, i.e. the form invariance of the WEYL equation (720) or (721) follows then directly without detour on the spinorial notation (714).

As can be easily seen by insertion, the PAULI matrices $\boldsymbol{\sigma}_{\alpha}$, $\alpha = 1, 2, 3$ have the following algebraic properties:

$$\left.\begin{aligned} \boldsymbol{\sigma}_1\boldsymbol{\sigma}_2 &= i\,\boldsymbol{\sigma}_3\,,\\ \boldsymbol{\sigma}_2\boldsymbol{\sigma}_3 &= i\,\boldsymbol{\sigma}_1\,, \qquad \boldsymbol{\sigma}_{\alpha}\boldsymbol{\sigma}_{\beta} + \boldsymbol{\sigma}_{\beta}\boldsymbol{\sigma}_{\alpha} = 2\,\delta_{\alpha\beta}\,\mathbf{1}\,, \quad \alpha, \beta = 1, 2, 3\,.\\ \boldsymbol{\sigma}_3\boldsymbol{\sigma}_1 &= i\,\boldsymbol{\sigma}_2\,, \end{aligned}\right\} \tag{726}$$

There we have used the KRONECKER symbol with

$$\delta_{11} = \delta_{22} = \delta_{33} = 1\,, \quad \delta_{12} = \delta_{21} = \delta_{13} = \delta_{31} = \delta_{23} = \delta_{32} = 0.$$

With Eqs. (726), one verifies the relation:

$$\left(\mathbf{1}\,\partial_0 + \boldsymbol{\sigma}_1\partial_1 + \boldsymbol{\sigma}_2\partial_2 + \boldsymbol{\sigma}_3\partial_3\right)\left(\mathbf{1}\,\partial_0 - \boldsymbol{\sigma}_1\partial_1 - \boldsymbol{\sigma}_2\partial_2 - \boldsymbol{\sigma}_3\partial_3\right) = \Box \tag{727}$$

with the invariant wave operator $\Box$, and the LAPLACE operator $\boldsymbol{\Delta}$,

$$\Box := \eta^{ik}\partial_i\,\partial_k = \partial_0^2 - \partial_1^2 - \partial_2^2 - \partial_3^2 \equiv \frac{1}{c^2}\frac{\partial^2}{\partial t^2} - \boldsymbol{\Delta}\,. \tag{728}$$

The two-component spinor field φ then satisfies with the WEYL equation (721) also the wave equation

$$\Box\,\varphi \equiv \left(\partial_0^2 - \partial_1^2 - \partial_2^2 - \partial_3^2\right)\varphi \equiv \frac{1}{c^2}\frac{\partial^2\varphi}{\partial t^2} - \Delta\,\varphi = 0\,. \tag{729}$$

Particles connected with the spinor field φ then are moving with the speed of light. The physical background of the WEYL equation leads to quantum theory as will be discussed in connection with the DIRAC equation, Sects. 4.2 and 5–7. There we shall see, that the PAULI matrices describe a so-called spin, a kind of inner angular momentum with unchanging value $1/2$, also called eigen angular momentum. It has all algebraic properties of an angular momentum, but cannot be interpreted with an orbital motion of the particle. The WEYL equation can therefore describe particles with vanishing rest mass, but with spin $1/2$ as for example massless neutrinos. However, this description is not completely straightforward. With the WEYL equation and the DIRAC equation, to be discussed in the next subsection, there began a very productive development in the history of the theory of elementary particles. We cannot follow this way within our book. But we now understand why we describe these particles with *complex* fields, namely since they are elements of a complex representation space $\mathbb{U}_2$ belonging to *complex* matrices $\mathbf{C}$.

We remark that the sign in the definition (728) of the wave operator with the signature $(+,-,-,-)$ used in this book, differs from the historical convention $\frac{\partial^2}{\partial x^2} + \frac{\partial^2}{\partial y^2} + \frac{\partial^2}{\partial z^2} - \frac{\partial^2}{c^2\partial t^2}$, but this is of no influence for the homogeneous wave equation connected with particles with vanishing rest mass. We have to take care of the definition in the following for equations describing particles with rest masses.

4.2 DIRAC *Equation*

We consider a spinor field $\varphi_a(x^i)$ from the representation space $\mathbb{U}_2$ and a further spinor field $\chi_{\dot a}(x^i)$ in the space $\mathbb{U}_{\dot 2}$. We are looking for equations describing the evolution of both fields. Since both representation spaces are independent of each other, we need a physical constant for getting a connection between both fields. This constant is denoted by κ. As most simple linear partial differential equation for determining both fields, satisfying the requirement of high symmetry in the form, we get an equation system denoted after P. A. M. DIRAC:

$$\left.\begin{aligned} \partial_{a\dot b}\,\varphi^a &= \kappa\,\chi_{\dot b}\,, \\ \partial_{a\dot b}\,\chi^{\dot b} &= -\kappa\,\varphi_a\,. \end{aligned}\right\} \qquad \text{DIRAC equation} \tag{730}$$

The minus sign seems to destroy the symmetry. However, it results only from the antisymmetry of the metric (690). We write the spinor of the derivatives by means of the metric contravariant,

$$\partial^{d\dot f} = g^{da}\,g^{\dot f\dot b}\partial_{a\dot b}\,. \tag{731}$$

Then we get with Eq. (691)

$$
\begin{aligned}
\partial_{a\dot{b}}\,\chi^{\dot{b}} &= -\kappa\,\varphi_a\,,\\
g^{da}\,\partial_{a\dot{b}}\,\chi^{\dot{b}} &= -\,g^{da}\,\kappa\,\varphi_a\,,\\
g^{da}\,\partial_{a\dot{b}}\,\delta^{\dot{b}}_{\dot{c}}\,\chi^{\dot{c}} &= -\,\kappa\,\varphi^d\,,\\
g^{da}\,\partial_{a\dot{b}}\,g^{\dot{b}\dot{e}}\,g_{\dot{e}\dot{c}}\,\chi^{\dot{c}} &= -\,\kappa\,\varphi^d\,,\\
g^{da}\,g^{\dot{b}\dot{e}}\,\partial_{a\dot{b}}\,\chi_{\dot{c}} &= -\,\kappa\,\varphi^d\,,\\
-g^{da}\,g^{\dot{e}\dot{b}}\,\partial_{a\dot{b}}\,\chi_{\dot{e}} &= -\,\kappa\,\varphi^d\,,\\
\partial^{d\dot{e}}\,\chi_{\dot{e}} &= \kappa\,\varphi^d\,.
\end{aligned}
$$

Hence we can write the DIRAC equation (730) also as

$$
\left.\begin{aligned}
\partial_{a\dot{b}}\,\varphi^a &= \kappa\,\chi_{\dot{b}}\,,\\
\partial^{a\dot{b}}\,\chi_{\dot{b}} &= \kappa\,\varphi^a\,.
\end{aligned}\right\}\quad \text{DIRAC equation} \tag{732}
$$

Below we will determine the constant κ by the connection of the DIRAC equation with the KLEIN–GORDON equation, and we shall see, that the DIRAC equation describes particles with non-vanishing rest mass.

By the covariant form, Eqs. (730) or (732), of the DIRAC equation, its validity in all inertial systems is guaranteed.

As for the WEYL equation, we shall now rewrite the DIRAC equation. Using Eqs. (696), (707), and (708) we find from Eqs. (730) or also from (732) the following equation, when we rewrite all terms on the spinor components $\varphi_1,\ \varphi_2,\ \chi^{\dot{1}}$ and $\chi^{\dot{2}}$, and reverse the order of both Eqs. (730) or (732)

$$
\left.\begin{aligned}
\partial_{1\dot{1}}\,\chi^{\dot{1}} + \partial_{1\dot{2}}\,\chi^{\dot{2}} &= -\kappa\,\varphi_1\,,\\
\partial_{2\dot{1}}\,\chi^{\dot{1}} + \partial_{2\dot{2}}\,\chi^{\dot{2}} &= -\kappa\,\varphi_2\,,\\
\partial_{2\dot{2}}\,\varphi_1 - \partial_{1\dot{2}}\,\varphi_2 &= -\kappa\,\chi^{\dot{1}}\,,\\
\partial_{1\dot{1}}\,\varphi_2 - \partial_{2\dot{1}}\,\varphi_1 &= -\kappa\,\chi^{\dot{2}}
\end{aligned}\right\}
$$

and then

$$
\left.\begin{aligned}
-\,\partial_1\,\chi^{\dot{2}} + i\partial_2\,\chi^{\dot{2}} - \partial_0\,\chi^{\dot{1}} - \partial_3\,\chi^{\dot{1}} &= \kappa\,\varphi_1\,,\\
-\,\partial_0\,\chi^{\dot{2}} + \partial_3\,\chi^{\dot{2}} - \partial_1\,\chi^{\dot{1}} - i\partial_2\,\chi^{\dot{1}} &= \kappa\,\varphi_2\,,\\
+\,\partial_1\,\varphi_2 - i\partial_2\,\varphi_2 - \partial_0\,\varphi_1 + \partial_3\,\varphi_1 &= \kappa\,\chi^{\dot{1}}\,,\\
-\,\partial_0\,\varphi_2 - \partial_3\,\varphi_2 + \partial_1\,\varphi_1 + i\partial_2\,\varphi_1 &= \kappa\,\chi^{\dot{2}}\,.
\end{aligned}\right\}\quad \text{DIRAC equation} \tag{733}
$$

This form of the DIRAC equation holds true in each inertial system, since it holds true there also Eqs. (730) and (732).

From both elementary spinors $\varphi_a(x^i)$ and $\chi^{\dot{a}}(x^i)$ we form the so-called DIRAC spinor $\psi_u(x^i),\ u = 1, 2, 3, 4,\ i = 0, 1, 2, 3,$

$$\psi = \begin{pmatrix} \psi_1 \\ \psi_2 \\ \psi_3 \\ \psi_4 \end{pmatrix} := \begin{pmatrix} \varphi_1 \\ \varphi_2 \\ \chi^{\dot{1}} \\ \chi^{\dot{2}} \end{pmatrix}, \quad \psi^T = \left(\varphi_1, \varphi_2, \chi^{\dot{1}}, \chi^{\dot{2}}\right). \tag{734}$$

We remark the four components of the DIRAC spinor $\psi_u(x^i)$, $u = 1, 2, 3, 4$, do *not* form a vector.

The covariance of the DIRAC equation (733), its form invariance in each inertial system, can be formulated as follows:

$$\left(\eta^{ik}\,\gamma_i\,\partial_k - \kappa\right)\psi_u = 0\,, \quad i, k = 0, 1, 2, 3\,, \quad u = 1, 2, 3, 4\,. \quad \text{DIRAC equation} \tag{735}$$

There we introduced the so-called DIRAC matrices γ_i, that can also be written in terms of the PAULI matrices,

$$\left.\begin{aligned}
\gamma_0 &= \begin{pmatrix} 0 & 0 & -1 & 0 \\ 0 & 0 & 0 & -1 \\ -1 & 0 & 0 & 0 \\ 0 & -1 & 0 & 0 \end{pmatrix} = \begin{pmatrix} \mathbf{0} & -\mathbf{1} \\ -\mathbf{1} & \mathbf{0} \end{pmatrix}, \\
\gamma_1 &= \begin{pmatrix} 0 & 0 & 0 & -1 \\ 0 & 0 & -1 & 0 \\ 0 & 1 & 0 & 0 \\ 1 & 0 & 0 & 0 \end{pmatrix} = \begin{pmatrix} \mathbf{0} & \boldsymbol{\sigma}_1 \\ -\boldsymbol{\sigma}_1 & \mathbf{0} \end{pmatrix}, \\
\gamma_2 &= \begin{pmatrix} 0 & 0 & 0 & -i \\ 0 & 0 & i & 0 \\ 0 & i & 0 & 0 \\ -i & 0 & 0 & 0 \end{pmatrix} = \begin{pmatrix} \mathbf{0} & \boldsymbol{\sigma}_2 \\ -\boldsymbol{\sigma}_2 & \mathbf{0} \end{pmatrix}, \\
\gamma_3 &= \begin{pmatrix} 0 & 0 & -1 & 0 \\ 0 & 0 & 0 & 1 \\ 1 & 0 & 0 & 0 \\ 0 & -1 & 0 & 0 \end{pmatrix} = \begin{pmatrix} \mathbf{0} & \boldsymbol{\sigma}_3 \\ -\boldsymbol{\sigma}_3 & \mathbf{0} \end{pmatrix}.
\end{aligned}\right\} \quad \text{DIRAC matrices} \tag{736}$$

We write the DIRAC matrices with a contravariant index $\gamma^i := \eta^{ik}\,\gamma_k$, and we get then the DIRAC equation in the form

$$\left(\gamma^i\,\partial_i - \kappa\right)\psi_u = 0\,, \quad i, = 0, 1, 2, 3\,, \quad u = 1, 2, 3, 4\,. \quad \text{DIRAC equation} \tag{737}$$

From the covariant spinorial form (733) of the DIRAC equation, that can be rewritten in each inertial system in the form (735), there follows the numerical invariance of the DIRAC matrices.

As for the WEYL equation, the numerical invariance of the operator

$$\eta^{ik}\,\gamma_i\,\partial_k\,, \quad i, k = 0, 1, 2, 3, \ \text{numerical invariant operator} \tag{738}$$

i.e. the numerical invariance of the DIRAC matrices, is not sufficient to write these matrices γ_i, $i = 0, 1, 2, 3$, with a covariant vector index and treat them as MINKOWSKI vectors as for example the partial derivatives ∂_k, $k = 0, 1, 2, 3$.

The transformation laws for the DIRAC matrices γ_i for LORENTZ transformations, that guarantee their numerical invariance, follow analogous to the procedure extensively presented for the PAULI matrices $\boldsymbol{\sigma}_i$, providing Eq. (725). We can therefore be quite brief.

We write the DIRAC equation with the spinor ψ as in Eq. (734) in two reference systems, belonging to dashed and undashed coordinates:

$$\gamma^i \, \partial_i \, \psi = \kappa \, \psi \quad \longleftrightarrow \quad \gamma^{i'} \, \partial_{i'} \, \psi' = \kappa \, \psi' \, . \tag{739}$$

In ψ now φ_a is a covariant spinor in $\mathbb{U}_2$, and $\chi^{\dot{b}}$ a contravariant spinor in $\mathbb{U}_{\dot{2}}$, hence

$$\left.\begin{array}{lll} \varphi_{a'} = C^a_{a'} \varphi_a & \longleftrightarrow & \boldsymbol{\varphi}' = \mathbf{C} \, \boldsymbol{\varphi} \, , \\ \chi^{\dot{b}'} = C^{\dot{b}'}_{\dot{b}} \, \chi^{\dot{b}} & \longleftrightarrow & \boldsymbol{\chi}' = (\bar{\mathbf{C}}^T)^{-1} \boldsymbol{\chi} \, . \end{array}\right\} \tag{740}$$

With a four-dimensional matrix $\mathbf{C}_4$, we write for the transformation of ψ

$$\psi' = \mathbf{C}_4 \, \psi \, , \quad \text{where} \quad \mathbf{C}_4 = \begin{pmatrix} \mathbf{C} & \mathbf{0} \\ \mathbf{0} & (\bar{\mathbf{C}}^T)^{-1} \end{pmatrix} . \tag{741}$$

From the transformed DIRAC equation in Eq. (739), it follows

$$\gamma^{i'} \, \frac{\partial x^i}{\partial x^{i'}} \, \partial_i \, \mathbf{C}_4 \, \psi = \kappa \, \mathbf{C}_4 \, \psi \, , \quad \text{hence}$$

$$\left(\mathbf{C}_4{}^{-1} \gamma^{i'} \, \frac{\partial x^i}{\partial x^{i'}} \, \mathbf{C}_4 \right) \partial_i \, \psi = \kappa \, \psi \quad \text{with} \quad \mathbf{C}_4{}^{-1} = \begin{pmatrix} \mathbf{C}^{-1} & \mathbf{0} \\ \mathbf{0} & \bar{\mathbf{C}}^T \end{pmatrix} . \tag{742}$$

The transformation law for the DIRAC matrices, guaranteeing its numerical invariance, reads than (see also explicitly our examples, Eqs. (672) and (680))

$$\gamma^i = \mathbf{C}_4{}^{-1} \gamma^{i'} \, \frac{\partial x^i}{\partial x^{i'}} \, \mathbf{C}_4 \quad \longleftrightarrow \quad \gamma^{i'} = \mathbf{C}_4 \, \gamma^i \, \frac{\partial x^{i'}}{\partial x^i} \, \mathbf{C}_4{}^{-1} \; : \; \gamma^{i'} \equiv \gamma^i \, . \tag{743}$$

Still we write down the following matrix relations that can be verified by means of the PAULI matrices or simply by direct evaluation:

$$\gamma_i \, \gamma_k + \gamma_k \, \gamma_i = 2 \, \eta_{ik} \, \mathbf{1} \, . \tag{744}$$

Then we find using the symmetry of $\partial_i \partial_k$

$$\begin{aligned}(\gamma^i\,\partial_i+\kappa)\,(\gamma^k\,\partial_k-\kappa) &= \gamma^i\,\partial_i\,\gamma^k\,\partial_k-\kappa^2+\gamma^i\,\partial_i\,\kappa-\kappa\gamma^k\,\partial_k\,,\\ &= \gamma^i\gamma^k\,\partial_i\,\partial_k-\kappa^2\,,\\ &= \frac{1}{2}\,(\gamma_i\gamma_k+\gamma_k\gamma_i)\,\partial_i\,\partial_k-\kappa^2\,,\\ &= \eta^{ik}\,\partial_i\,\partial_k-\kappa^2\,.\end{aligned}$$

Using the wave operator (728), then it follows by applying the operator $(\gamma^k\,\partial_k+\kappa)$ on the left-hand side of the DIRAC equation (737) that the DIRAC field ψ fulfils the so-called KLEIN–GORDON equation

$$(\eta^{ik}\,\partial_i\,\partial_k-\kappa^2)\,\psi \;=\; (\Box-\kappa^2)\,\psi \;=\; 0\,. \quad \text{KLEIN–GORDON equation} \qquad (745)$$

The constant κ^2 in the KLEIN–GORDON equation is now determined by the speed of light c, the PLANCK constant $\hbar=h/2\pi$ and the rest mass m_o of the particle

$$\kappa^2=-\frac{m_o^2\,c^2}{\hbar^2}\,. \qquad (746)$$

We emphasise again on the signature $(+,-,-,-)$ and resulting definition (728) of the wave operators, that leads to a negative sign for the constant κ^2 in Eq. (746). The constant κ itself then must have the imaginary unit i or $-i$ as factor, for getting no imaginary rest masses. Here the convention is

$$\kappa=-i\,\frac{m_o\,c}{\hbar}\,. \qquad (747)$$

The positive sign then appears in the later introduced Eq. (832) for the adjunct spinor. Since only products of both quantities are experimentally relevant, we could also choose the sign inversely.

The DIRAC equation describes particles with rest mass m_o, and we shall see, that such particles have the spin 1/2, for example, electrons, neutrons and protons. We remark once again that the connection of the DIRAC equation with particles, described by this equation, belongs to the field of elementary particle theory, much beyond the framework of our book.

We have shown up to now only that the DIRAC equation describes physical relations fulfilling the principle of relativity. We shall sketch in the following, that the principles of quantum theory lead to a reasonable description of the relativistic particle dynamics with the DIRAC equation, including also a physical interpretation of the DIRAC equation.

5 Physical Background of the DIRAC Equation

5.1 Remembering Quantum Mechanics

When we try to understand the behaviour of particles in atomic or subatomic scales, we need to replace the laws of NEWTONian mechanics by the principles of quantum theory. The complete quantum theory is now formulated axiomatically, and its consequences are verified with highest precision in different experimental set-ups, similar as the relativity theory, both the special theory within the range of its applicability and the general theory, EINSTEIN's gravitation theory. The student planning to become a theoretical physicist should careful study the concepts of these theories in order to be prepared for understanding the present and possibly in future extended theories of elementary particles. It seems that in particular the quantum theory represents the most radical revolution in our understanding of physical reality since NEWTON. Attempts to modify quantum theory or relativity theory in its basic statements are now the theme of such people that also intend to construct a perpetuum mobile, a Punch and Judy show. For advanced students we also refer to the explicit falsification of a classical interpretation of quantum mechanics with the help of BELL's inequality, cf. e.g. Saturai (1993).

Here we restrict ourselves only on a short remembering of the fascinating conception of quantum theory, the theory on the behaviour of particles in the submicroscopic range of our world, and we refer for a more complete discussion on the relevant textbooks, some of which are given in the references.

The states of physical systems are given in quantum theory by elements $|\psi>,\ |\chi>, \ldots$ of a HILBERT (Fig. 1) space $\mathbb{H}$ with a complex scalar product $<\psi|\chi> = c \in \mathbb{C}$, that is assigned to the system. There

$$<\psi|\chi> = \overline{<\chi|\psi>} \ . \tag{748}$$

The measurable quantities of the system, the so-called observables $A, B, \ldots$, are then operators in $\mathbb{H}$, and in particular linear HERMITEan operators $\mathbf{A}, \mathbf{B}, \ldots$ An operator $\mathbf{A}$ is HERMITEan, when the scalar product $<\psi\,|\,\mathbf{A}\,\chi>$ for arbitrary elements $|\psi>,\ |\chi> \in \mathbb{H}$ fulfils the equation

$$<\psi|\mathbf{A}\chi> = <\mathbf{A}\psi|\chi> \ . \tag{749}$$

The equation

$$\mathbf{A}\,|\,\psi> = a\,|\,\psi> \tag{750}$$

defines the eigenvalues a of an operator. It holds:

The eigenvalues of HERMITE*an operators are real.*

Formally, this follows immediately from

$< \psi \,|\, \mathbf{A}\,\psi > = < \psi \,|\, a\,\psi > = a \;< \psi \,|\, \psi >$ and
$< \psi \,|\, \mathbf{A}\,\psi > = < \mathbf{A}\,\psi \,|\, \psi > = < a\,\psi \,|\, \psi > = \overline{a} \;< \psi \,|\, \psi >$, hence
$\overline{a} = a$.

In quantum theory, the eigenvalues a of a HERMITEan operator **A** represent the physically possible measureable values of an observable A, that belongs to the operator **A**.

While the elements of $\mathbb{H}$, which are denoted as state vectors $|\,\psi >, \;\; |\,\chi >, \ldots$ of the system, in general are complex quantities, the measured values of the observables are real quantities due to the HERMITEcity of the operators belonging to the physical quantities.

If there exists a denumerable basis $|\,\varphi_i >, \; i = 1, 2, \ldots$ of the HILBERT space $\mathbb{H}$, then $\mathbb{H}$ is called *separable*. Each state vector $|\,\psi >$ of the system can then be decomposed by the series

$$|\,\psi > = \sum_{i=1}^{\infty} c_i \;|\,\varphi_i > \quad . \tag{751}$$

Now we take for simplicity that the eigenvectors of an HERMITE an operator **A**, belonging to the observable A, form the basis $|\,\alpha_i >, \; i = 1, 2, \ldots$ of the HILBERT space according to Eq. (751), and we assume in addition that the basis vectors are normalised and pairwise orthogonal, cp. Appendix B.1.

Hence we get for the operator **A** the eigenvalue equation

$$\mathbf{A}\,|\,\alpha_i > = a_i \;|\,\alpha_i > \;, \quad i = 1, 2, \ldots \tag{752}$$

with

$$< \alpha_i \,|\, \alpha_k > \; = \; \delta_{ik} \; . \tag{753}$$

The series for the state vector of the systems in the basis $|\,\alpha_i >$ then reads

$$|\,\psi > = \sum_{i=1}^{\infty} c_i \;|\,\alpha_i > \quad . \tag{754}$$

In a measurement of the observable A of the system, there appear only the values a_i .

A central postulate of quantum theory reads as follows:

$$\left.\begin{array}{l} \textit{The probability to measure an eigen value } a_i \\ \textit{for the observable } A \textit{ of a state } |\,\psi > \\ \textit{of the system amounts } |c_i|^2 \,. \end{array}\right\} \tag{755}$$

If the system is in an eigenstate of the operator **A**, say, e.g. in the state $|\,\alpha_1 >$, hence $|\,\psi > \; = \; |\,\alpha_1 >$, then it follows from Eq. (754), that $c_1 = 1$, while all other coefficients c_i vanish. In measuring the observable A one finds then, and only then, with certainty the eigenvalues a_1.

Fig. 1 DAVID HILBERT, * Königsberg 23.1.1862, † Göttingen 14.2.1943.

The *expectation value* $< \mathbf{A} >$ of an operator $\mathbf{A}$ for the observable A in state $|\,\psi >$ of the system is defined as

$$< \mathbf{A} > := < \psi \,|\, \mathbf{A}\, \psi > \ . \tag{756}$$

We get with Eqs. (749)–(753)

$$\begin{aligned} < \psi \,|\, \mathbf{A}\, \psi > &= < \sum_{i=1}^{\infty} c_i\, \alpha_i \,|\, \mathbf{A} \sum_{k=1}^{\infty} c_k\, \alpha_k > \\ &= \sum_{i=1}^{\infty} \sum_{k=1}^{\infty} < c_i\, \alpha_i \,|\, c_k\, a_k\, \alpha_k > \\ &= \sum_{i=1}^{\infty} \sum_{k=1}^{\infty} \overline{c}_i\, c_k\, a_k\, \delta_{ik} \ , \end{aligned}$$

hence

$$< \mathbf{A} > = \sum_{i=1}^{\infty} |c_i|^2\, a_i \ . \tag{757}$$

The expectation value of the observable A in state $|\,\psi >$ of the system represents a mean value of a large number of measurements of the observable A, when all measurements are done on one and the same state $|\,\psi >$ of the system.

Here we underline that due to the statistical interpretation of quantum theory (755), we can make only probability predictions for measuring an observable of a physical system being in a state $|\,\psi >$, unless the system is just in an eigenstate of this observable, where measurements provide a fixed or sharp value.

We have to take care in our formalism, that for the description of physical systems a separable HILBERT space is in general not sufficient. Then we have to use a continuous, not countable basis.

We define the basis of the HILBERT space again by the eigenvectors of a HERMITEan operator $\mathbf{K}$, now with continuous eigenvalues k:

$$\mathbf{K}\,|\,\varphi_k > = k\,|\,\varphi_k > \quad \text{mit} \quad k \in \mathbb{R} \ , \tag{758}$$

the basis vectors $|\,\varphi_k >$ with its continuous real eigenvalues k in $\mathbb{R}$ are normalised with the δ-function, cp. Appendix B.3:

$$< \varphi_k \,|\, \varphi_{k'} > = \delta(k - k') \ . \tag{759}$$

For the decomposition of a state vector $|\,\psi >$ into a set of basis vectors $|\,\varphi_k >$, we get then instead of the summation in Eq. (751) an integration,

$$|\,\psi > = \int dk\, c(k)\,|\,\varphi_k > \quad . \tag{760}$$

In this case, $|c(k)|^2 dk$ gives the probability to measure for an observable K a value between k and $k + dk$, if the system is in state $| \psi >$. The quantity $c(k)$ is called probability amplitude, and $|c(k)|^2$ represents a probability density.

One meets also the case that the basis vectors of $\mathbb{H}$ have both a discrete and a continuous range. Then the decomposition of the state vector into this basis includes both a summation and an integration,

$$| \psi > = \sum_i \int dk \, c_{ik} \, | \varphi_{ik} > \quad . \tag{761}$$

Now we consider the commutator

$$[\, \mathbf{A}, \mathbf{B} \,] := \mathbf{AB} - \mathbf{BA} \tag{762}$$

of two operators **A** and **B**. The theory of operators then tells us, that only when the commutator vanishes, both operators have a common eigenvector system, otherwise not. One can easily illustrate this statement for the case that the operators are finite matrices.

Hence for $[\, \mathbf{A}, \mathbf{B} \,] = 0$, there exist states of the systems being both eigenvector of operator **A** and also eigenvector of operator **B**. Measurements of observables A and B provide then in such a state for both observables simultaneously definite sharp values.

In the following, we consider as physical system a particle situated in a certain field as a physical system.

A fundamental law of quantum theory, modifying basically the classical notion of motion and particle orbit, represents HEISENBERG*'s* (Fig. 2) *uncertainty relation*. In the language of observables in HILBERT space this law can be formulated as follows:

For operators $\mathbf{X}_1, \mathbf{X}_2, \mathbf{X}_3$ representing the measurements of the space position of the system in the three orthogonal directions x, y, z and the corresponding operators of momentum measurements $\mathbf{P}_1, \mathbf{P}_2, \mathbf{P}_3$, there hold the following commutation relations:

$$\left. \begin{array}{l} [\, \mathbf{X}_\alpha \, , \, \mathbf{X}_\beta \,] = \mathbf{0} \, , \\ [\, \mathbf{P}_\alpha \, , \, \mathbf{P}_\beta \,] \ = \mathbf{0} \, , \\ [\, \mathbf{X}_\alpha \, , \, \mathbf{P}_\beta \,] = i \, \hbar \, \delta_{\alpha\beta} \, \mathbf{1} \, , \end{array} \right\} \quad \alpha, \beta = 1, 2, 3 \ . \tag{763}$$

Equation (763) then means the following:

There are states of the systems, in which all three spatial coordinates provide sharp values, and there exist other states, where the three components of momentum give simultaneously sharp values. Likewise, there are states with a sharp position in x-direction, and simultaneously a sharp momentum in y-direction, etc.

But there are no states where the spatial and momentum measurements of the same coordinate deliver simultaneously sharp values.

We write just for simplicity the operator for space measurements **X** and the momentum operator **P**, belonging to the same coordinate, without index. From the expectation values $< \mathbf{X} >$ and $< \mathbf{P} >$ we form the operators

Fig. 2 WERNER HEISENBERG, * Würzburg 5.12.1901, † München 1.2.1976.

$$\triangle \mathbf{X} = \mathbf{X} - < \mathbf{X} > \quad \text{and} \quad \triangle \mathbf{P} = \mathbf{P} - < \mathbf{P} > , \tag{764}$$

describing the deviation of the mean expectation value of a measurement. The expectation values

$$(\Delta X)^2 :=< (\Delta \mathbf{X})^2 > \quad \text{and} \quad (\Delta P)^2 :=< (\Delta \mathbf{X})^2 > \tag{765}$$

are then measures for the scatter of the corresponding space- and momentum measurements.

From Eq. (763) there follows HEISENBERG's uncertainty relation in the form

$$(\Delta X)^2 \, (\Delta P)^2 \geq \frac{\hbar^2}{4}\,, \tag{766}$$

or simpler in taking the square root

$$\Delta x \ \Delta p \ \geq \ \frac{\hbar}{2}\,. \tag{767}$$

Therefore, we can prepare the state of the system, here representing a particle, so that a measurement of the position x can be made arbitrary sharp, so that $\Delta x \longrightarrow 0$, simply by pushing the system more and more in an eigenstate of the space operator. But this has the unavoidable consequence, that the measurement of the momentum in this state becomes more and more uncertain, hence that $\Delta p \longrightarrow \infty$, in such a way that the product of both uncertainty ranges fulfils Eqs. (766) or (767).

The motion of a particle can therefore be described by the observables space and momentum, but these quantities cannot have simultaneously sharp values. The classical understanding of the orbit of a particle as a continuous curve in $\mathbb{R}_3$ must be abandoned for the description of dynamics in atomic and subatomic range.

The experience of our everyday life proves to be a bad advisor for recognising the submicroscopic world. We remark to this point the following: While the beginner may have difficulties to understand that the outcome of space measurement, e.g. for the motion of a slow neutron passing a double slit, only consists in probability statements, he will have no difficulty to grasp that an unstable elementary particle will decay after a time span that is likewise only statistically determined.

The commutation relations (763) are part of a fundamental principle of quantum theory. This principle says, in which way NEWTON's laws of mechanics must be changed to get the rules of quantum theory, but also what remains intact from NEWTON's mechanics. To this aim, one writes the laws of mechanics with so-called POISSON brackets, a formalism introduced in standard textbooks of mechanics. The principle of quantum mechanics then reads:

If the classical laws of mechanics are expressed in POISSON *brackets, the classically observable quantities* $A, B, \ldots$ *are to be replaced by* HERMITEan *operators* $\mathbf{A}, \mathbf{B}, \ldots$, *and the* POISSON *bracket has to be replaced by the commutator between these operators, precisely:*

$$[A, B]_{\text{POISSON}\,bracket} = K \quad \longrightarrow \quad \frac{1}{i\,\hbar}\,[\mathbf{A}, \mathbf{B}]_{commutator} = \mathbf{K}\,. \tag{768}$$

This principle sets the basis for the principally new insights into the particle dynamics in submicroscopic range. But it shows also that the algebraic structures in formulating mechanical laws in POISSON brackets are preserved in the transition to quantum theory.[3]

In this way, from the law $[X, P]_{\text{POISSON}\,bracket} = 1$ follows the third Eq. (763).

With the HAMILTON function H of the system, representing in general its total classical energy, the equation for the temporal change of the observable A in the formalisms of POISSON brackets[4] reads

$$\frac{d}{dt}A = [A, H]_{\text{POISSON}\,bracket} \,,$$

it leads to the quantum theoretical equation

$$\frac{d}{dt}\mathbf{A} = \frac{1}{i\,\hbar}\,[\mathbf{A}, \mathbf{H}]_{commutator} \,. \tag{769}$$

Here $\mathbf{H}$ is now the HAMILTON operator, belonging to the energy as observable of the systems.

With respect to the description of the temporal evolution, there is a peculiarity in quantum theory. The measurable quantities of a physical system, its eigenvalues of HERMITEan operators, are formed from scalar products as seen from the scalar multiplication of Eq. (750) with $|\,\psi >$,
$a \;=< \psi\,|\,\mathbf{A}\,\psi > \big/ < \psi\,|\,\psi >$.

This expression does not change, when a time-dependent HERMITEan operator $\mathbf{Z}(t)$ acts on the state vector,
$|\,\psi > \longrightarrow \widetilde{|\,\psi >} = \exp[-i\mathbf{Z}(t)]\,|\,\psi >$,
and instead the operators obtain the transformation
$\mathbf{A} \longrightarrow \widetilde{\mathbf{A}} = \exp[-i\mathbf{Z}(t)]\,\mathbf{A}\exp[i\mathbf{Z}(t)]$.
Then it holds

$$\begin{aligned} < \widetilde{\psi}\,|\,\widetilde{\mathbf{A}}\widetilde{\psi} > &\longrightarrow \; < \exp[-i\mathbf{Z}(t)]\,\psi\,|\,\exp[-i\mathbf{Z}(t)]\,\mathbf{A}\exp[i\mathbf{Z}(t)]\,\exp[-i\mathbf{Z}(t)]\,\psi > \\ &= \; < \psi\,|\,\exp[i\mathbf{Z}(t)]\exp[-i\mathbf{Z}(t)]\,\mathbf{A}\exp[i\mathbf{Z}(t)]\,\exp[-i\mathbf{Z}(t)]\,\psi > \\ &= \; < \psi\,|\,\mathbf{A}\,\psi > \,. \end{aligned}$$

According to Eq. (769), the temporal evolution of the systems is described by a temporal change of HERMITEan operators. The state vector $|\psi >$ of the physical systems then remains constant in time. This description is called HEISENBERG representation.

With a suitable operator $\mathbf{Z}(t)$, one can transform the whole time dependence also on the temporal change of the state vectors, leaving the HERMITEan operators for the observables temporal constant, without any change in the outcome of measurements of the system. This is called SCHRÖDINGER (Fig. 3) representation. For interacting systems, one transforms only a part of the time dependence on the operators, so that

[3] A more specific discussion of this procedure is contained, e.g. in the textbook of Fick (1968).

[4] Here the observable A should be explicitly time independent, i.e. the measuring process for this observable does not change temporally, $\partial A/\partial t = 0$.

the state vector changes temporarily only over the interaction term, leading to the interaction picture. We advise the reader for details on the relevant textbooks.

If we disregard for simplicity any explicit time dependence and take as initial time of the evolution of the system the time $t = 0$, so the transition from the HEISENBERG formulation to the SCHRÖDINGER formulation is given by the operator

$$\mathbf{Z}(t) = \mathbf{H}/\hbar \, . \tag{770}$$

For distinction we introduce the index H for state vectors and operators in the HEISENBERG representation, and for the SCHRÖDINGER representation the index S, then it holds for the HAMILTON operator the connection

$$\mathbf{H}_S = \exp[\mathbf{H}\,t/(i\,\hbar)]\,\mathbf{H}_H\,\exp[-\mathbf{H}\,t/(i\,\hbar)] = \mathbf{H}_H = \mathbf{H}\,, \tag{771}$$

so that the index can be omitted in this case.

In the HEISENBERG representation, we write with Eq. (769)

$$\left.\begin{aligned} \frac{d}{dt}\mathbf{A}_H &= \frac{1}{i\,\hbar}\left(\mathbf{A}_H\,\mathbf{H} - \mathbf{H}\,\mathbf{A}_H\right), \\ \frac{d}{dt}\,|\,\psi>_H &= 0\,. \end{aligned}\right\} \quad \text{HEISENBERG representation} \tag{772}$$

The SCHRÖDINGER representation is

$$\left.\begin{aligned} \mathbf{A}_S &= \exp[\mathbf{H}\,t/(i\,\hbar)]\,\mathbf{A}_H\,\exp[-\mathbf{H}\,t/(i\,\hbar)]\,, \\ |\,\psi>_S &= \exp[\mathbf{H}\,t/(i\,\hbar)]\,|\,\psi>_H\,. \end{aligned}\right\} \tag{773}$$

Then it follows

$$\begin{aligned} \frac{d}{dt}\mathbf{A}_S &= \frac{\mathbf{H}}{i\,\hbar}\mathbf{A}_S - \mathbf{A}_S\frac{\mathbf{H}}{i\,\hbar} + \exp[\mathbf{H}\,t/(i\,\hbar)]\left(\mathbf{A}_H\,\mathbf{H} - \mathbf{H}\,\mathbf{A}_H\right)\exp[-\mathbf{H}\,t/(i\,\hbar)] \\ &= \frac{1}{i\,\hbar}\left(\mathbf{H}\,\mathbf{A}_S - \mathbf{A}_S\,\mathbf{H} + \mathbf{A}_S\,\mathbf{H} - \mathbf{H}\,\mathbf{A}_S\right) = 0 \quad \text{as well as} \\ \frac{d}{dt}\,|\,\psi>_S &= \frac{\mathbf{H}}{i\,\hbar}\exp[\mathbf{H}\,t/(i\,\hbar)]|\,\psi>_H \end{aligned}$$

and therefore

$$\left.\begin{aligned} \frac{d}{dt}\mathbf{A}_S &= 0\,, \\ \frac{d}{dt}\,|\,\psi>_S &= \frac{1}{i\,\hbar}\mathbf{H}\,|\,\psi>_S\,. \end{aligned}\right\} \quad \text{SCHRÖDINGER representation} \tag{774}$$

The equations for the temporal evolution of the state vector in the SCHRÖDINGER representation is called SCHRÖDINGER *equation.*

Now we want to specify our considerations to the HILBERT space of, e.g. a particle in a field, and we take to this aim the eigenfunctions of the HERMITEan space operators $\mathbf{X}_S$ with continuous eigenvalues $x \in \mathbb{R}$. As the index indicates, we remain in the

SCHRÖDINGER representation. The operators are then constant in time, and therefore also its eigenvectors. Analogous to Eqs. (758), we write

$$\mathbf{X}_S \mid \varphi_x >_S = x \mid \varphi_x >_S \quad \text{with} \quad x \in \mathbb{R} \ . \tag{775}$$

The eigenvalues x are possible outcomes of measurements of the position of the particle. Additional discrete eigenvalues for characterising the state of the system are for simplicity disregarded.

The basis vectors $\mid \varphi_x >_S$ are normalised with the δ-function. Since this normalisation is independent of the chosen representation, we can omit the index,

$$< \varphi_x \mid \varphi_{x'} > = \delta(x - x') \ . \tag{776}$$

The evolution of a state vector $\mid \psi >_S(x, t)$ of the system describes its complete time dependence. We decompose it for the basis vectors $\mid \varphi_x >_S$

$$\mid \psi >_S = \int dx \, \psi(x, t) \mid \varphi_x >_S \quad . \tag{777}$$

Now $|\psi(x, t)|^2 dx$ gives the probability that a position measurement of the particle in state $\mid \psi >_S$ at time t leads to a value between x and $x + dx$. From the probability amplitudes, the complex functions $\psi(x, t)$, we then form the probability density of the position measurement by $\psi(x, t)^* \psi(x, t) \equiv |\psi(x, t)|^2$. The temporal evolution of the system is now contained in this complex function, the probability amplitude $\psi(x, t)$, also denoted as wave function.

Now we derive the SCHRÖDINGER equation in the spatial representation, i.e. as differential equation for the function $\psi(x, t)$.

With Eqs. (775) and (777), we have initially

$$\begin{aligned} \mid \chi >_S := \mathbf{X}_S \mid \psi >_S = \int dx \, \chi(x, t) \mid \varphi_x >_S &= \mathbf{X}_S \int dx \, \psi(x, t) \mid \varphi_x >_S \\ &= \int dx \, \psi(x, t) \, \mathbf{X}_S \mid \varphi_x >_S \\ &= \int dx \, \psi(x, t) \, x \mid \varphi_x >_S \ , \end{aligned}$$

hence

$$\chi(x, t) = x \, \psi(x, t) \ . \tag{778}$$

For the spatial representation of operators in the SCHRÖDINGER representation, we omit the indices. For Eq. (778), we can write

$$\mathbf{X} \, \psi(x, t) = x \ \psi(x, t) \ , \tag{779}$$

hence

$$\mathbf{X} = x \ . \tag{780}$$

The spatial form of the momentum operator in the SCHRÖDINGER representation follows then from the third commutation relation (763). For an explicit derivation, we recommend the textbook by Fick (1968). For us, it is sufficient to verify the result. It holds

$$\mathbf{P} = \frac{\hbar}{i}\,\frac{\partial}{\partial x}\,. \tag{781}$$

For an arbitrary complex function $\psi(x, t)$ there holds

$$\begin{aligned}\big(\mathbf{X}\,\mathbf{P} - \mathbf{P}\,\mathbf{X}\big)\,\psi(x,t) &= \left(x\,\frac{\hbar}{i}\,\frac{\partial}{\partial x} - \frac{\hbar}{i}\,\frac{\partial}{\partial x}\,x\right)\psi(x,t)\\ &= x\,\frac{\hbar}{i}\,\frac{\partial\psi(x,t)}{\partial x} - \frac{\hbar}{i}\,\frac{\partial}{\partial x}\Big(x\,\psi(x,t)\Big)\\ &= x\,\frac{\hbar}{i}\,\frac{\partial\psi(x,t)}{\partial x} - \frac{\hbar}{i}\,\frac{\partial x}{\partial x}\,\psi(x,t) - \frac{\hbar}{i}\,x\,\frac{\partial\psi(x,t)}{\partial x}\\ &= -\frac{\hbar}{i}\,\psi(x,t)\,,\end{aligned}$$

hence the relation required by the commutator (763) reads

$$\big(\mathbf{X}\,\mathbf{P} - \mathbf{P}\,\mathbf{X}\big) = i\,\hbar\,\mathbf{1}\,.$$

At first, we consider a non-relativistic particle of mass m, moving in the potential $V(x)$. From the classical HAMILTON function of the particle, the energy written in variables position x and momentum p, we construct the HAMILTON operator **H**, the energy observable of the system. Using the preceding considerations, we get **H** in the spatial form of the SCHRÖDINGER representation, when we replace position and momentum in the HAMILTON function H of classical mechanics by the corresponding operators (780) and (781), hence

$$H(p,x) = \frac{p^2}{2m} + V(x) \quad\longrightarrow\quad \mathbf{H} = \frac{\mathbf{P}^2}{2m} + V(\mathbf{X}) = -\frac{\hbar^2}{2m}\,\frac{\partial^2}{\partial x^2} + V(x)\,. \tag{782}$$

Here $V(\mathbf{X})$ is an operator function defined by the corresponding TAYLOR series, s. Problem 21.

In the spatial form of the SCHRÖDINGER representation, we get the SCHRÖDINGER equation (774) in a form generally used in quantum mechanical calculations,

$$-\frac{\hbar^2}{2m}\,\frac{\partial^2\psi(x,t)}{\partial x^2} + V(x)\,\psi(x,t) = -\frac{\hbar}{i}\,\frac{\partial\psi(x,t)}{\partial t}\,. \qquad \text{SCHRÖDINGER equation in one dimension} \tag{783}$$

Here we wrote the equation for simplicity in one dimension. With $\mathbf{x}$ for (x, y, z), the three-dimensional generalisation reads

$$-\frac{\hbar^2}{2m}\,\Delta\,\psi(\mathbf{x},t) + V(\mathbf{x})\,\psi(\mathbf{x},t) = -\frac{\hbar}{i}\,\frac{\partial\psi(\mathbf{x},t)}{\partial t}\,. \qquad \text{SCHRÖDINGER equation in three dimensions} \tag{784}$$

with the LAPLACE operator Δ, see Eq. (315).

Then the SCHRÖDINGER equation for a free particle reads

$$-\frac{\hbar^2}{2m}\,\Delta\,\psi(\mathbf{x},t) = -\frac{\hbar}{i}\,\frac{\partial\,\psi(\mathbf{x},t)}{\partial t}\ . \qquad \text{SCHRÖDINGER equation for a free particle} \tag{785}$$

From EHRENFEST's theorem there follow again the expectation values of physical observables and the classical equations of motion, s. our references.

We summarise: There are two possibilities *in formulating* the laws of quantum mechanics.

In the *picture*, one can choose how to distribute the *deterministic* time dependence of the physical system on the time dependence of the operators or of the state vectors. In the *representation*, one fixes a certain basis of the HILBERT space. Here we have chosen the spatial description in the SCHRÖDINGER picture.

A particular illustrative example concerns the representation of angular momentum in quantum theory. Its relativistic treatment in the framework of the DIRAC equation will then lead directly to a fundamental discovery, the phenomenon of the spin of elementary particles, resisting any attempt of a classical explanation.

5.1.1 Angular Momentum

For a particle in classical mechanics at position $\mathbf{x} = (x,\ y,\ z)$ and with momentum $\mathbf{p} = (p_x,\ p_y,\ p_z)$, the angular momentum $\mathbf{l} = (l_x,\ l_y,\ l_z)$ is defined by the vector product from $\mathbf{x}$ and $\mathbf{p}$, cp. Appendix B.1,

$$\left.\begin{aligned} &\mathbf{l} = \mathbf{x} \times \mathbf{p}\,, \quad \text{hence} \\ &l_x = y\,p_z - z\,p_y\,, \\ &l_y = z\,p_x - x\,p_z\,, \\ &l_z = x\,p_y - y\,p_z\,. \end{aligned}\right\} \tag{786}$$

Consequently in quantum mechanics, the components of the observable 'angular momentum' is defined by the procedure, that the angular momentum operators $\mathbf{L}_x, \mathbf{L}_y, \mathbf{L}_z$ are defined by the position- and momentum operators,

$$\left.\begin{aligned} \mathbf{L}_x &= \mathbf{Y}\,\mathbf{P}_z - \mathbf{Z}\,\mathbf{P}_y\,, \\ \mathbf{L}_y &= \mathbf{Z}\,\mathbf{P}_x - \mathbf{X}\,\mathbf{P}_z\,, \\ \mathbf{L}_z &= \mathbf{X}\,\mathbf{P}_y - \mathbf{Y}\,\mathbf{P}_x\,. \end{aligned}\right\} \tag{787}$$

With respect to the following generalisation, we will call these operators in Eq. (787) also *orbital angular momentum* operators.

By means of the commutation relations (763) for the position and momentum operators, one verifies the following commutation relations of the angular momentum operators (the general definition of the LEVI- CIVITA symbol $\epsilon_{\alpha\beta\gamma}$ s. Appendix B.1),

Fig. 3 ERWIN SCHRÖDINGER, * Wien 12.8.1887, † Wien 4.1.1961.

$$\left.\begin{aligned} &[\mathbf{L}_\alpha\,,\mathbf{L}_\beta] = i\,\hbar\,\epsilon_{\alpha\beta\gamma}\mathbf{L}_\gamma\,, \quad \text{hence} \\ &\mathbf{L}_x\,\mathbf{L}_y - \mathbf{L}_y\,\mathbf{L}_x = i\,\hbar\,\mathbf{L}_z\,, \\ &\mathbf{L}_y\,\mathbf{L}_z - \mathbf{L}_z\,\mathbf{L}_y = i\,\hbar\,\mathbf{L}_x\,, \\ &\mathbf{L}_z\,\mathbf{L}_x - \mathbf{L}_x\,\mathbf{L}_z = i\,\hbar\,\mathbf{L}_y\,. \end{aligned}\right\} \tag{788}$$

The three angular momentum operators $\mathbf{L}_x$, $\mathbf{L}_y$, $\mathbf{L}_z$ do not commute. Therefore, a particle cannot have sharp values for any two components of the angular momentum.

The square of the angular momentum $\mathbf{L}^2$ is defined by

$$\mathbf{L}^2 := \mathbf{L}_x^2 + \mathbf{L}_y^2 + \mathbf{L}_z^2\,. \tag{789}$$

Here the simple application of Eqs. (788) leads to the commutation relations

$$[\mathbf{L}^2\,,\mathbf{L}_a] = 0\,, \quad a = 1, 2, 3\,. \tag{790}$$

The square of the angular momentum commutes with each of its components. Hence, there are states of the system with sharp values of the operator $\mathbf{L}^2$ and simultaneously of sharp values of its z-component $\mathbf{L}_z$ as example.

For getting the possible values of $\mathbf{L}^2$ and $\mathbf{L}_z$, we have to solve simultaneously the eigenvalue equations of these operators. We write

$$\left.\begin{aligned} \mathbf{L}^2\,|\,\psi > &= \ell^2\,|\,\psi >\,, \\ \mathbf{L}_z\,|\,\psi > &= \ell_z\,|\,\psi >\,. \end{aligned}\right\} \tag{791}$$

We have derived the commutation relations (788) from the definition of the orbital angular momentum (787). For determining the eigenvalues in Eq. (791) we can choose the spatial form and replace the orbital angular momentum operators $\mathbf{X}$ by x, $\mathbf{P}_x$ by $\frac{\hbar}{i}\frac{\partial}{\partial x}$, and correspondingly for the y- and z-components, then there result from Eq. (791) partial differential equations that deliver the looking for eigenvalues. Since we shall derive these values below again in another algebraic way, we refer the reader here to the relevant textbooks, and we only give the solutions. The eigenvalues can be written as

$$\left.\begin{aligned} \ell^2 &= \hbar^2\,l(l+1)\,, \quad l = 1, 2, 3, \dots\,, \\ \ell_z &= \hbar\,\big(0\,,\,\pm 1\,,\,\pm 2\,,\,\dots \pm l\big)\,. \end{aligned}\right\} \tag{792}$$

This solution represents a typical quantum mechanical eigenvalue spectrum: For large quantum numbers l, we can write for the amount of the angular momentum simply $\ell \simeq \hbar\,l$. Due to the smallness of $\hbar$, one can assert that ℓ changes practically continuously as known from classical mechanics, and the same holds true for the z-component. Only for small integer values l, there result essential deviations from the classical picture, discussed intensively in the cited textbooks.

One can derive the eigenvalues of $\mathbf{L}^2$ and $\mathbf{L}_z$ also on another way, without reference on the angular momentum operators, Eq. (787). We shall see that the full information on the eigenvalues of angular momentum operators is already contained in the commutation relations (788), and there we shall make an important discovery.

For deriving the eigenvalues admitted by the commutation relations (788), we use a procedure that is of great significance in quantum theory.

We write the eigenvalue equations (791) in the form

$$\left.\begin{aligned} \mathbf{L}^2\,\psi_{jm} &= \hbar^2\,j\,(j+1)\,\psi_{jm}\,,\\ \mathbf{L}_z\,\psi_{jm} &= \hbar\,m\,\psi_{jm}\,. \end{aligned}\right\} \tag{793}$$

Till now we have found with Eq. (792) the solutions $j = l$ with positive integers l, and $m = 0\,,\,\pm 1\,,\,\pm 2\,,\,\ldots \pm l$ for the orbital angular momenta.

Now we define the so-called creation- and annihilation operators $\mathbf{L}^+$ and $\mathbf{L}^-$:

$$\left.\begin{aligned} \mathbf{L}^+ &:= \mathbf{L}_x + i\,\mathbf{L}_y\,,\\ \mathbf{L}^- &:= \mathbf{L}_x - i\,\mathbf{L}_y\,. \end{aligned}\right\} \tag{794}$$

Since the angular momentum operators are HERMITEan, the creation operator $\mathbf{L}^+$ is HERMITEan adjungated to $\mathbf{L}^-$ because of the imaginary unit in the second summand, and it holds also the inverse, so that

$$\left.\begin{aligned} <\psi\,|\,\mathbf{L}^+\,\chi> &= <\mathbf{L}^-\,\psi\,|\,\chi>\,,\\ <\psi\,|\,\mathbf{L}^-\,\chi> &= <\mathbf{L}^+\,\psi\,|\,\chi>\,. \end{aligned}\right\} \tag{795}$$

Under consideration of Eqs. (788) and (790), one proofs the following equations

$$\left.\begin{aligned} \mathbf{L}^+\,\mathbf{L}^- &= \mathbf{L}^2 - \mathbf{L}_z^2 + \hbar\,\mathbf{L}_z\,,\\ \mathbf{L}^-\,\mathbf{L}^+ &= \mathbf{L}^2 - \mathbf{L}_z^2 - \hbar\,\mathbf{L}_z\,, \end{aligned}\right\} \tag{796}$$

hence

$$\mathbf{L}^+\,\mathbf{L}^- \;-\; \mathbf{L}^-\,\mathbf{L}^+ = 2\,\hbar\,\mathbf{L}_z\,, \tag{797}$$

and with Eq. (788) one gets easily

$$\left.\begin{aligned} \mathbf{L}^+\,\mathbf{L}_z \;-\; \mathbf{L}_z\,\mathbf{L}^+ &= -\,\hbar\,\mathbf{L}^+\,,\\ \mathbf{L}^-\,\mathbf{L}_z \;-\; \mathbf{L}_z\,\mathbf{L}^- &= \;\hbar\,\mathbf{L}^-\,. \end{aligned}\right\} \tag{798}$$

Now we use Eq. (793) together with Eq. (798) to get the two following equations for the operators $\mathbf{L}^+$ and $\mathbf{L}^-$, which show that the 'creation' and 'annihilation' operators increase or decrease the quantum number m of the state vector ψ_{jm} by unity, that justifies its names:

$$\left.\begin{aligned} \mathbf{L}_z\left(\mathbf{L}^+\,\psi_{jm}\right) &= -\,\hbar\,(m+1)\left(\mathbf{L}^+\,\psi_{jm}\right),\\ \mathbf{L}_z\left(\mathbf{L}^-\,\psi_{jm}\right) &= -\,\hbar\,(m-1)\left(\mathbf{L}^-\,\psi_{jm}\right). \end{aligned}\right\} \tag{799}$$

Hence, the operators $\mathbf{L}^+$ and $\mathbf{L}^-$ create from an eigenstate ψ_{jm} of $\mathbf{L}_z$ again an eigenstate with an $\hbar$ larger or smaller eigenvalue. Can the eigenvalues of $\mathbf{L}_z$ become arbitrary large or end this procedure at a fixed angular momentum?

To answer this question, we form the expectation values for the eigenfunctions $\mathbf{L}^+ \psi_{jm}$ and $\mathbf{L}^- \psi_{jm}$. Using Eqs. (793), (795) and (796), we find

$$\left.\begin{aligned} 0 \leq < \mathbf{L}^+ \psi \,|\, \mathbf{L}^+ \psi > &= < \mathbf{L}^- \mathbf{L}^+ \psi \,|\, \psi > \\ &= < \left(\mathbf{L}^2 - \mathbf{L}_z^2 - \mathbf{L}_z\right) \psi \,|\, \psi > , \\ &= \hbar^2 \left(j(j+1) - m^2 - m\right) < \psi \,|\, \psi > \end{aligned}\right\} \tag{800}$$

and likewise

$$\left.\begin{aligned} 0 \leq < \mathbf{L}^- \psi \,|\, \mathbf{L}^- \psi > &= < \mathbf{L}^+ \mathbf{L}^- \psi \,|\, \psi > \\ &= < \left(\mathbf{L}^2 - \mathbf{L}_z^2 + \mathbf{L}_z\right) \psi \,|\, \psi > \\ &= \hbar^2 \left(j(j+1) - m^2 + m\right) < \psi \,|\, \psi > . \end{aligned}\right\} \tag{801}$$

Applying $\mathbf{L}^+$ on ψ_{jm} the eigenvalue m grows by unity, while it decreases by the application of $\mathbf{L}^-$ by unity. Now since $< \psi \,|\, \psi >$ cannot vanish, for fixed j, there must exist a largest eigenvalue m_+ and a smallest eigenvalue m_-, for which in the end the curved parenthesis in Eq. (800) or (801) vanishes

$$\left.\begin{aligned} j(j+1) - m_+^2 - m_+ = 0 \, , \\ j(j+1) - m_-^2 + m_- = 0 \, . \end{aligned}\right\} \tag{802}$$

This equation already determines the eigenvalue spectrum compatible with the commutation relations (788). It holds immediately

$$\left.\begin{aligned} m_+ &= j \, , \\ m_- &= -j \, , \end{aligned}\right\} \tag{803}$$

and because of the integer difference ± 1 of the eigen values m in applying the creation and annihilation operators on the eigenfunctions, the difference between m_+ and m_- must be also an integer, hence with Eq. (803)

$$m_+ - m_- = 2j \quad \text{integer, not negative.} \tag{804}$$

Possible eigenvalues for j and m are then

$$\left.\begin{aligned} &j : 0 \, , \frac{1}{2} \, , 1 \, , \frac{3}{2} \, , 2 \, , \frac{5}{2} \, , 3 \, , \ldots \, , \\ &m : -j \, , (-j+1) \, , \ldots \, , 0 \, , \ldots (j-1) \, , j \, . \end{aligned}\right\} \tag{805}$$

The integer values for j and the corresponding m-values reproduce the result found above in Eq. (792) for the orbital angular momentum. While there one has to solve partial differential equations, our result here follows from simple algebraic considerations. But we have found additional values for the angular momentum operator. According to Eq. (805), j can have also half-integer values. At least the commutation relations (788) admit this solution.

Which physical meaning have half-integer values for j ?

Here we get support from the PAULI matrices introduced in Eq. (717). By means of the in Eq. (726) given algebraic properties or also simply by direct evaluation, one gets the commutation relation

$$[\boldsymbol{\sigma}_\alpha, \boldsymbol{\sigma}_\beta] = 2i\,\epsilon_{\alpha\beta\gamma}\,\boldsymbol{\sigma}_\gamma \ . \tag{806}$$

Multiplying these equations with $\hbar^2/4$, we can define the spin operator $\mathcal{S}_\alpha$ by

$$\mathcal{S}_\alpha := \frac{\hbar}{2}\,\boldsymbol{\sigma}_\alpha \,. \tag{807}$$

In components, it is written as

$$\left.\begin{aligned} \mathcal{S}_x\,\mathcal{S}_y - \mathcal{S}_y\,\mathcal{S}_x &= i\,\hbar\,\mathcal{S}_z \,, \\ \mathcal{S}_y\,\mathcal{S}_z - \mathcal{S}_z\,\mathcal{S}_y &= i\,\hbar\,\mathcal{S}_x \,, \\ \mathcal{S}_z\,\mathcal{S}_x - \mathcal{S}_x\,\mathcal{S}_z &= i\,\hbar\,\mathcal{S}_y \,. \end{aligned}\right\} \tag{808}$$

The spin operators $\mathcal{S}_\alpha$, being HERMITEan as the PAULI matrices, fulfil the commutation relations (788) of angular momentum operators. Now it is simple to evaluate the eigenvalues of $\mathcal{S}^2$ and $\mathcal{S}_z$. Because of the properties (726), it follows after simple evaluation

$$\left.\mathcal{S}^2 \equiv \mathcal{S}_x^2 + \mathcal{S}_y^2 + \mathcal{S}_z^2 = \frac{3}{4}\,\hbar^2\,\mathbf{1}\,.\right\} \tag{809}$$

From the eigenvalue equations (793), now it follows

$$\left.\begin{aligned} \mathcal{S}^2\,\psi_{jm} &= \hbar^2\,j\,(j+1)\,\psi_{jm} \,, \\ \mathcal{S}_z\,\psi_{jm} &= \hbar\,m\,\psi_{jm} \,, \end{aligned}\right\} \tag{810}$$

hence taking into account that ψ_{jm} is a two-component spinor,

$$\left.\begin{aligned} \frac{3\,\hbar^2}{4}\begin{pmatrix} 1 & 0 \\ 0 & 1 \end{pmatrix}\begin{pmatrix} \varphi_1 \\ \varphi_2 \end{pmatrix} &= \hbar^2\,j\,(j+1)\begin{pmatrix} \varphi_1 \\ \varphi_2 \end{pmatrix} , \\ \frac{\hbar}{2}\begin{pmatrix} 1 & 0 \\ 0 & -1 \end{pmatrix}\begin{pmatrix} \varphi_1 \\ \varphi_2 \end{pmatrix} &= \hbar\,m\begin{pmatrix} \varphi_1 \\ \varphi_2 \end{pmatrix} . \end{aligned}\right\} \tag{811}$$

For the two simultaneous eigenvectors of this equation,

$$\psi_{\frac{1}{2},-\frac{1}{2}} = \begin{pmatrix} \varphi_1 \\ 0 \end{pmatrix} , \qquad \psi_{\frac{1}{2},+\frac{1}{2}} = \begin{pmatrix} 0 \\ \varphi_2 \end{pmatrix} , \tag{812}$$

one reads off the eigenvalues from Eq. (812)

$$\left.\begin{aligned} j &= \frac{1}{2} \,, \\ m_- &= -\frac{1}{2} \,, \quad m_+ = +\frac{1}{2} \,. \end{aligned}\right\} \tag{813}$$

Hence the spin operators $\mathcal{S}_\alpha$ belonging, for example, to the WEYL equation (721), describe an inner angular momentum property of the WEYL spinors (713), and we have seen, that we need both components of the spinor for both spin orientations. The observable of the WEYL spinors, the spin, is incompatible with any classical orbital motion of the particle. The spin belongs to those properties of elementary particles that can only be understood within the framework of quantum theory.

5.2 *Transition to the* DIRAC *Equation*

A particle with rest mass m_o and a velocity $\mathbf{u} = (u_x,\ u_y,\ u_z)$ has the relativistic momentum $\mathbf{p} = (p_x,\ p_y,\ p_z) = (m\,u_x,\ m\,u_y,\ m\,u_z)$ with the velocity-dependent mass $m = m_o/\sqrt{1-u^2/c^2}$. We write
$p^2 = \mathbf{p}\cdot\mathbf{p} = p_x^2 + p_y^2 + p_z^2$.

For the relativistic energy $E = m\,c^2$ of the particle, we can write, s. Chap. 9, Sect. 2, Eq. (451),

$$E = \sqrt{p^2\,c^2 + m_o^2\,c^4}\ . \tag{814}$$

When we apply the principles of quantum theory on the relativistic motion of particles with non-vanishing rest mass, we get the DIRAC equation and its physical interpretation, as we shall see in this subsection.

Here we will consider only free particles and refer for the interaction terms on the references. We see from Eq. (785), that we find the transition from classical mechanics to quantum mechanics in the now used spatial form in the SCHRÖDINGER representation by a quite simple procedure:

In classical mechanics, we write the energy E of a free particle of mass m with momentum $\mathbf{p}$

$$E = \frac{p^2}{2m}\ . \tag{815}$$

Then we replace the classical observables by operators

$$E \longrightarrow -\frac{\hbar}{i}\frac{\partial}{\partial t}\,, \quad p_x \longrightarrow \frac{\hbar}{i}\frac{\partial}{\partial x}\,, \quad p_y \longrightarrow \frac{\hbar}{i}\frac{\partial}{\partial y}\,, \quad p_z \longrightarrow \frac{\hbar}{i}\frac{\partial}{\partial z}\,. \tag{816}$$

There follows the following equation for the differential operators:

$$-\frac{\hbar^2}{2m}\,\Delta = -\frac{\hbar}{i}\frac{\partial}{\partial t}\ . \tag{817}$$

Using it for the probability amplitude $\psi(\mathbf{x}, t)$, we get the SCHRÖDINGER equation (785) for a free particle. Now we note an important peculiarity. This procedure gained from non-relativistic mechanics is already LORENTZ-invariant.

According to Eq. (350), the energy E and the momentum $\mathbf{p} = (p_x, p_y, p_z)$ of a particle are combined in the four-momentum p with the contravariant components p^k,

$$p^k = \left(m\,c\,,\, m\,u_x\,,\, m\,u_y\,,\, m\,u_z\right) = \left(\frac{E}{c}\,,\, p_x\,,\, p_y\,,\, p_z\right) \,. \tag{818}$$

Due to the signature $(+, -, -, -)$ (cp. our remarks in the footnote 2 in Chap. 9, Sect. 1.1, the covariant components p_k read

$$p_k = \left(m\,c\,, -m\,u_x\,,\, -m\,u_y\,,\, -m\,u_z\right) = \left(\frac{E}{c}\,, -p_x\,,\, -p_y\,,\, -p_z\right) \,. \tag{819}$$

With the coordinates $(ct, x, y, z) = x^k\ ,\ k = 0, 1, 2, 3\ ,$ we can formulate our procedure for the transition to quantum mechanics as follows:

In the equation for the energy expressed in terms of E and momenta $\mathbf{p}$ (and possibly also the coordinates), we replace the four-momenta by

$$p_k \quad \longrightarrow \quad -\frac{\hbar}{i}\frac{\partial}{\partial x^k}\,, \quad k = 0, 1, 2, 3 \quad , \tag{820}$$

and we create a differential equation for the probability amplitude $\psi(\mathbf{x}, t)$.

Now the framework is provided to describe the dynamics of particles in relativistic submicroscopic range, i.e. the transition to the relativistic quantum theory. We have to introduce the LORENTZ-invariant prescription (820) in the LORENTZ-invariant Eq. (814), and then to apply the resulting differential operations on the probability amplitude $\psi(\mathbf{x}, t)$. It follows when we divide Eq. (814) by c and change the order of summands under the square root,

$$-\frac{\hbar}{i}\frac{\partial\psi}{\partial x^0} = \sqrt{m_0^2\,c^2 \,-\, \hbar^2\,\Delta}\ \ \psi\,. \tag{821}$$

But with this equation, we are faced with a mystery. We do not know how the square root of the LAPLACE operator should be applied to the wave function $\psi(\mathbf{x}, t)$, hence how to interpret the differential equation (821).

We will get a well-understood result when we apply the differential operators of the left and right sides of Eq. (821) repeatedly on $\psi(\mathbf{x}, t)$. It follows, writing again $x^0 = c\,t$,

$$\begin{aligned}
\left(-\frac{\hbar}{i}\frac{\partial}{\partial x^0}\right)\left(-\frac{\hbar}{i}\frac{\partial}{\partial x^0}\right)\psi &= \left(\sqrt{m_0^2\,c^2 \,-\, \hbar^2\,\Delta}\ \right)\left(\sqrt{m_0^2\,c^2 \,-\, \hbar^2\,\Delta}\ \right)\psi\,,\\
-\hbar^2\,\frac{1}{c^2}\frac{\partial^2}{\partial t^2} &= \left(m_0^2\,c^2 \,-\, \hbar^2\,\Delta\right)\psi\,,\\
\left(\frac{\partial^2}{c^2\,\partial t^2} - \Delta\right)\psi &= -\frac{m_0^2\,c^2}{\hbar^2}\ \psi\,,
\end{aligned}$$

or in using D'ALEMBERT's wave operator $\Box$, cp. Eq. (728), or Eq. (316),

$$\left(\Box - \kappa^2\right)\psi = 0 \quad \text{with} \quad \kappa^2 = -\frac{m_o^2\,c^2}{\hbar^2} \ . \text{ KLEIN–GORDON equation} \qquad (822)$$

This is the KLEIN–GORDON equation for particles with rest mass m_o, already introduced in Eq. (745).

However, how can we understand Eq. (821)?

The left-hand side of Eq. (821) contains only the time derivative in first order. Equation (821) should form a LORENTZ-invariant relation, therefore also the right-hand side should contain only first derivatives for the spatial coordinates x^α. Can we replace Eq. (821) by a linear, LORENTZ-invariant differential equation of first order, so that, by its iterated application as above, there appears again the KLEIN–GORDON equation?

DIRAC's solution of this problem was one of the most remarkable achievements of physics, often not so easy to grasp for the beginner. But after the preparation in Chap. 9, Sects. 3 and 4 the solution should be suggestive.

With the in Eq. (734) introduced DIRAC spinor ψ and the DIRAC matrices γ_i, Eq. (736), the looking for differential equation is Eq. (737),

$$\left(\gamma^k\,\partial_k - \kappa\right)\psi = 0 \quad \text{with} \quad \kappa = -i\,\frac{m_o\,c}{\hbar}\,, \quad k = 0, 1, 2, 3 \ . \quad \text{DIRAC equation} \qquad (823)$$

Now we also know the physical interpretation of the DIRAC spinor ψ:

By complex conjugation and the transition to the transposed matrix, we create the HERMITEan adjungated DIRAC spinor ψ^+:

$$\psi^+ := \overline{\psi}^{\,T}\,, \qquad (824)$$

hence

$$\psi = \begin{pmatrix} \psi_1 \\ \psi_2 \\ \psi_3 \\ \psi_4 \end{pmatrix}, \quad \psi^+ = \left(\overline{\psi_1}\ ,\ \overline{\psi_2}\ ,\ \overline{\psi_3}\ ,\ \overline{\psi_4}\right). \qquad (825)$$

Then it is

$$\left(\psi^+\,\psi\right) dx^1 dx^2 dx^3 = \left(|\psi_1|^2 + |\psi_2|^2 + |\psi_3|^2 + |\psi_4|^2\right) dx^1 dx^2 dx^3$$

the probability density for finding the particle in a position measurement at time t in the space between x^1 and $x^1 + dx^1$, x^2 and $x^2 + dx^2$, and x^3 and $x^3 + dx^3$.

The probability density $\psi^+\psi$ can change in time within a spatial domain only, if it streams in other space domains. The probability density $\psi^+\psi$ must fulfil a continuity equation, s. Eqs. (469) and (527).

Indeed, from the DIRAC equation (823) we can derive with the so-called DIRAC current j^k

$$j^k = \psi^+ \gamma^0 \gamma^k \psi \tag{826}$$

a balance equation

$$\frac{\partial}{\partial x^k} j^k \equiv \frac{\partial}{\partial x^0} j^0 + \frac{\partial}{\partial x^1} j^1 + \frac{\partial}{\partial x^2} j^2 + \frac{\partial}{\partial x^3} j^3 = 0 . \tag{827}$$

For verifying the continuity equation (827), we form from the DIRAC equation (823) by HERMITEan adjunction at first the equivalent relation

$$\partial_k \psi^+ \gamma^{k+} - \overline{\kappa}\, \psi^+ = 0 ,$$

hence, since κ is purely imaginary,

$$\partial_k \psi^+ \gamma^{k+} + \kappa\, \psi^+ = 0 . \tag{828}$$

Now we get with Eq. (736) and with $\gamma^i := \eta^{ik} \gamma_k$

$$\left.\begin{aligned} \gamma^{0+} &= \gamma^0 , \\ \gamma^{\alpha +} &= -\gamma^\alpha , \quad \alpha = 1, 2, 3 , \end{aligned}\right\} \tag{829}$$

and we use Eq. (744),

$$\gamma^0 \gamma^k + \gamma^k \gamma^0 = 2\, \eta^{0k} ,$$

hence

$$\left.\begin{aligned} \gamma^0 \gamma^0 &= \mathbf{1} , \\ \gamma^0 \gamma^\alpha &= -\gamma^\alpha \gamma^0 . \end{aligned}\right\} \tag{830}$$

After multiplication of Eq. (828) from right with γ^0, we can evaluate, using Eqs. (829) and (830),

$$\begin{aligned} 0 &= \partial_k \psi^+ \gamma^{k+} \gamma^0 + \kappa\, \psi^+ \gamma^0 , \\ 0 &= \partial_0 \psi^+ \gamma^{0+} \gamma^0 + \partial_\alpha \psi^+ \gamma^{\alpha +} \gamma^0 + \kappa\, \psi^+ \gamma^0 , \\ 0 &= \partial_0 \psi^+ \gamma^0 \gamma^0 + \partial_\alpha \psi^+ \gamma^0 \gamma^\alpha + \kappa\, \psi^+ \gamma^0 , \\ 0 &= \partial_k \psi^+ \gamma^0 \gamma^k + \kappa\, \psi^+ \gamma^0 . \end{aligned}$$

Here we introduced the so-called DIRAC-adjoint spinor $\widetilde{\psi}$

$$\widetilde{\psi} := \psi^+ \gamma^0 . \tag{831}$$

For the adjoint spinor, the DIRAC equation takes the form

$$\partial_k \widetilde{\psi}\, \gamma^k + \kappa\, \widetilde{\psi} = 0\,, \tag{832}$$

and for the DIRAC current (826), we write

$$j^k = \widetilde{\psi}\, \gamma^k \psi\,. \tag{833}$$

The continuity equation now follows simply from

$$\begin{aligned}
\partial_k\, j^k &= \partial_k \left(\widetilde{\psi}\, \gamma^k\, \psi\right)\,,\\
&= \partial_k \left(\widetilde{\psi}\right)\, \gamma^k\, \psi + \widetilde{\psi}\, \gamma^k\, \partial_k \psi\,,\\
&= -\kappa\, \widetilde{\psi}\, \psi + \kappa\, \widetilde{\psi} \psi\,,\\
&= 0\,,
\end{aligned}$$

as claimed in Eq. (827).

The continuity equation (827) for the DIRAC current density j^k connects the spatial probability density j^0 of particles with rest mass m_o and with the probability current density j^α. This represents the first part of the physical interpretation of the DIRAC spinor ψ. Now we must still explain, why the DIRAC equation just describes particles with spin $\frac{1}{2}$.

After our introduction of the spin operators $\mathcal{S}_\alpha$ in Sect. 5.1.1 determined by the PAULI matrices (807), we can write down the corresponding spin operators S_α for the DIRAC-field. With definitions

$$\left.\begin{aligned}
\mathsf{S}_1 &= i\, \frac{\hbar}{2}\, \gamma_2 \gamma_3 = \frac{\hbar}{2} \begin{pmatrix} \sigma_1 & \mathbf{0} \\ \mathbf{0} & \sigma_1 \end{pmatrix},\\
\mathsf{S}_2 &= i\, \frac{\hbar}{2}\, \gamma_3 \gamma_1 = \frac{\hbar}{2} \begin{pmatrix} \sigma_2 & \mathbf{0} \\ \mathbf{0} & \sigma_2 \end{pmatrix},\\
\mathsf{S}_3 &= i\, \frac{\hbar}{2}\, \gamma_1 \gamma_2 = \frac{\hbar}{2} \begin{pmatrix} \sigma_3 & \mathbf{0} \\ \mathbf{0} & \sigma_3 \end{pmatrix}
\end{aligned}\right\} \tag{834}$$

one easily proofs, that the HERMITEan spin operators S_α of the DIRAC field fulfil the commutation relations (788) of the angular momentum operator,

$$\left.\begin{aligned}
\mathsf{S}_x\, \mathsf{S}_y - \mathsf{S}_y\, \mathsf{S}_x &= i\, \hbar \mathsf{S}_z\,,\\
\mathsf{S}_y\, \mathsf{S}_z - \mathsf{S}_z\, \mathsf{S}_y &= i\, \hbar \mathsf{S}_x\,,\\
\mathsf{S}_z\, \mathsf{S}_x - \mathsf{S}_x\, \mathsf{S}_z &= i\, \hbar \mathsf{S}_y\,.
\end{aligned}\right\} \tag{835}$$

All further steps can be easily presented after the preparation in Sect. 5.1.1. From the eigenvalue equation (793) it follows now

$$\left.\begin{aligned}
\mathsf{S}^2\, \psi_{jm} &= \hbar^2\, j\, (j+1)\, \psi_{jm}\,,\\
\mathsf{S}_z\, \psi_{jm} &= \hbar m\, \psi_{jm}
\end{aligned}\right\} \tag{836}$$

with

$$\mathsf{S}^2 = \frac{3\hbar^2}{4}\begin{pmatrix}1&0&0&0\\0&1&0&0\\0&0&1&0\\0&0&0&1\end{pmatrix}, \quad \mathsf{S}_z = \begin{pmatrix}-1&0&0&0\\0&1&0&0\\0&0&-1&0\\0&0&0&1\end{pmatrix}, \quad \psi_{jm} = \begin{pmatrix}\psi_1\\\psi_2\\\psi_3\\\psi_4\end{pmatrix}. \tag{837}$$

Because of the quasi-diagonal structure of the matrices in Eq. (837) the relation (807) is now doubly reproduced by Eq. (836). To the two simultaneously determined eigenvectors of Eq. (836),

$$\left.\psi_{\frac{1}{2},-\frac{1}{2}} = \begin{pmatrix}\psi_1\\0\\\psi_3\\0\end{pmatrix}, \quad \psi_{\frac{1}{2},\frac{1}{2}} = \begin{pmatrix}0\\\psi_2\\0\\\psi_4\end{pmatrix},\right\} \tag{838}$$

we get again the eigenvalues

$$\left.\begin{aligned} j &= \frac{1}{2}, \\ m_- &= -\frac{1}{2}, \quad m_+ = +\frac{1}{2}. \end{aligned}\right\} \tag{839}$$

This is the second part of the interpretation of the DIRAC spinor: It represents particles with rest mass m_o and spin $\frac{1}{2}$. Examples of particles described by the DIRAC equation are therefore electrons, neutrons or protons.

According to our derivation, the spin has the same algebraic properties as the orbital angular momentum. However, the spin is a pure quantum number of elementary particles, and it cannot be described by a rotation of the particle, neither in classical sense nor in the quantum theoretical description. A rotational motion should allow acceleration or braking. In the quantum theory of the orbital momentum, this happens by an increase or decrease of the quantum number as in Eq. (792). The spin of the electron or neutron has however always the value $j = \frac{1}{2}$; it cannot be changed. Its projections on a spatial axis only provide two values, $m_- = -\frac{1}{2}$ and $m_+ = +\frac{1}{2}$.

The incompatibility of the spin with an orbital motion has also direct experimental consequences. For example, the electron has also a magnetic dipole moment μ. By means of MAXWELL's equations, one can derive the ratio of the orbital motion of an electric charge and the thereby created magnetic moment, cp. Chap. 9, Sect. 3. The ratio of the magnetic moment μ to the spin of the electrons is however twice the value as for the orbital motion. This was measured in the famous EINSTEIN - DE HAAS experiment. We shall return to this point in the following Sect. 6.

With the derivation of the DIRAC equation, there began a far-reaching development in theoretical physics. Particles and fields became unified in a relativistic quantum field theory. The old classical space-time picture has been replaced by the notion of a physical vacuum, with spectacular measurable physical effects as for example the CASIMIR effect. It was predicted by the Dutch physicist CASIMIR in 1948, and verified with increasing accuracy, after 1997 with deviations below one percent. In this effect, two metal plates in vacuum suffer an attraction, since the electromagnetic modes of quantum vacuum fluctuations, that exist between the plates, are restricted by the boundary conditions on the plates, while outside of the plates all modes are possible. In this way, the stress tensor of the electromagnetic field, Eq. (603), causes a force on the plates, cp. e.g. Milton (2001).

6 Other Representations of the DIRAC Equation

The representation (733) of the DIRAC equation (732) with a set of γ matrices, Eq. (736), follows directly from the derivation from the spinor formalism.

The algebraic condition on the γ matrices to satisfy the relativistic energy equation (814) as well as allowing the transition to the quantum mechanical operators (816), are the matrix relations (744), here repeated,

$$\gamma_i \gamma_k + \gamma_k \gamma_i = 2\eta_{ik} \mathbf{1} . \tag{744'}$$

But this condition allows a large number of other representations of γ matrices in the DIRAC equation. A suitable selection of a representation can be favourable for the respective physical problem at hand, as can be seen below with the example of the non-relativistic approximation.

Now we present an algebraic method for producing just a couple of new sets of γ matrices.

From the two-dimensional PAULI matrices $\boldsymbol{\sigma}_\alpha$, $\alpha = 1, 2, 3$, Eq. (717), we form the four-dimensional matrices s_α and r_α

$$\left. \begin{aligned} \mathsf{s}_1 &= \begin{pmatrix} i\sigma_1 & \mathbf{0} \\ \mathbf{0} & i\sigma_1 \end{pmatrix}, & \mathsf{r}_1 &= \begin{pmatrix} \mathbf{0} & \mathbf{1} \\ \mathbf{1} & \mathbf{0} \end{pmatrix}, \\ \mathsf{s}_2 &= \begin{pmatrix} i\sigma_2 & \mathbf{0} \\ \mathbf{0} & i\sigma_2 \end{pmatrix}, & \mathsf{r}_2 &= \begin{pmatrix} \mathbf{0} & -i\mathbf{1} \\ i\mathbf{1} & \mathbf{0} \end{pmatrix}, \\ \mathsf{s}_3 &= \begin{pmatrix} i\sigma_3 & \mathbf{0} \\ \mathbf{0} & i\sigma_3 \end{pmatrix}, & \mathsf{r}_3 &= \begin{pmatrix} \mathbf{1} & \mathbf{0} \\ \mathbf{0} & -\mathbf{1} \end{pmatrix}. \end{aligned} \right\} \tag{840}$$

One verifies easily that the following sets of γ matrices fulfil the relations (744) as well as our original representation (717):

$$\left. \begin{aligned} &(a)\ \gamma_0 = \mathsf{r}_1 ,\ \gamma_\alpha = \mathsf{r}_2\, \mathsf{s}_\alpha , \\ &(b)\ \gamma_0 = \mathsf{r}_1 ,\ \gamma_\alpha = \mathsf{r}_3\, \mathsf{s}_\alpha , \\ &(c)\ \gamma_0 = \mathsf{r}_2 ,\ \gamma_\alpha = \mathsf{r}_3\, \mathsf{s}_\alpha , \\ &(d)\ \gamma_0 = \mathsf{r}_2 ,\ \gamma_\alpha = \mathsf{r}_1\, \mathsf{s}_\alpha , \\ &(e)\ \gamma_0 = \mathsf{r}_3 ,\ \gamma_\alpha = \mathsf{r}_1\, \mathsf{s}_\alpha , \\ &(f)\ \gamma_0 = \mathsf{r}_3 ,\ \gamma_\alpha = \mathsf{r}_2\, \mathsf{s}_\alpha , \end{aligned} \quad \alpha = 1, 2, 3 . \right\} \tag{841}$$

Each of the given sets of γ matrices defines also the DIRAC equation with identical physical consequences as our first representation (717). Still there exist other representations (e.g. the so-called MAJORANA representation) that will not be given here.

For many applications, it is advantageous to choose for γ matrices the so-called DIRAC representation (841 f) and to rewrite the DIRAC equation after multiplication with $\boldsymbol{\sigma}_0$ in the HAMILTON form, so that the DIRAC equation looks formally as the SCHRÖDINGER equation,

$$\left(\gamma^0\,\gamma^0\,\partial_0 + \gamma^0\,\gamma^1\,\partial_1 + \gamma^0\,\gamma^2\,\partial_2 + \gamma^0\,\gamma^3\,\partial_3\right)\psi = \kappa\,\gamma^0\,\psi\,, \tag{842}$$

where the γ matrices from Eq. (841 f) have to be inserted, which gives the following matrices after the explicit multiplication:

$$\left.\begin{aligned} \gamma_0 &= \begin{pmatrix} \mathbf{1} & \mathbf{0} \\ \mathbf{0} & -\mathbf{1} \end{pmatrix}, \\ \gamma_\nu &= \begin{pmatrix} \mathbf{0} & \boldsymbol{\sigma}_\nu \\ -\boldsymbol{\sigma}_\nu & \mathbf{0} \end{pmatrix}. \end{aligned}\right\} \tag{843}$$

with $\kappa = -i\,m_o\,c/\hbar$, $\gamma^0\;\gamma^0 = \mathbf{1}$ and the abbreviation

$$\gamma^0\,\gamma^\nu := \boldsymbol{\alpha}^\nu\,, \quad \nu = 1, 2, 3 \quad \text{and} \quad \gamma^0 := \boldsymbol{\beta}\,. \tag{844}$$

Then it follows after simple evaluation the DIRAC equation in the HAMILTON form

$$i\,\hbar\,\frac{\partial}{\partial t}\,\psi = \mathbf{H}_D\,\psi \qquad \text{HAMILTON form of the DIRAC equation} \tag{845}$$

with the HAMILTON operator

$$\mathbf{H}_D = c\,\frac{\hbar}{i}\,\frac{\partial}{\partial x^\nu}\,\boldsymbol{\alpha}^\nu + m_0\,c^2\,\boldsymbol{\beta}\,. \tag{846}$$

There one finds taking into account that $\gamma^0 = \gamma_0$, $\gamma^\nu = -\,\gamma_\nu$,

$$\left.\begin{aligned} \boldsymbol{\beta} &= \begin{pmatrix} \mathbf{1} & \mathbf{0} \\ \mathbf{0} & -\mathbf{1} \end{pmatrix}, \\ \boldsymbol{\alpha}_\nu &= \begin{pmatrix} \mathbf{0} & \boldsymbol{\sigma}_\nu \\ \boldsymbol{\sigma}_\nu & \mathbf{0} \end{pmatrix}. \end{aligned}\right\} \tag{847}$$

The DIRAC equation (845) then takes the form

$$i\,\hbar\,\frac{\partial}{\partial t}\,\psi = \left(c\,\boldsymbol{\alpha}^\nu\,\frac{\hbar}{i}\,\frac{\partial}{\partial x^\nu} + m_0\,c^2\,\boldsymbol{\beta}\right)\psi\,. \qquad \text{HAMILTON form of the DIRAC equation} \tag{848}$$

7 DIRAC Equation, SCHRÖDINGER Equation and PAULI Equation

The DIRAC equation represents a partial differential equation being of first order in all derivatives, and the wave function ψ has four complex-valued components. The SCHRÖDINGER equation is of first order in the time derivative, but of second order in spatial derivatives, and the wave function ψ has only one complex component. How does this fit together?

The DIRAC spinor ψ with γ matrices in DIRAC representation shall be written with two-component spinors φ and χ as

$$\psi = \begin{pmatrix} \psi_1 \\ \psi_2 \\ \psi_3 \\ \psi_4 \end{pmatrix} := \begin{pmatrix} \varphi \\ \chi \end{pmatrix}, \quad \psi^T = (\varphi^T, \chi^T) \tag{849}$$

with $\varphi = \begin{pmatrix} \psi_1 \\ \psi_2 \end{pmatrix}$ and $\chi = \begin{pmatrix} \psi_3 \\ \psi_4 \end{pmatrix}$.

We insert this into the DIRAC equation (848),

$$i\hbar \frac{\partial}{\partial t} \begin{pmatrix} \varphi \\ \chi \end{pmatrix} = \left(c\, \boldsymbol{\alpha}^\nu \frac{\hbar}{i} \frac{\partial}{\partial x^\nu} + m_0 c^2 \beta \right) \begin{pmatrix} \varphi \\ \chi \end{pmatrix}. \tag{850}$$

The matrices $\boldsymbol{\alpha}$ and β follow from Eq. (847),

$$\left. \begin{aligned} i\hbar \frac{\partial}{\partial t} \varphi &= c\, \boldsymbol{\sigma}^\nu \frac{\hbar}{i} \frac{\partial}{\partial x^\nu} \chi + m_0 c^2 \varphi, \\ i\hbar \frac{\partial}{\partial t} \chi &= c\, \boldsymbol{\sigma}^\nu \frac{\hbar}{i} \frac{\partial}{\partial x^\nu} \varphi - m_0 c^2 \chi. \end{aligned} \right\} \tag{851}$$

For getting the connection to the SCHRÖDINGER equation, we consider stationary states, i.e. we look for solutions of Eq. (851), belonging to fixed energy states E,

$$\left. \begin{aligned} \varphi(\mathbf{x}, t) &= \varphi(\mathbf{x}) \exp\left(-i \frac{E\,t}{\hbar} \right), \\ \chi(\mathbf{x}, t) &= \chi(\mathbf{x}) \exp\left(-i \frac{E\,t}{\hbar} \right). \end{aligned} \right\} \tag{852}$$

It follows

$$\left. \begin{aligned} E\,\varphi &= c\, \boldsymbol{\sigma}^\nu \frac{\hbar}{i} \frac{\partial}{\partial x^\nu} \chi + m_0 c^2 \varphi, \\ E\,\chi &= c\, \boldsymbol{\sigma}^\nu \frac{\hbar}{i} \frac{\partial}{\partial x^\nu} \varphi - m_0 c^2 \chi, \end{aligned} \right\} \tag{853}$$

hence

$$\left.\begin{aligned}(E - m_o c^2)\,\varphi &= c\,\boldsymbol{\sigma}^\nu \frac{\hbar}{i}\frac{\partial}{\partial x^\nu}\,\chi\,,\\ (E + m_o c^2)\,\chi &= c\,\boldsymbol{\sigma}^\nu \frac{\hbar}{i}\frac{\partial}{\partial x^\nu}\,\varphi\,.\end{aligned}\right\} \tag{854}$$

From the second Eq. (854) we find

$$\chi = \frac{c}{(E + m_o c^2)}\,\boldsymbol{\sigma}^\nu \frac{\hbar}{i}\frac{\partial}{\partial x^\nu}\,\varphi\,. \tag{855}$$

This equation allows a comparison of the order of magnitude of both fields. The part $\frac{\hbar}{i}\frac{\partial}{\partial x^\nu}$ is the momentum operator $\mathbf{p}$ belonging to the classical momentum $m\,\mathbf{v}$. For a particle with fixed momentum p there is

$$\chi = \frac{p\,c}{(E + m_o c^2)}\,\boldsymbol{\sigma}^\nu\,\varphi\,. \tag{856}$$

In the case of non-relativistic motion, we can write

$$\left.\begin{aligned}p &\approx m_o\,v\,,\\ E &\approx m_o\,c^2\end{aligned}\right\} \quad\longrightarrow\quad \frac{m_o\,v\,c}{E + m_o\,c^2} \approx \frac{v}{c}\,, \tag{857}$$

hence

$$\chi \approx \frac{1}{2}\frac{v}{c}\,\varphi\,. \qquad \text{Non-relativistic approximation} \tag{858}$$

Therefore in the non-relativistic case, $v \ll c$, one can neglect χ against φ. Then the four components of the relativistic DIRAC equation reduce to two components in the non-relativistic approximation.

The transition from the DIRAC equation to the SCHRÖDINGER equation can now be described as follows: We insert Eq. (855) in the first Eq. (854), and we get, still in the relativistic regime,

$$\left.\begin{aligned}E\,\varphi &= c\,\boldsymbol{\sigma}^\nu \frac{\hbar}{i}\frac{\partial}{\partial x^\nu}\,\frac{c}{(E + m_o c^2)}\,\boldsymbol{\sigma}^\nu \frac{\hbar}{i}\frac{\partial}{\partial x^\nu}\,\varphi + m_0\,c^2\,\varphi\,,\\ (E - m_0\,c^2)\,\varphi &= \frac{c^2}{(E + m_o c^2)}\,\boldsymbol{\sigma}^\nu \frac{\hbar}{i}\frac{\partial}{\partial x^\nu}\,\boldsymbol{\sigma}^\nu \frac{\hbar}{i}\frac{\partial}{\partial x^\nu}\,\varphi\end{aligned}\right\} \tag{859}$$

hence

$$\varphi = \frac{c^2}{(E - m_0\,c^2)(E + m_o c^2)}\,\boldsymbol{\sigma}^\nu \frac{\hbar}{i}\frac{\partial}{\partial x^\nu}\,\boldsymbol{\sigma}^\nu \frac{\hbar}{i}\frac{\partial}{\partial x^\nu}\,\varphi\,. \tag{860}$$

Now we denote the classical energy of a particle by ε. The non-relativistic approximation is then written as

$$E \approx m_o\,c^2 + \varepsilon\,, \quad \varepsilon \ll m_o\,c^2\,, \tag{861}$$

hence

$$\left.\begin{aligned}\frac{c^2}{(E - m_0\,c^2)(E + m_o\,c^2)} &= \frac{c^2}{E^2 - (m_o\,c^2)^2} \\ &\approx \frac{c^2}{(m_o\,c^2)^2 + 2\,\varepsilon\,m_o\,c^2 + \varepsilon^2 - (m_o\,c^2)^2} \\ &\approx \frac{c^2}{2\,\varepsilon\,m_o\,c^2} = \frac{1}{2\,\varepsilon\,m_o}\,.\end{aligned}\right\} \tag{862}$$

From Eq. (860) it follows then the non-relativistic approximation of φ. We write $\varphi \approx \phi$, neglect χ, and find

$$\varepsilon\,\phi = -\frac{\hbar^2}{2\,m_o}\,\boldsymbol{\sigma}^\nu\,\frac{\partial}{\partial x^\nu}\,\boldsymbol{\sigma}^\nu\,\frac{\partial}{\partial x^\nu}\,\phi\,. \qquad \text{Non-relativistic approximation of the DIRAC equation} \tag{863}$$

This is indeed already the SCHRÖDINGER equation for a free particle, and it appears twice, for both components of $\phi = \begin{pmatrix}\phi_1\\ \phi_2\end{pmatrix}$, that are decoupled in Eq. (863).

For making this obvious, we need an auxiliary formula for the PAULI matrices. If **a** and **b** denote two vectors, then it holds

$$\boldsymbol{\sigma}^\nu\,a_\nu\,\boldsymbol{\sigma}^\mu\,b_\mu = -a^\nu\,b_\nu - i\,\boldsymbol{\sigma}^\nu\,(\mathbf{a}\times\mathbf{b})_\nu\,, \tag{864}$$

as can be checked with Eq. (726) (the minus sign on the right-hand side results from the signature $(+, -, -, -)$). We evaluate the left-hand side and find after rearranging the terms

$$\begin{aligned}&(\boldsymbol{\sigma}^1\,a_1 + \boldsymbol{\sigma}^2\,a_2 + \boldsymbol{\sigma}^3\,a_3)(\boldsymbol{\sigma}^1\,b_1 + \boldsymbol{\sigma}^2\,b_2 + \boldsymbol{\sigma}^3\,b_3)\\ &\quad = \boldsymbol{\sigma}^1\,\boldsymbol{\sigma}^1\,a_1\,b_1 + \boldsymbol{\sigma}^2\,\boldsymbol{\sigma}^2\,a_2\,b_2 + \boldsymbol{\sigma}^3\,\boldsymbol{\sigma}^3\,a_3\,b_3\\ &\qquad - \boldsymbol{\sigma}^2\,\boldsymbol{\sigma}^3\,(a_2\,b_3 - a_3\,b_2) - \boldsymbol{\sigma}^3\,\boldsymbol{\sigma}^1\,(a_3\,b_1 - a_1\,b_3) - \boldsymbol{\sigma}^1\,\boldsymbol{\sigma}^2\,(a_1\,b_2 - a_2\,b_1)\\ &\quad = -a^\nu\,b_\nu - i\,\boldsymbol{\sigma}^\nu\,(\mathbf{a}\times\mathbf{b})_\cdot\,,\end{aligned}$$

what we wanted to show. If $\mathbf{b} = \lambda\,\mathbf{a}$, then the cross product in Eq. (864) vanishes.

With $a_\nu = b_\nu = \dfrac{\partial}{\partial x^\nu}$, we get with the LAPLACE operator $\boldsymbol{\Delta}$, s. Eq. (728),

$$\boldsymbol{\sigma}^\nu\,\frac{\partial}{\partial x^\nu}\,\boldsymbol{\sigma}^\mu\,\frac{\partial}{\partial x^\mu} = \frac{\partial}{\partial x^\nu}\,\frac{\partial}{\partial x^\mu}\,\delta^{\mu\nu} = \boldsymbol{\Delta}\,. \tag{865}$$

We insert Eq. (865) into Eq. (863) and get

$$\varepsilon\,\phi = -\frac{\hbar^2}{2\,m_o}\,\boldsymbol{\Delta}\,\phi\,. \tag{866}$$

This is nothing else than the SCHRÖDINGER equation (785) for a free particle, written for a stationary stage, and for each of the two components of

$$\phi = \begin{pmatrix} \phi_1 \\ \phi_2 \end{pmatrix}.$$

Is this duplication an unnecessary redundancy? This is only the first glance. Indeed it is the hint for the non-relativistic approximation of the interaction of the spin of charged particles with the electromagnetic field, it leads to the famous PAULI equation.

In mechanics, we have learned that the motion of a particle with charge e in an electromagnetic field can be included in the canonical formulation of dynamics, that the momentum $\mathbf{p}$ has to be replaced by the canonical momentum $\mathbf{p} - e\,\mathbf{A}$, and that the energy represented by the HAMILTON operator H must be supplemented by the term $e\,\varphi$, i.e. $H \longrightarrow H + e\,\varphi$. Here

$$A^k = (A^0,\ A^1,\ A^2,\ A^3) = \left(\frac{\varphi}{c},\ A_x,\ A_y,\ A_z\right) \tag{867}$$

denotes the four-potential of the electromagnetic field with the scalar potential φ and the vector potential $\mathbf{A} = (A_x,\ A_y,\ A_z)$, s. Chap. 9, Sect. 3.2.

This procedure for including the interaction of electrically charged particles with the electromagnetic field is called *gauge-invariant coupling*, and it plays an essential role in theoretical physics. This procedure is relativistically invariant, with the covariant formulation

$$p_k \quad \longrightarrow \quad p_k - e\,A_k\,. \qquad \text{Gauge-invariant coupling} \tag{868}$$

Now we investigate Eq. (820). In quantum theory, we include the interaction of the electromagnetic field by means of the transition

$$p_k = -\frac{\hbar}{i}\frac{\partial}{\partial x^k} \quad \longrightarrow \quad -\frac{\hbar}{i}\frac{\partial}{\partial x^k} - e\,A_k\,, \quad k = 0, 1, 2, 3 \qquad \text{Gauge-invariant coupling} \tag{869}$$

The procedure is also called gauge-invariant derivative, which can be written as

$$\frac{\partial}{\partial x^k} \quad \longrightarrow \quad \frac{\partial}{\partial x^k} + \frac{i\,e}{\hbar}\,A_k\,, \quad k = 0, 1, 2, 3 \qquad \text{Gauge-invariant coupling} \tag{870}$$

Here we take care of the signature, s. also Chap. 9, Sect. 3.2,

$$(A_0,\ A_1,\ A_2,\ A_3) = (\tfrac{1}{c}\,\varphi,\ -A_x,\ -A_y,\ -A_z).$$

Now we multiply the DIRAC equation (823) with $-\dfrac{\hbar}{i}$, and use Eq. (820) to get for a free particle

$$\left(\gamma^k\,p_k - m_o\,c\right)\,\psi = 0\,, \quad k = 0, 1, 2, 3\,. \qquad \text{DIRAC equation for a free particle} \tag{871}$$

The DIRAC equation for an electrically charged particle in an electromagnetic field then reads

$$\left[\gamma^k\left(-\frac{\hbar}{i}\frac{\partial}{\partial x^k}-e\,A_k\right)-m_o\,c\right]\psi=0\,. \quad \begin{array}{l}\text{DIRAC equation}\\ \text{for a charged particle}\\ \text{in an electromagnetic field}\end{array} \tag{872}$$

Here we insert the γ matrices in the DIRAC representation (843), and we get

$$\left[\begin{pmatrix}\mathbf{1}&\mathbf{0}\\ \mathbf{0}&-\mathbf{1}\end{pmatrix}\left(\frac{\hbar}{i}\frac{\partial}{\partial t}+e\,c\,A_0\right)+\begin{pmatrix}\mathbf{0}&-\sigma_\nu\\ \sigma_\nu&\mathbf{0}\end{pmatrix}\left(\frac{\hbar c}{i}\frac{\partial}{\partial x^\nu}+e\,c\,A_\nu\right)+m_0\,c^2\right]\psi=0\,. \tag{873}$$

We insert again $(A_0,\ A_1,\ A_2,\ A_3)=(\frac{1}{c}\,\varphi,\ -A_x,\ -A_y,\ -A_z)$.
Eq. (873) follows also from Eq. (851), when we substitute the gauge-invariant derivative with the result

$$\left.\begin{aligned}\left(i\,\hbar\,\frac{\partial}{\partial t}-e\,c\,A_0\right)\varphi&=\boldsymbol{\sigma}^\nu\left(\frac{\hbar c}{i}\frac{\partial}{\partial x^\nu}+e\,c\,A_\nu\right)\chi+m_0\,c^2\,\varphi\,,\\ \left(i\,\hbar\,\frac{\partial}{\partial t}-e\,c\,A_0\right)\chi&=\boldsymbol{\sigma}^\nu\left(\frac{\hbar c}{i}\frac{\partial}{\partial x^\nu}+e\,c\,A_\nu\right)\varphi-m_0\,c^2\,\chi\,.\end{aligned}\right\} \tag{874}$$

Here we consider stationary states, we eliminate χ, and we get as generalisation of Eq. (860) for φ

$$\left.\begin{aligned}\varphi&=f\,\boldsymbol{\sigma}^\nu\left(\frac{\hbar}{i}\frac{\partial}{\partial x^\nu}+e\,A_\nu\right)\boldsymbol{\sigma}^\mu\left(\frac{\hbar}{i}\frac{\partial}{\partial x^\mu}+e\,A_0\mu\right)\varphi\,,\quad\text{where}\\ f&=\frac{c^2}{\big((E-e\,c\,A_0)-m_0\,c^2\big)\big((E-e\,c\,A_0)+m_o\,c^2\big)}\end{aligned}\right\} \tag{875}$$

For deriving the non-relativistic approximation, we assume that the energy $e\,c\,A_0$ is of the order of the kinetic energy ε of the electron. With the steps in deriving Eq. (862), we get the non-relativistic approximation of the factor f

$$f \xrightarrow{\text{non-relativistic approximation}} \frac{1}{2\,m_o\,(\varepsilon-e\,c\,A_0)}\,. \tag{876}$$

In the non-relativistic approximation of the DIRAC equation (875), we can again neglect χ with respect to φ, and for φ we insert in Eq. (875) the expression (876) for f. It follows when we multiply the resulting equation with $(\varepsilon-e\,c\,A_0)$ from the denominator of Eq. (876), and if we use the two-component wave function $\boldsymbol{\phi}=\begin{pmatrix}\phi_1\\ \phi_2\end{pmatrix}$,

$$(\varepsilon-e\,c\,A_0)\,\boldsymbol{\phi}=\boldsymbol{\sigma}^\nu\left(\frac{\hbar}{i}\frac{\partial}{\partial x^\nu}+e\,A_\nu\right)\boldsymbol{\sigma}^\mu\left(\frac{\hbar}{i}\frac{\partial}{\partial x^\mu}+e\,A_0\mu\right)\boldsymbol{\phi}\,. \tag{877}$$

For converting this relation in a more intuitive form, we use Eq. (864), so that

$$(\varepsilon - e\,c\,A_0)\,\phi = \left[-\frac{\hbar^2}{2\,m_o}\left(\nabla + \frac{i\,e}{\hbar}\,\mathbf{A}\right)\cdot\left(\nabla + \frac{i\,e}{\hbar}\,\mathbf{A}\right) - \frac{e\,\hbar}{2\,m_o}\,\sigma^\nu(\nabla\times\mathbf{A})_\nu\right]\phi\,.$$

According to Eq. (807), the spin of the electron is given by $\mathcal{S}_\alpha := \frac{\hbar}{2}\,\sigma_\alpha\,.$ With rot$\mathbf{A} = \mathbf{B}$, we get the famous PAULI equation as non-relativistic approximation of the DIRAC equation:

$$(\varepsilon - e\,c\,A_0)\,\phi = \left[-\frac{\hbar^2}{2\,m_o}\left(\nabla + \frac{i\,e}{\hbar}\,\mathbf{A}\right)\cdot\left(\nabla + \frac{i\,e}{\hbar}\,\mathbf{A}\right) + \frac{e}{m_o}\,\boldsymbol{\mathcal{S}}\cdot\mathbf{B}\right]\phi\,,$$

PAULI equation (878)

written for stationary states. For the general case, we have to replace ε by $i\,\hbar\,\partial/\partial t$, hence

$$i\,\hbar\,\frac{\partial}{\partial t}\,\phi = \left[-\frac{\hbar^2}{2\,m_o}\left(\nabla + \frac{i\,e}{\hbar}\,\mathbf{A}\right)\cdot\left(\nabla + \frac{i\,e}{\hbar}\,\mathbf{A}\right) + \frac{e}{m_o}\,\boldsymbol{\mathcal{S}}\cdot\mathbf{B} + e\,c\,A_0\right]\phi\,. \quad (879)$$

This equation contains the important term $(e/m_o)\,\boldsymbol{\mathcal{S}}\cdot\mathbf{B}$, its interpretation will be next discussed. W. PAULI got this equation already in 1927, before P. A. M. DIRAC presented Eq. (872) in 1928.

It is obvious, that taking into account the electromagnetic interaction directly into the SCHRÖDINGER equation (785) according to the recipe Eq. (869) leads to the relation

$$(\varepsilon - e\,c\,A_0)\,\psi = \left[-\frac{\hbar^2}{2\,m_o}\left(\nabla + \frac{i\,e}{\hbar}\,\mathbf{A}\right)\cdot\left(\nabla + \frac{i\,e}{\hbar}\,\mathbf{A}\right)\right]\psi \quad (880)$$

with a one-component wave function ψ, and the term $(e/m_o)\,\boldsymbol{\mathcal{S}}\cdot\mathbf{B}$ is missing in relation to the PAULI equation.

Now we see the deeper meaning of the apparently redundant formulated SCHRÖDINGER equation (863). Without repeating the details of the derivation, it is obvious that taking into account the electromagnetic interaction in the SCHRÖDINGER equation (863) just delivers the PAULI equation (878) according to the procedure (869).

The expression $(e/m_o)\,\boldsymbol{\mathcal{S}}\cdot\mathbf{B}$ represents an energy term describing the interaction of the magnetic field strength $\mathbf{B}$ with the electron spin $\boldsymbol{\mathcal{S}}$.

For better understanding the meaning of this term, we remember again the classical electrodynamics.

A negatively charged mass m and a positively charged mass m displaced by the vector $\mathbf{a}$ form a dipole with the dipole moment

$$\mathbf{P} = e\,\mathbf{a}\,. \quad (881)$$

Since the total charge of the dipole is zero, it feels no force in an electric field $\mathbf{E}$, when we assume that the electric field strength $\mathbf{E}$ is constant over the extension of the dipole.

But the dipole can rotate. We put the origin of coordinates in the middle between the two charges, and call the radius vectors pointing to the positive and negative charges $+e$ and $-e$ by $\mathbf{x}_1$ and $\mathbf{x}_2$, respectively. The angular momentum of the dipole $\mathbf{L}$ is

$$\mathbf{L} = \mathbf{x}_1 \times \frac{d}{dt}\, m\, \mathbf{x}_1 + \mathbf{x}_2 \times \frac{d}{dt}\, m\, \mathbf{x}_2 \,. \tag{882}$$

The torque $\mathbf{D}$ on the dipole in the electric field $\mathbf{E}$ follows when we take into account that $\mathbf{a} \times \mathbf{a} = 0$, and assume that both charges can only rotate at a constant separation, hence $\mathbf{x}_1 = -\mathbf{x}_2 = \frac{1}{2}\mathbf{a}$ and $\frac{d}{dt}\mathbf{x}_1 = -\frac{d}{dt}\mathbf{x}_2$:

$$\begin{aligned} \mathbf{D} = \frac{d}{dt}\,\mathbf{L} &= \mathbf{x}_1 \times \frac{d^2}{dt^2}\, m\, \mathbf{x}_1 + \mathbf{x}_2 \times \frac{d^2}{dt^2}\, m\, \mathbf{x}_2 \\ &= \mathbf{x}_1 \times e\,\mathbf{E} + \mathbf{x}_2 \times (-e\,\mathbf{E}) \\ &= \frac{1}{2}\mathbf{a} \times e\,\mathbf{E} - \frac{1}{2}\mathbf{a} \times (-e\,\mathbf{E}) \\ &= e\,\mathbf{a}\times, \mathbf{E}\,, \end{aligned}$$

hence

$$\mathbf{D} = \mathbf{P} \times \mathbf{E}\,. \tag{883}$$

The amount D of the torque $\mathbf{D}$ depends on the angle ϑ between the dipole $\mathbf{P}$ and the field $\mathbf{E}$:

$$D = PE \sin\vartheta\,. \tag{884}$$

For $\vartheta = 0$, the torque of the dipole in the field vanishes. For rotating by an angle $d\,\vartheta$, we must exert a work $d\,A = D\,d\,\vartheta$. After a rotation by the angle $d\,\vartheta$, the dipole gets the energy W_{el}:

$$W_{el} = \int\limits_{-\frac{\pi}{2}}^{\vartheta} P\,E \sin\vartheta'\,d\,\vartheta' = P\,E\cos\vartheta\,, \tag{885}$$

or

$$W_{el} = \mathbf{P}\cdot\mathbf{E}\,. \tag{886}$$

In Problem 32, we will derive that the magnetic field, produced by a ring current I around an area $\mathbf{F}$ outside of the current corresponds to a magnetic moment $\mathbf{M}$,

$$\mathbf{M} = I\,\mathbf{F}\,. \tag{887}$$

So while we have no magnetic monopoles, we can produce magnetic dipole moments. Then it holds true analogous to the consideration above: On a magnetic moment $\mathbf{M}$ in a magnetic field $\mathbf{B}$ there acts a torque $\mathbf{D}$,

$$\mathbf{D} = \mathbf{M} \times \mathbf{B}\,, \tag{888}$$

and the magnetic moment has an energy W_{mag} in the field $\mathbf{B}$:

$$W_{mag} = \mathbf{M} \cdot \mathbf{B}\ . \tag{889}$$

We consider a mass m with electric charge e, orbiting on a circular path of radius r. Let ν be the number of circulations per second. Then the current in the time average over one orbit is

$$I = e\,\nu. \tag{890}$$

The corresponding magnetic moment has the value

$$\mathsf{M} = e\,\nu\,\pi\,r^2\,. \tag{891}$$

On the other hand, the orbiting mass has an angular momentum

$$\mathbf{L} = m\,\mathbf{x} \times d\,\mathbf{x}/dt\ . \tag{892}$$

For the circular motion it is $|\,\mathbf{x}| = r$. The velocity $d\,\mathbf{x}/dt$ is vertical to $\mathbf{x}$, and the amount of the angular momentum L follows easily, since ν orbits of length $2\,\pi\,r$ are traversed per second:

$$L = m\,r\,2\,\pi\,r\,\nu\ . \tag{893}$$

From Eqs. (891) and (893), it follows a ratio g of the magnetic moment to the angular momentum

$$g = \frac{M}{L} = \frac{e}{2\,m}\ . \tag{894}$$

The ratio g of the magnetic moment to the angular momentum is called *gyromagnetic factor*. The relation

$$\mathsf{M} = \frac{e}{2\,m_o}\mathbf{L} \tag{895}$$

is also called *magnetomechanical parallelism*, where we have used the mass m_o for comparison with the PAULI equation. In Problem 48 we shall present a more general derivation of the relation $\mathbf{D} = \frac{e}{2\,m_o}\mathbf{L} \times \mathbf{B}$, for the more theoretically interested reader, cp. also relevant textbooks on electrodynamics.

Now we can better understand the characteristic term $(e/m_o)\,\boldsymbol{S} \cdot \mathbf{B}$ in the PAULI equation (879). It describes that the electron has an inner magnetic moment $\mathbf{m}$ due to its spin that contributes to its interaction energy with the field $\mathbf{B}$:

$$W_m = \frac{e}{m_o} \, \boldsymbol{\mathcal{S}} \cdot \mathbf{B} = \mathbf{m} \cdot \mathbf{B} \ , \tag{896}$$

with

$$\mathbf{m} = \frac{e}{m_o} \, \boldsymbol{\mathcal{S}} \, . \tag{897}$$

Again we find the magnetomechanical parallelism. The gyromagnetic factor in Eq. (897) is however twice as large as in Eq. (895). This peculiarity is called *magnetomechanical anomaly*. It describes the important difference of the orbital angular momentum **L** as caused by an orbital motion, and the inner angular momentum described by the spin $\boldsymbol{\mathcal{S}}$. The spin cannot be reduced on any classical motion in space. We have here a direct proof, that the *inner* dynamics of particles in atomic or subatomic scales does not follow our classical comprehension, but must be described by quantum theory.

After our presentation of special relativity theory and quantum theory, we add here a short notice on IMMANUEL KANT, without entering into a philosophical discussion. KANT speaks of time as 'presupposed in all human experience' and of space as 'a form of intuition'.

The strict classical, hence non-relativistic and non-quantum-theoretical consideration of physics *creates* our *contemplation* of reality, or it has this done in the process of evolution. In our opinion, events take place in the three-dimensional Euclidean space, and the sequence of events represents an one-dimensional time. No one can recognise intuitively space and time in other ways. Because of the impossibility to escape from this classical thinking, it is so difficult to grasp our aim: To understand the physical reality mathematically better, that follows from the logical arguments of special relativity theory on space and time, and likewise from the statistical laws of quantum theory, and we get involved in apparent contradictions that require our full attention.

In the same degree, as we need abstract concepts to describe the experimental facts provided by nature, we deviate in our physical concepts from the direct contemplation. In quantum theory, this becomes especially striking. Then establishing contradictions between the physical predictions of special relativity theory or quantum theory and the above cited theses of KANT on space and time makes absolutely no sense, since we use actually the same terms, but we describe completely different objects. For grasp the physical space and the physical time, our intuition is insufficient.

We could also say it in the following way: In the beginning, we need space and time as poor opinions a priori for the imaginary world in our mind. In this connection, space and time may remain unchangeable in the sense of KANT. NEWTON has built his mechanics on this assumption, and he introduced a physical space and a physical time. The distinction to space and time as discussed by KANT is still not obvious, even if conceptual the important difference is already established. Now we saw that we could investigate the physical reality, and in particular the physical space and the physical time better and better, using new channels and instruments. Thereby we arrive at an essential conceptual distinction of the physical space-time from the notion of space and time as presupposed forms of contemplation. It is in no way different from or even in contradiction to KANT's notion, since KANT's definition is relevant only for the presupposed and unchangeable world of our intuition. The special relativity theory and the quantum theory are not in contradiction to KANT, they led us beyond NEWTON in the world of modern physics.

For a deeper reflexion of the basic notion of time, we recommend the books of the British physicist and philosopher J. BARBOUR, s. Barbour (1999, 2001).

Chapter 11
Electrodynamics in Exterior Calculus

MAXWELL's differential equations (503), presented in Sect. 3, Chap. 9, or MINSKOWSKI's equations (580), are indispensable to solve problems in electrodynamics, the scattering and bending of electromagnetic waves, the derivation of the field of moving charges and currents and so on. On the other hand, we have seen that important results, following in general from tedious calculations, can be obtained relatively simple from an investigation of the underlying algebraic structure. An example is given by the discussion of the angular momentum operators in quantum mechanics, s. Sect. 5.1.1, Chap. 10. In this chapter, we shall present a formulation of electrodynamics, from which the algebraic structure of these equations becomes especially obvious. Such a formulation of the theory that makes especially visible its structural peculiarities is especially important for theoretical investigations, for global problems and for extensions of the theory.

1 The Wedge Product

Vectors are tensors of first rank. We have shown in Sect. 2, Chap. 10, s. also Appendix B.1, that two vectors $\mathbf{u}$ and $\mathbf{v}$ can be combined in a tensor product, the simplest tensor of rank two $\mathbf{u}\,\mathbf{v}$, three vectors $\mathbf{u}$, $\mathbf{v}$, and $\mathbf{w}$ to a three-rank tensor $\mathbf{u}\,\mathbf{v}\,\mathbf{w}$, and so on; one also speaks of a 2-tensor or a 3-tensor. Since we shall introduce a further product below, in the following we write always the sign $\otimes$, for the tensor product as introduced in Eq. (652).

$$\mathbf{u}\,\mathbf{v} \equiv \mathbf{u} \otimes \mathbf{v}\ , \quad \mathbf{u}\,\mathbf{v}\,\mathbf{w} \equiv \mathbf{u} \otimes \mathbf{v} \otimes \mathbf{w}\ , \dots . \tag{898}$$

By means of the basis vectors $\mathbf{k}_i$, or the vectors of the dual basis $\mathbf{k}^i$ one can write a general two-rank tensor as

$$\mathbf{T} = T^{ij}\,\mathbf{k}_i \otimes \mathbf{k}_j \tag{899}$$

with the contravariant components T^{ij}, or as

H. Günther and V. Müller, *The Special Theory of Relativity*,
https://doi.org/10.1007/978-981-13-7783-9_11

$$\mathbf{T} = T_{ij}\, \mathbf{k}^i \otimes \mathbf{k}^j \tag{900}$$

with the covariant components T_{ij}.

For a three-rank tensor **T**, it holds analogously

$$\mathbf{T} = T^{ijl}\, \mathbf{k}_i \otimes \mathbf{k}_j \otimes \mathbf{k}_l \ , \tag{901}$$

with contravariant components T^{ijl}, or with the dual basis $\mathbf{k}^i$

$$\mathbf{T} = T_{ijl}\, \mathbf{k}^i \otimes \mathbf{k}^j \otimes \mathbf{k}^l \tag{902}$$

with the covariant components T_{ijl}, cp. also Appendix B.1.

Now we consider total antisymmetric tensors $\mathbf{T}^A$. In the simplest case, one gets such a tensor from two vectors **u** and **v** by

$$\mathbf{T}^A = \mathbf{u} \otimes \mathbf{v} - \mathbf{v} \otimes \mathbf{u} \ . \tag{903}$$

The expression on the right-hand side of Eq. (903) is called the exterior or *wedge* product of vectors **u** and **v**. One writes

$$\mathbf{u} \wedge \mathbf{v} := \mathbf{u} \otimes \mathbf{v} - \mathbf{v} \otimes \mathbf{u} \ . \tag{904}$$

The exterior product is defined by its linearity and its antisymmetry,

$$\left.\begin{aligned} &\mathbf{u} \wedge \mathbf{v} = -\mathbf{v} \wedge \mathbf{u} \ , \quad \text{and} \\ &\mathbf{u} \wedge \alpha\mathbf{v} = \alpha\mathbf{u} \wedge \mathbf{v} \ . \end{aligned}\right\} \tag{905}$$

The higher wedge products are also defined by the condition of linearity and total antisymmetry. For the outer product $\mathbf{a} \wedge \mathbf{b} \wedge \mathbf{c}$ of three vectors **a**, **b** and **c**, it holds then

$$\left.\begin{aligned} &\mathbf{a} \wedge \mathbf{b} \wedge \mathbf{c} = -\mathbf{b} \wedge \mathbf{a} \wedge \mathbf{c} = -\mathbf{c} \wedge \mathbf{b} \wedge \mathbf{a} = -\mathbf{a} \wedge \mathbf{c} \wedge \mathbf{b} \quad \text{and} \\ &\mathbf{a} \wedge \alpha\mathbf{b} \wedge \mathbf{c} = \mathbf{a} \wedge \mathbf{b} \wedge \alpha\mathbf{c} = \alpha\mathbf{a} \wedge \mathbf{b} \wedge \mathbf{c} \ . \end{aligned}\right\} \tag{906}$$

In terms of the ordinary tensor products, we get for the threefold wedge product

$$\left.\begin{aligned} \mathbf{a} \wedge \mathbf{b} \wedge \mathbf{c} = \ &\mathbf{a} \otimes \mathbf{b} \otimes \mathbf{c} + \mathbf{c} \otimes \mathbf{a} \otimes \mathbf{b} + \mathbf{b} \otimes \mathbf{c} \otimes \mathbf{a} \\ &-\mathbf{b} \otimes \mathbf{a} \otimes \mathbf{c} - \mathbf{a} \otimes \mathbf{c} \otimes \mathbf{b} - \mathbf{c} \otimes \mathbf{b} \otimes \mathbf{a} \ . \end{aligned}\right\} \tag{907}$$

From this exercise, one sees that one can write the wedge product formally as determinant,

$$\mathbf{u} \wedge \mathbf{v} = \begin{vmatrix} \mathbf{u}\,\mathbf{v} \\ \mathbf{u}\,\mathbf{v} \end{vmatrix} \ , \quad \mathbf{a} \wedge \mathbf{b} \wedge \mathbf{c} = \begin{vmatrix} \mathbf{a}\,\mathbf{b}\,\mathbf{c} \\ \mathbf{a}\,\mathbf{b}\,\mathbf{c} \\ \mathbf{a}\,\mathbf{b}\,\mathbf{c} \end{vmatrix} \ , \ \ldots \ , \tag{908}$$

but one has to carefully preserve the order of the tensorial factors in evaluating the determinant, on the first place a factor from the first row, on the second place a factor from the second row, etc.

From the connection with the structure of a determinant, it follows:

The wedge product is linear in each factor,

$$\mathbf{k}^1 \wedge \ldots \alpha\mathbf{k}^i \wedge \ldots = \alpha\,\mathbf{k}^1 \wedge \ldots \mathbf{k}^i \wedge \ldots \ . \tag{909}$$

The wedge product vanishes, when two factors are parallel,

$$\mathbf{k}^1 \wedge \ldots \mathbf{k}^i \wedge \mathbf{k}^j \wedge \ldots = 0 \quad \text{wenn} \quad \mathbf{k}^i = \alpha\, \mathbf{k}^j \, . \tag{910}$$

With the notion of tensors in components, it follows for the rank-2 tensor

$$\mathbf{T}^A = T^A_{ij}\, \mathbf{k}^i \otimes \mathbf{k}^j = \frac{1}{2}\left(T_{ij} - T_{ji}\right) \mathbf{k}^i \otimes \mathbf{k}^j = T_{ij}\, \frac{1}{2}\left(\mathbf{k}^i \otimes \mathbf{k}^j - \mathbf{k}^j \otimes \mathbf{k}^i\right) ,$$

hence

$$\mathbf{T}^A = \frac{1}{2}\, T_{ij}\, \mathbf{k}^i \wedge \mathbf{k}^j \, . \tag{911}$$

Likewise, we get a rank-3 tensor

$$\mathbf{T}^A = \frac{1}{3!}\, T_{ijl}\, \mathbf{k}^i \wedge \mathbf{k}^j \wedge \mathbf{k}^l := T^A_{ijl}\, \mathbf{k}^i \wedge \mathbf{k}^j \wedge \mathbf{k}^l \tag{912}$$

as well as analogous constructions for the totally antisymmetric tensors of higher rank.

There the dimension sets a limit on the rank of the antisymmetric tensors. An antisymmetric tensor of rank p is formed from the wedge product with p basis vectors: In an n-dimensional space, there vanishes therefore the antisymmetric tensors with order larger than n. Since there are only n different basis vectors, in the order $(n+1)$ there must exist in each summand at least two identical basis vectors, so that the tensors become zero.

We remark: The wedge product of two antisymmetric tensors is again an antisymmetric tensor, hence, e.g. for two rank-2 tensors $\mathbf{T}^A$ and $\mathbf{S}^A$

$$\mathbf{U}^A := \mathbf{T}^A \wedge \mathbf{S}^A = T^A_{ij}\, \mathbf{k}^i \wedge \mathbf{k}^j \wedge S^A_{pq}\, \mathbf{k}^p \wedge \mathbf{k}^q \tag{913}$$

hence

$$\mathbf{U}^A = U^A_{ijpq}\, \mathbf{k}^i \wedge \mathbf{k}^j \wedge \mathbf{k}^p \wedge \mathbf{k}^q \quad \text{with} \quad U^A_{ijpq} = T^A_{ij}\, S^A_{pq} \, . \tag{914}$$

In the simplest case, the wedge product of a rank-2 tensor $\mathbf{T}^A$ with a vector $\mathbf{u} = u_r\, \mathbf{k}^r$ forms a rank-3 antisymmetric tensor $\mathbf{W}^A$

$$\mathbf{W}^A := \mathbf{T}^A \wedge \mathbf{u} = T^A_{ij}\, u_r\, \mathbf{k}^i \wedge \mathbf{k}^j \wedge \mathbf{k}^r = W^A_{ijr}\, \mathbf{k}^i \wedge \mathbf{k}^j \wedge \mathbf{k}^r \, . \tag{915}$$

2 Differential Forms

The coordinates $x^1, x^2, \ldots, x^n$ in an n-dimensional space with a metric g_{ik} are normally written with contravariant indices, hence also its differentials $dx^1, dx^2, \ldots, dx^n$.[1]

[1]To the covariant differentials defined by $dx_i := g_{ik} dx^k$ there exist in general no global covariant coordinates x_i. However, they exist in the case that the metric g_{ik} originates from the unit matrix δ_{ik} ($\delta_{ik} = 1$ for $i = k$ and zero else) by a coordinate transformation, i.e. if $g_{ik} = (\partial X^r/\partial x^i)(\partial X^s/\partial x^k)\delta_{rs}$ with Cartesian coordinates $X^r \equiv X_r$ in an Euclidean space with metric δ_{ik}.

A linear differential form,[2] also called PFAFF form, is simply the expression

$$\omega = \sum_j f_j \, dx^j = f_1 \, dx^1 + f_2 \, dx^2 + \ldots + f_n \, dx^n \ , \tag{916}$$

where the factors f_j are in general functions of the coordinates,

$$f_j = f_j(x^i) \ . \tag{917}$$

Analogous to the vectors and tensors of arbitrary rank, one constructs an algebra of forms. We remember: A tensor of rank 0 is a scalar, in general, a scalar function $f = f(x^i)$. A tensor of rank 1 is a vector $\mathbf{u}$ that can be decomposed into a basis $\mathbf{k}^i \ , \ i = 1, 2, \ldots, n$, of n-dimensional space according to $\mathbf{u} = u_i \, \mathbf{k}^i$. Accordingly, a tensor of second rank can be decomposed as $\mathbf{U} = U_{ij} \, \mathbf{k}^i \, \mathbf{k}^i$, etc.

A basis $\boldsymbol{\alpha}^i \ , \ i = 1, 2, \ldots, n$, for the p-forms in an n-dimensional space is formed by the coordinate differentials $d\mathbf{x}^i$,

$$\boldsymbol{\alpha}^i := \boldsymbol{dx}^i \ , \quad i = 1, 2, \ldots, n \ . \tag{918}$$

A function $F(x^i)$ is called 0-form.

A linear differential form (916) is now called an 1-form $\boldsymbol{a}$ with a decomposition in the basis $\boldsymbol{\alpha}_i$,

$$\boldsymbol{a} = \sum_j a_j \, \boldsymbol{\alpha}^j = a_1 \, \boldsymbol{\alpha}^1 + a_2 \, \boldsymbol{\alpha}^2 + \ldots + a_n \, \boldsymbol{\alpha}^n \ , \tag{919}$$

where the factors a_j are in general functions of coordinates,

$$a_j = a_j(x^i) \ . \tag{920}$$

A differential form $\boldsymbol{A}$ of rank two, also simply called a 2-form, is then defined by

$$\boldsymbol{A} = \sum_{j<l} A_{j\,l} \, \boldsymbol{\alpha}^j \wedge \boldsymbol{\alpha}^l = A_{12} \, \boldsymbol{\alpha}^1 \wedge \boldsymbol{\alpha}^2 + A_{13} \, \boldsymbol{\alpha}^1 \wedge \boldsymbol{\alpha}^3 + \ldots + A_{n-1\,n} \, \boldsymbol{\alpha}^{n-1} \wedge \boldsymbol{\alpha}^n \ , \tag{921}$$

$$A_{j\,l} = A_{j\,l}(x^i) \ . \tag{922}$$

One can also sum over all indices without the restriction $j < l$. However, due to the antisymmetry of the wedge product, one has to add the factor 1/2.

$$\boldsymbol{A} = \sum_{j,l} \frac{1}{2} \, A_{j\,l} \, \boldsymbol{\alpha}^j \wedge \boldsymbol{\alpha}^l \quad \text{with} \quad A_{j\,l} = - \, A_{l\,j} \ . \tag{923}$$

The elements $A_{j\,l}$ here form also components of an antisymmetric tensor as in Eq. (911). By Eqs. (921) or (923), we express in the formalism of forms, what else would read

$$\sum_{j,l} A_{j\,l} \left(dx^j \, dx^l - dx^l \, dx^j \right) \quad \text{with} \quad A_{j\,l} = - \, A_{l\,j} \ .$$

[2] In the older algebraic language, a form represents any homogeneous polynom. Now it is used only for linear differential forms. Forms of higher order are defined by the condition of antisymmetry.

A form of rank p, also called p-form, $\boldsymbol{A}$ (we use for simplicity the same letter $\boldsymbol{A}$), is then given by

$$\left.\begin{array}{l} \boldsymbol{A} = \sum\limits_{j_1,\ldots j_p} \frac{1}{p\,!} A_{j_1,\ldots j_p}\, \alpha^{j_1} \wedge \ldots \wedge \alpha^{j_p} \\ \quad \text{with a totally antisymmetric tensor } A_{j_1,\ldots j_p}\,, \end{array}\right\} \qquad p - \text{form} \quad (924)$$

or one restricts the number of summands and then needs no factor $1/p\,!$,

$$\boldsymbol{A} \;=\; \sum_{j_1<j_2<\ldots<j_p} A_{j_1,\ldots j_p}\, \alpha^{j_1} \wedge \ldots \wedge \alpha^{j_p}\;. \qquad p - \text{Form} \quad (925)$$

The wedge product of a p-form $\boldsymbol{A}$ with a q-Form $\boldsymbol{B}$ is a $(p+q)$-form $\boldsymbol{C}$

$$\boldsymbol{C} = \boldsymbol{A} \wedge \boldsymbol{B} = \left(\sum_{i_1<i_2<\ldots<i_p} A_{i_1,\ldots i_p}\, \alpha^{i_1} \wedge \ldots \wedge \alpha^{i_p} \right) \wedge \left(\sum_{j_1<j_2<\ldots<j_q} B_{j_1,\ldots j_q}\, \alpha^{j_1} \wedge \ldots \wedge \alpha^{j_q} \right). \qquad (926)$$

Clearly, it is

$$\boldsymbol{C} = \boldsymbol{0} \quad \text{for} \quad p+q>n\,, \qquad (927)$$

since in the wedge product, then there appear always two identical factors.

We want to discuss Eq. (926) not in full generality, but explain its meaning with two examples.

1.

$$\begin{aligned} \boldsymbol{A} &= A_{12}\, \boldsymbol{dx}^1 \wedge \boldsymbol{dx}^2 + A_{24}\, \mathbf{d}x^2 \wedge \boldsymbol{dx}^4\,, \\ \boldsymbol{B} &= B_{13}\, \boldsymbol{dx}^1 \wedge \boldsymbol{dx}^3 + B_{23}\, \boldsymbol{dx}^2 \wedge \boldsymbol{dx}^3\,. \end{aligned}$$

It follows for $\boldsymbol{C} = \boldsymbol{A} \wedge \boldsymbol{B}$

$$\begin{aligned} \boldsymbol{C} &= \left(A_{12}\, \boldsymbol{dx}^1 \wedge \boldsymbol{dx}^2 + A_{24}\, \mathbf{d}x^2 \wedge \boldsymbol{dx}^4\right) \wedge \left(B_{13}\, \mathbf{d}x^1 \wedge \boldsymbol{dx}^3 + B_{23}\, \boldsymbol{dx}^2 \wedge \boldsymbol{dx}^3\right) \\ &= A_{12}\, B_{13}\, \boldsymbol{dx}^1 \wedge \mathbf{d}x^2 \wedge \boldsymbol{dx}^1 \wedge \boldsymbol{dx}^3 + A_{12}\, B_{23}\, \boldsymbol{dx}^1 \wedge \boldsymbol{dx}^2 \wedge \mathbf{d}x^2 \wedge \boldsymbol{dx}^3 \\ &\quad + A_{24}\, B_{13}\, \boldsymbol{dx}^2 \wedge \boldsymbol{dx}^4 \wedge \boldsymbol{dx}^1 \wedge \boldsymbol{dx}^3 + A_{24}\, B_{23}\, \boldsymbol{dx}^2 \wedge \boldsymbol{dx}^4 \wedge \boldsymbol{dx}^2 \wedge \boldsymbol{dx}^3 \\ &= A_{24}\, B_{13}\, \boldsymbol{dx}^2 \wedge \mathbf{d}x^4 \wedge \boldsymbol{dx}^1 \wedge \boldsymbol{dx}^3 \\ &= -\,A_{24}\, B_{13}\, \boldsymbol{dx}^1 \wedge \boldsymbol{dx}^2 \wedge \mathbf{d}x^3 \wedge \boldsymbol{dx}^4\,, \end{aligned}$$

since the other summands with two equal factors vanish. For the last equality, we have used a different order with taking care of the antisymmetry of the wedge product.

2.

$$\begin{aligned} \boldsymbol{u} &= u_0\, \boldsymbol{dx}^0 + u_3\, \boldsymbol{dx}^3\,, \\ \boldsymbol{F} &= F_{12}\, \boldsymbol{dx}^1 \wedge \boldsymbol{dx}^2\,. \end{aligned}$$

It follows for $\boldsymbol{C} = \boldsymbol{u} \wedge \boldsymbol{F}$

$$\begin{aligned} \mathbf{C} &= \left(u_0\, \boldsymbol{dx}^0 + u_3\, \boldsymbol{dx}^3\right) \wedge F_{12}\, \boldsymbol{dx}^1 \wedge \boldsymbol{dx}^2 \\ &= u_0\, F_{12}\, \boldsymbol{dx}^0 \wedge \boldsymbol{dx}^1 \wedge \boldsymbol{dx}^2 + u_3\, F_{12}\, \boldsymbol{dx}^3 \wedge \boldsymbol{dx}^1 \wedge \boldsymbol{dx}^2 \\ &= u_0\, F_{12}\, \boldsymbol{dx}^0 \wedge \mathbf{d}x^1 \wedge \boldsymbol{dx}^2 + u_3\, F_{12}\, \boldsymbol{dx}^1 \wedge \boldsymbol{dx}^2 \wedge \boldsymbol{dx}^3\,. \end{aligned}$$

Now we have to explain some conventions for operations between forms $\boldsymbol{A}, \boldsymbol{u}$ and tensors $\mathbf{A}$, or vectors $\mathbf{u}$. They must be compatible with the definition of the wedge product (904). For the basis forms α^i and the basis vectors $\mathbf{k}^i$ or $\mathbf{k}_i$ there should hold the following elementary contraction formulas:

$$\left.\begin{array}{l} (\alpha^i, \mathbf{k}_j) = \delta^i_j\,, \quad (\alpha^i \wedge \alpha^j, \mathbf{k}_l) = \frac{1}{2}(\alpha^i\,\delta^j_l - \alpha^j\,\delta^i_l)\,, \\ (\alpha^i \wedge \alpha^j, \mathbf{k}_r \wedge \mathbf{k}_s) = \delta^{ij}_{rs} = \begin{cases} +1\,, & (i\ j) \text{ is an even permutation of } (r\ s), \\ -1\,, & (i\ j) \text{ is an uneven permutation of } (r\ s), \\ 0 & \text{else}\,. \end{cases} \end{array}\right\} \tag{928}$$

Then the contraction of a 1-form $\boldsymbol{u} = u_i\,\alpha^i$ with a vector $\mathbf{v} = v^j\,\mathbf{k}_j$ reads

$$(\boldsymbol{u}, \mathbf{v}) = (u_i\,\alpha^i, v^j\,\mathbf{k}_j) = u_i\,v^i\,. \tag{929}$$

The contraction of a 2-form $\boldsymbol{F} = \frac{1}{2}F_{ij}\,\alpha^i \wedge \alpha^j$ with a vector $\mathbf{v} = v^j\,\mathbf{k}_j$ leads to a 1-form,

$$\begin{aligned}(\boldsymbol{F}, \mathbf{v}) &= \frac{1}{2}\,(F_{ij}\,\alpha^i \wedge \alpha^j, v^l\mathbf{k}_l) = \frac{1}{2}\,F_{ij}\,v^l\left(\alpha^i \wedge \alpha^j, \mathbf{k}_l - \alpha^j \wedge \alpha^i, \mathbf{k}_l\right) \\ &= \frac{1}{2}\,F_{ij}\,v^l\left(\alpha^i\,\delta^j_l - \alpha^j\,\delta^i_l\right),\end{aligned}$$

hence with the antisymmetry of F_{ij}

$$(\boldsymbol{F}, \mathbf{v}) = F_{ij}\,v^j\,\alpha^i\,. \tag{930}$$

The contraction of a 2-form $\boldsymbol{F} = \frac{1}{2}\,F_{ij}\,\alpha^i \wedge \alpha^j$ with an antisymmetric tensor $\mathbf{T}^A = \frac{1}{2}\,T^{rs}\,\mathbf{k}_r \wedge \mathbf{k}_s$, cp. (911), results in a scalar,

$$(\boldsymbol{F}, \mathbf{T}^A) = \frac{1}{2}\cdot\frac{1}{2}\,F_{ij}\,T^{rs}\,(\alpha^i \wedge \alpha^j, \mathbf{k}_r \wedge \mathbf{k}_s) = \frac{1}{4}\,F_{ij}\,T^{rs}\,\delta^{ij}_{rs}\,,$$

hence with the antisymmetry of F_{ij} and T_{ij}

$$(\boldsymbol{F}, \mathbf{T}^A) = \sum_{i<j} F_{ij}\,T^{ij}\,. \tag{931}$$

The norm of a p-form is defined by

$$||\boldsymbol{F}|| = \sum_{j_1<j_2<\ldots<j_p} F_{j_1 j_2 \ldots j_p}\,F^{j_1 j_2 \ldots j_p}\,. \tag{932}$$

For the norm of the p-form $\boldsymbol{F}$, we get from Eq. (932) with Eq. (532),

$$\begin{aligned}||\boldsymbol{F}|| &= F_{01}\,F^{01} + F_{02}\,F^{02} + F_{03}\,F^{03} + F_{12}\,F^{12} + F_{13}\,F^{13} + F_{23}\,F^{23} \\ &= -\frac{1}{c^2}\,\left(E_x^2 + E_y^2 + E_z^2\right) + \left(B_x^2 + B_y^2 + B_z^2\right)\end{aligned}$$

hence

$$||\boldsymbol{F}|| = \mathbf{B}^2 - \frac{1}{c^2}\,\mathbf{E}^2\,. \tag{933}$$

We shall not need other formulas.

Now we define the external derivative.

The total differential dF of a function $F = F(x^i)$ reads with $\partial F/\partial x^j \equiv F_{,j} := f_j$

$$dF = \sum_j \frac{\partial F}{\partial x^j}\, dx^j = F,_j\, dx^j = f_j\, dx^j \ .$$

We can write it in the outer calculus as 1-form

$$\boldsymbol{d}F = \sum_j \frac{\partial F}{\partial x^j}\, \boldsymbol{d}x^j = \sum_j f_j\, \boldsymbol{d}x^j = f_j\, \boldsymbol{\alpha}^j \ , \tag{934}$$

and we call $\boldsymbol{d}F$ the outer derivative of the 0-form F.

In general with the outer derivative $\boldsymbol{d}$, we create from a p-form $\boldsymbol{A}$, given by $\boldsymbol{A} = A_{j_1,\ldots,j_p}\, \boldsymbol{\alpha}^{j_1} \wedge \ldots \wedge \boldsymbol{\alpha}^{j_p}$ a $(p+1)$-form $\boldsymbol{d}\,\boldsymbol{A}$ with the prescription

$$\left.\begin{aligned} \boldsymbol{d}\,\boldsymbol{A} &= \frac{\partial A_{j_1,\ldots,j_p}}{\partial x^r}\, \boldsymbol{d}x^r \wedge \boldsymbol{\alpha}^{j_1} \wedge \ldots \wedge \boldsymbol{\alpha}^{j_p} \\ &= \frac{\partial A_{j_1,\ldots,j_p}}{\partial x^r}\, \boldsymbol{\alpha}^r \wedge \boldsymbol{\alpha}^{j_1} \wedge \ldots \wedge \boldsymbol{\alpha}^{j_p}. \end{aligned}\right\} \tag{935}$$

We illustrate this construction with an example. From the 2-form $\boldsymbol{A}$,

$$\boldsymbol{A} = f(x^1, x^3)\, \boldsymbol{d}x^0 \wedge \boldsymbol{d}x^1 + g(x^2, x^3)\, \boldsymbol{d}x^1 \wedge \boldsymbol{d}x^2 \, , \tag{936}$$

we evaluate the 3-form $\boldsymbol{d}\boldsymbol{A}$ after the prescription (935)

$$\begin{aligned} \boldsymbol{d}\boldsymbol{A} &= \boldsymbol{d}f(x^1, x^3) \wedge \boldsymbol{d}x^0 \wedge \boldsymbol{d}x^1 + \boldsymbol{d}g(x^2, x^3) \wedge \boldsymbol{d}x^1 \wedge \boldsymbol{d}x^2 \, , \\ &= \left(\frac{\partial f}{\partial x^1}\, \boldsymbol{d}x^1 + \frac{\partial f}{\partial x^3}\, \boldsymbol{d}x^3\right) \wedge \boldsymbol{d}x^0 \wedge \boldsymbol{d}x^1 + \left(\frac{\partial g}{\partial x^2}\, \boldsymbol{d}x^2 + \frac{\partial g}{\partial x^3}\, \boldsymbol{d}x^3\right) \wedge \boldsymbol{d}x^1 \wedge \boldsymbol{d}x^2 \\ &= \frac{\partial f}{\partial x^1}\, \boldsymbol{d}x^1 \wedge \boldsymbol{d}x^0 \wedge \boldsymbol{d}x^1 + \frac{\partial f}{\partial x^3}\, \boldsymbol{d}x^3 \wedge \boldsymbol{d}x^0 \wedge \boldsymbol{d}x^1 \\ &\quad + \frac{\partial g}{\partial x^2}\, \boldsymbol{d}x^2 \wedge \boldsymbol{d}x^1 \wedge \boldsymbol{d}x^2 + \frac{\partial g}{\partial x^3}\, \boldsymbol{d}x^3 \wedge \boldsymbol{d}x^1 \wedge \boldsymbol{d}x^2 \\ &= \frac{\partial f}{\partial x^3}\, \boldsymbol{d}x^3 \wedge \boldsymbol{d}x^0 \wedge \boldsymbol{d}x^1 + \frac{\partial g}{\partial x^3}\, \boldsymbol{d}x^3 \wedge \boldsymbol{d}x^1 \wedge \boldsymbol{d}x^2 \ , \end{aligned}$$

hence

$$\boldsymbol{d}\boldsymbol{A} = \frac{\partial f}{\partial x^3}\, \boldsymbol{d}x^0 \wedge \boldsymbol{d}x^1 \wedge \boldsymbol{d}x^2 + \frac{\partial g}{\partial x^3}\, \boldsymbol{d}x^1 \wedge \boldsymbol{d}x^2 \wedge \boldsymbol{d}x^3 \, . \tag{937}$$

Applying twice the operation $\boldsymbol{d}$, one gets zero, since all summands contain at least two equal factors in the wedge products,

$$\boldsymbol{d}\,\boldsymbol{d}\boldsymbol{A} = 0 \, , \tag{938}$$

and it holds the LEIBNIZ product rule

$$\boldsymbol{d}\,(\boldsymbol{A} \wedge \boldsymbol{B}) = \boldsymbol{d}\boldsymbol{A} \wedge \boldsymbol{A} + \boldsymbol{A} \wedge \boldsymbol{d}\boldsymbol{B} \, . \tag{939}$$

We remark again that the notation $\boldsymbol{d}x^i$ and $\boldsymbol{\alpha}^i$ can be used alternatively, see Eq. (918).

One defines the following:

A k-form $\boldsymbol{A}$ *is called closed, if it fulfils the equation* $\boldsymbol{d}\,\boldsymbol{A} = 0$.

For example, the external derivative of each closed form is zero because of Eq. (938).

The form (936) is closed, when the functions f and g do not depend on x^3.

Furthermore, one defines the following:

A k-form $\boldsymbol{A}$ *is called exact, when there exists a* $(k+1)$*-form* $\boldsymbol{B}$*, so that the equation* $\boldsymbol{A} = \boldsymbol{d}\,\boldsymbol{B}$ *if fulfilled.*

Due to Eq. (938) each exact form is closed.

The inversion of the statement is not trivial, since the closure of a form is a local statement: One can check it by differentiation. The statement of exactness requires to find a form $\boldsymbol{B}$, only knowing that its outer derivative $\boldsymbol{d}\,\boldsymbol{B} = \boldsymbol{A}$. Determining **B** is therefore a problem of integration that can become difficult for spaces with complex topology. Topological questions, representing a strength of the formalism, cannot be covered within this book. We recommend to the relevant literature. We only remark that spaces that are topologically contractible, as the full $\mathbb{R}_n$, allow the inverse statement, and there each closed form is also exact.

For each k-form $\boldsymbol{A}$ in an n-dimensional space with metric g_{ik}, a dual form $*\boldsymbol{A}$ of degree $(n-k)$ is defined by the following construction:

$$\left.\begin{aligned} \boldsymbol{A} \quad &= \sum_{j_1<\ldots<j_k} a_{j_1\ldots j_k}\, \boldsymbol{dx}^{j_1} \wedge \ldots \wedge \boldsymbol{dx}^{j_k} \,, \\ *\boldsymbol{A} \quad &= \sum_{l_1<\ldots<k_k} b_{l_1\ldots l_{n-k}}\, \boldsymbol{dx}^{l_1} \wedge \ldots \wedge \boldsymbol{dx}^{l_{n-k}} \\ \text{with} \quad & \\ b_{l_1\ldots l_{n-k}} &= \sum_{j_1<\ldots<j_k} a^{j_1\ldots j_k} \epsilon_{j_1\ldots j_k\, l_1\ldots l_{n-k}} \,, \quad a^{j_1\ldots j_k} = a_{l_1\ldots l_k}\, g^{j_1 l_1} \ldots g^{j_k l_k} \,. \end{aligned}\right\} \tag{940}$$

Here $\epsilon_{j_1\ldots j_k\, l_1\ldots l_{n-k}}$ denotes the totally antisymmetric tensor with components $+1$, -1 or 0, s. Eq. (1635). An example for this $*$-operation will be shown in the following discussion of the MAXWELL equations in outer calculus.

Now we introduce the integration: Let be $x^i = x^i(t)$ the parameter form of a curve in space. The integral of an 1-form $\boldsymbol{u} = u_i\, \boldsymbol{dx}^i$ over this curve between parameter values t_1 and t_2 defines the line integral,

$$\int_{t_1}^{t_2} \boldsymbol{u} := \int_{t_1}^{t_2} u_i \frac{d\,x^i}{dt}\, dt \,. \tag{941}$$

For the special case that $u_i = \partial f / \partial x^i$, it is

$$\int_{t_1}^{t_2} \boldsymbol{u} := \int_{t_1}^{t_2} \frac{\partial f}{\partial x^i} \frac{d\,x^i}{dt}\, dt = f\left(x^i(t_2)\right) - f\left(x^i(t_1)\right) \,.$$

Let be $\boldsymbol{C} = \frac{1}{2}\, C_{ij}\, \boldsymbol{\alpha}^i \wedge \boldsymbol{\alpha}^j$ a 2-form. We consider the parameter representation $x^i = x^i(u, v)$ of an area S.

The 2-form is expressed by the forms $\boldsymbol{du}$ and $\boldsymbol{dv}$ of the area,

$$\begin{aligned} \boldsymbol{C} &= \frac{1}{2}\, C_{ij}\, \boldsymbol{dx}^i \wedge \boldsymbol{dx}^j \\ &= \frac{1}{2}\, C_{ij} \left(\frac{\partial x^i}{\partial u}\, \boldsymbol{du} + \frac{\partial x^i}{\partial v}\, \boldsymbol{dv} \right) \wedge \left(\frac{\partial x^j}{\partial u}\, \boldsymbol{du} + \frac{\partial x^j}{\partial v}\, \boldsymbol{dv} \right), \end{aligned}$$

hence

$$\boldsymbol{C} = \frac{1}{2}\, C_{ij} \left(\frac{\partial x^i}{\partial u} \frac{\partial x^j}{\partial v} - \frac{\partial x^i}{\partial v} \frac{\partial x^j}{\partial u} \right) \boldsymbol{d}u \wedge \boldsymbol{d}v. \tag{942}$$

The twofold integral of $\boldsymbol{C}$ over this area is then given by

$$\iint\limits_S \boldsymbol{C} := \iint\limits_S \frac{1}{2}\, C_{ij} \left(\frac{\partial x^i}{\partial u} \frac{\partial x^j}{\partial v} - \frac{\partial x^i}{\partial v} \frac{\partial x^j}{\partial u} \right) du\, dv\,, \tag{943}$$

where the area integral has to be evaluated according to standard mathematics.

At this point we show how the outer calculus is suitable for formulating important theorems very compact and without need of a special coordinate representation. It is $\boldsymbol{C} = \boldsymbol{d}\,\boldsymbol{u}$, hence

$$\boldsymbol{C} = \boldsymbol{d}\,\boldsymbol{u} = \left(\frac{\partial u_i}{\partial x^k} - \frac{\partial u_k}{\partial x^i} \right) \boldsymbol{d}x^k \wedge \boldsymbol{d}x^i = \left(\frac{\partial u_k}{\partial x^i} - \frac{\partial u_i}{\partial x^k} \right) \boldsymbol{d}x^i \wedge \boldsymbol{d}x^k. \tag{944}$$

Then for the *vector* $\mathbf{u}$, it follows $(\operatorname{rot}\mathbf{u})_{ik} = \left(\frac{\partial u_k}{\partial x^i} - \frac{\partial u_i}{\partial x^k} \right)$.

The integration of Eq. (944) over the area S with the boundary ∂S provides STOKES's law in the form

$$\iint\limits_S \boldsymbol{d}\,\boldsymbol{u} = \oint\limits_{\partial S} \boldsymbol{u}\,. \tag{945}$$

3 MAXWELL's Equations

In the following, we shall be careful with respect to the denomination and distinction of vectors in three- and four-dimensional space as introduced in Sect. 2.1.2, Chap. 9. We denote with $\mathbf{A}$ a three-dimensional vector, but A for a vector in MINKOWSKI space.

In MINKOWSKI space with the signature $(+1, -1, -1, -1)$, we consider the electromagnetic field strength tensor F_{ik}, s. Eq. (532), using the SI-system, the connected 2-form $\boldsymbol{F}$,

$$\boldsymbol{F} := \sum_{i<k} F_{ik}\, \boldsymbol{d}x^i \wedge \boldsymbol{d}x^k\,,\quad i,k = 0,1,2,3,\quad x^0 = ct,\ x^1 = x,\ x^2 = y,\ x^3 = z\,. \tag{946}$$

We can write down the 2-form $\boldsymbol{F}$ as

$$\begin{aligned} \boldsymbol{F} := &\ F_{01}\, \boldsymbol{d}x^0 \wedge \boldsymbol{d}x^1 + F_{02}\, \boldsymbol{d}x^0 \wedge \boldsymbol{d}x^2 + F_{03}\, \boldsymbol{d}x^0 \wedge \boldsymbol{d}x^3 \\ &+ F_{12}\, \boldsymbol{d}x^1 \wedge \boldsymbol{d}x^2 + F_{13}\, \boldsymbol{d}x^1 \wedge \boldsymbol{d}x^3 + F_{23}\, \boldsymbol{d}x^2 \wedge \boldsymbol{d}x^3\,, \\ = &\ \frac{E_x}{c}\, c\,\boldsymbol{d}t \wedge \boldsymbol{d}x + \frac{E_y}{c}\, c\,\boldsymbol{d}t \wedge \boldsymbol{d}y + \frac{E_z}{c}\, c\,\boldsymbol{d}t \wedge \boldsymbol{d}z \\ &- B_z\, \boldsymbol{d}x \wedge \boldsymbol{d}y + B_y\, \boldsymbol{d}x \wedge \boldsymbol{d}z - B_x\, \boldsymbol{d}y \wedge \boldsymbol{d}z\,, \end{aligned}$$

hence

$$\left.\begin{aligned} \boldsymbol{F} = &\ E_x\, \boldsymbol{d}t \wedge \boldsymbol{d}x + E_y\, \boldsymbol{d}t \wedge \boldsymbol{d}y + E_z\, \boldsymbol{d}t \wedge \boldsymbol{d}z \\ &- B_z\, \boldsymbol{d}x \wedge \boldsymbol{d}y + B_y\, \boldsymbol{d}x \wedge \boldsymbol{d}z - B_x\, \boldsymbol{d}y \wedge \boldsymbol{d}z. \end{aligned}\right\} \tag{947}$$

Now we define the 2-form $*\boldsymbol{F}$ after the description (940),

$$*\boldsymbol{F} := \sum_{i<k} *F_{ik}\, \boldsymbol{d}x^i \wedge \boldsymbol{d}x^k \ , \ i,k = 0,1,2,3, \quad x^0 = ct,\ x^1 = x,\ x^2 = y,\ x^3 = z \ , \tag{948}$$

where $*F_{ik}$ are given by the terms $b_{\dots}$ in Eq. (940), hence

$$* \ F_{ik} := \sum_{rs} F^{rs} \epsilon_{ikrs} \tag{949}$$

with the totally antisymmetric tensor ϵ_{ikrs}, s. Appendix B.1, hence

$$* \ F_{01} = F^{23},\ *F_{02} = -F^{13},\ *F_{03} = F^{12},\ *F_{12} = F^{03},\ *F_{13} = -F^{02},\ *F_{23} = F^{01}, \tag{950}$$

and with the contravariant components F^{ik} defined in Eq. (532),

$$\left.\begin{aligned} *F_{01} &= -B_x \ , \ *F_{02} = -B_y \ , \ *F_{03} = -B_z \ , \\ *F_{12} &= -\frac{E_z}{c} \ , \ *F_{13} = \frac{E_y}{c} \ , \quad *F_{23} = -\frac{E_x}{c} \ . \end{aligned}\right\} \tag{951}$$

We get the 2-form $*\boldsymbol{F}$

$$\left.\begin{aligned} *\boldsymbol{F} = &-B_x\, c\,\boldsymbol{d}t \wedge \boldsymbol{d}x - B_y\, c\,\boldsymbol{d}t \wedge \boldsymbol{d}y - B_z\, c\,\boldsymbol{d}t \wedge \boldsymbol{d}z \\ &-\frac{E_z}{c}\,\boldsymbol{d}x \wedge \boldsymbol{d}y + \frac{E_y}{c}\,\boldsymbol{d}x \wedge \boldsymbol{d}z - \frac{E_x}{c}\,\boldsymbol{d}y \wedge \boldsymbol{d}z \ . \end{aligned}\right\} \tag{952}$$

With the four-current density $j^i = \rho_e\, u^i$, Eq. (526), for the invariant rest charge density ρ_e and the four-velocity u^i, one defines the current density 1-form $\boldsymbol{J}$,

$$\boldsymbol{J} := j_i\, \boldsymbol{d}x^i = j_0\, \boldsymbol{d}x^0 + j_1\, \boldsymbol{d}x^1 + j_2\, \boldsymbol{d}x^2 + j_3\, \boldsymbol{d}x^3 \ . \tag{953}$$

From this 1-form $\boldsymbol{J}$, we now form in standard way the form $*\boldsymbol{J}$. Here, we take care of Eq. (526), and we consider the *vector* $j_i = (\rho\, c\ , \ -\rho\, v_x\ , \ -\rho\, v_y\ , \ -\rho\, v_z)$, and for signature $(+1, -1, -1, -1)$ the contravariant components $j^i = (\rho\, c\ , \ +\rho\, v_x\ , \ +\rho\, v_y\ , \ +\rho\, v_z)$, where $\rho = \rho_e/\sqrt{1 - u^2/c^2}$. We find for the adjunct form

$$\begin{aligned} *\boldsymbol{J} = \quad &\rho c \ \ \boldsymbol{d}x \wedge \boldsymbol{d}y \wedge \boldsymbol{d}z \\ &-\rho\, v_x\ \boldsymbol{d}y \wedge \boldsymbol{d}z \wedge c\,\boldsymbol{d}t \\ &+\rho\, v_y\ \boldsymbol{d}x \wedge \boldsymbol{d}z \wedge c\,\boldsymbol{d}t \\ &-\rho\, v_z\ \boldsymbol{d}x \wedge \boldsymbol{d}y \wedge c\,\boldsymbol{d}t \ . \end{aligned} \tag{954}$$

We shall now demonstrate that MAXWELL's equations in vacuum in SI-units have the following compact form:

$$\left.\begin{aligned} \boldsymbol{d}\boldsymbol{F} \quad &= 0 \ , \\ \boldsymbol{d} * \boldsymbol{F} &= -\mu_o * \boldsymbol{J} \ . \end{aligned}\right\} \tag{955}$$

We consider the rules for forming the outer derivative, as for example Eq. (936) for $\boldsymbol{d}\boldsymbol{A}$ from the form $\boldsymbol{A}$ after prescription (935). For abbreviation, we write $\partial_x f$ for $\partial f/\partial_x$ and find after simple reordering of the terms in the first form in Eq. (955),

$$\left.\begin{aligned} \boldsymbol{d}\boldsymbol{F} = &-(\partial_x B_x + \partial_y B_y + \partial_z B_z)\,\boldsymbol{d}x \wedge \boldsymbol{d}y \wedge \boldsymbol{d}z \\ &-(\partial_y E_z - \partial_z E_y + \partial_t B_x)\,\boldsymbol{d}y \wedge \boldsymbol{d}z \wedge \boldsymbol{d}t \\ &-(\partial_z E_x - \partial_x E_z + \partial_t B_y)\,\boldsymbol{d}z \wedge \boldsymbol{d}x \wedge \boldsymbol{d}t \\ &-(\partial_x E_y - \partial_y E_x + \partial_t B_z)\,\boldsymbol{d}x \wedge \boldsymbol{d}y \wedge \boldsymbol{d}t \ . \end{aligned}\right\} \tag{956}$$

This form vanishes exactly when all terms in parenthesis are zero. This is nothing than the MAXWELL equations (503a),

$$\boldsymbol{d}\boldsymbol{F} = 0 \quad \longleftrightarrow \quad \mathrm{rot}\mathbf{E} + \frac{\partial \mathbf{B}}{\partial t} = 0 \ \text{ and } \ \mathrm{div}\,\mathbf{B} = 0\,. \tag{957}$$

With the expression (952), we now form $\boldsymbol{d} * \boldsymbol{F}$. Again we consider the prescription of the outer differentiation, and we get after simple reordering

$$\left.\begin{aligned} \boldsymbol{d} * \boldsymbol{F} = & -\frac{1}{c}\,(\partial_x E_x + \partial_y E_y + \partial_z E_z)\, \boldsymbol{dx} \wedge \boldsymbol{dy} \wedge \boldsymbol{dz} \\ & + (\partial_y B_z - \partial_z B_y - \frac{1}{c^2}\,\partial_t E_x)\, \boldsymbol{dy} \wedge \boldsymbol{dz} \wedge \boldsymbol{cdt} \\ & - (\partial_z B_x - \partial_x B_z - \frac{1}{c^2}\,\partial_t E_y)\, \boldsymbol{dx} \wedge \boldsymbol{dz} \wedge \boldsymbol{cdt} \\ & + (\partial_x B_y - \partial_y B_x - \frac{1}{c^2}\,\partial_t E_z)\, \boldsymbol{dx} \wedge \boldsymbol{dy} \wedge \boldsymbol{cdt}\,. \end{aligned}\right\} \tag{958}$$

Now it is clear that the second equation (955) is identical with the second group of MAXWELL's equations,

$$\boldsymbol{d} * \boldsymbol{F} = -\,\mu_o * \boldsymbol{J} \quad \longleftrightarrow \quad \mathrm{rot}\mathbf{H} - \frac{\partial \mathbf{D}}{\partial t} = \rho\,\mathbf{v} \ \text{ and } \ \mathrm{div}\,\mathbf{D} = \rho\,. \tag{959}$$

Here, we have to take into account that in vacuum, the case here exclusively considered, it holds $\mathbf{B} = \mu_o\mathbf{H}$, $\mathbf{D} = \varepsilon_o\mathbf{E}$ and $1/c^2 = \varepsilon_o\,\mu_o$.

Still we remark: The sign in Eq. (959) follows from the signature $(+1, -1, -1, -1)$, which is just the inverse of the sign in older texts. While this has no influence on the definition of $*F_{ik}$ in Eq. (940), where the metric appears twice, it is important for the definition of $*J_{ijk}$, since the metric appears in Eq. (940) three times as factor.

With the three-dimensional vector $\mathbf{j} = \rho\,\mathbf{u}$, $\rho = \rho_e/\gamma$ (invariant rest charge density ρ_e) and the vector of four-velocity $\mathsf{u} = (c/\gamma, \mathbf{u}/\gamma)$, $\gamma = \sqrt{1 - u^2/c^2}$, it follows for the four-vector of current density J,

$$\mathsf{J} = \rho_e\,\mathsf{u} \quad \text{or} \quad j_i = \rho_e\,u_i\ . \tag{960}$$

By contraction of the 2-Form $\boldsymbol{F}$ of the electromagnetic field strength (947) with the current density four-vector J, the 1-form of LORENTZ force density $\boldsymbol{k}$ (here we use for distinction the letter k)

$$\begin{aligned} \boldsymbol{k} := (\boldsymbol{F}, \mathsf{J}) = & \ F_{ik}\, j^k\, \boldsymbol{\alpha}^i \\ = & \left(F_{01}\, j^1 + F_{02}\, j^2 + F_{03}\, j^3\right) \boldsymbol{\alpha}^0 \\ + & \left(F_{10}\, j^0 + F_{12}\, j^2 + F_{13}\, j^3\right) \boldsymbol{\alpha}^1 \\ + & \left(F_{20}\, j^0 + F_{21}\, j^1 + F_{23}\, j^3\right) \boldsymbol{\alpha}^2 \\ + & \left(F_{30}\, j^0 + F_{31}\, j^1 + F_{32}\, j^2\right) \boldsymbol{\alpha}^3\ , \end{aligned}$$

hence, after replacement of F_{ik} by Eq. (532), and j^i by Eq. (526), and furthermore Eq. (918) for $\boldsymbol{\alpha}^i$,

$$\boldsymbol{k} = -\,\rho\Big(\mathbf{E}\cdot\mathbf{u}\Big)\,\boldsymbol{dt} \;-\; \rho\Big(E_\nu + (\mathbf{u}\times\mathbf{B})_\nu\Big)\,\boldsymbol{dx}^\nu\ ,\ \ \nu = 1, 2, 3\ . \tag{961}$$

This result coincides with our expression for the LORENTZ force density in Eq. (536).

Now we present furthermore MINKOWSKI's equations of electrodynamics for media in motion within the outer calculus, Eqs. (580).

For Eq. (571a), we write without any change

$$\boldsymbol{dF} = 0 \quad \longleftrightarrow \quad \partial_i F_{kl} + \partial_k F_{li} + \partial_i F_{kl} = 0 \ . \tag{962}$$

In addition to the 2-form $\boldsymbol{F}$, we define the 2-form $\boldsymbol{H}$

$$\boldsymbol{H} := \sum_{i<k} H_{ik}\, \boldsymbol{dx}^i \wedge \boldsymbol{dx}^k \ , \quad i,k = 0,1,2,3, \quad x^0 = ct,\ x^1 = x,\ x^2 = y,\ x^3 = z \ . \tag{963}$$

In detail, this 2-Form $\boldsymbol{H}$ is given with the tensor H_{ik} in Eq. (571),

$$\begin{aligned} \boldsymbol{H} := \ & H_{01}\, \boldsymbol{dx}^0 \wedge \boldsymbol{dx}^1 + H_{02}\, \boldsymbol{dx}^0 \wedge \boldsymbol{dx}^2 + H_{03}\, \boldsymbol{dx}^0 \wedge \boldsymbol{dx}^3 \\ & + H_{12}\, \boldsymbol{dx}^1 \wedge \boldsymbol{dx}^2 + H_{13}\, \boldsymbol{dx}^1 \wedge \boldsymbol{dx}^3 + H_{23}\, \boldsymbol{dx}^2 \wedge \boldsymbol{dx}^3 \ , \end{aligned}$$

hence

$$\begin{aligned} \boldsymbol{H} = \ & D_x\, c^2\, \boldsymbol{dt} \wedge \boldsymbol{dx} + D_y\, c^2\, \boldsymbol{dt} \wedge \boldsymbol{dy} + D_z\, c^2\, \boldsymbol{dt} \wedge \boldsymbol{dz} \\ & - H_z\, \boldsymbol{dx} \wedge \boldsymbol{dy} + H_y\, \boldsymbol{dx} \wedge \boldsymbol{dz} - H_x\, \boldsymbol{dy} \wedge \boldsymbol{dz} \ . \end{aligned} \tag{964}$$

We define the 2-form $*\boldsymbol{H}$ after prescription (940),

$$*\boldsymbol{H} := \sum_{i<k} *F_{ik}\, \boldsymbol{dx}^i \wedge \boldsymbol{dx}^k \ , \quad i,k = 0,1,2,3, \quad x^0 = ct,\ x^1 = x,\ x^2 = y,\ x^3 = z \ , \tag{965}$$

where $*H_{ik}$ contains the terms $b_{...}$ in Eq. (940), hence

$$* \ H_{ik} := \sum_{rs} H^{rs}\, \epsilon_{ikrs} \tag{966}$$

with the totally antisymmetric tensor ϵ_{ikrs}, s. Appendix B.1, hence

$$* \ H_{01} = H^{23},\ *H_{02} = -H^{13},\ *H_{03} = H^{12},\ *H_{12} = H^{03},\ *H_{13} = -H^{02},\ *H_{23} = H^{01}. \tag{967}$$

With the contravariant components H^{ik} given in Eq. (571),

$$\left.\begin{aligned} *H_{01} &= -H_x \ , & *H_{02} &= -H_y \ , & *H_{03} &= -H_z \ , \\ *H_{12} &= -D_z\, c^2 \ , & *H_{13} &= D_y\, c^2 \ , & *H_{23} &= -D_x\, c^2 \ , \end{aligned}\right\} \tag{968}$$

we get for the 2-form $*\boldsymbol{H}$

$$\left.\begin{aligned} *\boldsymbol{H} = & -H_x\, c\, \boldsymbol{dt} \wedge \boldsymbol{dx} - H_y\, c\, \boldsymbol{dt} \wedge \boldsymbol{dy} - H_z\, c\, \boldsymbol{dt} \wedge \boldsymbol{dz} \\ & - D_z\, c^2\, \boldsymbol{dx} \wedge \boldsymbol{dy} + D_y\, c^2\, \boldsymbol{dx} \wedge \boldsymbol{dz} - D_x\, c^2\, \boldsymbol{dy} \wedge \boldsymbol{dz} \ . \end{aligned}\right\} \tag{969}$$

The current density now includes in addition to the convective current density $\mathbf{j}$ with

$$\mathbf{j} = \rho\, \mathbf{u} \ , \tag{970}$$

representing the complete current in vacuum considered before, the additional conduction current density $\boldsymbol{\iota}$ because of the conductivity σ of the medium. The electric field strength $\mathbf{E}$ for an observer at rest with respect to the medium fulfils the relation due to experiments

$$\imath = \sigma \mathbf{E} \ . \tag{971}$$

We add here an important property: Only in the local rest system, with the current density there is not connected a charge density, i.e. its fourth component vanishes. This statement is, however, not invariant. In an arbitrary reference system, a current density is also connected with a conduction current, contributing to the current density ϱ of the total current, given in Eq. (983). We write for the total current

$$J^i = (\varrho c, J_x, J_y, J_z) \tag{972}$$

with a charge density ϱ, being in general different from the prescribed charge density ρ in Eq. (970) of the convective current $\mathbf{j}$.

The four-current density is then, instead of Eqs. (953) and (954),

$$\boldsymbol{J} := J_i \, \boldsymbol{d}x^i = J_0 \, \boldsymbol{d}x^0 + J_1 \, \boldsymbol{d}x^1 + J_2 \, \boldsymbol{d}x^2 + J_3 \, \boldsymbol{d}x^3 \ . \tag{973}$$

From the 1-form $\boldsymbol{J}$ defined in Eq. (973), now we form $*\boldsymbol{J}$ in standard way,

$$\begin{aligned} *\boldsymbol{J} = \ & \varrho c \ \boldsymbol{d}x \wedge \boldsymbol{d}y \wedge \boldsymbol{d}z \\ & - J_x \, \boldsymbol{d}y \wedge \boldsymbol{d}z \wedge c\,\boldsymbol{d}t \\ & + J_y \, \boldsymbol{d}x \wedge \boldsymbol{d}z \wedge c\,\boldsymbol{d}t \\ & - J_z \, \boldsymbol{d}x \wedge \boldsymbol{d}y \wedge c\,\boldsymbol{d}t \ . \end{aligned} \tag{974}$$

With expression (969), we form now $\boldsymbol{d} * \boldsymbol{F}$. Again we take care of the rules of outer differentiation, and we get after some rearrangements

$$\left. \begin{aligned} \boldsymbol{d} * \boldsymbol{H} = & -c \, (\partial_x D_x + \partial_y D_y + \partial_z D_z) \, \boldsymbol{d}x \wedge \boldsymbol{d}y \wedge \boldsymbol{d}z \\ & + (\partial_y B_z - \partial_z B_y - \partial_t D_x) \, \boldsymbol{d}y \wedge \boldsymbol{d}z \wedge c\boldsymbol{d}t \\ & - (\partial_z B_x - \partial_x B_z - \partial_t D_y) \, \boldsymbol{d}x \wedge \boldsymbol{d}z \wedge c\boldsymbol{d}t \\ & + (\partial_x B_y - \partial_y B_x - \partial_t D_z) \, \boldsymbol{d}x \wedge \boldsymbol{d}y \wedge c\boldsymbol{d}t \ . \end{aligned} \right\} \tag{975}$$

The second part of Eq. (580b) can now be written as

$$\boldsymbol{d} * \boldsymbol{H} = - * \boldsymbol{J} \quad \longleftrightarrow \quad \mathrm{rot}\mathbf{H} - \frac{\partial \mathbf{D}}{\partial t} = \mathbf{J} \ \text{and} \ \mathrm{div}\,\mathbf{D} = \varrho \ . \tag{976}$$

For Eq. (503c) now it holds, instead of the above in Eq. (961) given 1-form

$$\boldsymbol{k} = -\varrho \Big(\mathbf{E} \cdot \mathbf{u} \Big) \, \boldsymbol{d}t - \Big(\varrho \, E_a + (\mathbf{J} \times \mathbf{B})_a \Big) \, \boldsymbol{d}x^a \ , \quad a = 1, 2, 3 \ . \tag{977}$$

Still we have to transform Eqs. (580d), (580e) and (580f), describing the properties of the medium in the language of forms.

With the 1-form $\boldsymbol{w}$ of the velocity of the medium, belonging to the four-vector w,

$$\boldsymbol{w} = - w_i * \boldsymbol{\alpha}^i \ , \quad \mathsf{w} = w_i \, \mathbf{k}^i \ , \tag{978}$$

there follows with the rules of contraction

$$(\boldsymbol{H}, \mathsf{w}) = H_{ik} \, w^k \, \boldsymbol{\alpha}^i \ . \tag{979}$$

For Eq. (580d), we then write

$$(\boldsymbol{H}, \mathsf{w}) = \varepsilon \, (\boldsymbol{F}, \mathsf{w}) \quad \longleftrightarrow \quad H_{ik} \, w^k = \varepsilon \, F_{ik} \, w^k \ . \tag{980}$$

The cyclic sum in Eq. (580e) leads to the total antisymmetry, easily described in the language of forms,

$$\boldsymbol{F} \wedge \boldsymbol{w} = \mu \boldsymbol{H} \wedge \boldsymbol{w} \longleftrightarrow F_{ij}\, w_k + F_{jk}\, w_i + F_{ki}\, w_j = \mu \left(H_{ij}\, w_k + H_{jk}\, w_i + H_{ki}\, w_j \right). \tag{981}$$

Finally, the connection of current density and field strength tensor, Eq. (580f), is given by the 1-form $\boldsymbol{\iota}$ for the current density

$$\boldsymbol{\iota} = -\sigma\,(\boldsymbol{F}, \mathsf{w}) \quad \longleftrightarrow \quad \iota_i = -\sigma\, F_{ik}\, w^k \,. \tag{982}$$

Here $\iota^i = J^i - j^i$ is the conduction current density $j^i = \rho_e\, u^i$ a convective in general prescribed current density, and J^i denotes the density, of the total current.

According to Eq. (982), also the conduction current has in general a non-vanishing fourth component, and it contributes therefore with $\iota_0 := c\,\rho_\iota$ to the total charge density,

$$\iota_0 := c\,\rho_\iota = -\,\sigma\, F_{0\alpha}\, w^\alpha = \frac{\sigma}{c\,\gamma_w}\, \mathbf{E} \cdot \mathbf{w} \,. \tag{983}$$

We collect the previous equations, leading to the MAXWELL equations of media in motion (580), now in the outer calculus

$$\left.\begin{aligned}
&(a)\; d\boldsymbol{F} = 0\,,\\
&(b)\; d * \boldsymbol{H} = -\, * \boldsymbol{J}\,,\\
&(c)\; \boldsymbol{k} = -\,\varrho \left(\mathbf{E} \cdot \mathbf{u}\right) dt - \left(\varrho\, E_\nu + (\mathbf{J} \times \mathbf{B})_\nu\right) dx^\nu\\
&(d)\; (\boldsymbol{H}, \mathsf{w}) = \varepsilon\,(\boldsymbol{F}, \mathsf{w})\,,\\
&(e)\; \boldsymbol{F} \wedge \boldsymbol{w} = \mu\, \boldsymbol{H} \wedge \boldsymbol{w}\,,\\
&(f)\; \boldsymbol{\iota} = -\sigma\,(\boldsymbol{F}, \mathsf{w})\,,\\
&(g)\; \varepsilon = \varepsilon_o\, \varepsilon_r\,,\; (h)\; \mu = \mu_o\, \mu_r\,,\; (i)\; J^i = j^i + \iota^i
\end{aligned}\right\} \text{MAXWELL equations of media in motion in outer calculus} \tag{984}$$

Chapter 12
A Lattice Model of Relativistic Space-Time

Why does a clock in motion be delayed? Why does a bar in motion become contracted?

All matter known to us is subject to the laws of Special Relativity Theory, based on its axiomatics either on the universal constancy of the speed of light, or, in our equivalent procedure, on the length contraction and the time dilatation, Eqs. (99) and (100). Axioms are simplest rules or laws that cannot be further reduced on more basic concepts. The question for a reason of the LORENTZ contraction or a justification of the time dilatation in our procedure seems therefore without chance. What should be the basis for explaining the origin of the length contraction, if in principle all length measurements are subject to this contraction? Therefore, we shall not be able to answer the question for the reason of the changing measuring rods and clocks in the relativistic space-time.

To escape this conflict, we shall now start from the idea that the reason for the length contraction and the time dilatation is based already on the structure of our physical vacuum. We shall show which kind of structure *could be* characteristic for our world, so that with necessity occur relativistic phenomena. Based on such a model, we will discuss the possible concepts of a physical vacuum, without to claim that it looks so in reality.

1 The Lattice Model

The lattice structures, as they appear in nature as crystalline solids, represent a model for reducing the effects of length contraction and time dilatation on properties of the basic atomic lattice. The subject of this section was intensively discussed in

Electronic supplementary material The online version of this chapter (https://doi.org/10.1007/978-981-13-7783-9_12) contains supplementary material, which is available to authorized users.

H. Günther and V. Müller, *The Special Theory of Relativity*,
https://doi.org/10.1007/978-981-13-7783-9_12

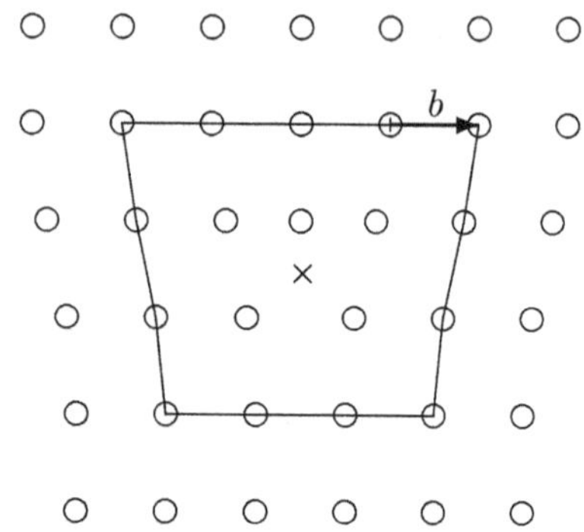

Fig. 1 Schematic representation of an edge dislocation seen from above with the BURGERS vector **b**. The direction vector **t** of the dislocation is directed vertically out of the picture plane. Only the position of the dislocation line marked by a cross is measurable.

GÜNTHER (2000). There is also a detailed list of references to the questions touched here, s. GÜNTHER (2000).

The significant point is the following: The ideal lattice with a totally regular symmetric ordering of its basic building blocks is assumed to model our physical vacuum. The ideal lattice with its unrestricted symmetric structure of its building blocks can model our physical vacuum. Only the perturbations in this lattice, the deviations from the ideal structure, represent the 'particles' or 'masses' that we shall investigate with respect to its relativistic properties. *'Mass is form'* tells C. F. v. WEIZSÄCKER. The atoms or molecules of the basic lattice principally cannot be observed by such 'particles'.

Among the different structure disturbances, we shall be concerned with so-called dislocations. They represent boundaries of lattice planes ending in the crystal, but not the incomplete lattice plane. Only the boundary line with direction vector **t** is an experimentally measurable quantity. We always assume crystals of infinite extensions. An idealised infinitely long dislocation line may be represented with the x-axis. As seen from Fig. 1, the dislocation is a purely geometrical place that cannot be identified with a position of the lattice atoms. A dislocation line in the crystal lattice can move within the lattice. The atoms of the lattice remain mainly fixed at its places. Only the disturbance line or the bending at the dislocation line can move over large distances in the crystal, the lattice atoms not.

Here, we do not consider the motion of a dislocation line as a whole. Subject of our considerations will be small local deviations from the straight dislocation line allowed by the equilibrium of the crystal structure and described by the function $q = q(x, t)$ of the vertical deviation of the dislocation on the x-axis from the ground state.

We shall derive the equation describing these deviations $q = q(x, t)$ of the dislocation line.

Only small parts of the dislocation line are able to move under the action of local forces. We divide the dislocation line into small pieces of length Δx_α. Each of them has an inertia Δm_α with respect to the lattice in the same way as lattice atoms have an inertia with respect to the physical space. Besides that, the line elements Δx_α are coupled pairwise elastically in the same way as parts of a strained string. The motion of the dislocation line is then given by the following model.

The dislocations form a linear chain of elastically coupled masses Δm_α at positions q_α. The motion of these relative to the regular lattice defined inertial masses[1] is determined by NEWTONian equations of motion, where the inertial system of mechanics is replaced by the rest system of the crystal lattice:

$$\frac{d}{dt}\left(\Delta m_\alpha \frac{d}{dt} q_\alpha(t)\right) = \sum_{b \neq \alpha} F_{b\alpha} \ , \tag{985}$$

$$F_{b\alpha} = -F_{\alpha b} \ . \tag{986}$$

Here, it is assumed that the variable q describes an elongation vertical to the dislocation line. The forces $F_{b\alpha}$ on the element Δx_α are decomposed as

$$\sum_{b \neq \alpha} F_{b\alpha} = F_{\alpha-1\,\alpha} + F_{\alpha+1\,\alpha} + \sum_g F_{g\alpha} \tag{987}$$

in the forces from the left and from the right side $F_{\alpha-1\,\alpha}$ and $F_{\alpha+1\,\alpha}$ of neighbouring elements, and the forces $F_{g\alpha}$, imposed by the other lattice atoms on the dislocation masses Δm_α.

For small elongations q_α, the lattice acts on the dislocation as an elastic spring, repelling the dislocation mass Δm_α with a directional constant D_m back to the equilibrium position $q_\alpha = 0$. For larger q_α, this force decreases, and it vanishes at the middle position of the dislocation at half of the lattice separation at $q_\alpha = a/2$ between two equilibrium positions $q_\alpha = 0$ and $q_\alpha = a$. Here a is the lattice constant.

We approximate the form of this force of the lattice atoms by a sine function and write

$$\sum_g F_{g\alpha} = -D_m \sin\left(\frac{2\pi}{a} q_\alpha(t)\right) \ . \tag{988}$$

The force $F_{\alpha+1\,\alpha}$ is equal to the *stress* acting from right on the element at q_α, called $\tau(q_\alpha, t)$. Also, $F_{\alpha-1\,\alpha} = -F_{\alpha\,\alpha-1} = -\tau(q_{\alpha-1}, t)$ is the stress acting from left, so that

$$F_{\alpha-1\,\alpha} + F_{\alpha+1\,\alpha} = -\tau(q_{\alpha-1}, t) + \tau(q_\alpha, t) \ . \tag{989}$$

Substituting the discrete quantities by a continuum

$$\left.\begin{aligned} \Delta m_\alpha &\longrightarrow \rho_o \Delta x \ , \\ q_\alpha(t) &\longrightarrow q(x, t) \ , \\ D_m &\longrightarrow D\Delta x \end{aligned}\right\} \quad \text{Dislocation line as one-dimensional continuum} \tag{990}$$

with the effective quasi-mass density ρ_o of the dislocation line and the directional constant per length unit D of the lattice.

With $q_\alpha \longrightarrow x$ and $q_{\alpha-1} \longrightarrow x - \Delta x$, it is

$$-\tau(q_{\alpha-1}, t) + \tau(q_\alpha, t) \longrightarrow -\tau(x - \Delta x, t) + \tau(x, t) \ ,$$

[1] Such inertial masses and the related particles only defined *with respect to the lattice* are called quasi-masses and quasi-particles. These names distinguish them from normal particles and masses in the physical space-time.

hence from Eq. (989) we get

$$F_{\alpha-1\,\alpha} + F_{\alpha+1\,\alpha} \longrightarrow \tau(x-\Delta x, t) + \tau(x,t)$$

and, restricting to the first term of the TAYLOR expansion

$$F_{\alpha-1\,\alpha} + F_{\alpha+1\,\alpha} \longrightarrow \Big(\tau(x,t) - \frac{\partial\tau(x,t)}{\partial x}\Delta x\Big) + \tau(x,t) = \frac{\partial\tau(x,t)}{\partial x}\Delta x\ . \quad (991)$$

For the force acting from the lattice (988), we get with Eq. (990)

$$\sum_g F_{g\,\alpha} = -D_m \sin\left(\frac{2\pi}{a}q_\alpha(t)\right) \longrightarrow -D\sin\left(\frac{2\pi}{a}q(x,t)\right)\Delta x\ . \quad (992)$$

We suppose *three basic assumptions of the linearised elasticity theory*:

1. In the NEWTONian inertia term the total time derivatives are replaced simply by the partial time derivatives.
2. The mass density is assumed to be constant.
3. There holds true HOOKE′ s law, i.e., the stress is directly proportional to the relative strain.

Applied on the quasi-particles of our linear chain of dislocation masses this reads as

$$\left.\begin{aligned} &1.\ \frac{d}{dt}\big(\Delta m_\alpha \frac{d}{dt}q_\alpha\big) \longrightarrow \rho_o \Delta x\,\frac{\partial}{\partial t}\frac{\partial q(x,t)}{\partial t}\ ,\\ &2.\ \rho_o = \text{const}\ ,\\ &3.\ \tau(x,t) = \sigma\,\frac{\partial q(x,t)}{\partial x}\ . \end{aligned}\right\}\ \text{Transition to the linear elasticity theory} \quad (993)$$

Here, the derivative $\partial q/\partial x$ describes the relative elongation vertical to the unperturbed dislocation line, the x-axis. The modulus of elasticity σ is called *line tension* of dislocations, a quantity being experimentally accessible by energy measurements.

Taking the relations (991), (992) and (993) into account in Eq. (985), we get after short calculation that the elongation q of the dislocation line fulfils the so-called *sine-*GORDON*equation*,

$$\left.\begin{aligned} &\frac{\partial^2}{\partial x^2}q(x,t) - \frac{1}{c_o^2}\frac{\partial^2}{\partial t^2}q(x,t) = \frac{D}{\sigma}\sin\big(\frac{2\pi}{a}q(x,t)\big)\ ,\\ &c_o = \sqrt{\frac{\sigma}{\rho_o}}\ . \end{aligned}\right\}\ \text{sine-GORDON equation of a dislocation} \quad (994)$$

The quantity $c_o = \sqrt{\sigma/\varrho}$ with an effective mass density ϱ of the dislocation line represents a limiting velocity for the propagation of disturbances along the dislocation line. Its value lies in the order of magnitude of the sound velocity.

Equation (994) was first established by Seeger (1949). We have given a simplified derivation of Eq. (994), s. also GÜNTHER (2000).

In the following, we use the dimensionless elongation $q(x,t)$, connected with the physical elongation on the lattice $q = q(x,t)$ by

$$q := \frac{2\pi}{a} q \,. \tag{995}$$

From Eq. (994), then we get the *sine*-GORDON *equation* in the form

$$\frac{\partial^2}{\partial x^2} q(x,t) - \frac{1}{c_o^2}\frac{\partial^2}{\partial t^2} q(x,t) = \frac{1}{\lambda_o^2} \sin\left(q(x,t)\right) . \quad \text{sine-GORDON equation} \tag{996}$$

Here

$$\lambda_o = \sqrt{\frac{a}{2\pi}\frac{\sigma}{D}} \tag{997}$$

is a characteristic length in the lattice, defined by the lattice constant a, the line tension σ of dislocation and the directional constant D belonging to collective atomic forces in the lattice.

It is essential that the lattice structure produces in the field theoretical description a nonlinearity, the term $\sin q$. Without the forces (988) from the lattice atoms in the surrounding of the dislocation line, the transverse elongation $q = q(x,t)$ would be described simply by a wave equation that results from the laws of NEWTONian mechanics

$$\frac{\partial^2}{\partial x^2} q(x,t) - \frac{1}{c_o^2}\frac{\partial^2}{\partial t^2} q(x,t) = 0 \,. \tag{998}$$

Solutions of Eq. (998) with two fixed points describe vibrations of a strained string. These so-called *strings* play nowadays an important role in the physics of elementary particles.

The solutions $q = q(x,t)$ of the sine-GORDON equation describe the energetically possible physically stable forms of dislocation lines near to the straight dislocation line $q = 0$, here coinciding with the x-axis. Now there are known many of such solutions. The corresponding line forms are typical for many real lattices. The motions in these dislocation lines describe the so-called microplastic deformations of the lattice.

Now we are looking for such physical stable line forms that may be suitable for providing a normal measure of a standard length, and for physically stable oscillations whose period could represent a normal clock.

The so-called kink solution of the sine-GORDON equation (996) reads, s. Fig. 2,

$$q_o^I(x) = 4 \arctan \exp\left[\frac{x}{\lambda_o}\right] . \quad \text{Localised kink solution} \tag{999}$$

The geometrical extension of the kink defines a length normal L_o in the lattice with middle point $x = 0$. It defines the distance over which the dislocation line changes from one equilibrium line $q = 0$ to the next one at $q = \pi$, see Fig. 2,

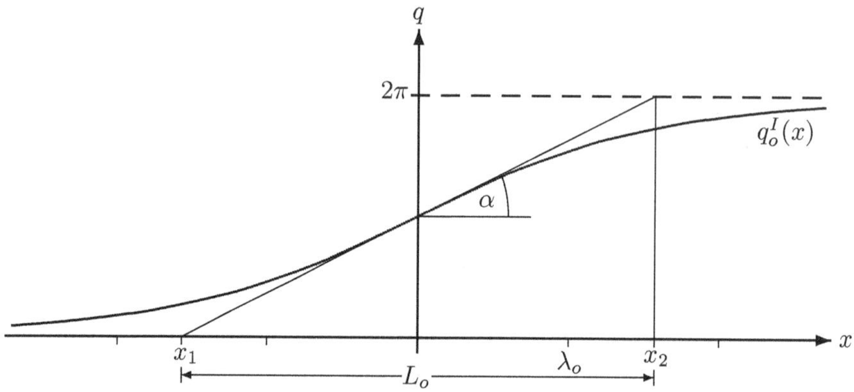

Fig. 2 Definition of a natural length standard L_o from the kink solution (999). The function $q_o^I(x)$ has a turning point at $x = 0$, $q_o^I(0) = \pi$. The tangent there has a slope $\tan\alpha = 2\pi/L_o = 2/\lambda_o$ that cuts the x-axis at $x_1 = -\pi\lambda_o/2$ and the asymptote $q = 2\pi$ (dashed line) at $x_2 = +\pi\lambda_o/2$, $y = 2\pi$. On the x-axis, we can define a length standard $L_o = x_2 - x_1 = \pi\lambda_o$.

$$L_o = \pi\,\lambda_o\ . \qquad \text{Natural length normal in a lattice} \qquad (1000)$$

The length normal of the kink can be fixed at each place $x = b$ in the lattice. The kink solution at place $x = b$ follows from Eq. (999) in replacing x by $x - b$, s. also Fig. 4,

$$q_o^I(x-b) = 4\arctan\exp\Big[\frac{x-b}{\lambda_o}\Big]\ . \qquad \text{Kink solution localised at } x = b \qquad (1001)$$

Obviously, this is also a solution of the sine-Gordon equation as Eq. (999), so that we define the same length normal at $x = b$ as above at $x = 0$. The kink exists as line form in the lattice, and it defines a general natural length unit L_o, which can be used as reference for all separations of structures in the lattice—in the same way as the separations of the orange-red spectral line in the spectrum of the Krypton isotope ^{86}Kr used in Chap. 1, Sect. 1 in our physical space.

The so-called breather solution of the sine-Gordon equation (996) is given by, s. Fig. 3,

$$q_o^{III}(x,t) = 4\arctan\frac{\sin(2\pi\,t/T_o)}{\cosh\big(x/(\sqrt{2}\,\lambda_o)\big)}\ . \qquad \text{Localised breather solution} \qquad (1002)$$

Here T_o is a period defined by properties of the crystal lattice,

$$T_o = \sqrt{2}\,2\pi\,\frac{\lambda_o}{c_o}\ . \qquad \text{Natural period in a lattice} \qquad (1003)$$

This breather solution is fixed with its centre in the lattice, here at $x = 0$, and it exists there as stable oscillating structure with fixed line form, s. Fig. 3.

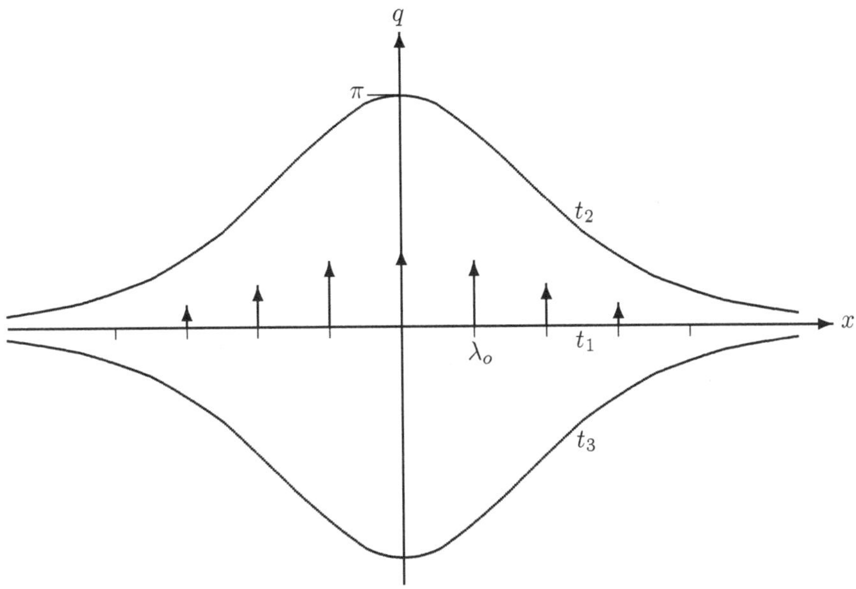

Fig. 3 Definition of a natural period T_o from the breather solution (1002). Shown are the oscillation states at $t_1 = 0$, $t_2 = T_o/4$ and $t_3 = 3T_o/4$. The arrows indicate the velocities of the dislocation elements at transition through the zero line.

The solution (1002) defines a normal clock U_o at place $x = 0$ in the lattice, which counts the number of periods T_o of the oscillating line $q_o^{III}(0, t)$, hence with Eq. (1002) for $x = 0$, $\cosh 0 = 1$,

$$q_o^{III}(0, t) = 4 \arctan \left[\sin(2\pi\, t / T_o) \right] . \tag{1004}$$

Such a clock, defined by an oscillating line, can exist at each place $x = b$ of the lattice.

The localised breather solution at place $x = b$ can be obtained from Eq. (1002), in replacing x by $x - b$,

$$q_o^{III}(x - b, t) = 4 \arctan \frac{\sin(2\pi\, t / T_o)}{\cosh\left((x - b)/(\sqrt{2}\,\lambda_o)\right)} . \qquad \text{Breather solution localised at } x = b \tag{1005}$$

Obviously, this is also a solution of the sine-GORDON equation as Eq. (1002). When we set $x = b$ in Eq. (1005), we get also an oscillating line (1004), defining at each place a normal clock. With this clock U_o, we can measure all time intervals in the lattice, as in our physical space all times reduces on the oscillations of a transition in the spectrum of the Caesium isotope ^{133}Cs, Chap. 1, Sect. 1.

If we now do not fix the kink and breather solutions at one and the same place, but we let them run with their line form along the x-axis with a constant velocity v, we notice a surprise.

We replace in the kink solution (999) at first x by $x - vt$ with a constant v and get a function propagating with velocity v to the right,

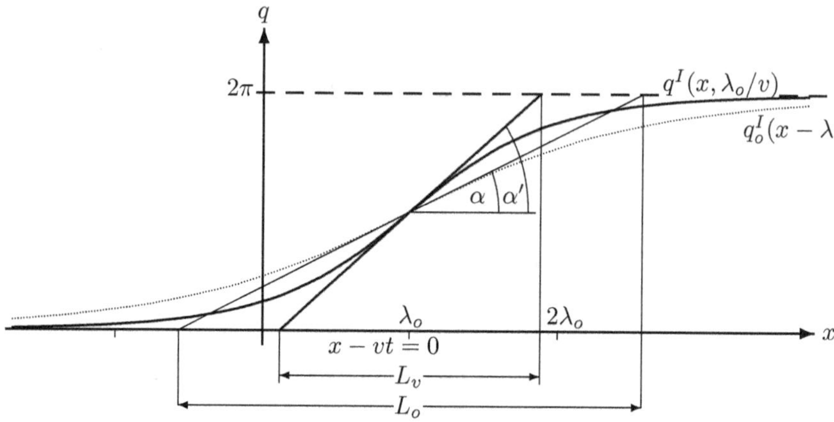

Fig. 4 Contraction of the moving kink. The picture shows the kink $q^I(x,t)$, Eq. (1008), at time $t = \lambda_o/v$ moving with a velocity $v = 0.8c_o$, hence $\gamma_o = 1.67$, so that $L_v = 0.6L_o$. For comparison, we also show a resting kink at $x = \lambda_o$, the solution $q_o^I(x - \lambda_o)$, Eq. (1001), with $b = \lambda_o$.

$$\tilde{q}_o^I(x - vt) = 4 \arctan \exp\left[\frac{x - vt}{\lambda_o}\right] . \tag{1006}$$

The important point is now that Eq. (1006) does *not* describe a possible line form in our lattice. The function (1006) is not a solution of the sine-GORDON equation, as can be simply verified by differentiation. We conclude

It is impossible to move our length normal without change of its length.

We get the same result with our normal clocks in the lattice. If we replace in the breather solution (1002) x by $x - vt$ with a constant v, we find a structure propagating to the right with velocity v

$$\tilde{q}_o^{III}(x,t) = 4 \ \arctan \frac{\sin\left[2\pi t/T_o\right]}{\cosh\left[(x - vt)/\sqrt{2}\lambda_o\right]} . \tag{1007}$$

Again this function is not a solution of the sine-GORDON equation, and we conclude

It is impossible to move the normal clock without change of its period.

Now we provide the solution of the sine-GORDON equation that describes the situation with a kink (999) moving with constant velocity v in x-direction. This solution $q^I(x,t)$, representing a moving measuring rod, reads

$$q^I(x,t) = 4 \arctan \exp\left[\frac{x - vt}{\lambda_v}\right], \quad \lambda_v = \frac{\lambda_o}{\gamma_o} . \qquad \text{Kink solution in motion} \tag{1008}$$

Here $\gamma_o := 1/\sqrt{1 - v^2/c_o^2}$.

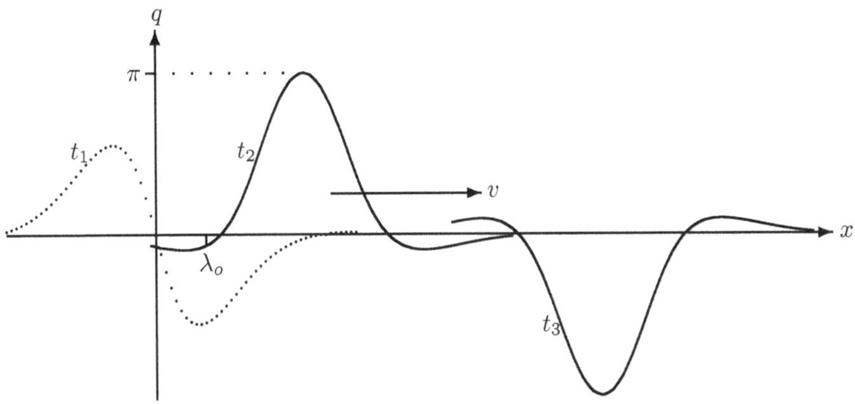

Fig. 5 Breather solution $q^{III}(x,t)$, Eq. (1010), moving with velocity $v = 0.8c_o$, hence $\gamma_o = 1.67$, at time $t_1 = 0$, $t_2 = T_v/4$ and $t_3 = 3T_v/4$. Here $T_v = \gamma_o T_o$ is the period of the moving breather. Please observe the wave-like changes of the lines compared with the localised oscillation in Fig. 3, originating from the definition of simultaneity in the crystal, cp. GÜNTHER (2000).

With $L_v = \pi\lambda_v$ we get from Eqs. (999), (1000) and (1008) immediately:

The length L_v of the measuring rod in motion is contracted with respect to the length L_o of the rod at rest in the lattice:

$$L_v = L_o\sqrt{1 - \frac{v^2}{c_o^2}}\,. \qquad \text{‘Lorentz contraction’} \qquad (1009)$$

The notation in quotation marked ‘Lorentz contraction’ indicates that here we are *not* concerned with the speed of light.

We add that this length contraction—in distinction to the situation in our physical space-time—can be directly measured.

Furthermore, we find the function $q^{III}(x,t)$ that describes the situation of an oscillating breather solution (1002) propagating with constant v along the x-axis, hence that our normal clock is in motion in the lattice with this velocity, s. Fig. 5,

$$q^{III}(x,t) = 4\arctan\frac{\sin[2\pi\gamma_o^2\,(t - xv/c_o^2)/\,T_v]}{\cosh[(x - v\,t)/\sqrt{2}\,\lambda_v]}\,, \quad T_v = \gamma_o\,T_o\,. \qquad \text{Breather solution in motion} \qquad (1010)$$

A moving breather represents a clock U_v, propagation with respect to the lattice. We can count the oscillations of this line, e.g. at $x = vt$. Which period shows this clock? To this aim, we insert into Eq. (1010) $x = v\,t$ and find the function shown in Fig. 6,

$$q^{III}(vt,t) = 4\arctan\left[\sin(2\pi\,t/\,T_v)\right]\,. \qquad (1011)$$

The period T_v of the breather clock U_v is then due to Eqs. (1004), (1010) and (1011) $T_v = \gamma_o\,T_o$.

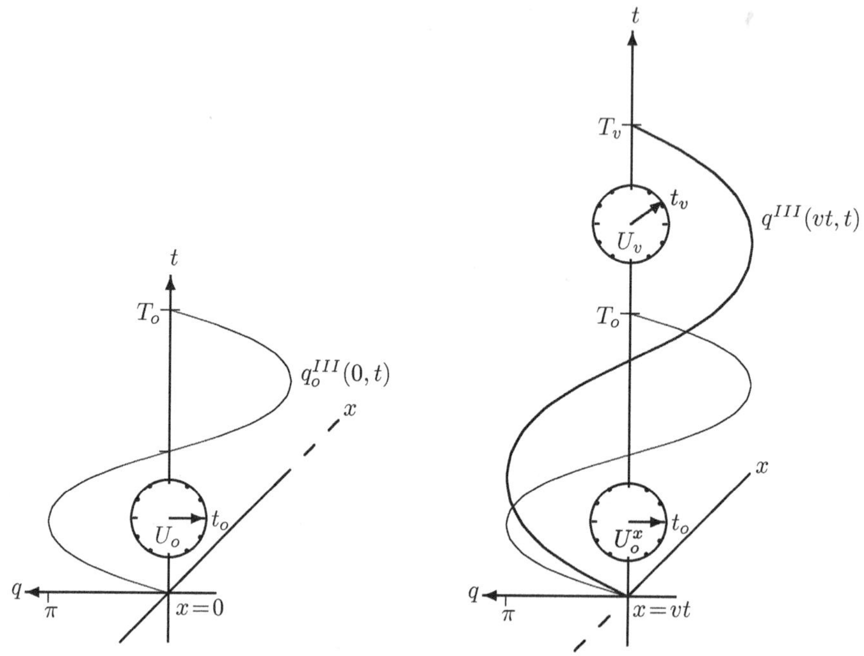

Fig. 6 Measurement of the 'time dilatation'. The pointer of the breather clock U_v in motion with respect to the lattice counts the oscillations shown by the thick line, the function $q^{III}(vt, t) = 4 \arctan \sin(2\pi\, t/T_v)$, Eq. (1011), for $x = v\,t$. It stays behind the pointer of the breather clock U_o^x, at rest in the lattice, counting the oscillations of the thin line of function $q_o^{III}(0, t) = 4 \arctan \sin(2\pi\, t/T_o)$, Eqs. (1004) or (1005) for $x = 0$ and $x = v\,t$. We have chosen the case that the clock U_v moves along the x-axis with constant velocity $v = 0.8\, c_o$. Then it is $\gamma_o = 1.67$ and $T_v = 1.67\, T_o$, and for $t_o = 15$ we get $t_v = 15 \cdot 0.6 = 9$, as shown.

The period T_v of a clock U_v in motion is retarded with respect to the clock U_o^x resting in the lattice:

$$T_v = \frac{T_o}{\sqrt{1 - v^2/c_o^2}}\,. \qquad \text{'Time dilatation'} \qquad (1012)$$

The pointer of a clock counts the oscillations; hence, the pointer of the moving clock stays behind the resting one, s. Fig. 6:

The pointer position t_v of a clock U_v in motion with respect to the lattice stays behind with respect to the resting clock U_o^x:

$$t_v = t_o \sqrt{1 - \frac{v^2}{c_o^2}}\,. \qquad \text{'Time dilatation'} \qquad (1013)$$

We remark that the reciprocity theorem (61), is fulfiled with Eqs. (1009) and (1012).

Here, the reference system Σ_o is defined by the crystal lattice. Not the constituents of the lattice are the objects of our consideration, but the internal structures of the crystal, i.e. structures where the crystal deviates from an ideal lattice. They have a typical geometrical extension with respect to the lattice. The atoms or molecules of the ideal lattice form the 'space'. The imperfections are the 'particles' in the space, cp. the footnote in Sect. 1. With the kink and the breather, we have selected certain measures for length and time. This is reasonable since in a real crystal, imperfections exist in a large number.

The space-time model of our considerations concerns a new 'internal world' of these imperfections. We take up here a theoretical conception developed by H. v. HELMHOLTZ (reprint 1971) in the year around 1870, cf. also Biagioli (2016), and also used by A. EINSTEIN (1955) in connection with the 'internal observer'[2] of our world. The single atoms of the lattice cannot be observed with these structures. They cannot be seen by the internal observer. The lattice has only the function of a vacuum, of the space where the structures are in motion, of a space with physical properties becoming manifests by the lattice structure.

With Eqs. (1009) and (1012) or (1013), we have found the physical postulates of a relativistic space-time on the background of the crystal lattice of atoms or molecules, the content of Eqs. (99) and (100) or (101). The only difference consists in the limiting velocity: The speed of light c was replaced by the limiting velocity c_o of microplastic deformations. Here, we are even better than in normal space-time: While Eqs. (99)–(101) are based only on precision measurements, either direct or over the measurements of the speed of light, the solutions Eqs. (1009), (1012) and (1013) are justified physically via the sine-GORDON equation (996) by the *basic* elastic interactions of the atoms of the underlying lattice. With the sine-GORDON equation, we have at hand a kind of *'elementary particle theory'* for the imperfections of the lattice.

Now we can go one step further. Until now we have always based our considerations on a certain reference system, the rest system of the crystal lattice, where we derived the sine-GORDON equation. This system was denoted by Σ_o. We restrict us on one spatial dimension, the x-axis distinguished by the dislocation line. The limiting velocity has in both directions of the x-axis the same value c_o, there it holds isotropy.

Now we ask whether the in Σ_o registered events $E(x, t)$ can be described, when we are in motion with respect to the lattice, hence if we are moving as internal observers with the moving kink or the moving breather. This reference system is called Σ'.

---

[2]We mean EINSTEIN's 'flat beings', living on 'spherical surfaces', on a two-dimensional surface with a non-Euclidean geometry imbedded in a normal three-dimensional space, the beings do nothing grasp from the third dimension. In connection with the General Relativity Theory, cp. Chap. 13, EINSTEIN demonstrated the restriction of the *three-dimensional* world of our experience, where we are also internal observers.

At first, we have to fix that we take in Σ' the same normal rods and clocks as the inner observer in Σ_o, namely, the kink and breather solutions of the sine-GORDON equation delivered by the crystal. The means for measurements are also of the same kind as the objects to be measured. In this way, we have a situation formed analogously to the physical space-time. Only by the repetition of many such normal rods, we can also assign in Σ' to each event its spatial coordinate x' as measured distance.

Now it seems quite obvious that the kink solution, resting in the crystal, seems longer than the moving kink in Σ' as shown in Fig. 4. For the observer in Σ_o, this is true without doubt, but not for the inner observer in Σ'. It cannot make a well-defined statement on the length of an in Σ_o resting, but with respect to Σ' moving kink, until he has *defined* the simultaneity in Σ'. Here, there appears the same situation as during the discussion of events from a moving reference system Σ' of our physical space-time in Chap. 1, Sect. 4.1.

Again we have the freedom of *definition* of simultaneity. Due to our restriction on lattice-based measuring rods and clocks with its well-defined physical properties by the lattice and Eqs. (1009), (1012), (1013), we can take over all conclusions there verbatim, only replacing the speed of light c by the limiting velocity c_o of microplastic deformations. This is the whole difference. Due to the advantage resulting from a symmetry, we here decide for a definition of simultaneity in Σ' on the basis of the elementary relativity principle as formulated in Chap. 2, Sect. 2. To each event, there are assigned space-time coordinates both in $\Sigma_o(x,t)$ and in $\Sigma'(x',t')$. For lattice-based measuring rods and clocks, we can now repeat all conclusions from Chap. 4, Sects. 3 and 4. It follows:

The *definition* of simultaneity according to the *elementary relativity principle* on the lattice requires the 'LORENTZ transformation':

$$\left.\begin{aligned} x' &= \frac{x - v\,t}{\sqrt{1 - v^2/c_o^2}}\,, & & x = \frac{x' + v\,t'}{\sqrt{1 - v^2/c_o^2}}\,, \\ & & \longleftrightarrow & \\ t' &= \frac{t - xv/c_o^2}{\sqrt{1 - v^2/c_o^2}}\,, & & t = \frac{t' + x'v/c_o^2}{\sqrt{1 - v^2/c_o^2}}\,. \end{aligned}\right\} \quad \begin{array}{l}\text{'LORENTZ transformation'}\\ \text{Internal observation}\\ \text{of imperfections}\\ \text{in a lattice}\end{array} \qquad (1014)$$

In consequence, there holds also 'EINSTEIN's addition theorem' (106), at with c_o instead of the speed of light c :

$$u = \frac{u' + v}{1 + u'\,v/c_o^2} \longleftrightarrow u' = \frac{u - v}{1 - u\,v/c_o^2}\,. \quad \begin{array}{l}\text{'EINSTEIN's addition theorem'}\\ \text{of velocities}\end{array} \qquad (1015)$$

It follows that the limiting velocity c_o of microplastic deformations has now in all reference systems Σ_o and Σ' a unique value:

$$c_o' = c_o\,. \qquad (1016)$$

This relation completes our lattice model of Special Relativity Theory.

Still, we discuss in which way also here in the crystal lattice all reference systems are equivalent.

In system Σ', the sine-GORDON equation holds true as in Σ_o, since displacements q transverse to the x-axis are invariant with respect to the 'LORENTZ transformation', $q' = \mathrm{q}$, hence

$$\frac{\partial^2}{\partial x'^2}\mathrm{q}(x',t') - \frac{1}{c_o^2}\frac{\partial^2}{\partial t'^2}\mathrm{q}(x',t') = \frac{1}{\lambda_o^2}\sin\left(\mathrm{q}(x',t')\right) . \tag{1017}$$

The observer resting in Σ' gets in consequence the same manifold of solutions as the observer in Σ_o. Each line form, measured in Σ_o, exists also among a registered line forms in Σ' and vice versa.

When we observe one and the same line from both reference systems, then the effects of Special Relativity Theory become visible. The in Σ' resting kink is the standard unit length shown in Fig. 2,

$$q_o^I(x') = 4\arctan\exp\left[\frac{x'}{\lambda_o}\right] . \qquad \text{Resting kink solution in } \Sigma' \tag{1018}$$

It is seen by an in Σ_o resting observer as moving kink, the solution (1008) shown in Fig. 4 as thick line. One needs to insert only the 'LORENTZ transformation' (1014) into Eq. (1018) to get Eq. (1008).

Likewise the in Σ' localised breather solution is shown as there resting normal clock in Fig. 3,

$$q_o^{III}(x',t') = 4\arctan\frac{\sin(2\pi\, t'/T_o)}{\cosh\left(x'/(\sqrt{2}\,\lambda_o)\right)} , \qquad \text{Localised breather solution in} \Sigma' \tag{1019}$$

seen by an in Σ_o resting observer as moving breather, namely, the solution (1010) as also shown in Fig. 5. Again we only need to insert the 'LORENTZ transformation' (1014) into Eq. (1019) to get (1010).

With the 'LORENTZ transformation' (1014), all reference systems Σ_o and Σ' are standing on the same footing. The crystal lattice resting in Σ_o is not distinguished. If one agrees to perform all measurements exclusively with the lattice-based length normals and clocks, so the defining crystal lattice cannot be determined anymore, *the crystal is absent*. To be more precise, we state that for the lattice we cannot determine a state of motion with the so-defined measuring rods and clocks.

The crystal is the space. It is the same as in 1920 A. EINSTEIN stated with respect to the aether, cp. Einstein (1923):

"The next position which it was possible to take up in face of this state of things appeared to be the following. The ether does not exist at all. …More careful reflection teaches us; however, that the special theory of relativity does not compel us to deny ether. …To deny ether is ultimately to assume that empty space has no physical qualities whatever. The fundamental fact of mechanics does not harmonise with this view."

The crystal provides a model of EINSTEIN's aether.

We shall finish our excursion on a lattice model of the Special Relativity Theory with a few vague hypothetical considerations of the structure of our physical space-time.

The solid body represents in a certain degree a minimal edition of the Special Relativity Theory, with a limiting velocity c_o, much smaller than the speed of light $c_o \ll c$. For avoiding any misunderstanding, this is once more explicitly stated.

The effects of EINSTEIN's Relativity Theory, caused by the finite speed of light, have nothing to do with the elastic interactions of a crystalline solid body. Nevertheless, one can take this lattice model as a serious hint for the possibility that our physical space has a discrete structure, some kind of a physically defined lattice structure. The constituents of such a lattice cannot be any kind of inertial masses of our space, neither of atoms or of elementary particles.

As lattice constant of our physical space-time, one could imagine the tiny PLANCK length $l_P = \sqrt{\hbar G/c^3} = 1.6 \cdot 10^{-33}$cm, according to our present understanding an absolute limit for measuring spatial distances. It is defined by the combination of PLANCK's constant $h = 6.63 \cdot 10^{-34}$ N m s, or $\hbar = h/2\pi$, NEWTON's gravitational constant $G = 6,67 \cdot 10^{-11}$ N m^2 kg^{-2} and the speed of light $c = 3 \cdot 10^8$ m s^{-1}. Such a space lattice with purely geometrically defined constituents could be a reasonable background for the Special Relativity Theory.

Possible consequences from such a model could be directly derived. In a solid body, the relativity is lost when we proceed into ranges of the lattice constant a, imposing correspondingly large energies. If the kink velocity is near to the limiting velocity c_o, so that its 'LORENTZ contracted' length becomes smaller than the lattice constant a . This is physically impossible. The assumption of the linear elasticity theory Eq. (993) is not fulfilled anymore. The sine-GORDON equation is not valid anymore, and with it the properties of its kink and breather solutions, its 'LORENTZ contraction' and 'time dilatation'. Also, the equality of all reference systems, resulting from the elementary relativity principle, would not be fulfilled. It is easy to understand that the lattice separation a now leads to a distinguished reference system Σ_o, in which the crystal is at rest: Σ_o is the system with the maximum lattice separation a. Relativistic conditions are obtained when we restrict ourself on distances A much larger than a, $A \gg a$. Then the reference system Σ_o is in no way distinguished.

Applied to our physical space-time, one could then ask the question whether a breaking of the LORENTZ symmetry is to be expected when we proceed to separations in the range of the PLANCK length l_P. However, our experiments are still far remote from this scale. For an elementary particle with rest mass m_o, its COMPTON length $\lambda = \hbar/mc$ provides a characteristic length scale. In coming close to the speed of light, m can become in principle arbitrary large. We get $\lambda = l_P$, when m equals the so-called PLANCK mass $m_P = \sqrt{\hbar c/f} = 2.2 \cdot 10^{-5}$ g. Taking a proton with a rest mass $m_o = 1.7 \cdot 10^{-24}$ g at a velocity v. For reaching $\gamma = m_P/m_o = 2.210^{-5}/1.7 \cdot 10^{-24} = 1.3 \cdot 10^{19}$, we would need a proton velocity $v_p = c(1 - 3 \cdot 10^{-39})$, or a proton energy of $2.4 \cdot 10^{18}$ GeV for getting a COMPTON length on scale of the PLANCK length. For this extreme approach to the speed of light, the model of a solid lattice for our space-time would lead to a violation of the relativity principle.

For a photon with a wavelength $\lambda \approx l_P$, we get a period $T_P = l_o/c = \sqrt{G\hbar/c^5} = 0.5 \cdot 10^{-44}$ s, the so-called PLANCK time, corresponding to a frequency $\nu_p = 1,9 \cdot 10^{43}$ Hz. In comparison with this frequency, the frequencies of very hard RÖNTGEN rays (X rays), at 10^{25} Hz, are extremely small. Again we could expect to come in the range of a violation of the LORENTZ symmetry for frequencies of the PLANCK scale. Clearly, such a discrete structure of our physical space-time could also appear at length scales larger than the PLANCK length. The expectations to measure such effect would be much better.

We should stress at this place that an approach to scales near the PLANCK length or the PLANCK time requires the inclusion of quantum physics and gravitation. It can be expected that the analogue of a classical solid body would requires a significant modification.

However, the idea of a discrete structure of our space could require a different sight on our test experiments of Special Relativity Theory. If the 'aether' is nothing else than a discrete vacuum structure, then we could not find a state of motion of this aether by definition, since all experimental objects are just structures within this aether. A velocity of an aether would be excluded from the beginning. With respect to the reputed measurement of an aether wind by D. C. MILLER in 1921, we refer to the famous comment of EINSTEIN, cited above, Chap. 4.

Further considerations require experiments looking for a dispersion of electromagnetic waves, a dependence of the speed of light from the frequency. When our space had indeed a discrete structure, then such a dispersion would be expected for sufficient small wavelengths. As a mathematical consequence, this follows from the strict validity of Eq. (985) for the linear chain, if we do not go over to the continuum approximation (990). We remark that in this model a vacuum dispersion would be expected from a purely classical point of view. In practice, such experiments could fail due to the smallness of the lattice constant and the corresponding high frequency of electromagnetic waves. Also, there is the objection that classical considerations may be modified by quantum effects and the action of gravity. We can expect that the universe as giant laboratory could present some surprises in this direction, cp. Chap. 13, Sect. 5.

Maybe we get sometimes an elementary lattice structure that directly provides the LORENTZ postulates. The photon would then propagate as elementary structure deformation through the lattice in a similar way as sound through a crystalline solid body. This could revive the old credo by MACH: 'Light should be as sound. Sound might be as light'. Only the lattice for photons would be our vacuum, while the lattice for the sound consists of the atoms of the crystal. This is the difference.

We further remark that a particle concept as presented here by means of dislocations in a lattice fulfils automatically important quantum physical axioms, s. GÜNTHER (2000). Both the principle of the identity of particles, the indistinguishability in particle collisions, the creation and annihilation of particles, or the interference of single particles in passing periodic barriers follow in elementary way with necessity from this model. The picture of an oscillating string gets its resemblance in the notions of the string theory of elementary particles, there however in higher dimensions then the here considered four dimensions of space-time.

2 A Clock Paradox

A. EINSTEIN has derived his Special Relativity Theory completely from the postulate that the speed of light c is a natural constant, having in each inertial system one and the same numerical value, $c = 299\,792\,458\ \mathrm{ms}^{-1}$. This implies in EINSTEIN's procedure also the synchronisation of clocks with this speed of light. All other effects are an inevitable consequence. A moving clock runs slower, or more precisely: We construct a number of identical clocks. Some are distributed in the inertial system $\Sigma_o(x,t)$ with coordinates (x,t) and synchronised along the x-axis. U_o^x should be this clock that in Σ_o is at rest at position x. Now we take one further clock U_v^o and carry them with constant velocity v along the series of clocks resting in Σ_o. The clock U_v^o should be at rest in inertial system $\Sigma'(x',t')$ with coordinates (x',t'). We compare the pointer position $t_v = t'$ of the in Σ_o moving clock U_v^o, with the pointer positions t of all the clocks U_o^x that the moving clock just passes by. Then we find[3]

$$t' = t\sqrt{1 - v^2/c^2}\,. \tag{1020}$$

At least this is the claim of EINSTEIN's Special Relativity Theory. The pointer of the moving clock stays behind. The only physical constant in this equation is the speed of light c. It is of no concern with which clock we try to verify experimentally this formula, and we have only to measure time sufficiently accurate.

Whether we move a good old mechanical pocket watch with velocity v or a modern Caesium clock based on atomic oscillations, always we will find in the experiment the factor $\sqrt{1 - v^2/c^2}$. The time dilatation is in EINSTEIN's words the 'experimentum crucis' of Special Relativity Theory, Chap. 4, Sect. 2. The technical construction of the clock does not appear in the formula for the time dilatation.[4]

Each precise oscillating system can realise a clock to measure EINSTEIN's time dilatation (1020). Therefore, we can also perform the test with the in Sect. 1, introduced breather clocks. Also, there we have found that the pointer of the moving breather clock stays behind the pointer of the resting breather clocks, consider Eq. (1013), namely, $t_v = t\sqrt{1 - v^2/c_o^2}$. Now this is just a time dilatation as described above, but the numerical value is completely different. Here, there appears *not* the speed of light c, but the critical velocity c_o of the sine-GORDON equation (996). It is of the order of magnitude of the sound velocity, in general, many powers of ten below the speed of light, $c_o \ll c$. EINSTEIN's theory, however, claims that in the equation for the time dilatation only the speed of light should appear, without exception for all clocks. This is the essential content of EINSTEIN's theory. Did we show with our model that the Special Relativity Theory is not applicable without exceptions, maybe somewhere wrong?

The resolution of the paradox is discussed in Problem 47.

---

[3]There it holds the standard initial condition: When clock U_v^o with pointer position $t' = 0$ passes by at $x = 0$, we assume that there the in Σ_o resting clock U_o^o also shows the time $t = 0$.

[4]Clocks, driven by the Earth gravity, pendulum clocks, hourglasses and water clocks should be excluded at this place. Gravitation cannot be included within the framework of Special Relativity Theory, and it was neglected in all our previous considerations. Some insights in a relativistic gravitation theory will be given in Chap. 13. For a pendulum clock moving with large velocity v, the influence of the gravity of Earth becomes smaller and smaller, if it is removed from Earth, maybe coming under the influence of the moon, other planets, the Sun or the gravity of the stars of our Milky Way.

Chapter 13
EINSTEIN'S General Relativity Theory

The final chapter is devoted to a short introduction to EINSTEIN's famous gravitation theory, the General Relativity Theory (GRT), and its experimental tests and applications. This theory now has been developed into a fundament for understanding of cosmic phenomena when large masses and densities are involved. EINSTEIN's gravity is a natural generalisation of NEWTON's gravitation theory. While NEWTON's gravity is sufficient for many cosmic phenomena, so-called post-NEWTONian effects with slight modifications from the classical theory become important in celestial mechanics of our planetary system and in other stellar systems. In modern time, however, not only fine deviations of the NEWTONian behaviour are observed. Extreme effects of time dilatation are found as gravitational redshifts. Light deflection from straight lines leads to the phenomenon of gravitational lensing, which provides some very spectacular observations including implications for the distribution of dark matter in the universe. We shall discuss in the following section that a relativistic gravitation theory requires a generalisation of the special-relativistic space-time concept. Therefore, it seems important to cover these ideas in the last part of the book. We cannot provide always complete derivations. Here, we hint for details at the relevant textbooks of Weinberg (1972), Misner, Thorne, Wheeler (1973), Wald (1984), Carroll (2004), Schutz (2009), to list a few recent ones. The discussion must be based on RIEMANN geometry, and an introduction can be found in Ibragimov (2015).

1 Gravitation by NEWTON and EINSTEIN

A fundamental property of the gravitation theory of NEWTON, and also of EINSTEIN, is the observation that all heavy bodies are falling with the same velocity. But this is not obvious. In NEWTON's second law, there appears the inertial mass m_i,

$$\frac{d^2 m_i \mathbf{r}}{dt^2} = \mathbf{F}. \tag{1021}$$

H. Günther and V. Müller, *The Special Theory of Relativity*,
https://doi.org/10.1007/978-981-13-7783-9_13

Here, any force can appear on the right-hand side. But when we consider the motion of the body with mass m_i in the gravitational field of a second mass M, there appears the mass of the body under consideration as heavy mass m_h,

$$\mathbf{F} = -G\, M\, m_h\, \frac{\mathbf{r}}{r^3}\,. \tag{1022}$$

This is the famous NEWTONian gravitation law, the radius vector $\mathbf{r}$ points from the source of the gravitational field M to the body under consideration. The motion is described by Eq. (1021). Here, we have considered point particles. Extended bodies can be understood as ensembles of point particles.

NEWTON's gravitational constant G has the value (cp. Li et al. 2018)

$$G = (6.6743 \pm 0.0002) \cdot 10^{-11}\, \mathrm{N\, m^2\, kg^{-2}}\,. \tag{1023}$$

We assume in the following that the source of the gravitational field, the mass M, is much larger than the mass of the point mass whose motion is investigated. Identifying the force $\mathbf{F}$ in Eq. (1021) with the force of the heavy mass $\mathbf{F}$, we get an equation of motion that leads to the identical orbits $\mathbf{r}(t)$ for different bodies, when the two masses m_i and m_h coincide.

The two logically separated masses in NEWTON's mechanics and gravitation theory should fulfil

$$m_i = m_h\,. \tag{1024}$$

The universal law of equally falling bodies was already known to GALILEI, but it remained unexplained within classical mechanics. However, it led EINSTEIN to the most important principle at the basis of his relativistic gravitation theory, his *equivalence principle*:

The inertial mass m_i and the heavy mass m_h of a body are identical under all conditions and also for different inner binding forces.

Sometimes, it is denoted as *weak equivalence principle* since it concerns the equivalence of the inertial mass m_i to the mass m_h in the force law Eq. (1022). The equal orbits of all bodies whatsoever constitution is generalised in EINSTEIN's equivalence principle:

It is locally indistinguishable whether a body feels an acceleration or whether it is sitting in a gravitational field.

If we still add the result from Chap. 7, Sect. 1, the equivalence of mass and energy (186), cp. also EINSTEIN's derivation (199), then it is

Heavy mass $=$ inertial mass $=$ energie$/c^2$.

It should be remarked that both masses m and M in the force law (1022) are standing on the same footing.[1] Since the inner gravitational binding energy contributes to the gravitating mass of the body itself, the identification of all heavy masses with its inertial masses is called *strong equivalence principle*. In the following, we omit any index on the masses. As source of gravitation, there remains only the energy that we translate to a mass by means of the speed of light. Gravity acts on all masses in the world, both on celestial objects and on elementary particles and also on antiparticles, even on particles without rest mass, and hence on light. Gravitation can principally not be shielded because of the positivity of the energy or of the gravitating masses. For a discussion of the dark energy, or more historically on EINSTEIN's cosmological term, s. later, Eq. (1057), and Sect. 5. The unique sign of the gravitating masses distinguishes gravitation from all other interactions, in particular, also from electromagnetic fields where one observes interaction with positively and negatively charged particles. We return to the assumption that the mass m is much smaller than the source of the gravitational field M, in principle, arbitrary small. One speaks of a test particle motion that is under the influence of M. But this can be also an arbitrary distribution of masses.

NEWTON's gravitational force $\mathbf{F}$ is according to Eq. (1022) conservative, since

$$\begin{aligned} \mathbf{F}_x &= -\frac{\partial}{\partial x}\left(-\frac{G\,M}{\sqrt{x^2+y^2+z^2}}\,m\right) = -\frac{1}{2}G\,M\,m\,\frac{2\,x}{\sqrt{x^2+y^2+z^2}^{\,3}} \\ &= -G\,M\,\frac{x}{\sqrt{x^2+y^2+z^2}^{\,3}}\,m\,, \end{aligned}$$

and similar for the other components, hence

$$\mathbf{F} = -\mathrm{grad}\,V \quad \text{with} \quad V = -\frac{G\,M}{r}\,m\,, \tag{1025}$$

The large mass M can now be replaced by a mass distribution with a continuous density $\varrho(\mathbf{x}')$. The position $\mathbf{x}'$ is called source point. The potential $\varphi = \varphi(\mathbf{x})$ of this mass distribution is defined by

$$\varphi(\mathbf{x}) = -G\int \frac{\varrho(\mathbf{x}-\mathbf{x}')}{\sqrt{(\mathbf{x}-\mathbf{x}')^2}}\,d^3\mathbf{x}'\,. \tag{1026}$$

If a mass m sits at place $\mathbf{x}$, then the force $\mathbf{F}$ of the continuously distributed masses on m is described by the gradient of $\varphi(x)$,

---

[1] By NEWTON's third axiom, Eq. (1022) holds also for the second body M, and it enters the corresponding equation of motion. So we have to solve in principle a two- (or many-) body problem, that only for the case of two bodies reduces to a one-body problem with a reduced mass $m_{red} = 1/(m^{-1} + M^{-1})$ and an equation of motion for the relative distance.

$$\mathbf{F} = -m \operatorname{grad} \varphi(\mathbf{x}) = G\, m \operatorname{grad} \int \frac{\varrho(\mathbf{x}-\mathbf{x}')}{\sqrt{(\mathbf{x}-\mathbf{x}')^2}}\, d^3\mathbf{x}' \ . \tag{1027}$$

We verify this equation for a point-like mass M at the coordinate origin $\mathbf{x} = 0$. For the continuous mass distribution, we use DIRAC's three-dimensional δ-function, s. Appendix B.3,

$$\varrho(\mathbf{x}-\mathbf{x}') = M\, \delta(x)\, \delta(y)\, \delta(z) \ , \tag{1028}$$

and we write $\sqrt{(\mathbf{x}-\mathbf{x}')^2} = r$,

$$\mathbf{F} = G\, m\, M \operatorname{grad} \int \frac{\delta(\mathbf{x}')}{\sqrt{(\mathbf{x}-\mathbf{x}')^2}}\, d^3\mathbf{x}' = G\, m\, M \operatorname{grad} \frac{1}{r}$$

equivalent to Eq. (1022).

With NEWTON's second axiom (140), NEWTON's gravitation theory consists of the two equations:

$$\left.\begin{aligned} \Delta\varphi &= 4\pi G\, \varrho \ , \\ \frac{d^2\mathbf{x}}{dt^2} &= \operatorname{grad} \varphi(\mathbf{x}) \ . \end{aligned}\right\} \quad \text{NEWTON's gravitation law} \tag{1029}$$

NEWTON's gravitation theory was verified with high precision in many investigations of celestial mechanics, beginning with the derivation of KEPLER's laws of planetary motion as solutions of the one-body problem. We shall discuss it in Sect. 3, where we shall also investigate experimental tests of EINSTEIN's General Relativity Theory and its NEWTONian limit. Stronger effects are investigated in Sects. 4 and 5 describing the application of relativistic gravitation theory on astrophysical phenomena.

EINSTEIN was motivated in the invention of this new theory of gravitation by the great success of Special Relativity Theory, and the impossibility to incorporate special-relativistic principles within the framework of NEWTON's theory. Especially, it seems insufficient that in NEWTON's law (1029) gravitational actions appear instantaneously, the force law (1022) describes an action over the distance, without time delay. Attempts to incorporate retardation into NEWTON's law failed, basically due to the equivalence principle as described above. It let EINSTEIN to the grandiose invention of a geometrical description of the gravitational field, where the equivalence principle is fulfiled automatically.

To get a hint on the nature of the gravitational field in a relativistic theory, we consider a simple thought experiment: We consider a disk in space. The geometry on the disk might be Euclidean; hence, for right-angular triangles there holds the law of PYTHAGORAS, and also equivalently the ratio of circumference U of the disk to its diameter d is given by $\pi = U/d$.

Now we set the disk in rotation, and we consider an inertial system Σ_o, with origin at rest at the centrum of the disk. An element ds of the circumference is then in longitudinal motion with velocity v with respect to Σ_o, while a radial element dr suffers only an orthogonal motion. The element ds then suffers a LORENTZ contraction, but not the radial element dr. In consequence on the rotating disk, we find $U < \pi d$, the geometry on the disk is not Euclidean anymore.

Until now we have assumed that the space is always Euclidean, with the sum of inner angles of triangles of 180° or π in arc scale. With KANT's a priori,[2] we assumed in the description of motion a Euclidean space. The rotation of the disk leads to an accelerated motion for all points besides the centre. We consider a small surrounding **U** of ds of the boundary of the rotating disk. The geometry in **U** is not Euclidean. According to EINSTEIN's equivalence principle, an observer in **U** cannot decide whether he is in an accelerated motion or in a gravitational field. Each acceleration of the circular motion of ds on the rotating disk creates a change in the geometry in **U** that is equivalent to a gravitational field. As we describe in Appendix B.1, the geometry of the space is completely specified by a *metric tensor* **g**, or in components g_{ik} that itself depends on the coordinates x_l. Knowing the principles of Special Relativity Theory we must go on to suppose that the metric tensor has to generalise the four-dimensional MINKOWSKI space with metric of the four-dimensional MINKOWSKI tensor η of Eq. (1623), Appendix B.1.

The quantity describing gravitational field is then a four-dimensional symmetric metric tensor **g** with the limit η for a vanishing gravitational field,

$$g_{ik} = \begin{pmatrix} g_{00} & g_{01} & g_{02} & g_{03} \\ g_{10} & g_{11} & g_{12} & g_{13} \\ g_{20} & g_{21} & g_{22} & g_{23} \\ g_{30} & g_{31} & g_{32} & g_{33} \end{pmatrix} \xrightarrow{\text{vanishing gravitation}} \eta_{ik} = \begin{pmatrix} 1 & 0 & 0 & 0 \\ 0 & -1 & 0 & 0 \\ 0 & 0 & -1 & 0 \\ 0 & 0 & 0 & -1 \end{pmatrix}. \tag{1030}$$

In General Relativity Theory, the metric tensor **g** replaces NEWTON's gravitational potential φ of Eq. (1025). It is essential that the gravitational field is not described by a scalar quantity, but by a four-dimensional tensor depending on the space-time coordinates x^i. The notion of EINSTEIN's equivalence principle means that we cannot distinguish locally whether a body is under the influence of a gravitational field or it is sitting in an accelerated reference system. Connected with this property is the possibility to choose a reference system, in which the body does not feel any gravitational field. In mathematical term, this means we can choose a coordinate transformation $x^{i'}(x^k)$ that translates the general metric $g_{ik}(x^l)$ into the special-relativistic form of Eq. (1029). However, this is possible only locally, i.e. not over finite regions of space-time. The translation to the MINKOWSKI form is also possible

[2] See Chap. 10, Sect. 7, where we noted that for KANT, space and time have an another meaning as our physical space-time accessible to measurements with rods and clocks.

along finite but not closed curves. This statement that locally the space-time can be brought to the form of the MINKOWSKI metric is the most general expression of the equivalence principle:

Locally also in a gravitational field, there hold the laws of Special Relativity Theory.

Formally, this local inertial system without any gravity effect, as it is free falling in the gravitational field, is produced by diagonalisation of the symmetric metric tensor g_{ik}, and then changing the scales of length and time that the eigenvalues of the tensor have values ± 1. The perception that the metric provides a full description of the gravitational field and that the law of the same orbits of bodies of different compositions also holds with respect to the inclusion of bodies having different inner gravitational binding energies is called EINSTEIN's equivalence principle.

The notion that we need a tensor field with the freedom to choose arbitrary coordinate transformations $x^{i'}(x^k)$, following from the equivalence principle, is a very satisfying property of EINSTEIN's General Relativity Theory, but at the same time it requires a careful discussion for distinguishing true gravitational fields on the one hand, and on the other hand, complicated metric tensors g_{ik} that arise from the special-relativistic MINKOWSKI metric (1030) by a coordinate transformation. We shall employ this coordinate freedom for special symmetric applications, but otherwise we have to characterise the coordinate-independent properties of the metric tensor **g**.

The introduction of a general metric tensor g_{ik} and of arbitrary coordinate transformation $x^{i'}(x^k)$ is a large step in the evolution of our understanding of space and time. We speak of a general RIEMANN space with curved coordinate systems and arbitrary moving observers.

The physical meaning of the four-dimensional tensors **g** is the description of separation between events in space-time, the *invariant line element*

$$ds^2 = g_{ik}(x^l)dx^i dx^k \qquad \text{Invariant line element} \qquad (1031)$$

for two events $A(t, x, y, z)$ and $B(t + dt, x + dx, y + dy, z + dz)$. We call the quantity ds^2 invariant separation, since the quantity ds^2 does not change when we go over from one coordinate system to any other one. We have only to transform both the coordinate differentials dx^i as contravariant vectors, and the elements of the metric tensor g_{ik} as covariant tensor of second rank,

$$dx^{i'} = \frac{\partial x^{i'}}{\partial x^i} dx^i \qquad\qquad g_{i'k'} = \frac{\partial x^i}{\partial x^{i'}} \frac{\partial x^k}{\partial x^{k'}} g_{ik}, \qquad (1032)$$

where we assume as always EINSTEIN's summation convention. In our exposition, the metric is always symmetric corresponding to Eq. (1031) as bilinear quadratic form in the coordinate differentials dx^i and dx^k. According to Eq. (1032), the covariant tensor g_{ik} transforms with the inverse transformation matrix $\partial x^i/\partial x^{i'}$ of the transformation of contravariant vectors or tensors. Therefore, we have the inverse metric g^{kl}, defined by

$$g_{ik}g^{kl} = \delta_i^l \,, \tag{1033}$$

as contravariant tensor, cp. Problem 49.

The invariant separation between two events A and B vanishes, when two events are connected in such a way that a light signal emitted at A just arrives at B. This is the generalisation of the signal propagation with the speed of light in MINKOWSKI space-time, cp. Chap. 9, Sect. 1. Then, we call this separation $ds = 0$ an element of the light cone connecting these two events. When the separation is larger than zero, $ds^2 > 0$, then a massive body can travel from one event to the other.

The physical meaning of the metric is its property to describe the separation between in the physical space-time also in cases, when the space-time is curved, or in other words in terms of the equivalence principle, in situations when we have heavy masses creating a gravitational field.

First, we consider the situation of the run of time on a body, moving in curved space-time. Inserting into Eq. (1031) the condition $dx^1 = dx^2 = dx^2 = 0$, we find

$$ds^2 = c^2 d\tau^2 = g_{00} dx^0 dx^0 \,. \tag{1034}$$

Taking the square root, we find for the run of time of the body, also called eigen-time τ that $d\tau = ds/c = (1/c)\sqrt{g_{00}}dx^0$, or after integration along the path

$$\tau = \frac{1}{c}\int \sqrt{g_{00}} dx^0 \qquad \text{Eigen time in curved space-time} \tag{1035}$$

Comparison with the relation between a general metric and the MINKOWSKI metric (1030) demonstrates that the time τ is really the eigentime measured by a clock that the body carries with it along the arbitrary path,

For discussing length measurements in general curved space-time, we decompose the general metric (1031) into spatial and time components,

$$ds^2 = g_{00}\,(dx^0)^2 + g_{0a}\,dx^0\,dx^a + g_{ab}\,dx^a\,dx^b \,, \tag{1036}$$

where the first indices of the alphabet a, b, c run only over the spatial coordinates $1, 2, 3$ as abbreviation for $x^1 = x, x^2 = y, x^3 = z$.

Now we consider the measurement of distances between bodies A and B, moving on different world lines through space-time, cp. Fig. 1. A light signal is emitted from event A to event B and immediately sent back to A'. For light propagation, the invariant separation vanishes between the events A and B as well as between B and A'. We get the two intervals of coordinate time run by solving the quadratic equation (1036) under the condition $ds^2 = 0$ for $dx^0_{(1)}$ and $dx^0_{(2)}$ (the index in parenthesis is not a space-time index, but means different moments of coordinate time at A).

$$dx^o_{(1,2)} = \frac{1}{g_{oo}}\left(-g_{oa}dx^a \mp \sqrt{(g_{oa}g_{ob} - g_{oo}g_{ab})dx^a dx^b}\right).$$

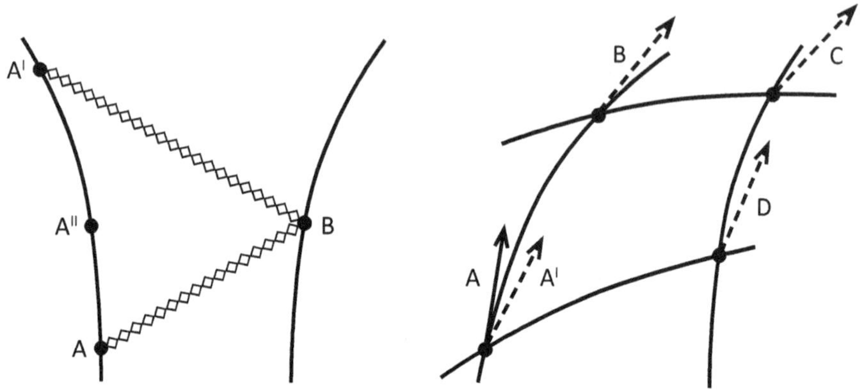

Fig. 1 World lines (solid lines) illustrate the meaning of the metric and parallel transport. Left: Signal propagation between two orbits A and B. Right: Parallel transport of a vector from A to B with inclusion of the affinities. Circular transport from A over B, C, and D back to A' leads to an angle deficit measuring the curvature.

The time interval between events A and A' (on the same world line) is obviously

$$dx^o_{(2)} - dx^o_{(1)} = \frac{2}{g_{oo}}\sqrt{(g_{oa}g_{ob} - g_{oo}g_{ab})dx^a dx^b}\,.$$

The eigentime between these events A and A' is after Eq. (1035)

$$d\tau = \frac{\sqrt{g_{oo}}}{c}(dx^o_{(2)} - dx^o_{(1)}).$$

The observer on orbit A will call as separation dl between events A and B the half of the eigentime difference between sending and receiving a signal to B multiplied by the speed of light. We get

$$dl = \frac{c\,d\tau}{2} = \sqrt{\left(\frac{g_{oa}g_{ob}}{g_{oo}} - g_{ab}\right)dx^a dx^b}\,.$$

Taking the square, we find the spatial metric in the curved space-time

$$dl^2 = \left(\frac{g_{oa}g_{ob}}{g_{oo}} - g_{ab}\right)dx^a dx^b\,. \qquad (1037)$$

Obviously, the mixed space-time components of the metric g_{oa} and the time–time components g_{oo} have an influence on the measurement of spatial separations in a general curved space-time.

The discussion of signal exchange between world lines A and B also allows a discussion of the definition of simultaneity that generalises our treatment in Special Relativity Theory. We consider again the situation of Fig. 1, the left-hand side. Obviously, the observer A will *define* the event A'' as simultaneously to event B at the coordinate time

$$x^o + \frac{1}{2}(dx^o_{(2)} - dx^o_{(1)}) = x^o - \frac{g_{oa}}{g_{oo}} dx^a .$$

We remark that the minus sign in this expression is a result of the chosen signature of the metric. The synchronisation of clocks along these lines is obviously only defined over infinitesimal regions of space-time. We can continue it along arbitrary special curves, but in general it is forbidden for closed curves, since coming back to the original world line would in general provide a different coordinate time of this observer. An exception is the case of vanishing of mixed space-time components of the metric, $g_{0a} = 0$. When we choose in addition in such a space-time at each point the eigentime, Eq. (1035), as x^0-coordinate, we get a metric of the form

$$ds^2 = d\tau^2 - g_{ab} dx^a dx^b , \tag{1038}$$

where the three spaces $\tau =$ const. represent a universal simultaneity. We call it *synchronous* or GAUSSian metric. We learn of such metrics in discussing the infall into a black hole in Sect. 3 and in cosmology, Sect. 5.

Now we go on in the discussion of a space-time with a general metric. We consider the right panel of Fig. 1. If the coordinates are arbitrary and the space-time metric is curved, we need special care in transporting vectors to different positions in space-time. We compare a vector field A^i at a certain point with its value at a neighbouring point where it takes the value B^i. For evaluating the change between both values, we consider the quantity $D A^i = B^i - A^i$ which is given by

$$DA^i = dA^i + \Gamma^i_{kl} A^k dx^l . \quad \text{Covariant differential} \tag{1039}$$

Here dA^i is the change of the vector field between the two positions, and the additional term comes from the curvature of space-time, describing the change of the contravariant base vector from point to point. The quantities Γ^i_{kl} are called *connection coefficients* or CHRISTOFFEL symbols that are constructed just in the way to lead to a *covariant derivative* DA^i, with coordinate transformation laws as given in Eq. (1032). In contrast to DA^i, the quantities dA^i as difference of vector components at different places in space-time and the quantity $\Gamma^i_{kl} A^k dx^l$ do not form vectors. This is obvious especially for the last term, since the CHRISTOFFEL symbols can be transformed to be zero at an arbitrary point in the curved space-time. To prove this, one goes over to the local MINKOWSKI form of the metric which always exists due to the equivalence principle. We generalise Eq. (1036) to the differential of covariant vectors V_i:

$$DA_i = dA_i - \Gamma^k_{il} A_k dx^l ,$$

where the negative sign arises from the requirement that the scalar product of a covariant and a contravariant vector $V_i\, W^i$ must form a scalar, which immediately provides the last relation.

Now we get the *definition* of the CHRISTOFFEL symbols in terms of the metric by the condition that $Dg_{ik} = 0$. Note that here for the covariant differential of a tensor, we have to apply the rules with the additional connection coefficients in Eq. (1039) for each index separately,

$$Dg_{ik} = dg_{ik} - \Gamma^l_{mk} g_{il} dx^m - \Gamma^l_{im} g_{lk} dx^m \, . \tag{1040}$$

Setting this expression to zero, we get an expression for the CHRISTOFFEL symbols in terms of the first partial derivatives of the metric,

$$\Gamma^i_{kl} = \frac{1}{2} g^{im} \left(\partial_k \, g_{lm} + \partial_l \, g_{mk} - \partial_m \, g_{kl} \right) \quad . \qquad \text{Christoffel symbols} \tag{1041}$$

We call the connection, derived in this way from the metric, as metric-compatible affinities which are characteristic for the RIEMANNian (Fig. 2) geometry. There, the metric completely describes the geometry of space-time. It is possible to introduce more general geometries, where in addition to Eq. (1041), there appears an intrinsic antisymmetric term in the affinities, describing a property of space-time called torsion. We shall not consider this mathematical possibility since it would lead to more general gravitation theories than EINSTEIN's General Relativity Theory that are not supported by experiments.[3]

Now it also becomes obvious that the CHRISTOFFEL symbols do not form a tensor since we can always use a coordinate transformation to convert the metric around a special point to the MINKOWSKI form. There then the affinities vanish identically, $\Gamma^i_{kl} = 0$.

When we divide the expression for the covariant derivative, Eq. (1024), and the corresponding one for a covariant vector by the coordinate differentials dx^i, we get the defining equations for the covariant derivatives in curved space-time.

For a covariant vector V^i and a contravariant vector V_i we write

$$\left.\begin{aligned} \frac{D\,V^i}{D\,x^k} &\equiv V^i_{\;;k} := \frac{\partial\,V^i}{\partial\,x^k} + \Gamma^i_{kl}\,V^l \, , \\ \frac{D\,V_i}{D\,x^k} &\equiv V_{i;\,k} := \frac{\partial\,V_i}{\partial\,x^k} - \Gamma^l_{ki}\,V_l \, . \end{aligned}\right\} \qquad \text{Covariant derivative} \tag{1042}$$

The covariantly derived vectors (1042) form a tensor of second rank in the same way, as in Euclidean space with Cartesian coordinates the partial derivatives of a vector form a tensor of second rank.

Closely connected with the covariant derivative is the problem of motion of test bodies in a given gravitational field. Here we give the direct answer to this problem

[3]Notice that the theory of elasticity for a crystalline solid with dislocations yields a model for both a metric and a torsion, s. Günther (1981).

Fig. 2 GEORG FRIEDRICH BERNHARD RIEMANN, *Dannenberg (Elbe) 19.9.1826, †Selasca (Italien) 20.7.1866.

in using EINSTEIN's equivalence principle. We discuss the justification by EINSTEIN's gravitational field equation in Sect. 2 and investigate it in more detail in Problem 50. In direct generalisation of NEWTON's first law of mechanics (141), bodies move on straight lines, if no forces are acting on them, we postulate that structureless bodies in a general gravitational field described by a metric **g** move on orbits that are as close as possible to straight lines. We call these orbits *geodesics* that are characterised by the property that the four-velocity of the body u^i is parallelly transported along this path, i.e. the covariant derivative of u^i for the eigentime vanishes,

$$\frac{Du^i}{d\tau} = \frac{du^i}{d\tau} + \Gamma^i_{kl} u^k\, u^l = 0. \qquad \text{Geodesic equation for massive bodies} \qquad (1043)$$

This equation becomes a nonlinear differential equation of second order for the path of the body $x^i(\tau)$, if we insert the definition of the four-velocity $u^i = dx^i/d\tau$. Its solution describes the extremum path of massive bodies in a given gravitational field. For massless particles as photons, we take the wave vector of the photon k^i orthogonal to its wavefronts and use a parameter σ for its propagation, then we have the propagation of photons or of electromagnetic waves as

$$\frac{Dk^i}{d\sigma} = \frac{dk^i}{d\sigma} + \Gamma^i_{kl} k^k\, k^l = 0. \qquad \text{Geodesic equation for photons} \qquad (1044)$$

The solution of this equation fixes also the parameter σ used for propagating the wave vector along the path. It is called *affine parameter* as it is only determined up to linear transformations. The wave vector k^i is a null vector, $k^i\, k_i = 0$, so that the light propagation is also described by the requirement of the vanishing of the proper distance along the path $ds = 0$.

The general description of the gravitational field by the metric **g**, which directs the motion of massive and massless particles via the geodesic equation, implies that the metric deviates from the MINKOWSKI metric. We speak from a *curvature* of the space-time. Intuitively, we can imagine a curvature in two-dimensional space as the uneven surface of a landscape or of water surfaces in the ocean. Even if this intuition is difficult for a four-dimensional space-time, we can extract two generic expectations from this simple picture: First, we expect that parallel lines will be deformed in such a way that its separation will change along its orbit, somewhere even crossing in the past or future. The second picture concerns a parallel transport of vectors along a closed orbit, cp. Fig. 1, the right panel. The evaluation provides a tensor of fourth rank, with one contravariant and three covariant indices, called RIEMANN–CHRISTOFFEL curvature tensor $R_{ijk}{}^l$, often simply denoted as *curvature tensor*, given by

$$R_{ijk}{}^l := \partial_i\, \Gamma^l{}_{jk} - \partial_j\, \Gamma^l{}_{ik} + \Gamma^m_{ik}\, \Gamma^l{}_{mj} - \Gamma^m_{jk}\, \Gamma^l{}_{mi}\,. \qquad \text{Curvature tensor} \qquad (1045)$$

According to our intuitive description above, the curvature tensor tells us by which amount an arbitrary vector A^i (see Fig. 1) changes during a parallel transport around a closed area element df^{ij} (the element is formed as antisymmetric product of two base vectors in i- and j-directions)

$$\Delta A^l = \oint \delta A^l = \oint \Gamma^l_{ik} A^k dx^i = \int\int df^{ij} \delta_j (\Gamma^l_{ik} A^k) .$$

For the change from the curve to the area integral, we used Stocke's law, cp. Appendix B.2. In differential form, it follows the definition (1045) of the curvature tensor

$$\Delta A^l = \frac{1}{2} df^{ij} R_{ijk}{}^l A^k . \tag{1046}$$

Related to the parallel transport of a vector around an infinitesimal area element of space-time as definition of the curvature tensor, one finds the commutator of the covariant derivatives of an arbitrary vector field A^l,

$$A^l_{;ij} - A^l_{;ji} = R_{kij}{}^l A^j , \tag{1047}$$

and the covariant derivative is not commutative as the partial derivative in Cartesian coordinates, as can be seen from the definition (1042). Furthermore, we consider a bundle of geodesics $x^i(\tau, \lambda)$, where λ is a parameter counting the geodesics. Then we form the vector of the four-velocity $t^i = x^i{}_{,\tau}$ as tangent vectors on the geodesics (the comma means the partial derivative $x^i{}_{,\tau} = \partial x^i / \partial \tau$) and the separation or *deviation vector* between geodesics $q^i = x^i{}_{,\lambda}$. Then, it follows that the transport of the change of the separation vector along the geodesics equals to the change of the tangent vectors projected in transverse directions,

$$q^k \frac{Dt^i}{dx^k} = t^k \frac{Dq^i}{dx^k} . \tag{1048}$$

This compatibility relation of first derivatives of the geodesic tangent and deviation vector again follows from the definition of the covariant derivative. However, the second derivative of the deviation vector is again given by the curvature tensor, and hence parallel geodesics are bent to approach to each other either in the past or in the future.

$$\frac{D^2 q^i}{d\tau} = R_{jkl}{}^i t^j t^k q^l , \qquad \text{Geodesic deviation} \tag{1049}$$

This equation is also called *equation of geodesic deviation.*

Finally, we give a formula for evaluating the curvature tensor with four covariant indices that follows straightforward from the definition Eq. (1045):

$$\begin{aligned} R_{ijkl} &= g_{lm} R_{ijk}{}^m \\ &= \frac{1}{2}\left(\frac{\partial g_{il}}{\partial x_j \partial x_k} + \frac{\partial g_{jk}}{\partial x_i \partial x_l} - \frac{\partial g_{ik}}{\partial x_j \partial x_l} + \frac{\partial g_{jl}}{\partial x_i \partial x_k}\right) + g_{mn}\left(\Gamma^m_{il}\Gamma^n_{jk} - \Gamma^m_{ik}\Gamma^n_{jl}\right) \end{aligned} \tag{1050}$$

From this equation, it becomes obvious that the curvature tensor is antisymmetric in the first two and the last two indices. The easy to remember reason is the property that the vector difference along the closed curve in Fig. 1 changes its sign in reversing the direction of the path. Furthermore, there vanishes the sum of three components of the curvature tensor with cyclic permutations of the last three indices. Summarising it holds

$$R_{ijkl} = -R_{jikl} = R_{ijlk} = R_{klij} \; , \qquad R_{ijkl} + R_{iklj} + R_{iljk} = 0 \; . \tag{1051}$$

The last equation must be postulated for one index i only, and then it follows by the former permutations for all four values. By counting one finds that the curvature tensor as rank-4-tensor has altogether 20 independent components.

Further one can also prove with the definition of the affinities that the curvature tensor fulfils the so-called BIANCHI *identity*

$$R^m{}_{ijk;l} + R^m{}_{ikl;j} + R^m{}_{ilj;k} = 0. \tag{1052}$$

In this relation, the semicolon means the covariant derivative, so it is indeed a relation fulfiled for the third derivatives of the metric. We will use this identity to derive the equations of motion as consequence of EINSTEIN's gravitational field equations.

From the full curvature tensor, we form the so-called RICCI tensor R_{ik} and the RICCI or curvature scalar R by contraction of two upper and lower indices,

$$R_{ik} := R_{ilk}{}^{l} \qquad \text{RICCI tensor} \tag{1053}$$

and

$$R := g^{ik} \, R_{ik}. \qquad \text{Curvature scalar} \tag{1054}$$

Now combining both tensors, we form the so-called EINSTEIN tensor E_{ik} that provides the basis for EINSTEIN's gravitational field equations

$$E_{ik} := R_{ik} - \frac{1}{2} \, g_{ik} \, R \; . \qquad \text{EINSTEIN tensor} \tag{1055}$$

Obviously, the EINSTEIN tensor is a symmetric tensor that contains quasi-linear second derivatives of the metric and nonlinear combinations of first derivatives arising from the products of affinities in the definition of the curvature tensor. We can form mixed and contravariant components of the EINSTEIN tensor,

$$E_i^k = E_{il} \, g^{lk} \quad \text{and} \quad E^{ik} = E_{lm} \, g^{il} \, g^{km} \; .$$

Then, it follows from the BIANCHI identity for the curvature tensor, Eq. 1052, that the covariant divergence, i.e. the contraction of the covariant derivative of the EINSTEIN tensor over one of its indices, vanishes:

$$\frac{DE^{ik}}{Dx^k} = E^{ik}_{\;;k} = \frac{\partial E^{ik}}{\partial x^k} + \Gamma^i_{kl} \, E^{kl} + \Gamma^k_{kl} \, E^{il} \equiv 0 \, , \qquad \text{BIANCHI identity} \tag{1056}$$

which is the main reason for the special construction of the EINSTEIN tensor from the RICCI tensor and curvature scalar. The idea of EINSTEIN was that all matter represents a source of the curvature of space-time. Therefore, he postulated his gravitational field equations, here already written with the additional cosmological term Λ, that becomes important in Sect. 5,

$$E_{ik} = \frac{8\pi G}{c^4} T_{ik} + \Lambda g_{ik} \, .$$ EINSTEIN'S gravitational field equations (1057)

Simultaneously with EINSTEIN the mathematician DAVID HILBERT has written down these field equations based on a field theoretical variational principle with the LAGRANGE density $\sqrt{-g}R$ where g is the determinant of the metric $\mathbf{g}$ and R denotes the curvature scalar.

On the right-hand side of EINSTEIN's equations, there stands the source of the gravitational field, the energy–momentum tensor T_{ik} defined with the following components:

$$\mathbf{T} = \begin{pmatrix} \epsilon & \frac{1}{c}\mathbf{s} \\ c\,\mathbf{q} & \mathbf{t} \end{pmatrix} .$$ Energy momentum tensor (1058)

Here $\epsilon = \rho c^2$ denotes the energy density, $\mathbf{s}$ is the current density of the energy, $\mathbf{q}$ is the momentum density, and $\mathbf{t}$ is the three-dimensional stress tensor. EINSTEIN's energy–mass equivalence means that $\mathbf{s} = c^2\,\mathbf{q}$. With the EINSTEIN tensor, also the energy–momentum tensor is symmetric. The gravitational field equations of General Relativity Theory are therefore ten equations for the ten tensor components g_{ik}. Because of the arbitrary coordinate choice in RIEMANNian geometry, we have at our disposal four free coordinate transformations. Therefore, only six equations remain to determine the general metric. The BIANCHI identity, Eq. (1056), imposes on the other hand four conditions on the EINSTEIN tensor that implies that also the covariant divergence of the energy–momentum tensor must vanish. As discussed in Sect. 2, this is just the requirement of local energy–momentum conservation, known from Special Relativity Theory. Sometimes, we fulfil the coordinate freedom in imposing additional conditions on the metric for simplifying calculations. An often chosen form is the so-called harmonic coordinates, fulfiling

$$\frac{\partial}{\partial x^r} g^{ir} = 0 \, .$$ Coordinate condition (1059)

In most applications, we can suppose an energy–momentum tensor for an ideal fluid, which is characterised by an energy density ϵ, vanishing momentum and energy current densities, and an isotropic PASCAL pressure p, where the three-dimensional stress–tensor $\mathbf{t}$ has only diagonal elements, i.e. we have

$$T^{ik} = \begin{pmatrix} \epsilon & 0 & 0 & 0 \\ 0 & p & 0 & 0 \\ 0 & 0 & p & 0 \\ 0 & 0 & 0 & p \end{pmatrix} = \epsilon u^i u^k + p(u^i u^k - g^{ik})$$ Ideal fluid (1060)

The four-velocity u^i denotes the flow vector of fluid elements of the matter under consideration. Note that the term $(u^i u^k - g^{ik})$ is a projection tensor for the direction orthogonal to the fluid streams. The simplest case is so-called *dust-like matter*, where the energy density is given by the rest mass density ρ_0, i.e. $\epsilon = \rho_0 c^2$, and $p = 0$.

We use this energy–momentum tensor, to derive the NEWTONian approximation of GRT, and to justify the proportional constant $8\,\pi\, G/c^4$ in the field Eqs. (1057).

In this case, the metric g_{ik} represents a small deviation from the MINKOWSKI metric and we write

$$g_{ik} = \eta_{ik} + h_{ik}\ . \tag{1061}$$

We suppose that the perturbations h_{ik} are much smaller than unity (the components of the MINKOWSKI metric), and we neglect all products of these terms in the evaluation of the curvature tensor,

$$|h_{ik}| \ll 1\ . \qquad \text{Linear approximation} \tag{1062}$$

The trace of the energy–momentum tensor T_{ik} is given by the rest mass density multiplied with c^2 and we get

$$\begin{aligned}
g^{ik} E_{ik} \quad &\longrightarrow \text{lin}\ \eta^{ik} R_{ik} - \tfrac{1}{2}\eta^{ik}\eta_{ik} R &&= \frac{8\,\pi\, G}{c^4}\, \eta^{ik} T_{ik}\ ,\\
& R - 2\,R &&= \frac{8\,\pi\, G}{c^4}\, \rho_0\, c^2\ ,\\
& R &&= -\frac{8\,\pi\, G}{c^2}\, \rho_0\ ,\\
R_{ik} - \frac{1}{2} g_{ik} R &\longrightarrow \text{lin}\ R_{ik} + \frac{1}{2}\,\eta_{ik}\, \frac{8\,\pi\, G}{c^4}\, \rho_0\, c^2 &&= \frac{8\,\pi\, G}{c^4}\, T_{ik}\ ,\\
& R_{ik} &&= \frac{8\,\pi\, G}{c^4}\left(T_{ik} - \frac{1}{2}\,\eta_{ik}\, T\right)
\end{aligned}$$

and therefore in linear approximation

$$R_{00} = \frac{4\,\pi\, G}{c^2}\, \rho_0\ . \tag{1063}$$

For the comparison with NEWTON's theory, we omit in Eq. (1063) the derivative $\dfrac{\partial}{\partial x_o} = \dfrac{1}{c}\dfrac{\partial}{\partial t}$ in view of the factor $\dfrac{1}{c}$ and also all nonlinear terms. This means we look for the static linear approximation of Eq. (1063). We take into account the definition of the CHRISTOFFEL symbols (1041) and find with Eqs. (1045) and (1053)

$$\left.\begin{aligned}
R_{00} &= \partial_a\, \Gamma^a_{00} \qquad = \frac{1}{2}\partial_a\, \eta^{ai}\,(\partial_0 g_{0i} + \partial_0 g_{i0} - \partial_i g_{00})\\
&= \frac{1}{2}\partial_a\, \eta^{ab}(-\partial_b g_{00})\\
&= \frac{1}{2}\Delta g_{00}\ . \qquad\qquad a, b = 1, 2, 3\ .
\end{aligned}\right\}$$

With Eq. (1061) we write $g_{00} = 1 + h_{00}$, and it follows

$$\Delta h_{00} = \frac{8\,\pi\, G}{c^2}\, \rho_0\ . \tag{1064}$$

With the identification of h_{00} with the gravitational potential Φ and taking the rest mass density ρ_0 as density in NEWTON's theory,

$$h_{00} = \frac{2}{c^2}\,\Phi \tag{1065}$$

we get the LAPLACE Eq. (1029) of NEWTON's gravitation theory.

2 Properties of EINSTEIN's Equations

The field equations of General Relativity Theory are at first ten nonlinear partial differential equations for the metric g_{ik}. Due to the four BIANCHI identities (1056), only six of the equations are independent. It is then required that the covariant divergence of the energy–momentum tensor T_{ik} on the right-hand side of Eq. (1057) vanishes identically,

$$\frac{DT^{ik}}{Dx^k} \equiv \frac{\partial T^{ik}}{\partial x^k} + \Gamma^i_{kl} T^{kl} + \Gamma^k_{kl} T^{il} = 0\,. \tag{1066}$$

This represents a translation of the energy–momentum conservation of the Special Relativity Theory to curved space-time, and it is a condition for the solutions of the EINSTEIN equations.

In the GRT, the gravitational field is described by geometry. On the one hand, the mass distribution, or more general, the energy–momentum tensor determines the metric g_{ik} of the RIEMANN space. A *small structureless mass* moves in the field of other *large masses* on geodesics, Eq. (1043, 1044), s. also Problem 50. The geodesics of massive bodies describe in lowest order approximation for small velocities and weak fields the motion according to NEWTON's second axiom in the classical space-time under the influence of the classical gravitational force, cp. Problem 51. This principle of motion on geodesics has been added to the field equations of GRT in the same way as one adds in NEWTON's theory the second axiom to the independent POISSON equation.

Only by the end of 1930th, it became finally clear that this additional postulate is unnecessary. EINSTEIN, INFELD and HOFFMANN presented an algorithm with which they showed that the equations of GRT can only be solved, when a given mass moves in a certain way: The *correct* motion of the masses are the condition that the field equations are solvable. To this end, one evaluates the field equations in a series expansion for approximation in terms of one over the speed of light $1/c$. Then the classical mechanics follows in lowest order. Here, it becomes clear that the motion of the masses according to NEWTON's mechanics is the condition that there is a solution to the field equation in next order. Parallel to the authors above, FOCK has presented a similar procedure for getting the equations of motions starting from Eq. (1066).

In Problem 51, Eq. (1593), it is derived how the formula for the covariant divergence can be simplified. There we considered a three-dimensional space $g \equiv \det g > 0$. EINSTEIN's equations generalise the MINKOWSKI space with $g \equiv \eta < 0$. Instead of the expression $\sqrt{g}$ in Eq. (1593), now there must appear $\sqrt{-g}$. The energy–momentum tensor T^{ik} fulfils equivalent to Eq. (1066) the relation

$$\frac{DT^{ik}}{Dx^k} \equiv T^{ik}_{\ ;k} = \frac{1}{\sqrt{-g}} \frac{\partial}{\partial x^k} \left(\sqrt{-g}\, T^{ik}\right) + \Gamma^i_{kl}\, T^{kl} = 0 . \tag{1067}$$

Now we consider the energy–momentum tensor for a point particle. We multiply it with $\sqrt{-g}$ and get the following condition for the motion of the particle in the gravitational field described by the metric $\mathbf{g}$ with the CHRISTOFFEL symbols Γ^i_{kl},

$$\frac{\partial}{\partial x^k} \left(\sqrt{-g}\, T^{ik}\right) + \sqrt{-g}\Gamma^i_{kl} T^{kl} = 0 . \tag{1068}$$

The integration of this equation over a volume including the particle under consideration then determines its motion in the gravitational field. We shall demonstrate how the motion of a test particle in lowest order approximation for weak gravitational fields and slow motion leads to NEWTON's equation of motion (1029) in a gravitational field.

A weak gravitational field means that the deviations of the metric g_{ik} from the MINKOWSKI metric η_{ik} in Eq. (1030) are small. These differences are written in terms of powers of $1/c^2$,

$$g_{ik} = \eta_{ik} + O\left(\frac{1}{c^2}\right) . \tag{1069}$$

In particular, it holds according to Eq. (1065), where we only wrote the first correction term,

$$g_{00} = 1 + \frac{2}{c^2}\,\Phi . \tag{1070}$$

The test particle has a rest mass m_0, and its motion is given by

$$\mathbf{x} = \mathbf{x}_0(t) , \quad \mathbf{u} = \frac{d\mathbf{x}_0(t)}{dt} . \tag{1071}$$

A point-like mass m_0 at position $P(x_0, y_0, z_0)$ can be described by a spatial mass density ρ_0,

$$\rho_0 = \rho_0(x, y, z) = m_0\, \delta(x - x_0)\, \delta(y - y_0)\, \delta(z - z_0) \ . \text{ Point-like mass density} \tag{1072}$$

Integration of Eq. (1072) over a region $K(P_0)$ containing the point P_0 then leads to

$$\int\limits_{K(P_0)} \rho_0(x, y, z)\, dx\, dy\, dz = m_0 \ , \tag{1073}$$

using the properties of the δ function. The energy–momentum tensor of the particle is then written as

$$T^{ik} = \rho_0 \, u^i \, u^k \quad \text{with} \quad u^i = \frac{dx^i}{d\tau}\,, \quad \tau = \frac{ds}{c}\,. \tag{1074}$$

Here τ is the eigentime. This is the time that is measured by an observer moving with the particle. When in first approximation the gravitational field is neglected, we can use the special-relativistic time dilatation (24), $d\tau = dt\sqrt{1 - v^2/c^2}$.

With indices $a, b, \ldots$ for the coordinates 1, 2, 3 we evaluate in Eq. (1068) the components T^{ak}, employ Eq. (1074), and integrate over a finite volume $K(P_o)$ containing the particle,

$$\int\limits_{K(P_0)} \frac{\partial}{\partial x^k}\left(\sqrt{-g}\,\rho_0\,u^a\,u^k\right) dxdydz + \int\limits_{K(P_0)} \sqrt{-g}\,\Gamma^a_{kl}\rho_0\,u^k\,u^l dxdydz = 0\,. \tag{1075}$$

In the first integral, the three terms with $k = 1, 2, 3$ form a divergence that can be translated into a surface integral using GAUSS law. The surface integral, however, vanishes since only in the interior $\rho_o \neq 0$, hence

$$\int\limits_{K(P_0)} \frac{\partial}{\partial x^k}\left(\sqrt{-g}\,\rho_0\,u^a\,u^k\right) dxdydz = \int\limits_{K(P_0)} \frac{\partial}{\partial x^0}\left(\sqrt{-g}\,\rho_0\,u^a\,u^0\right) dxdydz\,. \tag{1076}$$

In the lowest order, we get for the first term in Eq. (1075), when we observe that in this approximation $u^0 = c$, $dx^0 = cdt$,

$$\int\limits_{K(P_0)} \frac{\partial}{\partial x^k}\left(\sqrt{-g}\,\rho_0\,u^a\,u^k\right) dxdydz \;\underset{\text{lin}}{\longrightarrow}\; m_0 \frac{du^a}{dt}\,. \tag{1077}$$

This is nothing else than NEWTON's inertia term.

In the second integral of Eq. (1075), we have to take into account the factors $u^k\,u^l$ in the linear approximation only the terms $k = l = 0$ since $u^a u^b \ll u^0 u^0$. Furthermore, we replace $\sqrt{-g}$ by unity. In the CHRISTOFFEL symbols, there remain only the components

$$\Gamma^a_{00} = \frac{1}{2} g^{ab}(-g_{00,b} + g_{b0,0} + g_{0b,0})\,.$$

In linear approximation, we have to consider only the terms $g^{11} = g^{22} = g^{33} \underset{\text{lin}}{\longrightarrow} -1$, so that

$$\Gamma^a_{00} = \frac{1}{2}(g_{00,a} - g_{a0,0} - g_{0a,0}).$$

The non-diagonal terms of g_{ik} can be neglected against g_{00}, and we get

$$\Gamma^a_{00} \;\underset{\text{lin}}{\longrightarrow}\; \frac{1}{2} g_{00,a}\,. \tag{1078}$$

With Eq. (1065) and approximating u^0 by c, we obtain finally

$$\int\limits_{K(P_0)} \sqrt{-g}\,\Gamma^a_{kl}\rho_0\,u^k\,u^l dxdydz \;\underset{lin}{\longrightarrow}\; m_0 \frac{\partial}{\partial x^a}\Phi\,. \tag{1079}$$

Together we get from Eqs. (1075), (1077), and (1079), bringing the second integral on the right-hand side,

$$m_0 \frac{du^a}{dt} = -\, m_0 \frac{\partial \Phi}{\partial x^a} \quad \text{or} \quad m_0 \frac{d\mathbf{u}}{dt} = \mathbf{F} \,, \quad \mathbf{u} = (u^1, u^2, u^3) \,. \tag{1080}$$

These are the NEWTONian equations of motion (1029) (taking into account Eq. (139)), that are derived as a consequence of the field equations of EINSTEIN's General Relativity Theory. See also Problem 51, where the relation between the equations for a geodesic line and NEWTON's equation of motion is considered. In particular, one finds that the inertial mass does not appear in Eq. (1080), i.e. we come back to GALILEI's principle: *All bodies are falling at the same rate.*

3 Spherically Symmetric Gravitational Fields

3.1 SCHWARZSCHILD *Solution*

Spherically symmetric matter distributions play an important role in the universe, since the gravitational force is isotropic. In case of missing influence of rotation or of magnetic fields, in nature matter forms spherical matter distributions as equilibrium configurations. These are, for example, planets, for which the dominance of its inner gravity is a matter of definition, stars with central nuclear processes as energy reservoir, stars stabilised by degenerated matter as white dwarfs and neutron stars and finally black holes.

An ansatz for a spherically symmetric metric follows naturally from a generalisation of the MINKOWSKI metric in spherical coordinates, s. Problem 49:

$$ds^2 = e^{\nu} c^2 dt^2 - e^{\lambda} dr^2 - r^2 (d\theta^2 + \sin^2\theta d\phi^2). \tag{1081}$$

There are two free functions $\nu(t, r)$ and $\lambda(t, r)$ that depend on the coordinate time t and the radial coordinate r, which is fixed in such a way that the area of the sphere around the origin at radius r amounts $4\pi r^2$. Furthermore, we require that far from the source $r \to \infty$, the functions tend to zero, i.e. the metric goes over into the MINKOWSKI metric.

From the mixed components of EINSTEIN's equations (for derivation s. Problem 52, we write the partial derivatives in the following as $\partial\lambda/\partial t = \partial_t \lambda$):

$$E^t_r = \frac{\partial_t \lambda}{2r} e^{-\nu} = \frac{8\pi G}{c^4} T^t_r \,. \tag{1082}$$

When no energy currents $\mathbf{s}$ exist, so in particular in vacuum, we find the function λ is time-independent. The time–time and the diagonal spatial components of the field equations read

$$E_t^t = \left(\frac{\partial_r \lambda}{r} - \frac{1}{r^2}\right) e^{-\lambda} + \frac{1}{r^2} = \frac{8\pi G}{c^4} T_t^t \ , \tag{1083}$$

$$E_r^r = -\left(\frac{\partial_r \nu}{r} + \frac{1}{r^2}\right) e^{-\lambda} + \frac{1}{r^2} = \frac{8\pi G}{c^4} T_r^r \ . \tag{1084}$$

We consider at first the case of vacuum, i.e. we solve the equations for the range outside of a spherically symmetric source of the gravitational field. Subtraction of both equations leads to the condition

$$\partial_r \nu + \partial_r \lambda = 0 \ . \tag{1085}$$

Then the sum $\nu + \lambda$ is a pure function of time. This function can included in a redefinition of the time coordinate, so that the sum vanishes, $\nu + \lambda = 0$. Therefore, with the function λ, there is also ν independent on the time. We have proven the important BIRKHOFF *theorem*, telling that spherically symmetric vacuum fields are static, when there is supposed that the metric goes over to the MINKOWSKI form at large distances to the source, and the coordinate t is time-like. Spherically, symmetric gravitational fields cannot represent time-dependent gravitational waves.

With $\nu + \lambda = 0$, Eqs. (1083) and (1084) are equivalent, and they lead to an ordinary differential equation

$$r\frac{d\nu}{dr} + 1 = e^{-\nu} \ \rightarrow e^{\nu} = 1 + \frac{\text{const}}{r} \ .$$

with solution

$$g_{tt} = e^{\nu} = 1 - \frac{2\Phi}{c^2} = 1 - \frac{2GM}{rc^2} \ . \tag{1086}$$

The (tt)-component of the metric is determined by the NEWTONian potential in the outside region of a spherically symmetric mass distribution. The mass M represents the gravitating mass known from NEWTON's theory. The so-called SCHWARZSCHILD radius r_g or *gravitational radius* is given by const $= r_g = 2GM/c^2$. The spherically symmetric solution of the EINSTEIN equations (1057) reads

$$ds^2 = \left(1 - \frac{r_g}{r}\right) c^2 dt^2 - \frac{1}{1 - \frac{r_g}{r}} dr^2 - r^2 (d\theta^2 + \sin^2\theta d\phi^2) \ . \tag{1087}$$

This is the famous SCHWARZSCHILD solution (SCHWARZSCHILD, 1916) that describes the outer region of a non-rotating mass M. The mass is the only parameter in the solution, and all spherically symmetric gravitational fields in vacuum are described by a member of this solution family. Most important is the singularity of the solution at the SCHWARZSCHILD radius $r = r_g$. There the g_{tt}-component of the metric vanishes, and the component g_{rr} goes to infinity.

In general, the g_{tt}-component of the metric describes the rate of the time run with respect to a remote observer ($r \rightarrow \infty$): The eigentime $d\tau$ of an observer in the SCHWARZSCHILD field differs from the time that a distant observer determines as SCHWARZSCHILD time dt; it holds $d\tau = \sqrt{g_{tt}}\, dt$. The eigentime τ of the body in the gravitational field runs more slowly, and it runs infinitely slowly for approximating to the SCHWARZSCHILD radius. The consequence is that all signals from the SCHWARZSCHILD radius are infinitely stretched in the distant time t, or the signals are infinitely redshifted. A signal, sent out from the observer in the gravitational field, proceeds with the speed of light, i.e. it holds $ds = 0$. Therefore, the distant observer gets notice from the motion of an observer approaching the SCHWARZSCHILD area only at stretched definite time, $cdt = dr/(1 - r_g/r)$. This effect is called *gravitational red shift*. A consequence of this time dilatation will be discussed in Sects. 3.2 and 3.4.

Not only the run of time is delayed near to the SCHWARZSCHILD radius, but also the spatial geometry is distorted: The radius r determines the surface of a sphere around the mass. However, the radial separation $dl = dr/\sqrt{1 - r_g/r}$ is larger than dr, and the space has a positive curvature, analogous to the surface of a sphere in two-dimensional space.

A simple evaluation proves that the singularity of the spatial metric is a consequence of the chosen coordinates: Let us define a coordinate ρ by $r = \rho[1 + r/(4\rho)^2]$. Then the SCHWARZSCHILD metric transforms to a form, where the spatial diagonal elements of the metric are equal; we call this isotropic coordinates,

$$ds^2 = \left(\frac{1 - \frac{r_g}{4\rho}}{1 + \frac{r_g}{4\rho}}\right)^2 c^2 dt^2 - \left(1 + \frac{r_g}{4\rho}\right)^4 (d\rho^2 + \rho^2 d\omega^2)\,, \tag{1088}$$

where we denoted the unchanged angular term of the metric by $d\omega^2 = d\theta^2 + \sin^2\theta d\phi^2$. Obviously, we get a metric that is regular at the SCHWARZSCHILD radius $\rho = r_g/4$. For the understanding, it is important to note that the g_{tt} component remains zero. Hence, the most important physical consequence remains unchanged: At the SCHWARZSCHILD area, the time runs infinitely slowly as seen by a distant observer. We shall see in Sect. 4, that the SCHWARZSCHILD solution can be extended to the region $0 < r \leq r_g$, as long as we are still outside of the mass distribution, e.g. for a point mass. However, to this aim, we have to change the coordinates again, since in the central range inside the SCHWARZSCHILD radius, there do not exist static spherically symmetric solutions.

As vacuum solution around spherically symmetric stars, the SCHWARZSCHILD radius has no direct meaning, since all stellar radii are much larger than the SCHWARZSCHILD radius. For the Earth, it amounts about 1.2 cm, and for the Sun slightly below 3 km. The last value is only about two parts in a million of the solar radius. But nevertheless, the SCHWARZSCHILD geometry describes an influence on the run of time and the spatial geometry determining the motion of test bodies around objects as our Sun.

We use the equations of geodesics, Eqs. (1043) and (1044) for the evaluation, where we introduce a parameter μ, coinciding with the eigentime τ for particles with rest mass, and for rest massless particle representing the affine parameter σ.

In both cases, the derivative for the parameter is denoted by an overdot over the corresponding variable. We take the general metric (1081) with the relation $\lambda = -\nu$ derived above, and we denote with $\nu' = d\nu/dr$ the spatial derivative of the function $\nu(r)$. There are four components in the equation of geodesics for the four-velocity $(u^i) = (c\dot{t}, \dot{r}, \dot{\theta}, \dot{\phi})$:

$$\ddot{t} = -\nu' \dot{t}\, \dot{r} \,, \tag{1089}$$

$$\ddot{r} = -\frac{1}{2} c^2 \nu' e^{2\nu} \dot{t}^2 + \frac{1}{2} \nu' \dot{r}^2 + r\, e^{\nu} \dot{\theta}^2 + r \sin^2\theta\, e^{\nu} \dot{\phi}^2 \,, \tag{1090}$$

$$\ddot{\theta} = -\frac{2}{r} \dot{r}\, \dot{\theta} + \sin\theta \cos\theta\, \dot{\phi}^2 \,, \tag{1091}$$

$$\ddot{\phi} = -\frac{2}{r} \dot{r}\, \dot{\phi} - 2 \cot\theta\, \dot{\theta}\dot{\phi} \,. \tag{1092}$$

Equation (1091) can be immediately solved by $\theta = \pi/2 = \text{const.}$ The test body moves in a plane around the spherically symmetric mass distribution with the plane passing through its centre. The second angular Eq. (1092) can then be written as $d(r^2\dot{\phi})/d\lambda = 0$; hence $r^2\dot{\phi} = l = \text{const.}$ This corresponds to a property of the classical Kepler problem in Newtonian mechanics, the conservation of the specific (measured with respect to a unite mass) angular momentum l. This results from the spherical symmetry of the metric. The first component of the equation of geodesics (1089) can be once integrated, and we can write

$$(\ln \dot{t} + \nu)^{\cdot} = 0 \quad \Rightarrow \quad \dot{t} e^{\nu} = f = \text{const.} \tag{1093}$$

This relation implies the energy conservation for the motion in the static space-time. After multiplication with $2\dot{r}\, e^{-\nu}$, the second component (1090) leads to

$$\frac{d}{d\mu}\left(\dot{r}^2 e^{-\nu} + \frac{l^2}{r^2} - c^2 f^2 e^{-\nu}\right) = 0. \tag{1094}$$

The terms in parenthesis represent the square of the four-velocity of the particle multiplied with c^2, the corresponding integration constant (written as $-k$) determines its normalisation, $k = c^2$ for particles with rest mass and $k = 0$ for photons. The motion of the test mass is already determined by its conserved quantities energy and angular momentum. After inserting the Schwarzschild metric $e^{\nu} = 1 - 2GM/(rc^2)$, we get

$$\frac{\dot{r}^2}{2} - k\frac{GM}{c^2 r} + \frac{l^2}{2r^2} - \frac{l^2 GM}{c^2 r^3} = \frac{c^2 f^2 - k}{2} = \mathcal{E}. \tag{1095}$$

The equations of motion have the same form as the Newtonian equations of motion in an effective potential

$$V_{eff} = -k\frac{GM}{c^2r} + \frac{l^2}{2r^2} - \frac{l^2GM}{c^2r^3} \tag{1096}$$

known from the treatment of the KEPLER problem, but with an additional term $\propto r^{-3}$ caused by the GRT. The term $\mathcal{E}$ in Eq. (1095) represents the total energy of the particle. In Fig. 3, there is shown the potential for particles with non-vanishing rest mass by solid lines, and for photons by dashed lines.

Obviously, the orbits of test particles in the SCHWARZSCHILD metric show a significant different behaviour compared to the classical KEPLER problem in NEWTON's mechanics. Both for particles with rest mass and for photons, the effective potential becomes negative at approach to the centre. The influence of the centrifugal term l^2/r^2 is surmounted by the r^{-3}-term, when the particle approaches to the centre. Particles with rest mass fall even for $\mathcal{E} = 0$ (hence with zero velocity far away from the centre) on the origin $r = 0$, when the angular momentum is below a critical value, $l < 4cr_g$. For values of the angular momentum $l > \sqrt{3}cr_g$, there are minima and maxima in the effective potential, corresponding to stable and unstable circular orbits. In the limit $l = \sqrt{3}cr_g$, there is a *last stable circular orbit* at radius $r = 3r_g$. For smaller radii, there does not exist any circular orbit, and the particle must fall onto the central mass. The specific binding energy of this orbit amounts to 0.06. A particle on this orbit is gravitationally bound by an amount of 6% of its rest mass.

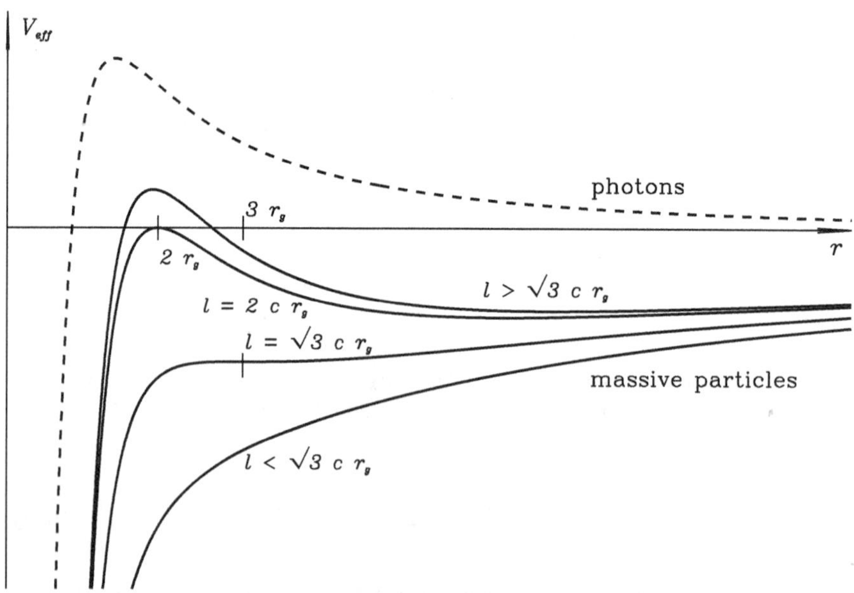

Fig. 3 Effective potential for the motion of particles with finite rest mass (solid lines) with specific angular momentum smaller, equal, and larger than the critical value $l_{crit} = \sqrt{3}cr_g$, and for photons (dashed line).

This amount of energy is available for emission, when it gets in the influence of the gravitating mass. Such a behaviour characterises the inflow of matter onto black holes, which is observed in different phenomena in the universe, the energy release during mass exchange in double stars with a black hole, or in supermassive black holes in the central region of galaxies.

The special case of a radial infall onto the mass centre can be easily solved, since it leads to a vanishing angular momentum $l = 0$: The equation of motion then reads as $\dot{r}^2 - r_g/r = \mathcal{E}$, and under the assumption of a vanishing velocity at the initial radius r_0 follows $\dot{r}^2 = r_g/r - r_g/r_0$ with the parametric solution

$$r = \frac{r_0}{2}(1 - \cos\eta) \quad \text{and} \quad \tau = \frac{r_0}{2c}\sqrt{\frac{r_0}{r_g}}(\eta - \sin\eta) \ , \tag{1097}$$

a parameter range $\eta = [0, \pi]$ with arrival at the centre for $\eta = \pi$ and therefore an infall time $\tau_{inf} = (\pi r_0/2c)\sqrt{r_0/r_g}$. Obviously, the infall occurs far from the Schwarzschild radius just as the normal free fall, relatively slowly. But near to the Schwarzschild radius the time is given approximately by a few light travel times, when measured by a clock carried by the infalling observer.

For photons, there do not exist stable circular orbits. For energies below a critical value, photons will be captured by the central mass. For larger energies, the orbits reach a minimal separation, and then they will move away from the central mass. The orbits of photons are curved in three-space, near to the critical energy, the photons can circulate a few times around the central mass before they leave again the attracting mass. The direction of scattering is sensitively dependent on the angular momentum.

It is interesting that the Schwarzschild radius does not play a special role in the effective potential. Particles and photons can freely pass through the surface $r = r_g$ in direction of the mass centre. This question is further discussed in Sect. 3.4.

3.2 *Classical Tests of General Relativity*

The General Relativity Theory (GRT) becomes important for the description of the motion of particles in the field of larger masses, and for velocities that are not anymore small with respect to the speed of light. This concerns mostly astronomical phenomena, for example, the motion of planets in the solar system, the motion of stars in double-star systems, the dynamics of stellar clusters, and the motion of stars in galaxies. Einstein's theory has crucial influence on the evolution of the whole universe and the origin and evolution of cosmic structures as galaxies and galaxy clusters. After its creation, GRT was considered as mathematical complicated gravitation theory with high elegance, but small empirical foundation. The last points concern especially the classical tests of GRT for the motion of planets in the solar system, and the observation of them by light propagation. Today, the situation has been strongly changed, already in the range of weak gravitational fields due to the high precision of our measuring devices. Also, satellite experiments have contributed to an improvement of the observational basis. The GRT is now extremely well justified by experiments.

Pendulum Experiments and Torsion Scale

As basis of GRT, we have discussed in Sect. 1 EINSTEIN's equivalence principle. Especially, the weak equivalence principle, the equality of heavy and inertial mass, has been already discussed before EINSTEIN as a significant property of gravity. Already, GALILEI and NEWTON have performed experiments with pendulums with weights of different materials to test the independence of the frequency on the kind of material. The accuracy of these experiments reaches about one promille. By the end of nineteenth century, the physicist EÖTVÖS invented his famous torsion scale, where the ratio of centrifugal force and gravitational force of Earth was measured with high accuracy, of about 10^{-8}, and there was no dependency on the composition of the studied bodies. With still higher accuracy, DICKE 1964, BRAGINSKY 1972, SHAPIRO 1976 and ADELSBERGER 1999 increased the accuracy of the proof of the equivalence principle on relative values of 10^{-12} up to 10^{-13}. So they could test the influence of the composition and different binding energies of molecules for their reaction on gravitational forces. Due to the fundamental meaning of the equivalence principle, the satellite missions STEP (Satellite Test of the Equivalence Principle), MICROSCOPE and GALILEO GALILEI are under construction with the aim to test the accuracy of the equivalence principle up to a relative factor of 10^{-18}. Preliminary tests are performed in the 110 m tower for gravity experiments in Bremen, Germany. Considering the free fall of material of different compositions, the free path of satellites in the Earth gravitational field independent of its constituents approves the validity of the equivalence principle.

Gravitational Redshift

A direct consequence of GRT, and in particular of the equivalence principle, is the dependence of the run time of a clock from the local gravitational potential: The eigentime τ of an experimenter is connected with the coordinate time t of a given reference system by $d\tau = \sqrt{g_{tt}}\, dt$. Correspondingly, the clocks in the SCHWARZSCHILD geometry will run as

$$d\tau = dt \cdot \sqrt{1 - r_g/r} \tag{1098}$$

hence slower than a clock far away in the gravitational-free space with the there defined time t. At first, this effect was planned to test in the spectra of the Sun—to this aim, there was built the so-called EINSTEIN tower of the Astrophysikalisches Observatorium Potsdam, see Fig. 4—however without success since the turbulent plasma motions on the surface of the Sun are larger than the redshift of the observed spectral lines.

Fig. 4 Solar observatory EINSTEIN tower of the Leibniz-Institut für Astrophysik Potsdam.

A measurement of this effect was performed successfully by ADAMS in 1925 by a measurement of the gravitational redshift in the spectrum of the compact companion of Sirius, called Sirius B. This star is a white dwarf, s. Sect. 3.3. The redshift of the spectral lines was determined with an accuracy of $\Delta\lambda/\lambda = 2 \cdot 10^{-4}$. In 1959, POUND and REBKA also determined the frequency and wavelength shift by an experiment on Earth in the gravitational field leading to an acceleration $g = GM/r_E^2$ with the Earth radius r_E. In the MÖSSBAUER spectroscopy, it is possible to produce and measure very sharp spectral lines, so that a redshift of $\Delta\lambda/\lambda = gh/c^2 = 2,5 \cdot 10^{-15}$ could be established with a one-percent accuracy.

Today, the gravitational redshift plays an important role in orbital parameters of artificial satellites around the Earth and for the Global Positioning System (GPS), where atomic clocks on a couple of satellites are used for sending precise time signals. From the relative arrival time on Earth, it is possible to measure the position with an accuracy of a few metres. For an orbital height of the GPS satellites of 20 000 km, the run-time differences amount about 100 ns. It is important to take into account both special-relativistic and general-relativistic effects. They would cause position errors of about 100 m. The GRT is included in the programming of the GPS, each user therefore performs tests of the Relativity Theory. The gravitative redshift was the first prediction or classical test of EINSTEIN's theory established on the basis of the equivalence principle. Therefore, it is crucial for the GRT until the presence.

Perihel Shift

Two further classical tests of the GRT concern the motion of planets around the Sun, and the light propagation in the SCHWARZSCHILD field of the Sun. We derived the orbit of test bodies by means of the effective potential derived in Sect. 3.1. Now at first we consider particles with rest mass ($k = 1$) and use as new variable the inverse radius $u = 1/r$. Then we can write the conservation of angular momentum as $\dot{\phi} = lu^2$. This relation is taken for changing the time derivative $\dot{r}$ into a derivative of u for the angle ϕ ($u' = du/d\phi$), $\dot{r} = -\dot{u}/u^2 = -u'\dot{\phi}/u^2 = -lu'$. We get a relation for u'^2 that can be derived for ϕ:

$$l^2u'^2 = f^2 - (1 - r_g u)(1 + l^2u^2) \quad \Rightarrow \quad u'' + u = \frac{r_g}{2l^2} + \frac{3}{2}r_g u^2 \,. \tag{1099}$$

In NEWTON's approximation, the second term $\propto u^2$ on the right-hand side is missing, and we get an ordinary oscillator equation with the solution u_0 (the index means the zeroth NEWTONian approximation) as harmonic function $u_0 = r_g(1 + e\cos\phi)/2l^2$, i.e. $r = p/(1 + e\cos\phi)$ with the integration constant e. This constant represents for $e < 1$ the eccentricity of a KEPLER ellipse with parameter $p = 2l^2/r_g$ and the large semi-axis $a = p/(1 - e^2)$. For $e \geq 1$, there follow analogously unbound orbits on a parabola or a hyperbola.

The relativistic approximation is now determined as perturbation $u = u_0 + u_1$, where we use NEWTON's solution in the relativistic correction term $\propto u^2$:

$$\begin{aligned} u_1'' + u_1 &= \frac{r_g}{2l^2} + \frac{3}{2}r_g u_0^2 \quad \Rightarrow \\ u_1 &= \frac{3r_g^3}{8l^4}\left[1 + e\phi\sin\phi + e^2\left(\frac{1}{2} - \frac{1}{6}\cos 2\phi\right)\right]. \end{aligned}$$

This solution contains the term $\propto e^2$ with periodic orbital perturbations that are very small. The term $\propto e$, however, grows proportional with ϕ during each circulation of the planet around the Sun. We call such an additive effect a secular change. It follows

$$\begin{aligned} u = u_0 + u_1 &= \frac{r_g}{2l^2}\left[1 + e\left(\cos\phi + \frac{3r_g^2}{4l^2}\phi\sin\phi\right)\right] \\ &\approx \frac{r_g}{2l^2}\left[1 + e\cos\left\{\left(1 + \frac{3r_g^2}{4l^2}\right)\phi\right\}\right], \end{aligned} \tag{1100}$$

where we used in the last step a TAYLOR expansion. The angle change during each orbit amounts to

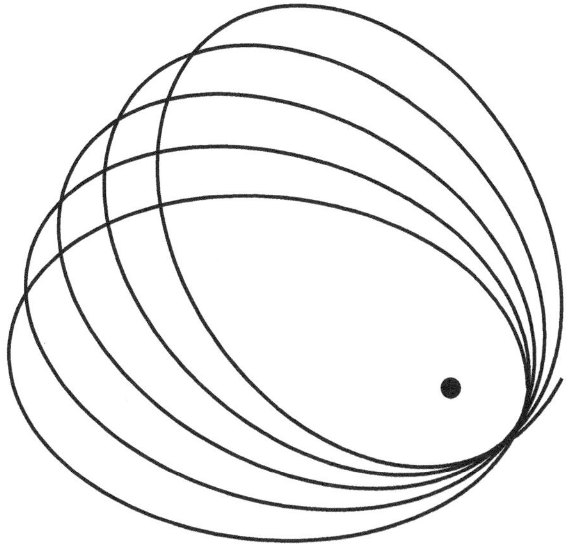

Fig. 5 Rotation of the orbit of a test body moving around the mass centre at the thick black dot, sitting in one pole of the ellipse.

$$\Delta\phi_p = 2\pi\frac{3r_g^2}{4l^2} = 3\pi\frac{r_g}{p} = \frac{6\pi GM}{ac^2(1-e^2)}\,. \tag{1101}$$

This is the famous rotation of the KEPLER ellipse in the SCHWARZSCHILD field, denoted historically as shift of perihelion, as shift of the position of the nearest point on the orbit with respect to the Sun (Fig. 5). This shift amounts 43 arcseconds in 100 years for the planet nearest to the Sun, for Mercury, for Venus 8 arcseconds and for the Earth 5 arcseconds. The last two values are very difficult to measure. The effect of the perihelion shift was already known by the end of nineteenth century, and it remained at this time without explanation. The explanation by the GRT, just derived immediately after the completion of the theory, was considered by EINSTEIN as one of his most important successes. It should be added here that this effect is very tiny, and it is found only after taking into account the orbital influence of the other planets in the solar system on Mercury, being a factor 10 larger than the GRT effect.

In 1974, HULSE and TAYLOR observed the double stars PSR 1913+16 consisting of a pulsar (a neutron star, see Sect. 3.3) and a compact companion. The pulsar emits periodically in the radio range, forming a natural clock. For the double-star system, a rotation of the periastron by 4.2 grad per year and still other effects of GRT were measured with high precision. This stellar system (there are now known at least two similar objects) has been developed to a real test laboratory for EINSTEIN's gravitation theory. We will further discuss this system in Sect. 4.

Light Deflection

The third classical EINSTEIN effect is connected with the light propagation in the SCHWARZSCHILD space-time. Hence, we use the equation for light-like geodesics. The form of the orbit follows in zeroth order without the u^2-term:

$$u'' + u = \frac{3}{2} r_g u^2 \quad \Rightarrow \quad u_0 \equiv \frac{1}{r_0} = \frac{1}{b} \sin \phi . \tag{1102}$$

This is the equation of a straight line with an impact parameter b against the mass centre. The first relativistic approximation then leads to a differential equation of second order as above

$$u_1'' + u_1 = \frac{3r_g}{2b^2} \sin^2 \phi \quad \Rightarrow \quad u = u_0 + u_1 = \frac{1}{b} \sin \phi + \frac{r_g}{2b^2}(1 + \cos \phi)^2 . \tag{1103}$$

The photon approaches the mass centre for parameter $\phi = \pi$ with transverse velocity $u_1 = 0$. For $\phi \to 0$, the orbit is given by

$$0 = \frac{1}{b} \sin \phi + \frac{r_g}{2b^2}(1 + \cos \phi)^2 . \tag{1104}$$

We approximate the term $(1 + \cos \phi) \approx 2$ for $\phi \to 0$, and get a deflection angle (here we approximate for small deflection angles)

$$\phi_a \approx \sin \phi_a = \frac{2r_g}{b} = \frac{4GM}{bc^2} . \tag{1105}$$

The photon suffers a small deflection ϕ_a. This spectacular prediction of EINSTEIN's theory and its approval led to a wide attention for the new relativistic gravitation theory.

For stars observed near to the solar disk, the deflection angle amounts to about 1.75 arcseconds, which represents a small value, of the same order as the apparent resolution of stellar positions in the Earth atmosphere. According to this deflection, the positions of stars appear to be more remote from the solar disk. This effect was at first measured during a total solar eclipse in 1919 by A. EDDINGTON. Today, the astrometric accuracy for measuring stellar position increased enormously, so that the light deflection can also be measured at larger angular separations from the Sun. The light deflection leads to the phenomenon of gravitational lensing that will be discussed in Sect. 5.2.

Time Delay and Precession of Gyroscopes

An additional effect, analogous to the light deflection, represents the run-time delay of light signals in the SCHWARZSCHILD space-time due to the curvature of space and the gravitational time dilatation. The gravitational field acts like an optical thick medium with refraction index $n = g_{tt}/c^2$, leading to a change of the light propagation by $\delta v = (n-1)c$ and to a time delay on the light propagation path from source 1 to the detector 2,

$$\Delta t = \int_1^2 \frac{2GM}{c^3} dl \,. \tag{1106}$$

This time delay has been proposed by SHAPIRO in 1964 as fourth EINSTEIN effect. It was repeatedly observed, for example, in the observation of reflections of radar signals from the surface of Mercury and Venus. The effect amounts at maximum 10^{-4} s in the solar system. All measurements confirm completely EINSTEIN's theory.

A further effect of the GRT concerns the direct approval of the space curvature in the spherically symmetric space-time. For the introduction of the curvature tensor (1046), it was discussed that the parallel transport of a vector along a closed path let an angle deficit open, cp. Fig. 1, right panel. The experiment Gravity Probe B (GP-B) realised in 2004 such a path with a satellite on a polar orbit around the Earth in a height of 650 km. The satellite carries a gyroscope with a stable parallel-transported angular momentum vector that changes its direction during each orbit by a small amount. This so-called geodesic precision amounts about 6.6 arcseconds per year. It was confirmed with about 1% accuracy.

The rotation of Earth distorts and tracks the local inertial system and causes a small precession of the gyroscope by about 0.054 arcseconds per year, a consequence of the GRT known as THIRRING-LENSE effect. Also, this precession was measured by the satellite GP-B with about 5% accuracy, confirming the predictions of GRT. Within the solar system, EINSTEIN's theory is now well confirmed by a wide range of effects.

3.3 Relativistic Stars

The General Relativity Theory has influence on the explanation of the structure and the stability of compact stars after the exhaustion of its inner energy supply. We consider spherically symmetric stars. There, we derive the so-called inner SCHWARZSCHILD metric from 1916 as solution of EINSTEIN's equations for a source as ideal fluid with energy–momentum tensor

$$T_{ik} = \rho c^2 u_i u_k + p (u_i u_k - g_{ik}) \,. \tag{1107}$$

We assume at first a constant density ρ, which leads to an analytical solution. The pressure p follows from the field equations as seen during the evaluation. We are looking for static solutions, i.e. we assume a time-independent metric of the form of Eq. (1081). The four-velocity has only a time component u_t. Later, we shall also discuss more realistic stellar models with pressure fulfiling an equation of state $p = p(\rho)$ for cold degenerate matter. It represents a so-called FERMI fluid with pressure caused by PAULI's exclusion principle. It characterises fermions, either electrons or neutrons, that dominate the pressure in compact stars. However, already the SCHWARZSCHILD solution shows the essential physical effect of GRT, the restricted compactness of relativistic stars.

At first, we employ the dynamical equations $T^{ik}_{\ ;k} = 0$. It takes the form $(\rho c^2 + p)\Gamma^i_{\ tt} u^t u^t - \partial_r p g^{ir} = 0$. We write $\partial_r p = p'$ and $\partial_r \nu = \nu'$; only the derivative for r is essential, and we get the energy conservation law $p' + (\rho c^2 + p)\nu'/2 = 0$. We can integrate this relation for a constant density, leading to the relation $p + \rho c^2 = c_1 \exp(-\nu/2)$ with a constant c_1.

We use the mixed co- and contravariant $\binom{t}{t}$-component of EINSTEIN's equation as

$$(\lambda' r - 1)e^{-\lambda} + 1 = -\partial_r(re^{-\Lambda}) + 1 = 8\pi G\rho r^2 c^2 \quad \Rightarrow \quad e^{-\lambda} = 1 - \frac{2Gm}{rc^2}\,, \tag{1108}$$

where we have introduced the mass function $m(r) := 4\pi \int_0^r \rho(r')r'^2 dr'$. This looks analogous to the classical mass definition in spherical coordinates. However, we have to observe that we use in this integral SCHWARZSCHILD's radial coordinate r for the volume element, not the invariant volume element of the SCHWARZSCHILD geometry. In this mass definition, the gravitational binding energy of the mass density is already included by the form of the integral.

The mixed $\binom{r}{r}$-component of the field Eq. (1083) is now subtracted from the $\binom{t}{t}$-component,

$$8\pi G(\rho c^2 + p) = 2e^{-\lambda}\frac{\nu' + \lambda'}{2r} \quad \Rightarrow \quad \sigma^{3/2}\partial_r(\sigma^{-1/2}e^{\nu}) = 4\pi G\rho c_1 r\,, \tag{1109}$$

where we have used at first the solution (1108) for λ and the energy conservation law $\rho c^2 + p$. Furthermore, we introduced the function $\sigma(r) = 1 - 8\pi G\rho r^2/c^2$. The linear differential Eq. (1109) can be integrated in closed form, leading to the second metric component and the pressure and providing a second integration constant c_2,

$$e^{\nu/2} = \frac{3c_1}{2\rho c^2} - c_2\sigma^{1/2} \quad \text{and} \quad p = c_1 e^{-\nu/2} - \rho c^2\,. \tag{1110}$$

The integration constants follow by the continuous transition to the outer SCHWARZSCHILD solution Eq. (1087). The boundary of the star $r = r_0$ follows from the condition of a vanishing pressure, $p(r_0) = 0$. The continuity of the (tt)-component of the metric leads to the relation $\exp(\nu) = 1 - r_g/r_0$ at radius $r = r_0$ with the SCHWARZSCHILD radius r_g of the outer SCHWARZSCHILD metric. We use the abbreviation for the scaled radius $x = r/r_0$ and the ratio of the gravitational radius and stellar radius $x_g = r_g/r_0$ (a value near unity means a compact star). This leads to the constants $c_1 = \rho c^2 \sqrt{1 - x_g}$ and $c_2 = 1/2$. Now the complete inner SCHWARZSCHILD solution (1916) reads

$$ds^2 = \left(\frac{3}{2}\sqrt{1 - x_g} - \frac{1}{2}\sqrt{1 - x_g x^2}\right) c^2 dt^2 - \frac{dr^2}{1 - x_g x^2} - r^2 d\omega^2 , \quad (1111)$$

$$p = \rho c^2 \frac{3\sqrt{1 - x_g x^2} - \sqrt{1 - x_g}}{\sqrt{1 - x_g} - \sqrt{1 - x_g x^2}} .$$

The central pressure grows monotonically with increasing x_g, corresponding to an increase of compactness. The pressure becomes infinite for $3\sqrt{1 - x_g} \to 1$, i.e. it must hold $x_g < 8/9$. The stellar radius fulfils therefore the condition $r_0 > 9r_g/8$. This is a very important result: The GRT restricts the compactness of stars. Later, we will see that this also leads to limited masses of relativistic stars.

In the case of a general equation of state $P(\rho)$ and a variable density, there are no closed analytical solutions of GRT inside of stars. However, one can proceed similarly as in the example above. We can solve the radial metric component (1108) with the same mass function $m(r)$. Inserting this definition in the $\binom{r}{r}$-component of the metric leads to a simple expression for the radial derivative of the function $\nu(r)$, determining the time–time component of the metric,

$$\frac{d\nu}{dr} = \frac{G}{c^2} \frac{m + 4\pi p r^3/c^2}{2r^2 (1 - 2Gm/rc^2)} . \quad (1112)$$

When we insert the dynamical equation $dp = -(\rho c^2 + p)d\nu/2$ into Eq. (1112), we get a differential equation for the pressure very similar to the normal NEWTONian equation of inner stellar structure,

$$\frac{dp}{dr} = -(\rho + p/c^2) \frac{G(m + 4\pi p r^3/c^2)}{r^2(1 - 2Gm/rc^2)} . \quad (1113)$$

It was derived in 1939 by TOLMAN, OPPENHEIMER and VOLKOFF and denoted as *TOV-equation.* It forms the basis for evaluating the structure of relativistic stars as white dwarfs or neutron stars. Already the first numerical integration led to stellar models with maximum masses of 1.4 solar masses. The existence of an upper mass limit is a consequence of the relativistic gravitation theory. It is interesting that the equation of stellar structure (1113) does not contain any explicit metric components or connection coefficients. It generalises the equation of stellar structure in NEWTONian theory. However, the pressure contributes to the passive mass density according to the term $\rho + p/c^2$. It contributes also to the active gravitational mass by $m + 4\pi p r^3/c^2$. Most important is the factor in the denominator $1 - 2Gm(r)/rc^2$ that increases the pressure gradient and leads finally to the instability of high-mass compact stars.

In compact stars with high central densities and pressures, the degeneration pressure of electrons e plays a dominant role. For still much higher densities, the electrons are pressed into the protons p. The matter then consists mainly of neutrons n, formed in the reaction $e + p \rightarrow n + \nu_e$, the inverse to the normal weak β-decay of neutrons. During the formation of such compact stars, the emitted neutrinos ν_e are playing an important role. They can lead to the loss of outer mass layers in Supernovae explosions.

Both electrons and neutrons are fermions. They can occupy each quantum state only one times (two times if the two spin orientations are extra considered). This PAULI principle leads to very simple equations of state, when the thermal motion of the basic constituents can be neglected. We speak of cold femionic matter which becomes important after exhaustion of the nuclear fusion processes inside stars. Per phase space cell d^3xd^3p/h^3 (h is PLANCK's constant), there can sit two fermions. Then adding particles in a volume element leads with necessity to an increase of the momentum. In the end, all states up to a maximum momentum p_F, the FERMI momentum and the corresponding FERMI energy $E_F = \sqrt{m^2 + p_F^2}$ are occupied. Then there follows the particle density n, the pressure p and the energy density ρc^2 (the sum of rest mass and kinetic energy density) after integration over the momenta:

$$n = \frac{2}{h^3}\int_0^{p_F} 4\pi p^2 dp = \frac{8\pi}{3}\frac{p_F^3}{h^3} = \frac{x^3}{3\pi^2\lambda^3}\,, \tag{1114}$$

$$p = \frac{2}{3h^3}\int_0^{p_F} \frac{p^2}{\sqrt{p^2+m^2}} 4\pi p^2 dp = \frac{m\,c^2}{\lambda^3}\Phi(x)\,, \tag{1115}$$

$$\rho c^2 = \frac{2}{h^3}\int_0^{p_F} \sqrt{p^2+m^2} 4\pi p^2 dp = \frac{m\,c^2}{\lambda^3}\Psi(x)\,. \tag{1116}$$

In these equations, we used the relative FERMI momentum $x = p_F/(mc)$ and the COMPTON wavelength $\lambda = \hbar/mc$. The two elementary functions coming from the integrals are given by

$$\Phi = \frac{1}{8\pi^2}\left[x\sqrt{1+x^2}(2x^2/3 - 1) + \ln(x + \sqrt{1+x^2})\right]\,,$$
$$\Psi = \frac{1}{8\pi^2}\left[x\sqrt{1+x^2}(1 + 2x^2) - \ln(x + \sqrt{1+x^2})\right]\,.$$

In case of non-relativistic degenerate matter with $x << 1$, i.e. the FERMI momentum is much smaller than the rest mass energy, it follows $\Phi \approx x^5/(15\pi^2)$, $\Psi \approx x^3/(3\pi^2)$, hence $p = K\rho_0^{5/3}$ (ρ_0 is the rest mass density, K is a constant, see below). Beyond the FERMI limit, the relativistic approximation becomes important, $\Phi \approx x^4/(12\pi^2)$, $\Psi \approx x^4/(3\pi^2)$, and therefore $p = K\rho_0^{4/3}$ and $p = \rho c^2/3$.

For a degenerate electron fluid, the rest mass density is dominated by protons, with $\rho_0 = \mu_e m_p n_e = 10^9 \mu_e x^3$ kg/m^3, hence about a factor 10 000 larger than the central density of the Sun. The composition of the matter enters by the specific weight per electron μ_e in units of the proton mass m_p. The constants in the equation of state are given by natural constants and have values of $K = 10^7/\mu_e^{5/3}$ Pa in the non-relativistic case, and $K = 10^{10}/\mu_e^{4/3}$ Pa in the relativistic case, when the density is given in kg/m^3. For a degenerate fluid of neutrons, there follows similarly $\rho_0 = m_n n_n = 6 \cdot 10^{18} x^3$ kg/m^3 and constants $K = 5 \cdot 10^3$ Pa and $K = 10^{10}$ Pa in non-relativistic and relativistic case, respectively.

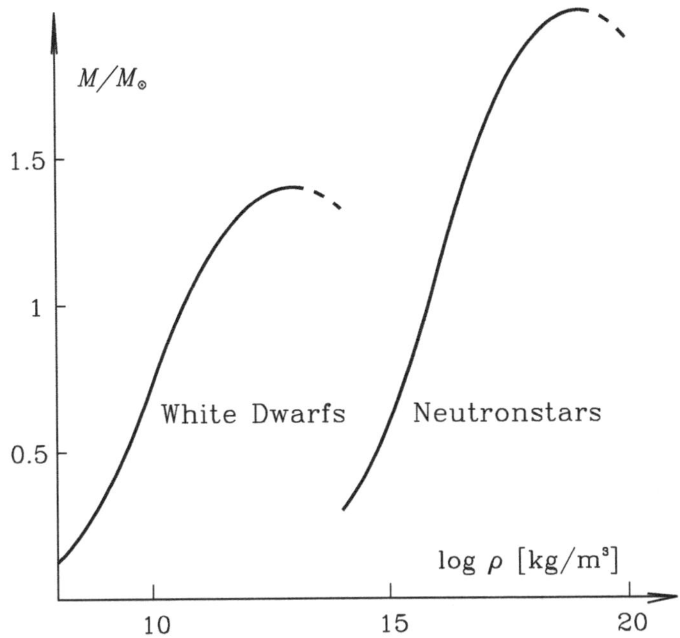

Fig. 6 Mass range of stable relativistic stars as function of the central density (solid lines), at left for electron degeneracy leading to white dwarfs, at right degeneracy by neutrons, leading to neutron stars.

The stellar models composed of degenerated electrons have central densities between 10^8 and 10^{13} kg/m^3 and stellar masses up to about 1.4 solar masses. The stellar radius has values of about 10 000 km. From the equation of state in the non-relativistic limit, it follows immediately an inverse relation between radius R and mass M, $R \propto M^{-1/3}$, i.e. the stellar radius falls off weakly with increasing central density and total mass. In Fig. 6, there is drawn on the left-hand side the mass range of stars with degenerated electrons. Beyond a limiting mass of about 1.4 solar masses, the so-called CHANDRASEKHAR *limit*, the mass of equilibrium configurations falls off again. But the objects are not in equilibrium anymore, since for a small contraction, *less* mass can be stabilised from the degeneration pressure. The star will collapse.

Stars stabilised by degenerated electrons are denoted as *white dwarfs*. They are observed with about 10 magnitudes below the normal main sequence of hydrogen burning stars. Its radiation is pure thermal and stems from the original star with nuclear fusion and contraction as energy source. During the further cooling, the white dwarfs will become fainter despite its high thermal capacity, and it will get finally invisible, without much changing its structure. Due to its compactness, the relativistic effects on the spectral lines from its surface have an order of magnitude of 10^{-4}, by which the gravitational redshift of the Sirius companion could be measured already shortly after creation of the GRT, see Sect. 3.2.

Still more extreme objects represent the *neutron stars*, with central densities between 10^{13} and 10^{19} kg/m^3. These densities are higher than the density in normal atomic nuclei. Therefore, for its understanding, we have to take into account the strong nuclear forces between the neutrons. The equation of state of an ideal FERMI fluid of neutrons leads in the same way as for white dwarfs to equilibrium configurations with an upper limit. For realistic calculations including nuclear forces, this limit amounts about 2 to 3 solar masses. For an increase of the central density, the pressure can again only stabilise a smaller mass, and the configuration becomes unstable. Neutron stars also have radii decreasing with increasing mass, with values between 30 and 10 km. In principle, they represent some kind of gigantic atomic nucleus.

Neutron stars are observed as *pulsars*, when they are in fast rotation and host a strong magnetic field. Due to the rotation, this field acts as a radiating electromagnetic dipole. Pulsars are radio sources with periodic signals stemming from active regions on its surface (probably connected with the magnetic poles) from which radiation is emitted in cone-like structures like in a lighthouse. Such a radiation mechanism is suggested by the observed slight slowdown of the pulse period observed in many pulsars. The energy corresponding to this radiation energy stems therefore from the rotation energy of the neutron star.

Neutron stars are extremely strange objects: Its rotation period is sometimes smaller than a second. There are also microsecond pulsars, probably young objects. Due to the large number of died mass-rich stars in the Milky Way, we expect the existence of a large number of neutron stars. But probably we cannot observe many of them, either since the signals do not reach the Earth due to the tightness of the radiation cones, or since they became to slow rotators due to its age.

Corresponding to its small radius, we expect strong relativistic effects of the order of 10 to 20% in neutron stars. An especially interesting object, a pulsar in a double-star system with a second invisible companion, has been described in Sect. 3.2.

We provide still a simple estimate of the masses and radii of stars with degenerate matter. The CHANDRASEKHAR limit follows from the transition from the non-relativistic to the relativistic approximation of the equation of state. When we use the FERMI momentum $x = 1$, so it follows an estimate of the maximum particle number and mass of the stars and a minimum radius of (here we take the nucleon mass m_N both for protons and neutrons)

$$
\begin{aligned}
N_{max} &= \left(\frac{\hbar c}{G m_N^2}\right)^{3/2} \approx 2 \cdot 10^{57} \, , \\
M_{max} &= m_N N_{max} \approx 1.4 M_\odot \, , \quad R_{min} \approx \frac{\hbar}{mc}\left(\frac{\hbar c}{G m_N^2}\right)^{1/3} \, , \qquad (1117)
\end{aligned}
$$

where in the last case for white dwarfs, we have to use $m = m_e$, and for neutron stars $m = m_n$. In this estimate, there appears the constant $G m_N^2/\hbar c \approx 6 \cdot 10^{-39}$, the fine structure constant of gravitation as generalisation of the analogous quantity in electrodynamics. However, it is smaller by many orders of magnitude than the last one, because of the weakness of the gravitational force between two nucleons.

3.4 Black Holes

The discussion of the SCHWARZSCHILD solution in Sect. 3.2 considered mainly the space outside of the SCHWARZSCHILD radius. A full understanding of this solution of GRT, however, requires a complete exhaustion of the range of its applicability. For describing the behaviour in approaching the SCHWARZSCHILD radius, we introduce new coordinates t' and r' instead of t and r thereby following LEMAÎTRE (1933),

$$ct' = ct + 2\sqrt{rr_g} + r_g \ln \frac{\sqrt{r} - \sqrt{r_g}}{\sqrt{r} + \sqrt{r_g}} ,$$

$$r = r_g^{1/3} \left[\frac{2}{3}(r' - ct') \right]^{2/3} . \tag{1118}$$

Then, it follows with the abbreviation $X = r_g/r$ for the differentials

$$cdt' = cdt + \frac{\sqrt{X}}{1 - X} dr \quad , \quad dr' = cdt + \frac{dr}{\sqrt{X}(1 - X)} , \tag{1119}$$

and we get the SCHWARZSCHILD metric in so-called GAUSS or synchronised coordinates (with metric component $g_{00} = 1$ and the mixed space-time components are zero, $g_{0a} = 0$, time-like orbits $x^a = const.$ are automatically geodesice in this metric, $a = 1, 2, 3$):

$$ds^2 = c^2 dt'^2 - \frac{dr'^2}{r/r_g} - r^2 d\omega^2 \tag{1120}$$

with the function $r(r', t')$ from Eq. (1118). Now we have a time-dependent metric, with observers resting in these coordinates falling onto the mass centre from infinity. The content of this form of the metric follows from the structure of the lightcones given at each point by the condition $ds = 0$. We restrict us on radial light paths ($d\omega = 0$) what leads to the differential equation for the light cones $dr'/dt' = \pm c\, r/r_g$. Outside of the radius r_g, i.e. for $r > r_g$, the areas $r =$ const. are inside the light cone, hence they are time-like. At the SCHWARZSCHILD radius, the area $r = r_g$ is part of the light cone. Inside of the SCHWARZSCHILD radius, the light cones shrink and point to the centre. The area $r = r_g$ is a null-area. Geodesics of massive particles and photons can only cross this area in direction of the centre $r \to 0$: We denote the SCHWARZSCHILD *area* as a *horizon* that does not allow escape of radiation or of matter. The objects described by this metric are denoted as *black holes*.

We expect to find such objects as end products of the evolution of massive stars, when its mass exceeds the limit of the CHANDRASEKHAR mass for degenerate neutron stars as described in Sect. 3.3. Then the stars collapse and vanish inside of the SCHWARZSCHILD radius. However, as seen from the outside, the time lapse becomes infinite when approaching the sphere $r = r_g$, cp. Eq. (1098) for the gravitational redshift. The collapsing star seems to be frozen at the SCHWARZSCHILD sphere.

A complete description of the SCHWARZSCHILD solution is possible in the so-called KRUSKAL- - SZEKERES coordinates (1963). To this aim, we define a new radial coordinate

$$r_* = r + r_g \ln\left(\frac{r}{r_g} - 1\right) . \tag{1121}$$

With this transformation, we introduce null coordinates $u = ct + r_*$ (an advanced light variable) and $v = ct - r_*$ (retarded light variable), which leads to a non-diagonal metric

$$ds^2 = (1 - X)dudv - r^2 d\omega^2 , X = r_g/r. \tag{1122}$$

With the further transformation

$$w = \frac{1}{2}\left[\exp\left(\frac{u}{2r_g}\right) - \exp\left(\frac{v}{2r_g}\right)\right] \quad z = \frac{1}{2}\left[\exp\left(\frac{u}{2r_g}\right) + \exp\left(\frac{v}{2r_g}\right)\right], \tag{1123}$$

we get a new form of the SCHWARZSCHILD metric,

$$ds^2 = \frac{4r_g^3}{r} \exp\left(-\frac{r}{r_g}\right)(dw^2 - dz^2) - r^2 d\omega^2 . \tag{1124}$$

The new time coordinate w and the radial coordinate z describe the maximum extension of the SCHWARZSCHILD metric. Radial null geodesics are represented as diagonal lines in the w-z-plane of Fig. 7; hence, they represent elements of the light cones as in the MINKOWSKI space-time. The first w-z-part of the KRUSKAL–SZEKERES representation of the SCHWARZSCHILD metric is proportional (conformal) to the space-time of Special Relativity, cp. Eq. (1124). The diagonals $w = \pm z$ represent the SCHWARZSCHILD sphere, where only the part right from the line $w = -z$ describes the complete black hole. The straight lines through the origin, $w = kz$ with $|k| < 1$ and constant are lines of constant SCHWARZSCHILD time t, and especially the diagonal $w = z$ belongs to $t \to \infty$. A distant observer can only observe after infinite time, how an object falls through the SCHWARZSCHILD radius. All hyperbola in the picture represent hyper-areas of the SCHWARZSCHILD radial coordinate $r =$ const. The hyperbolas around the w-axis are hyper-surfaces inside of the SCHWARZSCHILD radius $r < r_g$, and hyperbolas around the z-axis hyper-surfaces outside of the SCHWARZSCHILD radius.

The thick line in Fig. 7, running from below right upwards, represents the world line of a test body infalling into the black hole. For example, it could represent the surface of a collapsing star. In the last case, the complete range of the diagram left of this line until the diagonal $w = -z$ would be fictitious; it has to be replaced by the corresponding inner metric of the star. Obviously, the world line, corresponding to the thick line, describes the infall onto the black hole, that reaches the curvature singularity $r = 0$ (upper double thick line) within a finite time. For the distant ob-

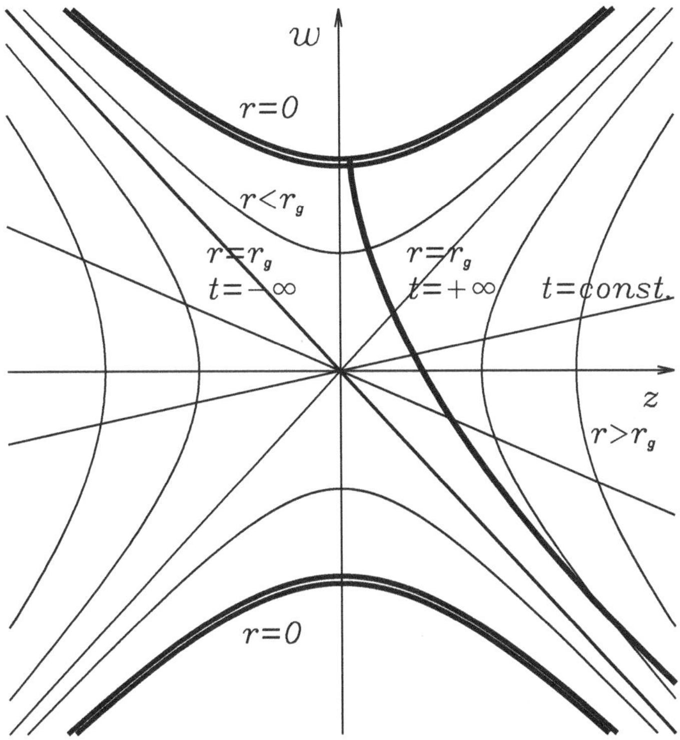

Fig. 7 Complete space-time geometry of the SCHWARZSCHILD–KRUSKAL–SZEKERES solution represented with time-like coordinates w and radial spatial coordinates z, angular coordinates are suppressed.

server, this world line seems to end on the SCHWARZSCHILD surface $r = r_g$ at $t \to \infty$; all signals from this area are infinitely redshifted; the body seems to vanish for the remote observer. For an outside observer, the collapse of stars seems to end at the SCHWARZSCHILD surface. We speak of a *collapsar*.

An unproven hypothesis, called *cosmic censorship*, claims that all singularities, also in more general than spherically symmetric gravitational fields, are hidden beyond such horizons. The classical space-time breaks down and becomes unapplicable for $r \to 0$, what becomes obvious, when evaluating the complete RIEMANN–CHRISTOFFEL curvature tensor which has in all coordinate systems diverging components.[4]

Now we consider the left part of the KRUSKAL–SZEKERES metric: When the left part $w < -z$ of the diagram would represent a real object in nature, it should show exactly the inverse properties of the black holes: World lines and light signals could only cross the SCHWARZSCHILD surface $w = -z$ away from the centre to the outside, escaping into an asymptotically flat space-time left in the diagram. Such objects are denoted as hypothetical *white holes*, but it must be observed that the regions $r \to \infty$

[4]We form so-called curvature invariants $R_{iklm}R^{iklm}$ that diverge as scalars in all coordinates.

left and right from the diagram represent not necessarily the same flat space-time region. However, there are not known any processes in nature that would lead to the formation of white holes analogously to the formation of black holes in a gravitational collapse. There are also not any observational hints on such strange objects.

4 Gravitational Waves

EINSTEIN's gravitation theory is a field theory with a finite propagation speed of its actions. This is quite different from NEWTON's theory, describing an immediate action of one body to all others. As we have seen in Sect. 1, the action of the gravitation is described by a deformation of the space-time geometry caused by heavy masses. The propagation of such deformations occurs with the speed of light c. This is best demonstrated with the propagation and the action of gravitational waves. We consider here in this section only weak fields, even if it can be easily shown that there exist exact solutions of the EINSTEIN equations describing also strong gravitational waves.

We come back to the ansatz of Eq. (1061) for the metric

$$g_{ik} = \eta_{ik} + h_{ik} \,, \quad |h_{ik}| << 1 \tag{1125}$$

with weak tensor deviations h_{ik} from the MINKOWSKI metric. Outside of matter, i.e. in vacuum, the EINSTEIN equations require the vanishing of the RICCI tensor

$$R_{ik} = \eta^{lm} R_{ilkm} = \frac{1}{2}(\eta^{lm} h_{ik,lm} + h^{l}{}_{i,lk} + h^{m}{}_{k,im} - h_{,ik}) = 0 \,. \tag{1126}$$

There we have neglected all products of CHRISTOFFEL symbols as nonlinear terms. Using the coordinate condition (1059) and the condition of the traceless of the *metric perturbation* h_{ik}, i.e. also $h = 0$, it follows that only the first term in Eq. (1126) remains, and this is already the wave equation for the metric perturbation h_{ik}. Hence, we see that EINSTEIN's equations lead in linear approximation to a wave equation that has plane gravitational waves as solution,

$$h_{ik} = a_{ik} \exp(i \, \mathbf{x} \, \mathbf{k}) \,. \tag{1127}$$

Here, the amplitudes a_{ik} and the wave number $k = 2\pi/\lambda$ corresponding to the wavelength λ are constants. Inserting this expression into the wave equation leads to the condition $\eta^{lm} k_l k_m = 0$, and hence the wave number vector is a *null vector*. The gravitational waves propagate with the speed of light. The imposed coordinate condition, the vanishing divergence of h_{ik}, proves that $a^{l}{}_{m} k_l = 0$. This means the deformation of the flat metric and its amplitude tensor are orthogonal to the propagation direction determined by the wave vector. Gravitational waves therefore represent *transverse* waves with traceless amplitudes. The coordinate conditions correspond to the LORENZ gauge in the description of electromagnetic waves. This gauge can be always choosen due to the free coordinate choice in GRT.

We consider for illustration a wave propagating in x-direction, so the wave number vector has as null vector only components in time- and in x-directions (with the amount $k = \omega/c$ and the circular frequency $\omega = 2\pi\nu$), $\nu = 1/T$ is the frequency to the period T, and components of the amplitude tensor a_{ik} only in y- and z-directions. Because of the symmetry and traceless of a_{ik}, only the components $a_{zz} = -a_{yy}$ and $a_{yz} = a_{zy}$ are different from zero,

$$a_{ik} = \begin{pmatrix} 0 & 0 & 0 & 0 \\ 0 & 0 & 0 & 0 \\ 0 & 0 & a_{yy} & a_{yz} \\ 0 & 0 & a_{yz} & -a_{yy} \end{pmatrix} . \qquad (1128)$$

The gravitational wave modifies the spatial separation of spatial neighbouring points,

$$dl^2 = dx^2 + (1 + h_{yy})dy^2 + (1 - h_{yy})dz^2 + 2h_{yz}dydz . \qquad (1129)$$

We consider a ring of test bodies in the plane orthogonal to the propagation directions. This ring will be deformed periodically as illustrated in Fig. 8. The two independent oscillations determine the two modes of gravitational waves, or its *polarisations*. As seen from the reaction of the test bodies, this is not a dipole radiation as in the electromagnetic case, but connected with a contraction in one direction, the body suffers always a dilatation in the orthogonal direction. This represents the principle way how gravitational waves can be measured. It is important to remark that these tensor waves require at least two test bodies for its measurement. A single point will not be moved, and only the *relative deviation* of neighbouring points is changing. To its description, we use the deviation equation, Eq. (1049) of neighbouring geodesics and the evaluation of the curvature tensor belonging to the wave (1127).

Before discussing actual measurements, we evaluate at first the emission of gravitational waves by the motion of large masses. To this aim, we have to solve the wave equation for the metric h_{ik} with inclusion of sources,

$$\Box h_{ik} = \frac{8\pi G}{c^4} T_{ik} \quad \Rightarrow \quad h_{ik} = -\frac{4G}{c^4} \int \frac{T_{ik}(x', t - |x - x'|/c)}{|x - x'|} d^3x' . \qquad (1130)$$

For evaluation of the emission of waves, we have to solve this equation approximately in the *near zone* $|x - x'|/c << t$,

$$T_{ik}(x', t - |x - x'|/c) \approx T_{ik}(x', t) - \partial_t T_{ik}(x', t)|x - x'|/c \pm \ldots . \qquad (1131)$$

For sources with slowly moving masses, the rest mass energy $T_{00} \approx \rho_0 c^2$ is dominant. In correspondence to the power in a series expansion for the inverse speed of light

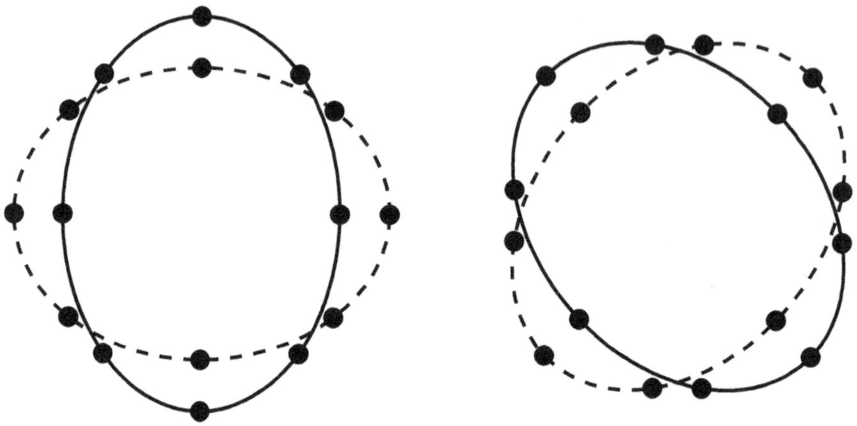

Fig. 8 Position changes of free test bodies during a passage of a gravitational wave. The left and right panels show the two polarisation states.

c^{-n}, we find in this case only in seventh order a term contributing to the emission of gravitational waves,

$$h_{00}^{(7)} = \frac{2G}{15c^7} \frac{d^5 Q_{ab}}{dt^5} x_a x_b \,, \tag{1132}$$

with the quadrupol moment of the matter distribution $Q_{ab} = 3I_{ab} - I_{aa}$ which consists of traceless components of the inertia tensor $I_{ab} = \int \rho_0(x') x'_a x'_b d^3x'$. The lower order evolution terms determine the static NEWTONian potential ($h_{00}^{(2)}$) and its static relativistic corrections $h_{00}^{(4)}$, or they vanish for odd powers due to the conservation of the total mass and the centre of mass of the source of the gravitational waves. As back-reaction for the emission of gravitational waves, there follows from the equation of geodesics, Eq. (1043), with definition of a damping potential,

$$\frac{d^2x^a}{dt^2} = -\frac{1}{2} c^2 h_{00,a}^{(7)} = \Phi_{,a}^{damp} \,, \tag{1133}$$

and a lost of energy by emission of

$$\frac{dE}{dt} = -\int \rho \Phi_{,a}^{damp} v^a d^3x = -\frac{G}{45c^5} \left\langle \frac{d^3 Q_{ab}}{dt^3} \frac{d^3 Q^{ab}}{dt^3} \right\rangle \tag{1134}$$

(v denotes the velocity of the mass elements of the source). This is EINSTEIN's famous quadrupol formula of 1921; it describes the emission of energy by gravitational waves and a corresponding energy lost of the source. The square bracket means the temporal mean. In case of a double-star system of stellar masses M_1 and M_2, a large semi-axis a and an eccentricity of the orbit e, there follows

$$\frac{dE}{dt} = -\frac{32G}{5c^5}\frac{M_1^2 M_2^2 (M_1 + M_2)}{a^5} f(e) \quad \text{with} \tag{1135}$$
$$f(e) = \left(1 + \frac{73e^2}{24} + \frac{37e^4}{96}\right) / (1 - e^2)^{1/2} .$$

This function is strongly growing with increasing eccentricity. The wave emission, however, reduces this eccentricity. According to the prediction of the theory, in the already in Sect. 3.2 mentioned double pulsar PSR 1913+16, in 1983 the emission of gravitational waves has been proven. There the orbital period P decreases with a rate $\dot{P} = 2 \cdot 10^{-12}$ corresponding to an energy lost described by Eq. (1135). Observed over a period of a few years, this represents the first indirect approval of the existence of gravitational waves.

The evaluation of the signal form of the emission of gravitational waves is an important problem in the research on gravitational waves, since it allows the identification of weak signals in a large background of other disturbations of the measuring device. Higher order post-MINKOWSKI calculations improve the estimates of the gravitational radiation damping and energy emission from compact binary double-star systems, BUONANNO and DAMOUR (1999). In general, the problem requires a numerical solution of EINSTEIN's equations, since a series expansion breaks down for strong gravitational fields.

As explained above, the emission of gravitational waves requires strong anisotropic motions of large masses, as realised in the anisotropic collapse of mass-rich stars in supernovae explosions, or else, the merging of tightly bound double stars containing compact relativistic stars and/or black holes. There the relevant masses have an order of magnitude of about $10\,M_\odot$, and the relevant frequencies $\nu \approx (kr_g/c)^{-1}$, with the SCHWARZSCHILD radius r_g, and a constant effectiveness $k \approx 10$, depending on the detailed processes and the geometry of the process. We expect high-frequency waves with $\nu \approx 10\,\text{kHz}$. Low-frequency gravitational waves are expected for the infall of stellar masses onto supermassive black holes of $(10^6 \ldots 10^8)\,M_\odot$ as observed in the nuclei of galaxies. Then the frequencies amount to $\nu \approx 0.1\,\text{Hz}$.

An estimate of conversion of ten percent of the rest mass energy in energy of the gravitational waves during such merging processes in a neighbouring galaxy about 10 million parsecs (10 Mpc) away, one half of the distance to the neighbouring Virgo cluster of galaxies, leads for a supposed isotropy of the emission to a power of about one Watt per square metre, $I_{gw} \approx 10\,\text{W/m}^2$, and a relative amplitude h of gravitational waves

$$h \approx \left(\frac{8\pi G}{c^3}\frac{I_{gw}}{\nu^2}\right)^{1/2} . \tag{1136}$$

On free test masses as in Fig. 8 this amplitude leads to relative changes in the separation of the masses of $\Delta l / l \approx h \approx 10^{-21}$. For long-wave gravitational waves, this signal is larger by a factor 10^5. There are, however, only expected signals of sources being by a factor ten times more distant to expect a sufficient signal rate of a few events per year.

Despite the small expected signal, in our example and for a separation of test bodies of 10 km, the length changes from passing a gravitational wave amounts only a distance of 10^{-17} m, being hence two orders of magnitude smaller than an atomic nucleus. There are running various experiments to measure these gravitational wave signals with devices analogous to the MICHELSON interferometer discussed in Chap. 4, Sect. 1. Improved laser interferometers work with free test masses with vacuum radiation cavities along the interferometer arms to improve the sensitivity. In 2015, the first gravitational wave signal was measured by the LIGO collaboration (Laser Interferometer Gravitational Observatory) simultaneously at two widely separated detectors in USA, which stem from the coalescence of two black holes, each of about 30 solar masses, where in the process a rest mass energy of about three solar masses was emitted as gravitational waves (ABOTT et al., 2016). There the signal stems from a source at about 400 Mpc distance, and the value is still quite uncertain since no counterpart of light or radio emission could be detected. Meanwhile, also the signal from the merging of compact neutron stars has been observed.

The proof of the emission of gravitational waves in double pulsars and in mergers of neutron stars and black holes represents a major triumph of EINSTEIN's prediction from 1915. It allowed a test of the GRT for very strong gravitational fields.

5 Cosmology and Gravitational Lensing

One of the most exciting subjects in natural sciences is the exploration of the world as a whole, the geometry of the large-scale matter distribution in the universe, and its origin and evolution. Long time, this was a subject strongly influenced by philosophical prejudice. However, with the General Theory of Relativity, there opened new ways to tackle cosmological questions. EINSTEIN's (1952) introduced his cosmological principle that the universe is homogeneous and isotropic on large scales. On this basis, he constructed a spherically closed, finite but unbounded and static world model. To this aim, he had to add a so-called cosmological term Λg_{ik} into his field equations, Eq. (1057). This term describes an universal repulsion also denoted as vacuum energy or *dark energy* that is necessary for the static model to compensate for the gravitational attraction of all normal matter.

A new paradigm was required for accepting the idea of a cosmological model with a time dependence in the metric. A first step is due to de Sitter (1916) with a solution of the field equations of GRT only with the Lambda term Λg_{ik} as source, but no matter. It is obtained as an imbedding of the four-dimensional space-time into a flat five-dimensional pseudo-Euclidean space. Depending on the choice of coordinates for the projection on a four-dimensional space-time, the solution seems to describe a static space-time with gravitational redshift, or an exponentially expanding world model, according to our present understanding important for the inflationary stage in the early universe, and for the late-time asymptotics of our cosmological standard model. In two seminal papers, the first one entitled 'On the curvature of space', Friedmann (1922, 1924) presented genuine time-dependent world models. Already in the first of these papers, FRIEDMANN speculated that his model could lead to a *'world period'* of 10 billion years, a value coming very close to the actual estimate of the age of the universe of 13.8 billion years. Later Lemaître (1927) reproduced the work of FRIEDMANN that got only little attention. LEMAÎTRA contributed much to the propagation of a world model with a temporal beginning, a big bang. Now they are known as FRIEDMANN–LEMAÎTRE models.

5.1 FRIEDMANN–LEMAÎTRE *Models*

Significant influence for the acceptance of the new understanding of an evolutionary cosmos had two at first sight-independent astronomical observations: SLIPHER could gain spectra of extended objects, so-called nebular flecks on the sky, now identified as galaxies comparable to our Milky Way. There he noted that the majority of the spectra shows a significant redshift that speaks via the DOPPLER effect for a positive radial velocity of these objects; they seem to escape from us. Later in the 1920, the astronomer HUBBLE could identify single bright stars belonging to these nebular flecks, and he concluded from the comparison with the brightness of stars of the same type in our Milky Way that these nebular objects are not part of our stellar system, but are much more distant away. With modern values, the distance till the boundary of our Milky Way is about 15 kpc (the unit parsec—pc—corresponds to about 3.26 light years). In comparison, the distance to our neighbouring galaxy, the Andromeda nebula M31, has a separation of 780 kpc. Comparing the observed distances d of the galaxies and their radial velocity v_{rad}, HUBBLE found a proportionality, his famous HUBBLE relation[5] from 1929 with the HUBBLE constant H_0:

$$v_{rad} = H_0\, d \quad \text{with} \quad H_0 = (70 \pm 3)\text{km/s/Mpc} . \tag{1137}$$

There we have given the modern value of the proportional constant. HUBBLE found in his first paper still a value of about eight times larger. The inaccuracy stems essentially from the systematics included in the measurement of the distances. The constant has the dimension of an inverse of time, and the inverse of the constant then gives an estimate of the age of the universe, where the exact value depends on the composition of the constituents of the different matter forms.

We observe the radial velocities v_{rad} as *cosmological redshifts* z in the spectra of galaxies,

[5]Now sometimes it is also called HUBBLE–LEMAÎTRE relation, with the HUBBLE–LEMAÎTRE constant.

$$z = \frac{\Delta\lambda}{\lambda} = \frac{\lambda_{beob} - \lambda_{em}}{\lambda_{em}} = \frac{v_{rad}}{c} , \tag{1138}$$

where the last relation between the redshift z and the radial velocity v_{rad} as DOPPLER effect is true only for velocities small against the speed of light. For larger redshifts, we interpret the redshift as relation of the world radius at measurement and at emission in the picture of an evolution cosmos. The redshift then becomes a measure of the distances of the relevant objects, and at the same time of a determination of the age of the universe, when the light was emitted. Now there are observed galaxies with redshifts up to about $z \approx 10$. The age of the universe at emission amounts about 500 million years, about three percent of the present age of the universe.

In this subsection, we discuss world models of the GRT with the supposition of EINSTEIN's cosmological principle, i.e. the assumption of homogeneous and isotropic space-like hyper-surfaces at constant cosmic time. Then each space point sees identical properties of the universe, and we are not sitting at an exceptional place in the universe. Extensions to describe inhomogeneities are then the subject of the next two subsections. Geometrically, we describe the space-time as a direct product of a one-dimensional time t and a homogeneous and isotropic space with a spatial metric being derived from a projection from a four-dimensional space,

$$d\sigma^2 = dr^2 + r^2 d\omega^2 + dx_4^2 , \qquad \text{angular part} \quad d\omega^2 = d\theta^2 + \sin^2\theta d\phi^2 . \tag{1139}$$

This is a Euclidean 4-space with an auxiliary coordinate x_4. We eliminate this coordinate in selecting a unite sphere with $k = 1$, or a pseudo-sphere with $k = -1$, by the requirement $r^2 + x_4^2 = k$. Then it holds for the spatial geometry

$$d\sigma^2 = \frac{dr^2}{1 - kr^2} + r^2 d\omega^2 .$$

This 3-geometry—determined by our symmetry requirements—is multiplied with a time-dependent world radius $R^2(t)$ that now also carries the length unit, leaving r dimensionless and restricted to the range $0 \le r < 1$ for $k = 1$, corresponding to a finite world model,

$$ds^2 = c^2 dt^2 - R^2(t)\left(\frac{dr^2}{1 - kr^2} + r^2 d\omega^2\right) = c^2 dt^2 - R^2(t)\left(d\chi^2 + S_k^2(\chi) d\omega^2\right) . \tag{1140}$$

This metric is called ROBERTSON–WALKER metric after the authors of a detailed mathematical justification based on the cosmological principle. In the second part of the equation, we have integrated the radial coordinate r to a dimensionless length χ

$$\chi = \int_0^r \frac{dr'}{1 - kr^2} = \begin{cases} \arcsin r \, , & \chi > r \quad \ldots \quad k = 1 \, , \\ \text{Arsinh}\, r \, , & \chi < r \quad \ldots \quad k = -1 \, . \end{cases} \tag{1141}$$

The case $k = 1$ corresponds to a spherically or positively curved finite word model $0 \leq \chi < \pi$ $(0 \leq r < 1)$, and the case $k = -1$ to an infinite, hyperbolic (negative) curved world model. The world radius $R(t)$ is normalised with the present value R_0 and written as $R(t) = R_0 a(t)$, and the *scale factor* $a(t)$ with present value equal to unity. Then the redshift defined as above in Eq. (1138) is given by the inverse of the scale factor, $1 + z = 1/a(t)$. The case $k = 0$ corresponds to a spatially flat world model. Then there does not exist a world radius, and the radial coordinate keeps the dimension of a length. The function $S_k(\chi)$ then reads

$$S_k = \begin{cases} \sin\chi & \ldots \quad k = 1\,, \\ \chi & \ldots \quad k = 0\,, \\ \sinh\chi & \ldots \quad k = -1\,. \end{cases} \tag{1142}$$

From the time–time and the radial–radial components of the EINSTEIN tensor (for derivation, see Problem 52)

$$E_t^t = 3\frac{\dot{R}^2}{R^2} + 3\frac{kc^2}{R^2}, \qquad E_r^r = 2\frac{\ddot{R}^2}{R^2} + \frac{\dot{R}^2}{R^2} + \frac{kc^2}{R^2}\,, \tag{1143}$$

an energy–momentum tensor of an ideal fluid with mass density ρ and the pressure p follow the two FRIEDMANN–LEMAÎTRE equations

$$\begin{aligned} \frac{\dot{R}^2}{R^2} + \frac{kc^2}{R^2} &= \frac{8\pi G}{3}\rho + \frac{\Lambda c^2}{3}\,, \\ \frac{\ddot{R}}{R} &= -\frac{4\pi G}{3}\left(\rho + 3\frac{p}{c^2}\right) + \frac{\Lambda c^2}{3}\,. \end{aligned} \tag{1144}$$

Both equations are compatible, when the conservation law, following from the BIANCHI identity (1056), is fulfiled

$$\dot{\rho} = -3H\left(\rho + 3\frac{p}{c^2}\right)\,. \tag{1145}$$

This relation describes the local energy conservation in the expanding universe, the increase or decrease of the energy density by the work done against the pressure of the ideal fluid. In the special case of a pressureless fluid $p = 0$, we speak of dust, or in modern language, of the dominant cold dark matter (CDM). But it holds also for homogeneous distributions of stars with slow peculiar velocities that are collisionless due to its large separation. One can integrate Eq. (1145) leading to $\rho \propto a^{-3}$. In the early phase of cosmological evolution for high densities, the radiation density dominates with an equation of stage $p_r = \rho_r c^2/3$, and an evolution law $\rho_r \propto a^{-4}$. Here, the index r means radiation.

The first FRIEDMANN–LEMAÎTRE equation defines a special so-called critical density ρ_{crit} for a vanishing curvature and vanishing cosmological term,

$$\rho_{crit} = 3H^2/(8\pi G) = 0.92 \cdot 10^{-26}\text{kg m}^{-3} = 1.4 \cdot 10^{11} M_\odot \text{ Mpc}^{-3}. \quad (1146)$$

The matter content of the universe is then characterised by a density parameter $\Omega_m = \rho_m/\rho_{crit}$ (and for radiation analogously Ω_r). Furthermore, we introduce parameters for the curvature $\Omega_k = -k/(H^2 R^2)$ and for the cosmological constant $\Omega_\Lambda = \Lambda c^2/(3H^2)$. Then the FRIEDMANN equation looks much simpler. We get an in general time-dependent expansion rate

$$H^2 = \frac{\dot{R}^2}{R^2} = H_0^2(\Omega_m^0 a^{-3} + \Omega_k^0 a^{-2} + \Omega_\Lambda^0 + \Omega_r^0 a^{-4}) \equiv H_0^2 E^2(a). \quad (1147)$$

The critical density depends on the cosmic time, and also all density parameters. On the right-hand side, we have used the present values with an index 0, and the dependences on the scale factor follow from the conservation law (1145). From the definition of the present density parameters, it follows that the sum of density parameters equals *unity*. The same conclusion follows clearly at each moment of the cosmic evolution for density parameters evaluated at the corresponding time.

The function $E(a)$ appears in the evaluation of different cosmological quantities, as cosmological distances or in the integral for determining the age of the Universe, the time since the big bang,

$$t_0 = \frac{c}{H_0}\int_0^a \frac{da'}{a'E(a')} . \qquad \text{Age of the Universe} \qquad (1148)$$

An explicit solution of the integral is derived in Problem 54.

If the cosmological constant can be neglected, we get the solutions of the FRIEDMANN–LEMAÎTRE equations in a quite easy way:

For vanishing curvature constant $k = 0$ and dust-like matter (pressure $p = 0$ as typical for the present evolution state of the universe), there follows a simple power law

$$a \propto t^{2/3} . \quad (1149)$$

This solution is called EINSTEIN–DE SITTER model. It was long a favourite model; however, it is not confirmed by observations.

In case of a positive curved space, there follows with the so-called conformal time $d\eta = dt/a(t)$ and a dash for the derivative for η

$$a'^2 = H_0^2\left[\Omega_m^0 a + (1 - \Omega_m^0)a^2\right] \Rightarrow \eta = H_0 \int \frac{da}{\sqrt{\Omega_m^0 a + (1 - \Omega_m^0)a^2}} . \quad (1150)$$

The integral can be solved elementary, leading to a parametric solution with the dimensionless parameter $\theta = \eta/H_0$, cp. the equation for the gravitational collapse, Eq. (1097):

$$a = \frac{\Omega_m^0}{2(\Omega_m^0 - 1)}(1 - \cos\theta) ,$$
$$t = \frac{\Omega_m^0}{2H_0(\Omega_m^0 - 1)^{3/2}}(\theta - \sin\theta) . \qquad (1151)$$

This solution describes a cosmological model with a finite life time, it begins with a big bang and a time dependence as solution (1149), it is time-symmetric around a maximum scale factor at $\theta = \pi$, and it ends in a 'big crunch' at $\theta = 2\pi$, when scale factor and world radius turn to zero. The maximum is reached for a scale factor $a_m = \Omega_m^0/(\Omega_m^0 - 1)$ at a time $\theta_m = \pi/(2H_0) \cdot \Omega_m^0/(\Omega_m^0 - 1)^{3/2}$.

A formally similar solution follows for the open cosmological models without a cosmological constant:

$$a = \frac{\Omega_m^0}{2(1 - \Omega_m^0)}(\cosh\theta - 1) ,$$
$$t = \frac{\Omega_m^0}{2H_0(1 - \Omega_m^0)^{3/2}}(\sinh\theta - \theta) . \qquad (1152)$$

This model also starts with a big bang and a time behaviour as the EINSTEIN–DE SITTER model; however, it expands forever, and there is no re-collapse.

The second FRIEDMANN equation leads to a determination of the so-called deceleration parameter $q_0 = -a\ddot{a}/\dot{a}^2|_{t_0}$, with present value $q_0 = -\Omega_\Lambda^0 + \Omega_m^0/2 + \Omega_r^0$. The deceleration parameter[6] appears in the second term of the TAYLOR expansion of the time dependence of the scale factor:

$$a(t) = 1 + H_0(t - t_0) - \frac{1}{2}q_0 H_0^2 (t - t_0)^2 + \ldots \qquad (1153)$$

Modern cosmological observations lead to well-determined values of the cosmological density parameters,

$$\Omega_m^0 = 0.3 \pm 0.02, \quad \Omega_\Lambda^0 = 0.7 \pm 0.03, \quad |\Omega_k^0| < 0.01, \quad \Omega_r^0 \approx 5 \cdot 10^{-5} . \qquad (1154)$$

These *cosmological parameters* characterise our cosmological standard model together with the HUBBLE constant (1137) . It describes at present time-accelerated cosmic expansion with $q_0 = -0.55$. This became clear in 1998 with the analysis of

[6]Sometimes, q_0 is also called acceleration parameter. However, its sign is defined in such a way that for the dominance of normal dust-like matter, its value is positive, and it describes a deceleration of the cosmic expansion due to the positivity of gravitational attractions.

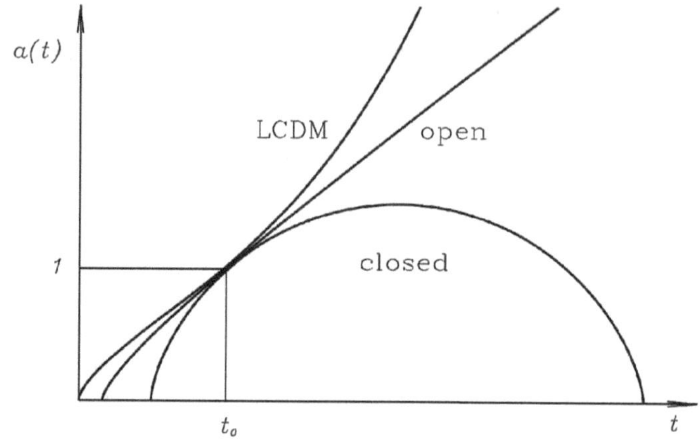

Fig. 9 Evolution of the scale factor $a(t)$ as function of the cosmic time t for a closed FRIEDMANN model with $k = 1$, an open model with $k = -1$, and the standard model LCDM with a flat Euclidean spatial geometry.

the data of supernovae surveys. The HUBBLE relation of supernovae shows larger distances at redshifts between $z = 0.3$ and 2 than the free expansion law (MILNE model, $a \propto t$), well described by a negative deceleration parameter, hence an *acceleration of the cosmic expansion*, which is caused by the cosmological term, also known as *dark energy*, cp. e.g. AMENDOLA and TSUJIKAWA (2010).

The matter density Ω_m^0 includes both *cold dark matter* (CDM) with a density parameter $\Omega_{CDM}^0 \approx 0.25$, while the baryonic matter, representing stars and both hot and cold diluted gas in and around galaxies, only amounts to $\Omega_b^0 = 0.045 \pm 0.002$. This cosmological model is called after the dominant matter constituents *LCDM-Model*, a model with a *Lambda-Term* Λ in EINSTEIN's equations (1057), and with the second important contribution of CDM, s. Eq. (1154).

In particular by the analysis of the primordial inhomogeneities in the cosmological background radiation, s. Sect. 5.3, there follows that the curvature parameter of the cosmological standard model is negligible. The upper limit is presently about $|\Omega_k^0| < 0.01$. The solution of the FRIEDMANN equation without curvature term can be found explicitly, see Problem 54. It follows an expansion law

$$a(t) = \left(\frac{\Omega_m}{\Omega_\lambda}\right) \sinh^{2/3}\left(\frac{3H_0\Omega_m^{1/2}}{2}t\right) . \tag{1155}$$

The age of the universe follows with the cosmological parameters of Eq. (1154), and it is about 13.8 billion years.

Figure 9 shows the three typical cosmological evolutions following from the FRIEDMANN–LEMAÎTRE equation.

The cosmological standard model is confirmed by a large number of observations in the deep universe. We shall especially discuss the application of gravitational lenses on cosmological problems in Sect. 5.2. The large number of galaxy surveys and its interpretation in cosmological structure formation processes are discussed in Sect. 5.3. We shall only consider processes after the decoupling of the cosmic background radiation, $z \approx 1100$, i.e. 380 000 years after the big bang.

In early stages of the cosmic evolution, the radiation density ρ_r becomes dynamically dominant, even if its present value is very small,[7] $\Omega_r^0 \approx 5 \cdot 10^{-5}$. However, the radiation density grows according to the conservation law (1145) for $a \to 0$ as $\propto a^{-4}$, and the cold dark and baryonic matter only as $\propto a^{-3}$. The photons are compressed as the CDM-matter $\propto a^{-3}$, but they get a *blueshift* $\nu = c/\lambda \propto a^{-1}$, i.e. they gain energy in following the cosmic evolution backwards. In the radiation era, neither the cold dark matter nor the cosmological constant play a role, and we get the simple power law for the scale factor

$$a \propto t^{1/2} \,. \tag{1156}$$

During the approach to the big bang, the temperature of the cosmic matter increases more and more, apparently without bound. At first, all baryonic matter will be ionised by the photons from the background radiation. In still earlier stages, there are possible high-energy reactions between the nuclei and elementary particles. During the first 10 min, free protons and electrons came into interaction and produced the first light nuclei of deuterium, tritium, helium and lithium. A dominant helium abundance (besides hydrogen) of 24% mass content of all baryons prove this hot evolutionary stage of the universe at a temperature of a billion Kelvin. This stage is called *big bang nucleosynthesis*.

In still earlier stages, the nucleons, protons and neutrons are broken up into quarks and gluons, and all energetically allowed elementary particles are produced in particle collisions. Relativistic particles (the rest mass is negligible with respect to the kinetic energy) dominate the evolution. We suspect a special phase transition in the first 10^{-35} s after the big bang, called *cosmic inflation*, during which phase the separation of the strong nuclear force and the electroweak interaction had a comparable interaction strength, and the different forces could eventually arose from a single unified elementary particle theory. The inflationary phase is probably caused by a scalar field on the right side of EINSTEIN's equation (1057).

This phase is extremely interesting with respect to the explanation of a few coincidences in the early values of the cosmological parameters. On the other hand, we expect zero-point fluctuations in the scalar field during this phase that may be the seed of structures in the cosmic background radiation, and in the large-scale structures in the distribution of galaxies and dark matter that shall be discussed in Sect. 5.3.

---

[7] The numerical value follows from STEFAN–BOLTZMANN's law and the very precise measurement of the temperature of the cosmic background radiation.

5.2 Gravitational Lensing

The light deflection of Sect. 3.2, the historically most spectacular effect of the GRT, led to the hypothesis of multi-images of distant cosmic objects by foreground deflectors, the so-called gravitational lensing effect. It is interesting that the basic relations for the lensing effect can be found already in EINSTEIN's notebook from 1912; hence, still before the final equations of GRT have been found. Different astronomers have put forward similar ideas, and EINSTEIN has published an article on this hypothesis, but he estimated that the probability to observe this effect is very small. The deflected object and the lens must be sitting at very nearby (projected) sky positions to find significant deflection angles.

Let in Fig. 10 the paper plane be defined by the observer B, the lens D and the source S, with separations d_D between observer and lens, d_S between observer and source, and d_{DS} between lens and source. More precisely, these are angular diameter distances that result from the ratio of the tangential proper separation in the ROBERTSON–WALKER metric and the observed angular separation on the sky. Due to the expansion of the universe, i.e. the curvature of the space-time, these distances are smaller than the formally Euclidean separation. In particular, there are differences in the addition, $d_S \neq d_D + d_{DS}$.

In comparison to these in general very large length scales, the light deflection happens in a very tight range around the lens. For its description, we can therefore assume a deflection only in the plane orthogonal to the line-of-sight, the so-called lens plane. In optics, this corresponds to the approximation of thin lenses.

In the transversal lens plane, we call the separation of the true bend light ray from the lens ξ as shown in Fig. 10.[8] We define the angle between the rays from the source

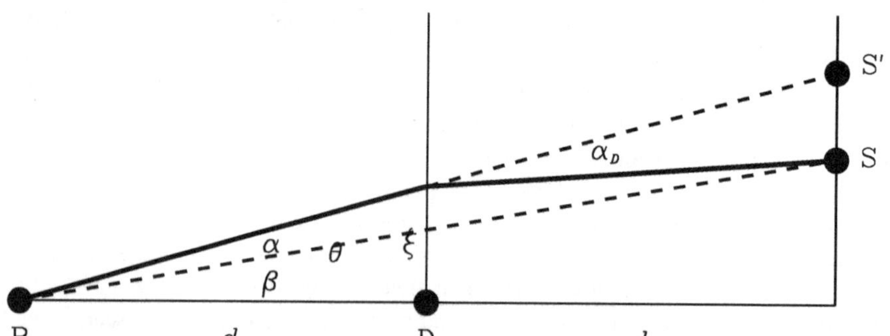

Fig. 10 Geometric configuration between observer B, a lens (deflector) D and a source S that is observed at the apparent position S' under the angle separation θ from the lens due to the light deflection. In the picture, we assume schematically a bending of the light path only at the deflector plane, in reality, this happened more gradually.

[8]In the general case of a more complex, not spherically symmetric lens, this is a two-dimensional vector in the lens plane.

S and the lens D as β, the angle between picture S' and lens D as θ, and the angle between the rays to the source S and picture S' as α. The deflection angle α_D is the angle between S and S' as seen from the lens plane.

It is immediately clear that for small angles, the deflection angle as seen from the observer is given by $\alpha = \alpha_D \cdot d_{DS}/d_S$. The deflection angle α_D itself is dependent on the matter distribution in the lens plane, and on the viewing angle θ of the vector ξ in the lens—or image plane, $\theta = \xi/d_D$.

With the dependence of the apparent deflection angle $\alpha(\theta)$ on the separation of the image from the lens θ, it follows the so-called *lens equation* as relation between the true, but unobservable object position in case on no deflector, β and the observed image position θ,

$$\beta = \theta - \alpha(\theta) \quad \text{and} \quad \mu = \frac{\theta d\theta}{\beta d\beta} = \frac{\theta}{\beta \frac{d\beta}{d\theta}}. \tag{1157}$$

Directly from the deflection equation, there follows the second relation for the *magnification* μ as ratio of the observed and the theoretical image extension without deflector. The lens equation is in general a nonlinear equation for the position β of the object without deflector as function of the observed image position θ. The inversion then provides the image position θ from a given configuration observer—lens—image. For a tight alignment of lens and object there may result more than one solution, i.e. more pictures of a single source. This is the so-called *strong gravitational lens effect.* When the alignment is not so good, or the lens mass is not big enough, the source in only weakly deformed, and we speak of a *weak lensing effect.*

For a point lens, the deflection angle is given by $\alpha_D = 2r_s/\xi = 2r_g/d_D\theta$ (r_g is the gravitational or SCHWARZSCHILD radius), hence

$$\alpha = \frac{2r_s}{\theta}\frac{d_{DS}}{d_D\, d_S} = \frac{\theta_E^2}{\theta} \quad \Rightarrow \beta = \theta - \frac{\theta_E^2}{\theta} \quad \text{with} \quad \theta_E = \sqrt{\frac{r_g\, d_{DS}}{d_D\, d_S}} \tag{1158}$$

with two solutions of the quadratic equation

$$\theta = \left(\beta \pm \sqrt{\beta^2 + 4\theta_E^2}\right)/2\,, \tag{1159}$$

i.e. exactly two images. Here, we have introduced the so-called EINSTEIN radius θ_E. It is given by the square root of the radio of the SCHWARZSCHILD radius of the lens and the effective angular diameter distance $d_D\, d_S/d_{DS}$, i.e. for stellar deflectors (say of mass of one solar mass) and an effective separation of 1 kpc for the near zone in our Milky Way, the EINSTEIN radius is about 10^{-6} degree or 2 mas (milli-arcseconds). For the mass of a galaxy of 10^{12} solar masses as lens and an effective separation of 100 Mpc (Megaparsec), the EINSTEIN radius has values of a few arcseconds.

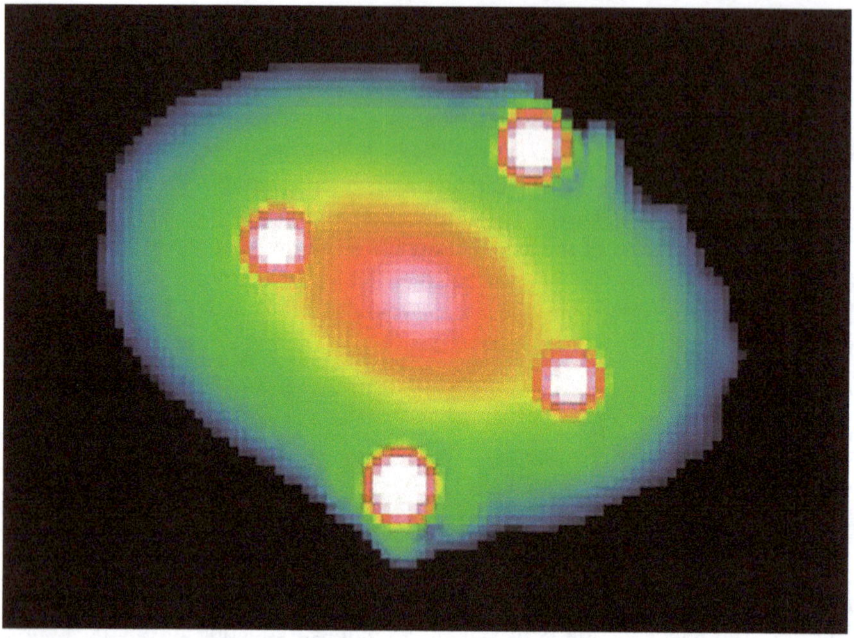

Fig. 11 Fourfold image of the quasars Q 2237 + 030—called EINSTEIN *cross*—taken from the Hubble-Space-Telescope within the CASTLES-survey.

The solution (1159) describing point mass lenses shows immediately that the separation of both images is always larger than the EINSTEIN radius, but this radius provides the order of magnitude, when two images can be observed. Then it follows that for stellar masses as lenses, the two images cannot be separated, but there can be observations of the magnification. Photometric observations looking for sharp brightness variations can lead to the detection of the so-called *micro-lensing*, see below. If galaxies act as lenses, the image separation of a few arcseconds is accessible to even ground-based telescopes. The first gravitational lens has been observed in 1979, the quasar Q 0957 + 561 A, B (the name encodes the sky coordinates right ascension and declination of the object, the notations A and B mean the two images). For this object, the angular separation amounts about 6 arcseconds. In this system, we observe also a time delay of the signals of about 400 days, becoming measurable due to the temporally variable source brightness. The time shift is a measure of the different light paths, providing a direct measurement of the SHAPIRO effect discussed in Sect. 3.2. Now there are detected many systems with double and multiple images of quasars, cp. the CASTLES programme.[9] The four images of Q 2237 + 030 shown in Fig. 11 have angular separations between one and two arcseconds. The galaxy in the centre overshines a weak fifth central image. Fourfold images arise from a deviation from axial symmetry of the lens, or the contribution of more than one lensing mass.

[9]Cfa-Arizona Space Telescope Lense Survey, s. https://www.cfa.harvard.edu/castles.

From second Eq. (1157), it follows for the image magnification μ of the point mass lens

$$\mu = \frac{1}{2} \pm \frac{u^2+2}{2u\sqrt{u^2+4}} \quad \text{with} \quad u = \frac{\beta}{\theta_e}\,, \tag{1160}$$

where the negative sign means a magnification with inverted parity of the image, and from a source picture, we get a mirror image. The magnification diverges for $u \to 0$, hence for an exact alignment of observer, lens and source. Then, the image appears exactly on the EINSTEIN radius as ring image. Since the lens is never an exact point source, the picture will only get a finite magnification, but it can have quite large values. For $u = 1$, i.e. when the unlensed picture is on the EINSTEIN radius, still we have magnifications by factors 1.17 and –0.67. Special observational programmes are under way to measure the image magnification of stars by other stars, or also of dark objects, in the Milky Way, and in particular in its bulge, the central thickening of the otherwise very flat stellar disk. There are many results, but no hints of a large abundance of massive dark objects as candidates of the dark matter.

For the description of the gravitational lens effects of galaxies or of quasars by galaxies, we must consider more general matter distributions, beyond the point mass lenses. The extended mass density distribution $\rho(\xi, z)$ of the deflector is described in cylinder coordinates, where z is lying along the line-of-sight, and ξ a vector in the lens plane. Since the z-range of the action of the lens is very small in comparison to the distances to the source and the observer, we can use the approximation of a thin lens and project the mass density on the lens plane, leading to an area mass density $\sigma(\xi) = \int \rho(\xi, z)dz$. We get a deflection angle

$$\alpha_D(\xi) = \frac{4G}{c^2} \int \frac{(\xi - \xi')\sigma(\xi')}{|\xi - \xi'|^2} d^2\xi'\,. \tag{1161}$$

Again it holds the lens equation (1157) which can be used to determine the two-component source vector β corresponding to the image vector θ in the lens plane. It follows a critical area density

$$\Sigma_{crit} = \frac{c^2}{4\pi G} \frac{D_s}{D_d \cdot D_{ds}} \approx 4,5 \text{ kg m}^{-3} \tag{1162}$$

as condition for the appearance of multi-images. This mass density is reached especially in mass-rich galaxies and in galaxy clusters. There, very interesting and complex image configurations are observed that in general lead to the possibility to invert the lens equation and to derive the surface mass density of the lens.

The meaning of the gravitational lens effects goes much beyond the verification of light deflection predicted by the GRT. All masses in the universe contribute to the lensing. Therefore, with the deflection effect, we get in a unique way a picture of the *total mass distribution* in the universe, hence also a measurement of the distribution of the dark beside the bright baryonic matter. Using the gravitational lens effect, we can employ the magnification to look deep into the universe in special sky regions with foreground mass assemblies as bright galaxy clusters. This has been applied to study very early proto-galaxies and galaxy structures formed soon after the big bang. Finally, the time delay measurements allow derivation of the length of the light path from the source, and then together with redshift measurements new independent clues in the HUBBLE relation (1137) and the value of the HUBBLE constant as most important quantity characterising our cosmological standard model.

5.3 *Cosmological Structure Formation*

The picture of an evolution cosmos developed originally by FRIEDMANN, LEMAÎTRE and HUBBLE allows a completely new sight on all structure formation processes in the universe. It starts with a hot dense early evolution stage following the initial *big bang*. From this phase, there remains a relict radiation of about 3 K that still contains faint fluctuations in the temperature that result from inhomogeneities in the density, the gravitational potential and the velocities. The relative temperature fluctuations are shown in Fig. 12: they have an amplitude of about 10^{-5}. Otherwise, the background looks isotropic, up to a dipole anisotropy resulting from the peculiar motion of the solar system in the Milky Way, and mostly from the motion of the Milky Way as a whole. The isotropy of the background radiation represents an important support for EINSTEIN's postulate of the cosmological principle.

Figure 12 shows the background radiation of the ESA-satellite Planck[10] with measurements during the time span 2009 till 2012. The data are foreground cleaned from emission of gas and dust in our own Milky Way that can be subtracted due to the different spectral signatures and the various used measuring bands. The statistical analysis strongly improved the knowledge of the values of the cosmological parameters, s. Eq. (1154). Among the fluctuations on the different scales, there are especially strong inhomogeneities on scales of about one degree, on a scale of 1/360 of a full circle. This maximum is caused by the combined action of potential and velocity fluctuations on the scale of the cosmic sound horizon, the maximum distance, signals can travel in the time since the big bang in the hot cosmic plasma. Besides this first maximum, there are found a couple of minor oscillations at smaller angles as overtones.

The scale of one degree corresponds to a comoving length of 150 Mpc. By the coupling of the baryonic matter to the radiation before decoupling of the background radiation, this scale is also depicted in the large-scale distribution of galaxies, which is, for example, seen in the data from the Sloan digital sky survey (SDSS),[11] s. Fig. 13. We denote them as *baryonic acoustic oscillations (BAO)*. We have used these oscillations for constraining the behaviour of the cosmic repulsion, whether it

[10]European Space Agency, s. http://www.esa.int/Our_Activities/Space_Science/Planck.
[11]http://www.sdss.org.

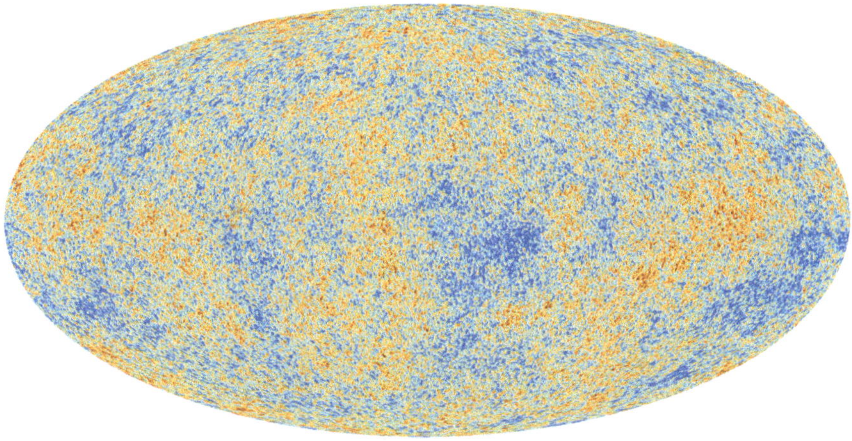

Fig. 12 Temperature fluctuations of the cosmic background radiation obtained with the Planck satellite. The inhomogeneities appear on a level of about $\Delta T/T$ of 10^{-5}, hence with an amplitude of a few ten micro-Kelvin.

stems from a cosmological term or a variable dark energy, cp. for example Wagner et al. (2008).

The origin of the faint fluctuations in the background radiation are quantum effects in the inflationary phase of the early cosmological evolution. Quantum fluctuations exist with necessity due to the zero-point fluctuations of the inflation scalar field due to HEISENBERG's uncertainty relation. With a series decomposition of the metric inhomogeneities, MÜLLER, SCHMIDT, STAROBINSKY (1992) have shown that the inflationary phase is an attractor solution among the general inhomogeneous cosmological models in GRT. The distribution of inhomogeneities over different length scales determines a distribution of inhomogeneities seen later in the large-scale matter distribution of the universe.

The inhomogeneities are described in the cosmic density field ρ,

$$\delta = \frac{\rho - \langle\rho\rangle}{\langle\rho\rangle}. \tag{1163}$$

These fluctuations in baryons and the dark matter get larger due to fluctuations in the gravitational potential connected with the density inhomogeneities: The matter leaves the potential hills and flows into the potential wells. This process is called *gravitational instability*. It is described by the peculiar NEWTONian gravitational potentials as perturbations of the background cosmological model, given by the FRIEDMANN–LEMAÎTRE equation. Combining the force equation for particles as perturbations, or EULER's equation for a cosmic fluid together with the POISSON equation for the gravitational potential, we get an evolution equation for the inhomogeneities. In the stage of small fluctuations, we get in linear approximation an oscillator equation for the density contrast δ,

$$\ddot{\delta} + 2H\dot{\delta} = 4\pi G\langle\rho\rangle\delta = \frac{3}{2}\Omega_m H^2\delta \,. \tag{1164}$$

Here, there enters the Hubble constant H as damping term on the left-hand side, and the density parameter Ω_m on the right-hand side. It appears by replacing the matter density ρ with the critical density and the density contrast. The amplification of the density contrast $\ddot{\delta}$ is caused by the gravitational action on the right-hand side, but damped by the action of the expansion. There is an increasing δ_+ and a decreasing solution δ_-,

$$\delta_+ \propto H(t)\int_0^a \frac{da}{a^3H^3} = E(z)\int_z^\infty \frac{(1+z')dz'}{E^3(z')} \quad \text{and} \quad \delta_- \propto H(t)\,. \tag{1165}$$

In the integral, there appears the energy function $E^2 = \Omega_m(1+z)^3 + \Omega_\Lambda$ from the FRIEDMANN–LEMAÎTRE equation (1147), where we only have taken the two significant matter contributions during the phase of structure formation. The solutions are simple power laws in the case of a spatially flat model without a cosmological term, the EINSTEIN–DE SITTER model (1156),

$$\delta_+ \propto a(t) \propto t^{2/3} \quad \text{and}\ \delta_- \propto t^{-1}\,. \tag{1166}$$

The cosmological term later leads to an end of the increase of linear perturbations due to the cosmic repulsion, however, only on scales where it dominates.

We can proceed over the linear approximation. Then, typically there appears an anisotropic collapse at first to flat structures, so-called ZEL'DOVICH pancakes, denoted after the inventor of the theory of the nonlinear gravitational instability, the ZEL'DOVICH solution. These flat structures decay into one-dimensional filaments. All structures on different scales are interwoven, and one finds characteristic density knots with large overdensities, called *halos of cold dark matter*, where baryons can condense and start to form stars in evolving galaxies. The stars in galaxies enlighten the cosmos and reveal the hierarchy of structures, see, e.g. Fig. 13. There are many attempts underway to reconstruct the different structure elements, pancakes, filaments and knots in the dark matter distribution, inferred from the galaxy halo distribution, see, e.g. MUNOS-CUARTAS, MÜLLER, FORERO-ROMANO (2011).

In the halos, there is contained a typical admixture of the baryonic gas. This is at first heated due to the compression in the gravitational collapse (and by radiation from neighbouring objects), but then it can cool through thermal radiation, forming stars and galaxies in the central regions of the large dark matter halos. The first such young galaxies with stars are expected at redshifts $z \approx 10 \ldots 20$. By secondary matter infall, the interaction and the merging of such young Galaxien, the whole diversity of galaxies forms as observed in the universe.

The galaxies stand not alone but form a huge network of large structures. An example is shown in Fig. 4, a slice through the presently largest galaxy redshift

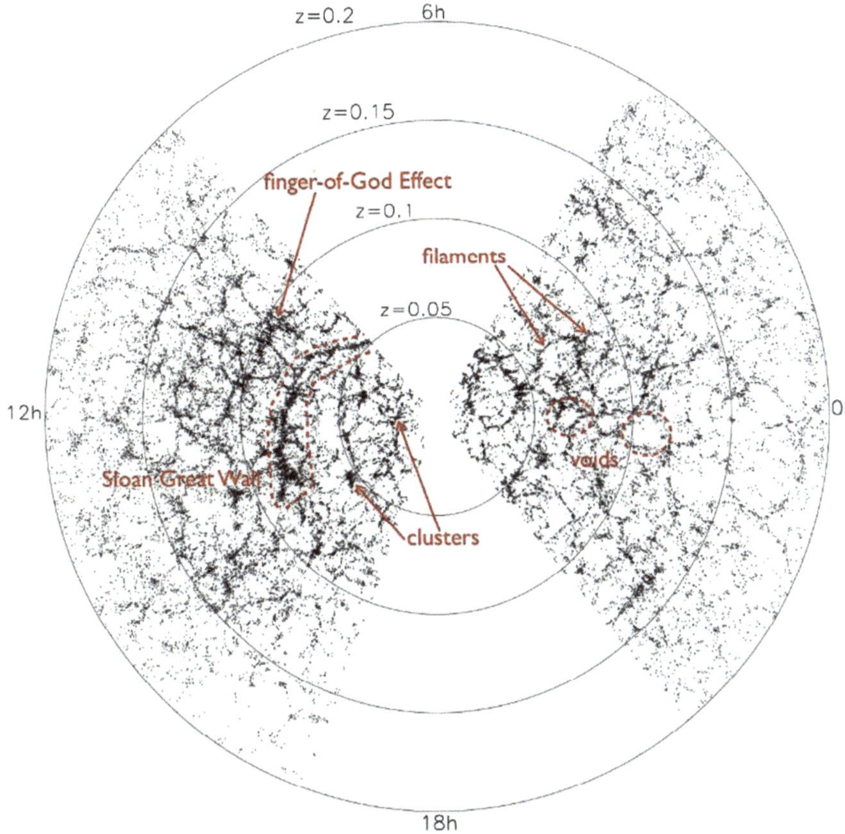

Fig. 13 Galaxy structures from a slice through the SDSS with indication of typical large structures. The numbers on the outer circle give the right ascension of the data, and the declination ranges over 5^0 around the sky equator, projected onto the picture plane. The redshift values z denote the depth of the survey between $z = 0.05$ and $z = 0.2$, providing the distance of the observer in the centre.

catalogue, the SDSS. It covers about one-quarter of the sky. In such cone pictures, the observer is in the centre of the concentric circles.

The nonlinear gravitational collapse leads to a variety of cosmic structures. Quite significant are the giant cosmic voids, e.g. Müller et al. (2000), that are, however, not completely empty, but are occupied with faint galaxies and a background of dark matter and gas of lower than average density. There exist furthermore a large number of faint galaxies, clustered around bright galaxies. Stellar systems appear typically at first as spiral galaxies, and later due to interactions as elliptical galaxies. Galaxies have a large mass range between 10^6 up to 10^{13} solar masses including dark matter. Galaxies are typically clustered; they form groups of galaxies and the spectacular galaxy clusters containing thousands of members and having masses up to 10^{15} solar masses. Finally, there exist fragile structures that are only very weakly bounded, with overdensity of order unity, called *superclusters of galaxies* that are enclosing the cosmic voids. The mechanism of cosmic structure formation represents a universal phenomenon that characterises our cosmological model beginning from the inflation stage after the big bang up to the largest presently observed structures. Detailed simulations on supercomputers allow us to follow this process with increasing accuracy.

With EINSTEIN's General Relativity Theory, we have an excellent basis for describing our observed universe.

Appendix A
Problems and Solutions

The problems help to understand the main text. Some exercises provide additional material extending the content of the book. We give the corresponding assignment to the relevant Sections after formulating the different problems.

Problem 1 *For a bar at rest, we measure a length l_o. The same bar might have velocity v with respect to the system Σ_o. Show how you can measure the length l_v of the moving bar with a time measurement, if you know its velocity v.*

Chapter 1, Section 3
The bar has a length l_o, when it rests in the system of the measurement. Now we consider the case that it is in motion with velocity v with respect to the system Σ_o. Then the end points have coordinates $x_1 = x_1(t)$ and $x_2 = x_2(t)$, with the condition $v = dx_1/dt = dx_2/dt$. In Σ_o one knows the isotropic speed of light c, that is used to synchronise the clocks in Σ_o. We are looking for the length $l_v = x_2(t) - x_1(t)$ of the bar defined as difference of the coordinates of the end points at the same time t. To this aim we send a light signal forth and back from one end, s. Fig. A.1.

At time $t = 0$, we send off a light signal from the left end point $x_1(0)$, which moves in Σ_o as $x(t) = x_1(0) + c\,t$ in direction to the right end point. The end point removes from the signal with the velocity v. According to Eq. (8) the light front *approaches* to the right end point with the positive velocity $c - v$ and arrives there at time $t_\rightarrow = l_v/(c - v)$. After reflection, it has in system Σ_o the velocity $dx/dt = -c$, so that it *approaches* to the left end point with the positive velocity $c + v$. For the way back, the signal needs the time $t_\leftarrow = l_v/(c + v)$ and arrives there at time t_v,

$$t_v = t_\rightarrow + t_\leftarrow = \frac{l_v}{c - v} + \frac{l_v}{c + v} = \frac{2c\,l_v}{c^2 - v^2} = \frac{2l_v}{c}\,\frac{1}{1 - v^2/c^2}\ , \qquad \text{hence}$$

$$l_v = \frac{t_v\,c}{2}\left(1 - \frac{v^2}{c^2}\right)\ . \tag{1167}$$

The run time of the signal can be determined directly by the pointer position t_v of a clock resting in Σ_o, where the left end of the bar $x_1(t)$ sits at time t_v, hence

H. Günther and V. Müller, *The Special Theory of Relativity*,
https://doi.org/10.1007/978-981-13-7783-9

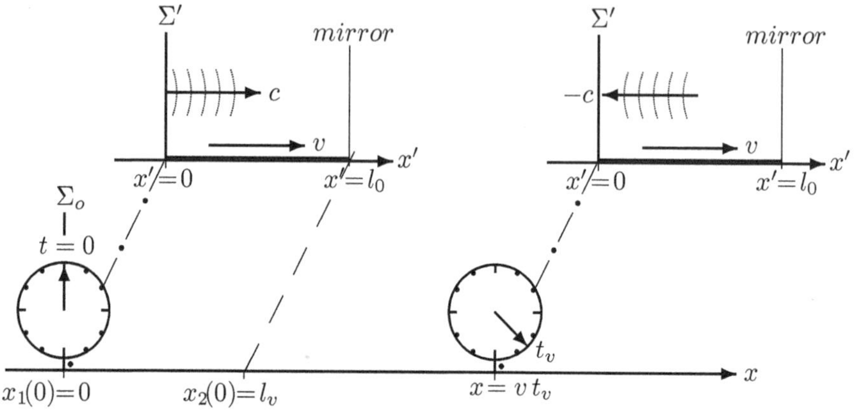

Fig. A.1 A bar with rest length l_o is in motion with constant v in x-direction of system Σ_o. Its length $l_v = x_2(t) - x_1(t)$ in this system is defined as the difference of end point coordinates at a time t. At time $t = 0$ in Σ_o, a light signal is emitted at the left end of the bar with coordinate $x = 0$. After its reflexion at the right end of the bar, it gets back to the left end, now being at the coordinate $x = v\,t_v$. We get the length l_v of the bar in Σ_o by Eq. (1167). Both dashed-dotted lines connect points belonging to the same event.

at $x = v\,t_v$. The measurements of this time t_v now determine, whether and how the length l_v depends on its velocity v in Σ_o.

Problem 2 *If the postulate of homogeneity and isotropy holds for a linear synchronisation (20), then the coordinate transformations (16) are linear.*

Chapter 1, Sections 4, 5
The general coordinate transformation (16) reads with Eqs. (18) and (19)

$$x' = f(x - vt, v)\,, \quad t' = f_4(x, t, v)\,, \text{ where } \; f_4(x, 0, v) = \Omega(x, v)\;. \tag{1168}$$

As introduced in Chap. 1, Sect. 5, a bar might be at rest on the x'-axis with its left end point in the origin of system Σ'. Then the rest length l_o is measured as coordinate of its right end pointes, $x'_1 = 0\,,\; x'_2 = l_o$.

To determine the length l_v of the bar moving in system Σ_o with velocity v, we need the positions of its end points at one and the same time in Σ_o, say, e.g. also for $t = 0$, s. Chap. 1, Fig. 10.

Due to the initial conditions (11), it follows for the left end point from ($x'_1 = 0,\ t' = 0$) also ($x_1 = 0,\ t = 0$). For the right end point, it follows from Eq. (1168) with $t = 0$ and $x' = x'_2 = l_o$, that $l_o = x'_2 = f(x_2, v)$.

Now it is $l_v = x_2 - x_1 = x_2$. Due to the postulate (21), the quotient l_o/l_v cannot depend on the coordinates (x, t), but only on the velocity v,

$$\frac{l_o}{l_v} = \frac{f(x_2, v)}{x_2} = k(v)\;. \tag{1169}$$

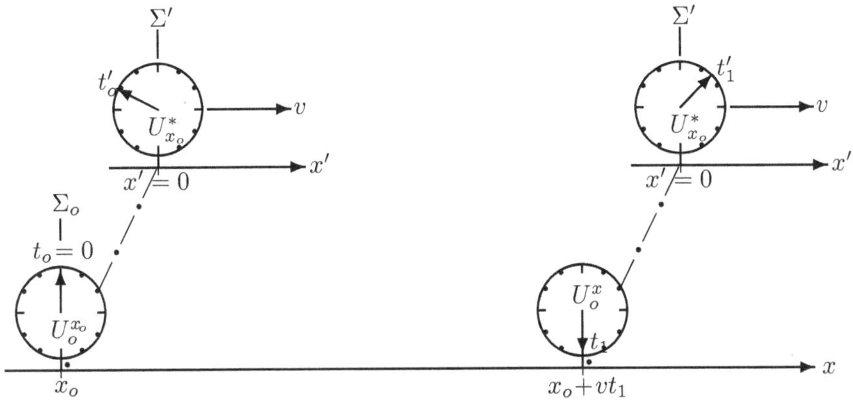

Fig. A.2 The pointer positions of an in Σ' resting clock $U^*_{x_o}$ are compared with the pointer positions of the in Σ_o resting clocks U^x_o, when it just passes by. The dashed-dotted lines connect points representing the same event.

The argument x_2 is arbitrary; hence, the function f is linear,

$$f(x - vt, v) = k(v)(x - vt) \ . \tag{1170}$$

Among the clocks resting in Σ', there might be $U^*_{x_o}$ the clock that for $t_o = 0$ in Σ_o just passes by at position x_o. Because of Eq. (1168) $U^*_{x_o}$ has then the pointer position t'_o,

$$t = t_o = 0 \ , \quad U^*_{x_o}\colon\ t'_o = f_4(x_o, 0, v) = \Omega(x_o, v) \, . \tag{1171}$$

After the time $t = t_1$, the clock $U^*_{x_o}$ sits at position $x = x_o + v\,t_1$ in Σ_o s. Fig. A.2. By Eq. (1168), it holds for time t'_1

$$t = t_1 \ , \quad U^*_{x_o}\colon\ t'_1 = f_4(x_o + v\,t_1, t_1, v) \, . \tag{1172}$$

The pointer of an in Σ' resting clock $U^*_{x_o}$ proceeds also by the difference

$$t' = t'_1 - t'_o = f_4(x_o + vt_1, t_1, v) - \Omega(x_o, v) \, , \tag{1173}$$

while in system Σ_o, the time difference t follows from the comparison of the resting clocks at positions x_o and $x = x_o + v\,t_1$,

$$t = t_1 - t_o = t_1 \ . \tag{1174}$$

By the postulate (21), the quotient t'/t cannot depend on the coordinates (x, t), but only on the velocity v,

$$\frac{t'}{t} = \frac{f_4(x_o + vt_1, t_1, v) - \Omega(x_o, v)}{t_1} = Q(v) \ , \tag{1175}$$

hence we can write for arbitrary values of x_o simply x,

$$f_4(x + vt, t, v) = \Omega(x, v) + Q(v)\,t \tag{1176}$$

with an in general nonlinear synchronisation, defined by the function Ω, Eq. (19).

If we restrict us on an according to Eq. (20) linear synchronisation to adopt the definition of simultaneity to the homogeneous description of space-time, hence $\Omega(x, v) = \theta(v)\,x$, then it follows from $f_4(x + vt, t, v) = \theta(v)\,x + Q(v)\,t$ after renaming the variables[1] the linearity of the function f_4 from Eq. (1168)

$$f_4(x, t, v) = \theta(v)\,x + q(v)\,t\ . \tag{1177}$$

Due to Eqs. (1170) and (1177), we get the linearity of the coordinate transformations (1168) from the postulate of homogeneity (21), when we *define* the simultaneity in systems Σ' with a *linear* synchronisation function.

Problem 3 *Compare the* Galilei *transformation (69) with the result of* Michelson*'s experiment.*

Chapter 3, Sections 1, 2, Chapter 4, Section 1

In the Michelson experiment described in Chap. 4, Sect. 1, the observer sits in the reference system Σ_o, and the interferometer moves with system Σ' at velocity v with respect to Σ_o, Chap. 4, Fig. 3. According to Eq. (80) we derive in Σ_o a run-time difference $\Delta t = t_2 - t_1$ of the light signal along both legs of the interferometer with length l_2 and l_1. Here we assumed l_1 in the direction of motion, and l_2 orthogonal to it, i.e. l_1 should depend on the velocity v, but not l_2 because of Eq. (13).

Now we test the dependence of the length change from the direction of motion by rotating the interferometer by an angle $\pi/2$. Again we evaluate the run-time difference $\Delta t_{\frac{\pi}{2}}$ as observed from Σ_o and find the result Eq. (81). Now it is interferometer leg l_2, lying in the direction of motion which could depend on the velocity v, but not l_1. In Eq. (82) we have derived both run-time differences, and there we have put both lengths l_1 and l_2 out of the parenthesis. That is, we have supposed that the length of an interferometer leg does *not* depend on the velocity v with respect to Σ_o. Equation (82) is based on the physical assumption Eq. (65) of the Galilei transformation (69). As shown in Chap. 4, Sect. 1, the outcome of the Michelson experiment is in contradiction with Eq. (82). This means that the Galilei transformation is in contradiction to the Michelson experiment, a precision experiment, suitable to measure the in v/c quadratic terms.

From an observation in an other reference system, the theoretical prediction of the Galilei transformation cannot change since there it holds $t' = t$, Eq. (69). The measured time difference is independent of the state of motion of the interferometer.

Problem 4 *A flying object might be equipped with precision clocks. It might start on the equator in a height of* 1000 m, *circulate around the Earth in western direction and take one star day for completing the circle. Compare the time in the on board*

[1] We can write $f_4(\overline{x} + v\,t, t, v) = \theta(v)\,\overline{x} + Q(v)\,t$. With the abbreviation $\overline{x} = x - v\,t$, we get then $f_4(x, t, v) = \theta(v)\,x - \theta(v)\,v\,t + Q(v)\,t = \theta(v)\,x + q(v)\,t$.

clock with the clock on the Earth after arrival. Does the result change when the flying object moves for a flight in the same height and time duration, but in eastern direction?

Remark: The General Relativity Theory tells us that locally gravitational fields and the corresponding accelerated motions change the run time of clocks, an effect that is of the same order as the special-relativistic time dilatation in this experiment. However, here the gravitational effects are disregarded, cp. Chap. 13, Sect. 3.2.

Chapter 1, Section 2, Chapter 4, Section 2, Chapter 8, Section 8

The clocks are situated in accelerated reference systems, but for small time intervals dt' there holds Eq. (97) describing the special-relativistic time dilatation for the temporal inertial systems Σ', having the velocity $\mathbf{v}(t)$ with respect to the inertial system Σ_o, i.e. $dt' = \sqrt{1 - \mathbf{v}\cdot\mathbf{v}/c^2}\, dt$ with the in Σ_o measured time interval dt. For finite times, this equation must be integrated

$$(t_2' - t_1') = \int_{t_1}^{t_2} \sqrt{1 - \frac{\mathbf{v}\cdot\mathbf{v}}{c^2}}\, dt = \sqrt{1 - \frac{v^2}{c^2}}\,(t_2 - t_1) \quad \text{for} \quad |\mathbf{v}| = \text{const}\,. \qquad (1178)$$

We neglect the minor change of the velocity caused by the orbital motion of the Earth centre P around the Sun during the day, we can treat the rest system of P with high precision as inertial system Σ_o.

A stationary clock on the equator is at each time t in an other inertial system. Our laboratory system Σ' has a velocity of constant amount v but changing direction with respect to Σ_o.

The time τ of Earth rotation is 4 min less than 24 hours, just the difference between a stellar and a solar day. Hence $\tau = (24\cdot 60\cdot 60 - 4\cdot 60)\,\text{s} = 86\,160\,\text{s}$. Because of the orbital motion of the Earth around the Sun, the Earth must fulfil still 4 min further eigenrotation, for having the Sun again at the same position. For a precise definition, we refer to astronomy textbooks. We take the Earth radius at the equator $R = 6\,378\,163\text{m}$, then we get for the angular velocity of the Earth, i.e. for the angle in arc scale the Earth rotates per second a value

$$\omega = (2\pi)/\tau\, \text{s}^{-1} = 0.000072924\ \text{s}^{-1}\,. \quad \text{Angular velocity of the Earth} \qquad (1179)$$

Then we get for the velocity v_L of the laboratory at equator with respect to Σ_o and the corresponding LORENTZ factor γ_L[2]

$$\left.\begin{aligned} v_L &= \omega R = 465.1\,\text{ms}^{-1}\,, \\ c &= 299\,792\,458\,\text{ms}^{-1}\,, \\ \gamma_L &= \frac{1}{\sqrt{1 - v_L^2/c^2}} \approx 1 + 1.2034\cdot 10^{-12}\,. \end{aligned}\right\} \quad \begin{array}{l}\text{Laboratory}\\ \text{on the equator}\end{array} \qquad (1180)$$

[2]The evaluation is not completely exact. $2\pi R$ is the length of an in Σ_o resting circle of radius R. By the rotation of the Earth with respect to Σ_o the surface elements in direction of motion are LORENTZ contracted as described in Chap. 4, Sect. 1. But this correction factor is of the order of magnitude of $1 - 1/2\cdot v_L^2/c^2 = 1 - 1/2\cdot 4,6^2\cdot 10^4/3^2\cdot 10^{16} \approx (1 - 1/2\cdot 10^{-12})$, hence it can be neglected within the measuring accuracy.

The flying object, moving to the west as seen from the rest system Σ_o at the centre of the Earth, moves with the same velocity to the west, as the Earth rotates to the east. The velocity of the flying object with respect to the inertial system Σ_o is then zero:

$$v_W = 0 \text{ ms}^{-1} . \qquad \text{Velocity of the flying object moving to the west with respect to } \Sigma_o \tag{1181}$$

The flight time of the west-directed moving object in Σ_o is one stellar day, hence

$$t_W = \tau = (24 \cdot 60 \cdot 60 - 4 \cdot 60)\,\text{s} = 86\,160\,\text{s} . \qquad \text{Flight time on the flying object going to the west} \tag{1182}$$

Furthermore, we denote with t_L the laboratory time on the equator. Then due to Eqs. (1178) and (1180)

$$t_L = \gamma_L\, t_W = (1 + 1.2034 \cdot 10^{-12}) \cdot 86\,160\,\text{s} . \qquad \text{Time in the laboratorium on the equator} \tag{1183}$$

The difference $t_w - t_L$ is then

$$t_W - t_L = (1 - \gamma_L)\, t_W = 1.2034 \cdot 10^{-12} \cdot 86\,160\,\text{s} = 103.7 \cdot 10^{-9}\,\text{s} . \tag{1184}$$

After landing the clock on the flying object will be advanced by 103.7 nanoseconds and not retarded as expected by a simple mind. The rotating Earth does not represent an inertial system. The observer in the laboratory on the equator is moving in an always changing inertial systems Σ', that leads him back to the original inertial system Σ_o after one rotation. It corresponds to the returning twin in the paradoxon, Chap. 8, Sect. 8.

An other situation occurs, when the flying object starts in eastern direction. As considered from Σ_o, at each time moment it sits in an other inertial system Σ''. Because of the flight hight of 1000 m we consider a circle K in the equatorial plane of $R_1 = (6\,378\,163 + 1000)\text{ m} = 6\,379\,163\text{m}$. With respect to Σ_o, the velocity v_K of a point Q_K amounts, cp. Eqs. (1178) and (1179),

$$v_K = \omega\, R_1 \ , \tag{1185}$$

$$v_K = 0.000072924 \cdot 6\,379\,163\,\text{ms}^{-1} = 465.2\,\text{ms}^{-1}. \qquad \text{Velocity of a point } Q_K \text{ with respect to } \Sigma_o \text{ in 1000 m heigth over the sea level} \tag{1186}$$

As seen from Σ_o, the motion of the eastern flying object on the circle K consists of two components, first of the rotation of the Earth, i.e. of a point fixed on the corotating circle Q_K with a velocity v_K with respect to Σ_o evaluated above, and from a motion of the flying object with respect to the with the Earth connected circle, that also amounts of a value v_K, but now with respect to the Earth.

The orbital velocity v_o of the flight object with respect to Σ_o is then[3]

$$v_E = 2\,v_K = 930.4\,\mathrm{ms}^{-1}\ . \qquad \text{Velocity of the eastern travelling flying object with respect to } \Sigma_o \tag{1187}$$

Then we evaluate the γ_E factor,

$$\gamma_E = \frac{1}{\sqrt{1 - v_E^2/c^2}} = \frac{1}{\sqrt{1 - 4\,v_K^2/c^2}} \approx 1 + 4.8158 \cdot 10^{-12}\ . \tag{1188}$$

The flight time t_O of the body going to the east is after Eq. (1178) by the factor $1/\gamma_E$ smaller then the in Σ_o measured time t_W coinciding with the time of the body flying to the west. Hence it is $t_E = t_W/\gamma_E$ and then

$$t_W - t_E = (1 - 1/\gamma_E)\,t = 4.8158 \cdot 10^{-12} \cdot 86\,160\,\mathrm{s} \approx 414.9 \cdot 10^{-9}\,\mathrm{s}\ . \tag{1189}$$

This is also the difference between the times measured on both flying objects. From Eq. (1184) and (1189) we get for $t_L - t_E = (t_W - t_E) - (t_W - t_L)$,

$$t_L - t_E = 311.2 \cdot 10^{-9}\,\mathrm{s}\ . \tag{1190}$$

According to Eq. (1190) the time t_O of the clock on the flying object on the eastern route is after landing by 311.2 nanoseconds retarded with respect to the time t_L, of the clock remaining fixed in the laboratory on the equator, since the time t_O is retarded by a larger amount with respect to time t_w, as time t_L is retarded with respect to t_w.

In reality, the performance and analysis of such experiments are quite expensive. For the general relativistic, i.e. the gravitative influence of the clock period, the precise flight hight becomes essential. Furthermore, the flight times must be measured precisely. J. C. Hafele and R. E. Keating (1971) performed this experiment with caesium atomic clocks on board of aircrafts, on circles around the Earth, which confirmed the time dilatation with high precision.

Problem 5 *We define a simultaneity according to the elementary relativity principle with the* Lorentz *synchronisation parameter (103) in supposing the postulates (99) and (101) in system Σ', than the value of the speed of light for system Σ' is also fixed. Evaluate without reference to the* Lorentz *transformation—and therefore without the addition theorem (106)—the velocity of the front of a light wave in Σ', for which the speed of light has the value c in system Σ_o.*

[3]To evaluate the velocity of the flying object $v_E = 2\,v_K$ with respect to Σ_o, we have to add the velocity v_K of the object with respect to the Earth to the velocity v_K of the rotating Earth with respect to Σ_o. Here in principle we had to apply Einstein's addition theorem (106), hence $\widetilde{v_E} = \frac{v_K + v_K}{1 + v_K^2/c^2} \approx v_E \cdot (1 - 2.4 \cdot 10^{-12})$. The correction of the γ_E-term can however be neglected.

Chapter 4, Sections 3, 4

We only use the postulates (99) and (101) of the LORENTZ contraction and the time dilatation in system Σ_o, and EINSTEIN's definition of simultaneity with the parameter (103) as consequence of the elementary relativity principle. For $t = 0$ in Σ_o we set the pointer of the in Σ' resting clock at t'

$$t' = -\gamma\, v\, x/c^2 \ . \tag{1191}$$

In Σ_o, the system Σ' has a velocity v in x-direction. We want to derive the value of the light propagation c' for an in Σ' resting observer, for which in Σ_o a value c was measured.

We take a bar resting on the x'-axis of Σ' as measuring length, with position of the left end point coinciding with the origin $x_1' = 0$ in Σ' as considered in Chap. 4, Sect. 2. The right end point is called x_2'. The bar has then in Σ' a rest length $l_o = \Delta x' = x_2' - x_1' = x_2'$, Fig. A.3.

At time t, we measure in Σ_o for both ends of the measuring bar the coordinates $x_1 = v\, t$ and x_2. The difference $\Delta x = x_2 - x_1$ of the at the same time measured coordinates is by definition the length of the bar $l_v = \Delta x = (x_2 - v\, t)$ moving in Σ_o with velocity v. According to the postulate (99), the moving line $l_v \equiv l'$ is shorter by the factor $1/\gamma$ with respect to the rest length l_o, $l' = l_o/\gamma$, hence

$$\Delta x' = \gamma\, \Delta x \ . \tag{1192}$$

At event O, the common coordinate origin with $(x_1 = 0, t_1 = 0)$ and $(x_1' = 0, t_1' = 0)$, a light signal is emitted in positive x'-direction. The run time $\Delta t_*'$ of the signal can then immediately be read off at clock U_v^* fixed at the end point of the measuring bar. It shows the pointer position $t_C' = \Delta t_*'$ at arrival of the signal. The observer in Σ' measures a signal velocity c'

$$c' = \frac{\Delta x'}{\Delta t_*'} = \frac{\gamma\, \Delta x}{\Delta t_*'} \ . \tag{1193}$$

Hence we need the pointer position $\Delta t_*'$. In Σ_o, we can observe that the pointer of the clock U_v^* for $x = \Delta x$ and $t = 0$ is at position t_B' according to the synchronisation prescription (1191),

$$t_B' = \frac{-\gamma\, \Delta x\, v}{c^2} \ . \tag{1194}$$

Furthermore, we measure in Σ_o for the travel time of the signal from emission to absorption at the ends of the measuring bar a time $\Delta t_\rightarrow$,

$$\Delta t_\rightarrow = \frac{\Delta x}{c - v} \ . \tag{1195}$$

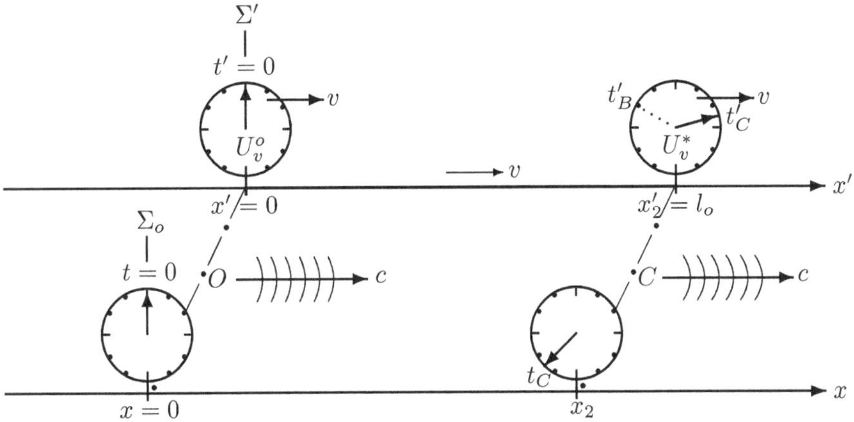

Fig. A.3 In Σ_o, we observe the measurement of the signal velocity in Σ', and we calculate an example. The light signal is emitted at event O with $x = x' = 0$ and $t = t' = 0$. Due to the definition of simultaneity according to the elementary relativity principle (1191) we observe at time $t = 0$ in Σ_o for the clock U_v^* at the end of the measuring line the pointer position t'_B. The arrival time of the signal on this clock t'_C is the travel time $\Delta t'_* = t'_C$ measured by the in Σ' resting observer, the velocity is therefore $c' = x'_2/t'_C$. In the picture, we have $v = 0.8\,c$, hence $\gamma = 1/\sqrt{1 - v^2/c^2} = 1.67$, and the clocks are gauged so that the time $\Delta t_o := 2x_2/c$ corresponds to a quarter position of the pointer. For a scale of 60 units on the watch face, it is then $\Delta t_o = 2x_2/c = 15$. Therefore in Chap. 4, Fig. 12, the pointer position shows $t'_B = t'(x_2, 0) = -x_2 v/c^2\gamma = -10$. The observer in Σ_o therefore states that the time span between t'_B and 0 on the clock at the end of the measuring bar does not count for the evaluation of the speed of light in Σ', even if from his point of view the light signal was already emitted. For the observer resting at the end of the measuring line in Σ', the signal is only emitted, when the pointer on his clock U_v^* stands at position 0. The pointer position t_C is evaluated according to Eq. (1195) with $v/c = 0.8$ to $t_C = \Delta x/(c - v) = \big(2x_2/c\big)\big(c/2(c - v)\big) = 15\big(1/2(1 - 0.8)\big) = 37.5$. Hence the observer resting in Σ_o finds, that the light signal transverses the in Σ' resting measuring bar with velocity $c - v = 0.2\,c$. The pointer position t'_C is found from Eq. (1197), $t'_C = \Delta t'_* = \big(2x_2/c\big)\big(\gamma/2\big) = 15/1.2 = 12.5$. Taking into account that $\Delta x = \Delta t_o\,c/2$, hence $x'_2 = \gamma\,\Delta x = \gamma\,\Delta t_o\,c/2 = 1.67 \cdot 15c/2 = 12.5\,c$, so it follows immediately that the in Σ' measured speed of light $c' = x'_2/t'_C = c$. The dashed-dotted lines again connect points describing the same event, here the events O and C.

Due to the time dilatation (101) the pointer on the moving clock U_v^* proceeds during the Σ_o-time only by $\Delta t'_{\rightarrow}$,

$$\Delta t'_{\rightarrow} = \frac{\Delta t_{\rightarrow}}{\gamma}\ . \tag{1196}$$

Altogether we find that the pointer of U_v^* at arrival of the signal shows a position $\Delta t'_* = t'_B + \Delta t'_{\rightarrow}$, hence with Eqs. (1194)–(1196)

$$\begin{aligned}\Delta t'_* &= \frac{-\gamma\,\Delta x\,v}{c^2} + \frac{\Delta x}{\gamma\,(c - v)} = \frac{\Delta x - \gamma^2\,\Delta x(c - v)\,v/c^2}{\gamma\,(c - v)} \\ &= \frac{\Delta x\,\gamma\,(1 - v^2/c^2 - v/c + v^2/c^2)}{(c - v)} = \frac{\Delta x\,\gamma\,(1 - v/c)}{(c - v)}\ ,\end{aligned}$$

$$\Delta t'_* = \frac{\Delta x \, \gamma}{c} \, . \tag{1197}$$

From Eqs. (1193) and (1197) it follows that the observer in Σ' measures the same value c for the propagation velocity of light as the observer in Σ_o,

$$c' = c \, . \tag{1198}$$

So the observer in Σ_o measures that the emitted light wave approaches the end of the measuring bar only with retarded velocity. However, this effect is compensated by the time dilatation and the synchronisation prescription Eq. (1191), s. Fig. A.3.

Problem 6 *Show by means of the* LORENTZ *transformation between the at first distinguished system Σ_o and two arbitrary inertial systems Σ' and Σ'', that the* LORENTZ *transformation between two arbitrary inertial systems has the same form.*

Chapter 4, Section 4

We denote with v and u the velocities of $\Sigma'(x', t')$ and $\Sigma''(x'', t'')$ in x-direction of $\Sigma_o(x, t)$, then we have the corresponding LORENTZ transformations, using the nomenclature of Eq. (102),

$$\left.\begin{array}{lll} x'' = \gamma_u \, (x - u\,t) \, , & x' = \gamma_v \, (x - v\,t) \, , & \longrightarrow \quad x = \gamma_v \, (x' + v\,t') \, , \\ t'' = \gamma_u \, (t - xu/c^2) \, , & t' = \gamma_v (t - xv/c^2) \, , & \qquad\;\; t = \gamma_v (t' + x'v/c^2) \, . \end{array}\right\} \tag{1199}$$

We insert the third column of Eq. (1199) into the first column and find

$$\begin{aligned} \frac{x''}{\gamma_u} &= x - u\,t = \gamma_v \, (x' + v\,t') - \gamma_v \left(u\,t' + \frac{u\,v\,x'}{c^2} \right) \, , \\ x'' &= \gamma_u \, \gamma_v \left[x'(1 - \tfrac{u\,v}{c^2}) - t'(u - v) \right] \, , \\ x'' &= \gamma_u \, \gamma_v (1 - u\,v/c^2) \left[x' - \tfrac{u - v}{1 - u\,v/c^2} \, t' \right] \, . \end{aligned}$$

Likewise it holds

$$\begin{aligned} t'' &= \gamma_u \, (t - x\,u/c^2) = \gamma_u \, \gamma_v \, (t' + x'\,v/c^2) - \tfrac{u}{c^2} \, \gamma_u \, \gamma_v \, (x' + v\,t') \, , \\ t'' &= \gamma_u \, \gamma_v \, (t' + \tfrac{v}{c^2} \, x' - \tfrac{u}{c^2} \, x' - \tfrac{u\,v\,t'}{c^2} \, t') \, , \\ t'' &= \gamma_u \, \gamma_v \, [t'(1 - \tfrac{u\,v}{c^2}) - \tfrac{x'}{c^2} (u - v)] \, , \\ t'' &= \gamma_u \, \gamma_v \, (1 - u\,v/c^2) [t' - \tfrac{x'}{c^2} \tfrac{u - v}{1 - u\,v/c^2}] \, . \end{aligned}$$

Due to Eq. (171) it is

$$\gamma_u \, \gamma_v \, (1 - u\,v/c^2) = \gamma_{u'} \quad , \; u' = \frac{u - v}{1 - u\,v/c^2} \, .$$

With $u' = (u - v) \Big/ (1 - u\,v/c^2)$ we find

$$
\begin{aligned}
\sqrt{1-\frac{u^2}{c^2}}\sqrt{1-\frac{v^2}{c^2}} &= (1-\frac{u\,v}{c^2})\sqrt{1-\frac{u'^2}{c^2}}\\
\sqrt{1-\frac{u^2}{c^2}}\sqrt{1-\frac{v^2}{c^2}} &= (1-\frac{u\,v}{c^2})\sqrt{1-\frac{1}{c^2}\left[(u-v)/(1-u\,v/c^2)\right]^2}\\
\left(1-\frac{u^2}{c^2}\right)\left(1-\frac{v^2}{c^2}\right) &= \left(1-\frac{u\,v}{c^2}\right)^2\left(1-\frac{1}{c^2}\left[(u-v)/(1-u\,v/c^2)\right]^2\right)\\
(c^2-u^2)\,(c^2-v^2) &= (c^2-u\,v)^2\left(1-\frac{1}{c^2}\left[(u-v)/(1-u\,v/c^2)\right]^2\right)\\
c^2\,(c^2-u^2)\,(c^2-v^2) &= (c^2-u\,v)^2\left(c^2-\left[(u-v)/(1-u\,v/c^2)\right]^2\right)\\
c^2\,(c^2-u^2)\,(c^2-v^2) &= c^2\,(c^2-u\,v)^2-c^4\,(u-v)^2\\
c^6+c^2\,u^2\,v^2-c^4\,u^2-c^4\,v^2 &= c^6-2\,c^4\,u\,v+c^2\,u^2\,v^2-c^4\,u^2+2\,c^4\,u\,v-c^4\,v^2\ ,
\end{aligned}
$$

hence

$$\gamma_u\,\gamma_v\,(1-u\,v/c^2)=\gamma_{u'}\quad ,\ u'=\frac{u-v}{1-u\,v/c^2}\ ,$$

and we get

$$
\begin{aligned}
x'' &= \gamma_{u'}\,(x'-u'\,t')\,, & & x'=\gamma_{u'}\,(x''+u'\,t'')\,,\\
& & \longleftrightarrow & \\
t'' &= \gamma_{u'}\,(t'-x'\,u'/c^2)\,, & & t'=\gamma_{u'}\,(t''+x''\,u'/c^2)\,.
\end{aligned}
\qquad (1200)
$$

Motion in x-direction
Special LORENTZ transformation for arbitrary inertial systems

Here u' is the velocity of system Σ'' measured from Σ', that results from velocities v and u of Σ' and Σ'', resp., both with respect to system Σ_o by EINSTEIN's addition theorem (106). Therefore, the LORENTZ transformation holds true between two arbitrary inertial systems.

Problem 7 *For an object moving with velocity $|u|\le c$ in Σ_o, it holds also $|u'|\le c$ in any other inertial system Σ'. Likewise for $c<|u|$ in Σ_o, it holds also $c<|u'|$ in Σ'. Remark: Starting from the inertial system Σ_o, all other inertial systems are represented by (real or virtual) bodies with $|v|<c$.*

Chapter 4, Section 3

We restrict us on velocities in x-direction of Σ_o. For system Σ' we measure in system Σ_o a velocity v with $0<v<c$.

Let u be the in Σ_o measured velocity of a body L with $0<u\le c$, hence

$$0<\frac{v}{c}<1\ ,\quad 0<\frac{u}{c}\le 1\ . \qquad (1201)$$

Now we form from velocities u and v quantities, for which we can apply EINSTEIN's addition theorem (106).

Because of (1201) we have

$$0 < 1 + \frac{v}{c} \quad \text{and} \quad 0 \le 1 - \frac{u}{c} \,. \tag{1202}$$

It follows

$$0 \le \left(1 + \frac{v}{c}\right)\left(1 - \frac{u}{c}\right) = 1 + \frac{v}{c} - \frac{u}{c} - \frac{u\,v}{c^2} \,,$$
$$\frac{u}{c} - \frac{v}{c} \le 1 - \frac{u\,v}{c^2} \,,$$
$$\frac{u/c - v/c}{1 - u\,v/c^2} \le 1 \,,$$

so that according to EINSTEIN's addition theorem (106), also for the velocity measured from Σ' with $u' \le c$ of the object L, our assertion is fulfilled,

$$u' = \frac{u - v}{1 - u\,v/c^2} \le c \;. \tag{1203}$$

Now we assume that a velocity w measured in Σ_o is a superluminal velocity, $0 < c < w$ or $-w < -c < 0$, i.e. $0 < c < |w|$, consequently

$$0 < \frac{v}{c} < 1 \;, \quad 0 < \frac{c}{|w|} < 1 \;. \tag{1204}$$

From Eq. (1204) it follows

$$0 < \left(1 + \frac{v}{c}\right)\left(1 - \frac{c}{|w|}\right) = 1 + \frac{v}{c} - \frac{c}{|w|} - \frac{v}{|w|} \,,$$

hence

$$c\left(1 - \frac{v\,|w|}{c^2}\right) < |w| - v$$

and therefore

$$c < \frac{|w| - v}{1 - v\,|w|/c^2} = |w'| \;, \tag{1205}$$

so that, with the addition theorem (106), we measure also in reference system Σ' a superluminal velocity

$$c < |w'| \;.$$

That means, if $0 < c < w$, then

$$c < \frac{w - v}{1 - v\,w/c^2} = w' \tag{1206}$$

and in case of negative velocity, $w < -c$

$$c < \frac{-w - v}{1 + v\,w/c^2} = -w' \;,$$

hence

$$w' = \frac{w+v}{1+v\,w/c^2} < -c\ . \tag{1207}$$

Hypothetical particles that move with super-light velocity, have this property in all inertial systems, they are called *tachyons*.[4] Their existence is not excluded by the special theory of relativity. Tachyons, however, will come in conflict with a fundamental experience, with *causality*.

From the systems Σ_o and Σ' with the velocity v with respect to Σ_o in the x-direction, two events $E_1(x_1, t_1) = E_1(x'_1, t'_1)$ and $E_2(x_2, t_2) = E_2(x'_2, t'_2)$ are observed. With the LORENTZ transformation (105) we find

$$\left(t'_2 - t'_1\right)\sqrt{1-v^2/c^2} = \left(t_2 - t_1\right) - \frac{v}{c^2}\,(x_2 - x_1) = (t_2 - t_1)\left[1 - \frac{v}{c^2}\,\frac{x_2 - x_1}{t_2 - t_1}\right]\,.$$

We denote by $u := (x_2 - x_1)/(t_2 - t_1)$ the velocity, with which the two events in Σ_o can be connected and write

$$\left(t'_2 - t'_1\right)\sqrt{1-v^2/c^2} = \left(t_2 - t_1\right)\left[1 - \frac{v\,u}{c^2}\right]\,. \tag{1208}$$

If u is the velocity of a 'normal particle' with $0 < u \le c$, a solid body or, e.g. a zero rest mass photons, which is sent to the event E_1 in Σ_o and there triggers the event E_2, so that E_2 is causally related to E_1, then it follows from Eq. (1208) because of the positive sign of the square bracket of $t_1 < t_2$ in Σ_o, also $t'_1 < t'_2$ in Σ' in accordance with our experience that the order of cause and effect, i.e. causality is independent of the state of motion of the observer. By circumventing the concept of causality, one also says that the two events are time-like related to each other.

Let u be the velocity of a tachyon, which we have denoted by w above, that is $0 < c < u \equiv w$. The tachyon is sent to the event E_1 in Σ_o, and it is supposed to trigger the event E_2, then we can set the velocity v of Σ' according to (1208) so that from $t_1 < t_2$ in Σ_o, now there follows $t'_2 < t'_1$ in Σ'. For this behaviour, we only have to choose a velocity v of Σ' according to $c^2/w < v < c$, which is always possible. According to Fig. A.4, c^2/w is the velocity of Σ', at which the velocity w of the tachyon, observed from Σ', changes its direction. For

$$1 < w\,v/c^2 \longrightarrow 1 - w\,v/c^2 < 0\ , \tag{1209}$$

follows $t'_2 < t'_1$ for $t_1 < t_2$. The two events E_1 and E_2, interpreted in Σ_o as cause and effect, would then change their order if we observe them from system Σ'. In this case, it is said that the events are space-like related.

One can easily solve this problem and exclude a violation of causality simply by saying, that tachyons do not exist. However, such an ad hoc hypothesis always represents a deficiency in theory. Note that tachyons only contradict

[4] According to a proposal of Feinberg (1967).

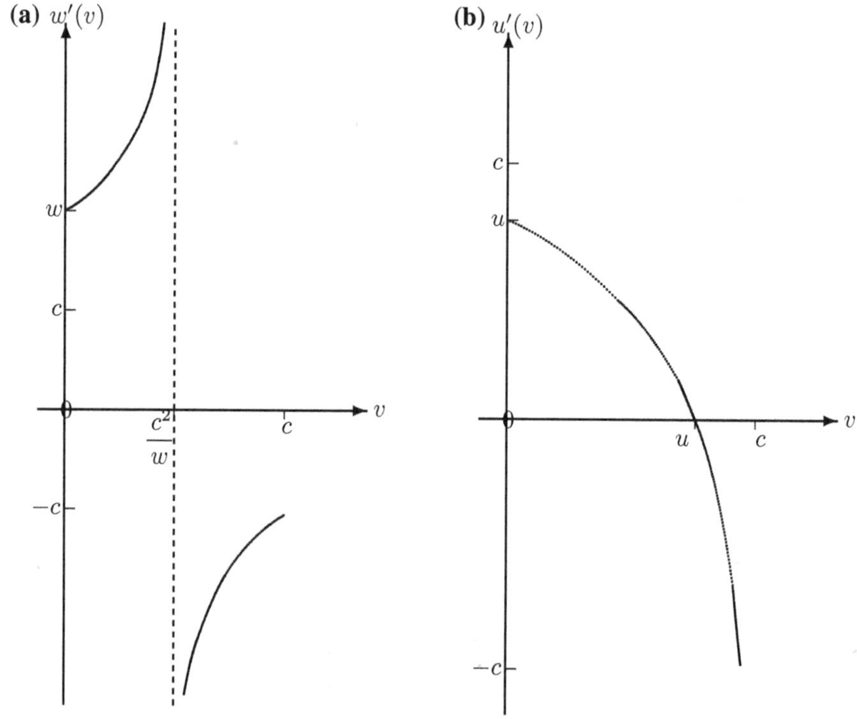

Fig. A.4 **a** Schematic representation of the function $w' = \frac{w-v}{1-v\,w/c^2}$, the tachyon velocity in system Σ', when the tachyon has the velocity w in system Σ_o. We denote with v the velocity of Σ' as measured from Σ_o. **b** Schematic representation of the function $u' = \frac{u-v}{1-v\,u/c^2}$, that is, the velocities of particles measured system Σ', while the particle velocity is u in system Σ_o. The velocity between both systems v cannot reach the speed of light c. The factor $\left[1 - \frac{v\,u}{c^2}\right]$ is always positive, while the factor $\left[1 - \frac{v\,w}{c^2}\right]$ changes its sign at $v = c^2/w$. This would lead to a causality violation if signals could be transmitted by tachyons.

our experience, if they are able to transmit signals and therefore energy, which we will discuss in more detail in Problem 9. There we investigate the physical consequences of collision processes of particles and hypothetical tachyons. It would be best, if we had a theory of elementary particles that excludes signal transmission by tachyons, that excludes the energy release of tachyons to particles, cp. TREDER (1974), LIEBSCHER (1975), GÜNTHER (1996). We formulate: Tachyons realise a correlation, but no interaction.

The existence of correlations, of mutually conditioned states over arbitrarily large distances without any interaction effect, is encountered in quantum mechanics.

Problem 8 *In this problem we investigate the question: We assume that tachyons exist. The physical parameter of a particle is its rest mass m_o. What is the physical parameter of a tachyon?*

Chapter 4, Section 4, Chapter 6, Section 3

We remind at first on the NOETHER theorem, cited on Chap. 4, Sect. 2.1, which leads to conservation laws of energy and momentum as consequence of the translation invariance of our space-time. We must secure in all considered collision processes these conservation laws.

Furthermore, we shall employ EINSTEIN's mass–energy equivalence,

$$E = m\,c^2 \ .$$

Here m denotes the inertial mass from NEWTON's mechanics, cp. Chap. 7, Sect. 1.

For distinction we shall denote in the following as 'normal' particles with rest mass m_o, particles with vanishing rest mass such as photons as light-like, and superluminal particles as tachyons.

As in Chap. 7, Sect. 1, we consider again a total inelastic collision of two particles, where we now admit that it could be tachyons.[5] We are again interested in the dependence of the particle mass m on its velocity u, i.e. we look for the function $f(u)$ in the relation

$$m = \xi_o\, f(u) \ . \tag{1210}$$

We suppose no restrictions for the velocity u and ξ_o is a parameter, not identified from the beginning with the rest mass.

In system Σ' we consider the total inelastic collision of two particles with velocities u_1', and $u_2' = -u_1'$, both are characterised by the same parameter ξ_o, but they have opposite momenta, so that the total momentum of the system vanishes.

The velocity of system Σ' measured in Σ_o is denoted by v.

The momentum conservation in Σ', regularly written as

$$\xi_o\, f(u_1')\, u_1' + \xi_o\, f(u_2')\,(u_2') \equiv \xi_o\, f(u_1')\, u_1' + \xi_o\, f(-u_1')\,(-u_1') = 0$$

with an uneven function $f(-u) = f(u)$, is now formulated with a more general ansatz, that is justified only after the following derivation:

$$\xi_o\, f(u_1')\, u_1' + \delta \cdot \xi_o\, f(-u_1')\,(-u_1') = 0 \ . \qquad \text{Momentum conservation in } \Sigma' \tag{1211}$$

Here we have introduced a factor δ in the mass parameter of the second particle, that takes values $+1$ or -1. It will be later determined. The relation (1211) is essential for including tachyonic particles.

According to our proposition of a total inelastic collision, both particles merge after the impact with a single particle with mass parameter M_o, staying at rest in system Σ' due to momentum conservation. For the equality of the total energy before and after the collision, we can then write in Σ'

[5] Particles with superluminal velocities had been excluded in Chap. 7, Sect. 1, implicitly by the assumption that the LORENTZ factor $\gamma = 1/\sqrt{1 - u^2/c^2}$ is real.

$$\xi_o\, f(u_1')\, c_o^2 + \delta \cdot \xi_o\, f(-u_1')\, c_o^2 = M_o\, f(0)\, c_o^2\ , \quad \text{Energy conservation in } \Sigma' \tag{1212}$$

so that

$$\xi_o \left[f(u_1') + \delta \cdot f(-u_1') \right] = M_o\ . \tag{1213}$$

In reference system Σ_o, the mass M_o has after the impact the velocity v of system Σ', since it is there at rest. The velocities of both particles before the impact are determined in Σ_o by the addition theorem of velocities (106),

$$u' = \frac{u - v}{1 - u\,/c^2} \longleftrightarrow u = \frac{u' + v}{1 + u'\,/c^2}\ , \quad \text{EINSTEIN's addition theorem of velocities}$$

where we insert in the second equation for the first particle u_1' in Σ', and for the second particle with mass parameter $\delta \cdot \xi_o$ the velocity $-u_1'$. The energy conservation law in system Σ_o is then written as

$$\xi_o\, f\Big(\frac{+u_1'}{1 + u_1'/c_o^2}\Big)\, c_o^2 + \delta \cdot \xi_o\, f\Big(\frac{-u_1'}{1 - u_1'/c_o^2}\Big)\, c_o^2 = M_o\, f(0)\, c_o^2\, . \quad \text{Energy conservation in } \Sigma_o \tag{1214}$$

From Eqs. (1213) and (1214) there follows the functional equation for $f = f(u)$,

$$f\left(\frac{+u'}{1 + u'/c_o^2}\right) + \delta \cdot f\left(\frac{-u'}{1 - u'/c_o^2}\right) = \left[f(u') + \delta \cdot f(-u') \right] \cdot f(0)\ . \tag{1215}$$

For the general solution of this equation we have to distinguish two cases:
(a) We assume both particles are normal particles, hence $|u'| < c_o \longrightarrow |u| < c_o$, s. Problem 7. In this case, we take $\delta = 1$, and the solution of the functional equation is given by the LORENTZ factor

$$f(u) = \frac{1}{\sqrt{1 - u^2/c_o^2}} \equiv \gamma\ , \quad |u| < c_o\ . \tag{1216}$$

Insertion of Eq. (1216) into (1215) provides

$$\begin{aligned}
&\frac{1}{\sqrt{1 - \frac{1}{c_o^2}\left(\frac{v + u'}{1 + v\,u'/c_o^2}\right)^2}} + \frac{1}{\sqrt{1 - \frac{1}{c_o^2}\left(\frac{v - u'}{1 - v\,u'/c_o^2}\right)^2}} \\
&= \frac{1 + v\,u'/c_o^2}{\sqrt{\left(1 + v\,u'/c_o^2\right)^2 - \frac{1}{c_o^2}\left(v + u'\right)^2}} + \frac{1 - v\,u'/c_o^2}{\sqrt{\left(1 - v\,u'/c_o^2\right)^2 - \frac{1}{c_o^2}\left(v - u'\right)^2}} \\
&= \frac{1 + v\,u'/c_o^2}{\sqrt{1 - v^2/c_o^2}\,\sqrt{1 - u'^2/c_o^2}} + \frac{1 - v\,u'/c_o^2}{\sqrt{1 - v^2/c_o^2}\,\sqrt{1 - u'^2/c_o^2}} \\
&= 2\,\frac{1}{\sqrt{1 - v^2/c_o^2}\,\sqrt{1 - u'^2/c_o^2}} = 2\, f(v)\, f(u')\ .
\end{aligned}$$

(We have to observe that for a 'normal' particle $\delta = 1$ and $f(u) = f(-u)$).
(b) We consider two tachyons, i.e. writing $u \equiv w$ with $|w'| > c_o \longrightarrow |w| > c_o$. For getting a solution of Eq. (1215), we must assume $\delta = -1$, and we have to complement our ansatz (1216),

$$f(u) = \left.\begin{array}{ll} \dfrac{1}{\sqrt{1-u^2/c_o^2}}\,, & \text{for} \quad |u| < c_o\,, \\ \dfrac{\operatorname{sign} w}{\sqrt{w^2/c_o^2 - 1}}\,, & \phantom{\text{for}} \quad |w| > c_o\,. \end{array}\right\} \tag{1216a}$$

Here we used sign $w = w/|w|$ in the second part of the solution, that might appear arbitrary. We return to this point in a minute, and we verify at first, that this function satisfies Eq. (1215), both for a positive and also a negative w. There we observe that $\delta = -1$. For the first particles, we assume without restriction of generality a velocity $w' > 0$ in Σ'. Then the $\operatorname{sign}(v + w') = +1$ and $\operatorname{sign}(v - w') = -1$. We find

$$\frac{\operatorname{sign}\dfrac{v+w'}{1+v\,w'/c_o^2}}{\sqrt{\dfrac{1}{c_o^2}\left(\dfrac{v+w'}{1+v\,w'/c_o^2}\right)^2 - 1}} + \delta\,\frac{\operatorname{sign}\dfrac{v-w'}{1-v\,w'/c_o^2}}{\sqrt{\dfrac{1}{c_o^2}\left(\dfrac{v-u'}{1-v\,w'/c_o^2}\right)^2 - 1}}$$

$$= \frac{\left[\operatorname{sign}\dfrac{v+w'}{1+v\,w'/c_o^2}\right](1+v\,w'/c_o^2)\left[\operatorname{sign}(1+v\,w'/c_o^2)\right]}{\sqrt{\dfrac{1}{c_o^2}(v+w')^2 - \left(1+v\,w'/c_o^2\right)^2}} +$$

$$+\,\delta\,\frac{\left[\operatorname{sign}\dfrac{v-w'}{1-v\,w'/c_o^2}\right](1-v\,w'/c_o^2)\left[\operatorname{sign}(1-v\,w'/c_o^2)\right]}{\sqrt{\dfrac{1}{c_o^2}(v-w')^2 - \left(1-v\,w'/c_o^2\right)^2}}$$

$$= \frac{\left[\operatorname{sign}(v+w')\right](1+v\,w'/c_o^2)}{\sqrt{1-v^2/c_o^2}\sqrt{w'^2/c_o^2-1}} + \delta\,\frac{\left[\operatorname{sign}(v-w')\right](1-v\,w'/c_o^2)}{\sqrt{1-v^2/c_o^2}\sqrt{w'^2/c_o^2-1}}$$

$$= \frac{1}{\sqrt{1-v^2/c_o^2}}\,\frac{1-\delta}{\sqrt{w'^2/c_o^2-1}} \quad = \quad f(v)\left[f(w') + \delta\, f(-w')\right].$$

For both tachyonic and normal particles, we find the following dependence of momenta and energies on its velocity:
(a) For the momentum and energy of a 'normal' massive particle with $u < c_o$ we write $\xi_o = m_o$and it holds

$$P = \frac{m_o}{\sqrt{1-u^2/c_o^2}}\,u\,, \quad E = \frac{m_o}{\sqrt{1-u^2/c_o^2}}\,c_o^2\,. \tag{1217}$$

The parameter ξ_o of this particle is called rest mass m_o, and it holds

$$P_o = 0 \ , \quad E_o = m_o \, c_o^2 \ . \tag{1217a}$$

(b) The general solution (1216a) now allows a second type of particles, also fulfilling the conservation laws of mechanics, the energy and momentum conservation. For the energy E^T and the momentum P^T of tachyonic particles, using again the letter of its velocity w, hence $u \equiv w > c_o$, we find now

$$P^T = \frac{m_*^T}{\text{sign}\, w \, \sqrt{w^2/c_o^2 - 1}} \, w \ , \quad E^T = \frac{m_*^T}{\text{sign}\, w \, \sqrt{w^2/c_o^2 - 1}} \, c_o^2 \ . \tag{1218}$$

Here we have again written $\xi_o := m_*^T$. This parameter m_*^T cannot be a rest mass. The physical meaning of m_*^T follows from the limit $w \longrightarrow \pm\infty$. As seen easily, it holds both for $w \longrightarrow +\infty$ and for $w \longrightarrow -\infty$

$$\lim_{u \to \pm\infty} P^T = P_\infty^T = m_*^T \, c_o \ , \quad \lim_{u \to \pm\infty} E^T = E_\infty^T = 0 \ . \tag{1218a}$$

If one takes the for tachyons excluded velocities $w < c_o$ formally in Eq. (1218), we get imaginary values for the mass of tachyons, hence an imaginary 'rest mass', required for getting momentum and energy real. Tachyons are therefore sometimes denoted as particles with imaginary rest mass. Such a statement is however misleading, as shall be demonstrated shortly.

The physical parameter of a tachyon is the quantity m_*^T. It replaces the rest mass m_o of normal massive particles. There is a basic difference between these two parameters as shall be shown further below.

The velocity w of a tachyon is defined using momentum and energy

$$w := \frac{c_o^2 \, P^T}{E^T} \ . \tag{1219}$$

The mass of the tachyon follows from the energy $E^T = m^T \, c_o^2$ can be both positive and negative. We could fulfil the conservation laws of momentum (1211) and energy (1214)—and therefore the functional equation (1215)—with the ansatz (1216a) only when we assume for the second particle $\delta = -1$, hence a negative mass. The factor sign$w$ in Eq. (1216a) would not be required for this statement. But this factor secures that we get for the momentum P^T of tachyons both for $w \longrightarrow +\infty$ as well as for $w \longrightarrow -\infty$ one and the same limit $m_*^T \, c_o$. This condition must be fulfilled for the following reason.

From the addition theorem (106) we read off directly: Both a tachyon observed in Σ' with velocity $w' = +\infty$, as well as the one with $w' = -\infty$ has in Σ_o one and the same velocity $w = c_o^2/v$, where v is the velocity of Σ' with respect to Σ_o. In Σ_o the tachyon has for both limits one and the same momentum. The tachyon cannot have two different limiting values of the momentum in Σ'. The factor δ is so required for the uniqueness of the tachyon momentum.

From Eq. (1218) for energy and momentum of a tachyon, there follows its physical property:

If the energy and the momentum of a tachyon in reference system Σ_o has the values E^T and P^T, and in system Σ', moving with velocity v with respect to Σ_o, the values $E^{T\prime}$ and $P^{T\prime}$, then it holds

$$\left.\begin{aligned} P^{T'} &= \frac{P^T - E^T/c_o^2}{\sqrt{1-v^2/c_o^2}} \,, \qquad E^{T'} = \frac{E^T - P^T}{\sqrt{1-v^2/c_o^2}} \\ &\text{with the reversed relation} \\ P^{T} &= \frac{P^{T'} + E^{T'}/c_o^2}{\sqrt{1-v^2/c_o^2}} \,, \qquad E^{T} = \frac{E^{T'} + P^{T'}}{\sqrt{1-v^2/c_o^2}} \,. \end{aligned}\right\} \tag{1220}$$

Equations (1220) hold true both for normal and tachyonic particles. They are characteristic for any relativistic particles. The proof of this relation can be given for tachyons only with the addition theorem of velocities (106). It is based on the fact that energy and momentum in Eq. (449), Chap. 9, Sect. 2, form a four-vector in MINKOWSKI space. We shall present the proof of the last Eq. (1220) explicitly. By

$$\nu := \frac{c_o^2}{w} \quad \longrightarrow \quad \sqrt{\frac{w^2}{c_o^2} - 1} = \frac{|w|}{c_o}\sqrt{1 - \frac{\nu^2}{c_o^2}} \tag{1221}$$

we assign to the velocities w and w' of tachyons in Σ_o and Σ' velocities ν and ν', that have amounts smaller than c_o. With the addition theorem (106) for w and w', there also holds the addition theorem for ν and ν', as can be easily shown (v is again the velocity of Σ' with respect to Σ_o),

$$\nu = \frac{\nu' + v}{1 + \nu' v/c_o^2} \quad \text{with the inversion} \quad \nu' = \frac{\nu - v}{1 - \nu v/c_o^2} \,. \tag{1222}$$

There follows by taking the square

$$\sqrt{1 - \frac{\nu^2}{c_o^2}} = \frac{\sqrt{1-v^2/c_o^2}\sqrt{1-\nu'^2/c_o^2}}{1 - v\nu'/c_o^2} \,, \tag{1223}$$

and we find with $\operatorname{sign} x = x/|x|$,

$$\begin{aligned} \frac{E^{T'} + v P^{T'}}{\sqrt{1-v^2/c_o^2}} &= \frac{\dfrac{m_*^T c_o^2}{\dfrac{w'}{|w'|}\dfrac{|w'|}{c_o}\sqrt{1-\nu'^2/c_o^2}} + v\,\dfrac{m_*^T w'}{\dfrac{w'}{|w'|}\dfrac{|w'|}{c_o}\sqrt{1-\nu'^2/c_o^2}}}{\sqrt{1-v^2/c_o^2}} \\ &= \frac{m_*^T c_o^2}{w'/c_o} \frac{1 + v w'/c_o^2}{\sqrt{1-v^2/c_o^2}\sqrt{1-\nu'^2/c_o^2}} \end{aligned}$$

$$= m_*^T\, c_o^2 \frac{\dfrac{\upsilon}{c_o} + \dfrac{\nu'}{c_o}}{\sqrt{1-\upsilon^2/c_o^2}\sqrt{1-\nu'^2/c_o^2}}$$

$$= m_*^T\, c_o^2 \frac{\dfrac{\nu}{c_o}\left(1 + \dfrac{\upsilon\,\nu'}{c_o^2}\right)}{\sqrt{1-\upsilon^2/c_o^2}\sqrt{1-\nu'^2/c_o^2}} \;=\; m_*^T\, c_o^2 \frac{\nu}{c_o} \frac{1}{\sqrt{1-\nu^2/c_o^2}}$$

$$= \frac{m_*^T\, c_o^2}{w/c_o \sqrt{1-\nu^2/c_o^2}}$$

and finally

$$\frac{E^{T'} + \upsilon\, P^{T'}}{\sqrt{1-\upsilon^2/c_o^2}} = \frac{m_*^T}{\operatorname{sign} w\,\sqrt{w^2/c_o^2 - 1}}\, c_o^2 = E^T\ .$$

On the same way there follows the prove for P^T.

Now we discuss that tachyons have quite strange properties.

The momentum of a tachyon has in all reference systems the same sign. This follows from Eq. (1220), when one chooses the system Σ_o in such a way, that the tachyon has limits (1218a): The energy is zero. The LORENTZ-factor $1/\sqrt{1-\upsilon^2/c_o^2}$ is always positive. Therefore $P^{T'}$ always has the same sign as P^T. From this property, it follows that again with the addition theorem (106) or (1222), the velocity w of a tachyon must not have the same sign as its momentum P^T. For positive ν' (hence also positive w') the quantity ν, and therefore also w gets negative, when one chooses $-\upsilon > \nu'$, as it is always possible by a choice of the reference system.

With two numerical examples, we shall illustrate the strange properties of the hypothetical tachyons.

An observer in Σ_o might observe a tachyon with velocity $w = \frac{5}{4}\, c_o$. He sends out an observer in the same direction with velocity $\upsilon = \frac{1}{2}\, c_o$. From Eq. (106) one determines the velocity w', that this observer measures for the tachyon,

$$w' = \frac{-\frac{1}{2}\, c_o + \frac{5}{4}\, c_o}{1 - \frac{c_o 5 c_o}{2\cdot 4 c_o^2}} = 2 c_o\ .$$

The observer, following the tachyon, measures a still larger velocity as the first one!

A kind of explanation for the behaviour is found in agreement with Eq. (1218): The observer following a removing particle must have a part of its kinetic energy. Due to Eq. (1218), the amount of energy of a tachyon becomes smaller, when its velocity increases. The tachyon loses energy when it becomes faster. For an unlimited increasing velocity, the tachyon loses all its energy.

Still more strange is the following situation. In Σ_o a tachyon has a velocity $w = 2c_o$. Now an observer follows the tachyon with velocity $= \frac{4}{5}\, c_o$. This observer finds for the tachyon a velocity

$$w' = \frac{-4/5 + 2c_o}{1 - 4c_o 2c_o/(5c_o)} = -2c_o \ .$$

Will the tachyon now approach to the observer that was following the tachyon? This would not be a strange situation for subluminal velocities $|u| < c_o$, for example, a normal overtaking of a car on a street. However, a tachyon cannot be overtaken since its velocity is always larger than c_o, and this velocity cannot be reached by the observer. But nevertheless, the velocity of the tachyon $w' = \Delta x'/\Delta t'$ becomes negative, while its velocity $w = \Delta x/\Delta t$ in Σ_o is positive. Now for a tachyon, the quantity Δx of the increase of its path cannot change its sign at the change of the reference system, the reason is the same as in the conservation of the sign of the momentum (and the proof follows the same lines as above). For a positive w and a positive Δx, w' can become negative only by a negative time interval $\Delta t'$! The tachyon will indeed not approach to the observer on Σ'. The measured velocity is negative, because the tachyon runs into the past! We always find that a tachyon runs into the past, when its velocity and momentum have opposite signs. If we keep the notion of velocity for a tachyon, we must accept that the tachyon moves into the future or into the past, in dependence on the reference system. From this property there results a violation of the causality, that we have already discussed in Problem 7.

With the sign of the velocity w of a tachyon, there changes also the sign of its energy E^T according to Eq. (1218), and then also by definition, the sign of its mass $m^T = E^T/c_o^2$.

We find an explanation of all these peculiarities from the following considerations. We stay in one and the same reference system Σ_o, and we consider a reflection of its coordinates. That means we describe all positions by new spatial coordinate $\overline{x}$ but preserving its timeline

$$\overline{x} = -x \ , \quad \overline{t} = t \ . \tag{1224}$$

By definition, the velocity $\overline{w}$ (as for all particles) reads in the new coordinates

$$\overline{w} = \frac{d\overline{x}}{d\overline{t}} = -w \ . \tag{1225}$$

Furthermore, the momentum is a spatial vector. Its components change its sign in the reflection

$$\overline{P}^T = -P^T \ . \tag{1226}$$

Due to $\overline{w}/\text{sign}\,\overline{w} = w/\text{sign}\,w$ it follows from Eq. (1218), that the tachyon parameter m_*^T also changes its sign during the reflection, for preserving Eq. (1226),

$$\overline{m}_*^T = -m_*^T \ . \tag{1227}$$

From these considerations we conclude:

The Tachyon parameter m_*^T *is a spatial vector.*

This is the key for understanding the tachyonic properties. The quantity m_*^T has nothing in common with the rest mass m_o of a normal massive particle. The particle parameter m_o, the rest mass, is a spatial scalar that does not change by a change of its spatial coordinates. However, m_*^T, multiplied with the constant quantity c_o, is a characteristic value of the momentum of the tachyon in that reference system, in which its energy is zero. The sometimes proposed classification of tachyons as particles with imaginary rest mass is misleading.

The quantity m_*^T *is the momentum parameter of the tachyon.*

While the rest mass m_o of a normal massive particle is always positive, the momentum parameter m_*^T can become both positive as well as negative. In our (three-dimensional) relativity theory we have two types of tachyons, such with positive and others with negative momentum parameter. When the direction of the x-axis changes its sign, also the momentum parameter of the tachyon changes its sign.

For all these properties of tachyons, which have common transformation relations (1220) with normal particles, its kinematic properties are basically different. There is a simple mathematical framework. In the language of MINKOWSKI geometry, all particles are represented by vectors in the space-time, s. Chap. 9, Sect. 2. The normal massive particles are described by time-like vectors. This has as consequence the existence of a distinguished reference system, in which only the time-like component $P_o = E_o/c_o = m_o c_o$ is different from zero. In this way, we define for all normal particles its rest mass m_o independent of the reference system. Tachyons are then described by space-like vectors with the consequence of a distinguished reference system, in which the time-like component vanishes. In our three-dimensional relativity theory, there remains for the in this way distinguished spatial momentum vector P_∞^T only one single component, for which we found in Eq. (1218a) $P_\infty^T = m_+^T c_o$. Because c_o is a constant, the spatial vector property is transferred to m_*^T, for which in the one-dimensional case only remains Eq. (1227) for a reflection (1226). With respect to the mathematical description of tachyons within the framework of MINKOWSKI space we refer to LIEBSCHER (2005).

Problem 9 *Which physical consequences follow for collision processes of normal particles and hypothetic tachyons?*

Chapter 4, Section 4, Chapter 7, Section 1

The pure replacement of the temporal order in the observation of two events E_1 and E_2 in different reference systems does not lead to a violation of causality. Only if one event E_2 is, e.g. caused by the other one E_1, then always E_2 must occur later than E_1, independent of the reference system, where both events are observed. An inversion in its temporal order, the inversion of the order of cause and action, would violate our understanding of the physical world.

The pure existence of tachyons as particles with energy and momentum does not represent the difficulty. The crucial question is, therefore, whether one can transmit signals with tachyons, and this means basically:

Can we transfer energy with tachyons? When this is not possible, which notion can we get from tachyons?

In full generality, we will discuss at first the elastic collision of a normal massive particle, with subluminal velocity, with a tachyon.

A particle with rest mass m_o might collide elastically with a tachyon having the momentum parameter m_*^T. The quantities after the collision will be denoted with an overbar. The velocities of the particle before and after the collision might be u and $\overline{u}$, and the corresponding velocities of the tachyon w and $\overline{w}$. The momenta and energies of the particle and the tachyon are given by expressions (1217) and (1218). The elasticity of the collision means that the LORENTZ-invariant parameters, the mass parameter m_o, the rest mass of the particle, and the momentum parameter m_*^T of the tachyon remain unchanged; there do not appear other particles or tachyons. For the energies and momenta before and after the collision, there holds

$$\left.\begin{aligned} \frac{m_o\,u}{\sqrt{1-u^2/c_o^2}} + \frac{m_*^T\,w}{\operatorname{sgn} w\sqrt{w^2/c_o^2-1}} &= \frac{m_o\overline{u}}{\sqrt{1-\overline{u}^2/c_o^2}} + \frac{m_*^T\,\overline{w}}{\operatorname{sgn}\overline{w}\sqrt{\overline{w}^2/c_o^2-1}}, \quad \text{Momentum} \\ \frac{m_o\,c_o^2}{\sqrt{1-u^2/c_o^2}} + \frac{m_*^T\,c_o^2}{\operatorname{sgn} w\sqrt{w^2/c_o^2-1}} &= \frac{m_o\,c_o^2}{\sqrt{1-\overline{u}^2/c_o^2}} + \frac{m_*^T\,c_o^2}{\operatorname{sgn}\overline{w}\sqrt{\overline{w}^2/c_o^2-1}}. \quad \text{Energy} \end{aligned}\right\} \tag{1228}$$

The tachyon velocities w and $\overline{w}$ are now replaced using ν and $\overline{\nu}$, Eq. (1221), by $w = c_o^2/\nu$ and $\overline{w} = c_o^2/\overline{\nu}$. The conservation laws (1228) then read

$$\left.\begin{aligned} \frac{m_o\,u}{\sqrt{1-u^2/c_o^2}} + \frac{m_*^T\,c_o}{\sqrt{1-\nu^2/c_o^2}} &= \frac{m_o\,\overline{u}}{\sqrt{1-\overline{u}^2/c_o^2}} + \frac{m_*^T\,c_o}{\sqrt{1-\overline{\nu}^2/c_o^2}}, \quad \text{Momentum} \\ \frac{m_o\,c_o^2}{\sqrt{1-u^2/c_o^2}} + \frac{m_*^T\,c_o\,\nu}{\sqrt{1-\nu^2/c_o^2}} &= \frac{m_o\,c_o^2}{\sqrt{1-\overline{u}^2/c_o^2}} + \frac{m_*^T\,c_o\,\overline{\nu}}{\sqrt{1-\overline{\nu}^2/c_o^2}}. \quad \text{Energy} \end{aligned}\right\} \tag{1229}$$

To simplify Eq. (1229) we use the fact that we can choose freely the reference system $\widetilde{\Sigma}$, in which we shall describe the collision process. Furthermore, we can choose suitably the positive x-axis. In general, we can assume that the particle is at rest before the collision in the laboratory system, in which we shall describe the collision with the tachyon, and that the momentum of the tachyon points into the direction of the positive x-axis, the momentum parameter m_*^T is then positive. From m_*^T and the rest mass m_o, then we form a positive parameter μ,

$$\mu := \frac{m_o}{m_*^T}\ . \tag{1230}$$

We remark once again the curiosity of the kinematics of tachyons: Even if the momentum of the tachyon remains positive, the sign of its velocity can change.

From Eq. (1229) we get after division by m_*^T,

$$\overline{\Sigma}: \quad \left. \begin{aligned} \frac{c_o}{\sqrt{1-\nu^2/c_o^2}} &= \frac{\mu\,\overline{u}}{\sqrt{1-\overline{u}^2/c_o^2}} + \frac{c_o}{\sqrt{1-\overline{\nu}^2/c_o^2}} \;, \quad \text{Momentum} \\ \mu\, c_o^2 + \frac{c_o\,\nu}{\sqrt{1-\nu^2/c_o^2}} &= \frac{\mu\, c_o^2}{\sqrt{1-\overline{u}^2/c_o^2}} + \frac{c_o\,\overline{\nu}}{\sqrt{1-\overline{\nu}^2/c_o^2}} \;. \quad \text{Energy} \end{aligned} \right\} \tag{1231}$$

For a better representation of the possible solutions, we introduce hyperbolic functions

$$\left. \begin{aligned} &\frac{\overline{\nu}}{c_o} = \tanh\overline{\alpha}, \quad \frac{\nu}{c_o} = \tanh\alpha \;, \\ &\frac{\overline{u}}{c_o} = \tanh\overline{\beta}. \end{aligned} \right\} \tag{1232}$$

For $u = 0$ we find $u/c_o = \tanh\beta = 0$ and also $\beta = 0$.

With $1/\sqrt{1-\tanh^2 x} = \cosh x,\ \tanh x/\sqrt{1-\tanh^2 x} = \sinh x$, it follows finally instead of Eq. (1231) after small calculation

$$\left. \begin{aligned} \mu \sinh\overline{\beta} &= \cosh\alpha - \cosh\overline{\alpha} \;, \\ \mu \cosh\overline{\beta} &= \sinh\alpha - \sinh\overline{\alpha} + \mu \;. \end{aligned} \right\} \tag{1233}$$

Since $0 < \mu$, Eq. (1230), $0 < \cosh\overline{\beta} - 1$ and the monotony of the hyperbolic sinus,

$$\sinh\overline{\alpha} < \sinh\alpha \quad \longleftrightarrow \quad \overline{\alpha} < \alpha \;, \tag{1234}$$

there follows from the second Eq. (1233) with

$$0 \le \sinh\alpha - \sinh\overline{\alpha} = \mu\,[\cosh\overline{\beta} - 1] \tag{1235}$$

the condition

$$\overline{\alpha} \le \alpha \;. \tag{1236}$$

With $\cosh^2\overline{\beta} - \sinh^2\overline{\beta} = 1$, we eliminate the velocity $\overline{u} = c_o \tanh\overline{\beta}$ of the particle after the collision from Eq. (1233), and we find with the addition theorem $\cosh(\alpha - \overline{\alpha}) = \cosh\alpha\cosh\overline{\alpha} - \sinh\alpha\sinh\overline{\alpha}$ after a short evaluation

$$0 \le \frac{1}{\mu}[\cosh(\alpha - \overline{\alpha})] = \sinh\overline{\alpha} - \sinh\alpha \;. \tag{1237}$$

With the same argument as above now it follows

$$\alpha \le \overline{\alpha} \;. \tag{1238}$$

Equations (1236) and (1238) require

$$\overline{\alpha} = \alpha \quad \longleftrightarrow \quad \overline{w} = w \ . \qquad \text{No signal transmission by the tachyon} \tag{1239}$$

The tachyon has to pass by without change of its velocity. There cannot be transmitted any energy, and therefore also no signal. Causality cannot be violated in this way:

By means of elastic collisions, no causality violation can be found.

When tachyon do not transmit energy, how can we get notice from them? Until now we have restricted ourself to elastic particle–tachyon collisions. Inelastic collisions change the situation completely, cp. LIEBSCHER (1975) and GÜNTHER (2000).

As initial state, we consider a resting particle of mass m_o. We are looking for a final state, when this particle has emitted a tachyon of velocity $\overline{w}$ without change of its rest mass, and itself it gets a velocity $\overline{u}$ different from zero.

It is easy to understand, that a resting particle cannot emit a normal particle without change of its rest mass due to the energy conservation, since the emitted particle has a positive energy $E_o = m_o\, c_o^2$. A particle at rest cannot provide energy without change of its rest mass. For tachyons, this is different, since they have negative energy.

With the same notation as above now it follows from the conservation of energy and momentum instead of Eq. (1233),

$$\left.\begin{aligned} \mu \sinh\overline{\beta} &= -\cosh\overline{\alpha} \ , \\ \mu \cosh\overline{\beta} &= -\sinh\overline{\alpha} + \mu \ . \end{aligned}\right\} \qquad \text{No tachyon before the collision} \tag{1240}$$

Here there is missing only the incoming tachyon with respect to Eq. (1233). From Eq. (1240), we get in eliminating $\overline{\beta}$ after a short evaluation

$$\begin{aligned} \sinh\overline{\alpha} &= -\frac{1}{2\mu} \ , \\ \frac{\overline{v}}{c_o} &= \tanh\overline{\alpha} = \frac{\sinh\overline{\alpha}}{\sqrt{1+\sinh^2\overline{\alpha}}} = -\frac{1}{2\mu}\frac{1}{\sqrt{1+1/4\mu^2}} \ , \\ \frac{\overline{w}}{c_o} &= \frac{c_o}{\overline{v}} = -\sqrt{4\mu^2+1} \ , \\ \overline{w} &= -c_o\,\frac{\sqrt{m_*^{T\,2}+4m_o^2}}{m_*^T} \end{aligned} \tag{1241}$$

and then from Eq. (1240)

$$\begin{aligned} \sinh\overline{\beta} &= -\frac{1}{\mu}\cosh\overline{\alpha} = -\frac{1}{\mu}\sqrt{1+\sinh^2\overline{\alpha}} = -\frac{\sqrt{1+4\mu^2}}{2\mu^2} \ , \\ \frac{\overline{u}}{c_o} &= \tanh\overline{\beta} = \frac{\sinh\overline{\beta}}{\sqrt{1+\sinh^2\overline{\beta}}} = -\frac{\sqrt{1+4\mu^2}}{2\mu^2}\frac{1}{\sqrt{1+(1+4\mu^2)/(4\mu^4)}} \\ &= -\frac{\sqrt{1+4\mu^2}}{\sqrt{(1+2\mu^2)^2}} \ , \end{aligned}$$

$$\overline{u} = -c_o \frac{\sqrt{m_*^{T\,2} + 4m_o^2}}{1 + 2m_o^2} \quad . \tag{1242}$$

Equations (1240) have a non-trivial solution $\overline{u}$ for a given rest mass m_o of the particle and for each arbitrary momentum parameter m_*^T of the emitted tachyon after the collision. Therefore, particles could emit arbitrarily many tachyons and could change its velocity without outer influence—and without change of its rest mass m_o. A terrible result! Has someone ever observed that a particle is set into motion by emitting tachyons?

Indeed, the emission of particles by tachyons is compatible with the conservation laws. When we omit in Eq. (1233) only the particle before the collision, then it follows

$$\left.\begin{aligned} \mu \sinh\overline{\beta} &= \cosh\alpha - \cosh\overline{\alpha} \ , \\ \mu \cosh\overline{\beta} &= \sinh\alpha - \sinh\overline{\alpha} \ . \end{aligned}\right\} \qquad \text{No particle before the collision} \tag{1243}$$

From $0 \leq \mu\cosh\overline{\beta} = \sinh\alpha - \sinh\overline{\alpha}$ and the monotony of the hyperbolic sinus there follows at first $\overline{\alpha} \leq \alpha$. With $\cosh^2\overline{\beta} - \sinh^2\overline{\beta} = 1$ we eliminate $\overline{\beta}$ by taking the square and subtract both equations (1243), and we find after a short evaluation

$$\overline{\alpha} = \alpha - \operatorname{arcosh}\big(1 + \frac{\mu^2}{2}\big) \ . \tag{1244}$$

Accordingly an arbitrary incoming tachyon (i.e. α and m_*^T are arbitrarily fixed) can emit a particle with arbitrary rest mass $m_o = \mu\, m_*^T$. The tachyon velocity $\overline{w}$ after the emission follows from Eq. (1244), taking into account $c_o/\overline{w} = \overline{v}/c_o = \tanh\overline{\alpha}$, and the velocity $\overline{u}$ of the emitted particle follows from Eqs. (1243) and (1244), $\overline{u}/c_o = \tanh\overline{\beta}$.

We aim to hint on a strange behaviour of our hypothetic tachyon physics. We consider the case of two totally inelastically colliding tachyons, so that the tachyons are annihilated, and for fulfilling the energy conservation, we expect the creation of a particle of rest mass M_o.

In system Σ_o, we consider two tachyons. One of them has a positive velocity $0 < c < w_1 = w$ and the momentum parameter m_*^T. The momentum parameter of the second tachyon with negative velocity $w_2 = -w < -c < 0$ is then $\delta \cdot m_*^T = -m_*^T$. By use of Eq. (1218), we find now for the total momentum P_G^T of the system of two tachyons with momenta P_1^T and P_2^T

$$\left.\begin{aligned} P_G^T &= P_1^T + P_2^T \\ &= \frac{m_*^T}{\operatorname{sign} w_1 \sqrt{w_1^2/c^2 - 1}}\, w_1 + \frac{\delta \cdot m_*^T}{\operatorname{sign} w_2 \sqrt{w_2^2/c^2 - 1}}\, w_2 \\ &= \frac{m_*^T}{(+1)\sqrt{w^2/c^2 - 1}}\, w - \frac{m_*^T}{(-1)\sqrt{w^2/c^2 - 1}}\, (-w) \\ &= 0 \ . \end{aligned}\right\} \tag{1245}$$

For the total energy E_G^T of the system of two tachyons with momenta E_1^T and E_2^T it follows

$$\left.\begin{aligned} E_G^T &= E_1^T + E_2^T \\ &= \frac{m_*^T}{\operatorname{sign} w_1 \sqrt{w_1^2/c^2 - 1}}\, c^2 + \frac{\delta \cdot m_*^T}{\operatorname{sign} w_2 \sqrt{w_2^2/c^2 - 1}}\, c^2 \\ &= \frac{m_*^T}{(+1) \sqrt{w^2/c^2 - 1}}\, c^2 - \frac{m_*^T}{(-1) \sqrt{w^2/c^2 - 1}}\, c^2 \\ &= \frac{2m_*^T\, c^2}{\sqrt{w^2/c^2 - 1}} \ . \end{aligned}\right\} \tag{1246}$$

As shown in Chap. 7, Sect. 1, we get from momentum conservation

$$\Sigma_o\colon\ \overline{M}\,\overline{U} = P_G^T = 0\ , \qquad \text{Momentum conservation in } \Sigma_o \tag{1247}$$

hence

$$\overline{U} = 0 \quad \longrightarrow \quad \overline{M}\{0\} = \overline{M}_o\ . \tag{1248}$$

Due to energy–mass equivalence, we find then with Eq. (1246)

$$M_o = \frac{2m_{*T}}{\sqrt{w^2/c_{o^2} - 1}}\ . \tag{1249}$$

For the totally inelastic collision of two tachyons, there appears a particle with rest mass M_o. If the tachyon velocity w goes to the speed of light c, we could observe in this process a creation of particles with arbitrary large rest masses. This particle creation from nothing would be compatible with the conservation of energy and momentum.

Once again we underline: Such processes would be only possible under the conservation laws of energy and momentum, when we suppose the physical existence of tachyons. But this does not mean that such processes are indeed a solution of the fundamental equations of the theory of elementary particles. In GÜNTHER (2000) we have provided an interpretation of tachyons, that could induce long-range correlations. The crystalline solid body provides also a model for this phenomenon.

Problem 10 *A body K might be in motion along the x-axis on the orbit $x = x(t)$, and in Σ_o there is determined a velocity $u = dx/dt$ and an acceleration $a = du/dt$. Which acceleration $a' = du'/dt'$ has the body in system Σ' for a velocity $u' = dx'/dt'$, when the system Σ' has a velocity v in x-direction with respect to Σ_o, and we apply the corresponding* LORENTZ *transformation?*

The body K is accelerated in its rest system with a constant amount of $a_o = 10\,g$ (g is the acceleration of the Earth on normal height zero). After which time T_o, an observer in Σ_o measures a velocity of $0.8\,c$ of body K, when it has started in Σ_o with zero velocity?

Which velocity would have the body after time T_o, when we would suppose this acceleration in classical space-time?

Chapter 3, Section 2, Chapter 4, Section 4

The body K has in Σ' the velocity $u' = dx'/dt'$ and the acceleration $a' = du'/dt'$; hence it holds, cp. Eqs. (105) and (106),

$$u' = \frac{dx'}{dt'} = \frac{dx'}{dt}\left(\frac{dt'}{dt}\right)^{-1} = \gamma\,(u-v)\left(\gamma(1-uv/c^2)\right)^{-1} = \frac{u-v}{1-uv/c^2}\ ,$$

$$a' = \frac{du'}{dt'} = \frac{d}{dt}\left(\frac{u-v}{1-uv/c^2}\right)\left(\frac{dt'}{dt}\right)^{-1}\ ,$$

$$= \frac{a\,(1-uv/c^2)+(u-v)\,av/c^2}{\left(1-uv/c^2\right)^2}\ \frac{1}{\gamma\,(1-uv/c^2)} = \frac{a\,(1-v^2/c^2)}{(1-uv/c^2)^2}\ \frac{1}{\gamma\,(1-uv/c^2)}\ , \quad \text{hence}$$

$$a' = \frac{a}{\gamma^3\,(1-uv/c^2)^3}\ . \tag{1250}$$

For small velocities, $v^2/c^2 \approx 0\,,\ uv/c^2 \approx 0$, there follows the independence of the acceleration from the reference system, i.e. the classical mechanics.

Now let be Σ' the temporal rest system of the body K that continuously changes,

$$u \equiv v \quad \longrightarrow \quad u' = \frac{u-v}{1-uv/c^2} = 0\ . \qquad \begin{array}{l}\Sigma' \text{ temporal rest system}\\ \text{of the body } K\end{array} \tag{1251}$$

For $u = v$, it follows from Eq. (1250) the ratio of acceleration a' in the temporal rest system to the acceleration a in the laboratory system Σ_o,

$$a = a'\,/\,\gamma^3\ . \tag{1252}$$

On Earth, we observe a smaller acceleration as from the rest system, and this effect strengthens with increasing velocity of the body K.

With a constant acceleration of $a' = a_o = 10\,g$ in the temporal rest system, we write for acceleration a in the laboratory system

$$a = \frac{du}{dt} = a_o\,\sqrt{1-u^2/c^2}^{\,3}\ .$$

Integration of the equation delivers the velocity $u = u(t)$ observed in Σ_o.

$$\int \frac{du}{\sqrt{1-u^2/c^2}^{\,3}} = \int a_o\,dt\ .$$

With the initial condition, that the body starts at rest for $t = 0$, hence $u(0) = 0$, it follows

$$\frac{u}{\sqrt{1-u^2/c^2}} = a_o\,t\ ,$$

or after resolving for u,

$$u = c\,\frac{a_o\,t}{\sqrt{c^2 + a_o^2\,t^2}}\ .$$

Even if the acceleration a_o stays constant, we observe in the laboratory system after a certain observation time only an approach to the speed of light, but it cannot be reached.

In our example with

$$a_o = 10\,g = 98.1\,\mathrm{ms}^{-2}\ ,\quad c = 3\cdot 10^8\,\mathrm{ms}^{-1}\quad \text{and}\quad u = 0.8\,c$$

we get for the time t_o the relation

$$\frac{0.8\,c}{\sqrt{1-0.64}} = a_o\,t_o\ ,$$

so with $a_o = 10\,g = 98.1\,\mathrm{ms}^{-2}$,

$$t_o = \frac{0.8}{0.6}\,\frac{c}{a_o} = \frac{4}{3}\,\frac{3\cdot 10^8}{98.1}\,\mathrm{s},\quad \text{so that}$$

$$t_o \approx 4.08\cdot 10^6\,\mathrm{s} = 47\ \text{days}\ 5\,\mathrm{h}\ 20\,\mathrm{min}\,.$$

We add: From the integration of $u = dx/dt$ there follows, that the observer on Earth would measure a distance of the body of about 600 billion km.

In classical mechanics, for a constant acceleration a_o, there is simply $u_{cl} = a_o\,t_o$, and therefore

$$u_{cl} = 98.1\cdot 4.08\cdot 10^6\,\mathrm{ms}^{-1} \approx 4\cdot 10^8\,\mathrm{ms}^{-1}\ .$$

This value already exceeds the speed of light by 33%.

Problem 11 *The values of the speed of light in classical and in relativistic space-time should be compared under the supposition of an in both cases defined absolute simultaneity.*

Chapter 1, Section 4, Chapter 3, Sections 1, 2, Chapter 4, Sections 3, 4, Chapter 8, Section 9

The starting point is a homogeneous and isotropic inertial system Σ_o with the value of the speed of light c given in Eq. (4). In systems Σ' with velocity v with respect to Σ_o, there might be defined an absolute simultaneity

$$\theta = \theta_a = 0\ . \qquad \text{Absolute simultaneity} \qquad (1253)$$

Now if $t_2 = t_1$ in Σ_o, then for $\theta = 0$ it follows from Eq. (22) also $t_2' = t_1'$ in Σ'.

In classical space-time we introduce the conventional simultaneity by Eq. (1253), which delivers with $k = q = 1$ the GALILEI transformation, Chap. 3, Sect. 2,

$$x' = x - v\,t\ ,\quad t' = t\ . \qquad \text{GALILEI transformation} \qquad (1254)$$

From $k = q = 1$, $\theta = 0$, there follows from Eq. (23), the classical addition theorem (70). Then it follows with $u = c$ in Σ_o for the speed of light c'_{cl} measured in Σ' the classical value

$$c'_{cl} = c - v\,. \tag{1255}$$

This result has been especially exciting for classical physics: With the absolute simultaneity, one introduces in the classical space-time the conventional simultaneity. The GALILEI transformations form a group. For the classical mechanics, there is then no inertial system distinguished among all others. With respect to the light propagation, the system Σ_o is distinguished by relation (1255). This could be explained in classical physics only by the introduction of a mechanical aether for the propagation of light analogous to the propagation of sound waves in air. Then electrodynamics would be reduced to mechanics. But such an aether could not be found by any experiment.

Equation (1255) represents a non-relativistic approximation.

In the relativistic space-time, one introduces with Eq. (1253) a non-conventional simultaneity, leading to an asymmetric mathematical description as done in Chap. 8, Sect. 9. From Eqs. (99) and (100), there follows from Eq. (1253), that then $k = 1/q = \gamma = 1/\sqrt{1 - v^2/c^2}$. From Eq. (22), and Eq. (1253) it follows then instead of the LORENTZ transformation the REICHENBACH transformation (298) from Chap. 8, Sect. 9,

$$x' = \gamma\,(x - v\,t)\,, \quad t' = t/\gamma\,. \qquad \text{REICHENBACH transformation} \tag{1256}$$

With $k = 1/q = 1/\gamma$, $\theta = 0$ we get the addition theorem (23), with $u = c$ in Σ_o for the speed of light $\widehat{c}\,'_{rel}$ in Σ' the relativistic value

$$\widehat{c}\,'_{rel} = \frac{c - v}{1 - v^2/c^2}\,, \tag{1257}$$

$$\widehat{c}\,'_{rel} \approx (c - v)\left(1 + \frac{v^2}{c^2}\right) = c'_{cl}\left(1 + \frac{v^2}{c^2}\right) \quad \text{for} \quad v \ll c\,. \tag{1258}$$

The classical value of the speed of light $c'_{cl} = c - v$ in Σ' differs by nonlinear terms in v/c from the relativistic value $\widehat{c}\,'_{rel}$, written with a dashed accent for marking the non-conventional synchronisation of clocks.

The unfamiliar value of the speed of light $\widehat{c}\,'_{rel}$ in Σ' is *produced* by the non-conventional gauging of clocks, i.e. by abandoning the elementary relativity. This value is not an approximation.

For the velocity of system Σ_o measured in Σ', we find from Eq. (24) instead of $-v$ the value $u'_o = -v/(1 - v^2/c^2)$, while Σ' has the velocity v in Σ_o. This is the

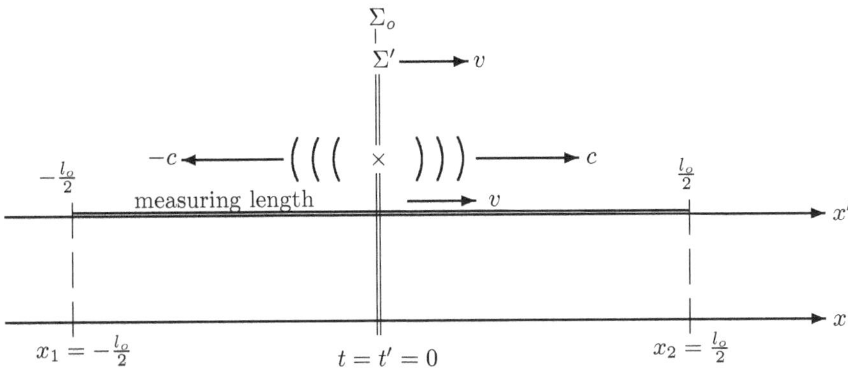

Fig. A.5 The measuring length is at rest in system Σ', that has a velocity v with respect to Σ_o. We investigate the difference between the linearised LORENTZ transformation and the GALILEI transformation for an in Σ' resting observer, which evaluates the arrival time of light signals at the end points of the measuring length in system Σ_o. We assume the postulates of Chap. 3, Sect. 1.

price for a description of the relativistic space-time with Eq. (1256) as consequence of an absolute simultaneity (1253).[6]

Problem 12 *We assume the classical space-time with Eqs. (65) and (66). A measuring distance of length l_o might be at rest in system Σ', which otherwise has a velocity v with respect to the laboratory system Σ_o as shown in Fig. A.5 and already discussed in Chap. 2, Sect. 1. At event $(x, t) = (x', t') = (0, 0)$ a light signal is emitted. Compare the arrival times of the signal at the end points of the rod as observed from two observers resting in Σ'. One of them should use the linearised* LORENTZ *transformation (131), and the other takes the* GALILEI *transformation (69).*

Chapter 2, Section 1, Chapter 3, Section 2, Chapter 4, Section 7

Because of Eq. (65), the measuring distance has in Σ_o also the length l_o: At time $t = 0$, the end points of the distance define in Σ_o the points $x_1 = -l_o/2$ and $x_2 = +l_o/2$ on the x-axis, s. Fig. A.5. Because of the isotropy of light propagation in Σ_o, the light signals get to these points at time $t = l_o/(2\,c)$, while the measuring distance has moved.

Both events, the arrival of the light signals at the right and left ends of the in Σ' resting measuring distance are denoted by $A(x_A, t_A) = A(x'_A, t'_A)$ and $B(x_B, t_B) = B(x'_B, t'_B)$. The right end point runs away from the light signal with velocity v and the left one approaches with velocity v. As in Chap. 2, Sect. 1, we find in Σ_o with Eq. (8), $t_A = l_o/\big(2(c+v)\big)$ and $t_B = l_o/\big(2(c-v)\big)$. The corresponding positions are

[6] Even if the REICHENBACH transformation (1256) is of no practical advantage in most cases, for some situations this equation can be indeed applied. Erroneous conclusions as, e.g. in the twin paradox or the problem of the length contraction, that results from an non-conventional use of the relativity of simultaneity, can be avoided in this way, s. Chap. 8, Sect. 9.

$$\Sigma_o : \left.\begin{aligned} x_A &= -\frac{l_o}{2} + v\,t_A\,, \\ x_B &= \frac{l_o}{2} + v\,t_B\,. \end{aligned}\right\} \tag{1259}$$

For the events A bzw. B, we find in Σ_o

$$\Sigma_o \left.\begin{aligned} A : x_A &= -\frac{l_o}{2} + v\frac{l_o}{2(c+v)} = -\frac{l_o}{2}\frac{c}{c+v}\,, \quad t_A = \frac{l_o}{2(c+v)}\,, \\ B : x_B &= \frac{l_o}{2} + v\frac{l_o}{2(c-v)} = \frac{l_o}{2}\frac{c}{c-v}\,, \quad t_B = \frac{l_o}{2(c-v)}\,. \end{aligned}\right\} \tag{1260}$$

For evaluation of t'_A and t'_B of events A and B for an in Σ' resting observer, we assume at first the linearised LORENTZ transformation (131). Therefore, we insert Eq. (1260) in the time transformation, the second line of Eq. (131), and we find

$$\Sigma' : \left.\begin{aligned} t'_A &= t_A - \frac{v\,x_A}{c^2} = \frac{l_o}{2(c+v)} + \frac{v}{c^2}\frac{l_o}{2}\frac{c}{c+v} = \frac{l_o}{2(c+v)}\left(1+\frac{v}{c}\right)\,, \\ t'_B &= t_B - \frac{v\,x_B}{c^2} = \frac{l_o}{2(c-v)} - \frac{v}{c^2}\frac{l_o}{2}\frac{c}{c-v} = \frac{l_o}{2(c-v)}\left(1-\frac{v}{c}\right)\,. \end{aligned}\right\} \tag{1261}$$

We notice: When we use the linearised LORENTZ transformation in the classical space-time, both events A and B, occurring according to Eq. (1260) in Σ_o at different times, seem to be simultaneously in Σ',

$$t'_A = t'_B = \frac{l_o}{2c}\,. \qquad \text{Linearised LORENTZ transformation simultaneity in } \Sigma' \tag{1262}$$

Now we compare this equality with the result that an in Σ' resting observer gets on the basis of the GALILEI transformation, which measures in each inertial system the same time. With equation $t' = t$ of the GALILEI transformation (69), we find from Eq. (1260) immediately

$$\left.\begin{aligned} t'_A &= t_A = \frac{l_o}{2(c+v)} = \frac{l_o}{2c}\frac{1}{1+v/c}\,, \\ t'_B &= t_B = \frac{l_o}{2(c-v)} = \frac{l_o}{2c}\frac{1}{1-v/c}\,. \end{aligned}\right\} \tag{1263}$$

Using the GALILEI transformation, the observer expects different arrival times of both signals at the end points of the in Σ' resting measuring rod, with the same time difference as in Σ_o already evaluated in Chap. 2, Sect. 1, Eq. (39),

$$t'_B - t'_A = \frac{l_o\,v}{(c^2 - v^2)}\,. \qquad \text{GALILEI transformation, unchanged time difference in } \Sigma' \text{ and } \Sigma_o \tag{1264}$$

In the classical space-time, i.e. for the GALILEI transformation, the events A and B are observed in Σ' not simultaneously, and this result is independent of the reference system.

Hence the linearised LORENTZ transformation and the GALILEI transformation differ in the *definition* of simultaneity for systems Σ'. The GALILEI transformation follows from a linearisation of the REICHENBACH transformation, s. Chap. 8, Sect. 9.

The linearised LORENTZ transformation however contains also the invariability of lengths and times as the GALILEI transformation, as discussed in Chap. 4, Sect. 7.

Problem 13 *We suppose the classical mechanics. Two bodies with equal masses* $m_1 = m_2 = m$ *might be approaching as seen from reference system* Σ_o *with equal but opposite oriented velocities* $\mathbf{u}_1 = (u, 0, 0)$, $\mathbf{u}_2 = (-u, 0, 0)$ *without the action of outer forces. What can you tell on their motion after its collision?*

Chapter 6, Section 2

For both bodies there hold Eq. (150) with $\mathbf{F}_a = 0$. Hence, there holds the momentum conservation (151). After the collision, the quantities are denoted with an overbar. Due to $m_1 = m_2 = m$ and $\mathbf{u}_1 = (u, 0, 0)$, $\mathbf{u}_2 = (-u, 0, 0)$, the total momentum vanishes. Therefore, if we restrict our calculation on the x-component,

$$P = mu_1 + mu_2 = mu - mu = 0 \quad \longrightarrow \quad \bar{P} = m\bar{u}_1 + m\bar{u}_2 = 0$$

and therefore

$$\left.\begin{aligned} u_1 &= -u_2 := u \ , \\ \bar{u}_1 &= -\bar{u}_2 := \bar{u} \ . \end{aligned}\right\} \tag{1265}$$

We denote the velocities generally with $w_1 = w_1(t)$ and $w_2 = w_2(t)$ with $w_1 = u_1$ before and $w_1 = \bar{u}_1$ after the collision, and likewise for w_2. We can write for Eq. (150)

$$m\frac{dw_1}{dt} = F_{21} \ , \quad m\frac{dw_2}{dt} = F_{12} \ . \tag{1266}$$

We multiply the first equation with w_1 and the second one with w_2, and we get by addition

$$m\left(w_1\frac{dw_1}{dt} + w_2\frac{dw_2}{dt}\right) = w_1 F_{21} + w_2 F_{12}$$

and therefore

$$\frac{m}{2}\frac{d}{dt}\left(w_1^2 + w_2^2\right) = w_1 F_{21} + w_2 F_{12} \ . \tag{1267}$$

With Eq. (149) it is $F_{21} = -F_{12} := F$, and with Eq. (1265) at each moment of time $w_1 = -w_2$, hence

$$\frac{m}{2}\frac{d}{dt}\left(w_1^2 + w_2^2\right) = 2Fu \ . \tag{1268}$$

On the right-hand side of Eq. (1268), there stands the power of the interaction forces during the collision process, acting during the collision within a small time interval δt around $t = 0$.

Taking $\Delta t > \delta t$, then the integration of Eq. (1268) delivers

$$\frac{m}{2}\int\limits_{-\Delta t}^{+\Delta t}\frac{d}{dt}(w_1^2 + w_2^2)dt = \frac{m}{2}\left[w_1^2 + w_2^2\right]_{-\Delta t}^{+\Delta t} = \int\limits_{-\Delta t}^{+\Delta t} 2Fudt \ .$$

With velocities u and $\bar{u}$ of particles before and after the collision, it follows

$$2\frac{m}{2}\bar{u}^2 - 2\frac{m}{2}u^2 = \int\limits_{-\Delta t}^{+\Delta t} 2Fudt := \Delta A \ . \tag{1269}$$

On the right-hand side of Eq. (1269) stands the work ΔA of the inner forces during the collision process, and on the left-hand side the difference of the kinetic energies before and after the collision.

For the case $\Delta A = 0$, we call the collision elastic. The interaction forces can absorb the kinetic energy of the colliding particles only for a short time, and they give them back thereafter completely. With Eqs. (1265), (1269) and $\Delta A = 0$, we get a solution of the collision problem,

$$\bar{u}_1 = -u \ , \quad \bar{u}_2 = u \ . \qquad \begin{array}{l}\text{Elastic collision}\\ \Delta A = 0\end{array} \tag{1270}$$

For $\Delta A \neq 0$ the collision is called inelastic. From Eqs. (1265), (1269), it follows

$$2\frac{m}{2}\bar{u}^2 = 2\frac{m}{2}u^2 + \Delta A \ . \qquad \begin{array}{l}\text{Inelastic collision}\\ \Delta A \neq 0\end{array} \tag{1271}$$

The proper physical process cannot be tackled purely from the conservation laws, in Eq.(1271) also non-mechanical energies can be included. As an example, we consider the totally inelastic collision with $2(m/2)\bar{u}^2 = 0$, hence

$$\bar{u} = 0 \quad \longrightarrow \quad \Delta A = -2\frac{m}{2}u^2 \ . \qquad \text{Totally inelastic collision} \tag{1272}$$

The inner forces cause that both particles lose their complete kinetic energy. It is open what happens with this energy. In the classical description of the collision process, only mechanical energy is included. There remains the possibility, that the work ΔA is transferred into the heating of the bodies. This is in general the standard model for the totally inelastic collision, and it can be observed, when two equal ideally plastic bodies collide, and later they remain at rest as combined somewhat warmer body. Equation (1272) can also be taken as a balance for a quite different process: Two particles collide, and it appears a new particle. Such processes of particle creation or annihilation cannot be covered by classical mechanics. These processes require

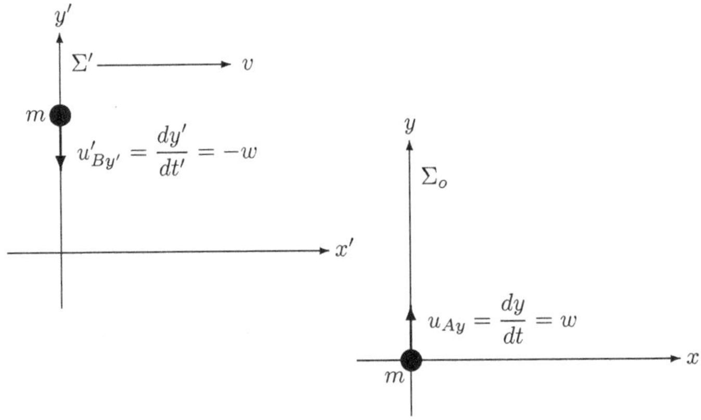

Fig. A.6 Schematic description of TOLMAN's thought experiment.

the application of relativistic mechanics and the inclusion of the rest masses into the transformation processes of energy, cp. Problems 17 and 18.

Problem 14 *Show with the* TOLMAN *experiment, that the mass m of a body under the supposition of the* GALILEI *transformation is independent of its velocity u. Derive the conservation of rest masses from momentum conservation in two different inertial systems for the totally inelastic collision.*

Chapter 6, Section 3, Chapter 7, Section 1

Since we know the result of the TOLMAN experiments in Chap. 6, Sect. 3 derived from the LORENTZ transformation, we have to take only the limit $\gamma = 1/\sqrt{1 - u^2/c^2} \longrightarrow 1$ in Eq. (167), and we get the result of the TOLMAN experiment on the basis of the GALILEI transformation, the independence of the mass m of bodies on its velocity, $m = m_o$.

Here we shall derive this result directly from the GALILEI transformation. To this aim, we consider an elastic collision of two ideally formed spherical balls A and B, as in Chap. 6, Sect. 3 and illustrated in Fig. A.6. Both balls have the same masses m. The ball A might have a velocity component in y-direction of Σ_o,

$$\Sigma_o : \quad \mathbf{u}_A = (dx/dt,\ dy/dt) = (u_{Ax},\ u_{Ay}) = (0,\ w).$$

The reference system Σ' has a velocity v in x-direction as seen from Σ_o. The ball B has in Σ' only a velocity component in direction of the negative y'-axis,

$$\Sigma' : \quad \mathbf{u}'_B = (dx'/dt',\ dy'/dt') = (u'_{Bx'},\ u'_{By'}) = (0,\ -w).$$

With the velocity $\mathbf{v} = (v, 0, 0)$ of Σ' in Σ_o and the velocity $\mathbf{u}' = (u'_x, u'_y, u'_z)$ of a body L in Σ', there follows from the GALILEI transformation the in Σ_o determined velocity $\mathbf{u} = (u_x, u_y, u_z)$ of L,

$$u_x = u'_x + v \ , \quad u_y = u'_y \ , \quad u_z = u'_z \ . \tag{1273}$$

For the velocity components of the balls A and B before the collision, we then get

$$\Sigma_o : \begin{array}{l} \mathbf{u}_A = (u_{Ax}, \, u_{Ay}) = (0, \ w) \ , \\ \mathbf{u}_B = (u_{Bx}, \, u_{By}) = (v, \ -w) \ . \end{array} \Bigg\} \quad \begin{array}{l}\text{Velocity components} \\ \text{before the collision}\end{array} \tag{1274}$$

The velocities v and w and positions of the spheres are chosen so that they collide in the moment, when the y'-axis exactly coincides with the y-axis, and the spheres are sitting on top of each other. The assumption of ideally smooth surface means, that at the collision no tangential forces are transmitted, i.e. no forces acting along the x-axis. In y-direction, the forces are described by the counteraction axiom, so that we can apply the momentum conservation law (145). After the collision, we denote the momenta and velocities by an overbar. As in Chap. 6, Sect. 3.1, we assume that the mass m could depend from the amount u of its velocity $\mathbf{u}$. We write $m = m\{u\}$, hence

$$\begin{array}{l} m\{u_A\} \ u_{Ax} + m\{u_B\} \ u_{Bx} = m\{\overline{u}_A\} \ \overline{u}_{Ax} + m\{\overline{u}_B\} \ \overline{u}_{Bx} \ , \\ m\{u_A\} \ u_{Ay} + m\{u_B\} \ u_{By} = m\{\overline{u}_A\} \ \overline{u}_{Ay} + m\{\overline{u}_B\} \ \overline{u}_{By} \ . \end{array} \Bigg\} \tag{1275}$$

Since no tangential forces appear, the velocities in x-direction remain unchanged. In y-direction, we denote the velocities after the collision $\overline{u}_{Ay} = \overline{w}_A$ and $\overline{u}_{By} = \overline{w}_B$. The velocity components after the collision are then

$$\begin{array}{ll} \Sigma_o : & \overline{\mathbf{u}}_A = (0, \ \overline{w}_A) \ , \\ \Sigma' : & \overline{\mathbf{u}}'_B = (\overline{u}'_{Bx'}, \, \overline{u}'_{By'}) = (0, \ \overline{w}_B). \end{array}$$

As considered from system Σ_o, we find the velocity components after the impact under consideration of unchanged velocities in x-direction as in Eq. (1274),

$$\Sigma_o : \begin{array}{l} \overline{\mathbf{u}}_A = (\overline{u}_{Ax}, \, \overline{u}_{Ay}) = (0, \ \overline{w}_A) \ , \\ \overline{\mathbf{u}}_B = (\overline{u}_{Bx}, \, \overline{u}_{By}) = (v, \ \overline{w}_B) \ . \end{array} \Bigg\} \quad \begin{array}{l}\text{Velocity components} \\ \text{after the collision}\end{array} \tag{1276}$$

The possible dependence of the mass m on the amount u of its velocity $\mathbf{u}$ in Σ_o is written as

$$\Sigma_o : \ m = m\{u^2\} \equiv m\{\mathbf{u}^2\} = m\{u_x^2 + u_y^2\} \ . \tag{1277}$$

With Eqs. (1274) and (1276) the x-component of the momentum balance (1274) reads

$$\Sigma_o : \ m\{v^2 + w^2\} \ v = m\{v^2 + \overline{w}_B^2\} \ v \ . \quad \begin{array}{l}x\text{-component} \\ \text{of momentum balance}\end{array} \tag{1278}$$

Assuming a monotonic function (1277), this equation can be fulfilled for arbitrary v only with $\overline{w}_B^2 = w^2$. After the collision, the second ball must run back in positive y-direction. The solution $\overline{w}_B = -w$ is therefore excluded, cp. also Chap. 6, Sect. 3, hence

$$\Sigma_o : \overline{w}_B = +w \ . \tag{1279}$$

The y-component of the momentum balance (1275) now reads with Eqs. (1274), (1276) and (1279)

$$m\{\mathbf{u}_A^2\}\, u_{Ay} + m\{\mathbf{u}_B^2\}\, u_{By} = m\{\overline{\mathbf{u}}_A^2\}\, \overline{u}_{Ay} + m\{\overline{\mathbf{u}}_B^2\}\, \overline{u}_{By},$$

hence

$$\Sigma_o : \left.\begin{aligned} & m\{w^2\}\ w - m\{v^2 + w^2\}\ w \\ & = m\{\overline{w}_A^2\}\ \overline{w}_A + m\{v^2 + w^2\}\ w \ . \end{aligned}\right\} \quad \begin{aligned} & y\text{-component} \\ & \text{of momentum balance} \end{aligned} \tag{1280}$$

Equation (1280) must hold true for arbitrary velocities v and w. We consider at first the limit $v \longrightarrow 0$, hence $\gamma \longrightarrow 1$,

$$0 = m\{\overline{w}_A^2\}\ \overline{w}_A + m\{w^2\}\ w \quad \text{for} \quad v \longrightarrow 0 \ . \tag{1281}$$

For $w \longrightarrow 0$ there follows from Eq. (1281) $m\{\overline{w}_A^2\}\ \overline{w}_A \longrightarrow 0$, and therefore, since the mass differs from zero, also $\overline{w}_A \longrightarrow 0$. With the limit

$$\lim_{\omega \to 0} \frac{\overline{w}_A}{w} = -\lim_{\omega \to 0} \frac{m\{w^2\}}{m\{\overline{w}_A^2\}} = -1$$

it follows from Eq. (1281)

$$\Sigma_o : \overline{w}_A = -w + O(w^2) \ . \tag{1282}$$

The quadratic correction terms $O(w^2)$ do not play any role and can be omitted in the following. Then we get from Eq. (1280)

$$2\,m\{v^2 + w^2\} = 2\,m\{w^2\} \ . \tag{1283}$$

This relation is the classical counterpart to Eq. (165), for the relativistic treatment. We write again

$$m\{0\} := m_o \ , \quad m\{v^2\} := m \ , \tag{1284}$$

and Eq. (1283) demonstrates with the limit $w \longrightarrow 0$ the independence of the mass m on its velocity for classical mechanics,

$$m = m_o \ . \qquad \begin{aligned} & \text{Classical space-time} \\ & \text{Constancy of mass} \end{aligned} \tag{1285}$$

Within the range of validity of the Galilei transformation, the mass of a body remains unchanged, just as the independence of the period of a moving clock.

Changing to the second part of the problem, we consider the experiment depicted in Chap. 7, Fig. 1. Two bodies of the same mass $m = m_o$ might approach against each other in reference system Σ_o with velocities of identical amount u. The motion is along the x-axis, hence the velocities are $\mathbf{u}_1 = (u, 0, 0)$, $\mathbf{u}_2 = (-u, 0, 0)$ and momenta $\mathbf{p}_1 = (m\,u, 0, 0)$, $\mathbf{p}_2 = (-m\,u, 0, 0)$. They collide inelastically, so that after the collision a single body remains with rest mass $\overline{M}$, the velocity $\overline{\mathbf{U}} = (\overline{U}, 0, 0)$ and the momentum $\overline{\mathbf{P}} = (\overline{M}\,\overline{U}, 0, 0)$. Outer forces $\mathbf{F}_a$ are excluded.

According to NEWTON's third axiom, Eq. (142) or equivalently Eq. (145), the momentum cannot change during the collision. In reference system Σ_o, there holds momentum conservation

$$\Sigma_o:\; P = m\,u + m\,(-u) = \overline{P} = \overline{M}\,\overline{U} = 0 \;, \qquad \text{Momentum conservation in } \Sigma_o \tag{1286}$$

hence

$$\overline{U} = 0 \;. \tag{1287}$$

Now we write down momentum conservation in the inertial system Σ', moving with velocity $v = u$ against the system Σ_o. The first body is then at rest in Σ', hence $u'_1 = 0$. The new body is at rest after the collision in Σ_o, hence it is $\overline{U}\,' = -u$ in Σ'. The velocity u'_2 of the second body before the collision is evaluated with the addition theorem (70), where we set u for the velocity v of Σ' and $-u$ instead of u for the velocity of the second body. Then we get in Σ'

$$\Sigma' :\; \left.\begin{array}{l} u'_1 = 0 \;,\quad u'_2 = -2u \;, \\ \overline{U}' = -u \;. \end{array}\right\} \tag{1288}$$

We take into account the independence of the mass on the velocity, Eq. (1285), and we find for the momentum balance in Σ', i.e. $P' = m_1 u'_1 + m_2 u'_2 = \overline{P}' = \overline{M}\,\overline{U}'$,

$$\Sigma' :\; P' = m\,u'_2 = \overline{P}' = \overline{M}\,\overline{U}' \;. \qquad \text{Momentum conservation in } \Sigma' \tag{1289}$$

We insert Eq. (1288) in Eq. (1289) and read off directly

$$\overline{M} = 2\,m \;, \tag{1290}$$

i.e. the mass $\overline{M} = \overline{M}_o$ after the collision is equal to the sum of masses $2\,m = 2\,m_o$ before the collision. Using the GALILEI transformation, the sum of rest masses is conserved in the collision.

Problem 15 *We make use of the* LORENTZ *transformation (105). We consider a water pot of mass $M = 215\,000\,l$ at temperature of $0\,°C$ and in addition a body at rest with mass m_o of one milligram. The rest mass energy of mass m_o might be transformed into heat that increases the temperature of the water. To which temperature, the water will be heated? For simplicity, we use as energy unit a kcal.*[7] *1 kcal is sufficient to increase 1 l water by 1 °C. The difference in the initial temperature should be neglected.*

[7]The unit calorie (cal) is not a SI-unit. The SI-unit of energy is Joule, 1 Joule (J) = 1 Newton metre. The conversion of 1 cal = 4.184 J $\longleftrightarrow$ 1 J = 0.239 cal has an uncertainty of about 1%.

Chapter 6, Section 3, Chapter 7, Section 1
In the pot, there is together a mass $M + m_o = 215\,000.000\,001$kg with a total energy E of

$$\left.\begin{array}{l} E = (M + m_o)\,c^2 = 215\,000.000\,001\,\text{kg} \cdot 299\,792\,458^2\,\text{m}^2\,\text{s}^{-2} \\ \quad = 1.93 \cdot 10^{22}\,\text{Joule}\ . \end{array}\right\} \qquad (1291)$$

The rest energy of one milligram is now converted completely into heat energy Q. We take into account that $1\,\text{cal} = 4.184\,\text{J} \longleftrightarrow 1\,\text{J} = 0.239\,\text{cal}$,

$$\left.\begin{array}{l} Q = m_o c^2 = 0.000\,001\,\text{kg} \cdot 299\,792\,458^2\,\text{m}^2\,\text{s}^{-2} = 8.987552 \cdot 10^{10}\,\text{Joule} \\ \quad = 8.99 \cdot 10^{10} \cdot 2.39 \cdot 10^{-4}\,\text{kcal} = 2.148 \cdot 10^7\,\text{kcal} \approx 215\,000 \cdot 100\,\text{kcal}\ . \end{array}\right\} \qquad (1292)$$

This energy would be sufficient to bring the pot with 215 0001 water from 0 °C to cooking. The total energy $E = (M + m_o)\,c^2$ of 215 0001 water and one milligram remains unchanged as the total mass $M + m_o$. The Special Relativity Theory now shows that the mass of 215 0001 water, i.e. the inertia of this amount of water at 100 °C, is increased by about one milligram with respect to the same amount of water (the same number of water molecules) at 0°C.

Problem 16 *Electrically charged $\pi^{\pm}$-mesons decay in its own rest system Σ' after a mean livetime of $\tau' = 2.60 \cdot 10^{-8}s$. The half lifetime T', after which one half of the mesons decayed, amounts $T' = \ln 2 \cdot \tau' = 1.804 \cdot 10^{-8}s$. Its rest energy is $m_o c^2 = 139.6 \cdot 10^6 eV$.*
Show that for an energy of $20 \cdot 10^9 eV$ $\pi^{\pm}$-mesons move almost with the speed of light, $v = 0.999\,975\,639\,c$. Determine the live time τ, the half live time T and the mean path length $l := v\,\tau$ before decay of these $\pi^{\pm}$-mesons in the laboratory system Σ_o. Compare the mean path length l with the length that an observer comoving with the $\pi^{\pm}$ mesons, i.e. in system Σ', would measure for the path of the laboratory Σ_o during the time τ'.

Chaper 3, Section 2, Chapter 4, Sections 1, 3
At first, we remind the reader in the notion of the lifetime, s. Fig. A.7.

The mean live time τ of a number $N_o = \sum \triangle N_i$ of unstable particles is defined as the mean value of the lifetime t_i of decaying particles multiplied with weights $\triangle N_i$, hence

$$\tau = \frac{\sum t_i \triangle N_i}{N_o}\,, \qquad \text{or for large particle numbers}$$

$$\tau = \frac{\int t(N)\,dN}{N_o}\,. \qquad (1293)$$

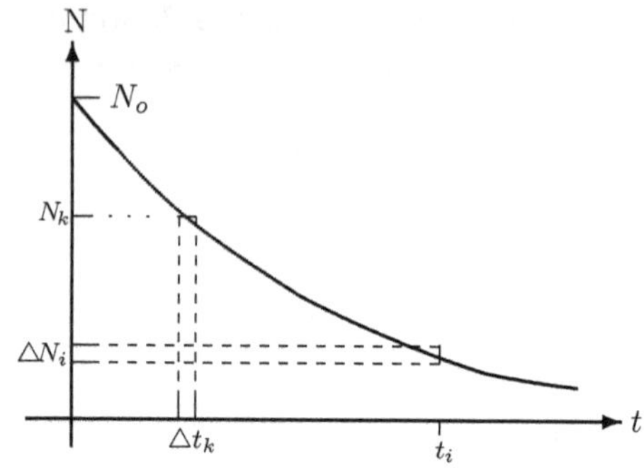

Fig. A.7 The definition of the mean lifetime of unstable particles (explanation in the text).

Since the integral is nothing else than the area below the curve of Fig. A.7, we can evaluate this area also as integral over the variable t, hence

$$\tau = \frac{\int N(t)\,dt}{N_o} \,. \tag{1294}$$

The radioactive decay follows an exponential law, therefore it follows[8]

$$N = N_o \, \exp\left[-\frac{t}{\tau}\right] \,. \tag{1295}$$

The integration of Eq. (1295) from $t = 0$ until $t = \infty$ just provides Eq. (1294),

$$\int_0^\infty N dt = N_o \int_0^\infty \exp\left[-\frac{t}{\tau}\right] dt = -\,N_o\,\tau\, \exp[-\frac{t}{\tau}\,]\,|_0^\infty \;=\; N_o\,\tau \;.$$

The inverse of the mean livetime is called decay constant λ,

$$N = N_o \, \exp[-\lambda\, t] \quad \text{with} \quad \lambda = \frac{1}{\tau} \,. \tag{1296}$$

According to Eq. (187) for the energy–mass equivalence, the ratio of the energy of the accelerated $\pi^{\pm}$-mesons to its rest energy is given by

$$\frac{mc^2}{m_o c^2} = \frac{200 \cdot 10^8}{139.6 \cdot 10^6} = \frac{1}{\sqrt{1 - v^2/c^2}} \,, \tag{1297}$$

and, when we use $c = 299\,792\,458\,\mathrm{ms}^{-1}$, Eq. (4), it follows

$$\left.\begin{aligned} \gamma &= 1/\sqrt{1 - \frac{v^2}{c^2}} = 143 \;, \\ v &= 0.999\,975\,6395\, c = 299\,785\,155\;\mathrm{ms}^{-1} \,. \end{aligned}\right\} \tag{1298}$$

---

[8]The halfe live time T follows from $N(T) := N_o/2 = N_o \, \exp[-\frac{T}{\tau}\,]$, $2 = \exp[\frac{T}{\tau})$, hence $T/\tau = \ln 2$.

The rest system Σ' of charged mesons moves with velocity v against the laboratory system Σ_o. Clocks in Σ' show the time t' which is smaller than the time t in the laboratory system, by the square root factor, i.e. by $1/\gamma$, according to Eq. (113), $t' = t/\gamma$. The live time τ' in the rest system of $\pi^{\pm}$-mesons and the half lifetime T' are t'-times. In the laboratory system the lifetime τ and the half lifetime T are by a factor γ larger,

$$\tau = \gamma\,\tau' = 0.000\,003\,7249\ \text{s}\ , \quad T = \ln 2 \cdot \tau = 0.000\,002\,582\ \text{s}\ . \tag{1299}$$

In the laboratory system, we measure a mean path length l

$$l = v\,\tau = 1117\ \text{m} = 1.117\ \text{km}\ . \tag{1300}$$

In other words, in the laboratory system, a half number of N $\pi^{\pm}$-mesons will still reach the target at a distance of $L = T\,v = 0.774\text{km}$.

As seen in system Σ', the system Σ_o has the velocity $-v$. Hence, the comoving observer with the $\pi^{\pm}$-mesons, i.e. in system Σ', measures a small distance l' transversed by the laboratory system Σ_o during time τ',

$$l' = |-v|\,\tau' = 7.7944\ \text{m}\ . \tag{1301}$$

l' is equal to the LORENTZ-contracted mean path length l, hence with Eq. (1298)

$$l' = l/\gamma = 0.006\,98 \cdot 1116.677\ \text{m} = 7.794\ \text{m}\ . \tag{1302}$$

Corresponding experiments are already performed some time ago, cp. Durbin et al. (1952) using π^{+}-mesons, and Burrowes et al. (1959) using K-mesons, cp. also Chap. 4, Sect. 6.

Problem 17 *Evaluate the amount of heat energy produced in the nuclear fusion of deuterium and tritium to helium in the reaction* $^2D + {}^3T \longrightarrow {}^4He + n$.

Chapter 6, Section 3, Chapter 7, Section 1

We use the value $m_n = 1.674\,928\,6 \cdot 10^{-27}\text{kg}$ for the rest mass of the neutron. We take the atomic mass unit u, equal to one twelfth of the mass of one atom of the carbon isotop ^{12}C, with value $u = 1.6605402 \cdot 10^{-27}\text{kg}$. We denote the relative atomic masses by m_r and the absolute values for the rest masses by $m_o = u\,m_r$, then we get for deuterium, tritium, helium and the neutron the values

$$\left.\begin{array}{lll} & m_r & m_o\,[10^{-27}\ \text{kg}] \\ ^2D & 2.014\,08 & 3.344\,460\,8\ , \\ ^3T & 3.01605 & 5.008\,272\,3\ , \\ ^4He & 4.002\,60 & 6.6464\,782\ , \\ n & 1.008\,664\,891 & 1.674\,928\,6\ . \end{array}\right\} \tag{1303}$$

In the reaction

$$^{2}D + {}^{3}T \longrightarrow {}^{4}He + n \tag{1304}$$

we get from Eq. (1303) for the sum of rest masses M_o and $\overline{M}_o$ before and after the reaction the values

$$\left.\begin{array}{l} M_o = 8.352\,733\,1 \cdot 10^{-27}\,\text{kg} \ , \\ \overline{M}_o = 8.321\,406\,8 \cdot 10^{-27}\,\text{kg} \ . \end{array}\right\} \quad \text{Rest masses before and after the reaction} \tag{1305}$$

The difference of rest masses in the nuclear reaction (1304) is then

$$\Delta M_o = M_o - \overline{M}_o = 0.0313\,263 \cdot 10^{-27}\,\text{kg} \ . \tag{1306}$$

During the reaction, the rest mass ΔM_o is transformed into the kinetic energy of the reaction products, the helium atom and the neutron. This energy can be transformed into heat energy Q of the environmental medium. The radiation energy of the Sun results from this process.

According to the mass–energy equivalence (186), the produced heat energy for a single atomic reaction (1304) is then

$$\begin{aligned} Q &= \Delta M_o\, c^2 = 0.031\,326\,3 \cdot 10^{-27}\text{kg} \cdot 2.997\,924\,58^2 \cdot 10^{16}\,\text{m}^2\text{s}^{-2} \\ &= 2.815\,467\,435\,5 \cdot 10^{-12}\,\text{J} \\ &= 2.815\,367\,435\,5 \cdot 10^{-12} \cdot 0.239\,\text{cal} \\ &= 2.815\,467\,435\,5 \cdot 10^{-12} \cdot 6.241\,506\,48 \cdot 10^{12}\,\text{MeV} \ , \end{aligned}$$

hence

$$Q \approx 17.57\,\text{MeV} = 0.673 \cdot 10^{-12}\,\text{cal} \ . \tag{1307}$$

For $6.022\,136\,7 \cdot 10^{23}$ reactions, i.e. for the fusion of one mol deuterium and on mol tritium to one mol helium, there react according to Eq. (1303) 2.014 08g deuterium and 3.016 05g tritium to 4.002 60g helium and 1.008 664 891g neutrons. From Eq. (1306), we get a molar mass defect of $\Delta M_{mol} = 0.0313\,263 \cdot 10^{-27} \cdot 6.0221\,367 \cdot 10^{23}$ kg,

$$\Delta M_{mol} \approx 0.019\,\text{g} \ . \quad \text{Molar mass defect} \tag{1308}$$

According to Eq. (1307), a huge heat energy is delivered during the fusion of only 4 g helium,

$$Q_{mol} \approx 0.673 \cdot 10^{-12} \cdot 6.02 \cdot 10^{23}\,\text{cal} \approx 4 \cdot 10^{8}\,\text{kcal} \ ,$$

or in other units

$$Q_{mol} = \Delta M_{mol} c^2 \text{J} = 19 \cdot 10^{-6} \cdot 9 \cdot 10^{16}\text{J} = 19 \cdot 10^{-6} \cdot 9 \cdot 10^{16} \cdot 239 \cdot 10^{-6}\text{kcal} \ ,$$

clearly again

$$Q_{mol} \approx 4 \cdot 10^8 \,\text{kcal} \ . \qquad \text{Molar fusion heat energy} \qquad (1309)$$

With the fusion energy of 1 g helium, one could heat the amount of one million litre water from 0 °C to boiling.

The mass of one proton $m_p = 1.672\,623\,1 \cdot 10^{-27}$ kg, corresponding to a relative atomic mass of $m_r = 1.007\,276\,487\,5$ is slightly different from the mass of the neutron. One gets immediately from Eq. (1303), that the rest mass of the helium nucleus is smaller than the sum of rest masses of its constituents, two protons and two neutrons. The protons or neutrons cannot leave the nucleus simply from energetic reasons. A helium nucleus is stable.

In general, one understands under a mass defect ΔW of an atomic nucleus the difference of its rest mass M_o and the sum of rest masses of its nuclear constituents, the sum of masses of protons m_p and neutrons m_n,

$$\Delta W = M_o - \sum m_p - \sum m_n \ . \qquad \text{Mass defect of atomic necleus} \qquad (1310)$$

If this mass defect is negative, the nucleus is stable. This stability is fulfilled for light atomic nuclei, in particular for helium. Some heavy atomic nuclei have $\Delta W > 0$, leading to natural radioactivity. The atomic nuclei decay spontaneously in stable or unstable subnuclei. Uranium ^{235}U has a positive mass defect of more than two neutron masses. Therefore, one can set free atomic energy by fission of uranium nuclei, a different process as the fusion of helium. The quantitative evaluation of the amount of energy becoming available is not so straightforward due to the large number of partly parallel processes compared to the case of nuclear fusion.

Problem 18 *Show that in a total inelastic collision with work done only by internal forces, a system cannot release energy due to the law of momentum conservation. Perform the proof for a relativistic space-time.*

Chapter 6, Section 3, Chapter 7, Section 1

We take the scalar product of Eq. (169), with the particle velocity $\mathbf{u}_a$ and get

$$\mathbf{u}_a \cdot \frac{d}{dt}\Big(\frac{m_{oa}}{\sqrt{1-u_a^2/c^2}}\,\mathbf{u}_a\Big) = \Big(\sum_{b=1}^{n} \mathbf{F}_{ba}\Big)\cdot \mathbf{u}_a + \mathbf{F}_a \cdot \mathbf{u}_a \ . \qquad (1311)$$

As in Eq. (175) we find

$$\frac{d}{dt}\frac{m_{oa}\,c^2}{\sqrt{1-u_a^2/c^2}} = \sum_{b=1}^{n} (\mathbf{F}_{ba}\cdot \mathbf{u}_a) + \mathbf{F}_a \cdot \mathbf{u}_a \ . \qquad (1312)$$

On the right-hand side of Eq. (1312) we get the power done by the forces $\mathbf{F}_{ba}$ and $\mathbf{F}_a$, i.e. the work done per time unit on the particles moving with velocity $\mathbf{u}_a$. For the total inelastic collision $\mathbf{F}_a = 0$. We sum in Eq. (1312) over the particles,

$$\frac{d}{dt}\sum_{a=1}^{n}\frac{m_{oa}\,c^2}{\sqrt{1-u_a^2/c^2}}=\sum_{a=1}^{n}\sum_{b=1}^{n}\left(\mathbf{F}_{ba}\cdot\mathbf{u}_a\right)\ . \tag{1313}$$

The collision process might take place during the small time interval δt around $t=0$. Now we use the assumption of a totally inelastic collision. Before the impact we have two particles, after the collision only one particle with rest mass $\overline{M}_o$. We denote all quantities after the impact with an overbar. Then we find in system Σ_o

$$\Sigma_o : \left.\begin{array}{l} u_1=-u_2:=-u\ ,\quad m_{o1}=m_{o2}=m_o\ , \\ \overline{U}=0\ ,\quad \overline{M}_o\ . \end{array}\right\} \tag{1314}$$

We assume $\Delta t>\delta t$, then integration of Eq. (1313) provides the work ΔA done by the internal forces during the collision

$$\Delta A=\int_{-\Delta t}^{+\Delta t}\sum_{a=1}^{n}\sum_{b=1}^{n}\left(\mathbf{F}_{ba}\cdot\mathbf{u}_a\right)dt=\int_{-\Delta t}^{+\Delta t}\frac{d}{dt}\sum_{a=1}^{n}\frac{m_{oa}\,c^2}{\sqrt{1-u_a^2/c^2}}\,dt\ ,$$

$$\Delta A=\left[\sum_{a=1}^{n}\frac{m_{oa}\,c^2}{\sqrt{1-u_a^2/c^2}}\right]_{-\Delta t}^{+\Delta t}\ .$$

Using Eq. (184) we conclude that the total work ΔA done by the inner forces vanishes

$$\Delta A=\overline{M}_o c^2-\frac{2m_o c^2}{\sqrt{1-u^2/c^2}}=0\ . \tag{1315}$$

Problem 19 *Show that one can measure the motion of the guiding medium within the framework of classical space-time from the validity of Eqs. (232), and (233), for the* DOPPLER *effect of moving sender or receiver.*

Chapter 8, Section 5

A sender S might emit per construction an eigenfrequency ν_S. A receiver E can measure the exact frequencies, e.g. by exciting resonances. There it is supposed that the signal period of sending or receiving systems does not depend on their velocities (classical space-time). The unknown rest system of the transmitting medium of waves is denoted by Σ_o. Waves travel there with velocity C. We consider two systems Σ' and Σ'', having velocities v_1 and v_2, respectively, with respect to Σ_o. Due to the unknown state of motion of Σ_o we do not know neither v_1 nor v_2, but the relative velocity v_R, that is equal for the classical space-time by Eq. (70), in each inertial system,

$$v_R=v_2-v_1\ . \qquad \text{Relative velocity between systems } \Sigma' \text{ and } \Sigma'' \tag{1316}$$

In systems Σ' and Σ'' we have the sender S and receiver E, respectively, cp. Fig. A.8 and Fig. A.9.

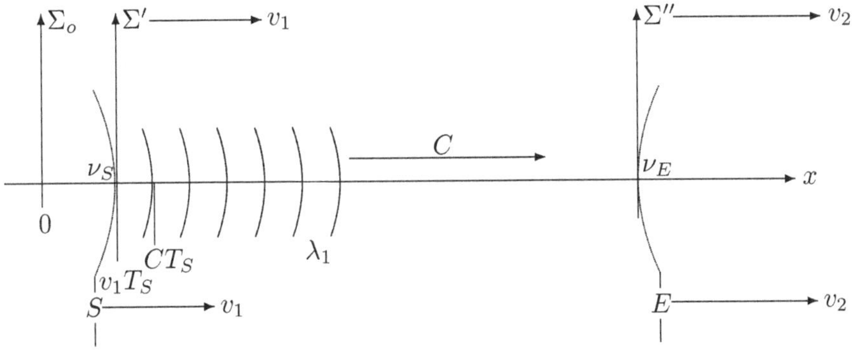

Fig. A.8 The sender *S* and, at the right side, the receiver *E* are at rest in systems Σ' and Σ'', with unknown velocities v_1 and v_2 with respect ot Σ_o. The propagation velocity of waves in Σ_o is C. Due to velocity v_1, in Σ_o the emitted waves have a wave length λ_1 according to Eq. (1317).

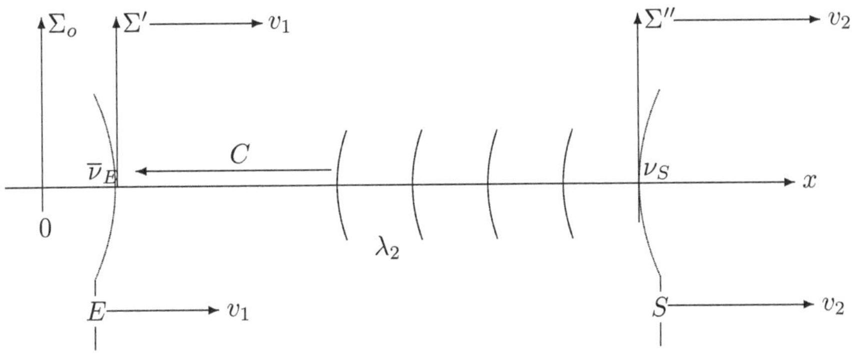

Fig. A.9 Now the receiver *E* and on the right of it the sender *S* are in rest systems Σ' and Σ'', with unknown velocities v_1 and v_2 with respect to Σ_o. Due to velocity v_2 in Σ_o the emitted waves now have a wavelength λ_2 given in Eq. (1319).

1. The sender S might be left from the receiver E, and resting in Σ'. The receiver is at rest in Σ'', Fig. A.8. We evaluate the frequency ν_E measured by the receiver in dependence of the unknown velocity v_1, that is measured by the receiver.

Due to its velocity v_1 in direction of signal propagation, the emitter produces a wave relative to the in Σ_o resting medium with wavelength λ_1 as given in Eq. (231),

$$\lambda_1 = (C - v_1)T_S = \frac{C - v_1}{\nu_S} \,. \tag{1317}$$

The receiver, moving with velocity v_2 in direction of signal propagation as shown in Fig. A.8, follows this wave with velocity $C - v_2$, hence it measures a frequency ν_E,

$$\nu_E = \frac{C - v_2}{\lambda_1} = \nu_S \frac{C - v_2}{C - v_1} = \nu_S \frac{C - v_R - v_1}{C - v_1} \ ,$$

hence

$$\nu_E = \nu_S \big(1 - \frac{v_R}{C - v_1}\big) \ . \qquad \text{Measured frequency } \nu_E \text{ according to the situation in Fig. 8} \qquad (1318)$$

The relative velocity v_R between receiver E and sender S can be measured. Unknown is however the velocity v_1 of the receiver E with respect to Σ_o.
2. We assume that sender and receiver both have identical properties. Both can then exchange their positions at arbitrary choice. Thereafter the sender S is right from the receiver E. The relative velocity v_R between sender and receiver is still $v_R = v_2 - v_1$, but now v_1 is the velocity of the in Σ' resting receiver E, and v_2 the velocity of the in Σ'' resting sender S, Fig. A.9. Because of its velocity v_2 with respect to Σ_o, the sender produces a wave in the medium propagating against the signal with wavelength λ_2, the value is given in Eq. (231), when we replace v by $-v_2$,

$$\lambda_2 = (C + v_2) T_S = \frac{C + v_2}{\nu_S} \ . \qquad (1319)$$

The receiver, moving as shown in Fig. A.9 with velocity v_1 in direction of signal propagation, i.e. this wave crosses the wavelength with velocity $C + v_1$, and therefore it determines a frequency $\overline{\nu}_E$

$$\overline{\nu}_E = \frac{C + v_1}{\lambda_2} = \nu_S \frac{C + v_1}{C + v_2} = \nu_S \frac{C + v_1 + v_R - v_R}{C + v_1 + v_R} \ ,$$

hence

$$\overline{\nu}_E = \nu_S \big(1 - \frac{v_R}{C + v_1 + v_R}\big) \ . \qquad \text{Measured frequency } \overline{\nu}_E \text{ for situation in Fig. 9} \qquad (1320)$$

For the considered constellation, an exchange of functions of sender and receiver leads to different frequencies (1318) and (1320). We add these frequencies, denote $v_1 = x$ and study to resulting function $f(x)$ with respect to extrema,

$$f(x) := \nu_E + \overline{\nu}_E = 2\nu_S - \nu_S\, v_R \big(\frac{1}{C - x} + \frac{1}{C + x + v_R}\big) = 2\nu_S - \nu_S\, v_R \frac{2C + v_R}{C^2 - x^2 + C v_R - x v_R} \ .$$

Under the condition that neither v_1 nor v_2 can reach the velocity C, it holds always that $2C + v_R \neq 0$, and we find from

$$f'(x) = -\nu_S\, v_R (2C + v_R) \frac{2x + v_R}{(C^2 - x^2 + C v_R - x v_R)^2} = 0$$

the solution

$$x = -\frac{v_R}{2} \; . \tag{1321}$$

After elementary calculation, it follows from

$$f''(x) = -\nu_S\, v_R(2C + v_R)\,\frac{2(C^2 - x^2 + C\,v_R - x\,v_R) + 2\,(2x + v_R)^2}{(C^2 - x^2 + C v_R - x v_R)^3}$$

that at $x = -v_R/2$

$$f''(-\frac{v_R}{2}) = -\frac{2^5 \nu_S\, v_R}{(2C + v_R)^3} < 0 \; . \tag{1322}$$

Hence, the function $f(x)$ has at $x = v_1 = -v_R/2$ a maximum.

Because of $v_R = v_2 - v_1 = v_2 + v_R/2$, we have $v_2 = v_R/2$. The sum of the measured frequencies $\nu_E + \overline{\nu}_E$ has therefore a maximum value when for fixed relative velocity v_R between the systems Σ' and Σ'', the velocity v_2 of system Σ'' against the medium is just $v_2 = v_R/2$, and the velocity of systems Σ' is $v_1 = -v_R/2$.

Keeping fixed the relative velocity v_R, we have only to change the velocities of sender and receiver so long, until we determine a maximum for the sum of frequencies $\nu_E + \overline{\nu}_E$. In practical set-ups, one could fix, e.g. the relative motion of sender and receiver on a carriage, and then we could vary the velocity of the carriage. In this way, it is possible to determine the state of motion of the carrier with the Doppler effect in case of validity of Eqs. (232) and (233), since in this case the Doppler effect depends on the absolute motion against the carrier medium.

Problem 20 *Evaluate from the invariance of phase ϕ the transversal* Doppler *effect for an in direction of negative y-axis incoming wave corresponding to the in Chap. 8, Sect. 7, considered situation.*

Chapter 8, Sections 5, 7, Chapter 9, Section 3.2.2
Independent of the physical postulate for space-time there is guaranteed the invariability of the in Σ' and Σ_o observed phases (260),

$$\phi' = \omega'\, t' - \mathbf{k}' \cdot \mathbf{x}' = \omega\, t - \mathbf{k} \cdot \mathbf{x} = \phi \; . \qquad \text{Phase invariance} \tag{1323}$$

Here we investigate a vertically incoming wave,

$$\mathbf{k} = (k_1,\, k_2,\, k_3) = (0,\, -k,\, 0) \; . \qquad \text{Vertical incoming wave} \tag{1324}$$

With Eq. (1323) it follows

$$\omega'\, t' - k_1' x' - k_2' y' - k_3' z' = \omega\, t + k y \; . \qquad \text{Phase invariance} \tag{1325}$$

As considered in Chap. 8, Sect. 7, the system Σ' should be in motion in system Σ_o with velocity $-v$.

At first, we consider the classical space-time. With the special GALILEI transformation

$$\left.\begin{aligned} x' &= x + v\,t, \\ t' &= t. \end{aligned}\right\} \qquad \text{Special GALILEI transformation} \qquad (1326)$$

there follows from Eq. (1325)

$$\omega' t - k_1'(x + vt) - k_2' y - k_3' z = \omega\,t + ky\ ,$$

hence

$$(\omega' - \omega - k_1' v)\,t - k_1'\,x - (k_2' + k)\,y - k_3'\,z = 0\ . \qquad (1327)$$

This can be fulfilled as identity in $x,\ y,\ z,\ t$ only for vanishing coefficients, and we get

$$\left.\begin{aligned} \mathbf{k}' &= (k_1',\ k_2',\ k_3') = (0,\ -k,\ 0)\ , \\ \omega' &= \omega\ . \end{aligned}\right\} \qquad (1328)$$

The first equation proofs in the Chap. 8, Sect. 7 discussed invariable direction of the wave normal. However, this does not mean no aberration.

We set $\omega' = 2\pi\nu_S$ and $\omega = 2\pi\nu_E$, then it follows from the second equation the already in Eq. (235), shown property, that in classical space-time there is no transversal DOPPLER effect.

Now we consider the relativistic space-time. With the special LORENTZ transformation, again with velocity $-v$ for Σ',

$$\left.\begin{aligned} x' &= \gamma\,(x + v\,t), \\ t' &= \gamma\,(t + vx/c^2)\ , \end{aligned}\right\} \qquad \text{Special LORENTZ transformation} \qquad (1329)$$

it follows from Eq. (1325)

$$\gamma\,\omega'\,(t + vx/c^2) - \gamma\,k_1'(x + vt) - k_2' y - k_3' z = \omega\,t + ky\ ,$$

hence

$$\big(\gamma\,\omega' - \gamma\,k_1'\,v - \omega\big)\,t - \big(\gamma\,k_1' - v\,\omega'/(c^2\gamma)\big)\,x - (k_2' + k)\,y - k_3'\,z = 0\ . \qquad (1330)$$

Again this is an identity in $x,\ y,\ z,\ t$ only for vanishing coefficients. Now this means however

$$\left.\begin{aligned} \mathbf{k}' &= (k_1',\ k_2',\ k_3') = (v\,\omega'/c^2,\ -k,\ 0)\ , \\ \omega' &= \omega/\gamma + k_1' v\ . \end{aligned}\right\} \qquad (1331)$$

Here we insert the expression for k_1' in the equation for ω' and observe again that $\omega = 2\pi\nu_E$ and $\omega' = 2\pi\nu_S$, then it follows with

$$\omega = \omega'/\gamma \tag{1332}$$

the validity of Eq. (243) for the transversale DOPPLER effect in the relativistic space-time.

From the ratio of the components of the vector $\mathbf{k}'$, it follows in this case again the aberration angle, since with $c\,k = \omega$ it holds

$$\tan\alpha' = \left|\frac{k'_x}{k'_y}\right| = \frac{v\,\omega'}{c^2 k} = \frac{\gamma\,v\,\omega}{c^2 k} = \frac{\gamma\,v}{c}\,, \tag{1333}$$

s. the discussion in Chap. 8, Sect. 7 and Problem 33.

Problem 21 *Evaluate from the* TAYLOR *series (352), for the* LORENTZ *transformation the rotation matrix* $\mathbf{D}_1$ *and the special* LORENTZ *transformation* $\mathbf{L}_1$ *from Eq. (346), if only* $\alpha_1 \neq 0$ *and* $\beta_1 \neq 0$ *is assumed.*

Chapter 9, Section 1

1. The series for $e^{-\alpha_1\,\mathbf{a}_1}$ reads (352)

$$\exp[-\alpha_1\,\mathbf{a}_1] = \mathbf{1} - \alpha_1\,\mathbf{a}_1 + \frac{1}{2}\,\alpha_1^2\,\mathbf{a}_1^2 - \frac{1}{3!}\,\alpha_1^3\,\mathbf{a}_1^3 + \frac{1}{4!}\,\alpha_1^4\,\mathbf{a}_1^4 - + \dots\,, \tag{1334}$$

or, since one can arbitrarily reorder the power series,

$$\left.\begin{aligned}\exp[-\alpha_1\,\mathbf{a}_1] = \mathbf{1} &+ \frac{1}{2}\,\alpha_1^2\,\mathbf{a}_1^2 + \frac{1}{4!}\,\alpha_1^4\,\mathbf{a}_1^4 + \dots\\ &- \alpha_1\,\mathbf{a}_1 - \frac{1}{3!}\,\alpha_1^3\,\mathbf{a}_1^3 - \dots\,.\end{aligned}\right\} \tag{1335}$$

One notes the diagonal structure according to Eq. (369) and the changing signs of the even powers $\mathbf{a}_1^{2n}$ and the relation $\mathbf{a}_1^3 = -\mathbf{a}_1$ for the uneven powers $\mathbf{a}_1^{2n+1}$, then we get with using the matrix $\mathbf{b}_1^2$ from Eq. (369) and the known series presentations

$$\begin{aligned}\cos\alpha &= 1 - \frac{1}{2}\,\alpha^2 + \frac{1}{4!}\,\alpha^4 - \dots\,,\\ \sin\alpha &= \alpha - \frac{1}{3!}\,\alpha^3 + \frac{1}{5!}\,\alpha^5 - \dots\,.\end{aligned}$$

There follows immediately the wanted result for the presentation of $\mathbf{D}_1$ from Eqs. (346), (348) and (369),

$$\exp[-\alpha_1\,\mathbf{a}_1] = \mathbf{b}_1^2 - \cos\alpha_1\,\mathbf{a}_1^2 - \sin\alpha_1\,\mathbf{a}_1 = \mathbf{D}_1\,. \tag{1336}$$

This expression also shows that the in Eq. (346) introduced parameter α_1 is the angle of spatial rotation.

2. For $e^{-\beta_1\,\mathbf{b}_1}$ we write the series (352)

$$\exp[-\beta_1\,\mathbf{b}_1] = \mathbf{1} - \beta_1\,\mathbf{b}_1 + \frac{1}{2}\,\beta_1^2\,\mathbf{b}_1^2 - \frac{1}{3!}\,\beta_1^3\,\mathbf{b}_1^3 + \frac{1}{4!}\,\beta_1^4\,\mathbf{b}_1^4 - + \dots\,, \tag{1337}$$

and again after reordering,

$$\left.\begin{aligned}\exp[-\beta_1\,\mathbf{b}_1] = \mathbf{1} + \frac{1}{2}\,\beta_1^2\,\mathbf{b}_1^2 + \frac{1}{4!}\,\beta_1^4\,\mathbf{b}_1^4\,+\,\dots\\ -\,\beta_1\,\mathbf{b}_1 - \frac{1}{3!}\,\beta_1^3\,\mathbf{b}_1^3\,-\,\dots\,.\end{aligned}\right\} \tag{1338}$$

We consider with Eq. (369) the diagonal structure and the same sign of the even powers $\mathbf{b}_1^{2n}$, and the relation $\mathbf{b}_1^3 = \mathbf{b}_1$ for the uneven powers $\mathbf{b}_1^{2n+1}$, and we remind the series presentations

$$\cosh\alpha = 1 + \frac{1}{2}\,\alpha^2 + \frac{1}{4!}\,\alpha^4\,+\,\dots\,,$$
$$\sinh\alpha = \alpha + \frac{1}{3!}\,\alpha^3 + \frac{1}{5!}\,\alpha^5\,+\,\dots\,.$$

For Eq. (1338) we can then write

$$\exp[-\beta_1\,\mathbf{b}_1] = \mathbf{1} - \mathbf{b}_1^2 + \cosh\beta_1\,\mathbf{b}_1^2 - \sinh\beta_1\,\mathbf{b}_1\;. \tag{1339}$$

Here we write again $\beta_1 = v_1/c$ and find with Eq. (337)

$$\exp[-\beta_1\,\mathbf{b}_1] = \mathbf{1} - \mathbf{b}_1^2 + \frac{1}{\sqrt{1-v^2/c^2}}\,\mathbf{b}_1^2 - \frac{v/c}{\sqrt{1-v^2/c^2}}\,\mathbf{b}_1\;. \tag{1340}$$

According to Eqs. (346) and (369) this is the representation of $\mathbf{L}_1$.

Problem 22 *We denote with $\mathbf{L}_{1,2}$ the* LORENTZ *transformation (382), producing from Σ_o the inertial system Σ'' with coordinates $\tilde{\mathbf{x}}''$. Show that the $\tilde{\mathbf{x}}''$ coordinate axes are seen from Σ_o as non-orthogonal.*

Chapter 8, Sections 3, 4, Chapter 9, Section 1
With the matrix $\mathbf{L}_{1,2}^{-1}$, explicitly written as the right matrix in Eq. (398) we evaluate from coordinates $\tilde{\mathbf{x}}''$ of an event in Σ'' the coordinates $\mathbf{x}$ in Σ_o,

$$\begin{pmatrix} x^0\\ x^1\\ x^2\\ x^3\end{pmatrix} = \begin{pmatrix} \gamma & \gamma\,\beta_1 & \gamma\,\beta_2 & 0\\ \gamma\,\beta_1 & 1+\frac{(\gamma-1)\beta_1^2}{\beta^2} & \frac{(\gamma-1)\beta_1\beta_2}{\beta^2} & 0\\ \gamma\,\beta_2 & \frac{(\gamma-1)\beta_2\beta_1}{\beta^2} & 1+\frac{(\gamma-1)\beta_2^2}{\beta^2} & 0\\ 0 & 0 & 0 & 1\end{pmatrix}\begin{pmatrix}\tilde{x}^{0''}\\ \tilde{x}^{1''}\\ \tilde{x}^{2''}\\ \tilde{x}^{3''}\end{pmatrix}\,. \tag{1341}$$

We consider an event $F(c\tilde{t}'', 0, \tilde{l}'', 0)$ on the $\tilde{y}''$-axis of Σ''. The corresponding coordinates in Σ_o are formed from (1341), setting on the right-hand side for $\tilde{\mathbf{x}}''$ the coordinates $F(c\tilde{t}'', 0, \tilde{l}'', 0)$,

$$F\left(ct_F = \tilde{l}''\tfrac{\beta_2}{\gamma} + \tfrac{c\tilde{t}''}{\gamma}\,,\; x_F = \tilde{l}''\tfrac{(1-\gamma)\beta_1\beta_2}{\beta^2\gamma} + c\tilde{t}''\tfrac{\beta_1}{\gamma}\,,\; y_F = \tilde{l}''(1+\tfrac{(1-\gamma)\beta_2^2}{\beta^2\gamma}) + c\tilde{t}''\tfrac{\beta_2}{\gamma}\,,\; z_E = 0\right).$$

The in Σ_o simultaneous positions of events O and F on the $\tilde{y}''$-axis are obtained by the condition $t_F = 0$,

$$c\tilde{t}'' = -\tilde{l}''\beta_2 \ .$$

After straightforward but simple calculation we get the result

$$\frac{y_F}{x_F} = \tan\eta_y = -\frac{\beta_1^2\gamma + \beta_2^2(1-\gamma) + \beta_2^2\beta_2}{\beta_1\beta_2(1-\gamma-\beta^2)} \ . \tag{1342}$$

Since $x_F/\tilde{l}'' < 0$, $y_F/\tilde{l}'' > 0$, this angle lies in the second quadrant, so that $\eta_y > \pi/2$.

Next we consider the events $O(0,0,0,0)$ and $E(c\tilde{t}'', \tilde{l}'', 0, 0)$ on the $\tilde{x}''$-axis of Σ''. In this case however, we find the direction of the x''-axis without calculation by remembering the THOMAS-precession, Chap. 8, Sect. 3. The looked for quantity is the angle α_3 that is there given by, s. Fig. 35

$$\tan\alpha_3 = \frac{(\gamma-1)\beta_1\beta_2}{\gamma\beta_1^2 + \beta_2^2} \ . \tag{1343}$$

Since $\alpha_3 < 0$, we find that the x''-axis of the system Σ'' is below the x-axis of Σ_o.

Summerising both results, the x''-axis (lying in the fourth quadrant) and the y''-axis (lying in the second quadrant) no longer are orthogonal.

Problem 23 *Explain the time dilatation observed in Σ_o and in Σ' from the geometrical representation of the* LORENTZ *transformation in the* MINKOWSKI *space analogous to the explanation of* LORENTZ *contraction in Chap. 9, Fig. 3.*

Chapter 9, Section 1

We consider the problem of Fig. A.10.

The intersection $F(x_F\ , ct_F)$ of the $c\,t'$-axis given by (423) with the gauge hyperbola (423) is given by equations

$$ct_F = \frac{c}{v}\,x_F\ , \quad \text{and} \quad x_F^2 - c^2\,t_F^2 = 1\ .$$

It follows

$$x_F = \frac{\gamma\,v}{c}\ , \quad ct_F = \gamma\ . \tag{1344}$$

This defines the points B and A on the axis. (A second intersection for negative $c\,t'$-values is disregarded.)

Since the point F is given by the invariant line element according to Eq. (1344) with

$$s^2 = c^2\,t_F^2 - x_F^2 = c^2\,t'^2_F - x'^2_F = 1$$

in Σ', the unit on the $c\,t'$-axis is fixed by $x'_F = 0$ at $c\,t'_F = 1$.

For describing time dilatation we have to compare the pointer positions of a moving clock with the pointer positions of two resting clocks, where the first just

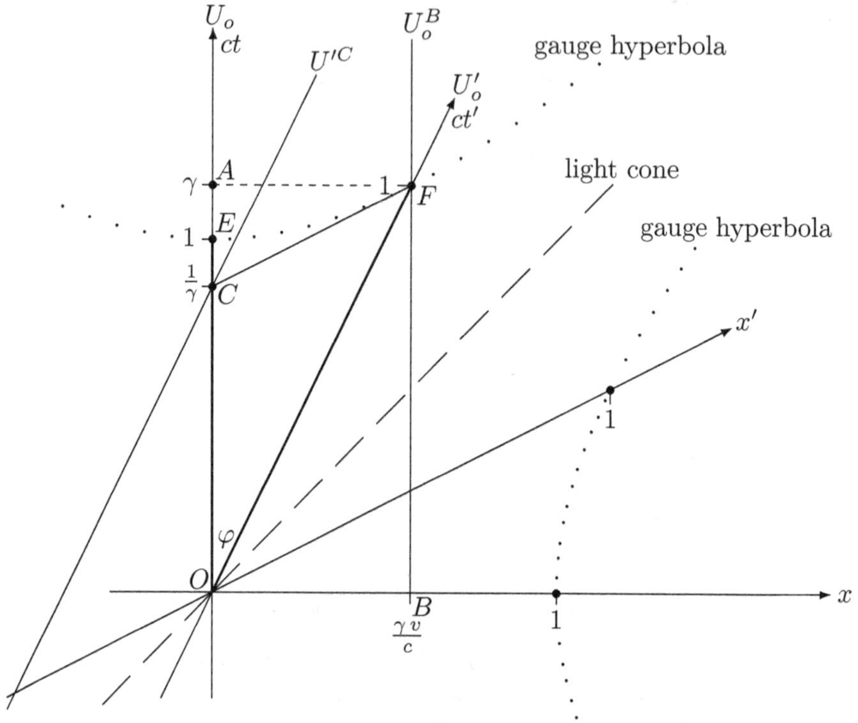

Fig. A.10 Special LORENTZ transformation and time dilatation. The length unit 1 is shown on the ct- and ct'-axis by the crossing of the dotted lines.

passes by. We remember, s. Chap. 9, Fig. 3, at length unit 1 on the ct-axis, the pointer of the clock shows $(1/c)\mathrm{s} \approx (1/3) \cdot 10^{-8}\mathrm{s}$ (SI-units in MINKOWSKI space).

The clocks U_o and U_o' are at rest at $x = 0$ in Σ_o and at $x' = 0$ in Σ', respectively, and its pointer positions are fixed on the common coordinate origin by the initial conditions

$$t = 0 \text{ and } t' = 0.$$

The world lines of the in Σ_o resting clocks are in the MINKOWSKI diagram the parallel lines to the $c\,t$-axis. To the pointer position 1 of clock U_o' is assigned by the gauge hyperbola (423) the point F, with coordinates Eq. (1344) in Σ_o. The world line of the at $x = \gamma\, v/c$ in Σ_o resting clock U_o^B cuts the world line of clock U_o' at point F, i.e. both clocks met there and can compare its pointer positions. The pointer position of U_o^B is the time coordinate of point F, i.e. the pointer of U_o^B shows the value γ, while the pointer of U_o' shows 1:

The clock U_o' moving in system Σ_o with velocity v is retarded by a factor $1/\gamma$ with respect to an in Σ_o resting clock.

The world lines of the in Σ' resting clocks are in the MINKOWSKI diagram parallel lines to the $c\,t'$-axis. The pointer position 1 of clock U_o corresponds to the point E given by the gauge hyperbola (423).

We determine the parallel line to the x'-axis going through the point F from

$$\frac{c\,t - c\,t_F}{x - x_F} = \frac{v}{c}\,,$$

hence

$$c\,t = \frac{v}{c}\,x + 1/\gamma\ . \qquad \text{Parallel line to } x'\text{-axis through } F \qquad (1345)$$

This line cuts the $c\,t$-axis, hence $x = 0$ in (1345), at the point C for $c\,t = 1/\gamma$. The parallel line to the $c\,t'$-axis (423) through the point C is the world line of an in Σ' synchronised clock U'^C, which therefore shows at event C with U'_o simultaneously the pointer position $t' = 1$. The pointer of the in Σ_o resting, but with respect to Σ' with velocity v moving clock U_o shows at the encounter with clock U'^C the time $t = 1/\gamma$. At the common origin O, both U_o and U'_o had the same initial value zero:

The clock U_o, moving in system Σ' with velocity v, runs by a factor $1/\gamma$ slower with respect to a clock resting in Σ'.

From the geometric representation of the Minkowski space, the time dilatation can be read off for all inertial systems.

Problem 24 *Show that the eigentime τ_E of a body moving with velocity $u = dx/dt$ along the x-axis in Σ_o,*

$$\tau_E = \int_0^{\tau_E} d\tau = \int_0^{t_E} \sqrt{1 - u^2(t)/c^2}\,dt = \int_0^{t'_E} \sqrt{1 - u'^2(t')/c^2}\,dt',$$

is a relativistic invariant parameter.

Chapter 1, Sections 5, Chapter 9, Section 1

We consider the world lines of the body from the event O till an arbitrary event E. There the origin O is chosen as start point, clearly no restriction of generality in the proof of the invariance. The inertial system Σ' moves along the x-axis of Σ_o with velocity v. For each point of the world line of the body, we can assume the special Lorentz transformation, Eqs. (105) or (339). The motion of the body $x = u(t)$ is inserted into Eq. (105), and we get immediately, omitting the y- and z-coordinates,

$$\left.\begin{aligned} x' &= \frac{x(t) - v\,t}{\sqrt{1 - v^2/c^2}} \\ t' &= \frac{t - x(t)\,v/c^2}{\sqrt{1 - v^2/c^2}}\ . \end{aligned}\right\} \qquad (1346)$$

For the evaluation of the integral

$$\int_0^{t'_E} \sqrt{1 - u'^2(t')/c^2}\,dt' \qquad (1347)$$

we substitute $t' = t'(t)$ from Eq. (1346),

$$\frac{dt'}{dt} = \frac{1 - u(t)\, v/c^2}{\sqrt{1 - v^2/c^2}} \quad \longleftrightarrow \quad \frac{dt}{dt'} = \frac{\sqrt{1 - v^2/c^2}}{1 - u(t)\, v/c^2} \tag{1348}$$

and therefore

$$u'(t') = \frac{dx'}{dt'} = \frac{dx'}{dt}\,\frac{dt}{dt'} = \frac{u(t) - v}{\sqrt{1 - v^2/c^2}}\,\frac{\sqrt{1 - v^2/c^2}}{1 - u(t)\, v/c^2} = \frac{u(t) - v}{1 - u(t)\, v/c^2}\ , \tag{1349}$$

where the second Eq. (1349) is nothing else than EINSTEIN's addition theorem of velocities (106).

After substitution the integration runs in the new variable t from 0 to t_E. We insert Eqs. (1348) and (1349) into (1347) and get

$$\begin{aligned}
\int_0^{t'_E} \sqrt{1 - u'^2(t')/c^2}\, dt' &= \int_0^{t_E} \sqrt{1 - \frac{1}{c^2}\,\frac{(u(t) - v)^2}{\left(1 - u(t)\, v/c^2\right)^2}}\;\frac{1 - u(t)\, v/c^2}{\sqrt{1 - v^2/c^2}}\, dt \\
&= \int_0^{t_E} \sqrt{\left(1 - u(t)\, v/c^2\right)^2 - \frac{1}{c^2}\,(u(t) - v)^2}\; dt \\
&= \int_0^{t_E} \sqrt{1 - 2\frac{uv}{c^2} + \frac{u^2\, v^2}{c^4} - \frac{u^2}{c^2} - \frac{v^2}{c^2} + 2\frac{u\, v}{c^2}}\;\frac{1}{\sqrt{1 - v^2/c^2}}\, dt \\
&= \int_0^{t_E} \sqrt{\left(1 - v^2/c^2\right)\left(1 - u^2(t)/c^2\right)}\;\frac{1}{\sqrt{1 - v^2/c^2}}\, dt\ ,
\end{aligned}$$

hence

$$\tau_E = \int_0^{\tau_E} d\tau = \int_0^{t'_E} \sqrt{1 - u'^2(t')/c^2}\, dt' = \int_0^{t_E} \sqrt{1 - u^2(t)/c^2}\, dt\ . \tag{1350}$$

The motion of the body from the origin with an arbitrary velocity up to an arbitrarily chosen event E leads to one and the same value of the covered eigentime τ_E, independent of the chosen inertial system as it should really be.

Problem 25 *Illustrate by means of the totally inelastic collision of two bodies of the same rest mass m_o the transformation of the kinetic energy of the bodies before the collision into an equivalent rest energy after the collision. Perform the evaluation by means of the covariant formalism in* MINKOWSKI *space for the centre-of-mass system of both bodies.*

Chapter 7, Section 1, Chapter 9, Section 2

In the inertial centre-of-mass system, we put the x-axis in direction of motion of the incoming masses m_o. After the collision, we denote the quantities by an overbar. A totally inelastic collision means that then there appears a single body of restmass $\overline{M}_o$. For the three-dimensional velocities, we can then write

$$\left.\begin{array}{ll}\text{Before the collision :} & \mathbf{u}_1 = -\mathbf{u}_2 = (u, 0, 0) \ , \\ \text{After the collision :} & \overline{\mathbf{U}} = (\overline{U}, 0, 0) \ . \end{array}\right\} \tag{1351}$$

We use the energy–momentum conservation (456), hence

$$p_1^i + p_2^i = \overline{P}^i \ . \tag{1352}$$

With rest masses m_o and $\overline{M}_o$, four velocities according to Eq. (431), and LORENTZ factors γ with indices u and $\overline{U}$, respectively, it holds

$$m_o\,\gamma_u\,u_1^i + m_o\,\gamma_u\,u_2^i = \overline{M}_o\,\gamma_{\overline{U}}\,\overline{U}^i \ . \tag{1353}$$

In the spatial components of this equation, for $i = 1, 2, 3$, there remains because of Eq.(1351) only the $i = 1$ component,

$$m_o\,\gamma_u\,u + m_o\,\gamma_u\,(-u) = \overline{M}_o\,\gamma_{\overline{U}}\,\overline{U} = 0 \ , \tag{1354}$$

hence

$$\overline{U} = 0 \quad \longrightarrow \quad \gamma_{\overline{U}} = 1 \ . \tag{1355}$$

We set in Eq. (1353) $i = 0$, so it follows with Eqs. (431) and (1355)

$$m_o\,\gamma_u\,c + m_o\,\gamma_u\,c = \overline{M}_o \ , \tag{1356}$$

hence our result from Chap. 6, Sect. 1, Eq. (184),

$$\overline{M}_o = 2\,\gamma_u\,m_o \ , \tag{1357}$$

and therefore the transformation of the kinetic energy of bodies before the collision into a corresponding amount of rest energy as in Eq. (185).

Problem 26 *An unstable particle of rest mass M_o might decay into two particles with rest masses $\overline{m}_{o1}$ and $\overline{m}_{o2}$. What can be inferred on the masses $\overline{m}_{o1}$ and $\overline{m}_{o2}$? Perform the evaluation in the covariant formalism.*

Chapter 9, Section 2
The problem is equivalent to an inelastic relativistic collision process. We use the inertial system, in which the decaying particle is at rest. Furthermore, we can orient x-axis along the direction of motion of the decay products. Then we get for the three-dimensional velocities

$$\left.\begin{array}{ll}\text{Before the collision :} & \mathbf{U} = (0, 0, 0) \ , \\ \text{After the collision :} & \begin{array}{l}\overline{\mathbf{u}}_1 = (\overline{u}_1, 0, 0) \ , \\ \overline{\mathbf{u}}_2 = (\overline{u}_2, 0, 0) \ . \end{array}\end{array}\right\} \tag{1358}$$

We use the energy–momentum conservation (456),

$$P^i = \overline{p}^i_1 + \overline{p}^i_2 \ . \tag{1359}$$

With rest masses $\overline{m}_{o1}$, $\overline{m}_{o2}$ and M_o and the four-velocity Eq. (431), where we specify the LORENTZ factors γ by the indices of the corresponding velocities $\overline{u}_1$ and $\overline{u}_2$, and since for $U = 0$ we have $\gamma_U = 1$, it holds

$$\overline{m}_{o1}\, \gamma_{\overline{u}_1}\, \overline{u}^i_1 + \overline{m}_{o2}\, \gamma_{\overline{u}_2}\, \overline{u}^i_2 = M_o\, U^i \ . \tag{1360}$$

Here we use the components $i = 1, 2, 3$. Because of Eq. (1358) there remains for $U = 0$ the relation

$$\overline{m}_{o1}\, \gamma_{\overline{u}_1}\, \overline{u}_1 + \overline{m}_{o2}\, \gamma_{\overline{u}_2}\, \overline{u}_2 = 0 \ . \tag{1361}$$

We set in Eq. (1360) $i = 0$, then it follows taking into account $U = 0$

$$\overline{m}_{o1}\, \gamma_{\overline{u}_1} + \overline{m}_{o2}\, \gamma_{\overline{u}_2} = M_o \ . \tag{1362}$$

We have found *two* Eqs. (1361) and (1362) for *four* unknown quantities $\overline{m}_{o1}$, $\overline{m}_{o2}$, $\overline{u}_1$, $\overline{u}_2$. Our system of equations is underdetermined. We cannot tell, which masses appear after the collision. We can only write these equations in a more elegant form. Using $v/c := \tanh\varphi$, it follows with Eq. (337)

$$\left.\begin{array}{l} \overline{m}_{o1}\ \sinh\varphi_1 + \overline{m}_{o2}\ \sinh\varphi_2 = 0 \ , \\ \overline{m}_{o1}\ \cosh\varphi_1 + \overline{m}_{o2}\ \cosh\varphi_2 = M_o \ . \end{array}\right\} \tag{1363}$$

These equations must be fulfilled for each particle decay. Whether a particle decays, into which particles and after which time can only be predicted by the theory of elementary particles.

Problem 27 *We consider once again the Problem 10: A body L might be in motion along the x-axis, and there would be measured an acceleration $a = d^2x_L/dt^2$ in system Σ_o. Which acceleration a' is measured in system Σ' that moves with velocity v in x-direction with respect to system Σ_o? Solve this problem in the covariant formalism for the four-vector of acceleration a^i.*

Chapter 9, Section 2
Due to the supposed one-dimensional motion of the body L along the x-axis in system Σ_o,

$$\mathbf{u} = (u, 0, 0) \ , \quad \mathbf{a} = (a, 0, 0) = \left(\frac{du}{dt}, 0, 0\right) \ , \tag{1364}$$

the four-vector a^i of acceleration (435) reads

$$a^i = \left(\gamma_u^4\, au/c\,, \gamma_u^2\, a + \gamma_u^4\, a\, u^2/c^2\,, 0,\, 0\right) = \left(\gamma_u^4\, a\, u/c, \gamma_u^4(a(1 - u^2/c^2) + a\, u^2/c^2)\,, 0,\, 0\right) \ .$$

Then it follows for the vector a^i in Σ_o

$$a^i = \left(\gamma_u^4\, a\, u/c\,,\, \gamma_u^4\, a\,,0\,,\,0\right)\ . \tag{1365}$$

In system Σ' there holds with

$$\mathbf{u}' = (u',\,0,\,0)\ \ ,\quad \mathbf{a}' = (a',\,0,\,0) = \left(\frac{du'}{dt'}\,,\,0,\,0\right) \tag{1366}$$

for the four-vector $a^{i'}$

$$a^{i'} = \left(\gamma_u^4\, a'\, u'/c\,,\,\gamma_u^4\, a'\,,0,\,0\right)\ . \tag{1367}$$

Then we find because of EINSTEIN's addition theorem (106) and with Eq. (171)

$$u' = \frac{u - v}{1 - u\,v/c^2}\ ,\quad \gamma_{u'} = \gamma_u\,\gamma_v\,(1 - u\,v/c^2)\ . \tag{1368}$$

Now the components of a^i transform as the components of x^i, s. Eq. (339). We are interested in the component $a^{1'}$. To avoid misinterpretations, we put on the factor γ of the transformation matrix the index v, and we get

$$a^{1'} = -\gamma_v\,(v/c)\,a^0 + \gamma_v\,a^1 = -\gamma_v(v/c)\,\gamma_u^4(a\,u/c) + \gamma_v\,a = a\,\gamma_v\,\gamma_u^4(1 - u\,v/c^2)\,.$$

Here we use on the left-hand side for $a^{1'}$ Eq (1367) and for γ_u' Eq. (1368),

$$\begin{aligned} a'\,\gamma_{u'}^4 &= a\,(1 - u\,v/c^2)\,\gamma_v\,\gamma_u^4\ ,\\ a' &= a\,(1 - u\,v/c^2)\,\gamma_v\,\gamma_u^4\frac{1}{\gamma_{u'}^4} = a\,(1 - u\,v/c^2)\gamma_v\,\gamma_u^4\,\frac{1}{(1 - u\,v/c^2)^4\gamma_u^4\,\gamma_v^4}\ ,\end{aligned}$$

and finally

$$a' = \frac{a}{\gamma_v^3\,(1 - u\,v/c^2)^3} \tag{1369}$$

in agreement with our result (1250).

Problem 28 *Evaluate the central elastic collision of two relativistic particles with rest masses m_{o1} and m_{o2}. Compare the result with the classical approximation for particle velocities $u \ll c$.*

Chapter 9, Section 2
We assume that both particles move along the x-axis of the reference system Σ_o. Our problem is one-dimensional. From the conservation law (456) of energy and momentum, there remains only the zeroth and first component, the others vanish identically. The total energy and the total momentum are denoted by E and P. The rest masses m_{o1} and m_{o2} remain constant for the elastic collision. With velocities

u_1, u_2 and $\overline{u}_1, \overline{u}_2$ of particles before and after the collision, we can write Eq. (456) as

$$\left.\begin{aligned} \frac{E}{c^2} &= \frac{m_{o1}}{\sqrt{1-u_1^2/c^2}} + \frac{m_{o2}}{\sqrt{1-u_2^2/c^2}} = \frac{m_{o1}}{\sqrt{1-\overline{u}_1^2/c^2}} + \frac{m_{o2}}{\sqrt{1-\overline{u}_2^2/c^2}} \,, \\ P &= \frac{m_{o1}u_1}{\sqrt{1-u_1^2/c^2}} + \frac{m_{o2}u_2}{\sqrt{1-u_2^2/c^2}} = \frac{m_{o1}\overline{u}_1}{\sqrt{1-\overline{u}_1^2/c^2}} + \frac{m_{o1}\overline{u}_2}{\sqrt{1-\overline{u}_2^2/c^2}} \,. \end{aligned}\right\} \tag{1370}$$

From the given initial velocities u_1 and u_2 we can derive the final velocities $\overline{u}_1$ and $\overline{u}_2$. For comparison, we show at first the limiting case of classical mechanics.

For $u \ll c$ the first non-vanishing, i.e. classical approximation of Eq. (1370) reads

$$\left.\begin{aligned} E_{cl} &= \frac{1}{2}m_{o1}u_1^2 + \frac{1}{2}m_{o2}u_2^2 = \frac{1}{2}m_{o1}\overline{u}_1^2 + \frac{1}{2}m_{o2}\overline{u}_2^2 \,, \\ P_{cl} &= m_{o1}u_1 + m_{o2}u_2 = m_{o1}\overline{u}_1 + m_{o1}\overline{u}_2 \,, \end{aligned} \quad u \ll c \,. \right\} \tag{1371}$$

These equations can be simply transformed into the relation

$$\begin{aligned} m_{o1}(u_1 - \overline{u}_1) &= m_{o2}(\overline{u}_2 - u_2) \,, \\ m_{o1}(u_1 + \overline{u}_1)(u_1 - \overline{u}_1) &= m_{o2}(\overline{u}_2 + u_2)(\overline{u}_2 - u_2) \,, \end{aligned} \quad u \ll c \,.$$

Then we find the trivial solution

$$\textbf{a)} \quad \overline{u}_1 = u_1 \,, \quad \overline{u}_2 = u_2 \,, \; u \ll c \,, \tag{1372}$$

i.e. the particles continue their path without interaction.

Or otherwise for $\overline{u}_1 - u_1 \neq 0$, it follows

$$\textbf{b)} \quad u_1 + \overline{u}_1 = u_2 + \overline{u}_2 \,, \; u \ll c \tag{1373}$$

and the final velocities $\overline{u}_1$ and $\overline{u}_2$ are obtained after a short evaluation,

$$\left.\begin{aligned} \overline{u}_1 &= \frac{P_{cl} + m_{o2}(u_2 - u_1)}{m_{o1} + m_{02}} \,, \\ \overline{u}_2 &= \frac{P_{cl} + m_{o1}(u_1 - u_2)}{m_{o1} + m_{o2}} \,, \end{aligned} \quad u \ll c \,. \right\} \tag{1374}$$

The full relativistic Eq. (1370) require a bit more tedious calculations due to the square root expressions. For simplification of Eq. (1370), we introduce the following abbreviations:

$$\left.\begin{aligned} &\mu := \frac{m_{o1}}{m_{o2}} \,, \quad \pi := \frac{P}{m_{o2}} \,, \\ &\beta_1 = \frac{u_1}{c} \,, \quad \beta_2 = \frac{u_2}{c} \,, \quad \overline{\beta}_1 = \frac{\overline{u}_1}{c} \,, \quad \overline{\beta}_2 = \frac{\overline{u}_2}{c} \,, \\ &\beta_1 := \tanh\varphi_1 \,, \; \beta_2 := \tanh\varphi_2 \,, \; \overline{\beta}_1 := \tanh\overline{\varphi}_1 \,, \; \overline{\beta}_2 := \tanh\overline{\varphi}_2 \,. \end{aligned}\right\} \tag{1375}$$

Because of $\cosh\varphi = 1/\sqrt{1-\tanh^2\varphi}$ and $\sinh\varphi = \tanh\varphi/\sqrt{1-\tanh^2\varphi}$ we have

$$\left.\begin{aligned} \cosh\varphi &= \frac{1}{\sqrt{1-u^2/c^2}} \ , \\ \sinh\varphi &= \frac{u/c}{\sqrt{1-u^2/c^2}} \ , \end{aligned}\right\} \tag{1376}$$

and we use the known properties of hyperbolic functions,

$$\left.\begin{aligned} &\cosh^2 x - \sinh^2 x = 1 \ , \\ &\sinh(x \pm y) = \sinh x \cosh y \pm \cosh x \sinh y \ , \\ &\cosh(x \pm y) = \cosh x \cosh y \pm \sinh x \sinh y \ . \end{aligned}\right\} \tag{1377}$$

After division by m_{o2} we get from Eq. (1370)

$$\left.\begin{aligned} \pi &= \mu \sinh\varphi_1 + \sinh\varphi_2 = \mu \sinh\overline{\varphi}_1 + \sinh\overline{\varphi}_2 \ , \\ \frac{E}{m_{o2}\, c^2} &= \mu \cosh\varphi_1 + \cosh\varphi_2 = \mu \cosh\overline{\varphi}_1 + \cosh\overline{\varphi}_2 \ . \end{aligned}\right\} \tag{1378}$$

With the first series terms of $\sinh x \approx x$, $\cosh x \approx 1 + (1/2)x^2$ and $\tanh x \approx x$ we get again the classical approximation (1371).

We take the square of both Eq. (1378) and subtract the first from the second one with the result

$$\begin{aligned} &\mu^2\cosh^2\varphi_1 + \cosh^2\varphi_2 + 2\mu\cosh\varphi_1\cosh\varphi_2 - \mu^2\sinh^2\varphi_1 - \sinh^2\varphi_2 - 2\mu\sinh\varphi_1\sinh\varphi_2 \\ &= \mu^2\cosh^2\overline{\varphi}_1 + \cosh^2\overline{\varphi}_2 + 2\mu\cosh\overline{\varphi}_1\cosh\overline{\varphi}_2 - \mu^2\sinh^2\overline{\varphi}_1 - \sinh^2\overline{\varphi}_2 - 2\mu\sinh\overline{\varphi}_1\sinh\overline{\varphi}_2 \ . \end{aligned}$$

With Eq. (1377), we find after some easy evaluations

$$\cosh(\varphi_1 - \varphi_2) = \cosh(\overline{\varphi}_1 - \overline{\varphi}_2) \ . \tag{1379}$$

Because of $\cosh(-x) = \cosh x$ the equation has two solutions,

$$(a) \quad \varphi_1 - \overline{\varphi}_1 = \varphi_2 - \overline{\varphi}_2 \tag{1380}$$

and

$$(b) \quad \varphi_1 + \overline{\varphi}_1 = \varphi_2 + \overline{\varphi}_2 \ . \tag{1381}$$

The physical meaning of a) follows from an approximation. We take in the first Eq. (1378) $\sinh x \approx x$ and replace with Eq. (1380) $\overline{\varphi}_1 = \varphi_1 + \overline{\varphi}_2 - \varphi_2$, so it follows after short calculation $(1+\mu)\varphi_2 = (1+\mu)\overline{\varphi}_2$. Since $(1+\mu) \neq 0$ we get $\varphi_2 = \overline{\varphi}_2$, so that also $\varphi_1 = \overline{\varphi}_1$ and therefore $\overline{u}_1 = u_1$ and $\overline{u}_2 = u_2$. This solution (1380) is therefore identical to the classical case (1372). The particles pass by without interaction.

The case b) must be the relativistic generalisation of solution (1373). With Eq. (1381) we set in the first Eq. (1378) $\overline{\varphi}_1 = \varphi_2 - \varphi_1 + \overline{\varphi}_2$, which leads to

$$\begin{aligned}
&\pi = \mu \sinh(\varphi_2 - \varphi_1 + \overline{\varphi}_2) + \sinh \overline{\varphi}_2 \ , \\
&\pi = \mu \sinh(\varphi_2 - \varphi_1) \cosh \overline{\varphi}_2 + \cosh(\varphi_2 - \varphi_1) \sinh \overline{\varphi}_2 + \sinh \overline{\varphi}_2 \ , \\
&\pi - \sinh \overline{\varphi}_2 [1 + \cosh(\varphi_2 - \varphi_1)] = \mu \sinh(\varphi_2 - \varphi_1) \cosh \overline{\varphi}_2 \ .
\end{aligned}$$

We take the square of the last equation,

$$\begin{aligned}
&\pi^2 + \sinh^2 \overline{\varphi}_2 [1 + \mu \cosh(\varphi_2 - \varphi_1)]^2 - 2\pi \sinh \varphi_2 [1 + \mu \cosh(\varphi_2 - \varphi_1)] \\
&\quad = \mu^2 \sinh^2(\varphi_2 - \varphi_1) + \mu^2 \sinh^2(\varphi_2 - \varphi_1) \sinh^2 \overline{\varphi}_2 \ , \\
&\sinh^2 \overline{\varphi}_2 \left([1 + \mu \cosh(\varphi_2 - \varphi_1)]^2 - \mu^2 \sinh^2(\varphi_2 - \varphi_1) \right) \\
&-2 \sinh \overline{\varphi}_2 \, \pi \, [1 + \mu \cosh(\varphi_2 - \varphi_1)] + \pi^2 - \mu^2 \sinh^2(\varphi_2 - \varphi_1) = 0 \ , \\
&\sinh^2 \overline{\varphi}_2 [1 + \mu^2 + 2\mu \cosh(\varphi_2 - \varphi_1)] \\
&-2 \sinh \overline{\varphi}_2 \, \pi \, [1 + \mu \cosh(\varphi_2 - \varphi_1)] + \pi^2 - \mu^2 \sinh^2(\varphi_2 - \varphi_1) = 0
\end{aligned}$$

and we find also a quadratic equation for $\sinh \overline{\varphi}_2$,

$$\sinh^2 \overline{\varphi}_2 - \sinh \overline{\varphi}_2 \frac{2\pi[1+\mu \cosh(\varphi_2-\varphi_1)]}{1+\mu^2+2\mu \cosh(\varphi_2-\varphi_1)} - \frac{\mu^2 \sinh^2(\varphi_2-\varphi_1)-\pi^2}{1+\mu^2+2\mu \cosh(\varphi_2-\varphi_1)} = 0 \, . \tag{1382}$$

Wir take the solution with the negative sign of the square root

$$\left.\begin{aligned}
\sinh \overline{\varphi}_2 = &\frac{\pi \, [1 + \mu \cosh(\varphi_2 - \varphi_1)]}{1 + \mu^2 + 2\mu \cosh(\varphi_2 - \varphi_1)} \\
&- \sqrt{\left[\frac{\pi \, [1 + \mu \cosh(\varphi_2 - \varphi_1)]}{1 + \mu^2 + 2\mu \cosh(\varphi_2 - \varphi_1)}\right]^2 + \frac{\mu^2 \sinh^2(\varphi_2 - \varphi_1) - \pi^2}{1 + \mu^2 + 2\mu \cosh(\varphi_2 - \varphi_1)}}
\end{aligned}\right\} \tag{1383}$$

The second solution with a positive sign of the square root can be refused for the following reason.

We consider the classical limit of small velocities, hence small arguments in the hyperbolic functions with $\sinh x \approx x$ and $\cosh x \approx 1$. Further, we assume without restriction of generality $\varphi_2 - \varphi_1 \geq 0$, so that $\sqrt{(\varphi_2 - \varphi_1)^2} = \varphi_2 - \varphi_1$.

For a positive sign of the square root, we then get

$$\begin{aligned}
\overline{\varphi}_2 &= \frac{\pi}{1 + \mu} + \sqrt{\frac{\pi^2}{(1 + \mu)^2} + \frac{\mu^2(\varphi_2 - \varphi_1)^2 - \pi^2}{(1 + \mu)^2}} \ , \\
&= \frac{\pi}{1 + \mu} + \frac{\mu(\varphi_2 - \varphi_1)}{1 + \mu} = \frac{\mu\varphi_1 + \varphi_2}{1 + \mu} + \frac{\mu(\varphi_2 - \varphi_1)}{1 + \mu} \\
&= \varphi_2 \ .
\end{aligned}$$

The particles move on with their initial velocities without interaction. This solution is already covered by Eq. (1380).

The solution (1383) become much simpler when we evaluate the collision in the inertial system, where the total momentum P vanishes. Hence we take as Σ_o the centre-of-mass system with $P = 0$ and therefore also $\pi = 0$. Without restriction, we assume again $\varphi_2 - \varphi_1 \geq 0$. With $\sinh(-x) = -\sinh x$ we get for Eq. (1383) in Σ_o

$$\sinh\overline{\varphi}_2 = \frac{\mu\sinh(\varphi_1 - \varphi_2)}{\sqrt{1+\mu^2+2\mu\cosh(\varphi_2-\varphi_1)}} \quad \text{for} \quad P = 0 \ . \tag{1384}$$

From Eq. (1381) then also $\overline{\varphi}_1$ is determined. The solution for $\overline{\varphi}_1$ looks more symmetric, when we repeat the same steps of the derivation. We put into the first Eq. (1378) from Eq. (1381) $\overline{\varphi}_2 = \varphi_1 - \varphi_2 + \overline{\varphi}_1$. The quadratic equation for $\sinh\overline{\varphi}_1$ has the solution

$$\left.\begin{aligned} \sinh\overline{\varphi}_1 = & \frac{\pi\,[\mu + \cosh(\varphi_2-\varphi_1)]}{1+\mu^2+2\mu\cosh(\varphi_2-\varphi_1)} \\ & + \sqrt{\left[\frac{\pi\,[\mu + \cosh(\varphi_2-\varphi_1)]}{1+\mu^2+2\mu\cosh(\varphi_2-\varphi_1)}\right]^2 + \frac{\sinh^2(\varphi_2-\varphi_1)-\pi^2}{1+\mu^2+2\mu\cosh(\varphi_2-\varphi_1)}}\ . \end{aligned}\right\} \tag{1385}$$

Here we must take the positive sign of the square root, since the negative sign leads under the assumption $\varphi_2 - \varphi_1 \geq 0$ again to the trivial solution $\overline{\varphi}_1 = \varphi_1$. In the centre-of-mass system of the colliding particles, we get from Eq. (1385) with $\pi = 0$

$$\sinh\overline{\varphi}_1 = \frac{\sinh(\varphi_2 - \varphi_1)}{\sqrt{1+\mu^2+2\mu\cosh(\varphi_2-\varphi_1)}} \ , \quad \text{for} \quad P = 0 \ . \tag{1386}$$

When all particle velocities are small, $u/c \ll 1$, we can use $\sinh x \approx x, \cosh x \approx 1$ and $\tanh x \approx x$. Then the relativistic solutions (1383) and (1385) go over into the classical solution (1374). The exact relativistic solutions (1383) to (1386) generalise the classical approximation (1374) valid for small particle velocities $u \ll c$. It is remarkable, that the relativistic solution in the centre-of-mass system (1384) and (1386) with $P = 0$, has a similar algebraic structure as the classical solutions (1374) for $P_{cl} = 0$.

Problem 29 *Verify by a rotation around the z-axis, that the three-dimensional vector property of the electric field strength* **E** *and the magnetic induction* **B** *is guaranteed by the transformation law of the tensor F_{ik} in* Minkowski *space.*

Chapter 9, Sections 1, 3

The tensor F_{ik} transforms according to Eq. (533), hence

$$F_{i'k'} = \frac{\partial x^i}{\partial x^{i'}}\frac{\partial x^k}{\partial x^{k'}} F_{ik} \ . \tag{1387}$$

The transformation matrix for a rotation around the z-axis can be read off from Eq. (346). There we have $x^{i'} = \mathbf{D}_{3}{}^{i'}_{i} x^i$ with the rotation angle α_3 around the z-axis. One can therefore perform in the evaluation of $F_{i'k'}$ the summation in Eq. (1387) with Eq. (532) for F_{ik} for each term separately. Now we use the matrix notation of Eq. (1387),

$$F_{i'k'} = \begin{pmatrix} 0 & \frac{E'_x}{c} & \frac{E'_y}{c} & \frac{E'_z}{c} \\ -\frac{E'_x}{c} & 0 & -B'_z & B'_y \\ -\frac{E'_y}{c} & B'_z & 0 & -B'_x \\ -\frac{E'_z}{c} & -B'_y & B'_x & 0 \end{pmatrix}$$
$$= \begin{pmatrix} 1 & 0 & 0 & 0 \\ 0 & \cos\alpha_3 & \sin\alpha_3 & 0 \\ 0 & -\sin\alpha_3 & \cos\alpha_3 & 0 \\ 0 & 0 & 0 & 1 \end{pmatrix} \begin{pmatrix} 0 & \frac{E_x}{c} & \frac{E_y}{c} & \frac{E_z}{c} \\ -\frac{E_x}{c} & 0 & -B_z & B_y \\ -\frac{E_y}{c} & B_z & 0 & -B_x \\ -\frac{E_z}{c} & -B_y & B_x & 0 \end{pmatrix} \begin{pmatrix} 1 & 0 & 0 & 0 \\ 0 & \cos\alpha_3 & \sin\alpha_3 & 0 \\ 0 & -\sin\alpha_3 & \cos\alpha_3 & 0 \\ 0 & 0 & 0 & 1 \end{pmatrix} .$$

Performing the multiplication of the matrices shows that the dashed components of $\mathbf{E}$ and $\mathbf{B}$ coincide indeed with the result of the three-dimensional vector calculus,

$$\left.\begin{aligned} \begin{pmatrix} B'_x \\ B'_y \\ B'_z \end{pmatrix} &= \begin{pmatrix} \cos\alpha_3 & \sin\alpha_3 & 0 \\ -\sin\alpha_3 & \cos\alpha_3 & 0 \\ 0 & 0 & 1 \end{pmatrix} \begin{pmatrix} B_x \\ B_y \\ B_z \end{pmatrix} , \\ \begin{pmatrix} E'_x \\ E'_y \\ E'_z \end{pmatrix} &= \begin{pmatrix} \cos\alpha_3 & \sin\alpha_3 & 0 \\ -\sin\alpha_3 & \cos\alpha_3 & 0 \\ 0 & 0 & 1 \end{pmatrix} \begin{pmatrix} E_x \\ E_y \\ E_z \end{pmatrix} . \end{aligned}\right\} \tag{1388}$$

Problem 30 *Derive* COULOMB *law from the* MAXWELL *equations.*

Chapter 9, Section 3

The problem deals with a static situation, so that all time derivatives vanish, $\partial/\partial t = 0$. At coordinate origin we consider a point charge e_1 at rest, described by a charge density ρ_1 as in Eq. (1729),

$$\rho_1(x, y, z) = e_1 \delta(x)\delta(y)\delta(z) . \qquad \text{Point charge at rest in the coordinate origin} \tag{1389}$$

The velocity of this charge is zero, $\mathbf{u} = 0$. In vacuum with $\varepsilon = \varepsilon_o$, $\mu = \mu_o$ there remain from the MAXWELL equations (503) only two equations besides $\mathbf{B} = 0$ and $\mathbf{H} = 0$

$$\text{rot}\,\mathbf{E} = 0 \tag{1390}$$

and

$$\text{div}\,\mathbf{E} = \frac{1}{\varepsilon_o} e_1 \delta(x)\delta(y)\delta(z) . \tag{1391}$$

Equation (1390) is fulfilled by the ansatz of a potential

$$\mathbf{E} = -\mathrm{grad}\,\varphi \ . \tag{1392}$$

Inserting into Eq. (1391) we get with Eq. (1668)

$$\Delta\varphi = -\frac{1}{\varepsilon_o}\, e_1\delta(x)\delta(y)\delta(z) \ . \tag{1393}$$

As solution of this equation, we find with Eq. (1734) the COULOMB potential

$$\varphi = \frac{1}{4\,\pi\,\varepsilon_o}\,\frac{e_1}{\sqrt{x^2+y^2+z^2}} \ . \qquad \text{COULOMB potential} \tag{1394}$$

We write $\mathbf{r} = (x, y, z)$ and find with

$$\left(\frac{\partial}{\partial x}, \frac{\partial}{\partial y}, \frac{\partial}{\partial z}\right)\frac{1}{r} = -\left(\frac{x}{r^3}, \frac{y}{r^3}, \frac{z}{r^3}\right)$$

from Eq. (1392) the COULOMB field of a point charge e_1,

$$\mathbf{E} = \frac{e_1}{4\,\pi\,\varepsilon_o}\,\frac{\mathbf{r}}{r^3} \ . \qquad \text{COULOMB field of a point charge } e_1 \tag{1395}$$

The force $\mathbf{F}$ on a second point charge e_2 at a position $\mathbf{r}_{12} = (x_o, y_o, z_o)$ with density $\rho_2(x, y, z) = e_2\delta(x - x_o)\delta(y - y_o)\delta(z - z_o)$ is found by integration over the LORENTZ force density (470) at $\mathbf{B} = 0$, hence

$$\mathbf{F} = \iiint \mathbf{E}\,\rho_2(x, y, z)dxdydz = \frac{e_1\,e_2}{4\,\pi\,\varepsilon_o}\iiint \frac{\mathbf{r}}{r^3}\,\delta(x-x_o)\delta(y-y_o)\delta(z-z_o)dxdydz$$

and therefore, with the property (1720) of the δ function, when we integrate over the coordinates (x_o, y_o, z_o),

$$\mathbf{F} = \frac{e_1\,e_2}{4\,\pi\,\varepsilon_o}\,\frac{\mathbf{r}_{12}}{r_{12}^3} \ . \qquad \text{COULOMB law} \tag{1396}$$

This is the COULOMB law in SI-units.

When we measure the charges in absolute units, then there vanishes the factor $1/4\pi\varepsilon_o$, hence $\mathbf{F} = \tilde{e}_1\tilde{e}_2\left(\mathbf{r}_{12}\right)/\left(r_{12}^3\right)$. This defines the unit of charge in the absolute unit system.

Problem 31 *Evaluate the electromagnetic field of a uniformly moving point charge by performing a* LORENTZ *transformation from the* COULOMB *field of a point charge at rest.*

Chapter 9, Sections 1, 3

The inertial system Σ' might have the velocity $\mathbf{v} = (v, 0, 0)$ in x-direction as measured in Σ_o. The systems Σ_o and Σ' are then connected by a LORENTZ transformation

(105). At coordinate origin of Σ' the point charge e is at rest, hence the charge density is $\rho' = \rho'(x', y', z')$, cp. Eq. (1729),

$$\rho'(x', y', z') = e\,\delta(x')\delta(y')\delta(z') \ . \qquad \text{Point charge at rest in } \Sigma' \qquad (1397)$$

The charge e is an invariant.

The electromagnetic field created from this charge and observed in Σ' is a pure Coulomb field, cp. (1395),

$$\left.\begin{array}{ll} \mathbf{E}' = \dfrac{e}{4\pi\varepsilon_o}\dfrac{\mathbf{r}'}{r'^3} \ , & \mathbf{r}' = (x', y', z') \ , \\ \mathbf{B}' = 0 \ , & r' = \sqrt{x'^2 + y'^2 + z'^2} \ . \end{array}\right\} \quad \text{Coulomb field of a point charge in } \Sigma' \qquad (1398)$$

The field of the charge e as measured in system Σ_o can be evaluated from the transformation law (533) of the field strength tensor, that is now needed in the form resolved for F_{ik}, since F'_{ik} is already known. To this aim, we have to replace only v by $-v$ and exchange in Eq. (534) the dashed and undashed components of the field strength tensor. The in Σ_o measured field strength $\mathbf{E}$ and $\mathbf{B}$ are then, when we set in addition $\mathbf{B}' = 0$,

$$\begin{pmatrix} 0 & \frac{E_x}{c} & \frac{E_y}{c} & \frac{E_z}{c} \\ -\frac{E_x}{c} & 0 & -B_z & B_y \\ -\frac{E_y}{c} & B_z & 0 & -B_x \\ -\frac{E_z}{c} & -B_y & B_x & 0 \end{pmatrix} = \begin{pmatrix} 0 & \frac{E'_x}{c} & \frac{\gamma E'_y}{c} & \frac{\gamma E'_z}{c} \\ -\frac{E'_x}{c} & 0 & -\frac{\beta\gamma E'_y}{c} & -\frac{\beta\gamma E'_z}{c} \\ -\frac{\gamma E'_y}{c} & \frac{\beta\gamma E'_y}{c} & 0 & 0 \\ -\frac{\gamma E'_z}{c} & \frac{\beta\gamma E'_z}{c} & 0 & 0 \end{pmatrix} . \qquad (1399)$$

The fields $\mathbf{E}$ and $\mathbf{B}$ measured in Σ_o have to be expressed in dependence on the corresponding coordinates (x, y, z, t). With the Lorentz transformation (105) we get

$$r' = \sqrt{x'^2 + y'^2 + z'^2} = \sqrt{\gamma^2\,(x - vt)^2 + y^2 + z^2} = \gamma\sqrt{(x - vt)^2 + (y^2 + z^2)/\gamma^2} \ ,$$

hence

$$r' = \gamma\sqrt{(x - vt)^2 + (y^2 + z^2)/\gamma^2} \ . \qquad (1400)$$

With Eqs. (1398)–(1400) we find for the field strengths $\mathbf{E}$ and $\mathbf{B}$,

$$\begin{pmatrix} E_x \\ E_y \\ E_z \end{pmatrix} = \begin{pmatrix} E'_x \\ \gamma E'_y \\ \gamma E'_z \end{pmatrix} = \frac{e}{4\pi\varepsilon_o}\begin{pmatrix} \frac{x'}{r'^3} \\ \frac{\gamma y'}{r'^3} \\ \frac{\gamma z'}{r'^3} \end{pmatrix} , \qquad \begin{pmatrix} B_x \\ B_y \\ B_z \end{pmatrix} = \begin{pmatrix} 0 \\ -\frac{\beta\gamma E'_z}{c} \\ \frac{\beta\gamma E'_y}{c} \end{pmatrix} ,$$

hence

$$\begin{pmatrix} E_x \\ E_y \\ E_z \end{pmatrix} = \frac{e}{4\pi\varepsilon_o\gamma^2} \begin{pmatrix} \dfrac{x - vt}{\sqrt{(x-vt)^2 + (y^2+z^2)/\gamma^2}^3} \\ \dfrac{y}{\sqrt{(x-vt)^2 + (y^2+z^2)/\gamma^2}^3} \\ \dfrac{z}{\sqrt{(x-vt)^2 + (y^2+z^2)/\gamma^2}^3} \end{pmatrix} \tag{1401}$$

and

$$\begin{pmatrix} B_x \\ B_y \\ B_z \end{pmatrix} = \frac{e}{4\pi\varepsilon_o} \frac{v}{\gamma^2 c^2} \begin{pmatrix} 0 \\ -\dfrac{z}{\sqrt{(x-vt)^2 + (y^2+z^2)/\gamma^2}^3} \\ \dfrac{y}{\sqrt{(x-vt)^2 + (y^2+z^2)/\gamma^2}^3} \end{pmatrix} \longrightarrow \mathbf{B} = \frac{1}{c^2}\mathbf{v}\cdot\mathbf{E} \,. \tag{1402}$$

We shall confirm this result by means of the evaluation of the electromagnetic potentials $\mathbf{A}$ and φ. With the ansatz for the potential (516), the potential form of the COULOMB field (1398) reads

$$\left.\begin{aligned} \mathbf{A}' &= 0 \,, \\ \varphi' &= \frac{e}{4\pi\varepsilon_o}\frac{1}{r'} \,. \end{aligned}\right\} \qquad \text{COULOMB potential of a point charge in } \Sigma' \tag{1403}$$

According to Eq. (529) the potentials represent a four-vector $A^{i'}$,

$$A^{i'} = \left(\frac{\varphi'}{c},\, A'_x,\, A'_y,\, A'_z\right) = \frac{e}{4\pi\varepsilon_o}\left(\frac{1}{c\,r'},\, 0,\, 0,\, 0\right) \,. \qquad \text{Four-potential of the COULOMB field in } \Sigma' \tag{1404}$$

We apply the LORENTZ transformation, Eqs. (105) or (339) on the four-potential A^i,

$$A^i = \left(\frac{\varphi}{c},\, A_x,\, A_y,\, A_z\right) \,,$$

hence

$$A^0 = \gamma(A^{0'} + \beta\, A^{1'}) \,, \quad A^1 = \gamma(A^{1'} + \beta\, A^{0'}) \,,$$

and get for the electromagnetic field observed in system Σ_o at first

$$A^i = \frac{e}{4\pi\varepsilon_o}\left(\frac{\gamma}{c\,r'},\, \frac{\beta\gamma}{c\,r'},\, 0,\, 0\right) \,.$$

We replace again r' by Eq. (1400) and find

$$A^i = \frac{e}{4\pi\varepsilon_o}\frac{1}{c}\left(\frac{1}{\sqrt{(x-vt)^2+(y^2+z^2)/\gamma^2}},\ \frac{v/c}{\sqrt{(x-vt)^2+(y^2+z^2)/\gamma^2}}\,0,\ 0\right) \tag{1405}$$

and therefore

$$\left.\begin{aligned} \mathbf{A} &= \frac{e}{4\pi\varepsilon_o}\frac{1}{c}\frac{1}{\sqrt{(x-vt)^2+(y^2+z^2)/\gamma^2}}(\beta,\ 0,\ 0)\ , \\ \varphi &= \frac{e}{4\pi\varepsilon_o}\frac{1}{\sqrt{(x-vt)^2+(y^2+z^2)/\gamma^2}}\ . \end{aligned}\right\}\quad \text{Four-potential in } \Sigma_o \tag{1406}$$

One easily verifies, that from $\mathbf{B} = \operatorname{rot}\mathbf{A}$ and $\mathbf{E} = -\operatorname{grad}\varphi - \partial\mathbf{A}/\partial t$ from these potentials there follows again the field strengths $\mathbf{E}$ and $\mathbf{B}$ in agreement with Eqs. (1401) and (1402).

Now we prove that the fields (1401) and (1402) found by a Lorentz transformation and describing a uniformly moving point charge in Σ_o also represent solutions of the Maxwell equations in vacuum, Eq. (503) for $\varepsilon = \varepsilon_o$ and $\mu = \mu_o$.

The homogeneous Maxwell equation (503a) are identically fulfilled by the ansatz of the potentials. Further there holds the Lorentz gauge (518). Since this equation holds true by Eq. (543) as vector equation in the Minkowski space in all inertial systems, when it is true in one of them, it is sufficient to prove its validity in Σ'. But this can be immediately seen from Eq. (1404),

$$\frac{\partial A^{i'}}{\partial x^{i'}} = \frac{1}{c}\frac{\partial}{\partial t'}\frac{1}{r'} = 0\ . \qquad \text{Lorentz gauge for Coulomb potential in } \Sigma' \tag{1407}$$

The second group of Maxwell equations (503b) is fulfilled, when we prove Eq. (521) for $\varepsilon = \varepsilon_o\,,\ \mu = \mu_o$.

To the point charge resting at coordinate origin of system Σ', i.e. with velocity vector $u^{i'} = (c, 0, 0, 0)$, there belongs according to Eq. (526) a four-vector $j^{i'}$ of current density,

$$j^{i'} = e\,\delta(x')\,\delta(y')\,\delta(z')\,(c\,, 0,\ 0,\ 0)\ . \tag{1408}$$

In Σ_o, the point charge is situated at time t at position $(x_o = v\,t,\ y_o = 0,\ z_o = 0)$. The corresponding four-vector of current density j^i follows in the simplest way from Eq. (1407) by a Lorentz transformation. With Eq. (105) or (339), we find at first

$$j^i = e\,\delta(x')\,\delta(y')\,\delta(z')\,(\gamma c\,,\ \gamma v\,,\ 0\,,\ 0)\ .$$

Here we insert for (x', y', z') the new variables (x, y, z, t) from Eq. (105) and take into account Eq. (1726), $\delta(a\,x) = (1/a)\delta(x)$, hence $\delta\big(\gamma\,(x - vt)\big) = \big(1/\gamma\big)\,\delta(x - vt)$. The current density j^i then reads in Σ_o

$$j^i = (c\,\rho\,,\, j_x\,,\, j_y\,,\, j_z) = \big(c\,e\,\delta(x - vt)\,\delta(y)\,\delta(z)\,,\, e\,v\,\delta(x - vt)\,\delta(y)\,\delta(z)\,,\, 0\,,\, 0\big) \tag{1409}$$

or

$$\left.\begin{aligned} &\mathbf{j} = \rho\,\mathbf{v} = e\,\delta(x - vt)\,\delta(y)\,\delta(z)\,\big(v\,,\, 0\,,\, 0\big)\ , \\ &\rho = e\,\delta(x - vt)\,\delta(y)\,\delta(z)\ . \end{aligned}\right\} \quad \begin{aligned}&\text{Uniformly moving}\\ &\text{point charge in } \Sigma_o\end{aligned} \tag{1410}$$

Now we have to show, that the potentials $\mathbf{A}$ and φ for $\varepsilon = \varepsilon_o\,,\ \mu = \mu_o$ fulfil Eq. (521). Because of the form of Eqs. (1406) and (1410) it is sufficient to prove the equation $\Box\varphi = (e/\varepsilon_o)\,\delta(x - vt)\,\delta(y)\,\delta(z)$. It holds from Eq. (1406)

$$\left.\begin{aligned} &\frac{1}{c^2}\frac{\partial^2}{\partial t^2}\,\varphi = \frac{v^2}{c^2}\frac{\partial^2}{\partial x^2}\,\varphi\ , \\ &\frac{1}{c^2}\frac{\partial^2}{\partial t^2}\,\varphi - \boldsymbol{\Delta}\varphi = -\left[\left(1 - \frac{v^2}{c^2}\right)\frac{\partial^2}{\partial x^2} + \frac{\partial^2}{\partial y^2} + \frac{\partial^2}{\partial z^2}\right]\varphi\ , \\ &\Box\varphi = -\left(\gamma^2\,\frac{\partial^2}{\partial x^2} + \frac{\partial^2}{\partial y^2} + \frac{\partial^2}{\partial z^2}\right)\varphi\ . \end{aligned}\right\} \tag{1411}$$

For the function φ we introduce as new variables,

$$\left.\begin{aligned} &u = x - vt\ ,\quad v = y/\gamma\ ,\quad w = z/\gamma\ , \\ &\frac{\partial}{\partial x} = \frac{\partial}{\partial u}\ ,\quad \frac{\partial}{\partial y} = \frac{1}{\gamma}\frac{\partial}{\partial v}\ ,\quad \frac{\partial}{\partial z} = \frac{1}{\gamma}\frac{\partial}{\partial w}\ , \\ &\varphi = \frac{e}{4\,\pi\,\varepsilon_o}\,\frac{1}{\sqrt{(x - vt)^2 + (y^2 + z^2)/\gamma^2}} = \frac{e}{4\,\pi\,\varepsilon_o}\,\frac{1}{\sqrt{u^2 + v^2 + w^2}}\ . \end{aligned}\right\} \tag{1412}$$

It follows

$$\left.\begin{aligned} &\left(\frac{1}{\gamma^2}\frac{\partial^2}{\partial x^2} + \frac{\partial^2}{\partial y^2} + \frac{\partial^2}{\partial z^2}\right)\frac{1}{\sqrt{(x - vt)^2 + (y^2 + z^2)/\gamma^2}} \\ &= \frac{1}{\gamma^2}\left(\frac{\partial^2}{\partial u^2} + \frac{\partial^2}{\partial v^2} + \frac{\partial^2}{\partial w^2}\right)\frac{1}{\sqrt{u^2 + v^2 + w^2}}\ . \end{aligned}\right\} \tag{1413}$$

Here we use Eq. (1734), and we find

$$\left(\frac{1}{\gamma^2}\frac{\partial^2}{\partial x^2} + \frac{\partial^2}{\partial y^2} + \frac{\partial^2}{\partial z^2}\right)\frac{1}{\sqrt{(x - vt)^2 + (y^2 + z^2)/\gamma^2}} = -4\,\pi\,\frac{1}{\gamma^2}\,\delta(u)\,\delta(v)\,\delta(w)\ . \tag{1414}$$

Now we change with Eq. (1412) to the old variables (x, y, z, t) and get finally with Eq. (1726), hence $\delta(y/\gamma) = \gamma\,\delta(y)$, $\delta(z/\gamma) = \gamma\,\delta(z)$,

$$\left.\begin{aligned} \Box\,\varphi &= -\left(\frac{1}{\gamma^2}\frac{\partial^2}{\partial x^2} + \frac{\partial^2}{\partial y^2} + \frac{\partial^2}{\partial z^2}\right)\frac{1}{\sqrt{(x-vt)^2 + (y^2+z^2)/\gamma^2}} \\ &= \frac{e}{\varepsilon_o}\,\delta(x-vt)\,\delta(y)\,\delta(z)\,. \end{aligned}\right\} \tag{1415}$$

For the vector potential the evaluation is completely equivalent, hence we have

$$\left.\begin{aligned} \Box\,\varphi &= \left[\frac{1}{c^2}\frac{\partial^2}{\partial t^2} - \left(\frac{\partial^2}{\partial x^2} + \frac{\partial^2}{\partial y^2} + \frac{\partial^2}{\partial z^2}\right)\right]\varphi = \frac{\rho}{\varepsilon_o}\,, \\ \Box\,\mathbf{A} &= \left[\frac{1}{c^2}\frac{\partial^2}{\partial t^2} - \left(\frac{\partial^2}{\partial x^2} + \frac{\partial^2}{\partial y^2} + \frac{\partial^2}{\partial z^2}\right)\right]\mathbf{A} = \mu_o\,\mathbf{j}\,. \end{aligned}\right\} \tag{1416}$$

Therefore, the electromagnetic field constructed by the LORENTZ transformation fulfils the MAXWELL equations, as claimed.

Problem 32 *For a circular current of strength I with a radius r, derive the magnetic field* **H** *for large separations R, $R \gg r$.*

Chapter 9, Section 3

Again w begin with electrostatics. A charge e_1 at position $\mathbf{x}_1$ produces the electrostatic potential φ_1, cp. (1394),

$$\varphi_1 = \frac{1}{4\pi\,\varepsilon_o}\frac{e_1}{|\mathbf{x}-\mathbf{x}_1|}\,, \qquad |\mathbf{x}-\mathbf{x}_1| := \sqrt{(x-x_1)^2 + (y-y_1)^2 + (z-z_1)^2}\,. \tag{1417}$$

Adding a second charge e_2 at position $\mathbf{x}_2$, the potential reads

$$\varphi = \varphi_1 + \varphi_2 = \frac{1}{4\pi\,\varepsilon_o}\left(\frac{e_1}{|\mathbf{x}-\mathbf{x}_1|} + \frac{e_2}{|\mathbf{x}-\mathbf{x}_2|}\right)\,. \tag{1418}$$

In case of a dipole, we have $e_1 = -\,e$ and $e_2 = +\,e$. The potential then reads

$$\varphi = \frac{1}{4\pi\,\varepsilon_o}\left(\frac{-e}{|\mathbf{x}-\mathbf{x}_1|} + \frac{e}{|\mathbf{x}-\mathbf{x}_2|}\right)\,,$$

hence

$$\varphi = \frac{1}{4\pi\,\varepsilon_o}\left(e\left(\frac{1}{|\mathbf{x}-\mathbf{x}_2|} - \frac{1}{|\mathbf{x}-\mathbf{x}_1|}\right)\right)\,. \tag{1419}$$

For the dipole, we write $\mathbf{x}_2 = \mathbf{x}_1 + \mathbf{a}$, and we consider the fields at position $\mathbf{x}$ with

$$|\mathbf{a}| \equiv |\mathbf{x}_2 - \mathbf{x}_1| \ll |\mathbf{x}| . \tag{1420}$$

The TAYLOR expansion leads to the approximation

$$\varphi \approx \frac{1}{4\pi\varepsilon_o}\left(e\,\frac{\mathbf{a}\cdot(\mathbf{x}-\mathbf{x}_1)}{|\mathbf{x}-\mathbf{x}_1|^3}\right) . \tag{1421}$$

We write with the dipole moment $\mathbf{P} = e\,\mathbf{a}$, and instead of the approximation an equality sign that can also be understood as redefinition of φ,

$$\varphi = \frac{1}{4\pi\varepsilon_o}\,\frac{\mathbf{P}\cdot(\mathbf{x}-\mathbf{x}_1)}{|\mathbf{x}-\mathbf{x}_1|^3} ,$$

hence

$$\varphi = \frac{1}{4\pi\varepsilon_o}\,\frac{\mathbf{P}}{|\mathbf{x}-\mathbf{x}_1|^2}\cos\left(\mathbf{P}, (\mathbf{x}-\mathbf{x}_1)\right) . \tag{1422}$$

We consider an area $\mathbf{F}$ uniformly covered with dipoles of constant momenta in direction of the area normal $\mathbf{n}$ and with a density $\mathbf{p}$. The corresponding potential follows by integration, where we replace in Eq. (1422) the variable $\mathbf{x}_1$ by the integration variable $\mathbf{x}'$,

$$\varphi(\mathbf{x}) = \frac{1}{4\pi\varepsilon_o}\,|\mathbf{p}|\iint\frac{\cos\left(\mathbf{n}(\mathbf{x}'), (\mathbf{x}-\mathbf{x}')\right)}{|\mathbf{x}-\mathbf{x}'|^2}\,dS' . \tag{1423}$$

We shall return to this formula in the evaluation of a magnetic field where we find an analogous equation. (The comma in the argument of the cos-function means that we take the angle between both vectors.)

We consider MAXWELL's vacuum equations. As source we consider the current density $\mathbf{j}$, later taken in the form of a line current. Electromagnetic waves are excluded as solutions. We only need the equations

$$\operatorname{rot}\mathbf{H} = \mathbf{j} , \qquad \operatorname{div}\mathbf{H} = 0 . \tag{1424}$$

The solution of these equations for a closed current distribution is a standard exercise in vector analysis, also found, e.g. in G. JOOS (1987).

With the ansatz

$$\mathbf{H} = \operatorname{rot}\mathbf{U} \tag{1425}$$

the second Eq. (1424) is identically fulfilled. The vector potential $\mathbf{U}(\mathbf{x})$ is also subject to the auxiliary condition

$$\operatorname{div}\mathbf{U} = 0 . \tag{1426}$$

We show that this is always possible: Assuming that div $\mathbf{U} = r(\mathbf{x}) \neq 0$. We determine a function $f(\mathbf{x})$ as solution of the equation

$$\text{div grad } f \equiv \Delta f = -r$$

and use in the following the new vector potential $\mathbf{U}'(\mathbf{x}) := \mathbf{U}(\mathbf{x}) + \text{grad } f$.

Since the rotation of a gradient always vanishes, it follows

$$\text{rot } \mathbf{U}' = \text{rot } \mathbf{U} = 0$$

as well as

$$\text{div } \mathbf{U}' = \text{div } \mathbf{U} + \text{div grad } f = r - r = 0\,.$$

We can therefore assume after a corresponding renaming, that the vector potential $\mathbf{U}(\mathbf{x})$ fulfils both Eqs. (1425) and (1426).

MAXWELL's Eq. (1425) then is reduced on a single equation for $\mathbf{U}$,

$$\text{rot } \mathbf{H} \;=\; \text{rot rot } \mathbf{U} \;=\; \text{grad div } \mathbf{U} + \Delta U \;=\; \mathbf{j}\,,$$

hence

$$\Delta U \;=\; \mathbf{j}\,. \tag{1427}$$

Now we write Eq. (1734) a bit more general,

$$\Delta \frac{1}{\sqrt{(x-x')^2 + (y-y')^2 + (z-z')^2}} \;=\; -4\pi\,\delta(x-x')\,\delta(y-y')\,\delta(z-z')\,. \tag{1428}$$

One gets immediately the solution of Eq. (1427),

$$\mathbf{U}(\mathbf{x}) \;=\; -\frac{1}{4\pi} \iiint \frac{\mathbf{j}(\mathbf{x}')}{|\mathbf{x}-\mathbf{x}'|}\, d^3x'\,. \tag{1429}$$

For a current along a line, we get

$$\iiint \frac{\mathbf{j}(\mathbf{x}')}{|\mathbf{x}-\mathbf{x}'|}\, d^3x' \;=\; \iiint \frac{\mathbf{j}(\mathbf{x}')\, d^2\mathbf{f}'}{|\mathbf{x}-\mathbf{x}'|} d\mathbf{s}'\,. \tag{1430}$$

Here we assumed that the volume element is decomposed in an area element $d^2\mathbf{f}'$ orthogonal to the current and the tangential element $d\mathbf{s}'$ along the current, hence $d^3x' = d^2\mathbf{f}' \cdot d\mathbf{s}'$.

For a constant current along a closed circular line J it follows

$$\iiint \frac{\mathbf{j}(\mathbf{x}')}{|\mathbf{x}-\mathbf{x}'|}\, d^3x' \;=\; J \oint \frac{d\mathbf{s}'}{|\mathbf{x}-\mathbf{x}'|}\,. \tag{1431}$$

Here we take for $\mathbf{x}'$ the circle of the current. The vector potential is then

$$\mathbf{U}(\mathbf{x}) = -\frac{J}{4\pi} \oint \frac{d\mathbf{s}'}{|\mathbf{x}-\mathbf{x}'|} \ . \tag{1432}$$

Before we derive a suitable expression for the magnetic field **H**, we derive a consequence of Stokes law, Eq. (1701), telling that for every vector field **V** there holds

$$\oint_{\partial S} \mathbf{V} \cdot d\mathbf{s} = \iint_S \mathrm{rot}\mathbf{V} \cdot d\mathbf{S} \ . \tag{1433}$$

We consider the special case $\mathbf{V} = h\,\mathbf{c}$ with a constant vector **c** and a function $h = h(\mathbf{x})$. It follows

$$\oint_{\partial S} h\ \mathbf{c} \cdot d\mathbf{s} = \iint_S \mathrm{rot}\,(h\,\mathbf{c}) \cdot d\mathbf{S} = \iint_S (\mathrm{grad}\,h \times \mathbf{c}) \cdot d\mathbf{S} \ ,$$

hence, putting the constant vector in front of the integral and taking into account, that **c** is an arbitrary vector,

$$\mathbf{c} \cdot \oint_{\partial S} h\, d\mathbf{s} = \mathbf{c} \cdot \iint_S (d\mathbf{S} \times \mathrm{grad}\,h) \ ,$$

and therefore

$$\oint_{\partial S} h\, d\mathbf{s} = \iint_S (d\mathbf{S} \times \mathrm{grad}\,h) \ . \tag{1434}$$

This relation is now applied to Eq. (1432) for **U**. We observe that in Eq. (1434) the derivative below the integral sign refers to the integration variable, denoted in the integral for the vector potential by a dash $'$. We write for the line integral $d\mathbf{S} = \mathbf{n}\, dS$, then it follows

$$\mathbf{U}(\mathbf{x}) = -\frac{J}{4\pi} \oint_{\partial S} \frac{d\mathbf{s}'}{|\mathbf{x}-\mathbf{x}'|} = -\frac{J}{4\pi} \iint_S \left(\mathbf{n}(\mathbf{x}') \times \mathrm{grad}' \frac{1}{|\mathbf{x}-\mathbf{x}'|} \right) dS' \tag{1435}$$

We observe that under the integral, the derivative of the difference $|\mathbf{x}-\mathbf{x}'|$ for $\mathbf{x}'$ can be replaced by the derivative for **x**. Then we find **H**

$$\left.\begin{aligned} \mathbf{H}(\mathbf{x}) = \mathrm{rot}\,\mathbf{U}(\mathbf{x}) &= -\frac{J}{4\pi}\, \mathrm{rot} \iint_S \left(\mathbf{n}(\mathbf{x}') \times \mathrm{grad}' \frac{1}{|\mathbf{x}-\mathbf{x}'|} \right) dS' \\ &= +\frac{J}{4\pi} \iint_S \mathrm{rot} \left(\mathbf{n}(\mathbf{x}') \times \mathrm{grad} \frac{1}{|\mathbf{x}-\mathbf{x}'|} \right) dS' \ . \end{aligned}\right\} \tag{1436}$$

For evaluation of the integrand, we use the formula

$$\mathbf{a} \times (\mathbf{b} \times \mathbf{c}) = \mathbf{b}\,(\mathbf{a} \cdot \mathbf{c}) - \mathbf{c}\,(\mathbf{a} \cdot \mathbf{b})\,,$$

and we must observe that all differentiations are covered, hence

$$\left.\begin{aligned} \text{rot}\left(\mathbf{n}(\mathbf{x}') \times \text{grad}\,\frac{1}{|\mathbf{x}-\mathbf{x}'|}\right) &\equiv \nabla \times \left(\mathbf{n}(\mathbf{x}') \times \nabla\,\frac{1}{|\mathbf{x}-\mathbf{x}'|}\right) \\ &= \mathbf{n}(\mathbf{x}')\,\boldsymbol{\Delta}\,\frac{1}{|\mathbf{x}-\mathbf{x}'|} - (\mathbf{n}(\mathbf{x}') \cdot \nabla)\,\nabla\,\frac{1}{|\mathbf{x}-\mathbf{x}'|} \\ &= 0 + \nabla\left(\mathbf{n}(\mathbf{x}') \cdot \nabla\,\frac{1}{|\mathbf{x}-\mathbf{x}'|}\right)\,. \end{aligned}\right\} \tag{1437}$$

Then we find for $\mathbf{H}$

$$\begin{aligned} \mathbf{H}(\mathbf{x}) &= \nabla\,\frac{J}{4\pi} \iint_S \mathbf{n}(\mathbf{x}') \cdot \nabla\,\frac{1}{|\mathbf{x}-\mathbf{x}'|}\,dS' \\ &= -\nabla\,\frac{J}{4\pi} \iint_S \frac{\mathbf{n}(\mathbf{x}') \cdot (\mathbf{x}-\mathbf{x}')}{|\mathbf{x}-\mathbf{x}'|^3}\,dS' \end{aligned}$$

and therefore

$$\mathbf{H}(\mathbf{x}) = -\nabla\,\frac{J}{4\pi} \iint_S \frac{\cos\big(\mathbf{n}(\mathbf{x}'), (\mathbf{x}-\mathbf{x}')\big)}{|\mathbf{x}-\mathbf{x}'|^2}\,dS'\,. \tag{1438}$$

For a comparison of Eq. (1438) with our dipole formula (1423) for the electric field, it has to be observed that we evaluate the electric field $\mathbf{E} = -\nabla\varphi$ from Eq. (1423), and the magnetic induction corresponding to $\mathbf{E}$ follows from Eq. (1438) by $\mathbf{B} = \mu_o\,\mathbf{H}$, hence

$$\left.\begin{aligned} \mathbf{B}(\mathbf{x}) &= -\nabla\,\frac{\mu_o}{4\pi}\,J \iint_S \frac{\cos\big(\mathbf{n}(\mathbf{x}'), (\mathbf{x}-\mathbf{x}')\big)}{|\mathbf{x}-\mathbf{x}'|^2}\,dS'\,, \\ \mathbf{E}(\mathbf{x}) &= -\nabla\,\frac{1}{4\pi\,\varepsilon_o}\,|\mathbf{p}| \iint_S \frac{\cos\big(\mathbf{n}(\mathbf{x}'), (\mathbf{x}-\mathbf{x}')\big)}{|\mathbf{x}-\mathbf{x}'|^2}\,dS'\,. \end{aligned}\right\} \tag{1439}$$

The constants ε_o and μ_o are defined inversely with respect to the corresponding fields $\mathbf{E}$ vs. $\mathbf{D}$ and $\mathbf{B}$ vs. $\mathbf{H}$, s. Chap. 9, Sect. 3.1.4. Hence the current J is equivalent to a density $\mathbf{m}$ of constant magnetic momenta. Therefore, the magnetic moment $\mathbf{M}$, produced by a current J around an area F is given by the following equation, when we still write the area as a vector:

$$\mathbf{M} = J\,\mathbf{F}\,, \tag{1440}$$

what was to be shown.

Problem 33 *Explain the classical approximation of aberration of light waves on the basis of electromagnetism.*

Chapter 8, Sections 6, 7, Chapter 9, Section 3.2

Now we discuss again the paradoxon of aberration from Chap. 8, Sect. 7. As there assumed, we consider a system Σ' moving in direction of the negative x-axis of Σ_o with velocity of amount $|v|$.

From Eq. (534), one can read off directly, in which components the electromagnetic field is decomposed when it is measured in system Σ_o as **E** and **B** and otherwise in system Σ' as $\mathbf{E}'$ and $\mathbf{B}'$. There it becomes obvious, that the general form of the decomposition is unchanged when we go over to the non-relativistic approximation, only taking into account terms linear in v/c. In Eq. (534) one can therefore set $\gamma \approx 1$. We assume for simplicity linear polarised waves, then we can fulfil the transversality Eq. (514) of electromagnetic waves, following from MAXWELL's equations, and the in Eq. (515) formulated relation $|\mathbf{E}_o| = c\,|\mathbf{B}_o|$ by $\mathbf{E} = (c\,F,\ 0,\ 0)$, $\mathbf{B} = (0,\ 0,\ F)$ and $\mathbf{k} = (0, -|k|, 0)$. We neglect terms of second and higher order in v/c, then it follows from Eq. (534) for the in Σ' observed fields $\mathbf{E}'$ and $\mathbf{B}'$, when we replace in Eq. (534) v by $-v$ for the direction of motion of system Σ',

$$\Sigma_o:\ \begin{aligned} \mathbf{E} &= (c\,F, \quad 0\ ,\ 0)\,, \\ \mathbf{B} &= (0, \quad 0\ ,\ F)\,, \\ \mathbf{k} &= (0, \ -|k|,\ 0)\,, \end{aligned} \quad \underset{\gamma\approx 1}{\longrightarrow} \quad \Sigma':\ \left.\begin{aligned} \mathbf{E}' &= (c\,F, \quad |v|\,F,\ 0)\,, \\ \mathbf{B}' &= (0, \quad 0, \quad F)\,, \\ \mathbf{k}' &= (\frac{|v|}{c}\,|k|,\ -|k|,\ 0\ )\,. \end{aligned}\right\} \tag{1441}$$

For getting the last equation, we have taken into account that the wave four-vector $(k^0,\ k^1,\ k^2,\ k^3) = (\frac{\omega}{c},\ k_x,\ k_y,\ k_z)$, s. Eq. (549), fulfils the same LORENTZ transformation Eq. (105) as $(x^0,\ x^1,\ x^2,\ x^3) = (c\,t,\ x,\ y,\ z)$, where we have put again $v = -|v|$ because of the direction of motion of system Σ'.

We choose the polarisation of the in system Σ_o in direction of the negative y-direction propagating wave as $\mathbf{E} = (0,\ 0,\ c\,F)$ and $\mathbf{B} = (-F,\ 0,\ 0)$, then it follows instead of Eq. (1441)

$$\Sigma_o:\ \begin{aligned} \mathbf{E} &= (\,0, \quad 0, \ -c\,F\,)\,, \\ \mathbf{B} &= (\,F, \quad 0, \quad 0\ )\,, \\ \mathbf{k} &= (\,0, -|k| \quad 0\ )\,, \end{aligned} \quad \underset{\gamma\approx 1}{\longrightarrow} \quad \Sigma':\ \left.\begin{aligned} \mathbf{E}' &= (0 \quad 0, \quad -c\,F\ )\,, \\ \mathbf{B}' &= (F, \quad \frac{|v|}{c}\,F, \quad 0\,)\,, \\ \mathbf{k}' &= (\frac{|v|}{c}\,|k|,\ -|k|, \quad 0\,)\,. \end{aligned}\right\} \tag{1442}$$

With the relations of the rotation of a vector $\mathbf{V} = (V_x,\ V_y,\ V_z)$ around the z-axis,

$$\begin{aligned} V'_x &= \ \ V_x \cos\alpha + \ V_y \sin\alpha\ , \\ V'_y &= -\,V_x \sin\alpha + V_y \cos\alpha\ , \end{aligned} \qquad V'_z = V_z\ ,$$

one gets immediately, that the vector $\mathbf{k}'$ is rotated by the angle α against the vector **k**, and in the first case (1441) the vector $\mathbf{E}'$, against the vector **E**, and in the second case (1442) the vector $\mathbf{B}'$ against the vector **B**, where $\tan\alpha = v/c$ with $v = -|v|$.

Then the in Σ' observed vectors $\mathbf{E}'$ and $\mathbf{B}'$ are now orthogonal to the propagation direction $\mathbf{k}'$, being inclined with respect to the original direction **k** by the angle α. In the considered non-relativistic approximation, the aberration is given by Eq. (251),

$$\tan\alpha \approx \alpha \approx \frac{v}{c} \ .$$

This conclusion can be inverted. When we know the property (514) of the transversality of electromagnetic waves, one can interpret the tilt of the telescope in system Σ' as follows: If one compares the electromagnetic field of one light wave in two reference systems Σ_o and Σ' being in motion with respect to each other, then it holds in agreement with Eqs. (1441) and (1442),[9]

$$\Sigma_o : \begin{array}{l} \mathbf{E}\,, \\ \mathbf{B}\,, \end{array} \longrightarrow \ \Sigma' : \begin{array}{l} \mathbf{E}' = \mathbf{E} + \mathbf{v} \times \mathbf{B}\,, \\ \mathbf{B}' = \mathbf{B} - \dfrac{\mathbf{v}}{c^2} \times \mathbf{E}\,. \end{array} \quad \begin{array}{l} \text{Transformation} \\ \text{of the electromagnetic field} \\ \text{classical approximation} \end{array} \tag{1443}$$

Equation (1443) is the non-relativistic approximation of Eq. (534).

One can read Eq. (1443) also as follows: The aberration demonstrates that we 'see' one and the same electromagnetic field in different inertial systems with different electric and magnetic components. There we remain still in the framework of classical physics. However this conclusion was not drawn in the classical, hence not relativistic electrodynamics. With Eq. (1443), the symmetry for the explanation of the experimentally given induction law is established. Only EINSTEIN has established the transformation laws of the electromagnetic field for the change of the reference system with the Special Relativity Theory, cp. Chap. 9, Sects. 3.1.3 and 3.2.1. The classical approximation of Eq. (1443) was thereby skipped.

Problem 34 *Verify with the charging of a plate capacitor, s. Chap. 9, Sect. 3.1.4, that the energy density of the electric field is given by the relation* $\upsilon_e = (1/2)\mathbf{D} \times \mathbf{E}$.

Chapter 9, Section 3.1

We consider an ideal plate capacitor. Between two equal parallel flat plates $\mathbf{A}_1$ and $\mathbf{A}_2$ of area $A_1 = A_2 = A$ and separation d, there might be a homogeneous material with dielectricity constant ε. The electric field $\mathbf{E}$ and the dielectric displacement vector $\mathbf{D}$ with amount D and E should be non-zero only in the volume $V = A\,d$ between the plates. We can assume that they are parallel to the normal vector $\mathbf{n}$ of the area $\mathbf{A}_1$. Hence it holds

$$\left. \begin{array}{l} \mathbf{A}_1 = A\,\mathbf{n}\ , \quad \mathbf{A}_2 = -A\,\mathbf{n}\ , \\ \left. \begin{array}{l} \mathbf{D} = D\,\mathbf{n}\ , \\ \mathbf{E} = \dfrac{1}{\varepsilon}\mathbf{D} = E\,\mathbf{n}\ , \end{array} \right\} \begin{array}{l} \text{between} \\ \text{the plates,} \end{array} \quad \left. \begin{array}{l} \mathbf{D} = 0\ , \\ \mathbf{E} = 0\ . \end{array} \right\} \text{else}\,. \end{array} \right\} \tag{1444}$$

On the areas $\mathbf{A}_1$ and $\mathbf{A}_2$ there are charges $+e_o$ and $-e_o$, respectively, homogeneously distributed. This defines the area charge density $\omega_o = e_o/A$.

[9]The asymmetric position of the speed of light in Eq. (1443) follows only from the SI-unit system. We consider the non-relativistic approximation despite the terms $1/c^2$, what can be verified easily with our example of Eqs. (1441) and (1442). In the absolute unit system, Eqs. (1443) are to be replaced by $\tilde{\mathbf{E}}' = \tilde{\mathbf{E}} + \frac{\mathbf{v}}{c} \times \tilde{\mathbf{B}}$ and $\tilde{\mathbf{B}}' = \mathbf{B} - \frac{\mathbf{v}}{c} \times \tilde{\mathbf{E}}$.

We consider a thin cuboid K_1 enclosing the area $\mathbf{A}_1$. Then it holds Eq. (486), the integral form of the differential Eq. (490), MAXWELL's Eq. (503),

$$\iint_{\partial K_1} \mathbf{D} \cdot d\mathbf{A} = \iiint_{K_1} \rho \, dxdydz \ . \tag{1445}$$

Since by definition only the limiting area ∂K_1 of K_1 lies within the capacitor, and only there the field $\mathbf{D}$ is different from zero, we get for the area integral with Eq. (1444)

$$\iint_{\partial K_1} \mathbf{D} \cdot d\mathbf{A} = \mathbf{D} \cdot \mathbf{A}_1 = D\, A \ . \tag{1446}$$

The volume integral is equal to the sum of the enclosed charges,

$$\iiint_{K_1} \rho \, dxdydz = e_o = \omega_o A \ . \tag{1447}$$

From Eqs. (1444) to (1447) it follows

$$\mathbf{D} = \omega \mathbf{n} \ , \quad \mathbf{E} = \frac{\omega_o}{\varepsilon} \mathbf{n} \ . \tag{1448}$$

A small charge de on $\mathbf{A}_2$ feels in the electric field $\mathbf{E}$ a force $d\mathbf{F} = \mathbf{E}\, de$. We put the x-axis in opposite direction of $\mathbf{n}$ with the starting point at $\mathbf{A}_2$, and we transport the charge de against the force of the electric field on the area $\mathbf{A}_1$, that means we have spent the work dW,

$$dW = de \int_0^d \mathbf{E} \cdot \mathbf{n}\, dx = de\, E\, d \ . \tag{1449}$$

The charge de is uniformly distributed on the area $\mathbf{A}_1$, the transport increases the charge area density by $d\omega = de/F$, so that

$$dW = E\, d\, F\, d\omega \ . \tag{1450}$$

The total work W, necessary for putting charges starting from zero charge area density until getting a charge area density ω_o, we get with Eq. (1448) by summation,

$$W = \int_0^{\omega_o} E\, d\, F\, d\omega = F\, d \int_0^{\omega_o} \frac{\omega}{\varepsilon}\, d\omega = \frac{1}{2} F\, d\, \frac{\omega_o^2}{\varepsilon} \ . \tag{1451}$$

With the volume $V = F\, d$ between plates of the capacitor and the fields $\mathbf{E}$ and $\mathbf{D}$ in Eq. (1448) we can write

$$W = \frac{1}{2}\, V\, \mathbf{D} \cdot \mathbf{E} \ . \tag{1452}$$

Already MAXWELL has put up the idea that this work is stored in the energy of the electric field. Due to the considered homogeneity, we find from Eq. (1452) an electric energy density $\upsilon_e = W/V$,

$$\upsilon_e = \frac{1}{2}\, \mathbf{D} \cdot \mathbf{E} \ . \qquad \text{Energy density of the electrid field} \tag{1453}$$

In the absolute unit system with $\tilde{\mathbf{D}} = \sqrt{\tilde{\varepsilon}_o/4\pi}\, \mathbf{D}$ and $\tilde{\mathbf{E}} = \sqrt{1/4\pi\tilde{\varepsilon}_o}\, \mathbf{E}$, cp. Eq. (581), we find an electric energy density $\tilde{\upsilon}_e$

$$\tilde{\upsilon}_e = \frac{1}{8\pi}\, \tilde{\mathbf{D}} \cdot \tilde{\mathbf{E}} \ . \tag{1454}$$

Problem 35 *In system Σ_o during a finite time span, a body might emit an amount of light corresponding to an energy $\Delta E/2$ in direction* **k**, *and also the same amount in opposite direction. The system Σ' might be in motion in x-direction of system Σ_o with velocity v. Determine the emitted light energy in system Σ'.*

Chapter 7, Section 2, Chapter 9, Section 3.3

In the previous Problem 34, we have determined the energy density υ_e of the electric field in a medium, Eq. (1453). With the corresponding term for the energy density υ_m of the magnetic field in a medium,

$$\upsilon_m = \frac{1}{2}\, \mathbf{H} \cdot \mathbf{B} \ , \qquad \text{Energy density of a magnetic field} \tag{1455}$$

we get the energy density υ_{em} of the electromagnetic field

$$\upsilon_{em} = \frac{1}{2}\, (\mathbf{D} \cdot \mathbf{E} + \mathbf{H} \cdot \mathbf{B}) \ . \qquad \text{Energy density of the electromagnetic field} \tag{1456}$$

We need the energy density υ_{vem} of the electromagnetic field in vacuum. With $\mathbf{D} = \varepsilon_o \mathbf{E}$, $\mathbf{H} = (1/\mu_o)\mathbf{B}$ and $c = 1/\sqrt{\varepsilon_o \mu_o}$, s. Eqs. (503), (506), we have

$$\begin{aligned} \upsilon_{vem} &= \frac{1}{2}\Big(\varepsilon_o\, \mathbf{E} \cdot \mathbf{E} + \frac{1}{\mu_o}\mathbf{B} \cdot \mathbf{B}\Big) \\ &= \frac{1}{2\,\mu_o}\Big(\varepsilon_o\, \mu_o\, \mathbf{E} \cdot \mathbf{E} + \mathbf{B} \cdot \mathbf{B}\Big) \ , \end{aligned}$$

hence with $\mathbf{E}^2$ for $\mathbf{E} \cdot \mathbf{E}$ and $\mathbf{B}^2$ for $\mathbf{B} \cdot \mathbf{B}$

$$\upsilon_{vem} = \frac{1}{2\,\mu_o}\Big(\frac{1}{c^2}\, \mathbf{E}^2 + \mathbf{B}^2\Big) \ . \qquad \text{Energy density of the electromagnetic field in vacuum} \tag{1457}$$

With the transformation formulae (581), one proves easily that this expression v_{vem} leads to the energy density $\tilde{v}$ in the absolute unit system, Eq. (603).

Equation (1457) is the starting point for EINSTEIN's considerations to the energy–mass equivalence, cp. Chap. 7, Sect. 2.

The amplitudes of a plane wave can be written according to Eq. (515) as

$$\frac{1}{c}\,|\mathbf{E}_o| = |\mathbf{B}_o| = \mathrm{A}\ . \qquad \text{Amplitude of a plane wave} \tag{1458}$$

It is

$$\frac{1}{T}\int_0^T \cos^2(\omega t)dt = \frac{1}{2}\ , \qquad \omega = \frac{2\pi}{T}\ . \tag{1459}$$

Using the field strength of Eq. (507), $\mathbf{E} = \mathbf{E}_o \cos(\omega\, t)$ and $\mathbf{B} = \mathbf{B}_o \cos(\omega\, t)$, we take the square and the temporal mean values $\overline{\mathbf{E}^2}$ and $\overline{\mathbf{B}^2}$ over one oscillation period T. From Eqs. (1457), (1458), and (1459), we get the energy density h of a plane wave by the mean value over one period T,

$$h = \frac{1}{2\,\mu_o}\left(\frac{1}{c^2}\,\overline{\mathbf{E}^2} + \overline{\mathbf{B}^2}\right) = \frac{1}{2\,\mu_o}\left(\frac{1}{c^2}\,\mathbf{E}_o^2\,\frac{1}{T}\int_0^T \cos^2(\omega t)dt\, + \mathbf{B}_o^2\,\frac{1}{T}\int_0^T \cos^2(\omega t)dt\right),$$

hence

$$h = \frac{1}{2\,\mu_o}\left(\frac{1}{2\,c^2}\,\mathbf{E}_o^2 + \frac{1}{2}\,\mathbf{B}_o^2\right) = \frac{1}{2\,\mu_o}\,\mathrm{A}^2\ . \qquad \text{Reference system } \Sigma_o \text{ average energy density of a plane wave} \tag{1460}$$

We consider a plane wave with wave vector in the x-y-plane and forming with the x-axis an angle α, hence because of the transversality of the wave, Eq. (514),

$$\mathbf{k} = k\,(\cos\alpha\,, \sin\alpha\,, 0)\ , \quad \mathbf{E}_o = c\,\mathrm{A}\,(-\sin\alpha\,, \cos\alpha\,, 0)\ , \quad \mathbf{B}_o = \mathrm{A}\,(0\,, 0\,, 1)\ . \tag{1461}$$

For the energy density h of the wave in reference system Σ_o, we get the expression (1460), independent on the angle α of the wave vector $\mathbf{k}$.

The inertial system Σ' might be in motion with velocity v with respect to the x-axis of Σ_o. With Eq. (1461) we read off from Eq. (534), that the fields $\mathbf{E}_o'$ and $\mathbf{B}_o'$ in Σ' have components,

$$\mathbf{E}_o' = c\,\mathrm{A}\,(-\sin\alpha\,,\ \gamma(\cos\alpha - \beta)\,,\ 0)\ , \quad \mathbf{B}_o' = \mathrm{A}\,(0\,,\ 0\,,\ \gamma(1 - \beta\cos\alpha))\ . \tag{1462}$$

Now one finds that

$$\sin^2\alpha + \gamma^2\,(\cos\alpha - \beta)^2 = \sin^2\alpha + \gamma^2\left(\cos^2\alpha - 2\beta\,\cos\alpha + \beta^2\right) = \gamma^2\,(1 - \beta\,\cos\alpha)^2\ .$$

Then we get the energy density h' of this wave in Σ'

$$h' = \frac{1}{2\,\mu_o}\left(\frac{1}{2\,c^2}\,\mathbf{E}_o^{2'} + \frac{1}{2}\,\mathbf{B}_o^{2'}\right) = \frac{1}{2\,\mu_o}\,\mathrm{A}^2\gamma^2\,(1 - \beta\cos\alpha)^2\,. \tag{1463}$$

Reference system Σ' : energy density of the plane wave

We consider a certain amount of light in form of our plane waves, enclosed in a volume K so that no light crosses its surface. To this aim, one can use a volume that moves as a whole with the plane wave, i.e. it has in each inertial system a velocity equal to the speed of light c.

At first, we assume that the plane wave moves in direction of the positive x-axis of Σ_o, so that $\mathbf{k} = (k, 0, 0)$. The system Σ' has a velocity v in x-direction of Σ_o. We compare the values K and K', giving the energies inside the considered volume in Σ_o and in Σ', resp. Here it is *not* a pure LORENTZ contraction, since there is no rest system for the considered volume. We shall discuss in Problem 37 the length l of a with the speed of light propagating length on the x-axis in Σ_o with respect to the value l' in Σ'.

For simplicity, we assume that the considered volume K in Σ_o is a sphere with unity radius 1, moving with the speed of light c along the x-axis. This sphere is defined by

$$\Sigma_o:\ \left.\begin{aligned} &(x - c\,t)^2 + y^2 + z^2 = 1\ ,\\ &K = \frac{4}{3}\,\pi\ . \end{aligned}\right\} \qquad \text{Sphere in } \Sigma_o \tag{1464}$$

We insert into Eq. (1464) the special LORENTZ transformation (105), and we get the relation for the surface of the considered volume in Σ',

$$\Sigma':\ \begin{aligned} &\left(\gamma\,(x' + v\,t') - \gamma\,(c\,t' + x'\beta)\right)^2 + y'^2 + z'^2 = 1\ ,\\ &\left(\gamma\,(1-\beta)\,x' - \gamma\,((1-\beta))\,c\,t'\right)^2 + y'^2 + z'^2 = 1\ . \end{aligned}$$

This is nothing else than an ellipsoid with semi-axes $a = \gamma/(1 - v/c)$, $b = 1$, $c = 1$ propagating in Σ' along the x'-axis with the speed of light, hence

$$\Sigma':\ \left.\begin{aligned} &\left(\gamma\,(1-\beta)\,(x' - c\,t')\right)^2 + y'^2 + z'^2 = 1\ ,\\ &K' = \frac{4}{3}\,\frac{\pi}{\gamma(1-\beta)}\ . \end{aligned}\right\} \qquad \text{Ellipsoid in } \Sigma' \tag{1465}$$

For the ratio of values K and K' of the considered volumes in Σ_0 and Σ', it follows

$$K' = \frac{1}{\gamma\,(1-\beta)}\,K\ . \tag{1466}$$

What is the ratio of the values K and K' for the volumes enclosing a fixed amount of light, when the plane wave propagates with an angle α with respect to the x-axis of Σ_o? The components of the wave vector are then $\mathbf{k} = k\,(\cos\alpha\,,\,\sin\alpha\,,\,0)$. The evaluation is given in the following Problem 36, leading to the result

$$K' = \frac{1}{\gamma\,(1-\beta\,\cos\alpha)}\,K\ . \qquad \text{Wave vector } \mathbf{k} = k\,(\cos\alpha\,,\,\sin\alpha\,,\,0) \text{ system } \Sigma_o \qquad (1467)$$

The same ratio holds true between two lengths propagating with the speed of light according to Eq. (1487) in Problem 37.

The total amount of light within the considered volume has in the temporal average over one period in Σ_o and in Σ' energy values $H = h\,K$ and $H' = h'\,K'$. With Eqs. (1460), (1463), (1464), and (1465) it follows

$$\left.\begin{aligned} H = \ h\,K &= \frac{2\,\pi}{3\,\mu_o}\,\mathrm{A}^2\ , \\ H' = h'\,K' &= \frac{2\,\pi}{3\,\mu_o}\,\mathrm{A}^2\,\gamma\,(1-\beta\cos\alpha)\ , \end{aligned}\right\} \qquad (1468)$$

hence

$$H' = \gamma\,(1-\beta\cos\alpha)\,H\ . \qquad \text{Ratio of energies of light in } \Sigma' \text{ and } \Sigma_o \qquad (1469)$$

Now a body might rest in system Σ_o and have an energy U_o. In a finite time interval the body may emit a light energy $\Delta E/2$ in direction of the wave vector $\mathbf{k} = k\big(\cos\alpha\,,\ \sin\alpha\,,\ 0\big)$ and an equal amount of energy $\Delta E/2$ in opposite direction with wave vector $\mathbf{k}_- = k\big(\cos(\alpha+\pi)\,,\ \sin(\alpha+\pi)\,,\ 0\big)$. We denote the quantities after the emission process with an overbar. The body remains at rest after emission. Its energy is then $\overline{U}_o$, and it holds according to the energy conservation

$$U_o = \overline{U}_o + \left[\frac{\Delta E}{2} + \frac{\Delta E}{2}\right]\ . \qquad \text{Energy conservation in } \Sigma_o \qquad (1470)$$

The system Σ' is in motion in x-direction of Σ_o with velocity v. As seen from Σ', the body has before and after the emission energy values U_v' and $\overline{U}'_v$. The energies of the emitted light can be evaluated with Eq. (1469), when we replace H by H' and $\Delta E/2$ by the in both directions emitted energies $\Delta E'_+/2$ and $\Delta E'_-/2$. In Σ' there holds the conservation law

$$\begin{aligned} U_v' &= \overline{U}'_v + \left[\frac{\Delta E'_+}{2} + \frac{\Delta E'_-}{2}\right] \\ &= \overline{U}'_v + \left[\gamma\,(1-\beta\cos\alpha)\,\frac{\Delta E}{2} + \gamma\,(1-\beta\cos(\alpha+\pi))\,\frac{\Delta E}{2}\right] \\ &= \overline{U}'_v + \left[\gamma\,(1-\beta\cos\alpha)\,\frac{\Delta E}{2} + \gamma\,(1+\beta\cos\alpha)\,\frac{\Delta E}{2}\right]\ , \end{aligned}$$

hence

$$U'_v = \overline{U}'_v + \gamma \left[\frac{\Delta E}{2} + \frac{\Delta E}{2} \right] . \qquad \text{Energy conservation in } \Sigma' \qquad (1471)$$

From Eqs. (1470) and (1471) it follows the relation used in Chap. 7, Sect. 2

$$\left(U'_v - U_o \right) - \left(\overline{U}'_v - \overline{U}_o \right) = [\gamma - 1] \; \Delta E \; . \qquad (1472)$$

Problem 36 *A plane wave with fixed amount of light might be enclosed by a comoving volume. In system Σ_o, the wave vector has components $\mathbf{k} = k\,(\cos\alpha\,,\ \sin\alpha\,,\ 0)$. The volume K might be a sphere of radius unity, $R = 1$, propagating in Σ_o in direction $\mathbf{k}$ with speed of light. The reference system Σ' has a velocity $\mathbf{v} = (v, 0, 0)$ with respect to Σ_o. Which value K' is measured in Σ' for the volume comoving with the wave?*

Chapter 4, Section 4, Chapter 7, Section 2

The equation for the sphere K propagating in Σ_o with the speed of light reads

$$\Sigma_o : \left. \begin{array}{l} (x - c\,t\cos\alpha)^2 + (y - c\,t\sin\alpha)^2 + z^2 = 1\,, \\ K = \dfrac{4}{3}\pi\ . \end{array} \right\} \qquad \text{Sphere of radius 1 in } \Sigma_o \qquad (1473)$$

We insert into Eq. (1473) the special Lorentz transformation (105), then we get the equation for the surface of the considered volume in system Σ',

$$\Sigma' : \; \left(\gamma\,(x' + v\,t') - \gamma\,(c\,t' + x'v/c)\,\cos\alpha\right)^2 + \left(y' - \gamma\,(c\,t' + x'v/c)\,\sin\alpha\right)^2 + z'^2 = 1\ .$$

We consider this equation at a fixed moment of time, taken for simplicity $t' = 0$, hence with $\beta = v/c$ it follows

$$\left(x'\,\gamma\,(1 - \beta\cos\alpha)\right)^2 + y'^2 + \left(x'\,\gamma\,\beta\sin\alpha\right)^2 - 2x'y'\,\gamma\,\beta\sin\alpha + z'^2 = 1\ .$$

After a simple evaluation, we can then write

$$\Sigma' : \left. \begin{array}{l} A\,x'^2 + y'^2 - B\,x'y' + z'^2 = 1 \\ \text{with} \\ A = \gamma^2\,(1 + \beta^2 - 2\beta\cos\alpha)\ , \quad B = 2\,\gamma\,\beta\sin\alpha\ . \end{array} \right\} \qquad (1474)$$

In reference system Σ', we perform a pure rotation around the z'-axis to get new coordinates (x'^*, y'^*, z'^*) where the bilinear term vanishes

$$\Sigma' : \left. \begin{array}{l} x' = x'^*\cos\psi - y'^*\sin\psi\,, \\ y' = x'^*\sin\psi + y'^*\cos\psi\,, \\ z' = z'^*\ . \end{array} \right\} \qquad (1475)$$

With Eq. (1475) we get from Eq. (1474),

$$\left.\begin{aligned}\Sigma' : &\left[A\cos^2\psi + \sin^2\psi - B\sin\psi\cos\psi\right]x'^{*2} \\ &+ \left[A\sin^2\psi + \cos^2\psi + B\sin\psi\cos\psi\right]y'^{*2} \\ &+ \left[-A\,2\sin\psi\cos\psi + 2\sin\psi\cos\psi - B(\cos^2\psi - \sin^2\psi)\right]x'^{*}y'^{*} + z'^{*2} = 1 .\end{aligned}\right\} \tag{1476}$$

With $2\sin\psi\cos\psi = \sin(2\psi)$, $\cos^2\psi - \sin^2\psi = \cos(2\psi)$ the bilinear term in $x'^{*}y'^{*}$, vanishes, when

$$\tan(2\psi) = \frac{B}{1-A} . \tag{1477}$$

From Eq. (1474) it follows

$$\frac{B}{1-A} = \frac{\gamma\sin\alpha}{\cos\alpha - \beta} . \tag{1478}$$

With

$$\tan(2\psi) = \frac{2\tan\psi}{1-\tan^2\psi} , \qquad C := \frac{1-A}{B} = \frac{\cos\alpha - \beta}{\gamma\sin\alpha} \tag{1479}$$

we get a quadratic equation for $\tan\psi$,

$$\tan^2\psi + 2C\tan\psi - 1 = 0 , \tag{1480}$$

with solution

$$\tan\psi = -C \pm \sqrt{C^2+1} . \tag{1481}$$

It follows with Eqs. (1478) and (1479)

$$\tan\psi = \gamma\,\frac{\beta - \cos\alpha \pm (1-\beta\cos\alpha)}{\sin\alpha} . \tag{1482}$$

Since for $\alpha \longrightarrow 0$ there must also follow $\psi \longrightarrow 0$, the negative sign in the numerator can be excluded, and we get finally

$$\tan\psi = \frac{\gamma(1+\beta)(1-\cos\alpha)}{\gamma\sin\alpha} . \tag{1483}$$

By a rotation with the angle ψ, the bilinear term in Eq. (1476) vanishes, and we have only to determine the coefficients of x'^{*2} and y'^{*2}. To this aim, we need the relations

$$\sin^2\psi = \frac{\tan^2\psi}{1+\tan^2\psi} , \qquad \cos^2\psi = \frac{1}{1+\tan^2\psi} , \qquad \sin\psi\cos\psi = \frac{\tan\psi}{1+\tan^2\psi} .$$

With Eq. (1483) for $\tan\psi$, the evaluation provides

$$\left.\begin{aligned} \sin^2\psi &= \frac{(1+\beta)(1-\cos\alpha)}{2(1-\beta\cos\alpha)}\,, \\ \cos^2\psi &= \frac{\sin^2\alpha}{2\gamma^2(1+\beta)(1-\cos\alpha)(1-\beta\cos\alpha)}\,, \\ \sin\psi\cos\psi &= \frac{\sin\alpha}{2\gamma(1-\beta\cos\alpha)}\,. \end{aligned}\right\} \tag{1484}$$

With Eqs. (1476) and (1484) we get after lengthy but straightforward evaluation, that can be also verified at the end by insertion,

$$\left.\begin{aligned} \left[A\cos^2\psi+\sin^2\psi-B\sin\psi\cos\psi\right] &= \frac{1-\beta\cos\alpha}{1+\beta}\,, \\ \left[A\sin^2\psi+\cos^2\psi+B\sin\psi\cos\psi\right] &= \frac{1-\beta\cos\alpha}{1-\beta}\,. \end{aligned}\right\} \tag{1485}$$

From these expressions, we obtain with Eq. (1476) the semi-main axes of an ellipsoid in the coordinates x^{i*} of the reference systems Σ'
$a=\sqrt{(1+\beta)/(1-\beta\cos\alpha)}$, $b=\sqrt{(1-\beta)/(1-\beta\cos\alpha)}$, $c=1$
and with a volume $K'=(4/3)\pi abc$,

$$\left.\begin{aligned} \Sigma':\ &\frac{x'^{*2}}{(1+\beta)/(1-\beta\cos\alpha)}+\frac{y'^{*2}}{(1-\beta)/(1-\beta\cos\alpha)}+z'^{*2}=1\,, \quad \text{Ellipsoid in } \Sigma' \\ &K'=\frac{4\pi}{3}\,\frac{1}{\gamma(1-\beta\cos\alpha)}\,. \end{aligned}\right\} \tag{1486}$$

In agreement with Eq. (1467) we read off from Eqs. (1473) and (1486),

$$K'=\frac{1}{\gamma(1-\beta\cos\alpha)}\,K\,. \tag{1487}$$

Problem 37 *The points x_1 and x_2 on the x-axis of system Σ_o define a distance $l=x_2-x_1$. Both points now move on paths $x_1(t)=c\,t$ and $x_2(t)=l+c\,t$ in positive x-direction with the speed of light, so that the distance l between both points is independent on time. Which distance is measured in reference system Σ' moving in x-direction of Σ_o with velocity v?*

Chapter 4, Section 4
We consider two events E_1 and E_2 in Σ_o

$$\left.\Sigma_o:\qquad E_1:\ \begin{aligned} &x_1=c\,t_1\,,\\ &t_1\,, \end{aligned}\qquad E_2:\ \begin{aligned} &x_2=l+c\,t_2\,,\\ &t_2\,. \end{aligned}\right\} \tag{1488}$$

We suppose E_1 and E_2 are simultaneously in Σ_o, then the difference of its space coordinates $x_2(t_1) - x_1(t_1)$ determines the distance l measured in Σ_o,

$$\Sigma_o: \quad l = x_2(t_1) - x_1(t_1) \ . \tag{1489}$$

For arbitrary times t_1 and t_2 both events have in Σ' the coordinates $E_1(x_1', t_1')$ and $E_2(x_2', t_2')$, hence with a special LORENTZ transformation (105) it follows

$$\left.\Sigma' \begin{array}{ll} E_1: & x_1' = \gamma\,(x_1 - vt_10) = \gamma\,(ct_1 - vt_1) = \gamma\,(c - v)\,t_1 \ , \\ & t_1' = \gamma\,(t_1 - x_1 v/c^2) = \gamma\,(t_1 - t_1 v/c) = \gamma\,(1 - v/c)\,t_1 \ , \\ E_2: & x_2' = \gamma\,(x_2 - vt_2) = \gamma\,(l + ct_2 - vt_2) = \gamma\,\left(l + (c - v)t_2\right) \ , \\ & t_2' = \gamma(t_2 - x_2 v/c^2) = \gamma(t_2 - lv/c^2 - t_2 v/c) = \gamma\left((1 - v/c)t_2 - lv/c^2\right) . \end{array}\right\} \tag{1490}$$

When we suppose that both events E_1 and E_2 are simultaneously in Σ', then the difference of its spatial coordinates is equal to the in Σ' measured distance l'. Simultaneity in Σ' means with Eq. (1490)

$$\left. t_2' = t_1': \quad \begin{array}{rl} \gamma\left((1 - v/c)\,t_2 - lv/c^2\right) = & \gamma(1 - v/c)\,t_1 \ , \\ \left(1 - \frac{v}{c}\right)t_2 - \frac{lv}{c^2} = & \left(1 - \frac{v}{c}\right)t_1 \ , \end{array} \quad \longrightarrow \quad t_2 = t_1 + \frac{l\,v/c}{c - v} \ . \right\} \tag{1491}$$

Equations (1490) and (1491) lead to the relation for l',

$$l' = x_2'(t_1') - x_1'(t_1') = \gamma\,\left(l + (c - v)t_1 + l\,v/c\right) - \gamma\,(c - v)t_1,$$

$$l' = \gamma\,(1 + v/c)\,l \tag{1492}$$

or with

$$\gamma(1 + v/c) = \frac{\sqrt{(1 + v/c)(1 + v/c)}}{\sqrt{(1 + v/c)(1 - v/c)}} = \frac{\sqrt{1 + v/c}}{\sqrt{1 - v/c}} = \frac{\sqrt{(1 + v/c)(1 - v/c)}}{\sqrt{(1 - v/c)(1 - v/c)}} = \frac{1}{\gamma\,(1 - v/c)}$$

hence

$$l' = \frac{1}{\gamma\,(1 - v/c)}\,l \ . \tag{1493}$$

This relation of lengths determination in both inertial systems Σ' and Σ_o is different from the LORENTZ contraction, Eq. (112), it depends also on the sign of the velocity v of system Σ' with respect to Σ_o. The difference results from the propagation of the length with the speed of light that does not allow a rest system for this length.

Problem 38 *Derive the equation of motion for a continuously distributed finite mass from* NEWTON*'s second axiom using Eq. (1711), i.e. derive the equation of motion for a mechanical continuum.*

Chapter 6, Section 1, Chapter 9, Section 3.2

In a material continuum, the velocity $\mathbf{u} = \mathbf{u}(x, y, z, t)$ of the mass elements depends on the position of the point $P(x, y, z)$ and on the time t. The rest mass density is defined similarly as the rest charge density ρ_e in Chap. 9, Sect. 3.2.1, Eq. (525), $\varrho_o = \Delta m_o/\Delta V_o$, where Δm_o is the rest mass in a small comoving volume ΔV_o. Then this mass density ϱ again depends on the velocity, cp. Eq. (525)

$$\varrho = \gamma\,\varrho_e\,, \quad \varrho_o = \frac{\Delta m_o}{\Delta V_o}\,, \quad \gamma_u = \frac{1}{\sqrt{1 - u^2/c^2}}\,. \qquad \text{Mass density } \varrho \text{ in motion and invariant rest mass density } \varrho_o \tag{1494}$$

We define with

$$\mathbf{g}_m = \frac{\Delta \mathbf{p}_m}{\Delta V} \qquad \text{Momentum density} \tag{1495}$$

the momentum density in a material continuum. Here $\Delta \mathbf{p}_m = U \Delta M$ is defined by the mass ΔM in the volume ΔV and the centre-of-mass velocity U.

In a reference system Σ_o, we consider a volume K, occupied by a certain part of the masses. During the motion the volume K changes its form and position with time, so that $K = K(t)$. In particular, the surface ∂K of K changes with velocity $\mathbf{u} = (dx/dt, dy/dt, dz/dt)$ of the mass elements on the boundary of K.

We determine the change of its momentum $\mathbf{p}_m = \iiint\limits_{K(t)} \mathbf{g}_m\, dxdydz$ in volume K.

From Eq. (1711) it follows

$$\frac{d\mathbf{p}_m}{dt} = \frac{d}{dt} \iiint\limits_{K(t)} \mathbf{g}_m\, dxdyz = \iint\limits_{\partial K(t)} \mathbf{g}_m\, \mathbf{u} \cdot d\mathbf{A} + \iiint\limits_{K(t)} \frac{\partial}{\partial t}\, \mathbf{g}_m\, dxdydz\ .$$

We use the GAUSS law (1703) for the first integral and get

$$\frac{d\mathbf{p}_m}{dt} = \iiint\limits_{K} \mathrm{Div}_2\,(\mathbf{g}_m\, \mathbf{u})\, dxdydz + \iiint\limits_{K(t)} \frac{\partial}{\partial t}\, \mathbf{g}_m\, dxdydz\ ,$$

hence

$$\frac{d\mathbf{p}_m}{dt} = \iiint\limits_{K(t)} \left(\mathrm{Div}_2\,(\mathbf{g}_m\, \mathbf{u}) + \frac{\partial}{\partial t}\, \mathbf{g}_m\right) dxdydz\ . \tag{1496}$$

The differential operator Div_2 means, cp. Eq. (1655), that the contraction in forming the divergence should be performed with respect to the vector index of $\mathbf{u}$. Then the x-component of Eq. (1496) reads

$$\frac{dp_x}{dt} = \iiint\limits_{K(t)} \left(\frac{\partial}{\partial x} (g_x\, u_x) + \frac{\partial}{\partial y} (g_x\, u_y) + \frac{\partial}{\partial z} (g_x\, u_z) + \frac{\partial}{\partial t}\, g_x \right) dxdydz \ . \tag{1497}$$

Similar to the decomposition of the forces in Eqs. (143) and (169), $\sum\limits_{b=1}^{n} \mathbf{F}_{ba} + \mathbf{F}_a$, we decompose in the continuum the force $\mathbf{F}_m$, on the mass elements in volume K, into two parts, a force $\mathbf{F}_k$, stemming from the force action of the boundary of the volume K and an outer force $\mathbf{F_m}$. The force $\mathbf{F}_k$ can be written as surface integral over the boundary ∂K, and the force $\mathbf{F_m}$ as volume integral over an outer force density $\mathbf{f}$,

$$\mathbf{F}_m = \iint\limits_{\partial K(t)} d\mathbf{F}_k + \iiint\limits_{K(t)} \mathbf{f}\ dxdydz \ .$$

The force $d\mathbf{F}_k$ is defined at a fixed moment of time. When the continuum is separated by an area $d\mathbf{A}$, then an observer measures on the surface element $d\mathbf{A}$ a force $d\mathbf{F}_k = \sigma \cdot d\mathbf{A}$ that is needed on the surface element to keep it at the given place. This defines a stress tensor $\boldsymbol{\sigma}$

$$\mathbf{F}_m = \iint\limits_{\partial K(t)} \boldsymbol{\sigma} \cdot d\mathbf{A} + \iiint\limits_{K(t)} \mathbf{f}\ dxdydz \ . \tag{1498}$$

We use the Gauss law and get

$$\mathbf{F}_m = \iiint\limits_{K(t)} (\mathrm{Div}_2\, \boldsymbol{\sigma} + \mathbf{f})\ dxdydz \ . \tag{1499}$$

With Eqs. (1496) and (1499) we find from Eq. (169) the second axiom of mechanics $d\mathbf{p}_m/dt = \mathbf{F}_m$ for a mechanical continuum

$$\iiint\limits_{K(t)} \left(\mathrm{Div}_2\, (\mathbf{g}_m\, \mathbf{u}) + \frac{\partial}{\partial t}\, \mathbf{g}_m\right) dxdydz = \iiint\limits_{K(t)} (\mathrm{Div}_2\, \boldsymbol{\sigma} + \mathbf{f})\ dxdydz \ .$$

This equation holds true for an arbitrary volume K only when

$$\mathrm{Div}_2\, (\mathbf{g}_m\, \mathbf{u}) + \frac{\partial}{\partial t}\, \mathbf{g}_m = \mathrm{Div}_2\, \boldsymbol{\sigma} + \mathbf{f} \ . \tag{1500}$$

The quantity $\mathbf{g}_m\, \mathbf{u}$ is the local material momentum current density, that contributes with the term $\mathrm{Div}_2\, (\mathbf{g}_m\, \mathbf{u})$ to the local momentum balance. The local area element at a fixed position defines a local stress tensor $\mathbf{t}_m$ as

$$\mathbf{t}_m := \mathbf{g}_m\, \mathbf{u} - \boldsymbol{\sigma} \ . \tag{1501}$$

Then we write the second axiom of a mechanical continuum under the action of an outer force density $\mathbf{f}$, e.g. the LORENTZ force density in the form

$$\frac{\partial}{\partial t}\,\mathbf{g}_m + \mathrm{Div}_2\,\mathbf{t}_m = \mathbf{f}\ . \qquad \text{Material continuum, Second axiom of mechanics} \qquad (1502)$$

When there are no outer forces $\mathbf{f}$, then the equilibrium condition for a material continuum reads

$$\frac{\partial}{\partial t}\,\mathbf{g}_m + \mathrm{Div}_2\,\mathbf{t}_m = 0\ . \qquad \text{Force-free continuum} \qquad (1503)$$

We have to point to a peculiarity of the relativistic mechanics of continua. Both in the relativistic and also the non-relativistic point mechanics, the momentum $\mathbf{p}_m$ of a body is equal to the product of its mass m and its velocity $\mathbf{u}$, $\mathbf{p}_m = m\,\mathbf{u}$. A similar relation between the momentum density $\mathbf{g}_m$ and the mass density ϱ of a mechanical continuum moving with velocity field $\mathbf{u}$, hence $\mathbf{g}_m = \varrho\,\mathbf{u}$ is in general true only as approximation for non-relativistic velocities. In a relativistic continuum the momentum density contains additional terms by the mechanical stresses $\boldsymbol{\sigma}$, so that the momentum density $\mathbf{g}_m$ is in general not parallel to the velocity $\mathbf{u}$. Only for $\boldsymbol{\sigma} = 0$, hence, e.g. for a laminar streaming, also for relativistic velocities there holds again $\mathbf{g}_m = \varrho\,\mathbf{u}$.

When no transformations of the rest masses take place as always in the classical case of small velocities, and also for laminar streamings in the relativistic case, then the temporal change of the local mass density ϱ and the momentum current density $\mathbf{g}_m = \varrho\,\mathbf{u}$ also fulfil a continuity equation in the same form as for electric charges,

$$\frac{\partial}{\partial t}\,\varrho + \mathrm{div}\,(\varrho\,\mathbf{u}) = 0\ . \qquad \text{Continuity equation for conservation of rest masses} \qquad (1504)$$

For $\boldsymbol{\sigma} = 0$ and $\mathbf{g}_m = \varrho\,\mathbf{u}$ we get the second axiom for a continuum (1500) in the form

$$\mathrm{div}_2\,(\varrho\,\mathbf{u}\,\mathbf{u}) + \frac{\partial}{\partial t}\,\mathbf{g}_m = \mathbf{f}\ . \qquad (1505)$$

For small velocities, we get the fundamental equation of classical elasticity theory in neglecting nonlinear terms from Eq. (1500)

$$\frac{\partial}{\partial t}\,\mathbf{g}_m = \mathrm{div}\,\boldsymbol{\sigma} + \mathbf{f}\ . \qquad \text{Classical elasticity theory } u \ll c\,,\ u^2 \approx 0 \qquad (1506)$$

Problem 39 *When a plane wave propagates along the negative y-axis of system Σ_o, hence $(\eta\,,\theta\,,\zeta) = (-\pi/2,\pi,\pi/2)$, so that $(\cos\eta,\cos\theta,\cos\zeta) = (0,-1,0)$, then we get with $\nu' = \nu_E$ and $\nu = \nu_S$ from the last Eq. (556) the relation $\nu_E = \nu_S/\sqrt{1-\beta^2}$, in contradiction to Eq. (561), for the transversal* DOPPLER*-effect. Where is the error?*

Chapter 8, Sections 5, 6, Chapter 9, Section 3.2
From $(\cos\eta, \cos\theta, \cos\zeta) = (0, -1, 0)$ we get the first and last relation from Eq. (556),

$$\nu' \cos\eta' = \frac{\nu\,\beta}{\sqrt{1-\beta^2}} \;, \qquad \nu' = \frac{\nu}{\sqrt{1-\beta^2}} \;. \tag{1507}$$

These equations are only true when $\cos\eta' = \beta$, hence $\eta' \neq \pi$. Therefore, in Σ' we have *not* a transversal observation. From sight of system Σ_o, transversal emitted waves are not anymore transversal, when they are observed in system Σ' as described in case of aberration, s. Chap. 8, Sect. 6.2.

Problem 40 *Let Σ_o be the rest system of a star. We observe the light, emitted from the star in direction of the negative y-axis. Our rest system Σ', the Earth, moves with velocity $v = 29\,783\,\mathrm{ms}^{-1}$ in direction of the negative x-axis of Σ_o. By which angle α' we have to tilt the telescope against the y'-axis, when we observe the star in the focus?*

Chapter 8, Sections 6, Chapter 9, Section 3.2
For observing a star, we have to orient the telescope in Σ' in direction $-\mathbf{k}'$, when the emitted waves have the wave vector $\mathbf{k}'$, s. Fig. A.11. Emission along the negative y-direction of Σ_o means $\eta = -\pi/2$, hence $\cos\eta = 0$. For the direction $\mathbf{k}'$ of waves on Earth, there holds with Eq. (562), $\cos\eta' = \beta$. Since the direction of $\mathbf{k}'$ has to be also in the fourth quadrant as that of $\mathbf{k}$, we get $\eta' = -\arccos\beta$. For the direction of the telescope against the x'-axis we have to add π, hence for the direction α' against the y'-axis only $\pi/2$. The solution for the angle α' is then $\alpha' = -\eta' + \pi/2$, so that $\sin\alpha' = \sin(-\eta' + \pi/2) = \sin(-\eta')\cos(\pi/2) + \cos(-\eta')\sin(\pi/2) = \cos\eta'$,

$$\sin\alpha' = \cos\eta' = \frac{v}{c} \;, \tag{1508}$$

With the velocity $v = 29\,783\,\mathrm{ms}^{-1}$ of the Earth against the stars and the speed of light $c = 299\,792\,458\,\mathrm{ms}^{-1}$ we get $\beta = v/c = 0.000\,099\,345$, hence $\sin\alpha' = \cos\eta' = 0.000\,099\,345$ and therefore $\alpha' = 0.000\,099\,345$ or $\alpha' = 0.005\,692\,049° = 0.341\,522\,953' = 20.49''$. The telescope must be tilted by $20.49''$ (20.49 arc seconds) against the y'-axis, in Fig. A.11 against the clock direction. This angle α' is called *aberration constant*,

$$\alpha' = 20.49'' \;. \qquad \text{Aberration constant} \tag{1509}$$

Problem 41 *Use the directions of the coordinate axes and verify, that Eq. (681), represents a rotation $\mathbf{D}(\mathbf{n}, \alpha)$ in the group of unimodular transformations C_2.*

Chapter 9, Sections 1, Chapter 10, Section 3
We prove that we get for $\mathbf{n} = (1, 0, 0)$ in the general formula (681) the representation of the rotation matrix $\mathbf{D}_1$. This matrix is already known from Eq. (346) in Chap. 9, Sect. 1.2. To recover it, we have to show that the matrix $\mathbf{C}(\mathbf{n}, \alpha_1) := \mathbf{C}_{\alpha_1}$ in Eq. (682) delivers the matrix $\mathbf{D}_1$. It is

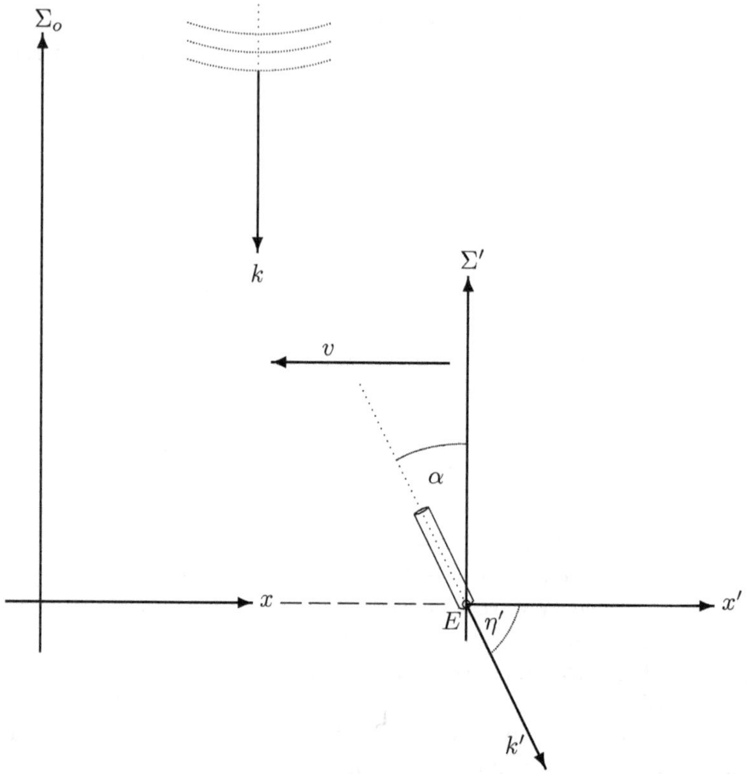

Fig. A.11 Schematic presentation of the aberration constant. In system Σ_o, a light wave propagates from above parallel to the y-axis. The system Σ' is in axis-parallel motion in direction of the negative x-axis of Σ_o with velocity amount v. That the light wave comes through the telescope to the ocular, the telescope resting in Σ' must be tilted by an angle α'. For the value of the orbital velocity of Earth around Sun, this angle is called aberration constant, with value given in Eq. (1509). For illustration, we have drawn the case of a telescope velocity of $v = 0.448\,c$ with respect to Σ_o, leading to an angle $\alpha' \approx 26.6°$.

$$\mathbf{D}_1 = \mathbf{T}\left(\mathbf{C}_{\alpha_1} \otimes \bar{\mathbf{C}}_{\alpha_1}\right)\mathbf{T}^{-1} \ . \tag{1510}$$

For $\mathbf{n} = (1, 0, 0)$ it follows from Eq. (681)

$$\mathbf{C}_{\alpha_1} = \cos\frac{\alpha_1}{2}\ \mathbf{1} + i\ \sin\frac{\alpha_1}{2}\boldsymbol{\sigma}_1 \ ,$$

hence with the Pauli matrix $\boldsymbol{\sigma}_1$ from Eq. (717),

$$\mathbf{C}_{\alpha_1} = \cos\frac{\alpha_1}{2}\begin{pmatrix} 1 & 0 \\ 0 & 1 \end{pmatrix} + i\ \sin\frac{\alpha_1}{2}\begin{pmatrix} 0 & 1 \\ 1 & 0 \end{pmatrix}$$

and therefore

$$\mathbf{C}_{\alpha_1} = \begin{pmatrix} \cos\frac{\alpha_1}{2} & i\,\sin\frac{\alpha_1}{2} \\ i\,\sin\frac{\alpha_1}{2} & \cos\frac{\alpha_1}{2} \end{pmatrix}, \quad \bar{\mathbf{C}}_{\alpha_1} = \begin{pmatrix} \cos\frac{\alpha_1}{2} & -i\,\sin\frac{\alpha_1}{2} \\ -i\,\sin\frac{\alpha_1}{2} & \cos\frac{\alpha_1}{2} \end{pmatrix}.$$

For the KRONECKER product of the two matrices we get

$$\mathbf{C}_{\alpha_1}\otimes\bar{\mathbf{C}}_{\alpha_1} = \begin{pmatrix} \cos\frac{\alpha_1}{2}\cos\frac{\alpha_1}{2} & -i\,\sin\frac{\alpha_1}{2}\cos\frac{\alpha_1}{2} & i\,\sin\frac{\alpha_1}{2}\cos\frac{\alpha_1}{2} & \sin\frac{\alpha_1}{2}\sin\frac{\alpha_1}{2} \\ -i\,\sin\frac{\alpha_1}{2}\cos\frac{\alpha_1}{2} & \cos\frac{\alpha_1}{2}\cos\frac{\alpha_1}{2} & \sin\frac{\alpha_1}{2}\sin\frac{\alpha_1}{2} & i\,\sin\frac{\alpha_1}{2}\cos\frac{\alpha_1}{2} \\ i\,\sin\frac{\alpha_1}{2}\cos\frac{\alpha_1}{2} & \sin\frac{\alpha_1}{2}\sin\frac{\alpha_1}{2} & \cos\frac{\alpha_1}{2}\cos\frac{\alpha_1}{2} & -i\,\sin\frac{\alpha_1}{2}\cos\frac{\alpha_1}{2} \\ \sin\frac{\alpha_1}{2}\sin\frac{\alpha_1}{2} & i\,\sin\frac{\alpha_1}{2}\cos\frac{\alpha_1}{2} & -i\,\sin\frac{\alpha_1}{2}\cos\frac{\alpha_1}{2} & \cos\frac{\alpha_1}{2}\cos\frac{\alpha_1}{2} \end{pmatrix}$$

$$= \begin{pmatrix} \cos^2\frac{\alpha_1}{2} & -\frac{i}{2}\sin\alpha_1 & \frac{i}{2}\sin\alpha_1 & \sin^2\frac{\alpha_1}{2} \\ -\frac{i}{2}\sin\alpha_1 & \cos^2\frac{\alpha_1}{2} & \sin^2\frac{\alpha_1}{2} & \frac{i}{2}\sin\alpha_1 \\ \frac{i}{2}\sin\alpha_1 & \sin^2\frac{\alpha_1}{2} & \cos^2\frac{\alpha_1}{2} & -\frac{i}{2}\sin\alpha_1 \\ \sin^2\frac{\alpha_1}{2} & \frac{i}{2}\sin\alpha_1 & -\frac{i}{2}\sin\alpha_1 & \cos^2\frac{\alpha_1}{2} \end{pmatrix}.$$

With the matrices $\mathbf{T}$ and $\mathbf{T}^{-1}$ in Eq. (654) we must form with Eq. (1510)

$$\frac{1}{2}\begin{pmatrix} 1 & 0 & 0 & 1 \\ 0 & -1 & -1 & 0 \\ 0 & -i & i & 0 \\ -1 & 0 & 0 & 1 \end{pmatrix} \begin{pmatrix} \cos^2\frac{\alpha_1}{2} & -\frac{i}{2}\sin\alpha_1 & \frac{i}{2}\sin\alpha_1 & \sin^2\frac{\alpha_1}{2} \\ -\frac{i}{2}\sin\alpha_1 & \cos^2\frac{\alpha_1}{2} & \sin^2\frac{\alpha_1}{2} & \frac{i}{2}\sin\alpha_1 \\ \frac{i}{2}\sin\alpha_1 & \sin^2\frac{\alpha_1}{2} & \cos^2\frac{\alpha_1}{2} & -\frac{i}{2}\sin\alpha_1 \\ \sin^2\frac{\alpha_1}{2} & \frac{i}{2}\sin\alpha_1 & -\frac{i}{2}\sin\alpha_1 & \cos^2\frac{\alpha_1}{2} \end{pmatrix} \begin{pmatrix} 1 & 0 & 0 & -1 \\ 0 & -1 & i & 0 \\ 0 & -1 & -i & 0 \\ 1 & 0 & 0 & 1 \end{pmatrix}.$$

Careful multiplication provides indeed the rotation matrix around the x-axis $\mathbf{D}_1$, Eq. (346),

$$\mathbf{D}_1 = \begin{pmatrix} 1 & 0 & 0 & 0 \\ 0 & 1 & 0 & 0 \\ 0 & 0 & \cos\alpha_1 & \sin\alpha_1 \\ 0 & 0 & -\sin\alpha_1 & \cos\alpha_1 \end{pmatrix}.$$

The evaluation for rotations around the y- and z-axis is clearly analogous. It can be easily repeated by the reader.

Problem 42 *Verify for the directions of coordinate axes that Eq. (683) represents a special* LORENTZ *transformation* $\mathbf{L}(v)$ *in the group of unimodular transformations* $\mathcal{C}_2$.

Chapter 9, Section 1, Chapter 10, Section 3

As in the previous problem, we prove here that we get for $\mathbf{n} = (1, 0, 0)$ from Eq. (684) a special LORENTZ transformation $\mathbf{L}_1$. This matrix is already known from Eq. (346) in Chap. 9, Sect. 1.2. To this aim, we show that $\mathbf{n} = (1, 0, 0)$ in Eq. (683) leads to the matrix $\mathbf{C}(\mathbf{n}, \varphi_1) := \mathbf{C}_{\varphi_1}$ in Eq. (684), the matrix $\mathbf{L}_1$. It is

$$\mathbf{L}_1 = \mathbf{T}\left(\mathbf{C}_{\varphi_1} \otimes \bar{\mathbf{C}}_{\varphi_1}\right)\mathbf{T}^{-1} \ . \tag{1511}$$

From $\mathbf{n} = (1, 0, 0)$ it follows from Eq. (683),

$$\mathbf{C}_{\varphi_1} = \cosh\frac{\varphi_1}{2}\ \mathbf{1} + \sinh\frac{\varphi_1}{2}\boldsymbol{\sigma}_1 \ ,$$

hence with the PAULI matrix $\boldsymbol{\sigma}_1$ from Eq. (717),

$$\mathbf{C}_{\varphi_1} = \cosh\frac{\varphi_1}{2}\begin{pmatrix} 1 & 0 \\ 0 & 1 \end{pmatrix} + \sinh\frac{\varphi_1}{2}\begin{pmatrix} 0 & 1 \\ 1 & 0 \end{pmatrix} \qquad \text{and therefore}$$

$$\mathbf{C}_{\varphi_1} = \bar{\mathbf{C}}_{\varphi_1} = \begin{pmatrix} \cosh\frac{\varphi_1}{2} & \sinh\frac{\varphi_1}{2} \\ \sinh\frac{\varphi_1}{2} & \cosh\frac{\varphi_1}{2} \end{pmatrix} .$$

For the KRONECKER product $\mathbf{C}_{\varphi_1} \otimes \bar{\mathbf{C}}_{\varphi_1}$, we get then

$$\mathbf{C}_{\alpha_1} \otimes \bar{\mathbf{C}}_{\alpha_1} = \begin{pmatrix} \cosh\frac{\varphi_1}{2}\cosh\frac{\varphi_1}{2} & \sinh\frac{\varphi_1}{2}\cosh\frac{\varphi_1}{2} & \sinh\frac{\varphi_1}{2}\cosh\frac{\varphi_1}{2} & \sinh\frac{\varphi_1}{2}\sinh\frac{\varphi_1}{2} \\ \sinh\frac{\varphi_1}{2}\cosh\frac{\varphi_1}{2} & \cosh\frac{\varphi_1}{2}\cosh\frac{\varphi_1}{2} & \sinh\frac{\varphi_1}{2}\sinh\frac{\varphi_1}{2} & \sinh\frac{\varphi_1}{2}\cosh\frac{\varphi_1}{2} \\ \sinh\frac{\varphi_1}{2}\cosh\frac{\varphi_1}{2} & \sinh\frac{\varphi_1}{2}\sinh\frac{\varphi_1}{2} & \cosh\frac{\varphi_1}{2}\cosh\frac{\varphi_1}{2} & \sinh\frac{\varphi_1}{2}\cosh\frac{\varphi_1}{2} \\ \sinh\frac{\varphi_1}{2}\sinh\frac{\varphi_1}{2} & \sinh\frac{\varphi_1}{2}\cosh\frac{\varphi_1}{2} & \sinh\frac{\varphi_1}{2}\cosh\frac{\varphi_1}{2} & \cosh\frac{\varphi_1}{2}\cosh\frac{\varphi_1}{2} \end{pmatrix}$$

$$= \begin{pmatrix} \cosh^2\frac{\varphi_1}{2} & \frac{1}{2}\sinh\varphi_1 & \frac{1}{2}\sinh\varphi_1 & \sinh^2\frac{\varphi_1}{2} \\ \frac{1}{2}\sinh\varphi_1 & \cosh^2\frac{\varphi_1}{2} & \sinh^2\frac{\varphi_1}{2} & \frac{1}{2}\sinh\varphi_1 \\ \frac{1}{2}\sinh\varphi_1 & \sinh^2\frac{\varphi_1}{2} & \cosh^2\frac{\varphi_1}{2} & \frac{1}{2}\sinh\varphi_1 \\ \sinh^2\frac{\varphi_1}{2} & \frac{1}{2}\sinh\varphi_1 & \frac{1}{2}\sinh\varphi_1 & \cosh^2\frac{\varphi_1}{2} \end{pmatrix} .$$

With matrices $\mathbf{T}$ and $\mathbf{T}^{-1}$ from Eq. (654) we must form according to Eq. (1511)

$$\frac{1}{2}\begin{pmatrix} 1 & 0 & 0 & 1 \\ 0 & -1 & -1 & 0 \\ 0 & -i & i & 0 \\ -1 & 0 & 0 & 1 \end{pmatrix}\begin{pmatrix} \cosh^2\frac{\varphi_1}{2} & \frac{1}{2}\sinh\varphi_1 & \frac{1}{2}\sinh\varphi_1 & \sinh^2\frac{\varphi_1}{2} \\ \frac{1}{2}\sinh\varphi_1 & \cosh^2\frac{\varphi_1}{2} & \sinh^2\frac{\varphi_1}{2} & \frac{1}{2}\sinh\varphi_1 \\ \frac{1}{2}\sinh\varphi_1 & \sinh^2\frac{\varphi_1}{2} & \cosh^2\frac{\varphi_1}{2} & \frac{1}{2}\sinh\varphi_1 \\ \sinh^2\frac{\varphi_1}{2} & \frac{1}{2}\sinh\varphi_1 & \frac{1}{2}\sinh\varphi_1 & \cosh^2\frac{\varphi_1}{2} \end{pmatrix}\begin{pmatrix} 1 & 0 & 0 & -1 \\ 0 & -1 & i & 0 \\ 0 & -1 & -i & 0 \\ 1 & 0 & 0 & 1 \end{pmatrix} .$$

The careful matrix multiplication provides indeed the special LORENTZ transformation $\mathbf{L}_1$ from Chap. 9, Sect. 1.2, Eq. (346),

$$\mathbf{L}_1 = \begin{pmatrix} \cosh\varphi_1 & -\sinh\varphi_1 & 0 & 0 \\ -\sinh\varphi_1 & \cosh\varphi_1 & 0 & 0 \\ 0 & 0 & 1 & 0 \\ 0 & 0 & 0 & 1 \end{pmatrix} = \begin{pmatrix} \gamma_1 & -\beta_1\gamma_1 & 0 & 0 \\ -\beta_1\gamma_1 & \gamma_1 & 0 & 0 \\ 0 & 0 & 1 & 0 \\ 0 & 0 & 0 & 1 \end{pmatrix}$$

with $\beta_1 = \dfrac{v_1}{c}$, what we wanted to show. The evaluation for rotations around the y- and z-axis proceeds analogously; it is left to the reader for an exercise in matrix calculus.

Problem 43 *In the relativistic space-time, we define after* W. Thirring *an absolute simultaneity. Instead of the* Lorentz *transformation there follows then the so-called* Reichenbach *transformation, given below, Eq. (1516). Two bodies L and P might have velocities* $u' = dx'_L/dt' = 0.4\,c$ *bzw.* $u' = dx'_P/dt' = 0.8\,c$ *in reference system* Σ'. *With respect to* Σ_o, *the system* Σ' *has the velocity* v. Σ'' *is the rest system of body L. Which velocity has body P, as seen from* Σ''?

Chapter 8, Section 9

We introduce shortly the background of the problem.

Here we refer as well to H. Reichenbach (1972), Sect. 19, p. 148: 'We could arrange the definition of simultaneity of a system K in such a manner that it leads to the same results as that of another system K' which is in motion relative to K;...'

With this in mind, we consider now an empirical determination of the relativistic space-time according to Chap. 2, Sects. 1–4. The physical postulates (99) and (100), s. also Chap. 1, Sect. 5, leading to the Lorentz contraction and the time dilatation, are maintained in the *initially* distinguished reference system Σ_o,

$$\Sigma_o: \quad \frac{l_v}{l_o} = \frac{1}{k} = \sqrt{1 - \frac{v^2}{c^2}}, \qquad \text{In } \Sigma_o \text{ the moving ruler is shortened.} \tag{1512}$$

$$\Sigma_o: \quad \frac{t_v}{t_o} = v\,\theta(v) + q(v) = \sqrt{1 - v^2/c^2}\;. \qquad \text{In } \Sigma_o \text{ the moving clock is delayed.} \tag{1513}$$

However, now we shall not require the elementary relativity principle for definition of simultaneity.

On the basis of the freely selectable synchronisation function in the coordinate transformation, Chap. 1, Sect. 4, we can realise Reichenbach's claim, Eq. (296), by the choice of parameters $\theta = \theta_a = 0$ in Eq. (1513) for synchronisation of clocks in systems Σ', illustrated in Fig. A.12.

Now we consider again the behaviour of moving and resting measuring rods and clocks, Chap. 4, Sect. 3, Eqs. (99) and (100), but we replace the definition of simultaneity. The Lorentz parameter $\theta_L = -v/(c^2\,\gamma)$, s. Eq. (103), leading to the Lorentz transformation, is replaced by an other definition $\theta_a = 0$. We change only the definition, but not the physics.

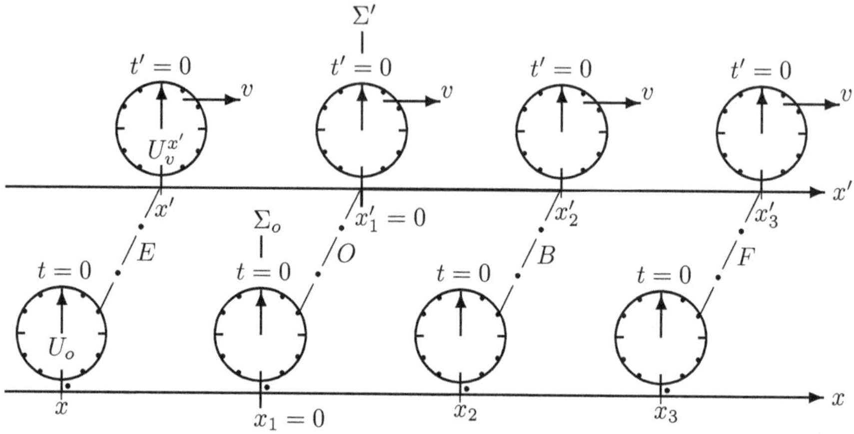

Fig. A.12 In systems Σ' we introduce an absolute simultaneity with the parameter $\theta_a(v) = 0$. At time $t = 0$ in Σ_o, in all inertial systems, we put the Σ'-clocks at position $t' = 0$. The dashed-dotted lines connect points in the picture representing the same events, here the events E, O, B and F.

$\theta_a = 0$ means a non-conventional synchronisation parameter of relativistic space-time.

For the validity of Eqs. (1512) and (1513), it is now essential, that $\theta_a = 0$ is required for the inertial system Σ_o.

From the general transformation law (103),

$$\left.\begin{array}{lcl} x' = k\,(x - v\,t)\,, & & x = \dfrac{q}{\Delta}\,x' + \dfrac{v\,k}{\Delta}\,t'\,, \\ & \longleftrightarrow & \\ t' = \theta\,x + q\,t\,, & & t = -\dfrac{\theta}{\Delta}\,x' + \dfrac{k}{\Delta}\,t'\,, \\ \Delta = k\,(v\,\theta + q)\,, & k = k(v)\,, \quad q = q(v)\,, & \theta = \theta(v)\,, \end{array}\right\} \text{General coordinate transformation} \tag{1514}$$

we get the transformation formula with

$$k = \frac{1}{\sqrt{1 - v^2/c^2}}\,, \quad q = \sqrt{1 - v^2/c^2}\,, \quad \theta = 0\,, \qquad \text{Relativistice space-time with absolute simultaneity} \tag{1515}$$

at first explicitly written down by W. Thirring, that we here present with its inversion and which was denoted by Günther (2000) as Reichenbach transformation,

$$\left.\begin{array}{lcl} x' = \dfrac{x - v\,t}{\sqrt{1 - v^2/c^2}}\,, & \longleftrightarrow & x = \dfrac{x'(1 - v^2/c^2) + t'\,v}{\sqrt{1 - v^2/c^2}}\,, \\ t' = t\,\sqrt{1 - v^2/c^2}\,, & & t = \dfrac{t'}{\sqrt{1 - v^2/c^2}}\,. \end{array}\right\} \text{Reichenbach transformation} \tag{1516}$$

It is to be kept in mind, that here v denotes the velocity of systems Σ' with respect to the distinguished system Σ_o. The system Σ' is arbitrary.

From the general linear coordinate transformation (103), there follows the general addition theorem (23), of velocities that we write down once more,

$$u' = k\,\frac{u - v}{\theta\, u + q} \quad \longleftrightarrow \quad u = \frac{q\, u' + v\, k}{-\theta\, u' + k}\,. \qquad \text{Addition theorem of velocities to the general transformation (1514)} \tag{1517}$$

With parameters k, q and θ of Eq. (1515), the transformation (1516) leads to the addition theorem

$$u' = \frac{u - v}{1 - v^2/c^2} \quad \longleftrightarrow \quad u = (1 - v^2/c^2)u' + v\,. \qquad \text{Addition theorem of velocities to the transformation (1516)} \tag{1518}$$

In the coordinate transformation (1514) and the addition theorem (1518), one notes the asymmetric mathematical *description* of the relativistic space-time. This follows from the arbitrary distinction of the reference systems Σ_o. A violation of the relativity principles cannot be inferred. EINSTEIN's relativity principle, the equivalence of all inertial systems, is here realised by the possibility to choose arbitrarily each inertial system as a mathematically distinct one.

Now we come back to the formulated problem.

When the body L has the velocity w' with respect to Σ', then it has according to Eq. (1518) in Σ_o the velocity

$$w = w'/\gamma_v^2 + v\,, \qquad \gamma_v = \frac{1}{\sqrt{1 - v^2/c^2}}\,. \tag{1519}$$

With respect to the body L representing the reference system $\Sigma''(x'', t'')$, that has a velocity w with respect to Σ_o, the REICHENBACH transformation (1516) reads

$$\left.\begin{aligned} x'' &= \gamma_w\,(x - w\,t)\,, \\ t'' &= \frac{t}{\gamma_w}\,, \qquad \gamma_w = \frac{1}{\sqrt{1 - w^2/c^2}}\,. \end{aligned}\right\} \tag{1520}$$

From Eqs. (1516) and (1520) we find the REICHENBACH transformation between Σ' and Σ''

$$\left.\begin{aligned} x'' &= \gamma_w \gamma_v \left(\frac{x'}{\gamma_v^2} + t'(v - w)\right)\,, \\ t'' &= t'\frac{\gamma_v}{\gamma_w}\,. \end{aligned}\right\} \tag{1521}$$

We look for the in Σ'' measured velocity $u'' = dx''_P/dt''$ of body P, for which we have a velocity $u' = dx'_P/dt'$ in Σ'. With Eq. (1521) we get

$$u'' = \frac{dx''_P}{dt''} = \frac{dx''_P}{dt'}\frac{dt'}{dt''} = \gamma_w \gamma_v \left(\frac{u'}{\gamma_v^2} + (v - w) \right) \frac{\gamma_w}{\gamma_v} ,$$

hence

$$u'' = \gamma_w^2 \left(\frac{u'}{\gamma_v^2} + (v - w) \right) . \tag{1522}$$

With Eq. (1518) it follows

$$u'' = (u' - w') \frac{\gamma_w^2}{\gamma_v^2} . \tag{1523}$$

For $v = 0$, hence $\gamma_v = 1$, we get that $\Sigma'(x', t')$ is identical with $\Sigma_o(x, t)$, and Eq. (1523) goes over to the addition theorem (1518), when there one replaces u' by u'', u by u', v by w', and $1 - v^2/c^2$ by $1/\gamma_v^2$.

The value for the wanted velocity u'' now depends on the velocity v of system Σ' with respect to Σ_o by the factors γ_w and γ_v.

As an example, we consider the case $v = 0.8\,c$, hence $\gamma_v = 1/\sqrt{1 - 0.64^2} = 1.7$. With $w' = 0.4\,c$ and $u' = 0.8\,c$ we get

$$u'' = 0.4 \frac{0.36}{1 - (0.4 \cdot 0.36 + 0.8)^2} \approx 0.4 \frac{0.36}{0.109} \approx 1.32\,c \quad \text{for} \quad v = 0.8\,c \ . \tag{1524}$$

We still add some remarks.

As seen from system Σ_o, the front of a light wave has the velocity $u = c$. The velocity of the inertial system Σ' might be v. The velocity of the light front measured from system Σ' is called c'. The three velocities c, c' and v are connected by REICHENBACH's addition theorem of velocities (1518), hence

$$c' = \frac{c - v}{1 - v^2/c^2} = c \frac{1 - v/c}{(1 - v/c)(1 + v/c)}$$

and therefore

$$c' = \frac{c}{1 + v/c} . \tag{1525}$$

Now the speed of light has values different from c in systems Σ', it also depends on the sign of v, Eq. (1525), i.e. from the propagation direction. Does this mean a violation of EINSTEIN's famous principle of the universal constancy of the speed of light?

By no means! H. REICHENBACH has stressed emphatically, that the principle of the universal constancy of the speed of light should be better formulated as follows: It is *possible*—and also H. POINCARÉ has mentioned it before 1900,[10] it may be of advantage—in *all* inertial systems to let run the clocks is such a way, that we thereafter can determine in *all* inertial systems the same speed of light c. This is only possible when all measuring rods and clocks fulfil the postulates of Eqs. (1512) and (1513). When we introduce an other than EINSTEIN's definition of simultaneity, we have to accept a change in the velocity of light propagation. This has no other foundation as the arbitrarily changed synchronisation of clocks in systems Σ', s. also the discussion in Chap. 2, Sect. 3 and Chap. 4, Sect. 7.

Problem 44 *A body L is in motion along the x-axis with an orbit $x = x(t)$ as seen in system Σ_o. There are measured velocity $u = dx/dt$ and acceleration $a = du/dt$. Which acceleration $a' = du'/dt'$ is measured for a velocity $u' = dx'/dt'$ of the body in system Σ', being in motion with velocity v in x-direction with respect to Σ_o, when we suppose the* REICHENBACH *transformation (1516)?*

Chapter 8, Section 9
In system Σ', the body L has a velocity $u' = dx'_L/dt'$ and an acceleration $a' = du'/dt'$, hence with Eq. (1516),

$$u' = \frac{dx'_L}{dt'} = \frac{dx'_L}{dt}\left(\frac{dt'}{dt}\right)^{-1} = \gamma\left(\frac{dx_L}{dt} - v\right)\gamma = \gamma^2(u - v) \ ,$$
$$a' = \frac{du'}{dt'} = \frac{du'}{dt}\left(\frac{dt'}{dt}\right)^{-1} = \gamma^2\frac{du}{dt}\gamma \ ,$$

and therefore—in distinction to the more complicated result in Eq. (1250), $a' = a\left(\gamma\left(1 - uv/c^2\right)\right)^{-3}$ in case of the LORENTZ transformation—

$$a' = \gamma^3 a \ . \tag{1526}$$

For small velocities, $v^2/c^2 \approx 0\ ,\ uv/c^2 \approx 0$, we get from Eq. (1526) the independence of the acceleration from the reference system, i.e. the result of classical mechanics.

Problem 45 *Derive the dependency of mass m of a body from its velocity under the assumption of the* REICHENBACH *transformation (1516).*

Chapter 8, Section 9
As shown in Fig. A.13, we consider two balls A and B that collide ideal elastically. Both balls have the same mass m.

In reference system Σ_o, the ball A has only a velocity component in y-direction, $u_{Ay} = dy/dt = w$, hence

$$\Sigma_o : \mathbf{u}_A = (u_{Ax}, u_{Ay}) = (0,\ w) \ . \tag{1527}$$

[10]Rev. Métaphys. Morale, **6** (1898), 1.

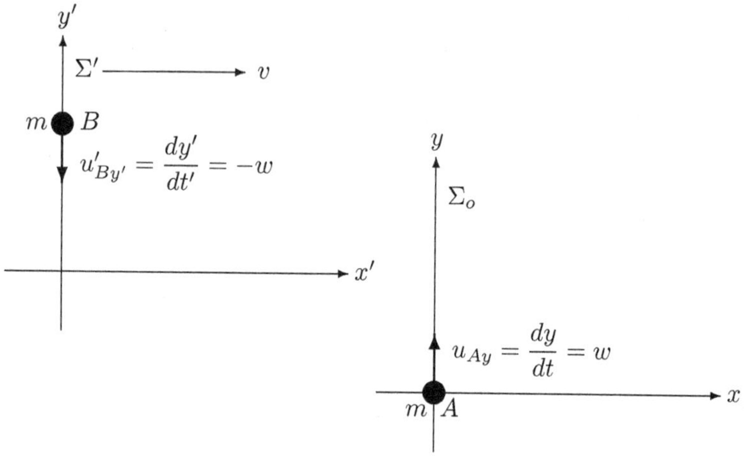

Fig. A.13 Schematic illustration of TOLMAN's thought experiment.

The reference system Σ' has a velocity v in x-direction as measured in Σ_o. The ball B has in Σ' only a velocity component in direction of the negative y'-axis, it is denoted by $u'_{By'} = dy'/dt' = -w$,

$$\Sigma' : \mathbf{u}'_B = (u'_{Bx'}, u'_{By'}) = (0, -w) . \tag{1528}$$

With velocity $\mathbf{v} = (v_1, 0, 0)$ of Σ' with respect to Σ_o, we generalise at first REICHENBACH's addition theorem of velocities (1518) for the case, that a body L in Σ' has an arbitrarily oriented velocity $\mathbf{u}' = (u'_x, u'_y, u'_z)$, while in Σ_o one determines a velocity $\mathbf{u} = (u_x, u_y, u_z)$,

$$\left.\begin{aligned} \Sigma_o \ : \ \mathbf{u} = (u_x, u_y, u_z) &= (\frac{dx}{dt}, \frac{dy}{dt}, \frac{dz}{dt}) , \\ \Sigma' \ : \ \mathbf{u}' = (u'_{x'}, u'_{y'}, u'_{z'}) &= (\frac{dx'}{dt'}, \frac{dy'}{dt'}, \frac{dz'}{dt'}) . \end{aligned}\right\} \tag{1529}$$

We write $\gamma_1 = 1/\sqrt{1 - v_1^2/c^2}$. The motion $[x = x(t),\ y = y(t),\ z = z(t)]$ and $[x' = x'(t'),\ y' = y'(t'),\ z' = z'(t')]$ of both balls are inserted into the REICHENBACH transformation (1516), and we find

$$\begin{aligned} \frac{dx}{dt} &= \frac{dx}{dt'}\frac{dt'}{dt} = \gamma_1 \left(\frac{dx'}{dt'} \frac{1}{\gamma_1^2} \frac{dt'}{dt} + v_1 \frac{dt'}{dt} \right) , \\ \frac{dy}{dt} &= \frac{dy}{dt'}\frac{dt'}{dt} = \frac{dy'}{dt'}\frac{dt'}{dt} , \\ \frac{dz}{dt} &= \frac{dz}{dt'}\frac{dt'}{dt} = \frac{dz'}{dt'}\frac{dt'}{dt} , \\ \frac{dt'}{dt} &= \frac{1}{\gamma_1} . \end{aligned}$$

As in case of the generalised EINSTEIN addition theorem (117), there follows a generalisation of REICHENBACH's addition theorems (1518),

$$\begin{aligned} u'_{x'} &= \gamma_1^2\,(u_x - v_1)\,, \\ u'_{y'} &= \gamma_1\, u_y\,, \\ u'_{z'} &= \gamma_1\, u_z\,, \end{aligned} \quad \longleftrightarrow \quad \left.\begin{aligned} u_x &= \frac{u'_{x'}}{\gamma_1^2} + v_1\,, \\ u_y &= \frac{u'_{y'}}{\gamma_1}\,, \\ u_z &= \frac{u'_{z'}}{\gamma_1}\,. \end{aligned}\right\} \quad \begin{array}{l} \text{REICHENBACH's} \\ \text{addition theorem with a} \\ \text{velocity } (v_1, 0, 0) \\ \text{of } \Sigma' \text{ with respect to } \Sigma_o\,, \\ \gamma_1 = 1/\sqrt{1-\beta_1^2}\,,\ \beta_2 = \frac{v_1}{c}\,. \end{array} \tag{1530}$$

For the velocity component $\mathbf{u}_B$ of ball B, we get in Σ_o with Eq. (1528) and the addition theorem (1530) the relations $u_{Bx} = u'_{Bx'}/\gamma^2 + v = v$ and $u_{By} = u'_{By'}/\gamma = -w/\gamma$, when we use there $v = v_1$ and $\gamma = \gamma_1$. Hence it holds in Σ_o,

$$\Sigma_o:\ \left.\begin{aligned} u_{Ax} &= 0\,, \quad u_{Bx} = v\,, \\ u_{Ay} &= w\,, \quad u_{By} = -\tfrac{w}{\gamma}\,. \end{aligned}\right\} \quad \begin{array}{l} \text{Velocity components} \\ \text{before the collision} \end{array} \tag{1531}$$

After the collision, we denote the momenta and velocities again by an overbar, and we write the momentum conservation

$$\left.\begin{aligned} m\, u_{Ax} + m\, u_{Bx} &= m\,\overline{u}_{Ax} + m\,\overline{u}_{Bx}\,, \\ m\, u_{Ay} + m\, u_{By} &= m\,\overline{u}_{Ay} + m\,\overline{u}_{By}\,. \end{aligned}\right\} \tag{1532}$$

Since there are no tangential forces during the collision, the velocities in x-direction will not change. In y-direction, we have the velocities $u_{Ay} = \overline{w}_A$ and $\overline{u}'_{By'} = \overline{w}_B$. As seen in system Σ_o, we find for the components after the collision as above,

$$\Sigma_o:\ \left.\begin{aligned} \overline{u}_{Ax} &= 0\,, \quad \overline{u}_{Bx} = v\,, \\ \overline{u}_{Ay} &= \overline{w}_A\,, \overline{u}_{By} = \tfrac{\overline{w}_B}{\gamma}\,. \end{aligned}\right\} \quad \begin{array}{l} \text{Velocity components} \\ \text{after the collision} \end{array} \tag{1533}$$

We remark that Eqs. (1531) and (1533), obtained on the basis of the REICHENBACH transformation are identical with the relations (152), and (155), derived in Chap. 6, Sect. 3 from the LORENTZ transformation. In system Σ_o, there holds the same momentum conservation. The further considerations can therefore be repeated verbatim to Chap. 6, Sect. 3. We get again the result (167),

$$\Sigma_o:\ m = \frac{m_o}{\sqrt{1-u^2/c^2}}\,. \quad \begin{array}{l} \text{REICHENBACH transformation} \\ \text{mass formula in } \Sigma_o \end{array} \tag{1534}$$

In distinction to the same Eq. (167), this equation holds only in system Σ_o. In this respect the result is almost trivial: Since m and m_o are both measured in system Σ_o for the moving and resting mass, resp., it would be strange if the result would depend on the way how the clocks are synchronised in the system Σ' being in motion with respect to Σ_o. On the other hand, we need this synchronisation in TOLMAN's thought experiment for deriving Eq. (1534), that depends via the addition theorem of velocities on this synchronisation. Without doubt, the formula (1534) is not applicable when we ask for the mass dependence on the velocity on the basis of the REICHENBACH transformation in systems Σ'. The equations of mechanics then lead to much more complex relations, as can be seen by direct substitution of the REICHENBACH transformation into the momentum balance.

Problem 46 *Consider a telescope at rest in system Σ' that is in motion along the negative x-direction of Σ_o with a velocity of amount v, Fig. A.14. Show in direct way that the run-time differences of light waves of the pathes ACO' and BO' coincide as seen from system Σ_o, when the telescope has a tilt angle α with* $\tan\alpha = v/c$ *against the y-axis in system Σ'.*

Chapter 8, Section 6

Since the telescope is in motion in negative x-direction with velocity v with respect to Σ_o, the right boundary ray must travel in horizontal direction a distance $a + v\,t$ and in vertical direction a distance b for getting from B to the point O', hence together with the PYTHAGORAS rule a distance $\sqrt{(a+v\,t)^2+b^2}$. Here, the time t is the run time for propagating this distance with the speed of light c in Σ_o,

$$ct = \sqrt{(a+v\,t)^2+b^2}\ . \tag{1535}$$

In order that both boundary rays overlay at O', also the left ray, in Σ_o must propagate with the speed of light from point A until the arrival at O' the same distance $c\,t$. The first part of the path until point C at the telescope has the length $2b$. The telescope approaches the left ray in Σ_o with the velocity v. Until the arrival at O', the left boundary ray must travel in vertical direction again the distance b, hence in horizontal direction the distance $a - v\,\Delta t$, hence altogether after the rule of PYTHAGORAS the distance $\sqrt{(a-v\,\Delta t)^2+b^2}$. Here Δt is not anymore the total run time t, but the time after subtraction of the time needed for the distance $2b$, hence $\Delta t = t - 2b/c$.

For the left boundary ray, we get

$$c\,t = 2b + \sqrt{\left(a - v\left(t - \frac{2b}{c}\right)\right)^2 + b^2}\ . \tag{1536}$$

For proving the equality of both run times (1535) and (1536), when the telescope in Σ_o is tilted by an angle α with $\tan\alpha = v/c$ against the y-axis, we introduce the following abbreviations:

$$\left.\begin{aligned} &u = ct\ , \qquad \beta = \frac{v}{c}\ , \qquad 1-\beta^2 = \frac{1}{\gamma^2}\ , \\ &a = \frac{a'}{\gamma} = \frac{L\cos\alpha'}{\gamma}\ ,\ b = b' = L\sin\alpha'\ . \end{aligned}\right\} \tag{1537}$$

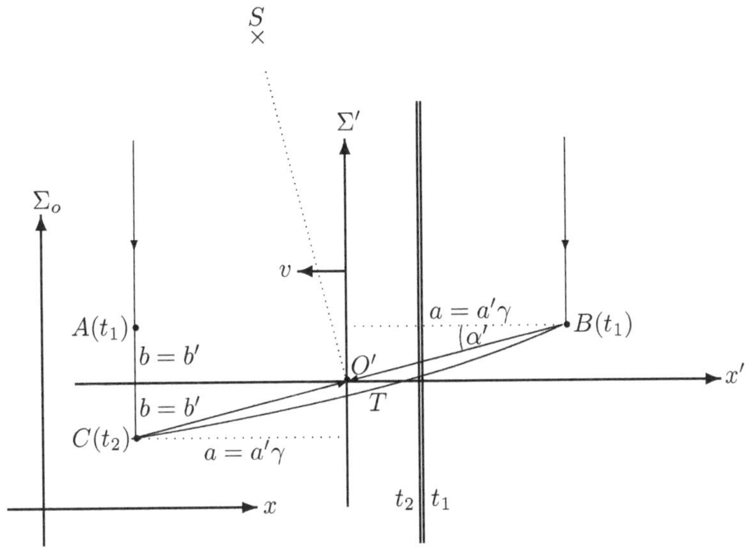

Fig. A.14 The telescope is at rest in Σ'. It has an opening diameter of $2L$ with projections $a' = L\cos\alpha'$ and $b' = L\sin\alpha'$ for the x'- and the y'-axis, so that $\tan\alpha' = b'/a'$. In Σ_o we have the lengths $a = a'/\gamma$ and $b = b'$ because of the LORENTZ-contraction in direction of motion. For the aberration angle α in Σ_o we find $\tan\alpha = v/c$. When the right boundary ray arrives at B, the left boundary ray is still at A. It should be remarked that in the figure we consider two different times t_1 and t_2.

With Eq. (1537) we get from Eq. (1535)

$$u = \sqrt{(\frac{L}{\gamma}\cos\alpha' + \beta u)^2 + L^2\sin^2\alpha'}\ . \tag{1538}$$

Taking the square of Eq. (1538) we get a quadratic equation for u, and we find after an easy evaluation the zero points

$$u_{1,2} = \gamma L(\beta\cos\alpha' \pm 1)\ . \tag{1539}$$

With Eq. (1537) we get from Eq. (1536)

$$u - 2L\sin\alpha' = \sqrt{\left(\frac{L}{\gamma}\cos\alpha' - \beta(u - 2L\sin\alpha')\right)^2 + L^2\sin^2\alpha'}\ . \tag{1540}$$

This leads to a quadratic equation in u

$$\left.\begin{aligned} &u^2/\gamma^2 - u2L(2\sin\alpha'/\gamma^2 - \tfrac{\beta}{\gamma}\cos\alpha') \\ &+ \tfrac{4L^2}{\gamma^2}\sin^2\alpha' - L^2(\tfrac{\cos^2\alpha'}{\gamma^2} + \sin^2\alpha') - 4L^2\tfrac{\beta}{\gamma}\cos\alpha'\sin\alpha' = 0\ . \end{aligned}\right\} \tag{1541}$$

The solution is

$$u_{1,2} = \gamma L\Big[\big(\frac{2}{\gamma}\sin\alpha' - \beta\cos\alpha'\big) \pm\sqrt{\Big(\frac{2}{\gamma}\sin\alpha' - \beta\cos\alpha'\Big)^2 - \frac{4}{\gamma^2}\sin^2\alpha' + \frac{1}{\gamma^2}\cos\alpha' + \sin^2\alpha' + 4\frac{\beta}{\gamma}\sin\alpha'\cos\alpha'}\ \Big] \tag{1542}$$

Evaluation of the terms under the square root results in the value 1, hence

$$u_{1,2} = \gamma L\left(\frac{2}{\gamma}\sin\alpha' - \beta\cos\alpha' \pm 1\right) . \tag{1543}$$

The run times following from Eqs. (1539) and (1543) for the left and right boundary ray coincide exactly when

$$\frac{1}{\gamma}\sin\alpha' = \beta\cos\alpha' . \tag{1544}$$

Therefore it holds the known Eqs. (254) or (255), for the aberration angle,

$$\Sigma' :\ \tan\alpha' = \beta\gamma \quad\longrightarrow\quad \sin\alpha' = \frac{v}{c} . \tag{1545}$$

Expressed with the in system Σ_o measured lengths a and b, i.e. for length in motion, we can write with Eq. (1537) in agreement with Eq. (255),

$$\Sigma_o :\ \tan\alpha = \frac{\gamma^2 b}{a} = \frac{v}{c} . \tag{1546}$$

Problem 47 *Solve the in Chap. 12, Sect. 2 formulated paradox.*

Chapter 12, Sections 1, 2

In a crystal, stable localised elongations from straight dislocation lines fulfil the so-called sine-GORDON equation (996), Certain solutions of this equation, the so-called 'breather', define stable oscillating states with a period determined by physical constants of the lattice.

We consider at first a breather oscillating at a fixed position $x = b$ in the crystal, Eq. (1005). As each uniformly oscillating system, we can consider the localised breather as normal clock U_o^b, that rests in the inertial system Σ_o. According to Eq. (1003), this clock has a period T_o

$$T_o = \frac{\sqrt{2}\,2\pi\,\lambda}{c_o} . \qquad \text{Period of the in } \Sigma_o \text{ resting clock } U_o^b \tag{1547}$$

Here $\lambda = a\,\sigma/(2\pi D)$, a denotes the lattice constant, σ the line tension of dislocations that can be determined by energy measurements, and D the restoring force between lattice atoms. Furthermore c_o is the limiting velocity of the propagation of disturbations along the dislocation line

$$c_o = \sqrt{\frac{\sigma}{\varrho_o}} \,, \qquad \text{Velocity limit} \qquad (1548)$$

cp. (994). The time t of clock U_o^b at position $x = b$ counts the number of oscillation periods T_o.

There are localised breathers, moving with constant velocity u as given in solutions (1010), that can be used for construction of clocks U_u^*, which have a period T_u^*,

$$T_u^* = \frac{T_o}{\sqrt{1 - u^2/c_o^2}} \,. \qquad \text{Period of the with velocity } u \text{ propagating clock } U_u^* \qquad (1549)$$

The time of clock U_u^* is denoted by t^*.

The clock U_u^{*o} sits at $x = 0$, and the pointer should show $t^* = 0$ when the pointer of the resting clock U_o^o, shows $t = 0$. This determines the initial conditions.

When we compare moving and resting clocks in Special Relativity Theory, we always use the property that pointer positions are inverse to the periods. The periods of the moving and resting breather are accordingly related with Eq. (1549) by the ratio $T_u^*/T_o = 1/\sqrt{1 - u^2/c_o^2}$. We conclude, that the pointer at the position t^* of the clock U_u^*, is related in the inverse ratio, when it passed by the clock at rest U_o^b at pointer position t. Hence after the equation for the time dilatation, it remains behind. Hence the time t^* measured on clock U_u^* at $b = v\,t$, is related to the time t of clock U_o^b by (cp. also Eq. (1013))

$$\frac{t^*}{t} = \sqrt{1 - u^2/c_o^2} \,. \qquad \text{'Time dilatation' with the limiting velocity } c_o \qquad (1550)$$

The velocity c_o is of the order of magnitude of the sound velocity, hence much smaller than the speed of light c. Due to $c_o \ll c$, the effect of 'time dilatation' as described by Eq. (1550) should be much larger than the effect in EINSTEIN's formula (1020).

The Special Relativity Theory stands and falls with EINSTEIN's Eq. (1020), of the time dilatation. Did we show, that EINSTEIN's theory does not hold true everywhere, possibly being even wrong?

In no way! - The key statement in the description of the time dilatation of the main text reads: 'We construct a number of absolutely identical clocks'. The breather clocks in inertial system Σ_o represent oscillating dislocation lines, whose geometrical centre rests with respect to the crystal. The moving breather clock, that leads to the time dilatation (1550), represents an oscillating line with centre having a velocity u with respect to the crystal. For the experimenter of special relativity theory, who will use crystals for construction of his clocks, the oscillating line forms the clock together with the whole crystal, and not only the oscillating line, as could be imaged by an 'internal observer' in the crystal as considered in Chap. 12, Sect. 1. For construction of a crystal clock, it would be necessary that the centre of the oscillating line remains at rest with respect to the crystal. A breather clock in motion with velocity v requires a motion together with the crystal i.e. an oscillating breather line that remains at rest in the crystal. When the oscillating line is in motion with respect to the crystal, then this is also a clock, but on the other hand a quite different one. We shall discuss this in the following in detail.

As each other precise oscillating system, we can at first use our crystal clock, more precisely our oscillating breather, which rests in the crystal, as standard clock U_o. The crystal might be at rest in the inertial system Σ_o. The time of clocks U_o^b at position $x = b$ is called t as usually.

When we move a crystal clock, hence a crystal with a breather clock resting in it, with a velocity v with respect to Σ_o, we have a moving clock U_v^o, resting in system Σ'.

We call the pointer position of U_v^o by t' and compare this clock with two in Σ_o resting crystal clocks, where we take the normal initial condition then it holds the EINSTEIN relation (1020),

$$t' = t\sqrt{1 - v^2/c^2}\,. \qquad (1551)$$

Now we consider a breather, moving with a velocity u in the crystal, that is at rest in system Σ_o. This is also a precise oscillating system that can be used as standard clock in system Σ_o. When we set the crystal at position b in Σ_o, we call this clock U_u^{*b}. The oscillation period of this clock is called T_u^*.

The duration of an oscillation period determines the gauging of the clock. The oscillation period of a quartz clock is smaller by many powers of ten than that of a clock with a balance spring. Both show the same time of 1 s, but in the quartz clock one has to count many more oscillations, till the second pointer proceeds by one unit.

Equation (1549) is therefore only a prescription for gauging the clocks U_o^b and U_u^{*b}![11]

When n oscillations of period T_o in the clock U_o^b correspond to one second, then we should gauge clock U_u^{*b} in such a way, that the pointer for seconds after n^* oscillations proceeds by one unit with

$$n^* = n\sqrt{1 - u^2/c_o^2}\,. \qquad \text{Gauge prescription for crystal clocks } U_u^* \text{ and } U_o \qquad (1552)$$

Also for these clocks, there holds without restriction the equation of EINSTEIN's time dilatation, Eq. (1551). We compare crystal breather clocks U_u^*, measuring time t, where the crystal is at rest in Σ_o, while a breather has a velocity u, in it, with breather clocks $U_u^{*\prime}$, measuring time t', that are moving with the velocity u, however in another crystal, being at rest in a system Σ', but in motion with velocity v, with respect to Σ_o.

We have overhasty concluded a revision of EINSTEIN's time dilatation by Eq. (1550). From Eq. (1549) it only follows the gauge prescription (1552) and no new version of time dilatation. There is no correction of EINSTEIN's formula. The paradox is solved.

[11] In Chap. 12, Sect. 1, we have provisionally considered a lattice as a model for the physical vacuum. The formulas used there should not be mixed with our considerations here in this problem.

Problem 48 *Determine the angular momentum produced by the orbital motion of a distribution of electrically charged masses m_A in a magnetic field* **B**. *There we assume a constant ratio $e/m = e_A/m_A$, and furthermore that the separations of the particles of its centre of mass remain finite.*

Chapter 9, Section 3, Chapter 10, Section 5

At first, we have to collect some facts from classical mechanics:

• For a system of N masses $m_A,\ A = 1, 2, \ldots, N$ at positions $\mathbf{x}_A$, we denote with $M = \sum_{A=1}^{N} m_A$ the total mass and with $\mathbf{X} = \sum_{A=1}^{N} m_A\ \mathbf{x}_A \Big/ M$ the centre of mass.

• The total momentum is then given by $\mathbf{P} = \sum_{A=1}^{N} m_A\, d\,\mathbf{x_A}/dt = M\, d\,\mathbf{X}/dt$.

• With the coordinates for the centre of mass $\mathbf{X}$, one writes for the positions of the masses

$\mathbf{x}_A := \mathbf{X} + \mathbf{x}^r_A$ with the relative positions $\mathbf{x}^r_A$ with respect to the centre of mass.

• To the momenta $\mathbf{p}_A = m_A\, d\,\mathbf{x}_A/dt$, there belongs the *angular momentum* $\mathbf{L}_A = m_A\ \mathbf{x}_A \times d\,\mathbf{x}_A/dt$.

By definition of the centre of mass, it holds for all times $\sum_{A=1}^{N} m_A\, \mathbf{x}^r_A = 0$, one finds the total angular momentum

$$\begin{aligned}
\mathbf{L} &:= \sum_{A=1}^{N} \left(m_A\ \mathbf{x}_A \times \frac{d\,\mathbf{x}_A}{dt} \right) \\
&= \sum_{A=1}^{N} \left(m_A\, (\mathbf{X} + \mathbf{x}^r_A) \times \frac{d}{dt}\, (\mathbf{X} + \mathbf{x}^r_A) \right) \\
&= \sum_{A=1}^{N} \left(m_A\, \mathbf{X} \times \frac{d\,\mathbf{X}}{dt} \right) + \sum_{A=1}^{N} \left(m_A\, \mathbf{X} \times \frac{d\,\mathbf{x}^r_A}{dt} \right) \\
&\quad + \sum_{A=1}^{N} \left(m_A\ \mathbf{x}^r_A \times \frac{d\,\mathbf{X}}{dt} \right) + \sum_{A=1}^{N} \left(m_A\ \mathbf{x}^r_A \times \frac{d\,\mathbf{x}^r_A}{dt} \right).
\end{aligned}$$

Here there vanishes the second and third term after the last equality sign because of the definition of the centre of mass.

The first term is the outer angular momentum that depends by definition on the position of the coordinate origin. Here we are interested in the *internal orbital momentum of the system* $\mathbf{L}^I$, given by the last term,

$$\mathbf{L}^I = \sum_{A=1}^{N} \left(m_A\ \mathbf{x}^r_A \times \frac{d\,\mathbf{x}^r_A}{dt} \right). \tag{1553}$$

The quantities with index I denote 'internal quantities', they have nothing to do with the internal characteristics as the spin. Here we are only concerned with the orbital motion.

When there act forces on the charged masses of the system, so there follows an internal torque $\mathbf{D}^I$

$$\mathbf{D}^I = \frac{d}{dt}\mathbf{L}^I = \frac{d}{dt}\sum_{A=1}^{N}\left(m_A\,\mathbf{x}_A^r \times \frac{d\,\mathbf{x}_A^r}{dt}\right)$$
$$= \sum_{A=1}^{N}\left(m_A\,\frac{d\,\mathbf{x}_A^r}{dt} \times \frac{d\,\mathbf{x}_A^r}{dt}\right) + \sum_{A=1}^{N}\left(\mathbf{x}_A^r \times \frac{d}{dt} m_A\,\frac{d\,\mathbf{x}_A^r}{dt}\right).$$

The first term on the right-hand side vanishes; hence, it remains

$$\mathbf{D}^I = \sum_{A=1}^{N}\left(\mathbf{x}_A^r \times \frac{d}{dt}\frac{d\,(m_A\,\mathbf{x}_A^r)}{dt}\right). \tag{1554}$$

Due to Eq. (473), a charge e_A, moving with velocity $d\,\mathbf{x}_A/dt$ in a magnetic field $\mathbf{B}$ feels the LORENTZ force $\mathbf{F} = e_A\,d\,\mathbf{x}_A/dt \times \mathbf{B}$. We insert this in the second factor of the cross product and get

$$\mathbf{D}^I = \sum_{A=1}^{N}\left(\mathbf{x}_A^r \times e_A\,\frac{d\,\mathbf{x}_A^r}{dt} \times \mathbf{B}\right). \tag{1555}$$

We use the supposed constancy of the charge–mass ratio, $e_A/m_A = e/m$, to get

$$\mathbf{D}^I = \frac{e}{m}\sum_{A=1}^{N} m_A\left(\mathbf{x}_A^r \times \left(\frac{d\,\mathbf{x}_A^r}{dt} \times \mathbf{B}\right)\right). \tag{1556}$$

For simplification, we take the derivative

$$\frac{d}{dt}\left(\sum_{A=1}^{N} m_A\,\mathbf{x}_A^r \times \left(\mathbf{x}_A^r \times \mathbf{B}\right)\right),$$

and we suppose a motion within a finite range of space. Such a sum will then have a vanishing derivative in the temporal average, since otherwise the relative coordinates would grow without limit. The temporal average is now indicated by an overbar on the expressions. We find

$$\left.\begin{aligned}\overline{\frac{d}{dt}\left(\sum_{A=1}^{N} m_A\,\mathbf{x}_A^r \times \left(\mathbf{x}_A^r \times \mathbf{B}\right)\right)} &= \overline{\sum_{A=1}^{N} m_A\frac{d\,\mathbf{x}_A^r}{dt} \times \left(\mathbf{x}_A^r \times \mathbf{B}\right)} \\ &\quad \overline{+\sum_{A=1}^{N} m_A\,\mathbf{x}_A^r \times \left(\frac{d\,\mathbf{x}_A^r}{dt} \times \mathbf{B}\right)} \\ &= 0\,.\end{aligned}\right\} \tag{1557}$$

Now we go over to the temporal average in Eq. (1556). We can write with Eq. (1557),

$$\overline{\mathbf{D}^I} = \left. \begin{aligned} &\overline{\frac{e}{2\,m}\sum_{A=1}^{N} m_A \left(\mathbf{x}_A^r \times \left(\frac{d\,\mathbf{x}_A^r}{dt} \times \mathbf{B} \right) - \frac{d\,\mathbf{x}_A^r}{dt} \times \left(\mathbf{x}_A^r \times \mathbf{B} \right) \right)} \\ &+ \overline{\frac{e}{2\,m}\sum_{A=1}^{N} m_A \frac{d\,\mathbf{x}_A^r}{dt} \times \left(\mathbf{x}_A^r \times \mathbf{B} \right)} \\ &+ \overline{\frac{e}{2\,m}\sum_{A=1}^{N} m_A\, \mathbf{x}_A^r \times \left(\frac{d\,\mathbf{x}_A^r}{dt} \times \mathbf{B} \right)} \ , \end{aligned} \right\}$$

hence

$$\overline{\mathbf{D}^I} = \overline{\frac{e}{2\,m}\sum_{A=1}^{N} m_A \left(\mathbf{x}_A^r \times \left(\frac{d\,\mathbf{x}_A^r}{dt} \times \mathbf{B} \right) - \frac{d\,\mathbf{x}_A^r}{dt} \times \left(\mathbf{x}_A^r \times \mathbf{B} \right) \right)} \ . \tag{1558}$$

With the identity $\mathbf{a} \times (\mathbf{b} \times \mathbf{c}) = \mathbf{b}\,(\mathbf{a} \times \mathbf{c}) - \mathbf{c}\,(\mathbf{a} \times \mathbf{b})$, one verifies that

$$\overline{\mathbf{D}^I} = \overline{\frac{e}{2\,m}\sum_{A=1}^{N} m_A \left(\mathbf{x}_A^r \times \frac{d\,\mathbf{x}_A^r}{dt} \right) \times \mathbf{B}} \ . \tag{1559}$$

Here the sum is nothing else than the internal angular momentum $\mathbf{L}^I$ of the system, cp. Eq. (1553). Then in omitting the overbar, it holds

$$\mathbf{D}^I = \frac{e}{2\,m}\,\mathbf{L} \times \mathbf{B} \ . \tag{1560}$$

Problem 49 *Determine the contravariant metric g^{ik} as inverse to the metric g_{ik}.*

Chapter 13, Section 1

For simplifying the notation, we consider here as example an Euclidean space in general coordinates. We take the metric g_{ab} of a three-dimensional space, with indices a, b running from 1 to 3. With the unit tensor $\boldsymbol{\delta}$ analogous to the metric $\boldsymbol{\eta}$ of the Minkowski space, the inverse metric g^{ab} is defined by

$$g_{ac}g^{cb} = \delta_i^k \quad \text{mit} \quad \delta_b^a = \begin{pmatrix} 1 & 0 & 0 \\ 0 & 1 & 0 \\ 0 & 0 & 1 \end{pmatrix} \ . \tag{1561}$$

Then it holds after the decomposition law for determinants

$$g\delta_a^b = g_{ac}G^{cb} \quad \text{with} \quad g := \det g_{ab} \ , \tag{1562}$$

where the vertical bars on a square matrix denote its determinant, and the elements G^{ab} are the adjunct elements to g_{ab},

$$\left.\begin{aligned} g_{ik} &= \begin{pmatrix} g_{11} & g_{12} & g_{13} \\ g_{21} & g_{22} & g_{23} \\ g_{31} & g_{32} & g_{33} \end{pmatrix} , \\ G^{11} &= \begin{vmatrix} g_{22} & g_{23} \\ g_{32} & g_{33} \end{vmatrix} , \quad G^{12} = \begin{vmatrix} g_{21} & g_{23} \\ g_{31} & g_{33} \end{vmatrix} , \; \ldots\ldots , \quad G^{33} = \begin{vmatrix} g_{11} & g_{12} \\ g_{21} & g_{22} \end{vmatrix} . \end{aligned}\right\} \tag{1563}$$

Hence it holds

$$g^{ab} = \frac{G^{ab}}{g} . \tag{1564}$$

The metric g_{ab} determines the separation ds between two neighbouring points x^a and $x^a + dx^a$.

As an example, we write down the metric in case of spherical coordinates r, θ, ϕ defined in terms of the Euclidean coordinates x, y, z,

$$\left.\begin{aligned} x &= r \sin\theta \cos\phi , \\ y &= r \sin\theta \sin\phi , \\ z &= r \cos\theta . \end{aligned}\right\} \tag{1565}$$

This leads to

$$ds^2 = dr^2 + r^2 d\theta^2 + r^2 \sin^2\theta \, d\phi^2 . \tag{1566}$$

The metric g_{ab} then reads

$$g_{ab} = \begin{pmatrix} 1 & 0 & 0 \\ 0 & r^2 & 0 \\ 0 & 0 & r^2 \sin^2\theta \end{pmatrix} \quad \longrightarrow \quad g^{ab} = \begin{pmatrix} 1 & 0 & 0 \\ 0 & \frac{1}{r^2} & 0 \\ 0 & 0 & \frac{1}{r^2 \sin^2\theta} \end{pmatrix} . \tag{1567}$$

Here we have the case of curvilinear coordinates in a plane Euclidean space. The curvature tensor vanishes. There is no gravitational field. Since the spherical coordinates are orthogonal, the non-diagonal components of the metric are zero.

Problem 50 *Form the covariant derivative of a vector V^i in arbitrary coordinates and derive the equation of geodesics.*

Chapter 13, Section 1

When we overly our space with arbitrary coordinates x^i, so we define coordinate lines, along which there changes only the value of one coordinate. For Cartesian coordinate in an Euclidean space, these lines are the mutual orthogonal x-, y- and z-axes. The geometrical meaning of the coordinates is fixed, when the separation ds between neighbouring points is determined by a metric g_{ab} in three dimensions, or by the metric g_{ik} in higher dimensions, i, k running from 1 to n. Here we are mainly concerned with the case $n = 4$.

$$ds^2 = g_{ik}dx^i dx^k \ , \tag{1568}$$

as in the case of spherical coordinates (1566). With the vector separation between both points $d\mathbf{x} = \mathbf{n}_a d\mathbf{x}^a$ it holds

$$ds^2 = d\mathbf{x} \cdot d\mathbf{x} = \mathbf{n}_i \cdot \mathbf{n}_k dx^i dx^k = g_{ik}dx^i dx^k \ , \tag{1569}$$

hence the metric represents the scalar product of the basis vectors,

$$\mathbf{n}_i \cdot \mathbf{n}_k = g_{ik} \ . \tag{1570}$$

The in this way introduced basis vectors point in the direction of the coordinate lines, they have lengths

$$\left.\begin{array}{l} |\mathbf{n}_1| = \sqrt{g_{11}} \ , \quad |\mathbf{n}_2| = \sqrt{g_{22}} \ , \quad |\mathbf{n}_3| = \sqrt{g_{33}} \ , \quad |\mathbf{n}_4| = \sqrt{g_{44}} \ , \\ \mathbf{n}_1 \cdot \mathbf{n}_2 = |\mathbf{n}_1||\mathbf{n}_2| \cos\left(\mathbf{n}_1 \cdot \mathbf{n}_2\right) \text{ , and so on.} \end{array}\right\} \tag{1571}$$

A change of coordinates, a coordinate transformation, means $x^i \longrightarrow x^{i'}$ accompanied by a change of the base vectors $\mathbf{n}_i \longrightarrow \mathbf{n}_{i'}$, leading to

$$d\mathbf{x} = \mathbf{n}_i dx^i = \mathbf{n}_{i'} dx^{i'} = \mathbf{n}_i \frac{\partial x^i}{\partial x^{i'}} dx^{i'} \tag{1572}$$

and therefore

$$\mathbf{n}_{i'} = \frac{\partial x^i}{\partial x^{i'}} \mathbf{n}_i \quad \longleftrightarrow \quad \mathbf{n}_i = \frac{\partial x^{i'}}{\partial x^i} \mathbf{n}_{i'} \ . \tag{1573}$$

With decomposition of vectors $\mathbf{V}$ into the basis vectors

$$\mathbf{V} = V^i \mathbf{n}_i = V^{i'} \mathbf{n}_{i'} = V^i \partial x^{i'} \partial x^i \mathbf{n}_{i'} = V^{i'} \frac{\partial x^i}{\partial x^{i'}} \mathbf{n}_i \tag{1574}$$

it follows the transformation law of vectors V^i for coordinate transformations,

$$V^{i'} = V^i \frac{\partial x^{i'}}{\partial x^i} \quad \longleftrightarrow \quad V^i = V^{i'} \frac{\partial x^i}{\partial x^{i'}} \ . \tag{1575}$$

The relation

$$V_i = g_{ik} V^k \tag{1576}$$

defines the covariant components V_i belonging to the contravariant components V^i of one and the same vector $\mathbf{V}$. Now it follows the equivalence

$$V_{i'} = V_i \frac{\partial x^i}{\partial x^{i'}} \quad \longleftrightarrow \quad V_i = V_{i'} \frac{\partial x^{i'}}{\partial x^i} \ . \tag{1577}$$

For tensors with more indices, the transformation follows in applying the same formula for each index separately. For a tensor T^{ik} and the metric g_{ik}, it follows then

$$\left.\begin{aligned} T^{i'k'} &= T^{ik}\frac{\partial x^{i'}}{\partial x^{i}}\frac{\partial x^{k'}}{\partial x^{k}}\,, \\ g_{i'k'} &= g_{ik}\frac{\partial x^{i}}{\partial x^{i'}}\frac{\partial x^{k}}{\partial x^{k'}}\,, \end{aligned}\right\} \tag{1578}$$

When we consider only linear coordinate transformations, i.e. transformations with constant coefficients $A^{i'}_{k}$,

$$x^{i'} = A^{i'}_{k}x^{k} \quad\longleftrightarrow\quad x^{k} = A^{k}_{i'}x^{i'}\,, \text{ hence } \frac{\partial x^{i'}}{\partial x^{i}} = A^{i'}_{i} \text{ with } A^{i'}_{k}A^{k}_{j'} = \delta^{i'}_{j'}\,, \tag{1579}$$

then it follows from the partial derivatives of Eq. (1577)

$$\frac{\partial V_{i'}}{\partial x^{k'}} := V_{i',k'} = \frac{\partial}{\partial x^{k'}}\left(V_{i}A^{i}_{i'}\right) = \frac{\partial V_{i}}{\partial x^{k}}\frac{\partial x^{k}}{\partial x^{k'}}A^{i}_{i'} = V_{i,k}A^{i}_{i'}A^{k}_{k'}\,, \quad \text{hence}$$

$$V_{i',k'} = V_{i,k}A^{i}_{i'}A^{k}_{k'} = V_{i',k'}\frac{\partial x^{i}}{\partial x^{i'}}\frac{\partial x^{k}}{\partial x^{k'}}\,. \tag{1580}$$

For linear coordinate transformations, the partial derivatives of a vector (a tensor of first rank) form a tensor of second rank. For a general coordinate transformation, we have to compare the components V^{i} of a vector field at neighbouring points, with changes both of the vector and also the basis vectors $\mathbf{n_i}$ of the coordinate system. We write

$$\left.\begin{aligned} \mathbf{V}(\mathbf{x}) &= V^{i}(\mathbf{x})\mathbf{n_i}(\mathbf{x})\,, \\ d\mathbf{V}(\mathbf{x}) &= \left(\frac{\partial V^{i}}{\partial x^{k}}\mathbf{n}_{i} + V^{i}\frac{\partial \mathbf{n}_{i}}{\partial x^{k}}\right)dx^{k}\,. \end{aligned}\right\} \tag{1581}$$

Since the basis vectors determine the metric g_{ik}, cp. Eq. (1570), $\mathbf{n}_{i}\cdot\mathbf{n}_{k} = g_{ik}$, we can evaluate its change from the metric. We write

$$\frac{\partial \mathbf{n}_{i}}{\partial x^{k}} = \Gamma^{l}_{ki}\mathbf{n}_{l}\,. \tag{1582}$$

For evaluation of the coefficients Γ^{l}_{ki}, called CHRISTOFFEL symbols or affinities, we form the first partial derivatives of Eq. (1570) for x^{k} and insert Eq. (1582) into (1570), then we get

$$\Gamma^{l}_{ki}\,g_{lm} + \Gamma^{l}_{km}\,g_{li} = g_{im,k}\,. \tag{1583}$$

We are looking for solutions with symmetric coefficients $\Gamma^{l}_{im} = \Gamma^{l}_{mi}$ (then the space has no torsion, see, e.g. RASCHEWSKI, 1995). For a special application to mechanics, cp. Günther (1981)). One verifies easily that the relation

$$\Gamma^{l}_{im}\, g_{lk} = \frac{1}{2}(g_{mk,i} + g_{ki,m} - g_{im,k}) \tag{1584}$$

fulfils Eq. (1583), and the CHRISTOFFEL symbols are given by

$$\Gamma^{i}_{kl} = \frac{1}{2}\, g^{im}\, \left(g_{km,l} + g_{ml,k} - g_{kl,m}\right) \ . \tag{1585}$$

For Eq. (1581) we can now write

$$d\mathbf{V}(\mathbf{x}) = \left(\frac{\partial V^i}{\partial x^k} + \Gamma^{i}_{kl} V^l\right) \mathbf{n}_i dx^k \ . \tag{1586}$$

We can interpret this result as follows: When a vector V^i at point P is parallel transported by dx^k to point Q, hence $d\mathbf{V} = 0$, then there it has components $V^i + dV^i$ with

$$dV^i = -\Gamma^{i}_{kl} V^l dx^k \ . \qquad \text{Vector } V^i \text{parallel transported} \tag{1587}$$

When the vector dx^i is parallel transported in its own direction, hence we replace in Eq. (1587) dx^i by V^i, then we get the differential equation for auto-parallel curves, the equation for *geodesics*. In Euclidean space, these are straight lines. When one requires

$$dx^i + \Gamma^{i}_{cl} dx^k dx^l = 0$$

for a curve $x^i = x^i(\tau)$ with parameter τ, then the equation for geodesics reads

$$\frac{d^2x^i}{d\tau^2} + \Gamma^{i}_{kl} \frac{dx^k}{d\tau}\frac{dx^l}{d\tau} = 0 \ . \qquad \text{Geodesic equations} \tag{1588}$$

The large circles on the surface of a globe are geodesics. The shortest way between two points on the globe are always segments of such large circles.

When we compare two vectors $V^i(x^k)$ at $P(x^k)$ and $V^i(x^k + dx^k)$ at $Q(x^k + dx^k)$, we have at first parallel transport $V^i(x^k)$ to Q, and we get from Eq. (1587) the vector $V^i + dV^i = V^i - \Gamma^{i}_{cl} V^k dx^l$. When we form the difference

$$V^i(x^k + dx^k) - \left(V^i(x^k) + dV^i\right) = V^i(x^k + dx^k) - \left(V^i(x^k) - \Gamma^{i}_{cl} V^k dx^l\right) \ ,$$

hence

$$DV^i \equiv V^i(x^r) + \frac{\partial V^i}{\partial x^k} dx^l - \left(V^i - \Gamma^{i}_{cl} V^k dx^l\right) \ . \tag{1589}$$

In this way, we define the covariant derivative DV^i/Dx^k that generalises the partial derivative in Cartesian coordinates,

$$\frac{DV^i}{Dx^k} = \frac{\partial V^i}{\partial x^k} + \Gamma^i_{cl} V^l \,. \tag{1590}$$

For denoting the covariant derivative, one often simply replaces the comma for the partial derivative by a semicolon. The covariant differentiation of a tensor of first rank, of a vector, produces a tensor of second rank. We show the following examples:

$$\left.\begin{aligned}
\frac{DV^i}{Dx^k} &\equiv V^i_{;\,k} &&= \frac{\partial V^i}{\partial x^k} + \Gamma^i_{cl} V^l \,,\\
\frac{DV_i}{Dx^k} &\equiv V_{i;\,k} &&= \frac{\partial V_i}{\partial x^k} - \Gamma^l_{ik} V_l \,,\\
\frac{Dg_{ik}}{Dx^l} &\equiv g_{ik\ l} &&= g_{ik,\,l} - \Gamma^m_{il}\, g_{km} - \Gamma^m_{cl}\, g_{im} = 0,\\
\frac{DV_i}{Dx^k} - \frac{DV_k}{Dx^i} &\equiv V_{i;\,k} - V_{k;\,i} &&= \frac{\partial V_i}{\partial x^k} - \frac{\partial V_k}{\partial x^i} \equiv V_{i,\,k} - V_{k,\,i} \,.
\end{aligned}\right\} \tag{1591}$$

From Eq. (1585) it follows by straightforward evaluation,

$$\Gamma^k_{ik} = \frac{1}{g}\frac{\partial g}{\partial x^i} \,, \tag{1592}$$

we get simple expressions for the expressions of the covariant divergence:

$$\left.\begin{aligned}
\frac{DV^k}{Dx^k} &\equiv V^k;_{k} = \frac{1}{\sqrt{g}}\frac{\partial}{\partial x^k}\left(\sqrt{g}\, A^k\right) \,,\\
\frac{DT^{ik}}{Dx^k} &\equiv T^{ik};_{k} = \frac{1}{\sqrt{g}}\frac{\partial}{\partial x^k}\left(\sqrt{g}\, T^{ik}\right) + \Gamma^i_{cl} T^{kl} \,.
\end{aligned}\right\} \tag{1593}$$

Problem 51 *Establish the connection between the* NEWTONian *equation of motion (1080) and the geodesic equations (1588) following from the field equations of General Relativity Theory (GRT).*

Chapter 13, Section 2

The differential equation for a geodesic line reads

$$\frac{d^2x^i}{d\tau^2} + \Gamma^i_{cl}\frac{dx^k}{d\tau}\frac{dx^l}{d\tau} = 0 \,. \tag{1594}$$

We are interested only in the spatial components $i = a = 1, 2, 3$ and look for expressions in lowest order of power in $1/c^2$. Then we can write $d\tau = dt$, so that from the products $(dx^k/d\tau)\,(dx^l/d\tau)$ only the expression $(dx^0/d\tau)\,(dx^0/d\tau) \longrightarrow c^2$ remains. Equation (1594) is then reduced to

$$\frac{d^2x^a}{dt^2} + \Gamma^a_{00} c^2 = 0 \,. \tag{1595}$$

With Eq. (1078), $\Gamma^a_{00} = g_{00,a}/2$, and Eq. (1065), $g_{00,a} = 2\Phi_{,a}/c^2$, we get

$$\frac{d^2x^a}{dt^2} + \Phi_{,a} = 0\,. \tag{1596}$$

This is Eq. (1080), what we wanted to show. The postulate of the motion of particles on geodesics was set up as an additional principle in the beginning of GRT. But in truth, it is already a consequence of EINSTEIN's field equations. To put it on the point again: The KEPLER orbits of the solar planets are a consequence of the GRT, i.e. these orbits are the zeroth approximation of the geodesics of a relativistic space-time, cp. Chap. 13, Sect. 3.2.

Problem 52 *Evaluate the components of the* EINSTEIN *tensor for a general spherically symmetric metric, Eq. (1081).*

Chapter 13, Section 3
For the diagonal metric, the inverse metric with contravariant components is also a diagonal metric with inverse components. Therefore, the CHRISTOFFEL symbols have only components different from zero, when at minimum two indices are equal. Therefore, there are up to the symmetry in the lower indices only 12 components to evaluate

$$\begin{aligned}
&\Gamma^t_{\ tt} = \partial_t\nu/2\,, \quad \Gamma^t_{\ tr} = \partial_r\nu/2\,, \quad \Gamma^r_{\ tt} = (\partial_r\nu/2)\exp(\nu-\lambda)\,,\\
&\Gamma^t_{\ rr} = (\partial_t\lambda/2)\exp(\lambda-\nu)\,, \quad \Gamma^r_{\ tr} = \partial_t\lambda/2\,, \quad \Gamma^r_{\ rr} = \partial_r\lambda/2\,,\\
&\Gamma^r_{\ \theta\theta} = -r\exp(-\lambda)\,, \quad \Gamma^r_{\ \phi\phi} = -r\sin^2\theta\exp(-\lambda)\,,\\
&\Gamma^\theta_{\ r\theta} = \Gamma^\phi_{\ r\phi} = 1/r\,, \quad \Gamma^\theta_{\ \phi\phi} = -\sin\theta\cos\theta\,, \quad \Gamma^\theta_{\ \theta\phi} = -\cot\theta\,.
\end{aligned} \tag{1597}$$

The following mixed components of the RIEMANN curvature tensor are needed:

$$\begin{aligned}
&R^{t\theta}_{\ \ t\theta} = R^{t\phi}_{\ \ t\phi} = -(\partial_r\nu/2r)\exp(-\lambda)\,,\\
&R^{r\theta}_{\ \ r\theta} = R^{t\phi}_{\ \ t\phi} = -(\partial_r\lambda/2r)\exp(-\lambda)\,,\\
&R^{\theta\phi}_{\ \ \theta\phi} = (1/r^2)(\exp(-\lambda)-1)\,.
\end{aligned} \tag{1598}$$

It follows for the EINSTEIN tensor with the definition of the RICCI tensor and the RICCI scalar

$$\begin{aligned}
&E^t_t = -R^{r\theta}_{\ \ r\theta} - R^{\theta\phi}_{\ \ \theta\phi} - R^{r\phi}_{\ \ r\phi} = (\partial_r\lambda/r - 1/r^2)\exp(-\lambda) + 1/r^2\,,\\
&E^r_r = -R^{t\theta}_{\ \ r\theta} - R^{\theta\phi}_{\ \ \theta\phi} - R^{t\phi}_{\ \ t\phi} = -(\partial_r\nu/r + 1/r^2)\exp(-\lambda) + 1/r^2\,,\\
&E^t_r = R^{t\theta}_{\ \ r\theta} + R^{t\phi}_{\ \ r\phi} = (\partial_t\lambda/r)\exp(-\nu)\,.
\end{aligned} \tag{1599}$$

Obviously, a time derivative appears only in the EINSTEIN tensor with mixed $\binom{t}{r}$-components.

Problem 53 *Evaluate the components of the* EINSTEIN *tensors for the cosmological* ROBERTSON-WALKER *metric, Eq. (1140).*

Chapter 13, Section 5

For the diagonal metric, the inverse metric with contravariant components is also a diagonal metric with inverse components. There are besides the symmetry in the lower indices 12 non-vanishing components

$$
\begin{aligned}
\Gamma^{t}{}_{rr} &= \frac{R\dot{R}}{1-kr^2}\,, \quad \Gamma^{t}{}_{\theta\theta} = R\dot{R}r^2\,, \quad \Gamma^{t}{}_{\phi\phi} = R\dot{R}r^2\sin^2\theta\,,\\
\Gamma^{t}{}_{rr} &= \Gamma^{t}{}_{\theta\theta} = \Gamma^{t}{}_{\phi\phi} = \frac{\dot{R}}{R}\,,\\
\Gamma^{r}{}_{rr} &= -\frac{kr}{1-kr^2}\,, \quad \Gamma^{r}{}_{\theta\theta} = (1-kr^2)r\,, \quad \Gamma^{r}{}_{\phi\phi} = (1-kr^2)r\sin^2\theta\,, \qquad (1600)\\
\Gamma^{\theta}{}_{r\theta} &= \Gamma^{\phi}{}_{r\phi} = \frac{1}{r}\,, \quad \Gamma^{\theta}{}_{\phi\phi} = \sin\theta\cos\theta\,, \quad \Gamma^{\phi}{}_{\theta\phi} = -\mathrm{ctg}\,\theta\,.
\end{aligned}
$$

We need five components of the RIEMANN curvature tensor,

$$
\begin{aligned}
E^t_t &= -R^{r\theta}{}_{r\theta} - R^{\theta\phi}{}_{\theta\phi} - R^{\phi r}{}_{\phi r}\,,\\
E^r_r &= -R^{t\theta}{}_{t\theta} - R^{\theta\phi}{}_{\theta\phi} - R^{\phi t}{}_{\phi t}\,. \qquad (1601)
\end{aligned}
$$

Insertion into the Eq. (1045) provides.[12]

$$
\begin{aligned}
R^{r\theta}{}_{r\theta} &= -\frac{\dot{R}^2}{R^2} - 3\frac{kc^2}{R^2}\,, \quad R^{\theta\phi}{}_{\theta\phi} = -\frac{\dot{R}^2}{R^2} - \frac{kc^2}{R^2}\,, \quad R^{\phi r}{}_{\phi r} = -\frac{\dot{R}^2}{R^2} + \frac{kc^2}{R^2}\,,\\
R^{\theta t}{}_{\theta t} &= R^{\phi t}{}_{\phi t} = -\frac{\ddot{R}}{R}\,. \qquad (1602)
\end{aligned}
$$

With Eq. (1601) there follow the two components of the EINSTEIN tensor in the form of Eq. (1143). Obviously the $\binom{t}{t}$-component of the EINSTEIN tensor contains only first time derivatives, hence this equation forms a first integral of the spatial field equations. It contains the curvature term kc^2/R^2. On the contrary, the spatial components contain the acceleration $\ddot{R}$ of the world radius, but not the spatial curvature.

Problem 54 *Solve the* FRIEDMANN- LEMAÎTRE *equation for dust-like matter and a cosmological constant, described by the density parameters* Ω_m *and* Ω_Λ.

Chapter 13, Section 5

The equation for the scale factor $a(t)$ reads

$$
\frac{\dot{a}^2}{a^2} = H_0^2(\Omega_m a^3 + \Omega_\Lambda)\,. \qquad (1603)
$$

In generalisation of the power law behaviour of solutions for equations without cosmological constant and with $\Omega_m = 1$ (EINSTEIN- DE- SITTER model, Eq. (1149), we use the ansatz $a(t) = [h(t)]^p$, leading to a relation

---

[12]Take care not to mix here the notion of the world radius $R(t)$ with the component of the RIEMANN tensor $R^{r\theta}{}_{r\theta}$ or similar components, and components of the RICCI tensor and RICCI scalar.

$$\dot{h}^2 = \frac{H_0^2}{p^2} \left(\Omega_m h^{2-3p} + \Omega_\Lambda h^2 \right), \tag{1604}$$

hence again suggesting $p = 2/3$. Going further we use the ansatz $h = h_0 \cdot \sinh(t/t_0)$ and find a solution with the addition theorem of hyperbolic functions

$$\frac{h_0^2}{t_0^2} \cosh^2\left(\frac{t}{t_0}\right) = \frac{9H_0\Omega_m}{4} \left[1 + \frac{\Omega_\Lambda}{\Omega_m} v_0^2 \sinh^2\left(\frac{t}{t_0}\right) \right] \tag{1605}$$

with constants $t_0 = 2/(3H_0\sqrt{\Omega_m})$, $v_0 = \sqrt{\Omega_m/\Omega_\Lambda}$ and therefore

$$a = \left(\frac{\Omega_m}{\Omega_\Lambda}\right)^{1/3} \sinh^{2/3}\left(\frac{3H_0\Omega_m^{1/2}}{2} t\right). \tag{1606}$$

Resolving this equation for t we get a simple analytic expression of the age of the universe in the LCDM model.

Appendix B
Mathematical Methods

B.1 Remembering of Tensor Calculus

In the n-dimensional linear space $\mathbb{R}^n$ each point $P(x^i) = P(x^{i'})$ is described by its coordinates x^i or else by $x^{i'}$, where both are connected by coordinate transformations,

$$x^{i'} = x^{i'}(x^i) \quad \longleftrightarrow \quad x^i = x^i(x^{i'}), \quad i,\, i' = 1,\, 2,\, \ldots\, n\,. \tag{1607}$$

The transformation matrix $(\partial x^{i'})/(\partial x^i)$ is called Jacobi *matrix* $J^{i'}_{\ i}$, (a portrait of G. J. JACOBI is available in the supplementary material to this book),

$$J^{i'}_{\ i} = \frac{\partial x^{i'}}{\partial x^i} \qquad \text{Jacobi matrix} \tag{1608}$$

with the determinant

$$J := \left|\partial x^{i'}/\partial x^i\right|\,. \qquad \text{Jacobi determinant} \tag{1609}$$

We denote rows with upper indices and columns with lower ones. The inverse Jacobi matrix, identified by the index with accent below, then reads

$$J^{i}_{\ i'} = \frac{\partial x^i}{\partial x^{i'}}\,, \qquad \text{Inverse Jacobi matrix} \tag{1610}$$

and it holds

$$J^{i}_{\ i'}\, J^{i'}_{\ k} = \frac{\partial x^i}{\partial x^{i'}}\, \frac{\partial x^{i'}}{\partial x^k} = \delta^i_k$$

with the Kronecker symbol δ^i_k as

$$\delta^i_k = \left\{\begin{matrix} 1 & & i = k \\ & \text{for} & \\ 0 & & i \neq k \end{matrix}\right\}\,, \quad \text{e. g. for } n = 3: \quad \delta^i_k = \begin{pmatrix} 1\,0\,0 \\ 0\,1\,0 \\ 0\,0\,1 \end{pmatrix}\,. \tag{1611}$$

H. Günther and V. Müller, *The Special Theory of Relativity*,
https://doi.org/10.1007/978-981-13-7783-9

Here and elsewhere we use EINSTEIN's sum convention: It is to be summed over equal lower and upper indices, $A^i B_i := \sum_{i=1}^{n} A^i B_i$. The same summation index can therefore appear only twice in each term. If the sum convention should not be applied, we set the corresponding indices in parenthesis: $A^{(i)}_{(i)}$ should mean no sum.

For the total differential, it follows from Eq. (1607)

$$dx^{i'} = \frac{\partial x^{i'}}{\partial x^i} dx^i \quad \longleftrightarrow \quad dx^i = \frac{\partial x^i}{\partial x^{i'}} dx^{i'} \ . \tag{1612}$$

For the partial derivative, it holds

$$\frac{\partial}{\partial x^{i'}} = \frac{\partial x^i}{\partial x^{i'}} \frac{\partial}{\partial x^i} \quad \longleftrightarrow \quad \frac{\partial}{\partial x^i} = \frac{\partial x^{i'}}{\partial x^i} \frac{\partial}{\partial x^{i'}} \ . \tag{1613}$$

This defines the tangent space at each point.

The basis vectors $\mathbf{k}_i$ of the tangent space are coupled to the coordinate lines, so that each coordinate transformation (1607) is accompanied by a change of the basis

$$\mathbf{k}_{i'} = \frac{\partial x^i}{\partial x^{i'}} \mathbf{k}_i \quad \longleftrightarrow \quad \mathbf{k}_i = \frac{\partial x^{i'}}{\partial x^i} \mathbf{k}_{i'} \ . \qquad \text{Basis vectors} \tag{1614}$$

The transformations (1612) and (1613) are called *contragredient*. They are provided by the inverse and transpose matrices.

To get $dx^{i'}$, we have to sum over the column index of the JACOBI matrix in Eq. (1612), and the dashed index appears above; for getting $\mathbf{k}_{i'}$, we have to sum in Eq. (1614) over the raw index of the inverse JACOBI matrix, and the dashed index appears below.

The metric tensor g_{ik} is defined by the scalar product of basis vectors $\mathbf{k}_i$ in the local tangent space,

$$g_{ik} := \mathbf{k}_i \cdot \mathbf{k}_k \quad \text{with} \quad g_{ik} = g_{ki} \ . \qquad \text{Metric} \tag{1615}$$

It will soon be obvious that the metric g_{ik} represents a tensor. We write

$$d\mathbf{x} := dx^i \mathbf{k}_i = dx^{i'} \mathbf{k}_{i'} \ . \tag{1616}$$

The vector $\mathbf{k}_i$ points in the direction of the coordinate lines x^i, and its length defines the length unit on this axis.

For two neighbouring points $P(x^i)$ and $Q(x^i + dx^i)$, we define the invariant separation

$$ds^2 := d\mathbf{x} \cdot d\mathbf{x} = dx^i \mathbf{k}_i \cdot dx^k \mathbf{k}_k = g_{ik} dx^i dx^k \ . \qquad \text{Invariant separation} \tag{1617}$$

ds^2 is called *invariant separation* in the space.

Let g^{ik} be the inverse matrix to g_{ik}, hence

$$g_{ir}g^{rk} := \delta_i^k \ . \tag{1618}$$

Then the inverse basis vectors $\mathbf{k}^i$ to the vectors $\mathbf{k}_i$ are defined by

$$\mathbf{k}^i = g^{ir}\mathbf{k}_r \ , \qquad \text{Inverse basis vectors} \tag{1619}$$

and from $\mathbf{k}^i \cdot \mathbf{k}^k = g^{ir}\mathbf{k}_r \cdot g^{ks}\mathbf{k}_s$, it follows with Eqs. (1615) and (1618)

$$g^{ik} := \mathbf{k}^i \cdot \mathbf{k}^k \quad \text{with} \quad g^{ik} = g^{ki} \ . \tag{1620}$$

Each vector $\mathbf{a}$ can then be written as

$$\mathbf{a} = a^i\mathbf{k}_i = a_i\mathbf{k}^i \ . \tag{1621}$$

Here the quantities a^i are called *contravariant components* and a_i *covariant components* of the vector $\mathbf{a}$.

The scalar product of two vectors $\mathbf{a}$ and $\mathbf{b}$ is then defined by

$$\mathbf{a} \cdot \mathbf{b} = a^i b^k g_{ik} = a^i b_i = a_i b^i \ . \qquad \text{Scalar product} \tag{1622}$$

In case of a positive definite metric, i.e. $ds^2 \geq 0$, where the value zero is possible only for coinciding points, the line element is nothing else than the square of the separation of both points P and Q.

In view of the MINKOWSKI space, we allow also indefinite metrics. Then ds^2 can be positive and negative, and we can have a vanishing separation $ds^2 = 0$ for two points, that do not coincide. The diagonal components of g_{ik}, the scalar product of the basis vectors with themselves, $\mathbf{k}_{(i)} \cdot \mathbf{k}_{(i)} = g_{(i)(i)}$, can be positive or negative. In the chosen signature of the space-time metric in this book, s. footnote 2 in Chap. 9, Sect. 1.1, a vector $\mathbf{a}$ with $\mathbf{a} \cdot \mathbf{a} < 0$ is called *spacelike*, one with $\mathbf{a} \cdot \mathbf{a} > 0$ *timelike*, and vectors with $\mathbf{a} \cdot \mathbf{a} = 0$, where not all components of $\mathbf{a}$ vanish, are called *lightlike* or simply *null vectors*. As basis vectors, they are only used in Chap. 13, Sect. 3.4 in discussing black holes in general relativity theory.

Of particular physical interest is the dimension $n = 4$, the MINKOWSKI space with one timelike and three spacelike base vectors. The pseudo-orthogonal and normalised metric is then denoted by η_{ik},

$$\eta_{ik} = \eta^{ik} = \begin{pmatrix} 1 & 0 & 0 & 0 \\ 0 & -1 & 0 & 0 \\ 0 & 0 & -1 & 0 \\ 0 & 0 & 0 & -1 \end{pmatrix} \ . \qquad \text{MINKOWSKI space metric tensor} \tag{1623}$$

We write this definition also shorter as

$$\eta_{ik} = (+1, -1, -1, -1) \ . \tag{1624}$$

The coordinate systems belonging to this metric (1623) are called *inertial systems* or LORENTZ *systems*.

For the vector calculus, the basis vectors are in principle superfluous. The components a^i or a_i of vectors $\mathbf{a} = a^i \mathbf{k}_i = a_i \mathbf{k}^i$ are collected as ensembles denoting contravariant or covariant vectors a^i and a_i, resp. Since we have defined the basis vectors by the coordinate lines, we get the transformation laws by a change of the basis characteristic for a vector, also required for the change of the vector components. Therefore we define:

The quantities a^i form a contravariant vector, when its components $a^{i'}$ change by a coordinate transformation (1607) after the same relations as the coordinate differentials in Eq. (1612),

$$a^{i'} = \frac{\partial x^{i'}}{\partial x^i} a^i \quad \longleftrightarrow \quad a^i = \frac{\partial x^i}{\partial x^{i'}} a^{i'} \ . \tag{1625}$$

The quantities a_i form a covariant vector, when they transform for a coordinate transformation (1607) after the same relations as the basis vectors in Eq. (1614),

$$a_{i'} = \frac{\partial x^i}{\partial x^{i'}} a_i \quad \longleftrightarrow \quad a_i = \frac{\partial x^{i'}}{\partial x^i} a_{i'} \ . \tag{1626}$$

Hence covariant and contravariant vectors transform as contragredient quantities, as explained after Eq. (1614).

Vectors are tensors of first rank. Tensors of second and higher rank are defined by respective transformation laws as products of the corresponding tensors.

The quantities t^{ik} form a contravariant tensor of second rank, when they transform for a coordinate transformation (1607) to quantities $t^{i'k'}$ as

$$t^{i'k'} = \frac{\partial x^{i'}}{\partial x^i} \frac{\partial x^{k'}}{\partial x^k} t^{ik} \quad \longleftrightarrow \quad t^{ik} = \frac{\partial x^i}{\partial x^{i'}} \frac{\partial x^k}{\partial x^{k'}} t^{i'k'} \ . \tag{1627}$$

A covariant tensor of second rank transforms as

$$t_{i'k'} = \frac{\partial x^i}{\partial x^{i'}} \frac{\partial x^k}{\partial x^{k'}} t_{ik} \quad \longleftrightarrow \quad t_{ik} = \frac{\partial x^{i'}}{\partial x^i} \frac{\partial x^{k'}}{\partial x^k} t_{i'k'} \ . \tag{1628}$$

In general, one defines n-fold contravariant and m-fold covariant tensors by the transformation laws

$$\begin{aligned} t^{i'_1 \dots i'_n}{}_{k'_1 \dots k'_m} &= \frac{\partial x^{i'_1}}{\partial x^i_1} \cdots \frac{\partial x^{i'_n}}{\partial x^i_n} \frac{\partial x^{k_1}}{\partial x^{k'_1}} \cdots \frac{\partial x^{k_m}}{\partial x^{k'_m}} t^{i_1 \dots i_n}{}_{k_1 \dots k_m} \ , \\ t^{i_1 \dots i_n}{}_{k_1 \dots k_m} &= \frac{\partial x^{i_1}}{\partial x^{i'_1}} \cdots \frac{\partial x^{i_n}}{\partial x^{i'_n}} \frac{\partial x^{k'_1}}{\partial x^{k_1}} \cdots \frac{\partial x^{k'_m}}{\partial x^{k_m}} t^{i'_1 \dots i'_n}{}_{k'_1 \dots k'_m} \ . \end{aligned} \tag{1629}$$

One has only to take care of the position of the indices, then the correct formula follows automatically. Contravariant indices are standing above, and covariant ones below. Coordinates have contravariant indices, and the partial derivates for the coordinates behave as covariant indices.

The summation over a co- and a contravariant index reduces the rank of the tensors by two. For example, from the tensor of third rank $T_i{}^k{}_l$, we get by $T_i{}^k{}_k = t_i$ a tensor of first rank, a vector. This procedure is called *contraction*.

The contraction of a rank one co- and contravariant tensor t^i_k leads to a tensor of rank zero, a scalar $t = t^i_i$.

The product of tensors leads to tensors where the ranks are added. For example, from a tensor of third rank, $T_i{}^k{}_l$ and a tensor of second rank $S^i{}_k$ we get a tensor of rank five,

$$V_i{}^k{}_l{}^p{}_q = T_i{}^k{}_l S^p{}_q .$$

From Eqs. (1615), (1618), (1611), and (1620) it follows:

The metric coefficients g_{ik} form a twofold covariant tensor, g^{ik} is a twofold contravariant tensor, and the KRONECKER symbol δ^i_k forms a rank one covariant and a rank-one contravariant tensor.

The KRONECKER symbol δ^i_k is a *numerically invariant tensor*, since

$$\frac{\partial x^{i'}}{\partial x^i}\frac{\partial x^k}{\partial x^{k'}}\delta^i_k = \frac{\partial x^{i'}}{\partial x^i}\frac{\partial x^i}{\partial x^{k'}} = \frac{\partial x^{i'}}{\partial x^{k'}} = \delta^{i'}_{k'} \ . \tag{1630}$$

By means of the metric tensor g_{ik}, now to each contravariant vector a^i, there are assigned the covariant components a_i and vice versa,

$$a^i = g^{ik} a_k \ , \quad a_i = g_{ik} a^k \ , \tag{1631}$$

so that we can speak again of the co- and contravariant components of one and the same vector $\mathbf{a}$, that we can represent by the respective basis vectors. This was the starting point of our presentation

$$\mathbf{a} = a^i \mathbf{k}_i = a_i \mathbf{k}^i \ .$$

On this definition, we shall return only occasionally, but we shall in general perform calculations with the components a^i or a_i.

As in Eq. (1631), we can represent arbitrary tensors with contra- or covariant components, including the metric tensor itself,

$$\left.\begin{aligned} T^i{}_k &= g^{ip} g_{kq} T_p{}^q \ , \\ T_i{}^k{}_l &= g_{ir} T^{rk}{}_l \ , \\ g^{ik} &= g^{ir} g^{ks} g_{rs} \ , \\ \delta^i_k &= g^{ir}\, g_{rk} \ . \end{aligned}\right\} \tag{1632}$$

There especially the KRONECKER symbol is the onefold co- and onefold contravariant representation of the metric tensor.

On the basis of the existence of a metric g_{ik} in our linear space, we can speak of a tensor $\mathbf{T}$ as a geometrical object and use its co- or contravariant components. For a rank three tensor $\mathbf{T}$, it holds, for example,

$$\mathbf{T} = T_i{}^k{}_l \mathbf{k}^i \mathbf{k}_k \mathbf{k}^l = T^{ik}{}_l \mathbf{k}_i \mathbf{k}_k \mathbf{k}^l \ .$$

The scalar product (1622) of two vectors a^i and b^i represents a contraction of the tensor $a^i b^k$ with the metric tensor g_{ik}.

Each tensor of second or higher rank can be decomposed with respect to each index pair into a symmetric and an antisymmetric part. We write

$$\left.\begin{aligned} T_{ik} &= T_{(ik)} + T_{[ik]} \ , \\ T_{(ik)} &:= \tfrac{1}{2}\left(T_{ik} + T_{ki}\right) \ , \\ T_{[ik]} &:= \tfrac{1}{2}\left(T_{ik} - T_{ki}\right) \ . \end{aligned}\right\} \tag{1633}$$

This decomposition is conserved at raising or lowering of indices with the metric tensor with the following rules: By the contraction of the decomposition (1633) with the metric tensor, we get

$$T_{ir} g^{rk} = \left[\tfrac{1}{2}\left(T_{ir} + T_{ri}\right) + \tfrac{1}{2}\left(T_{ir} - T_{ri}\right)\right] g^{rk},$$

hence

$$T_i{}^k = \frac{1}{2}\left(T_i{}^k + T^k{}_i\right) + \frac{1}{2}\left(T_i{}^k - T^k{}_i\right) \ . \tag{1634}$$

The indices have a fixed order in the tensor symbols, independent of characterisation as contravariant index standing above, or standing below as covariant index. If the tensor is symmetric in these indices, one can simply write the co- and contravariant indices one over the other, T_i^k, when $T_{ik} = T_{ki}$.

The LEVI- CIVITA symbol $\epsilon^{i_1 \dots i_n}$ is defined by

$$\epsilon^{i_1 \dots i_n} = \left.\begin{matrix} 1 & & i_1 \dots i_n = \text{even permutations of numbers } 1, \dots, n \\ -1 & \text{for} & i_1 \dots i_n = \text{odd permutation of numbers } 1, \dots, n \\ 0 & & \text{else} \end{matrix}\right\} . \tag{1635}$$

By the definition of a determinant it holds

$$\frac{\partial x^{i'_1}}{\partial x^{i_1}} \cdots \frac{\partial x^{i'_n}}{\partial x^{i_n}} \epsilon^{i_1 \dots i_n} = \left|\frac{\partial x^{i'}}{\partial x^{i}}\right| \epsilon^{i'_1 \dots i'_n} \ . \tag{1636}$$

If one defines the quantity $\epsilon^{i_1 \dots i_n}$ as *tensor density* of weight 1 according to the transformation law

$$\epsilon^{i'_1 \dots i'_n} = \left|\frac{\partial x^{i'}}{\partial x^{i}}\right|^{-1} \frac{\partial x^{i'_1}}{\partial x^{i_1}} \cdots \frac{\partial x^{i'_n}}{\partial x^{i_n}} \epsilon^{i_1 \dots i_n} \ , \tag{1637}$$

then $\epsilon^{i_1 \dots i_n}$ is numerically invariant.

The corresponding quantity $\epsilon_{i_1 \dots i_n}$ is defined as tensor density of weight -1, they are also numerically invariant during coordinate transformations,

$$\epsilon_{i'_1 \dots i'_n} = \left| \frac{\partial x^{i'}}{\partial x^i} \right|^{+1} \frac{\partial x^{i_1}}{\partial x^{i'_1}} \cdots \frac{\partial x^{i_n}}{\partial x^{i'_n}} \epsilon_{i_1 \dots i_n} \ . \tag{1638}$$

From the transformation law of a twofold covariant tensor (1628), one reads off immediately, that, e.g. the determinant of the metric tensor $g := \det(g_{ik})$ is a scalar density of weight 2,

$$g' = \left| \frac{\partial x^{i'}}{\partial x^i} \right|^{-2} g \ . \tag{1639}$$

When we take the square root of this equation, we must observe that g can be negative for an indefinite metric. In this case, we must multiply it with -1. For short notation, we can simply write $\sqrt{|g|}$ instead of $\sqrt{\text{sign}(|g|)\,|g|}$. Also the Jacobi determinant (1609) can become negative for a change of the relative orientations of the coordinates, e.g. for a reflection. Since the square root is by definition positive, we must still add the factor sign(J) in the definition,

$$\sqrt{|g'|} = \text{sign}(J) \left| \frac{\partial x^{i'}}{\partial x^i} \right|^{-1} \sqrt{|g|} \ . \tag{1640}$$

$\sqrt{|g|}$ represents the volume of the parallelepiped, formed by the basis vectors $\mathbf{k}_i$.

From Eqs. (1637) and (1640), it follows that the quantity $\epsilon^{i_1 \dots i_n}/\sqrt{|g|}$ has the transformation law of so-called *pseudotensors*,

$$\frac{1}{\sqrt{|g'|}} \epsilon^{i'_1 \dots i'_n} = \text{sign}(J) \frac{\partial x^{i'_1}}{\partial x^{i_1}} \cdots \frac{\partial x^{i'_n}}{\partial x^{i_n}} \frac{1}{\sqrt{|g|}} \epsilon^{i_1 \dots i_n} \ . \qquad \text{Pseudotensor} \tag{1641}$$

These quantities have the normal tensorial transformation laws, but get in addition a negative sign, when the orientation of coordinate axes changes.

For a three-dimensional space, $n = 3$, we provide the following formulae that can be verified by elementary insertion:

$$\left.\begin{aligned} \epsilon^{ikl}\,\epsilon_{pqr} &= \delta^i_p\,\delta^k_q\,\delta^l_r + \delta^i_q\,\delta^k_r\,\delta^l_p + \delta^i_r\,\delta^k_p\,\delta^l_q - \delta^i_q\,\delta^k_p\,\delta^l_r - \delta^i_p\,\delta^k_r\,\delta^l_q - \delta^i_r\,\delta^k_q\,\delta^l_p \ , \\ \epsilon^{rkl}\,\epsilon_{rpq} &= \delta^k_p \delta^l_q - \delta^k_q\,\delta^l_p \ , \\ \epsilon^{irs}\,\epsilon_{krs} &= 2\,\delta^i_k \ , \\ \epsilon^{ikl}\,\epsilon_{ikl} &= 6 \ . \end{aligned}\right\} \tag{1642}$$

With the pseudotensor $(1/\sqrt{|g|})\,\epsilon^{i_1 \dots i_n}$ we can assign to each tensor different pseudotensors, e.g. to the tensor T_{ik} a pseudotensor $\tau^{i_1 \dots i_{n-2}}$,

$$\mathcal{T}^{i_1 \dots i_{n-2}} = T_{rs} \frac{1}{\sqrt{|g|}} \epsilon^{i_1 \dots i_{n-2} rs} \,. \qquad \text{Pseudotensor} \qquad (1643)$$

In three-dimensional space, an important role is played by a pseudotensor of first level, a pseudovector ζ^i, formed from two vectors a_i and b_i,

$$\zeta^i = \frac{1}{\sqrt{|g|}} \epsilon^{irs} a_r b_s \,. \qquad \text{Pseudovektor for } n = 3 \qquad (1644)$$

Restricted on right-oriented Cartesian coordinates in three-dimensional Euclidean space, one calls the operation (1644) a *vector product* $\mathbf{a} \times \mathbf{b}$, formed from the two vectors $\mathbf{a}$ and $\mathbf{b}$. In this case, it holds $\sqrt{|g|} = 1$, and instead of ζ^i, we write c^i,

$$c^i = \epsilon^{irs} a_r b_s \quad \text{or} \quad \mathbf{c} = \mathbf{a} \times \mathbf{b} \,. \qquad \text{Vector product, } n = 3 \text{ Right-oriented Cartesian coordinates} \qquad (1645)$$

From Eq. (1645), it follows immediately $\mathbf{c} \times \mathbf{a} = \mathbf{c} \times \mathbf{b} = 0$, and there hold following formulas:

$$\left.\begin{array}{l} \mathbf{c} \times \left(\mathbf{a} \times \mathbf{b}\right) = \mathbf{a} \times \left(\mathbf{b} \times \mathbf{c}\right) = \mathbf{b} \times \left(\mathbf{c} \times \mathbf{a}\right) \,, \\ \mathbf{a} \times \left(\mathbf{b} \times \mathbf{c}\right) = \left(\mathbf{a} \times \mathbf{c}\right)\mathbf{b} - \left(\mathbf{a} \times \mathbf{b}\right)\mathbf{c} \,, \\ \left(\mathbf{a} \times \mathbf{b}\right) \times \left(\mathbf{c} \times \mathbf{d}\right) = \begin{vmatrix} \left(\mathbf{a} \times \mathbf{c}\right) & \left(\mathbf{a} \times \mathbf{d}\right) \\ \left(\mathbf{b} \times \mathbf{c}\right) & \left(\mathbf{b} \times \mathbf{d}\right) \end{vmatrix} \,, \\ |\mathbf{a} \times \mathbf{b}| = \sqrt{\left(\mathbf{a} \times \mathbf{a}\right)\left(\mathbf{b} \times \mathbf{b}\right) - \left(\mathbf{a} \times \mathbf{b}\right)^2} \,. \end{array}\right\} \qquad \text{Right-oriented Cartesian coordinates } n = 3 \qquad (1646)$$

We shall restrict us in a linear space on linear coordinate transformations. Then the Jacobi matrix $\mathbf{J}$, Eq. (1608), is constant, and also the basis vectors $\mathbf{k}_i$ are independent on the coordinates. On the basis vectors, we apply Schmidt's rule for producing an orthonormal basis. There to the basis vectors $\mathbf{k}_i$, we construct new basis vectors $\mathbf{e}_i$ according to the following scheme:

$$\left.\begin{array}{l} 1. \; \mathbf{e}_i^* = \mathbf{k}_i - \sum\limits_{k=1}^{i-1} (\mathbf{k}_i \times \mathbf{e}_k)\mathbf{e}_k \,, \quad i = 1, \dots, n \,, \\ 2. \; \mathbf{e}_i = \dfrac{\mathbf{e}_i^*}{\mathbf{e}_{(i)}^* \times \mathbf{e}_{(i)}^*} \,, \quad \text{if} \;\; \mathbf{e}_{(i)}^* \times \mathbf{e}_{(i)}^* \neq 0 \,. \end{array}\right\} \qquad (1647)$$

Beginning with $i = 1$, where there is no sum in the first step, one iterates the steps 1 and 2 for $i = 2$, $i = 3$, etc., until $i = n$. At the end, one gets normalised base vectors $\mathbf{e}_i$, that are mutually orthogonal. Then the metric tensor g_{ik} gets a diagonal form, in the main diagonal stand the numbers $+1$ or -1, all other components of g_{ik} are zero. When the condition $\mathbf{e}_{(i)}^* \cdot \mathbf{e}_{(i)}^* \neq 0$ is not fulfilled, e.g. at step $i = r$, hence $\mathbf{e}_{(r)}^* \cdot \mathbf{e}_{(r)}^* = 0$, we cannot construct the $r-$th basis vector $\mathbf{e}_r$ according to Eq. (1647), and the algorithm stops. This occurs when the vector $\mathbf{k}_r$ only differs from the subspace formed by the vectors $\mathbf{e}_1, \mathbf{e}_2 \dots \mathbf{e}_{r-1}$ by a light-like vector. It is then always possible to chose instead of the vectors $\mathbf{k}_r$ such a vector $\overline{\mathbf{k}}_r$, that differs from this subspace by a space-like or time-like vector, and then to continue the procedure.

Now we consider *tensor fields* and restrict us again on linear coordinates, the base vectors then remain constant. In this case, the partial derivatives of the tensor field can be evaluated by the partial derivatives of its co- or contravariant components.

The gradient Grad ϕ of a scalar field $\phi = \phi(x^i)$ is defined by

$$\mathrm{Grad}_k\, \phi(x^i) := \frac{\partial}{\partial x^k}\, \phi(x^i) = V_k(x^i) \ . \tag{1648}$$

This is a covariant vector field, since

$$V_{k'}(x^{i'}) := \frac{\partial}{\partial x^{k'}}\, \phi(x^i(x^{i'})) = \frac{\partial}{\partial x^k}\, \phi(x^i) \frac{\partial x^k}{\partial x^{k'}} = \frac{\partial x^k}{\partial x^{k'}}\, V_k(x^i) \ . \tag{1649}$$

The field $\mathrm{Grad}_k\, \phi(x^i)$ therefore fulfils the transformation laws of covariant vectors for transformations (1626) as claimed.

For simplification of the notation, the partial derivatives of a function $f = f(x^i)$ are often written as

$$\frac{\partial}{\partial x^k} f(x^i) := \partial_k f(x^i) := f(x^i)_{,k} \ . \tag{1650}$$

The quantities $\phi_{,k} = \partial_k \phi$ form a covariant vector field.

The gradient of a scalar function $f = f(x, y, z)$ in three-dimensional Euclidean space with Cartesian coordinates then reads

$$\mathrm{grad}\, f = (\partial_x f,\ \partial_y f,\ \partial_z f\,) \ . \qquad n = 3 \tag{1651}$$

The gradient of a rank-two tensor field $\mathbf{T}$ is given by

$$(\mathrm{Grad}\,\mathbf{T})_l := T_{ik,l} \ . \tag{1652}$$

From the proof of Eq. (1649), there follows immediately that in general, the partial derivative of an arbitrary tensor field raises its rank by one. For example, $\partial_l T_i{}^k$ is a twofold co- and a single contravariant tensor, and $U^{ik}{}_{r,p}$ a twofold co- and a twofold contravariant tensor.

The n-dimensional divergence $\mathrm{Div}\mathbf{V}$ of a vector field V^k is defined by

$$\mathrm{Div}\mathbf{V} = \mathrm{Div}_k V^k := \partial_k V_l g^{kl} = \partial_k V^k = V^k{}_{,k} = V_{l,k}\, g^{kl} \ . \tag{1653}$$

For forming the divergence, we have therefore at first to form a partial derivative, and then to contract the result with the metric.

The divergence of a three-vector $\mathbf{V} = (V_x, V_y, V_z)$ in an Euclidean space in Cartesian coordinates is written as

$$\mathrm{div}\,\mathbf{V} = \partial_x V_x + \partial_y V_y + \partial_z V_z \ . \qquad n = 3 \tag{1654}$$

The divergence of a vector is a scalar. For a tensor of higher order, forming the divergence is possible with respect to each index. Therefore we have to distinguish between different forms of divergences, that will lead to different results in particular for asymmetric tensors. Taking a tensor of second rankT_{kl}, we get as example

$$\left.\begin{aligned}(\mathrm{Div}_1\mathbf{T})_l &= T^i{}_{l}, {}_i = \partial_i(T_{kl}g^{ik}) \ , \\ (\mathrm{Div}_2\mathbf{T})_k &= T_k{}^l, {}_l = \partial_i(T_{kl}g^{il}) \ . \end{aligned}\right\} \tag{1655}$$

With the divergence, the rank of a tensor is in any case reduced by one due to the contraction with the metric tensor.

For the n-dimensional rotation, we have to distinguish two quantities: We discuss it at first with a vector field V_k and form from it a twofold covariant antisymmetric tensor that will be denoted as Rot$\mathbf{V}$,

$$(\mathrm{Rot}\mathbf{V})_{ik} := \partial_i V_k - \partial_k V_i = V_{k},_i - V_i,_k \ . \tag{1656}$$

In forming the rotation, we have at first to form partial derivatives, and then we subtract the resulting quantity with exchanged indices. To this antisymmetric rank-2 tensor, it is attributed a rank-$(n-2)$ pseudotensor rot$\mathbf{V}$ by means of the rank-n Levi- Civita pseudotensor (1641) $\frac{1}{\sqrt{|g|}}\epsilon^{iki_3\ldots i_n}$, that is denoted by a small 'rot' for distinction from Eq. (1656),

$$(\mathrm{rot}\mathbf{V})^{i_3\ldots i_n} := \frac{1}{2}(\mathrm{Rot}\mathbf{V})_{ik}\frac{1}{\sqrt{|g|}}\epsilon^{iki_3\ldots i_n} = \frac{1}{\sqrt{|g|}}\epsilon^{iki_3\ldots i_n}\,\partial_i V_k \ . \tag{1657}$$

The only distinction from a real tensor consists in the behaviour of the pseudotensor rot$\mathbf{V}$ at coordinate transformations containing a reflection. Then in addition the pseudotensors must be multiplied by -1. In the three-dimensional space, this becomes the known pseudovector rot$\mathbf{V}$

$$\mathrm{rot}\mathbf{V} = \frac{1}{\sqrt{|g|}}\epsilon^{ijk}\,\partial_j V_k\,\mathbf{k}_i \ . \qquad \mathrm{n} = 3 \tag{1658}$$

For right-oriented Cartesian coordinates (x, y, z), it is $\sqrt{|g|} = 1$, and Eq. (1658) leads to the known component definition of the rotation of a vector field

$$\mathrm{rot}\mathbf{V} = \left(\partial_2 V_3 - \partial_3 V_2\,,\ \partial_3 V_1 - \partial_1 V_3\,,\ \partial_1 V_2 - \partial_2 V_1\right). \qquad \begin{array}{l}\text{Right-oriented}\\ \text{Cartesian coordinates}\\ n=3\end{array} \tag{1659}$$

Using in Eq. (1656) $\mathbf{V} = \mathrm{Grad}\,\phi$, hence $V_k = \partial_k\phi$, so it follows

$$\mathrm{Rot\,Grad}\,\phi = 0 \quad \text{by} \quad \partial_i\partial_k\phi - \partial_k\partial_i\phi = 0 \tag{1660}$$

corresponding to the well-known three-dimensional relation

$$\mathrm{rot\ grad}\,\phi = 0 \ . \qquad n = 3 \tag{1661}$$

We form in Eq. (1657) a divergence with respect to an arbitrary index r and find

$$\frac{1}{\sqrt{|g|}}\,\epsilon^{iki_3\ldots r\ldots i_n}\,\partial_r\partial_i V_k = 0\,, \quad \text{i.e.} \quad \text{Div rot}\mathbf{V} = 0\,, \tag{1662}$$

since $\epsilon^{iki_3\ldots r\ldots i_n}$ is in i and r antisymmetric, and $\partial_r\partial_i V_k$ is in i and r symmetric, and the contraction of an antisymmetric tensor V^{ir} with a symmetric tensor S_{ir} vanishes, $V^{ir}S_{ir} = 0$. Equation (1662) is the n-dimensional generalisation of the known relation from three-dimensional vector analysis,

$$\text{div rot}\,\mathbf{V} = 0\,. \qquad n = 3 \tag{1663}$$

Still we define the scalar differential operator of second order $\mathbf{\Delta}_n$ that leaves unchanged the tensor character of a quantity. Applied to the twofold covariant tensor T_{ik}, it is for example

$$\mathbf{\Delta}_n T_{ik} := g^{pq}\frac{\partial}{\partial x^p}\frac{\partial}{\partial x^q}\,T_{ik} = T_{ik\,,pq}\,g^{pq}\,. \tag{1664}$$

In a three-dimensional space with positive definite metric, $\mathbf{\Delta}_3$ is given in Cartesian coordinates (x, y, z) by the LAPLACE *operator* $\mathbf{\Delta}_3 = \mathbf{\Delta}$,

$$\mathbf{\Delta} = \frac{\partial^2}{\partial x^2} + \frac{\partial^2}{\partial y^2} + \frac{\partial^2}{\partial z^2}\,. \qquad \textit{Laplace}\text{ operator} \tag{1665}$$

In the MINKOWSKI space with $n = 4$ we get for inertial systems, hence in LORENTZ coordinates (ct, x, y, z) with metric η_{ik} of Eq. (1623) from $\mathbf{\Delta}_4$ the D'ALEMBERT *wave operator* $\mathbf{\Delta}_4 = \Box$,

$$\Box = \frac{1}{c^2}\frac{\partial^2}{\partial t^2} - \frac{\partial^2}{\partial x^2} - \frac{\partial^2}{\partial y^2} - \frac{\partial^2}{\partial z^2}\,. \qquad \text{Wave operator} \tag{1666}$$

There are the following connections between the differential operators of second order $\mathbf{\Delta}_n$ and the differential operators of first order Grad, Div and Rot.

In Eq. (1653) one inserts $\mathbf{V} = \text{Grad}\,\phi$, hence $V_k = \partial_k\phi$, so it follows

$$\text{Div Grad}\,\phi = \partial_k\partial_l\phi g^{kl} = \phi_{,kl}\,g^{kl} = \mathbf{\Delta}_n\phi\,. \tag{1667}$$

In the three-dimensional space, this is the known relation

$$\text{div grad}\,\phi = \mathbf{\Delta}\phi\,. \qquad n = 3 \tag{1668}$$

With Eq. (1655) we form in Eq. (1656) the divergence with respect to the first index,

$$(\text{DivRot}\mathbf{V})_k = \partial_l(\text{Rot}\mathbf{V})_{ik}g^{li} = (\partial_l\partial_i V_k - \partial_l\partial_k V_i)g^{li} = V_{k\,,il}\,g^{li} - \partial_k(V_{i\,,l}\,g^{li})\,,$$

hence with Eqs. (1653), (1664) and (1648)

$$\text{Div Rot } \mathbf{V} = \boldsymbol{\Delta}_n \mathbf{V} - \text{Grad Div } \mathbf{V} \ , \tag{1669}$$

or in components

$$(\text{Div Rot } \mathbf{V})_k = \boldsymbol{\Delta}_n V_k - V^l{}_{,lk} \ . \tag{1670}$$

From the twofold application of Eq. (1659), we get the relation in three-dimensional vector analysis

$$\text{rot rot } \mathbf{V} = -\boldsymbol{\Delta}\mathbf{V} + \text{grad div } \mathbf{V} \ . \qquad n = 3 \tag{1671}$$

We remark, that the partial derivative of a tensor field for curvilinear coordinates, resulting from nonlinear coordinate transformations, do not produce in general tensors. For conserving the tensor characters, one has to replace the partial with the so-called covariant derivative.

The formation of the covariant derivative depends on the rank of the tensor field. Only for a scalar field, all previous relations remain intact, i.e. Eqs. (1648), (1649) hold true also for nonlinear coordinates.

Here we define only the covariant derivative of a vector field $\mathbf{V}$,

$$\left.\begin{aligned} \frac{D\,V^i}{D\,x^k} &= \frac{\partial\,V^i}{\partial\,x^k} + \Gamma^i_{kr}\,V^r \ , \\ \frac{D\,V_i}{D\,x^k} &= \frac{\partial\,V_i}{\partial\,x^k} - \Gamma^r_{ki}\,V_r \ . \end{aligned}\right\} \qquad \begin{aligned}&\text{Covariant derivative}\\ &\text{of a vector field}\end{aligned} \tag{1672}$$

The so-called CHRISTOFFEL symbols Γ^i_{kl} are given by the metric tensor and its first partial derivatives

$$\Gamma^i_{k\,l} = \frac{1}{2} g^{i\,r}\,(-\partial_r\,g_{k\,l} + \partial_k\,g_{l\,r} + \partial_l\,g_{r\,k}\} \ . \qquad \text{CHRISTOFFEL symbols} \tag{1673}$$

In Eq. (1656), and e.g. in the definition of the potential in electrodynamics, Eq. (530), s. also Eq. (587), the CHRISTOFFEL symbols are just cancelling.

The covariant derivatives have a fundamental meaning for EINSTEIN's General Relativity Theory, Chap. 13, Sects. 1 and 2. There the metric describes the gravitational field.

B.2 Integral Laws

We consider the transformation of the integration variables for the evaluation of the n-fold integral in an n-dimensional space, Eq. (1607), i.e. using instead of $x^1, \ldots, x^n$ the new variables $x^{1'}, \ldots, x^{n'}$, then it follows the general substitution equation,

$$\int_{B_n} \cdots \int f(x^i)\, dx^1 \cdots dx^n = \int_{B'_n} \cdots \int v(x^{i'}) \det\left(J^{i'}_{\ i}\right) dx^{1'} \cdots dx^{n'} \tag{1674}$$

with $v(x^{i'}) = f\left(x^i(x^{i'})\right)$, $J^{i'}_{\ i}$ is the JACOBI matrix (1608), and the integration domain B'_n in variables $x^{i'}$ follows by the substitution $x^{i'} = x^{i'}(x^i)$ from the range B_n.

For $n = 1$, Eq. (1674) is the substitution formula for one-dimensional integrals,

$$\int_a^b f(x)\, dx = \int_\alpha^\beta f\left(x(x')\right) \frac{dx}{dx'}\, dx' \quad , \quad \alpha = x'(a) \quad , \quad \beta = x'(b) \; . \tag{1675}$$

For $n = 2$ it follows correspondingly

$$\iint_S f(x, y)\, dx\, dy = \iint_{S'} f\left(x(x', y'), y(x', y')\right) \frac{\partial(x, y)}{\partial(x', y')}\, dx'\, dy' \; , \tag{1676}$$

where the integration range S' for variables (x', y') follows by substitution from the range S for variables (x, y).

The case $n = 3$ reads

$$\iiint_K f(x, y, z)\, dx\, dy\, dz = \iiint_{K'} v(x', y', z') \frac{\partial(x, y, z)}{\partial(x', y', z')}\, dx'\, dy'\, dz' \tag{1677}$$

with $v(x', y', z') = f\left(x(x', y', z'), y(x', y', z'), z(x', y', z')\right)$, and the range K' of variables (x', y', z') results by substitution of the range K for the variables (x, y, z). We provide two special cases explicitly:

1. Cylinder coordinates with ρ for x' and φ for y', hence

$$\left.\begin{aligned} x &= x' \cos y' = \rho \cos\varphi \; , \\ y &= x' \sin y' = \rho \sin\varphi \; , \\ z &= z' \; . \end{aligned}\right\} \quad \text{Cylinder coordinates} \tag{1678}$$

With Eq. (1678) it follows Eq. (1677),

$$\int_{x_1}^{x_2}\int_{y_1}^{y_2}\int_{z_1}^{z_2} f(x, y, z)\, dx\, dy\, dz = \int_{\rho_1}^{\rho_2}\int_{\varphi_1}^{\varphi_2}\int_{z_1}^{z_2} v(\rho, \varphi, z)\, \rho\, d\rho\, d\varphi\, dz \; , \tag{1679}$$

where the integration limits and the function $f(x, y, z)$ and $v(\rho, \varphi, z)$ are connected by Eq. (1678).

2. Spherical coordinates with r for x', ϑ for y' and φ for z', hence

$$\left.\begin{aligned} x &= x' \sin y' \cos z' = r \sin\vartheta \cos\varphi \; , \\ y &= x' \sin y' \sin z' = r \sin\vartheta \sin\varphi \; , \\ z &= x' \cos y' \qquad\quad = r \cos\vartheta \; . \end{aligned}\right\} \quad \text{Spherical coordinates} \tag{1680}$$

With Eq. (1680) it follows from Eq. (1677)

$$\int_{x_1}^{x_2}\int_{y_1}^{y_2}\int_{z_1}^{z_2} f(x, y, z)\, dx\, dy\, dz = \int_{r_1}^{r_2}\int_{\vartheta_1}^{\vartheta_2}\int_{\varphi_1}^{\varphi_2} v(r, \vartheta, \varphi)\, r^2 \sin\vartheta\, dr\, d\vartheta\, d\varphi\ , \tag{1681}$$

where functions $f(x, y, z)$ and $v(r, \vartheta, \varphi)$ and the integration limits are connected with Eq. (1680).

Now we provide the most important integral relations in the three-dimensional space.

With (x, y, z) we denote Cartesian coordinates. We consider in the following three functions: $P = P(x, y, z)$, $Q = Q(x, y, z)$ and $R = R(x, y, z)$.

The parametric representation of a curve k in space is given by three functions, depending on a parameter t,

$$k:\ x = \varphi(t)\,,\ y = \psi(t)\,,\ z = \chi(t)\,. \qquad \text{Curve in space} \tag{1682}$$

Each value of the parameter t corresponds to a point $C(t)$ on the curve. When the parameter t can be replaced by a variable, e.g. by x, so it is called an explicit representation of the curve

$$k:\ y = y(x)\,,\ z = z(x)\ .$$

The line element ds of the curve, the length ds between two neighbouring points $C(t)$ and $C(t + dt)$ on the curve, is

$$ds^2 = dx^2 + dy^2 + dz^2 = \big((d\varphi/dt)^2 + (d\psi/dt)^2 + (d\chi/dt)^2\big)\, dt^2\,. \qquad \text{Line element} \tag{1683}$$

For a function $P = P(x, y, z)$ defined on the curve, the following conception is introduced:

Curve integral of first kind

$$\left.\begin{aligned} I &= \int_{C_1}^{C_2} P(x,\ y,\ z)\, ds \\ &= \int_{t_1}^{t_2} P\big(x(t),\ y(t),\ z(t)\big)\sqrt{\Big(\frac{dx}{dt}\Big)^2 + \Big(\frac{dy}{dt}\Big)^2 + \Big(\frac{dz}{dt}\Big)^2}\ dt \\ &= \int_{x_1}^{x_2} P\big(x,\ y(x),\ z(x)\big)\sqrt{1 + \Big(\frac{dy}{dx}\Big)^2 + \Big(\frac{dz}{dx}\Big)^2}\ dx\ . \end{aligned}\right\} \tag{1684}$$

For a curve integral of second kind, the function $P = P(x, y, z)$, defined on the curve k, is not integrated with the line element, but with the projections dx, dy and dz on the coordinate axes. Considering again the points C_1 and C_2 on the curve k, then it is, e.g.

$$I = \int_{C_1}^{C_2} P(x,\ y,\ z)\, dx = \int_{t_1}^{t_2} P\big(x(t),\ y(t),\ z(t)\big)\, \frac{dx}{dt}\, dt = \int_{x_1}^{x_2} P\big(x,\ y(x),\ z(x)\big)\, dx$$

a curve integral of second kind. In the general case, there are needed three functions $P = P(x, y, z)$, $Q = Q(x, y, z)$ and $R = R(x, y, z)$, and it is defined:
General curve integral of second kind

$$\left.\begin{aligned} I &= \int_{C_1}^{C_2} P(x,\ y,\ z)\, dx + Q(x,\ y,\ z)\, dy + R(x,\ y,\ z) dz \\ &= \int_{t_1}^{t_2} \left[P\big(x(t), y(t), z(t)\big)\frac{dx}{dt} + Q\big(x(t), y(t), z(t)\big)\frac{dy}{dt} + R\big(x(t), y(t), z(t)\big)\frac{dz}{dt}\right] dt \\ &= \int_{x_1}^{x_2} \left[P\big(x, y(x), z(x)\big) + Q\big(x, y(x), z(x)\big)\frac{dy}{dx} + R\big(x, y(x), z(x)\big)\frac{dz}{dx}\right] dx\,. \end{aligned}\right\} \tag{1685}$$

The form of the curve integral depends therefore on the parameter used to represent the curve k.

The vectorial line element $d\mathbf{s} = (dx\,, dy\,, dz)$ connects the two neighbouring points $C(t) = \big(x(t)\,,\ y(t)\,,\ z(t)\big)$ and $C(t + dt) = \big(x(t + dt)\,,\ y(t + dt)\,,\ z(t + dt)\big)$.

Now we assume, that the functions $P(x, y, z)$, $Q(x, y, z)$ and $R(x, y, z)$ are the three components of a vector field $\mathbf{V}$, hence $\mathbf{V} = (P,\ Q,\ R)$.

The curve integral of second kind (1685) can then be written as

$$I = \int_{C_1}^{C_2} P(x,\ y,\ z)\, dx + Q(x,\ y,\ z)\, dy + R(x,\ y,\ z)\, dz = \int_{C_1}^{C_2} \mathbf{V}\cdot d\mathbf{s}\ . \tag{1686}$$

Let α, β and γ be the angles of the vector $d\mathbf{s}$ with the coordinate axes, then it follows from Eq. (1686) with $d\mathbf{s} = (\cos\alpha\,,\ \cos\beta\,,\ \cos\gamma)\, ds$ a connection between the curve integral of first and second kinds

$$\int_{C_1}^{C_2} P\, dx + Q\, dy + R\, dz = \int_{C_1}^{C_2} (P\ \cos\alpha + Q\ \cos\beta + R\ \cos\gamma)\ ds\ . \tag{1687}$$

The parameter representation of an area S in space is given by three functions that depend on two parameters u and v,

Fig. B.1 CARL FRIEDRICH GAUSS, *Braunschweig 30.4.1777, † Göttingen 23.2.1855.

$$S: x = \varphi(u, v), \quad y = \psi(u, v), \quad z = \chi(u, v). \qquad \text{Area in space} \quad (1688)$$

Each pair of parameters (u, v) corresponds to a point $C(u, v)$ on the area. If it is possible to eliminate the parameters u and v, and to replace them, e.g. by x and y, then we get an explicit representation of the area S,

$$S: z = z(x, y) .$$

With $x^1 = u$ and $x^2 = v$ we have introduced curvilinear coordinates on the area. The separation ds of two neighbouring points $C(u, v)$ and $C(u + du, v + dv)$ on the area are then given by

$$ds^2 = g_{ab}\, dx^a\, dx^b , \qquad \alpha, \beta = 1, 2 . \qquad (1689)$$

Here g_{ab} is the so-called inner metric of the area. From Eqs. (1614) and (1615), it follows

$$g_{ab} = \mathbf{k}_a \cdot \mathbf{k}_b . \qquad \text{Inner metric of the area } S \qquad (1690)$$

The vectors $\mathbf{k}_a$ have directions of the x^a-coordinate lines. Its length is connected with the separation of two neighbouring points on the x^1-coordinate line $|k_1| = ds/du$ and $|k_2| = ds/dv$.

The two vectors $\mathbf{k}_1\, du$ and $\mathbf{k}_2\, dv$ form a parallelogram. We define the area normal vector $d\mathbf{S}$ by the vector product of the two basis vectors,

$$d\mathbf{S} := \mathbf{k}_1 \times \mathbf{k}_2\, du\, dv . \qquad \text{Infinitesimal area vector} \qquad (1691)$$

The amount of the vector dS follows from Eq. (1646),

$$dS = \sqrt{g}\, du\, dv , \quad g = |g_{ab}| . \qquad (1692)$$

An arbitrary element of the area S can then be decomposed into these area elements. Given a function $F = F(x, y, z)$ defined on the area, one defines the area integral of first kind with the area element dS:

Area integral of first kind

$$I = \iint_S F(x, y, z)\, dS = \iint_S F\Big(x(u, v), y(u, v), z(u, v)\Big) \sqrt{g}\, du\, dv . \qquad (1693)$$

The area integral of second kind follows from the function $F = F(x, y, z)$ defined over the area S and not multiplied by the area elements, but by its projections on the coordinate planes xy, yz and zx. Taking again the integration range S, then it follows for a projection on the xy-plane the area integral of second kind

$$I = \iint\limits_S F\big(x,\ y,\ z(x,\ y)\big)\, dx\, dy \ . \tag{1694}$$

In the general case, one has to use three functions $P = P(x, y, z)$, $Q = Q(x, y, z)$ and $R = R(x, y, z)$, and then:

General area integral of second kind

$$\begin{aligned} I &= \iint\limits_S \Big(P\big(x(y,\ z),\ y,\ z\big)\, dy\, dz + Q\big(x,\ y(x,\ z),\ z\big)\, dz\, dx + R\big(x,\ y,\ z(x,\ y)\big)\, dx\, dy\Big) \\ &= \iint\limits_S \Big[P\big(x(u,\ v),\ y(u,\ v),\ z(u,\ v)\big) J_{yz} + Q\big(x(u,\ v),\ y(u,\ v),\ z(u,\ v)\big) J_{zx} + \\ &\qquad + R\big(x(u,\ v),\ y(u,\ v),\ z(u,\ v)\big) J_{xy}\Big]\, du\, dv \ . \end{aligned} \tag{1695}$$

Here J_{xy}, J_{yz} and J_{zx} are the determinants of the JACOBI matrices (1608), that result from the change in the integration variables in Eq. (1688) from $(x,\ y)$ to $(u,\ v)$, from $(y,\ z)$ to $(u,\ v)$ and from $(z,\ x)$ to $(u,\ v)$ according to the substitution formulae (1677). It should be remarked, that the order of variables is important due to the sign in the determinant, hence

$$J_{xy} = |\partial(x,\ y)/\partial(u,\ v)| \ , \quad J_{yz} = |\partial(y,\ z)/\partial(u,\ v)| \ , \quad J_{zx} = |\partial(z,\ x)/\partial(u,\ v)| \ . \tag{1696}$$

The infinitesimal area, that the vector area element $d\mathbf{S}$ represents, can be chosen so that its projections on the coordinate plane just give the area elements $dy\, dz$, $dz\, dx$ and $dx\, dy$, and $d\mathbf{S} = \big(dy\, dz,\ dz\, dx,\ dx\, dy\big)$.

We understand again the functions $P(x, y, z)$, $Q(x, y, z)$ and $R(x, y, z)$ as components of a vector field $\mathbf{V} = (P,\ Q,\ R)$.

The area integral of second kind (1695) can then be written as

$$I = \iint\limits_S \big(P(x,\ y,\ z)\, dy\, dz + Q(x,\ y,\ z)\, dz\, dx + R(x,\ y,\ z)\big)\, dx\, dy = \iint\limits_S \mathbf{V} \cdot d\mathbf{S} \ . \tag{1697}$$

Still we define the flux of the vector $\mathbf{V}$ through the area S,

$$\Phi := \iint\limits_S \mathbf{V} \cdot d\mathbf{S} \ . \qquad \text{Flux of the vector } \mathbf{V} \text{ through the area } S \tag{1698}$$

α, β and γ might be the angles of the vector $d\mathbf{S}$ with the coordinate axes, then it follows from Eq. (1697) with $d\mathbf{S} = (\cos\alpha\, ,\, \cos\beta\, ,\, \cos\gamma)\, dS$ a connection between the area integrals of first and second kind,

$$I = \iint\limits_S (P\,dy\,dz + Q\,dz\,dx + R\,dx\,dy = \iint\limits_S (P\cos\alpha + Q\cos\beta + R\cos\gamma)\,dS\,. \tag{1699}$$

We denote with ∂S the closed curve bounding the area S. Then it follows for the connection of the line integral (curve integral of second kind) over ∂S and the area integral over S, where the closure of the curve is symbolised by the circle in the integral sign,

STOKES *integral theorem*

$$\oint\limits_{\partial S} P\,dx + Q\,dy + R\,dz$$
$$= \iint\limits_S \left(\frac{\partial R}{\partial y} - \frac{\partial Q}{\partial z}\right) dy\,dz + \left(\frac{\partial P}{\partial z} - \frac{\partial R}{\partial x}\right) dz\,dx \left(\frac{\partial Q}{\partial x} - \frac{\partial P}{\partial y}\right) dx\,dy\,. \tag{1700}$$

With the vector $\mathbf{V} = (P, Q, R)$, we get the vector form of STOKES theorem

$$\oint\limits_{\partial S} \mathbf{V}\cdot d\mathbf{s} = \iint\limits_S \operatorname{rot}\mathbf{V}\cdot d\mathbf{S}\,. \qquad \text{STOKES integral theorem} \tag{1701}$$

Now we consider the spatial range K, and we denote its closed surface with ∂K. The threefold integral over the divergence of the vector $\mathbf{V} = (P, Q, R)$ then reads

GAUSS *integral theorem*

$$\iiint\limits_K \left(\frac{\partial P}{\partial x} + \frac{\partial Q}{\partial y} + \frac{\partial R}{\partial z}\right) dx\,dy\,dz = \iint\limits_{\partial K} P\,dydz + Q\,dzdx + R\,dxdy\,. \tag{1702}$$

With Eq. (1659) we get the vectorial form of GAUSS theorem, Fig. B.1,

$$\iiint\limits_K \operatorname{div}\mathbf{V}\,dxdydz = \iint\limits_{\partial K} \mathbf{V}\cdot d\mathbf{S}\,. \qquad \text{GAUSS integral theorem} \tag{1703}$$

The projection of the gradient of a function $f = f(x, y, z)$ on a unite vector $\mathbf{n}$ is called derivative of f in direction of this vector. It is

$$\frac{\partial f}{\partial n} := \operatorname{grad} f\cdot\mathbf{n} = \frac{\partial f}{\partial x}\cos\alpha + \frac{\partial f}{\partial y}\cos\beta + \frac{\partial f}{\partial z}\cos\gamma\,, \qquad \text{Directional derivative} \tag{1704}$$

where the vector $\mathbf{n} = (\cos\alpha\,, \cos\beta\,, \cos\gamma)$ forms with the coordinate axes the angles α, β and γ.

On a three-dimensional volume K with the surface ∂K, we consider two functions $P = P(x, y, z)$ and $Q = Q(x, y, z)$. The unit vector $\mathbf{n}$ might be orthogonal to ∂K and outside-directed. For the LAPLACE operator $\boldsymbol{\Delta}$, Eq. (1665), there follow the following integral theorems:

1st GREEN*'s theorem*

$$\iiint\limits_K P\Delta Q\,dx\,dy\,dz = \iint\limits_{\partial K} P\,\frac{\partial Q}{\partial n}\,dS - \iiint\limits_K \operatorname{grad}P \cdot \operatorname{grad}Q\,dx\,dy\,dz \tag{1705}$$

and after simple subtraction

2nd GREEN*'s theorem*

$$\iiint\limits_K (P\Delta Q - Q\Delta P)\,dx\,dy\,dz = \iint\limits_{\partial K} \left(P\,\frac{\partial Q}{\partial n} - Q\,\frac{\partial P}{\partial n}\right) dS\;. \tag{1706}$$

For integrals, depending on a parameter, we have the following relations. For simplicity, we suppose a continuously differentiable integrand f, that depends on the parameter t. The differentiation of the integral over f for the parameter can be performed below the integral. We formulate this theorem in a n-dimensional space with a continuously differentiable function $f = f(x^1\,, \ldots\,, x^n\,, t)$. Let B_n be a spatially fixed n-dimensional range B_n. Then the derivative of the n-fold integral over f is given by

$$\frac{d}{dt}\int\limits_{B_n}\!\cdots\!\int f(x^1, \ldots, x^n, t)\,dx^1\cdots dx^n = \int\limits_{B_n}\!\cdots\!\int \frac{\partial}{\partial t} f(x^1, \ldots, x^n, t)\,dx^1\cdots dx^n\,. \tag{1707}$$

For a one-dimensional integral with fixed integration limits, this reads

$$\frac{d}{dt}\int\limits_a^b f(x, t)\,dx = \int\limits_a^b \frac{\partial}{\partial t} f(x, t)\,dx\,. \tag{1708}$$

Now we consider in addition the case that the integration region depends also on the parameter t. The one-dimensional integral then reads

$$\frac{d}{dt}\int\limits_{a(t)}^{b(t)} f(x, t)\,dx = \int\limits_{a(t)}^{b(t)} \frac{\partial}{\partial t} f(x, t)\,dx - f(a(t), t)\frac{da}{dt} + f(b(t), t)\frac{db}{dt}\;. \tag{1709}$$

We assume for illustration, that the parameter t represents the time, then there appears in the last two summands of Eq. (1709) the product of the integrand and the velocity, with which the integration region *changes*. If f is a density, e.g. a mass density, then the total mass $F(t)$ between a and b can change by the amount of mass entering or leaving the region in time with $-f\,da/dt$ and $+f\,db/dt$.

The n-fold integral can be evaluated by a succession of n single integrations. Therefore we get a direct generalisation of relation (1709) on n-fold integrals. For two- and threefold integrals, this is given explicitly.

Might S be a region in the x-y-plane, bounded by a closed curve ∂S. The position and form of this range might depend on a parameter t, e.g. the time. The region $S = S(t)$ then changes with time, so that each point on the boundary curve ∂S can be assigned a velocity $\mathbf{u} = (u_x\,, u_y)$. For any function $f = f(x, y, t)$, then it holds

$$\frac{d}{dt}\iint\limits_{S(t)} f(x,y,t)\,dxdy = \iint\limits_{S(t)} \frac{\partial}{\partial t} f(x,y,t)\,dxdy + \oint\limits_{\partial S(t)} f(x,y,t)\,\mathbf{u}\cdot d\mathbf{n} \qquad (1710)$$

with the normal vector $d\mathbf{n} = (\cos\alpha\,, \cos\beta)\,ds$ on the boundary curve $\partial S(t)$, i.e. $\mathbf{n}$ forms angles α and β with the x- and y-axis, and ds is the line element of the curve. The line integral is performed in mathematically positive sense. Again we can interpret this relation as law of the change of the total mass $F(t)$, formed by the mass density f inside S, that changes by in- or outstreaming with velocity $\mathbf{u}$ through the boundary ∂S, since $f(x, y, t)\,\mathbf{u}\cdot d\mathbf{n}$ is the mass flowing through the boundary ds per time.

Now we consider a spatial volume K with the limiting surface ∂K. Position and form of the volume K should now also depend on time t, so that we can assign to each point on the surface ∂K a velocity $\mathbf{u} = (u_x\,, u_y\,, u_z)$. For any function $f = f(x, y, z, t)$, then there holds

$$\frac{d}{dt}\iiint\limits_{K(t)} f(x,y,z,t)\,dxdyz = \iiint\limits_{K(t)} \frac{\partial}{\partial t} f(x,y,t)\,dxdydz + \iint\limits_{\partial K(t))} f(x,y,z,t)\,\mathbf{u}\cdot d\mathbf{S} \qquad (1711)$$

with the vectorial area element $d\mathbf{S}$, so that for a mass density f, the quantity $f(x, y, z, t)\,\mathbf{u}\cdot d\mathbf{S}$ means the mass streaming through the surface element $d\mathbf{S}$ per time.

B.3 The δ-Function

For the description of singular distributions, as, e.g. for point masses, linear currents or area charges, Dirac introduced the δ-function. In mathematical terms, it represents a so-called distribution that can be understood as limit of continuous functions.[13]

[13]The mathematically interested reader is referred to the literature on distributions, e.g. Schwarz (1950).

With a positive parameter $\alpha > 0$, we define the following on the whole x-axis continuous functions:

$$\left.\begin{aligned} \Theta(x,\alpha) &:= \frac{1}{\pi}\int_0^\infty \frac{\sin(kx)}{k}\exp(-\alpha k)\,dk = \frac{1}{\pi}\arctan\frac{x}{\alpha}\;, \\ \delta(x,\alpha) &:= \frac{1}{\pi}\int_0^\infty \cos(kx)\exp(-\alpha k)\,dk = \frac{1}{\pi}\frac{\alpha}{\alpha^2+x^2}\;, \quad \alpha > 0\;, \end{aligned}\right\} \tag{1712}$$

so that

$$\delta(x,\alpha) = \frac{\partial}{\partial x}\Theta(x,\alpha)\;. \tag{1713}$$

In the theory of FOURIER integrals, one proves the following relation:

$$\left.\lim_{\alpha\to 0}\int_a^b f(x')\,\delta(x'-x,\alpha)dx' = \begin{matrix} f(x) \\ 0 \end{matrix} \;\text{ for }\; \begin{matrix} a < x < b\,, \\ x < a \text{ or } x > b\,. \end{matrix}\right\} \tag{1714}$$

The limits $\alpha \longrightarrow 0$ in Eq. (1712) read

$$\left.\begin{aligned} \lim_{\alpha\to 0}\Theta(x,\alpha) &= \begin{matrix} +\frac{1}{2} \\ -\frac{1}{2} \end{matrix} \;\text{ for }\; \begin{matrix} x > 0\,, \\ x < 0\,, \end{matrix}\;, \\ \lim_{\alpha\to 0}\delta(x,\alpha) &= \begin{matrix} 0 \\ \infty \end{matrix} \;\text{ for }\; \begin{matrix} x \neq 0\,, \\ x = 0\,. \end{matrix} \end{aligned}\right\} \tag{1715}$$

The *step function* $\Theta(x)$ is now defined by

$$\Theta(x) := \lim_{\alpha\to 0}\Theta(x,\alpha) \qquad \text{Step function} \tag{1716}$$

and DIRAC*'s* δ*-function* by

$$\delta(x) := \lim_{\alpha\to 0}\delta(x,\alpha)\;. \qquad \text{DIRAC's } \delta\text{-function} \tag{1717}$$

We note that the $\delta(x)$-function can also be represented by other limits, e.g.

$$\delta(x) = \lim_{K\to\infty}\delta(x,K) = \lim_{K\to\infty}\frac{1}{\pi}\frac{\sin(Kx)}{x}\;. \tag{1718}$$

Obviously $\delta(x)$ is not anymore a simple function in the sense of classical analysis. For a sufficient small parameter α, all properties of the δ-function are however arbitrary good realised by the continuous functions $\delta(x,\alpha)$, or for sufficient large K by the continuous functions $\delta(x,K)$. The following mathematical rules for

δ-function are then true in the strict sense of analysis of RIEMANNian integrals and differentiable functions, when always the δ-function is replaced by the continuous function $\delta(x,\alpha)$, and the Θ-function by the function $\Theta(x,\alpha)$. Only *after* performing all differentiations and integrations, one evaluates the limit $\alpha \longrightarrow 0$.[14] From Eqs. (1713) and (1714), there follows immediately

$$\delta(x) = \frac{d}{dx}\,\Theta(x) \tag{1719}$$

and[15]

$$\left.\int\limits_a^b f(x')\,\delta(x'-x)\,dx' = \begin{matrix} f(x) \\ 0 \end{matrix} \quad \text{for} \quad \begin{matrix} a < x < b\,, \\ x < a \text{ or } x > b\,. \end{matrix}\right\} \tag{1720}$$

In a short notation, Eq. (1720) is also taken as equation defining the δ-function. In an n-dimensional space, we write

$$\left.\begin{aligned} &\int\limits_{a_1}^{b_1}\ldots\int\limits_{a_n}^{b_n} f(x_1',\ldots,x_n')\,\delta(x_1'-x_1,\,\ldots,\,x_n'-x_n)\,dx_1'\ldots dx_n' \\ &= \begin{matrix} f(x_1,\,\ldots,\,x_n) \\ 0 \end{matrix} \quad \text{for} \quad \begin{matrix} a_1 < x_1 < b_1,\,\ldots,\,a_n < x_n < b_n\,, \\ x_1 < a_1 \text{ or } x_1 > b_1,\,\ldots,\,x_n < a_n \text{ or } x_n > b_n\,. \end{matrix} \end{aligned}\right\} \tag{1721}$$

Since n-fold integrals can be reduced on n simple integrals, the product of n one-dimensional δ-functions defines the n-dimensional δ-function,

$$\delta(x_1,\,\ldots,x_n) = \delta(x_1)\,\delta(x_2)\,\ldots\,\delta(x_n)\ . \tag{1722}$$

Furthermore, one gets the following formulas:

$$\delta(-x) = -\delta(x)\ , \tag{1723}$$

$$x\,\delta(x) = 0\ , \tag{1724}$$

$$f(x')\,\delta(x'-x) = f(x)\,\delta(x'-x)\ , \tag{1725}$$

$$\delta(ax) = \frac{\delta(x)}{|a|}\ . \tag{1726}$$

---

[14] Treating the $\delta(x)$-functions in mathematical strength as distributions, the notion of RIEMANN's integral is insufficient, and it must be replaced by the STIELTJES integral.

[15] If the point x is located on an integration limit, so the result depends on the kind of continuous functions used to define the δ-function.

If the function $\varphi(x)$ only has unique zero points, x_k , $k = 1, \ldots, n$, so that $\varphi(x_k) = 0$, but $(d/dx)\,\varphi(x_k) = \varphi'(x_k) \neq 0$, then it holds

$$\delta[\varphi(x)] = \sum_{k=1}^{n} \frac{\delta(x - x_k)}{|\varphi'(x_k)|} \ , \tag{1727}$$

hence, e.g.

$$\delta(x^2 - a^2) = \frac{\delta(x - a) + \delta(x + a)}{2\,|x|} \ . \tag{1728}$$

To a mass m_o sitting at point $P(x_o, y_o, z_o)$, we can assign a spatial mass density ϱ_o

$$\varrho_o = \varrho_o(x, y, z) = m_o\, \delta(x - x_o)\, \delta(y - y_o)\, \delta(z - z_o) \ . \tag{1729}$$

Point-like mass density

Integration of Eq. (1729) over the region $K(P_o)$ containing the point P_o leads directly to

$$\iiint\limits_{K(P_o)} \varrho_o(x, y, z)\, dx\, dy\, dz = m_o \ . \tag{1730}$$

A current along the z-axis with strength $J = J(z)$, can be described by a current density j

$$j = j(x, y, z) = J(z)\, \delta(x)\, \delta(y) \ . \tag{1731}$$

Linear current density

The integration in the x-y-plane over a region S, containing the origin, then leads to

$$\iint\limits_{S} j(x, y, z)\, dx\, dy = J(z) \ . \tag{1732}$$

With the δ-function, one can extend the range of validity of equations to singular points. It holds

$$\Delta_2 \ln \frac{1}{r_2} = \left(\frac{\partial^2}{\partial x^2} + \frac{\partial^2}{\partial y^2} \right) \ln \frac{1}{\sqrt{x^2 + y^2}} = -2\pi\, \delta(x)\, \delta(y) \ , \tag{1733}$$

$$\Delta \frac{1}{r} = \left(\frac{\partial^2}{\partial x^2} + \frac{\partial^2}{\partial y^2} + \frac{\partial^2}{\partial z^2} \right) \frac{1}{\sqrt{x^2 + y^2 + z^2}} = -4\pi\, \delta(x)\, \delta(y)\, \delta(z) \ . \tag{1734}$$

References

Abbott, B.P., et al.: Observation of gravitational waves from a binary black hole merger. Phys. Rev. Lett. **116**, 061102 (2016)
Alvänger, T., Farley, F.J.M., Kjellman, J., Wallin, J.: Test of the second postulate of special relativity in the GeV region. Phys. Lett. **12**, 260 (1964)
Arzeliès, H.: Relativistic Kinematics. Oxford, London, Edinburgh, New York, Toronto, Pergamon Press, Frankfurt (1966) (English translation of the French edition 1955)
Barbour, J.: The End of Time. University Press, Oxford (1999)
Barbour, J.: The Discovery of Dynamics: A Study from a Machian Point of View of the Discovery and the Structure of Dynamical Theories. University Press, Oxford (2001)
Becker, R.: Electromagnetic Fields and Interactions, Volume I: Electromagnetic Theory and Relativity. Dover Publications Inc., New York (1982)
Berzi, V., Gorini, V.: Reciprocity principle and the Lorentz transformation. J. Math. Phys. Phys. **10**, 1518 (1969)
Biagioli, F.: Space, Number and Geometry from Helmholtz to Cassirer. Springer International Publishing, Switzerland (2016)
Blanchet, L., Damour, T., Iyler, B.R., Will, C.M., Wiseman, A.G.: Gravitational-radiation damping of compact binary systems to second post-Newtonian order. Phys. Rev. Lett. **74**, 3515 (1995)
Bleyer, U., Gottlöber, S., Haubold, H.J., Hempelmann, A., Mücket, J.P., Müller, V., Stoll, D.: Zur Geschichte der Theorie der Lichtausbreitung (in German). Sterne **55**, 24 (1979)
Brillet, A., Hall, J.L.: Improved Laser Test of the Isotropy of Space. Phys. Rev. Lett. **42**, 549 (1979)
Brown, B.C., Masek, G.E., Maung, T., Miller, E.S., Ruderman, H., Vernon, W.: Experimental comparison of the velocities of eV (visible) and GeV electromagnetic radiation. Phys. Rev. Lett. **30**, 763 (1973)
Buonanno, A., Damour, T.: Effective one-body approach to general relativistic two-body dynamics. Phys. Rev. D **59**, 084006 (1999)
Burrowes, H.G., Caldwell, D.O., Frisch, D.H., Hill, D.H., Ritson, D.M., Schluter, R.A.: K-Meson-Nucleon total cross section from 0.6 to 2.0 BeV. Phys. Rev. Lett. **2**, 117 (1959)
Champeney, D.C., Isaak, G.R., Khan, A.M.: The Lifetimes of the π^+ and π^- mesons. Phys. Rev. **88**, 179 (1952)
Champeney, D., Isaak, G.R., Khan, A.M.: A time dilatation experiment based on the Mössbauer effect. Proc. Phys. Soc. (Lond.) **85**, 583 (1965)
DiSalle, R.: Conventionalism and the origins of the inertial frame concept. Philos. Sci. Assoc. **2**, 139 (1990)

H. Günther and V. Müller, *The Special Theory of Relativity*,
https://doi.org/10.1007/978-981-13-7783-9

de Sitter, W.: On Einstein's theory of gravitation and its astronomical consequences. Mon. Not. R. Astr. Soc. **77**, 155 (1916)

Durbin, R.P., Loar, H.H., Havens, W.W. jr.: The lifetimes of the π^+ and π^- mesons. Phys. Rev. **88**, 179 (1952)

Einstein, A.: Zur Elektrodynamik bewegter Körper. Ann. Phys. (Lpz.) **17**, 891, p. 1952. Printed in Lorentz, H.A. et al. (1905a)

Einstein, A.: Ist die Trägheit eines Körpers von seinem Energiegehalt abhängig? Ann. Phys. (Lpz.) **18**, 639, p. 1952. Printed in Lorentz, H.A. et al. (1905b)

Einstein, A.: Relativity. The Special and the General Theory. Methuen & Co. LTD. London 1955. English edition of: Über die spezielle and die allgemeine Relativitäts theorie. Braunschweig: Vieweg-Verlag 1916, Berlin, Heidelberg: Springer-Verlag, 24. Edition 2009

Einstein, A.: Kosmologische Betrachtungen zur allgemeinen Relativitätstheorie. Preuss. Akad. Wiss. Berlin, Sitzber. **142**, p. 1952. Printed in Lorentz, H.A. et al. (1917)

Einstein, A.: Ether and the Theory of Relativity. An Address delivered on May 5th, 1920, in the University of Leyden (1920)

Einstein, A.: Sidelights on Relativity. Methuen, London (1922)

Einstein, A.: The Meaning of Relativity. University Press, Princeton (1923)

Farley, F.J.M., Bailey, J., Brown, R.C.A., Griesch, M., Jostlein, H., v.d.Meer, S., Picasso, E., Tannenbaum, M.: The anomalous magnetic moment of the negative muon. Nuovo Cimento. **45**, 281 (1966)

Farley, F.J.M., Bailey, J., Picasso, E.: Is the special theory right or wrong?: Experimental verifications of the theory of relativity. Phys. Rev. **159**, 1089 (1967)

Feinberg, G.: Possibility of faster than light particles. Phys. Rev. **159**, 1089 (1967)

FitzGerald, G.F.: The ether and the Earth's atmosphere. Letter to the editor. Science **13**, 390 (1889)

Fock, V.: The Theory of Space, Time and Gravitation. Pergamon Press, Oxford (1964)

Frank, P., Rothe, H.: Zur Herleitung der Lorentztransformation. Phys. Zeitschr. **13**, 750 (1912). Erratum: p. 839

Freund, D.: Privat communication (2009)

Friedmann, A.: Über die Krümmung des Raumes. Z. Phys. **10**, 377 (1922)

Friedmann, A.: Über die Möglichkeit einer Welt mit konstanter negativer Krümmung des Raumes. Z. Phys. **21**, 377 (1924)

Gabrielse, G., Phillips, D., Quint, W., Kalinowsky, H.: Special Relativity and the Single Antiproton. Phys. Rev. Lett. **74**, 3544 (1995)

Giulini, G.: *Algebraic and Geometric Structures in Special Relativity*. In EHLERS, J., LÄMMERZAHL, C., Lecture Notes in Physics 45 (2006)

Grieser, R., Albrecht, F., Dickopf, S., Grieser, M., Habs, D., Huber, G., Klaft, I., Klein, R., Knobloch, P., Kühl, T., Merz, P., Schwalm, D.: A test of special relativity with storaged lithium ions. Appl. Phys. B **59**, 127 (1994)

Grieser, R., Grieser, M., Habs, D., Huber, G., Kühl, T., Merz, P., Schmidt, M., Schwalm, D., Webastian, V., Then, V.: Test of special relativity in an ion storage ring. Hyperfine I. **99**, 135 (1996)

Günther, H.: On the Geometrical Approach to the Theory of Structural Defects. Int. J. Engng. Sci. **19**, 1799 (1981)

Günther, H.: *The Crystalline Structure as a Basis for a Reversed Access to the Special Theory of Relativity*. phys. stat. sol. (b) **185** (1994) 335

Günther, H.: *Elementaryy Theory of Relativity – Lattice·Ether·Symmetry.* Aachen: Shaker-Verlag 2000. Revised English edition of: *Grenzgeschwindigkeiten and ihre Paradoxa. Gitter·Äther·Relativität.* Stuttgart·Leipzig: Teubner-Verlag 1996

Günther, H.: *On the aberration of waves in classical spacetime*. Astron. Nachr. **322**, 153 (2001)

Günther, H.: EINSTEIN*'s Ether - A Lattice Model of Relativistic Spacetime*. Micromaterials and Nanomaterials. A Tribute to ALBERT EINSTEIN. Lectures held at the Commemoration Symposium by Fraunhofer IZM Berlin. **5**, 38 (2006)

Günther, H.: *Spezielle Relativitätstheorie. Ein neuer Einstieg in* EINSTEINs *Welt.* Wiesbaden: Teubner-Verlag, 1. Aufl. 2007
Günther, H.: Die Spezielle Relativitätstheorie. Einsteins Welt in einer neuen Axiomatik, Springer Fachmedien (2013)
Günther, H.: *Starthilfe Relativitätstheorie. Ein neuer Einstieg in* EINSTEINs *Welt.* Wiesbaden: Vieweg+Teubner, 4. Auflage 2010
Günther, H.: *Bewegung in Raum und Zeit. Eine Einführung in die Physik.* Leipzig: Edition am Gutenbergplatz 2012
Günther, H., Müller, V.: Relativitätstheorie von A bis Z. Edition am Gutenbergplatz, Einsteins Spezielle Relativitätstheorie. Leipzig (2013)
Hafele, J.C., Keating, R.E.: Around-the-World Atomic Clocks: Predicted Relativistic Time Gains. Science **177**, 166 (1972)
Halliday, D., Resnick, R., Walker, J.: *Fundamentals of Physics.* Weinheim: John Wiley & Sons, Inc. 8th edition 2008
Haughan, M.P., Will, C.M.: Modern Tests of Special Relativity. Physics Today **40**(5), 69 (1987)
Hehl, F.W., Obukhov, Y.N.: *Foundations of Classical Electrodynamics.* Progress in Math. Phys. 33. Birkhäuser Verlag 2003
Hehl, F. W., Itin, Y., Obukhov, Y.N.: *On* KOTTLER*'s path: origin and evolution of the premetric program in gravity and in electrodynamics.* Int. J. Mod. Phys. D **25**, 1640016 (2016)
Hils, D., Hall, J.L.: Improved Kennedy-Thorndike experiment to test special relativity. Phys. Rev. Lett. **64**, 1697 (1990)
Ibragimov, N.H.: Tensors and Riemannian Geometry. De Gruyter, Berlin/Boston (2015)
Ignatowski, W.v.: *Einige allgemeine Bemerkungen zum Relativitätsprinzip.* Verh. Dt. Phys. Ges. *12*, 788 (1910)
Ives, H.J., Stilwell, G.J.: An Experimental Study of the Rate of a Moving Atomic Clock. Journ. Opt. Soc. **28**, 215 (1938)
Kant, I.: Critique of Pure Reason. University Press, Cambridge (1998)
Kennedy, R.J., Thorndike, E.M.: Experimental Establishment of the Relativity of Time. Phys. Rev. **42**, 400 (1932)
Lange, L.: *Über das Beharrungsgesetz.* Ber. d. Königl. Sächsischen Ges. d. Wiss. Leipzig. Mathematisch physikalische Klasse, **37**, 333 (1885)
Lemaître, G.: Un Universe homogène de masse constante et de rayon croissant rendant compte de la vitesse radiale des nèbuleuses extra-galactiques. Ann. Soc. Sc. Bruxelles **A47**, 49 (1927)
Lewis, G. N., Tolman, R. C.: *The Principles of Relativity, and Non-Newtonian Mechanics.* In: *Proceedings of the American Academy of Arts and Sci* **44**, 709 (1909). In EHLERS, J.,LÄMMERZAHL, C., Lecture Notes in Physics 45 (2006)
Li, Q., Xue, C., Liu, J.-F., et al.: Measurements of the gravitational constant using two independent methods. Nature **560**, 582 (2018)
Liebscher, D.-E., Brosche, B.: Aberration and relativity. Astron. Nachr. **319**, 309 (1998)
Liebscher, D.-E.: *The Geometry of Time.* Weinheim: Wiley-VCH Verlag GmbH & Co. KGaA 2005. English edition of: EINSTEINs *Relativitätstheorie and die Geometrien der Ebene.* Stuttgart, Leipzig, Wiesbaden: Teubner-Verlag 1999
Liebscher, D.-E.: Newton's first law and the existence of free tachyons. Ann. Phys. **32**, 363 (1975)
Lorentz, H.A.: De relatieve bewegung van de aarde en den aether. Versl. K. Ak. Amst. **1**, 74 (1892); Coll. Papers, ed. M. Nijhoff, **4**, 219 (1939)
Lorentz, H.A.: Electromagnetic phenomena in a system with any velocity smaller than that of light. Proc. Acad. Sc. Amst. **6**, 809 (1904)
Lorentz, H.A., Einstein, A., Minkowski, H., Weyl, H.: The Principle of Relativity. A Collection of Original Memoirs on the Special and General Relativity. Dover Publications, Inc., New York (1952). Republication of the 1923 translation of LORENTZ, H. A., EINSTEIN, A., MINKOWSKI, H. first published by Methuen and Company
Lorentz, H.A., Einstein, A., Minkowski, H.: *Das Relativitätsprinzip.* Stuttgart: Teubner-Verlag 1958, 1. Edition 1913

Miller, D.C.: The ether-drift experiment and the determination of the absolute motion of the Earth. Rev. Mod. Phys. **5**, 203 (1933)
Milton, K.A.: The Casimir Effect. World Scienrific, Singapore (2001)
Minkowski, H.: Raum und Zeit, for the English translation s. Lorentz, H.A. (1952)
Minkowski, H.: *Die Grundgleichungen für die elektromagnetischen Vorgänge in bewegten Körpern*. Nachr. von der Ges. der Wiss. zu Göttingen. 53 (1908)
Müller, V., Arbabi-Bidgoli, S., Einasto, J., Tucker, D.: Voids in the LCRS versus CDM models. Mon. Not. R. Astron. Soc. **325**, 280 (2000)
Müller, V., Schmidt, H.-J., Starobinsky, A.A.: Power law inflation as attractor solution for inhomogeneous cosmological models. Class. Quant. Grav. **7**, 1163 (1990)
Munoz-Cuartas, Müller, V., Forero-Romero, J.: Halo based reconstruction of the cosmic mass density field. Mon. Not. R. Astron. Soc. **417**, 1303 (2011)
Nolting, W.: Theoretical Physics 1 - Classical Mechanics. Springer, Berlin (2016)
Otting, G.: Der quadratische Dopplereffekt. Dissertation München. Phys. Z. **40**, 681 (1939)
Pais, A.: Subtle is the Lord .... The Science and the Life of Albert Einstein. Oxford University Press, Oxford (1982)
Poincaré, H.: La mesure du temps. Rev. Métaphys. Morale **6** (1898). English translation in The Foundation of Science. Science Press, New York (1913)
Poincaré, H.: The measure of time. The Value of Science. Reprinted in: The Foundation of Science. Science Press, New York (1913)
Rawls, J.M.: Pulsar data and the dispersion relation for light. Phys. Rev. D **5**, 487 (1972)
Rindler, W.: Essential Relativity. Springer, Berlin (1977)
Saturai, J.J.: Modern Quantum Mechanics. Addison-Wesley, Boston (1993)
Sands, M., Feynman, R., Leighton, R. B.: The Feynman Lectures on Physics. Vol. I: Mainly Mechanics, Radiation, and Heat; Vol. II: Mainly Electromagnetism and Matter. Addison-Wesley, Amsterdam (1970)
Schild, A.: *Lectures in General Relativity Theory*. in EHLERS, J.: *Relativity Theory and Astrophysics*. Amer. Math. Soc., Providence, R.I. 1976
Seeger, A.: Zur Dynamik von Versetzungen in Gitterreihen mit verschiedenen Gitter konstanten. Dipl. Arbeit. Techn, Hochschule Stuttgart (1949)
Shankland, R.S., McCukey, S.W., Leone, F.C., Kuerit, G.: New analysis of the interferometer observations of Dayton C. Miller. Rev. Mod. Phys. **27**, 167 (1955)
Sommerfeld, A.: Electrodynamics. Lectures on Theoretical Physics, Vol. III. Academic, New York (1952)
Taylor, E.F., Wheeler, J.A.: Spacetime Physics. Freeman, San Francisco (1966)
Thirring, W: A Course in Mathematical Physics 1 and 2. Classical Dynamical Systems and Classical Field theory. Springer, New York. Engl. Ed. (1992)
Tonnelat, M.-A.: Histoire du principe de relativité. Flammarion, Paris (1971)
Treder, H.-J.: Philosophische Probleme des physikalischen Raumes. Akademie-Verlag, Berlin (1974)
Wagner, C., Müller, V., Steinmetz, M.: Constraining dark energy via baryon acoustic oscillations in the (anisotropic) light-cone power spectrum. Astr. Astroph. **487**, 63 (2008)

Further Textbooks

Blochinzew, D.I.: Principles of Quantum Mechanics. Boston (1964)
Born, M.: Einstein's Theory of Relativity. Dover Pubkications Inc., New York (1962)
Carroll, S.: Spacetime and Geometry. Addison Wesley, San Francisco (2004)
Fick, E.: Einführung in die Grandlagen der Quantentheorie. Akademische Verlags gesellschaft Geest & Portig K.-G, Leipzig (1968)

French, A.P.: Special Relativity. Norton & Company Inc., M.I.T. (1968)
Goenner, H.F.: Einführung in die spezielle und allgemeine Relativitätstheorie. Spektrum Akademischer Verlag, Heidelberg (1996)
Gottlöber, S., Hauboldt, H.-J., Mücket, J.P., Müller, V.: Early Evolution of the Universe and Formation of Structure. Akademie-Verlag (1990)
Greiner, W.: Quantum Mechanics. Springer, Berlin (2001)
Günter, H., Müller, V.: EAGLE-Starthilfe Allgemeine Relativitätstheorie. Edition am Gutenbergplatz Leipzig (2015)
Helmholtz, H.V.: Philosophische Vorträge und Aufsätze. Akademie-Verlag, Berlin (1971)
Joos, G., Freeman, I.M.: Theoretical Physics. Dover Books on Physics, New York (1987)
Laue, M.V.: Die Relativitystheorie. Vieweg-Verlag, Braunschweig (1961)
Kelvin, Lord: Baltimore Lectures on Molecular Dynamics and the Wave Theory of Light. Clay and Sons, London (1904)
Ludwig, G.: Foundations of Quantum Mechanics. Springer, New York (1983)
Misner, C.W., Thorne, K.S., Wheeler, J.A.: Gravitation. W. H. Freeman and Company, New York (1973)
Møller, C.: The Theory of Relativity. At the Clarendon Press, Oxford (1952)
Papapetrou, A.: Lectures on General Relativity. D. Reidel Publishing Company, Dordrecht (1974)
Pauli, W.: Theory of Relativity. Dover Publications Inc., New York (1981)
Reichenbach, H.: The Philosophy of Space and Time. Dover Publikations, Inc., New York (1958). cf. also Kindle Edition 2003
Schutz, B.: A First Course in General Relativity. University Press, Cambridge (2009)
Schwabl, F.: Quantum Mechanics. Springer, Heidelberg (2007)
Schwarz, L.: Theory of Distributions. Hermann & Cie, Paris (1950/51)
Shankar, R.: Principles of Quantum Mechanics. Springer, Berlin·Heidelberg·New York (1994)
Tonnelat, M.-A.: The Principles of Electromagnetic Theory and of Relativity. Dordrecht Reidel Publishing Company. Springer, Holland (1966)
Wald, R.M.: General Relativity. University of Chicago Press, Chicago (1984)
Weinberg, S.: Gravitation and Cosmology. Wiley, New York (1972)
Weyl, H.: Space Time Matter. Dover Books on Physics, Methuen & Company (1952)

Index

H. Günther and V. Müller, *The Special Theory of Relativity*,
https://doi.org/10.1007/978-981-13-7783-9

T

FSC
www.fsc.org
MIX
Papier | Fördert
gute Waldnutzung
FSC® C083411

Kolibri 360 GmbH
Ferdinand-Jühlke-Straße 7
99095 Erfurt, Deutschland
produktsicherheit@kolibri360.de